PROGRESS IN

Molecular Biology and Translational Science

Volume 85

PROGRESS IN
Molecular Biology and Translational Science

Molecular Biology of RNA Processing and Decay in Prokaryotes

edited by

Ciaran Condon
University of Paris
Institut de Biologie Physico-Chimique
Paris, France

Volume 85

AMSTERDAM • BOSTON • HEIDELBERG • LONDON
NEW YORK • OXFORD • PARIS • SAN DIEGO
SAN FRANCISCO • SINGAPORE • SYDNEY • TOKYO
Academic Press is an imprint of Elsevier

Academic Press is an imprint of Elsevier
32 Jamestown Road, London, NW1 7BY, UK
Radarweg 29, PO Box 211, 1000 AE Amsterdam, The Netherlands
30 Corporate Drive, Suite 400, Burlington, MA 01803, USA
525 B Street, Suite 1900, San Diego, CA 92101-4495, USA

This book is printed on acid-free paper. ∞

Copyright © 2009, Elsevier Inc. All Rights Reserved

No part of this publication may be reproduced, stored in a retrieval system or transmitted in any form or by any means electronic, mechanical, photocopying, recording or otherwise without the prior written permission of the Publisher

Permissions may be sought directly from Elsevier's Science & Technology Rights Department in Oxford, UK: phone (+44) (0) 1865 843830; fax (+44) (0) 1865 853333; email: permissions@elsevier.com. Alternatively you can submit your request online by visiting the Elsevier web site at http://elsevier.com/locate/permissions, and selecting *Obtaining permission to use Elsevier material*

Notice
No responsibility is assumed by the publisher for any injury and/or damage to persons or property as a matter of products liability, negligence or otherwise, or from any use or operation of any methods, products, instructions or ideas contained in the material herein. Because of rapid advances in the medical sciences, in particular, independent verification of diagnoses and drug dosages should be made

Library of Congress Cataloging-in-Publication Data
A catalog record for this book is available from the Library of Congress

British Library Cataloguing in Publication Data
A catalogue record for this book is available from the British Library

ISBN: 978-0-12-374761-7
ISSN: 0079-6603

For information on all Academic Press publications
visit our website at elsevierdirect.com

Printed and Bound in the USA
09 10 11 12   10 9 8 7 6 5 4 3 2 1

Working together to grow
libraries in developing countries

www.elsevier.com | www.bookaid.org | www.sabre.org

ELSEVIER   BOOK AID International   Sabre Foundation

Contents

A Phylogenetic View of Bacterial Ribonucleases 1
Antoine Danchin

I. Functions for Life	2
II. Ribonucleases of the Paleome	4
III. Ribonucleases of the Cenome	26
IV. Provisional Conclusion	30
References	32

RNA Processing and Decay in Bacteriophage T4 43
Marc Uzan

I. Introduction	44
II. Part 1: Mechanisms and Regulation of mRNA Degradation During T4 Development	45
III. Part 2: Transfer RNA Processing	66
References	78

Endonucleolytic Initiation of mRNA Decay in *Escherichia coli*. 91
Agamemnon J. Carpousis, Ben F. Luisi, and Kenneth J. McDowall

I. Background	92
II. General Features of mRNA Degradation in *E. coli*	97
III. Evidence for Diverse Pathways of mRNA Degradation in *E. coli*	102
IV. The Structure of RNase E	106
V. Substrate Specificity and Catalytic Mechanism of RNase E	112
VI. Modulators and Adaptors of RNase E	118
VII. Other Endoribonucleases in *E. coli*	119
VIII. Phylogenetic Distribution of RNase E and The RNA Degradosome	123
References	125

Poly(A)-Assisted RNA Decay and Modulators of RNA Stability 137
Philippe Régnier and Eliane Hajnsdorf

    I. Introduction 138
   II. Regulation of Gene Expression in Response to
       Environmental Signals 139
  III. Polyadenylation and Poly(A)-Dependent Decay 142
  IV. Other Modulators of RNA Stability 158
    V. Conclusions and Open Questions 167
       References 169

The Role of 3'–5' Exoribonucleases in RNA Degradation 187
José M. Andrade, Vânia Pobre, Inês J. Silva, Susana Domingues, and Cecília M. Arraiano

    I. Introduction 188
   II. Polynucleotide Phosphorylase (PNPase) 193
  III. RNase II 200
  IV. RNase R 205
    V. Oligoribonuclease 210
  VI. Mechanisms of RNA Degradation 212
 VII. Concluding Remarks 217
       References 218

Messenger RNA Decay and Maturation in *Bacillus Subtilis* 231
David H. Bechhofer

    I. Introduction and Historical Perspective 232
   II. The 5' End 234
  III. Body of The Message 251
  IV. The 3' End 258
    V. Regulated mRNA Decay 264
  VI. Conclusion 267
       References 267

RNA Degradation in Archaea and Gram-Negative Bacteria Different from *Escherichia coli*........... 275

Elena Evguenieva-Hackenberg and Gabriele Klug

I. Introduction 276
II. RNA Degradation Mechanisms in Bacteria and Eukarya 277
III. Features of mRNA in Archaea 287
IV. Archaeal Proteins with Unexpected Endoribonucleolytic Activities 293
V. Archaeal Proteins with Similarities to Bacterial or Eukaryotic Proteins Involved in RNA Degradation—A Short Overview 299
VI. The Archaeal Exosome 300
VII. RNA Degradation in Archaea 308
VIII. Concluding Remarks 310
References 311

The Making of tRNAs and More – RNase P and tRNase Z............................ 319

Roland K. Hartmann, Markus Gößringer, Bettina Späth, Susan Fischer, and Anita Marchfelder

I. Introduction 320
II. Processing by RNase P 324
III. Removal of the tRNA 3′-Trailer 347
References 358

Maturation and Degradation of Ribosomal RNA in Bacteria............................ 369

Murray P. Deutscher

I. Introduction 369
II. Early Studies of rRNA Maturation and Degradation 370
III. Current Understanding of rRNA Processing 373
IV. Degradation of rRNA 383
References 388

RNA Polyadenylation and Decay in Mitochondria and Chloroplasts 393

Gadi Schuster and David Stern

I. Introduction	394
II. Polyadenylation of RNA	395
III. The Enzymes	399
IV. RNA Degradation and Polyadenylation in Chloroplasts	408
V. RNA Degradation and Polyadenylation in Mitochondria	409
VI. Conclusions and Perspectives	413
References	414

Killer and Protective Ribosomes 423

Marc Dreyfus

I. Introduction	424
II. Nucleolytic *Versus* Non-nucleolytic Inactivation of mRNAs	426
III. Translation–Degradation Interplay: Conceptual Ambiguities and Technical Caveats	430
IV. Killer Ribosomes	436
V. Protective Ribosomes	446
VI. Concluding Remarks	457
References	458

mRNA Interferases, Sequence-Specific Endoribonucleases from the Toxin–Antitoxin Systems 467

Yoshihiro Yamaguchi and Masayori Inouye

I. Introduction	468
II. MazF: An mRNA Interferase	471
III. MazF Homologues and Other mRNA Interferases	480
IV. Other Toxins in TA Systems	489
V. The Regulation of MazF Expression Under Stress Conditions	491
VI. Concluding Remarks	494
References	496
Index	501

A Phylogenetic View of Bacterial Ribonucleases

ANTOINE DANCHIN

Genetics of Bacterial Genomes, CNRS URA2171, Institut Pasteur, 28, rue du Docteur Roux, 75724 Paris Cedex 15, France

I. Functions for Life 2
II. Ribonucleases of the Paleome 4
   A. RNase P, a Ribozyme from the RNA Genome World? 4
   B. Energy-Dependent Degradation of RNA 8
   C. Hydrolytic Nucleases 17
III. Ribonucleases of the Cenome 26
   A. Exported Ribonucleases 28
   B. Plasmid and Phage Ribonucleases 29
IV. Provisional Conclusion 30
   References 32

A phylogenetic analysis of bacterial genomes shows them to comprise persistent genes, the "paleome" (Greek: palaios, ancient, reminiscent of the origin of life), associated with genes permitting development of life in a particular niche, the "cenome" (from koinos, common, a radical often used in ecology). Most ribonucleases belong to the former, demonstrating their central position in core life processes. These enzymes appear to have often (but not always) evolved through consistent scenarios, generally grouping bacteria into well-defined clades. The evolution of phosphorylases (which salvage energy) is particularly revealing, resulting in diverse complex structures whose function is to degrade RNA. The degradosome of the gamma-Proteobacteria is a paradigm of such complex structures that emphasizes the essential role of energy in degradative processes. The A+T-rich Firmicutes behave in a highly original manner, where many ribonucleases and related proteins coevolve as a group. The recent identification of novel activities in these organisms, stresses the (underestimated) importance of degradation of very short RNAs, as well as 5′–3′ degradative processes in Bacteria.

I. Functions for Life

The degradation of cellular components is usually viewed as part of a "cleaning" function, a maturation function, or as part of a regulatory cascade. In parallel, as can be seen from the huge amount of effort devoted to the study of transcription factors, chromatin structure and regulatory cascades at the transcription initiation level, the paradigm for the control of gene expression at this level implicitly assumes that mRNA molecules will turnover, and, in Bacteria, turnover fast (1). Curiously, however, the exact contribution of the degradation process to the ultimate level of the products of gene expression, namely proteins, has not been analyzed systematically. And despite widespread observations of discrepancies between mRNA levels and proteins levels in Bacteria, the role of RNA degradation processes in the control of gene expression has, to say the least, been overlooked.

"Nothing in biology makes sense except in the light of evolution" (2) and the multiple roles of ribonucleases (RNases) in Bacteria should be explored in this light. Condon and Putzer analyzed their phylogenetic distribution (3). Rather than take the same approach here, which would simply update their thorough work, we take a different stance: a functional view of RNases considered from the stand point of both the organization of the bacterial genome and the organization of metabolism. In this review, we will use observations based on the study of evolution to try to convey the message that degradative processes are specifically involved in essential steps that are not simply meant to support life, but required to perpetuate life.

Before proceeding, a cautionary word is necessary. The triplet variation/selection/amplification that drives evolution continuously provides niches for particular new functions to be invented. These functions are performed by chemical objects, which must be either recruited from previously existing objects (4, 5) or created *de novo*. As a consequence, and this is essential for the discussion that follows, there is not a one-to-one relationship between protein structure and function, so that, while many functions may be essential, and therefore ubiquitous, this does not have to be the case for the protein structures that support them. This means that our view of phylogeny is not necessarily one that stems from a unique origin, for example, LUCA. Particular circumstances that led to the invention or capture of a function will be borne witness in phylogenetic trees with different shapes, related to the various scenarios for the origin and evolution of life that we may be tempted to consider. However, because living organisms always derive from close ancestors, there is a tendency for organisms to stick to one object when it fulfills a given function. Hence, it is most likely that the structure/function relationship will hold within a particular clade. From time to time a jump, a discontinuity, will be observed, corresponding to the moment when a particular object was replaced by a new one. In extant organisms, we do not expect to see many situations where two objects with the same function are found in a single species. This is the famous "missing link" often observed in paleontology (i.e., the missing organism establishing the transition between two very different forms). In this context, we will try in some cases to separate between a function (a generic answer to a general need, e.g., double-stranded RNA degradation) and its corresponding functionalities (features associated with the function which provide the closest response to a specific need, for example, tagging for activity/inactivity, localization in the cell, specification of a function to a narrow range of substrates, etc.), trying to avoid making a direct connection between the connotations of the concept of function and that of evolution (i.e., trying to avoid teleological interpretations: function "for" a particular purpose) (6).

Finally, we need to add another, technical, caveat that must be taken into account. Most of the data we analyze here are derived from genome sequencing projects. These projects are experimental and are not error-free. This implies that, while the presence of a sequence can be taken as firmly established, the same cannot be said for the absence of a sequence. Before we draw important conclusions about a missing object, a very thorough investigation of the data leading to this conclusion is required. In many cases, the absence of a sequence should therefore be taken with a grain of salt.

The concept of an RNA genome world at the onset of the evolution of life places RNA-centered functions at the core of the bacterial genome organization. However, this concept requires an earlier step, omitted from most scenarios that led to the creation of nucleotides. While the prebiotic soup scenario remains popular in the mass media, in-depth analysis of the physico-chemical constraints for a medium in which life could be born supports a scenario where autocatalytic reproduction of a surface metabolism would play a major role (7–11). This scenario requires that surfaces be subsequently replaced by some sort of charged support. The common hypothesis is that RNA played this role. This implies that RNA was involved in the metabolic processes that led not only to its own synthesis, but also to that of amino acids, coenzymes, and nucleotides. Among extant RNAs involved in metabolism, transfer RNA (tRNA) has a special role. Not only is it involved in ribosomal protein synthesis, but it is also involved in a variety of reactions that are unrelated to this process, including, for example, synthesis of the heme precursor aminolevulinate, in a reaction that should not necessarily involve tRNA *per se* (11). This makes ancestral tRNA a candidate of choice in the transition between the mineral world to the RNA genome world. It is therefore expected that, very early on, tRNA was metabolized via ribozymes, typical of the enzymatic activities postulated to have developed in the RNA genome world. Remarkably, the organization of bacterial genomes is consistent with this scenario of the origin of life. Indeed, bacterial genomes are made up of two sets of genes, persistent genes, that tend to be shared by a majority of genomes (12), and genes that are highly variable and often present in only one strain of a given species. Persistent genes make up a core structure that is organized into a network reminiscent of this scenario, forming the *paleome*. Highly variable genes, directly related to the way bacteria occupy a particular niche, form the *cenome*, an unlimited collection of genes that are spread by horizontal gene transfer and (we do not have any indication yet about this possible process) sometimes perhaps created *de novo* (13) (Fig. 1). The conservation of syntenies in genomes can be used to substantiate functional inferences, as the genes belonging to the paleome and the genes belonging to the cenome are often clustered together, albeit via completely different molecular processes (14, 15).

RNases belonging to the paleome are more directly involved in the basic functions involved in sustaining and perpetuating life. The way they evolved will be discussed first in detail. RNases (often secreted) belonging to the cenome, which provides functions permitting the cell to explore its environment and scavenge nutrients, will be discussed in a later section of this review.

II. Ribonucleases of the Paleome

A. RNase P, a Ribozyme from the RNA Genome World?

Evolution from a surface metabolism creating coenzymes, lipids, and the basic building blocks (nucleotides in particular) to that of an RNA genome world suggests that the first macromolecular RNase activity could have been a ribozyme (16). With this in mind, it is natural to begin the list of these essential functions with RNase P, a ribozyme that is involved in processing tRNA and other RNAs, riboswitches in particular (17, 18).

Ribonuclease P (RNase P) is the most widespread ribonuclease (3). It is found in the vast majority of organisms so far examined (19). Two exceptions have been documented, *Aquifex aeolicus* (3, 20, 21) and *Nanoarchaeum equitans* (22). Although a recent article describes an RNase P-like activity in cell lysates of *A. aeolicus*, it has not been characterized and it is not yet possible to understand its relationship with known counterparts (23). In *N. equitans*, tRNAs lack a 5' leader and have therefore presumably been free to lose RNase P activity. Other similar examples may be revealed with time. RNase P is present within the subcellular compartments of eukaryotes known to synthesize tRNAs (24). A key enzyme involved in the processing of tRNA, it is responsible for the generation of the 5' termini of mature tRNA, via a specific endonucleolytic cleavage (19, 25). In Bacteria, RNase P is a ribonucleoprotein consisting of a large catalytic RNA (P RNA, 350–400 nucleotides long) coded

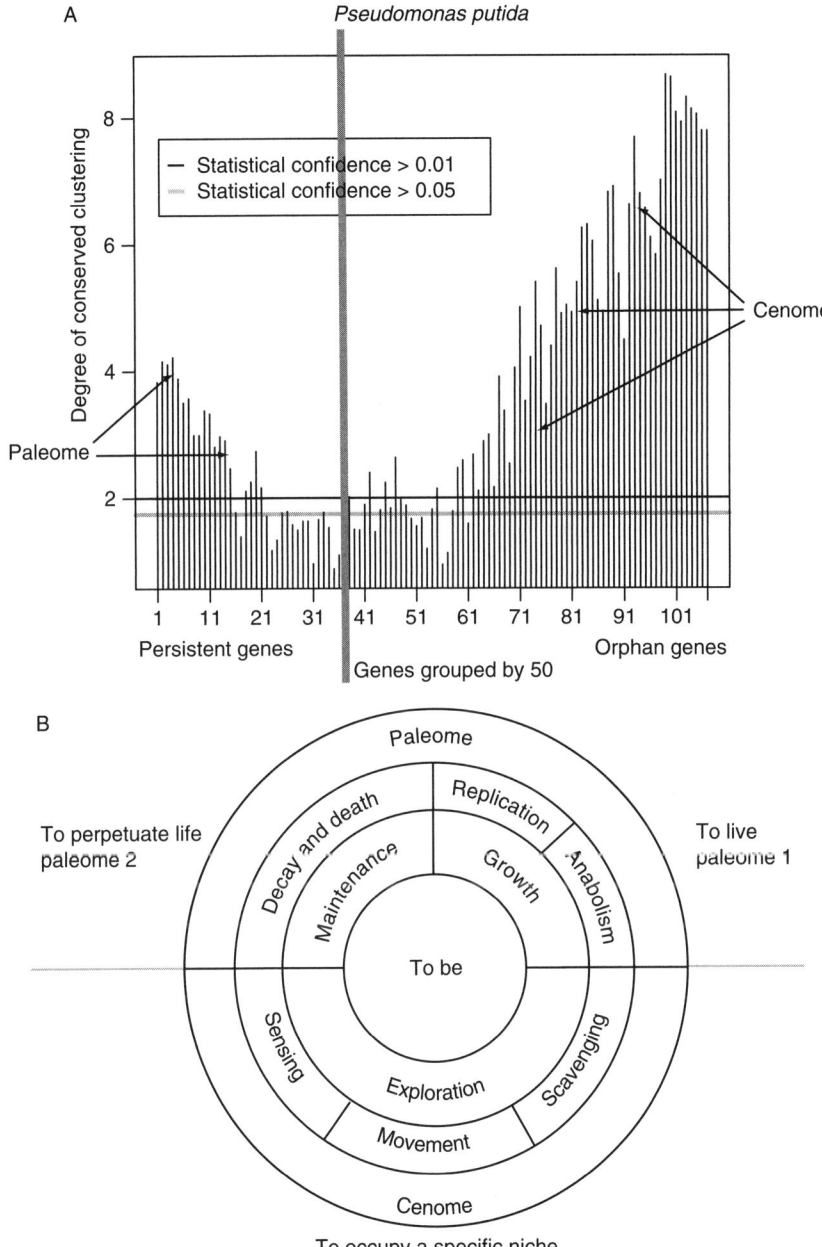

FIG. 1. (A) Organization of bacterial genomes illustrated by the conserved clustering of genes in *Pseudomonas putida* KT2440. Genes have been grouped together into groups of 50 genes as a function of their frequency in more than 150 bacterial genomes comprising more than 1500 genes

by the *rnpB* gene and a small protein subunit (approximately 13 kDa), encoded by the *rnpA* gene. The RNA component of RNase P, which derives from a precursor processed by another ribonuclease, RNase E/G (26) (see below), is active *in vitro* in the absence of its protein subunit, but this activity is too slow to permit growth of RnpA-deficient strains (27). This supports the idea that, during the transition from an RNA genome world to the present biological world, proteins, with their superior catalytic properties, progressively invaded the catalytic functions of cells, both by acting as enzymes themselves and by stabilizing and enhancing the catalytic efficiency of ribozymes.

Two types of bacterial P-RNAs exist, based on sequence alignments associated with the 3D structure of the ribozyme: type A (found in *Escherichia coli* and *Thermus thermophilus*, e.g.) and type B (at present, found exclusively in Firmicutes) (28, 29), which are functionally interchangeable *in vivo* (30). P-RNA contains two independent domains: the catalytic domain and the substrate binding domain (the S-domain) (31, 32). The S-domain recognizes and binds to the pre-tRNA substrate through the highly conserved T-psi loop of pre-tRNAs (33). Despite the invariant structure of the pre-tRNA substrate, the RNA components of type A and type B P-RNAs show characteristic differences in their secondary structure. The structure outside of the conserved structural core of P-RNAs is quite variable, as in group I intron RNAs (33). Conserved regions of the RNA have evolved by concerted substitution of alternative structures, rather than by insertion and deletion of helical elements that occur in the more variable regions of the RNA. Most of the unusual structural elements of type B P-RNAs of the Firmicutes have evolved independently in *Thermomicrobium roseum*, a member of the green nonsulfur Bacteria (34), possibly illustrating a case of convergent evolution (horizontal gene transfer can never be excluded, however). Interestingly, this phylogenetic dichotomy, which places Firmicutes in an original clade, is not paralleled by a particular feature of their tRNAs: the enzyme which modifies the uracil located in the T-psi loop of

---

(horizontal axis) and their tendency to remain clustered in many genomes is plotted on the vertical axis (drawn after a panel of Supplementary Fig. 1 in Ref. (13)). The horizontal line spanning the figure gives the statistical validation of this tendency for genes to remain clustered (p-value < 0.01 and p-value < 0.05). Genes that tend to be present in all genomes tend to remain clustered, while rare genes are also clustered together. This clustering process stems from completely different reasons as demonstrated in (15). The former class constitutes the paleome, while the latter makes up the cenome (197). (B) Genes constituting the paleome and the cenome code for functions that display consistent roles in the general processes defining life. The paleome is formed of two major parts. Paleome 1 makes life itself, coding for anabolism and replication, while Paleome 2 codes for maintenance functions that are required to perpetuate life (69). Most RNases are distributed among both parts of the paleome. The cenome comprises the rarer RNases that are used to scavenge RNA from the environment and perform all functions that would be useful for exploration, as well as those involved in managing horizontally transferred elements.

the tRNA to thymine, *trmA*, coding for S-adenosyl-L-methionine-dependent tRNA (uracil-54,C5)-MTase in most bacterial clades, is mutually exclusive with *trmFO*, coding for an enzyme using tetrahydrofolate instead. While TrmFO exists in most Firmicutes, it is also present in all alpha- and delta-Proteobacteria (except Rickettsiales in which the *trmFO* gene is missing), Deinococci, Cyanobacteria, Fusobacteria, Thermotogales, Acidobacteria, and in one *Actinobacterium* sp. (35).

When P-RNA sequence data from representatives of all bacterial clades was used (RNase P database http://www.mbio.ncsu.edu/RNaseP/home.html) to generate phylogenetic trees, the analysis repeatedly resulted in unstable tree topologies. This is not unexpected because of limited length and high degree of P-RNA variability, resulting in the introduction of gaps in the alignments. This limits the usefulness of the sequence comparison to the construction of phylogenies of narrow clades (32, 36). For example, P-RNA is conserved in plastids and it produces a stable phylogeny for Cyanobacteria, comparable to that of 16S rRNA (37). In the same way, P-RNA supports a phylogenetically consistent tree within the Chlamydiae (38).

In this context, it is most interesting to study bacterial clades that are outliers in the bacterial tree of life. In the case of Aquificales, this gave an interesting result: phylogenetic trees based on structure-based sequence alignment using the program fastDNAml did not corroborate an early origin for the Aquificales P-RNAs (36). With manual improvement of the alignments, the type A P-RNA trees grouped the *Aquificales* with the green sulfur bacteria, Cyanobacteria, and delta/epsilon-Proteobacteria, in general agreement with several protein-based phylogenies, but not with the usual 16S rRNA trees (39). The major discriminating feature is the lack of helix P18 in the *Persephonella marina* and *Sulfurihydrogenibium azorense* P-RNAs. This feature, shared with Archaea and Eukarya, thus seems to represent not an ancient, but a convergent trait (40).

Another group is particularly interesting to explore. Planctomycetes form a distinct clade of Bacteria. They display a number of unusual phenotypic features, including peptidoglycan-less cell walls, short 5S rRNA and an unlinked *rrn* operon organization (41). They have a compartmentalized cell structure, unique among Bacteria. In the P-RNA of *Gemmata* sp. isolates, an insert in helix P13, not found in any other member of the Bacteria, permits the construction of a tree that is consistent with the 16S rRNA tree (42).

In conclusion, the type A P-RNA tree appears to follow the standard 16S rRNA tree except for the Aquificales. The tree position of the Aquificales has been challenged recently, however, and it is now more consistent with that generated by P-RNA (39). It seems likely, however, that controversies will persist, as it is extremely difficult to establish robust phylogenies spanning the whole tree of life based on nucleic acids. Even ribosomal RNA itself is prone to vary, sometimes with significant differences within a same strain (43, 44). Nevertheless, a highly specific pattern is observed with the Firmicutes, which have type B P-RNAs, suggesting a particular pattern of evolution of this particular class of catalytic activities.

The protein component of active RNase P is encoded by rnpA gene. In general, this gene is located near the origin of replication (when it exists) in a phylogenetically conserved way and it belongs to the group of genes used to characterize the origin (45). As stated above, A. aeoliqus is a remarkable exception: all genes in the region are conserved, except that rnpA is missing (40). An RNase P protein knockdown strain provided direct proof that the rnpA gene is essential in B. subtilis and, by inference, in other bacteria (27). The strain conditionally expressing rnpA was used to screen for functional conservation among bacterial RnpA proteins from a representative spectrum of bacterial clades showing conserved function of RnpA proteins despite low sequence conservation. Even rnpA genes from psychrophilic and thermophilic bacteria rescued growth of B. subtilis rnpA mutants (30), despite a large difference in their P RNAs. A deletion analysis of B. subtilis RnpA defined the structural elements essential for bacterial RNase P protein function *in vivo*. Despite a low degree of sequence similarity the overall fold of the protein is conserved. In B. subtilis RnpA, a loop containing a cluster of acidic residues is thought to mediate RNA contacts via coordinated metal ions. Terminal extensions and insertions in this loop remained compatible with the RnpA function in B. subtilis. Some protein variants exhibit striking sequence extensions of unknown function, which may correlate with functional differences. Overall, the similar three-dimensional structures of protein subunits from unrelated bacteria with B-type (B. subtilis, Staphylococcus aureus) and A-type (Thermotoga maritima) P-RNAs suggested that structural and functional conservation may be a hallmark of bacterial RnpA proteins (46–48).

Interestingly, the corresponding phylogenetic tree of RnpA matches that of P-RNA, with Firmicutes being placed out of their commonly admitted place in the standard 16S phylogeny. This further indicates that this member of the paleome has followed a distinct evolutionary path. This feature will be further discussed below.

B. Energy-Dependent Degradation of RNA

Genes of the bacterial paleome can be divided into two more or less equal classes depending on whether or not they are essential to bacterial growth under laboratory conditions (Fig. 1B). Essential genes, well identified in E. coli (49) and B. subtilis (50), cannot be inactivated without loss of viability. In contrast, a second category of paleome genes can be inactivated when cells are grown in the laboratory. These genes correspond to a functional category of genes that are involved in maintenance and adaptation to rapid transitions under usual living conditions (12).

Degradation of weathered compounds is an essential step in the perpetuation of life. The enzymes responsible for this function can be assumed to have played an essential role in perpetuating life over the generations, by permitting the cell to use aged structures to generate young cells with correct products. If degradation were to be performed randomly, that is, without screening between functional and nonfunctional compounds (51), it is unlikely that this would be successful, given the limited number of molecules present in the small volume of a bacterium. Hence, the production of progeny requires a selective step in the choice of the molecules that need to be conserved, in particular those absolutely required for life. To be effective in eliminating deleterious molecules, while conserving those that are functional (even if slightly altered), this step should consume energy (52). In this context, it is important to analyze in detail the evolution of genes encoding proteins that salvage energy.

One evolutionarily conserved component of the general RNA degradation machinery that is remarkably poised to perform this essential energy-dependent function is the degradosome, which exists in the majority of Bacteria (53). A similar RNA degradation complex also exists in the Archaea and Eukarya, named the exosome (54). Interestingly, compartmentalization may play a major role in the organization of these structures. While the biochemical definition of the exosome shows that it has a simpler, less-evolved structure that cannot be split into active subcomponents, the *E. coli* degradosome is a protein complex with several core components that also play a role in the organization of the bacterial cell: polynucleotide phosphorylase (PNPase), ribonuclease E (RNase E), RhlB (an ATP-dependent helicase), and enolase (53, 55). As a case in point, suggesting association of hydrolytic enzymes to these structures, RNase II (Rrp44) has been found to be associated with the exosome in Eukarya (56). An association of the exosome with the cytoskeleton may be absent from Eukarya, as enolase (see below) is associated with the RNase E backbone in Bacteria (discussed in (57)), while in higher organisms enolase isoforms interact directly with microtubules and the centrosome (58–60), making it a core component of a cell's organization which differs considerably in Bacteria and in Eukarya. Several other proteins associate loosely with the degradosome: polyphosphate kinase (55, 61, 62), poly(A) polymerase (63), and ribosomal protein S1 (64, 65). We will discuss these components in the sections that follow.

It should be noted at this point that biochemical definitions of cell structures should be qualified, since even mild biochemical techniques tend to be extremely disruptive. As a matter of fact, it is very difficult to distinguish between a "contaminant" of a purified complex and a protein that may be transiently associated with it at some point in the cell life cycle. In this respect, it may be worth noting that the association of ribosomal protein S1 to the degradosome was predicted *in silico* well before this was demonstrated biochemically (*64, 65*). Furthermore, while systematically practiced by biochemists all over the world, breaking open a cell in the presence of dioxygen is extremely disruptive by nature. Any complex where a ferrous ion is not extremely tightly bound to its partner will simply vanish, whereas *in silico* analyses suggest that iron may play a much more important role than previously suspected (*13*), well in line with scenarios about the origin of life (*8, 10, 11, 13*). In this context, it may be relevant to take into account *in silico* studies that suggest that some objects are synthesized in the same zone in the cell and may therefore tend to interact (*66, 67*).

1. Phosphorylases

Degradation of polynucleotides by phosphorolysis is ubiquitous. Remarkably, phosphate is not only part of the backbone of nucleic acids and a variety of other building blocks of structural components of the cell (membranes, teichoic acids and the like), but it is also a major player in the management of energy (*68*). In this respect, it is worth noting that RNA degradation can proceed via the formation of energy-rich bonds: the energy present in phosphodiester bonds is recovered as the RNA molecules are degraded, by using phosphate for phosphorolysis instead of the ubiquitous water-mediated hydrolysis. Degradation by phosphorolysis produces nucleoside diphosphates that play a central role in the channeling of the cell's metabolism toward anabolism. Energy saving is of major importance when cells enter into stationary phase of growth, where many components of the cell age with a concomitant loss of energy synthesizing capacity, while the cell needs to produce a rejuvenated progeny when growth resumes. We have proposed that this requires a selective process which uses energy to discriminate between newly synthesized components and old ones ((*69*) and see below). The salvaging of energy by phosphorolysis of RNA would be admirably suited to this purpose, especially if polyphosphates, the product of another loosely associated component of the degradosome, could be used to restore the triphosphate levels of ATP and GTP when the cell reconstitutes a functional ATP synthase and/or glycolytic flux. In line with this conjecture, it has just been discovered that PNPase activity is allosterically regulated by ATP, preventing its action when the energy levels are high (*70*). It seems, therefore, quite remarkable that polyphosphate kinase is indeed associated with the degradosome.

A second property of nucleoside diphosphates is that they are almost ubiquitously preferred over ribonucleoside triphosphates for the biosynthesis of deoxyribonucleotides (there are only a few exceptions where NTPs can be used (*71*), but this is a backup of the normal NDP route). Surprisingly, the role of RNA phosphorolysis in generating the NDP precursors required for DNA synthesis, which might play a role in optimization of the relative distribution of DNA nucleotides to RNA (as remarked by Seymour Cohen as early as 1960 (72)), has only rarely been pointed out, despite its importance (64). Briefly, while the pyrimidine biosynthesis pathway goes through UDP (which, interestingly, should not enter DNA), CDP is not formed directly, creating a relative starvation for this precursor of dCDP, and consequently dCTP. This creates a general tendency for most genomes to increase their A+T content (73). The consequence of this organization of metabolism is that incorporation of cytosine into DNA results essentially from processes where CTP is recycled to CMP, unless CDP can be produced directly, which is exactly the outcome of RNA phosphorolysis (66).

Ribonuclease PH and PNPase, which descend from a common ancestor, are the two avatars of these ubiquitous phosphorolytic enzymes. PNPase is a structurally complex enzyme (74–77), that is certainly the product of a very intricate evolution scenario. It comprises five domains: two RPH-like domains (PNP1 and PNP2), one alpha helical domain, and two RNA binding domains (KH and S1 domains). We will discuss RNase PH before considering the evolution of PNPase, since the latter is organized around RPH domains. Except for the Mollicutes, Bacteria generally possess one or other of these enzymes, or both (3).

Ribonuclease PH. RNase PH (pre-tRNA phosphorylase, mistakenly referred to as tRNA nucleotidyltransferase in several UniProtKB (78) entries, probably because the reaction is reversible *in vitro* (79)) was discovered as an important enzyme involved in the maturation of the 3'-ends of tRNAs (80). The enzyme has a ring structure made up of six subunits, arranged as a trimer of dimers (81, 82). This structure allows the binding to tRNA precursors and the trimming of their 3' extremities (81, 83, 84). The CCA motif is then added to these extremities by tRNA nucleotidyltransferase (Cca), one nucleotide at a time (85). The RNase PH ring structure appears to be the model from which PNPase and the eukaryotic exosome have evolved, by domain duplication and fusion with a variety of other domains (86). Its strong sequence conservation in distantly related organisms argues in favor of a very constrained activity. Its function appears to sometimes overlap with that of PNPase, since some organisms have only one or other of these enzymes (3, 12). RNase PH also forms the core structure of the archaeal exosome.

Polynucleotide phosphorylase. PNPase is also a trimeric complex, forming a doughnut-shaped structure where the RNA-binding domains create a central pore where RNA enters. Each subunit contains five domains, including two RNase PH domains (PNP1 and PNP2), one alpha helical domain, one KH domain, and one S1 domain (77). Phylogenetic analysis suggests that PNPase was formed via a duplication of an ancestor of RNase PH. The current distribution of these domains on the tree of life suggests that the PNP domain predated the separation between Archaea, Eukarya, and Bacteria. PNP2 and RNase PH are more closely related to each other than either one is to PNP1, suggesting a functional specialization of PNP1 during evolution. The function of PNP1 is unclear, but it appears to synthesize guanosine 3'-diphosphate 5'-triphosphate (pppGpp) in *Streptomyces antibioticus* (87), while the phosphorolytic catalytic site is thought to be located within the PNP2 domain in both this organism and in *E. coli* PNPase. The pppGpp-related function should probably be explored further, as it provides a functional link with another component of the degradosome, polyphosphate kinase, and (p)ppGpp has been shown to control polyphosphate metabolism (88).

PNPase belongs to the degradosome in Proteobacteria, where it is associated with RNase E, the scaffold for the different components of this RNA degradation complex (53). It is also weakly associated with a variety of other enzymes such as Poly(A) polymerase, involved in nontemplated RNA synthesis, and CspE, an RNA binding factor that recognizes the sequence AAAUUU (65). Finally, the conserved codon usage bias of degradosome components suggests that other components, including other hydrolytic RNases, may be synthesized in the same region of the cell and be loosely associated with it (66, 67). The situation is different in Firmicutes, where there are generally few orthologs of RNase E, but this does not preclude the existence of a functional degradosome-like complex (see below). Indeed, it is remarkable that PNPase coevolves with enolase and the central enzymes of glycolysis in the Firmicutes, as it does in the Proteobacteria, suggesting that at least part of the degradosome may be conserved throughout the evolution of Bacteria (Table I). A domain analysis of

TABLE I
RNases and RNase-related Proteins in *B. subtilis* and in *E. coli* Identified by their Coevolutionary Pattern

	RNases and putative RNases	Coevolving accessory proteins
Coevolution with 16S rRNA (*B. subtilis*)	RnhB, Rnc, PnpA, Rnr	YpfD, SmpB, Eno, TpiA, Pgk, GapA, GlmM, Zwf
Coevolution with RNase J (*B. subtilis*)	RnmV, MrnC(YazC), YacP, YmdA, YhaM, YkzG, YlbM, YlmH, YloA, RnjB(YmfA), RnhC(YsgB), NrnA(YtqI), YusF, YybT	Cca, CshB(YqfR), YjbKLMN, YsnB GlcU, GlcT, GlcK, Pgi, GntK, CggR, FruR, UgtP
Coevolution with 16S rRNA (*E. coli*)	Rnr, Rnc, Pnp, RnhA	SpoT, RpsA, Hfq GlpK, Zwf, TpiA, Pgk, Eno
Coevolution with RNase E (*E. coli*)	Orn, Rng, Rnd	PcnB, Cca, RelA, Ppx, YhcM PykA, Glk

chloroplast PNPase revealed discrete functions in RNA degradation, polyadenylation, and sequence homology with exosome proteins (75). The evolution of PNPase and related proteins is therefore inseparable from the evolution of its various domains. A parallel situation is encountered with RNase E (see below).

The S1 domain. The S1 domain is typical of an ancestral RNA-binding domain. It gets its name from ribosomal protein S1 (RpsA in the gamma-Proteobacteria), which contains six similar sub-domains (S1 domains) and which interacts with the 30S subunit of the ribosome (89). RpsA is essential for ribosome binding site (RBS) recognition in these bacteria. In contrast, its homolog in Firmicutes, YpfD, while comprising four S1 domains, does not appear to bind to the upstream region of messenger RNAs to permit initiation of translation. The lack of this function explains why RBSs are highly conserved in these organisms, with the consequence that Firmicute genes are usually highly expressed in Proteobacteria (and often toxic for this reason (90)), making them better hosts for genetic engineering than Firmicutes.

Interestingly, the S1 protein of *E. coli* recognizes a pseudoknot in tmRNA (91), and YpfD has a similar function *in vitro* (92). This suggests an interesting role for this presumably primitive domain in the recognition or presentation RNA. The S1 domain is found in several other RNases beside PNPase, notably RNase II (93) and RNase E (94). RpsA is also required for the specificity of ribonuclease RegB from bacteriophage T4. It recognizes the core motif of the RBS, GGAG, acting as a presenting protein to the endonuclease (95, 96). This may account for the function of YpfD in Firmicutes, which could also play the role of presenting RNA to their still elusive RNA degradation apparatuses (66, 92). In summary, the S1 domain appears to be an RNA-presenting domain that has been recruited by several enzymes by acquisitive evolution. Ribosomal protein S1 in delta-Proteobacteria is particularly long, and coded in a different gene environment compared to other organisms. It also evolves along a similar tree to other proteins which appear to be specific to this clade (97). All these features suggest that the S1 domain is associated with sequence- or structure-specific RNA motifs that define specific properties of bacterial clades.

The K homology domain. The K homology (KH) domain is also an ancestral and widespread RNA-binding domain. It has been detected by sequence similarity searches in eukaryotic proteins, such as heterogeneous nuclear ribonucleoprotein K and ribosomal protein S3 (98). This motif is present in the EngA (YphC) GTPase of *B. subtilis* and *T. maritima*, involved in ribosomal 50S subunit assembly. It is present both in Era, a small GTP-binding protein essential for cell growth in *E. coli*, and RbfA, responsible for a step of ribosomal RNA maturation (99). It is also present in the RsmA (CsrA) protein which interacts with small regulatory RNAs (srRNAs, often named with the misleading "noncoding" tag, sncRNAs) in Proteobacteria, presumably providing specificity to the recognition of RNA substrates (100). It has also been proposed that the KH domain of PNPase plays a role in substrate recognition (101) and this may account for the fact that PNPase is a key player in degradation of srRNAs (102). Its pattern of evolution makes its function less clear than that of the S1 domain. This may be related to the observation that the proteins with a KH motif provide a rare example of protein domains that share significant sequence similarity in the motif regions, but possess globally distinct structures (98).

2. ATP-Dependent RNA Helicases

Some studies have shown that PNPase has a beta subunit in addition to its catalytic core. It was initially thought that this subunit was the enzyme enolase, but further work showed that it is in fact the ATP-dependent RNA helicase RhlB (103). More generally, ATP-dependent helicases are conserved components of the degradosome (104). Counterparts are also present in the exosomes of Eukarya, where an energy-dependent process is important for mRNA surveillance (105). The RNA helicases might be important for substrate recognition and recruitment, and for the activation of the catalytic activities of various types of degradosomes or exosomes, sometimes as isolated complexes (106). There are a considerable number of DEA[D/H]-box ATP-dependent RNA helicases involved in RNA folding, maintenance, and degradation in Bacteria. This is the case for the ribosome-associated cold-shock protein CsdA in *E. coli*, and CshA (YdbR) and CshB (YqfR) in *B. subtilis*. Interestingly, another putative DEAH-box ATP-dependent RNA helicase, HrpA, is involved in the processing of a fimbrial mRNA molecule in *E. coli* (107). Homologs of this protein are widely present in bacterial clades (but seemingly absent from the Firmicutes) and it appears to coevolve with RNase E.

This indicates that when investigating the evolution of ribonuclease activities, we should also explore in depth the many genes of unknown function that code for proteins having features related to the DEA(D/H)-box ATP-dependent helicases and which may in fact be involved in specific RNA processing. Because their activity depends on ATP hydrolysis, it is tempting to speculate that their role is similar to that of EF-Tu in translation, using energy to discriminate between substrates and degrade only those that are relevant via a kinetic proofreading mechanism (52, 108). Indeed it appears that the ATPase activity of RhlB is regulated allosterically (109) and that one role of the enzyme is to control RNA misfolding (110). Like other energy-dependent helicases, it could also be involved in the processes of surveillance and degradation of damaged RNA molecules. Targets might include stable RNAs and RNAs modified by a variety of chemical processes related to reactive oxygen species, errors in transcription or other modifications of RNA, often found in Eukarya, but poorly documented in Bacteria. Bacteria growing in cold or very cold conditions should be explored in priority, as it can be expected that RNA folding and stability at low temperature might interfere with smooth regulation (*111*). As this would require a full study in itself, this area falls outside of the scope of this review.

3. EVOLUTION OF THE ASSOCIATION OF ENERGY WITH RNA DEGRADATION: POLYPHOSPHATE SYNTHASE/KINASE AND ENOLASE

Poly(P), a ubiquitous source of energy, generally assumed to play the role of phosphate storage, is systematically associated with RNA degradation and in particular to the degradosome (*62*). Poly(P) has been detected in various amounts in all organisms tested to date (*112*). The pathways of its metabolism are not understood in detail, however, despite the extensive work of the late A. Kornberg and his co-workers, who demonstrated the remarkable features of poly(P)-deficient strains (*113*). Three biosynthetic pathways are known, but they apparently fail to account for the bulk of poly(P) accumulation. The major enzyme of poly(P) metabolism, polyphosphate kinase (PPK), processively and reversibly catalyzes phospho-transfer between ATP and poly(P). McMahon and coworkers, who analyzed the metagenome of sludge plants, determined a general tree of PPK evolution in the corresponding genome samples (*114*). Mycobacteria and Streptomycetes are present, as well as some Firmicutes (but *B. subtilis* and its neighbors, as well as Streptococci are absent). One also finds there some Cyanobacteria as well as alpha-Proteobacteria. Another class of processive polyphosphate kinases, PPK2, prefers GTP and is widely conserved in Bacteria (*115*). Many organisms have both PPK and PPK2 in their genomes (*D. radiodurans* and gamma-Proteobacteria such as *P. aeruginosa*, *Vibrio cholera*, etc.), a few have only PPK2, some only PPK (Enterobacteria, the *Bacillus cereus* complex, *Staphylococcus epidermidis*, and *S. haemolyticus*, but not *S. aureus*, etc.), and many have neither PPK nor PPK2 (*B. subtilis*, *Sulfolobus solfataricus*, *T. maritima*, etc.). The latter, however, have enzymes, namely PpnKA (YjbN) and PpnKB (YtdI), involved in the poly(P)-dependent biosynthesis of NADP (this activity has been biochemically established in *Mycobacterium tuberculosis* (*116*)) and might account for the missing (or a subunit of the missing) polyphosphate kinase. In *B. subtilis ppnKA* is essential and is in synteny with several genes involved in RNA metabolism or related activities. Interestingly, PpnKA coevolves with several RNase-related genes in Firmicutes, including RnmV, MrnC(YazC), YacP, YhaM, RnjB(YmfA), CshB (YqfR), RnhC(YsgB), NrnA(YtqI), YusF, YybT (Table I). Considerable work is needed to explore the link between poly(P) and RNA degradation further. It is clear, however, at this point that there is a strong functional association between this source of energy and some ribonuclease activities.

The second energy-managing component of the degradosome is enolase (53). Besides its structural role in the bacterial cytoskeleton via the RNase E scaffold ((117), see also (57)) the presence of enolase in the degradosome complex makes sense since, by providing phosphoenolpyruvate to nucleoside diphosphokinase (NDK), it permits the regeneration of GTP from the GDP produced by PNPase (64, 118). NDK exists in two forms in *E. coli*, in *Pseudomonas aeruginosa*, and in the Mycobacteria. The long cytoplasmic form of NDK binds to succinyl-CoA synthetase and exhibits a low substrate specificity, whereas the short membrane-bound form is strongly and specifically associated with pyruvate kinase, and synthesizes GTP from GDP and phosphoenolpyruvate (for a review see (119)). Enolase follows a complex phylogenetic pattern: while clearly separating Archaea and Eukarya from Bacteria, the fine distribution in bacterial clades looks somewhat erratic (120), as does that of various other components of the degradosome.

4. POLY(A) POLYMERASES

Two processes require nucleotide addition at the 3' end of RNAs, tRNA nucleotidyl transfer (which adds CCA, as already discussed earlier), and the addition of poly(A) stretches, which modifies a variety of RNAs and is involved in RNA turnover. tRNA nucleotidyl transferase is widely distributed. Interestingly its distribution appears to parallel that of a small RNA binding protein, Hfq, which is also widely present in Bacteria (for that matter, it is present in most Firmicutes, with the exception, perhaps, of the Streptococci). In this respect, an ancestral function of this protein might be interaction with tRNA and tRNA nucleotidyl transferase, more so than its role in increasing the processivity of poly(A) polymerase (121).

Many features of mRNA decay differ between the Bacteria and the Eukarya. Polyadenylation plays a role in the decay of some bacterial mRNAs, as well as in the quality control of stable RNA. However, while poly(A) polymerase is widely conserved, polyadenylation of mRNAs appears to serve a very different function in prokaryotes than its primary role in eukaryotes (122). In *E. coli*, the main polyadenylating enzyme is poly(A) polymerase I (PAP I), but the addition of 3' tails also occurs in the absence of PAP I via the synthetic activity of PNPase when the ADP levels are high enough, indicating a functional interaction with adenylate kinase. In *S. coelicolor*, PNPase is likely to be responsible for 3'-end poly(A) addition. In the same way, in *Synechocystis* and spinach chloroplast, 3'-end poly(A) addition is carried out by PNPase. *B. subtilis* lacks an identifiable PAP I homolog. However, analysis of 3'-tails revealed a similar pattern in wild-type and PNPase-deficient strains, indicating the existence of an alternative poly(A) polymerase activity in this organism (123). We have already stressed that *B. subtilis* and many Firmicutes possess an RNA degradation system that differs significantly from that of most Bacteria, with genes which tend to coevolve. It will therefore be of interest to explore whether one of the members of this set of genes of unknown function displays poly(A) synthesis activity (Table I).

C. Hydrolytic Nucleases

1. Hydrolytic Endonucleases

The chemical activity of water is extremely high in all biological systems. Therefore, the catabolism of macromolecules, nucleic acids in particular, employs hydrolysis reactions ubiquitously. While phosphorolysis salvages energy, hydrolysis is not energy saving, and as such it can only discriminate its substrates by passive selection (52), serving as a general scavenging activity, sometimes with associated regulatory properties. It is therefore not unexpected that the degradosome and the exosome have hydrolytic RNase activities (such as RNase E and probably others in the case of the degradosome) to complement the phosphorolytic activity of PNPase (66). The three major classes of RNases use polynucleotide hydrolysis for RNA degradation: those responsible or endonucleolytic attack, processive degradation from the 3′-end and degradation from the 5′-end. The first process usually takes into account sequence and/or structural properties of the RNA substrate, and, as such, it may be associated with helicases and exonucleases.

Exactly as in the case of PNPase and RNase PH, there exists a widespread core endonuclease framework, descending from a common ancestor, which has been characterized in both RNase E and RNase G. The RNase E/G family preferentially cleaves A/U-rich single-stranded regions of otherwise structured RNAs. Interestingly, as stated in the cautionary word at the beginning of this review, when this family is absent and replaced by members of the RNase J1/J2 family (see below), the AU-rich single stranded cleavage function is conserved; that is, this function is conserved even if the enzyme structure is not. RNase E/G members play an essential role in RNA maturation and decay. Our current understanding of these enzymes originates from studies of *E. coli* RNase G and RNase E, as well as studies in some isolated examples of particular species.

Ribonuclease G. RNase G is a remarkable endonuclease, in that it senses the monophosphate located at the 5′-end of the RNA to be cleaved (124). It is involved in the maturation of pre-16S rRNA and its evolution parallels that of 16S rRNA, suggesting that it is related to an ancestral form that gave rise to present day RNase G and E. RNase G is also involved in the degradation of the mRNA coding for enolase in *E. coli*, creating an interesting RNA degradation feedback loop (125). RNase G was previously named CafA, because its overproduction led to an alteration of the *c*ytoskeleton through formation of an *a*xial *f*ilament, suggesting that the protein is, functionally at least, associated with structuration of the cytoplasm (*126*, *127*). It may therefore interfere with the organization of the degradosome, possibly because of an interference with the RNase E scaffold, via some type of interaction with a component associated to the degradosome (*66*).

The core catalytic center of RNase G is highly similar to that of RNase E, so that both proteins are often classified within a single category, RNase E/G (*3*), as in *A. aeolicus* (*128*), but RNase G cannot fully complement an RNase E defect in *E. coli*, except with highly overproduced variants (*129*). It must be noted, however, that the number of cases where the actual function of the enzyme has been assayed is very limited. In one case where it has been studied, it was shown that the RNase E/G homolog (MycRne) from *M. tuberculosis* is a 5'-end-dependent endoribonuclease with some overlap with RNase E function, such as 5S ribosomal RNA processing (*130*). Since Mycobacteria also have RNase J1 (Rv2752c), an enzyme that processes 16S rRNA in *B. subtilis* (*131*), it will be interesting to see which enzyme catalyzes 16S rRNA maturation in this organism. Sequence analysis of the distribution of RNase G in genomes is somewhat inconsistent, suggesting that there has been either convergent evolution for its function, or that a variety of other proteins underwent acquisitive evolution of the corresponding function (see below). While this enzyme domain is frequently observed in Bacteria, it does not apparently have many counterparts in A+T-rich Firmicutes (but it has in some (*3*)), where ribosomal RNA maturation is, potentially at least, performed by structurally different enzymes (see below).

Ribonuclease E. In *E. coli*, RNase E is considered the major endoribonuclease responsible for the degradation and processing of mRNAs and srRNAs (*132*), as well as stable RNAs, in particular tRNA maturation (*133*). It is composed of three domains, an N-terminal catalytic region comprising an S1 domain (*94*), a central RNA-binding domain, and a C-terminal scaffold region responsible for binding of the associated proteins (*53*). Like RNase G, RNase E cleaves RNA internally, while its catalytic power may be determined by the 5' terminus of the substrate, even if this lies at a distance from the cutting site ((*134*), but see (*124*)). RNase E is present in most Bacteria, with the same core structure, but with a variety of extensions at the N- or C- terminus, sometimes both. It is absent however in many Firmicutes and in Mollicutes. Long extensions of RNase E permit it to become a component of the bacterial cytoskeleton (*117*). The sequence of the C-terminal half of RNase E is not highly conserved evolutionarily, suggesting that there may be some diversity among RNase E interactions with other components in different organisms. Furthermore, there could even be epigenetic variation in the polypeptide sequence due to mRNA sliding in the ribosome at a specific AGCU site (ribosomal hopping) as a function of the environment (*57*). As an example, the *Synechocystis* sp. RNase E homolog does not permit assembly of *E. coli* degradosome components (*135*). In *Streptomyces coelicolor*, RNase ES is a structurally shuffled RNase E homolog showing evolutionary conservation of functional RNase E-like enzymatic activity. This suggests the existence of degradosome-like complexes in at least some Gram-positive bacteria (*76*). As already indicated, there is no conserved counterpart of RNase E in *B. subtilis* and many Firmicutes (see section on the RnjA/B family below).

In *E. coli*, the activity of RNase E is regulated by several effectors, RraA (*136*) and RraB (*137*) in particular. These proteins appear to alter the composition of the degradosome and it is therefore interesting to analyze their corresponding phylogenies. Overall, it appears that the trio of RNase E, RraA, and RraB coevolves, and this could probably be used to infer the function of a protein belonging to the RNase E/G class in the absence of experimental data. There are a few instances where RraA and RraB seem to be both present, while an obvious RNase E/G counterpart is missing: *Arcobacter butzleri*, for example, representing the epsilon-Proteobacteria, and the representatives of the Thermales/Deinococcales, *Deinococcus radiodurans* and *T. thermophilus*. In the Firmicute *Geobacillus kaustophilus*, the situation is similar, while in *Bacillus pumilus* one observes a fairly divergent pair of these proteins. They may represent relics of an RNase E-type degradosome that was present in Firmicutes (it can be observed in *B. cereus* and partially in *Listeria innocua*) and subsequently displaced by a totally divergent degradosome structure (see discussion) that became a specific marker of the Firmicute genome organization. Alternatively, but perhaps less likely, they could represent attempts by the Gram-negative degradosome to invade an older degradosome structure present in the ancestors of the Firmicutes, by horizontal gene transfer with subsequent loss of the functionally duplicated elements. Finally, in *Rubrobacter xylanophilus*, there is a weak conservation of these regulators, with no obvious Rne-related sequence. Because the whole system is present in the other Actinobacteria, this could simply reflect a fast evolution of the structure in Rubrobacteridae. It should also be remembered that genome sequences are not 100% accurate, so that experiments would be needed to make a strong point from the apparent absence of this gene.

Ribonuclease M5. RNase M5 (RnmV, YabF) is a highly specialized endonuclease that appears to be specifically involved in maturation of 5S RNA in most Firmicutes (including Mollicutes) and Fusobacteria. It is absent elsewhere, except in the Spirochete *Borrelia burgdorferi* (*3*). This RNase belongs to the core class of proteins that coevolve in Firmicutes, consisting of RnmV, MrnC (YazC), YacP, YhaM, RnjB(YmfA), CshB(yqfR), RnhC(YsgB), NrnA (YtqI), YusF and YybT, and which seems to be a hallmark of this particular class of organisms (Table I).

Metallo-β-lactamases. The metallo-β-lactamase fold, known for its important role in resistance to penicillin-related antibiotics, is widely spread in bacterial proteomes. The corresponding protein fold is used in a large variety of hydrolytic reactions (*138*). In particular these proteins are often found as nucleases, acting either on DNA or on RNA (*139*). Specificity appears to come from a variety of domains associated to the hydrolytic alphabetabetaalpha-fold and its conserved catalytic residues. Two major families, without a very consistent phylogenetic distribution, play an important role in RNA degradation.

Rnz. RNase Z (ElaC, Trz, Rnz, YqjK in *B. subtilis*) is an endonuclease belonging to the metallo-beta-lactamase family (with a core HXHXDH motif). It is generally involved in tRNA maturation (*140*) and has been shown to process CCA-less tRNA precursors in *B. subtilis* (*141*). It is found in Firmicutes (not in Mollicutes), Cyanobacteria, Spirochetes and is fairly widespread elsewhere, including in some gamma-Proteobacteria. It is identical with RNase BN in *E. coli*, where all tRNA molecules are coded by a CCA-containing genes, making the function of this enzyme fairly enigmatic (*142*). It has been shown to play a significant role in mRNA decay, in conjunction with RNase E (*143*). RNase Z displays less identity, but is conserved over its whole length in Thermotogales (*140, 144*). In general, its distribution is not consistent with phylogenetic trees of proteins with related functions, suggesting widespread horizontal gene transfer (*140*).

RnjA/B. Interestingly, RNase Z shows some sequence similarity with a large class of RNases, RNase J1 (RnjA) and J2 (RnjB), which play the role of RNase E where it is absent, in particular in *B. subtilis* (*145*). RnjA (YkqC) has been found to be involved in 16S rRNA maturation (*131*). Remarkably, both RnjA and RnjB have an additional activity that of a long sought after 5′–3′ exonuclease (*146*). Both activities have been demonstrated in *T. thermophilus* (*147*). Most Firmicutes contain both enzymes. However, as for RNase E and G, their activities appear to be overlapping and it would be extremely interesting to perform a detailed study of the presence of the enzyme in one or several copies in different organisms. Homologs of RnjA/B are present also in Actinobacteria, Cholorflexi, Deinococcus/Thermus, Fibrobacteres/Acidobacteria, Cyanobacteria, and, remarkably, in some alpha- and epsilon-Proteobacteria. It is also present in a few gamma-Proteobacteria, and in several delta-Proteobacteria.

Others. Metallo-beta-lactamase superfamily proteins with related sequences are widespread, but their function is often unidentified: the putative *B. subtilis* metallo-dependent hydrolases YflN, YybB, YycJ, YqgX display some sequence similarity with the Rnz/RnjA/B class of RNases and tend to coevolve with the RnjA/B proteins. It will be interesting to investigate their possible role in specific RNA metabolism. Indeed mRNA turnover might be modulated by different RNases, providing an additional control to gene expression.

Ribonuclease III and ribonuclease MrnC. RNase III (Rnc, RanA, AbsB) is a ubiquitous endonuclease that participates in the maturation of ribosomal RNA from precursors. The protein belongs to the paleome, and it evolves following a tree that parallels that of the 16S rRNA. The structural fold of the protein is present in Eukarya as well, where proteins of this class have a major role in RNA-regulated processes. Its phylogeny in individual species and in the alpha-Proteobacteria has been explored in some detail (148, 149). There are many situations when mRNA is cleaved during the translation process, in particular when ribosomes are stalled by amino acid starvation or other physico-chemical transitions (such as rapid temperature up or downshift). It has been shown that, under such conditions mRNA molecules are often cleaved by ribonucleases (150) and that tmRNA is involved in rapid degradation of the truncated RNAs. A major role of the tmRNA is to extend translation of truncated proteins *in statu nascendi*, with a peptide tag that send them to the ClpXP degradation system (151). Furthermore, the tmRNA system facilitates the release of the truncated mRNA from the stalled ribosome and allows its rapid degradation by RNase R (see below) to prevent production of aberrant polypeptides (150, 152). The proteins responsible for cleavage of the stalled mRNAs have not yet been identified, but they are probably endonucleases associated with the ribosome (153).

The related protein MrnC (YazC), with a typical RNase III fold is involved in 23S RNA maturation in *B. subtilis* (154). It is present in Firmicutes, (not Mollicutes) and present in Cyanobacteria and Thermotogales, but with a fairly divergent sequence. It belongs to the group of coevolving RNases present in Firmicutes and highly specific to this bacterial family (Table I).

Miscellanea. The YjgH protein is a putative L-PSP (mRNA) endoribonuclease (PFAM PF01042), that is similar to ribonuclease UK114 from mouse (155) and is possibly responsible for the inhibition of translation by cleaving mRNA in *E. coli*. These RNases cleave phosphodiester bonds only in single-stranded RNA. YjgH coevolves with YjgI a putative oxido-reductase that may act on nucleotides. Enzymes involved in the degradation of RNA molecules containing modified nucleotides have not yet been explored. Curiously, YjgH has a fold (PDB entry 1PF5) that is similar to that of the YjgF/TdcF family of proteins, and that have been crystallized, but which do not yet have a well-identified function. The putative catalytic center residues are not conserved, however. An inference of functional from the sequence alone may be extremely misleading. The most likely counterparts of the protein can be found in alpha-Proteobacteria.

YmdA is an essential putative phosphohydrolase that is conserved among Firmicutes (156). It contains a HD/KH domain strongly suggestive of RNA binding (157). Its activity should therefore be explored as a priority, especially as it belongs to the group of proteins coevolving with other RNases in Firmicutes. Interestingly, it is also conserved in Planctomycetes, in delta/epsilon-Proteobacteria and in Thermotogales. In the same vein, YusF, which contains the Toprim domain found in RNase M5, belongs to the coevolving family of RNases associated with Firmicute. It is also possible, however, that this gene product is more functionally related to the DNA primases (157).

The B. subtilis YfkH protein is incorrectly annotated in many genomes as the tRNA-processing ribonuclease BN (158), due to a mis-identification of the gene associated with this activity (159). RNase BN was later shown to be encoded by the *elaC/rnz* gene and to be identical with RNase Z (above) (142). It should be noted that YfkH is a typical integral inner membrane protein (IIMP) (160), suggesting that its activity is membrane associated. Counterparts of YfkH appear to be almost ubiquitous (no obvious counterpart in Thermotogales, but this may be due to specific evolution of IIMPs at high temperature), making its function the more interesting to identify, even if it is unlikely to have ribonuclease activity.

Cleavage of RNA/DNA hybrids: Ribonucleases HI HII HIII. RNase H enzymes encode an essential function: they degrade the RNA–DNA hybrids formed during the priming of DNA replication, essentially on the lagging strand. There are three types of this RNase, named RNases HI, HII, and HIII. The molecular evolution of this family of enzymes, often present in multiple copies in genomes, has been used to examine the implications of functional redundancy for gene evolution (161). It appears that the RNase H group evolved in such a way that RNase HI and HIII are mutually exclusive despite appearing to have fairly similar substrates. In contrast, RNase HII, which coevolves with DnaA and PolA, as expected considering its function in degrading RNA in DNA–RNA hybrids, can coexist with either RNase HI or HIII. The latter, however, lacks the sequence corresponding to a basic protruding region of the *E. coli* RNase HI (162). RNase HI (RnhA) is present here and there in various clades or species, but is absent from most Firmicutes and the Thermotogales. In contrast, RNase HIII (RnhC) belongs to the family of RNases that coevolve in A + T-rich Firmicutes in a highly specific manner (Table I).

Finally, the protein family DUF458 (YkuK) is distantly related to RNase H (163). In sporulating Firmicutes, the protein coevolves with several proteins involved in sporulation/germination, and it is present in thermophilic organisms (A. aeolicus, T. tengcongensis, and Thermotogales) as well as in Clostridium acetobutylicum. It is not possible to know at this point whether this protein is involved in DNA or RNA metabolism.

2. HYDROLYTIC EXONUCLEASES

3′–5′ exonucleases.
Ribonuclease II and ribonuclease R. 3′–5′ exoribonucleases that processively hydrolyze single-stranded RNAs, generating 5′ mononucleotides and liberating short oligoribonucleotides (typically less than 5nts) as final degradation products, constitute a distinct functional class. RNase II and RNase R of *E. coli* are close kins, but they are probably involved in different activities. An analysis of the codon usage bias of the RNase II (Rnb) gene in *E. coli* shows that it may be synthesized in the same region of the cell and suggests that is functionally associated with the activity of the degradosome (*66, 67*). Its catalytic core is surrounded by three RNA-binding domains. There is a typical S1 domain at its C-terminus that is critical for RNA binding, suggesting some sort of screening mechanism for specific sequences. It is highly related to RNase R (Rnr, VacB, YjeC in *E. coli*), which has a similar degradative capacity, but may be involved in degradation of misfolded ribosomal RNA (*164*). Both enzymes appear to have overlapping activity and structural properties (the situation is somewhat similar to that of RNase E/G and RNase J1/2) (*165*). Like PNPase, RNase R can degrade structured RNAs (*164*). RNase R is associated to the degradosome in *Pseudomonas syringae* (*166*).

RNase II and R are so similar that, unless explicit experiments are performed to identify their activity, it is not yet possible to know whether a genome possesses either of the activities, or both, using sequence information only. That said, RNase R is generally longer than RNase II by about 200 residues, some 100 of which or more are located at the C-terminus. Their phylogenetic pattern parallels that of 16S rRNA. Both enzymes work best on poly(A) *in vitro* (*165*). However, while RNase R shortens RNA processively to di- and trinucleotides, RNase II becomes more distributive when the length of the substrate reaches ∼10 nucleotides, and it leaves an undigested core of 3–5 nucleotides (*56*). Thus, several types of short oligonucleotides ("nanoRNAs") accumulate in the cell as the consequence of the activity of these enzymes (see below). The activity of RNase II is modulated in *E. coli* by the Gmr (YciR) GGDEF diguanylate cyclase/phosphodiesterase protein (*167*). As this family of proteins is very frequent, it is difficult to know at present whether this type of RNase modulation is widespread. It should be noted, however (see below), that the degradation of cyclic-di-GMP generates the dinucleotide pGpG, which must ultimately be cleaved by a nuclease.

Ribonuclease D and ribonuclease T. RNase D (Rnd) (*168*) and RNase T (Rnt) (*169*) catalyze tRNA-end turnover. The former is present in Actinobacteria, Aquificales, Bacteroidetes/Chlorobi, Cyanobacteria, Planctomycetes, alpha-, gamma-, and some delta-Proteobacteria, and some organisms (alpha-Proteobacteria in particular) appear to have two variants, a long and a short form of the enzyme. In contrast RNase T is unambiguously present only in the Proteobacteria (but not in delta/epsilon-Proteobacteria). Both seem to be totally lacking in Firmicutes and in Thermotogales.

A putative exonuclease of the same family, KapD in *B. subtilis*, is similar to eukaryotic histone mRNA exonucleases. It is involved in the control of the KinA pathway to sporulation (UniProtKB/TrEMBL O24685). This could correspond to an effect on a specific RNA involved in the process. The fate of the many srRNAs that exist in Bacteria is extremely poorly known and it is expected that ribonucleases are involved in their turnover. However, nothing is known about the real substrates of this putative enzyme, and it is more or less randomly distributed in the bacterial phylogenetic tree (Bacilli and some Clostridiales, some Cyanobacteria, Planktomycetes, Alteromonadales, some *Pseudomonas* sp., *Campylobacter jejuni*, and Thermotogales).

5′–3′ exonucleases. An examination of the distribution of the 5′–3′ exonuclease PolA domain among 250 bacterial genomes showed that all Bacteria, but not Archaea, possess this domain (*170*). As far as RNA degradation was concerned, however, the situation was far from clear, until it was found that RNase J1 has 5′–3′ exonuclease activity in addition to its endonuclease activity (*146*). As a matter of fact, the difficulty in identifying this activity in Bacteria up to now may have come from the want of an extra step: the 5′–3′ exoribonuclease activity of RNase J1 only functions on a 5′ monophosphorylated RNAs. Thus, while exonucleolytic RNA degradation can begin from the 5′-end, this process requires an additional step, that of a 5′-triphosphatase, releasing to produce a 5′-monophosphorylated RNA molecule.

Messenger RNA 5′-pyrophosphatase. An investigation of such an activity was only recently undertaken in *E. coli*, with the identification of the RNA pyrophosphohydrolase RppH (YgdP, NudH), tentatively present (NUDIX hydrolases are ubiquitous) in most Proteobacteria, but not in delta-Proteobacteria, and present in *Leptospira interrogans*, but not in Thermotogales (*171*). The phylogenetic tree pattern of this protein is not that of the 16S rRNA. It appears to be absent from all other bacterial families (but other distant NUDIX hydrolases are systematically present). This requires experimental identification of the cognate function, which could be performed after recruitment of completely different structures. In addition to NUDIX hydrolases, many candidates could be found among the enzymes that release phosphate or pyrophosphate and recognize nucleotides, in particular adenylyl and guanylyl cyclases, which have never been assayed for this activity, despite their large number in some species (*172, 173*).

Putative ribonuclease YacP. Using sensitive sequence profile searches and contextual information associated with domain architectures and predicted operons, Anantharaman and Aravind identified a putative 5′–3′ exonuclease NYN domain that shares a common protein fold with two other previously characterized groups of nucleases, namely the PIN (PilT N-terminal) and FLAP/5′–3′ exonuclease superfamilies. The *Bacillus subtilis* protein YacP, which belongs to this family has been proposed to be involved in ribosomal RNA maturation because of its chromosomal context (*174*). It appears to be particular to Firmicutes and Cyanobacteria, being also present in Actinobacteria, *R. xylanophilus* and Chloroflexi, *Chloroflexus aurantiacus*. It belongs to the coevolving family of proteins that contains several Firmicute-specific RNases (Table I).

3. NanoRNases

At this point of our inventory it appears that we have identified all of the steps involved in RNA degradation except for one: as we have seen, processive exonucleases release nanoRNAs in the cell. This is potentially extremely toxic, both for replication and for transcription, especially because the open replication and transcription bubbles can accommodate oligonucleotides of as many as seven residues, with a strong effect of 5-mers (*175*). NanoRNA degradation is therefore an essential function.

In *E. coli* short oligoribonucleotides are degraded by a processive mechanism, after attack at a free 3′ hydroxyl group on single-stranded RNAs, releasing 5′ mononucleotides in a sequential manner (*176*). This is performed by a unique protein Orn (YjeR), that is inhibited by 3′,5′-adenosine-bisphosphate (pAp), a product of sulfate assimilation and 4-phosphopantetheine formation (*177, 178*). This protein is widespread in living organisms (it has a counterpart with the same role in Eukarya), but is absent from many bacterial clades. It is duplicated in some cases, as its gene may be carried by plasmids. This is the case in *Pseudomonas* plasmid pQBR103, where expression is induced by interaction with plants. Its origin is not known, but it is unlikely to come from the Proteobacteria (*179*). This work also revealed that the *orn* gene, while essential in *E. coli*, is not essential in *Pseudomonas putida* KT2440, suggesting that at least one other enzyme can complement the Orn defect.

In Firmicutes (including Mollicutes), an Orn counterpart has been identified as NrnA (YtqI) (*180*). Remarkably NrnA is not only a nanoRNase, but it also hydrolyzes pAp, again associating sulfur metabolism and RNA degradation. NrnA is not essential in *B. subtilis*, indicating that the function can be performed by other proteins as well. A 3′–5′ exonuclease of B. subtilis, YhaM, has been proposed to perform this function (181). However, this enzyme probably prefers deoxyribonucleotides as substrates (157). It may nevertheless contribute to degradation of short (but perhaps not very short) RNAs in this organism, as it hydrolyzes RNA molecules in vitro (180). Finally, RNase J1/J2 5′–3′ exonuclease activity is not size limited: it can go all the way to mononucleotides (146) and thus oligoribonuclease or nanoRNase function may not be strictly essential in organisms containing RNase J. We considerably lack information about the specificity of the various nanoRNases in terms of nucleotide sequence or length. This may be particularly important in some circumstances. Indeed, many Bacteria have genes coding for the so-called GGDEF cyclic di-GMP cyclases (182) and this molecule is hydrolyzed first by a phosphodiesterase, releasing pGpG (183), the fate of which has not been clearly established.

The distribution of Orn and NrnA does not span all organisms, and some (such as Actinobacteria) have both (180). This is interesting as this may reflect the invasion of the genome of the ancestor of Actinobacteria by one of them (possibly Orn, as Proteobacteria are diderm organisms, and more recent than monoderms (184)) creating a situation where the functional redundancy has not yet been eliminated. This suggests that there may be some advantage in having multiple proteins with nanoRNase activity. A role in the hydrolysis of specific oligonucleotides could be a reason for stabilizing redundancy. Remarkably several clades of Bacteria have neither Orn nor NrnA (the alpha-Proteobacteria in particular), which shows that some other protein has been recruited to perform this essential function (Fig. 2). This is also consistent with the observation that while *orn* is not essential in P. putida, one does not find obvious counterparts of either *nrnA* or *yhaM*, indicating that some other gene codes for this function.

Finally, there are many candidate exonucleases/phosphodiesterases in bacterial genomes, in particular those of the DHH/DHHA1 family (185) (Table II). It is however difficult, in the absence of experimental data, to predict whether they have DNA or RNA as substrates. The proteins listed in Table II do not have counterparts in P. putida, showing that, even if some are involved in nanoRNA degradation, there still remains other polypeptides to be discovered that are endowed with this activity.

III. Ribonucleases of the Cenome

RNA is ubiquitous. It is involved in central processes of the cell, which are encoded by the paleome. It is also an essential component of horizontal gene transfer, in particular via expression of bacteriophage functions. Besides its role

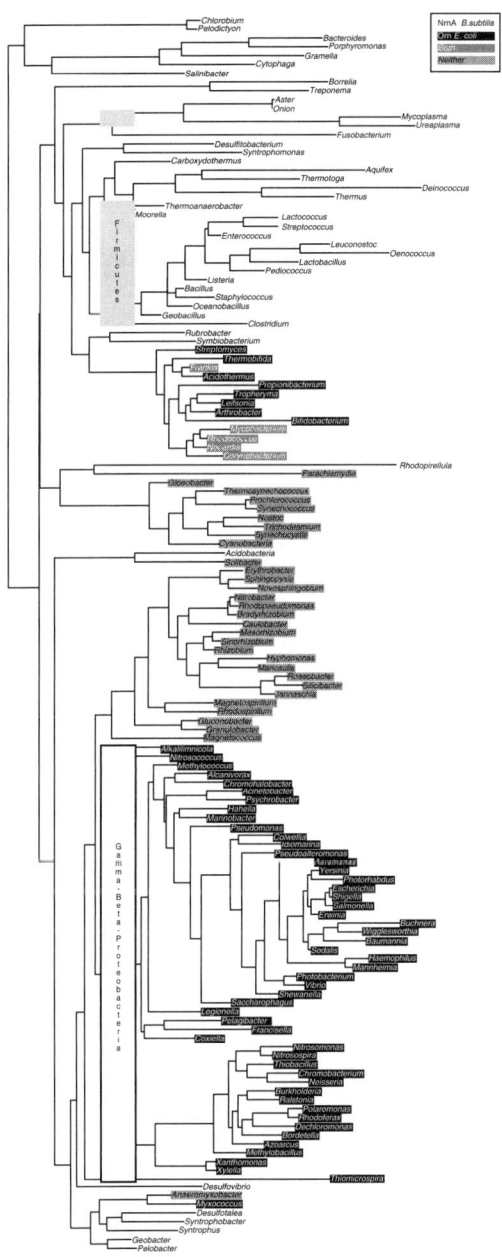

FIG. 2. Distribution of nanoRNases in the bacteria phylogenetic tree (drawn after Supplementary Fig. 1 from Ref. (180)). Firmicutes as well as bacteria growing at high temperature (Aquificales and Thermotogales) have the NrnA counterpart, while beta- and gamma-Proteobacteria have Orn. Actinobacteria appear to have both. Cyanobacteria and alpha-Proteobacteria have none, asking for identification of the protein structure performing the corresponding activity.

TABLE II
Proteins of the DHH/DHHA1 Family in some Organisms

B. subtilis	E. coli	A. baylyi	H. pylori	H. arsenicoxydans	B. quintana	C. glutamicum
AlaS	AlaS	AlaS	AlaS	AlaS	AlaS	AlaS
RecJ	RecJ	RecJ	RecJ	RecJ	RecJ	
PpaC						
NrnA						
YngD,			HP0425,			CGR_1812
YorK,			HP1042,			
YybT			HP1410			

The AlaS and RecJ proteins are highly conserved, while various members of this family are present in many Bacteria. Proteins where the function has not been yet identified are not homologous to each other.

as a template, RNA is also a nutrient supply. These specific roles, typical of what can be found in the occupation of a particular niche, are encoded by the highly variable part of the genome, the cenome. Functions related to plasmid and phage expression would be cytoplasmic when expressed under specific conditions (e.g., the onset of the lytic phase for bacteriophages), while functions related to scavenging are most likely to be exported, as they would most probably be highly deleterious in the cell, unless associated with specific inhibitors.

A. Exported Ribonucleases

RNase I (Rna, RnsA) of *E. coli* belongs to the RNase T2/S-RNase group of nonspecific endoribonucleases (*186*). It is an exported protein, absent from the majority of organisms, but with a counterpart present in individual species here and there, such as in the Cyanobacteria, for example, *Trichodesmium erythreum*. It is present in few species of the beta-Proteobacteria (*Chromobacterium violaceum* and *Thiobacillus denitrificans*). In the gamma-Proteobacteria, it is present in Pasteurellales, in some Enterobacteria: in *Shewanella oneidensis* and *S. baltica*, as well as in *Methylococcus capsulatus*, *Legionella pneumophila*, *Photobacterium profondum*. It is widespread in the alpha-Proteobacteria, where it appears to have evolved fast and this could be an indication that these bacteria are the original niche of the enzyme.

BaRNase (pronounced "barn-ase") is another exported RNase, with guanyl-specificity. It is present in some Firmicutes (*B. pumilus*, *C. acetobutylicum*) and considerably evolved in *Nocardia farcinica*, and in the Alteromonadales, *Marinobacter aquaeolei*. Identified in *Bacillus amyloliquifaciens*, it is not present in its close homolog *B. subtilis*. In Enterobacteria, it is present in

Yersinia pseudotuberculosis and *Y. pestis*, *Serratia proteomaculans* and *Photorhabdus luminescens*, that is, in members of the family that do not code for RNase I. It is also present in the delta-Proteobacteria, in *Geobacter metallidurens*. The ribonuclease activity of BaRNase is controlled by an inhibitor, barstar, which is present together with the protein in some organisms, but remains alone in others (such as *B. subtilis*) suggesting that the couple BaRNase/barstar evolved via sequential loss of the genes, with that coding for the ribonuclease lost before that of the inhibitor (*3*). This is reminiscent of the way restriction/modification systems evolve, with methylases being conserved when restriction has disappeared (*187*).

RNase Bsn (YurI) is another exported nuclease, present in a very limited number of species: some Firmicutes (*B. subtilis* and related species, *B. pumilus* and *Oceanobacillus iheyensis*, possibly in *G. kaustophilus*) and in a variety of Proteobacteria, where it is probably a DNase with limited similarity (*3*). It may also be present in *D. radiodurans*. Here again the distribution of the protein is consistent with repeated horizontal gene transfer. The actual specificity of the protein both in terms of sequence and of the nature of the substrate nucleic acid (RNA or DNA) cannot easily be predicted at this point, in the absence of complementary experimental data.

The YhcR protein is a large exported *B. subtilis* nonspecific endonuclease (it cleaves both DNA and RNA), with a unique domain structure (the OB-fold), typically found in the thermonuclease family. It is likely to be secreted by the twin-arginine secretion system, as it possesses a twin-arginine signal peptide, but it appears to have some activity in the cytoplasm as it can complement a multiple RNase defect (*188*). As a large protein, presumably with the same function, it is conserved in *B. pumilus* and *O. iheyensis*. It is not possible, however, in the absence of experimental data, to make inferences about the function of the many related proteins that have the same 5′-nucleotidase fold.

B. Plasmid and Phage Ribonucleases

RNA is essential for the expression of many horizontally transferred elements, phage and plasmids in particular. In the latter, suicidal toxin–antitoxin systems are often mediated by RNase activities, associated with a specific inhibitor (*189*). The *E. coli* ChpAI(MazE)-ChpAK(MazF) toxin–antitoxin system (*190*), for example, consists of an endoribonuclease (and its specific inhibitor) that cleaves target sequences at the 5′ side of A residues in 5′-NAC-3′ sequences (where N is preferentially U or A) (*191*). There are five such systems in *E. coli* K12, presumably acquired by horizontal gene transfer. This type of system, which can be considered "selfish," is also well poised to regulate apoptosis in bacteria, often in a ppGpp-dependent manner, to permit some cells in a population to survive in an unfavorable environment. The role of chromosomally encoded toxin–antitoxin systems in apoptosis is the subject of some controversy.

We have seen that ribosome stalling results in endonucleolytic cleavage, followed by tmRNA mediated degradation of the truncated mRNA and of the truncated proteins it encodes. Recent work in *E. coli* has shown that none of its toxin/antitoxin RNase systems is involved in this process (153). Several related systems have been also identified under different names, for example, Endo A, PemK (192, 193). The presence of these systems span a large part of the bacterial tree. As there is no straightforward counterpart yet identified in reference Bacteria growing at high temperature, it may be absent from these organisms. They appear also to be lacking in obligate intracellular parasites (194).

Many RNases could be coded by prophages. In *E. coli* RNase LS (RnlA, YfjN) is an endonuclease coded in the CP4-57 prophage region of the genome (195). A mutation in *rnlA* differentially reduced the decay rate of many *E. coli* mRNAs, while a 307-nucleotide fragment corresponding to an internal fragment of 23S rRNA accumulated to a high level (195). This protein appears to be almost unique to *E. coli*: a counterpart exists in *P. profondum*, and perhaps in *Desulfotalea psychrophila*. However, careful exploration of other proteomes suggests that the protein could be related to the putative metal-dependent phosphohydrolases present in many genomes. This could fit well with a phage origin, as these parasites evolve extremely fast. Several putative phage endonucleases such as YokF or YncB could act on DNA or RNA substrates. At this point we could not find enough experimental data to support a significant analysis, and we leave this type of activity for future work.

IV. Provisional Conclusion

Three classes of RNA molecules organize bacterial cell life. Stable RNAs (essentially ribosomal RNA and transfer RNA) form the core of the gene expression machinery. RNAs with an intermediate life time, small regulatory RNAs (srRNAs) and riboswitches are essential regulators whose fundamental importance was only recently uncovered. Finally messenger RNAs, usually with a short life time, are the intermediates between genes and their final protein products. All three cases-interact with specific RNases.

In the first class, we find RNases essential for shaping the RNAs into their final active form. The corresponding functions are therefore ubiquitous. However, several types of protein folds have been recruited to perform them: this is particularly true when one considers the two model bacteria *E. coli* and *B. subtilis*, which do not appear to share many of the enzymes performing the required functions. These functions form the core of the essential genes of the paleome (paleome 1) involved in RNA degradation (Fig. 1). Interestingly, more functions have been explicitly identified in *B. subtilis* than in *E. coli*. This suggests that a coevolution pattern of genes in the latter organism should be explored in more detail, in particular to make inferences about the putative function of genes still devoid of a biochemically identified function.

In the second class, we find RNases involved in maintenance, allowing active RNAs to persist in cells in an active form only. They are expected to be active on misfolded stable RNAs, but also on srRNAs and as controllers of riboswitch states (on/off). This functional category is extremely important during transitions from one type of environment to another (and this is particularly the case during the universal transitions from exponential growth to stationary phase, and from stationary phase to exponential growth). The actual experimental analysis of this function is particularly difficult as it is not obviously amenable to biochemical experiments (*153*). Here physiology, combined with *in silico* analysis of genomes, provides some much-wanted information. While these functions are obviously essential to perpetuate life, there are not strictly indispensable when the cell's environment is stable and provides most of the basic building blocks of the cell. We therefore expect these genes to belong to the second half of the paleome (paleome 2, Fig. 1). Also, as conditions are particularly constant for bacteria that have become obligate parasites, it is expected that many of the corresponding functions will be among those which are generally persistent, but often missing in obligate parasites.

Finally, gene expression is regulated in Bacteria by the combination of mRNA levels and mRNA translation control. The standard pervasive view is that mRNA levels are controlled at the level of transcription initiation; however, this is only meaningful if mRNA degradation is fast. Furthermore, the type of degradation is important, as the way ribosomes pull mRNA from its DNA template and translate it is essential. In this context, RNA degradation plays a major role in gene expression. This can be witnessed by the frequent discrepancy between proteome and transcriptome expression data (*196*).

Different evolution patterns are expected for these three classes of activities. In particular, regulation is probably a most variable process in the course of evolution. One expects therefore that mRNA degradation will follow different paths in different species and the most interesting phylogenetic studies will probably deal with specific bacterial clades, associated with specific niches. By contrast, management of stable RNAs is an essential process that will remain stable, until a major horizontal gene transfer will bring together several types of RNases that work in a concerted way. This will be a rare event and will concern only large bacterial families. In this context, the present study emphasized the particular position of Firmicutes: in these organisms, a completely original RNA degradation system coevolves, which tends to predict the existence of a degradosome-like complex which would comprise elements that differ considerably from the degradosome of the gamma-Proteobacteria. A most interesting feature of this system is that, as in the gamma-Proteobacteria, it would be associated with energy saving and managing processes (see enzymes of glycolysis in Table I), suggesting the existence of a novel function of major importance to the RNA maintenance and degradation process.

Acknowledgments

This work benefited from discussions with many colleagues, in particular in the networks set up by the European Union. I wish to thank in particular Eric Westhof for his comments on RNase P, and Undine Mechold and Ciaran Condon for suggestions. Analysis of the *B. subtilis* RNase genes derived from the resequencing and reannotation of the genome, supported by the latter and the EuroPathoGenomics Network of Excellence, grant LSHB-CT-2005-512061. Finally, the identification of gene sets essential to permit construction of stable synthetic bacteria has been supported by the PROBACTYS NEST European programme grant CT-2006-029104.

References

1. Alpers DH, Tomkins GM. The order of induction and deinduction of the enzymes of the lactose operon in *E. coli*. *Proc Natl Acad Sci USA* 1965;**53**:797–802.
2. Dobzhansky T. Biology, molecular and organismic. *Am Zool* 1964;**4**:443–52.
3. Condon C, Putzer H. The phylogenetic distribution of bacterial ribonucleases. *Nucleic Acids Res* 2002;**30**:5339–46.
4. Thompson LW, Krawiec S. Acquisitive evolution of ribitol dehydrogenase in *Klebsiella pneumoniae*. *J Bacteriol* 1983;**154**:1027–31.
5. Ashida H, Danchin A, Yokota A. Was photosynthetic RuBisCO recruited by acquisitive evolution from RuBisCO-like proteins involved in sulfur metabolism? *Res Microbiol* 2005;**156**:611–8.
6. Allen C, Bekoff M, and, Lauder G. Editors. *Nature's Purposes*. Cambridge, MA: MIT Press; 1998.
7. Bernal JD. *The physical basis of life*. London: Routledge and Kegan Paul; 1951.
8. Granick S. Speculations on the origin and evolution of photosynthesis. *Annals New York Acad Sci* 1957;**69**:292–308.
9. Cairns-Smith A. *Genetic takeover and the mineral origin of life*. Cambridge (UK): Cambridge University Press; 1982.
10. Wachtershauser G. Before enzymes and templates: Theory of surface metabolism. *Microbiol. Rev.* 1988;**52**:452–84.
11. Danchin A. Homeotopic transformation and the origin of translation. *Prog Biophys Mol Biol* 1989;**54**:81–6.
12. Fang G, Rocha E, Danchin A. How essential are nonessential genes? *Mol Biol Evol* 2005;**22**:2147–56.
13. Danchin A, Fang G, Noria S. The extant core bacterial proteome is an archive of the origin of life. *Proteomics* 2007;**7**:875–89.
14. Lawrence JG, Roth JR. Selfish operons: Horizontal transfer may drive the evolution of gene clusters. *Genetics* 1996;**143**:1843–60.

15. Fang G, Rocha EP, Danchin A. Persistence drives gene clustering in bacterial genomes. *BMC Genomics* 2008;**9**:4.
16. Chen X, Li N, Ellington AD. Ribozyme catalysis of metabolism in the RNA world. *Chem Biodivers* 2007;**4**:633–55.
17. Altman S, Wesolowski D, Guerrier-Takada C, Li Y. RNase P cleaves transient structures in some riboswitches. *Proc Natl Acad Sci USA* 2005;**102**:11284–9.
18. Ko JH, Altman S. OLE RNA, an RNA motif that is highly conserved in several extremophilic bacteria, is a substrate for and can be regulated by RNase P RNA. *Proc Natl Acad Sci USA* 2007;**104**:7815–20.
19. Altman S. Ribonuclease P: An enzyme with a catalytic RNA subunit. *Adv Enzymol Relat Areas Mol Biol* 1989;**62**:1–36.
20. Willkomm DK, Feltens R, Hartmann RK. tRNA maturation in *Aquifex aeolicus*. *Biochimie* 2002;**84**:713–22.
21. Willkomm DK, Minnerup J, Huttenhofer A, Hartmann RK. Experimental RNomics in *Aquifex aeolicus*: Identification of small non-coding RNAs and the putative 6S RNA homolog. *Nucleic Acids Res* 2005;**33**:1949–60.
22. Randau L, Schroder I, Soll D. Life without RNase P. *Nature* 2008;**453**:120–3.
23. Marszalkowski M, Willkomm DK, Hartmann RK. 5′-End maturation of tRNA in *Aquifex aeolicus*. *Biol Chem* 2008;**389**:395–403.
24. Frank DN, Pace NR. Ribonuclease P: Unity and diversity in a tRNA processing ribozyme. *Annu Rev Biochem* 1998;**67**:153–80.
25. Guerrier-Takada C, Lumelsky N, Altman S. Specific interactions in RNA enzyme-substrate complexes. *Science* 1989;**246**:1578–84.
26. Lundberg U, Altman S. Processing of the precursor to the catalytic RNA subunit of RNase P from *Escherichia coli*. *RNA* 1995;**1**:327–34.
27. Gossringer M, Kretschmer-Kazemi Far R, Hartmann RK. Analysis of RNase P protein (RnpA) expression in *Bacillus subtilis* utilizing strains with suppressible *rnpA* expression. *J Bacteriol* 2006;**188**:6816–23.
28. Waugh DS, Pace NR. Complementation of an RNase P RNA (*rnpB*) gene deletion in *Escherichia coli* by homologous genes from distantly related eubacteria. *J Bacteriol* 1990;**172**:6316–22.
29. Massire C, Jaeger L, Westhof E. Derivation of the three-dimensional architecture of bacterial ribonuclease P RNAs from comparative sequence analysis. *J Mol Biol* 1998;**279**:773–93.
30. Wegscheid B, Condon C, Hartmann RK. Type A and B RNase P RNAs are interchangeable *in vivo* despite substantial biophysical differences. *EMBO Rep* 2006;**7**:411–7.
31. Loria A, Pan T. Domain structure of the ribozyme from eubacterial ribonuclease P. *RNA* 1996;**2**:551–63.
32. Kazantsev AV, Pace NR. Bacterial RNase P: A new view of an ancient enzyme. *Nat Rev Microbiol* 2006;**4**:729–40.
33. Loria A, Pan T. Modular construction for function of a ribonucleoprotein enzyme: The catalytic domain of *Bacillus subtilis* RNase P complexed with *B. subtilis* RNase P protein. *Nucleic Acids Res* 2001;**29**:1892–7.
34. Haas ES, Brown JW. Evolutionary variation in bacterial RNase P RNAs. *Nucleic Acids Res* 1998;**26**:4093–9.
35. Urbonavicius J, Brochier-Armanet C, Skouloubris S, Myllykallio H, Grosjean H. *In vitro* detection of the enzymatic activity of folate-dependent tRNA (Uracil-54,-C5)-methyltransferase: evolutionary implications. *Methods Enzymol* 2007;**425**:103–19.
36. Pitulle C, Strehse C, Brown JW, Breitschwerdt EB. Investigation of the phylogenetic relationships within the genus *Bartonella* based on comparative sequence analysis of the *rnpB* gene, 16S rDNA and 23S rDNA. *Int J Syst Evol Microbiol* 2002;**52**:2075–80.

37. Honda D, Yokota A, Sugiyama J. Detection of seven major evolutionary lineages in cyanobacteria based on the 16S rRNA gene sequence analysis with new sequences of five marine Synechococcus strains. *J Mol Evol* 1999;**48**:723–39.
38. Herrmann B, Pettersson B, Everett KD, Mikkelsen NE, Kirsebom LA. Characterization of the rnpB gene and RNase P RNA in the order Chlamydiales. *Int J Syst Evol Microbiol* 2000;**50**(Pt 1)), 149–58.
39. Griffiths E, Gupta RS. Signature sequences in diverse proteins provide evidence for the late divergence of the Order Aquificales. *Int Microbiol* 2004;**7**:41–52.
40. Marszalkowski M, Teune JH, Steger G, Hartmann RK, Willkomm DK. Thermostable RNase P RNAs lacking P18 identified in the Aquificales. *RNA* 2006;**12**:1915–21.
41. Fuerst JA. The planctomycetes: Emerging models for microbial ecology, evolution and cell biology. *Microbiology* 1995;**141**(Pt 7), 1493–506.
42. Butler MK, Fuerst JA. Comparative analysis of ribonuclease P RNA of the planctomycetes. *Int J Syst Evol Microbiol* 2004;**54**:1333–44.
43. Cilia V, Lafay B, Christen R. Sequence heterogeneities among 16S ribosomal RNA sequences, and their effect on phylogenetic analyses at the species level. *Mol Biol Evol* 1996;**13**:451–61.
44. Acinas SG, Marcelino LA, Klepac-Ceraj V, Polz MF. Divergence and redundancy of 16S rRNA sequences in genomes with multiple rrn operons. *J Bacteriol* 2004;**186**:2629–35.
45. Zawilak A, Cebrat S, Mackiewicz P, Krol-Hulewicz A, Jakimowicz D, Messer W, et al. Identification of a putative chromosomal replication origin from *Helicobacter pylori* and its interaction with the initiator protein DnaA. *Nucleic Acids Res* 2001;**29**:2251–9.
46. Hall TA, Brown JW. The ribonuclease P family. *Methods Enzymol* 2001;**341**:56–77.
47. Kazantsev AV, Krivenko AA, Harrington DJ, Carter RJ, Holbrook SR, Adams PD, et al. High-resolution structure of RNase P protein from *Thermotoga maritima*. *Proc Natl Acad Sci USA* 2003;**100**:7497–502.
48. Niranjanakumari S, Day-Storms JJ, Ahmed M, Hsieh J, Zahler NH, Venters RA, et al. Probing the architecture of the B. subtilis RNase P holoenzyme active site by cross-linking and affinity cleavage. *RNA* 2007;**13**:521–35.
49. Baba T, Ara T, Hasegawa M, Takai Y, Okumura Y, Baba M, et al. Construction of *Escherichia coli* K-12 in-frame, single-gene knockout mutants: The Keio collection. *Mol Syst Biol* 2006;**2**2006.0008.
50. Kobayashi K, Ehrlich SD, Albertini A, Amati G, ersen KK, Arnaud M, et al. Essential *Bacillus subtilis* genes. *Proc Natl Acad Sci USA* 2003;**100**:4678–83.
51. Kimura M, Ota T. On some principles governing molecular evolution. *Proc Natl Acad Sci USA* 1974;**71**:2848–52.
52. Qian H. Reducing intrinsic biochemical noise in cells and its thermodynamic limit. *J Mol Biol* 2006;**362**:387–92.
53. Carpousis AJ. The RNA degradosome of *Escherichia coli*: An mRNA-degrading machine assembled on RNase E. *Annu Rev Microbiol* 2007;**61**:71–87.
54. Evguenieva-Hackenberg E, Walter P, Hochleitner E, Lottspeich F, Klug G. An exosome-like complex in *Sulfolobus solfataricus*. *EMBO Rep* 2003;**4**:889–93.
55. Rauhut R, Klug G. mRNA degradation in bacteria. *FEMS Microbiol Rev* 1999;**23**:353–70.
56. Frazao C, McVey CE, Amblar M, Barbas A, Vonrhein C, Arraiano CM, et al. Unravelling the dynamics of RNA degradation by ribonuclease II and its RNA-bound complex. *Nature* 2006;**443**:110–4.
57. Henaut A, Lisacek F, Nitschke P, Moszer I, Danchin A. Global analysis of genomic texts: The distribution of AGCT tetranucleotides in the *Escherichia coli* and *Bacillus subtilis* genomes predicts translational frameshifting and ribosomal hopping in several genes. *Electrophoresis* 1998;**19**:515–27.

58. Johnstone SA, Waisman DM, Rattner JB. Enolase is present at the centrosome of HeLa cells. *Exp Cell Res* 1992;**202**:458–63.
59. Xi JH, Bai F, McGaha R, Andley UP. Alpha-crystallin expression affects microtubule assembly and prevents their aggregation. *Faseb J* 2006;**20**:846–57.
60. Keller A, Peltzer J, Carpentier G, Horvath I, Olah J, Duchesnay A, et al. Interactions of enolase isoforms with tubulin and microtubules during myogenesis. *Biochim Biophys Acta* 2007;**1770**:919–26.
61. Kovacs L, Csanadi A, Megyeri K, Kaberdin VR, Miczak A. Mycobacterial RNase E-associated proteins. *Microbiol Immunol* 2005;**49**:1003–7.
62. Blum E, Py B, Carpousis AJ, Higgins CF. Polyphosphate kinase is a component of the *Escherichia coli* RNA degradosome. *Mol Microbiol* 1997;**26**:387–98.
63. Jasiecki J, Wegrzyn G. Localization of *Escherichia coli* poly(A) polymerase I in cellular membrane. *Biochem Biophys Res Commun* 2005;**329**:598–602.
64. Danchin A. Comparison between the *Escherichia coli* and *Bacillus subtilis* genomes suggests that a major function of polynucleotide phosphorylase is to synthesize CDP. *DNA Res* 1997;**4**:9–18.
65. Feng Y, Huang H, Liao J, Cohen SN. *Escherichia coli* poly(A)-binding proteins that interact with components of degradosomes or impede RNA decay mediated by polynucleotide phosphorylase and RNase E. *J Biol Chem* 2001;**276**:31651–6.
66. Nitschke P, Guerdoux-Jamet P, Chiapello H, Faroux G, Henaut C, Henaut A, et al. Indigo: A World-Wide-Web review of genomes and gene functions. *FEMS Microbiol Rev* 1998;**22**:207–27.
67. Bailly-Bechet M, Danchin A, Iqbal M, Marsili M, Vergassola M. Codon usage domains over bacterial chromosomes. *PLoS Comput Biol* 2006;**2**:e37.
68. Lipmann F. Metabolic generation and utilization of phosphate bond energy. In: Nord FF, Werkman CH, editors. *Advances in enzymology and related areas of molecular biology*, vol. 1. New York, NY: Fordham University; 2006. p. 99–162.
69. Danchin A. Natural selection and immortality. *Biogerontology* 2008; Aug 22 [Epub ahead of Print].
70. Del Favero M, Mazzantini E, Briani F, Zangrossi S, Tortora P, Dehò G. Regulation of *Escherichia coli* polynucleotide phosphorylase by ATP. *J Biol Chem* 2008;**283**:27355–9.
71. Ollagnier S, Mulliez E, Gaillard J, Eliasson R, Fontecave M, Reichard P. The anaerobic *Escherichia coli* ribonucleotide reductase. Subunit structure and iron sulfur center. *J Biol Chem* 1996;**271**:9410–6.
72. Cohen SS. A hypothesis on a possible competitive relation between DNA synthesis and protein synthesis. *Cancer Res* 1960;**20**:698–9.
73. Rocha EP, Danchin A. Base composition bias might result from competition for metabolic resources. *Trends Genet* 2002;**18**:291–4.
74. Symmons MF, Jones GH, Luisi BF. A duplicated fold is the structural basis for polynucleotide phosphorylase catalytic activity, processivity, and regulation. *Structure* 2000;**8**:1215–26.
75. Yehudai-Resheff S, Portnoy V, Yogev S, Adir N, Schuster G. Domain analysis of the chloroplast polynucleotide phosphorylase reveals discrete functions in RNA degradation, polyadenylation, and sequence homology with exosome proteins. *Plant Cell* 2003;**15**:2003–19.
76. Lee K, Cohen SN. A *Streptomyces coelicolor* functional orthologue of *Escherichia coli* RNase E shows shuffling of catalytic and PNPase-binding domains. *Mol Microbiol* 2003;**48**:349–60.
77. Bermudez-Cruz RM, Fernandez-Ramirez F, Kameyama-Kawabe L, Montanez C. Conserved domains in polynucleotide phosphorylase among eubacteria. *Biochimie* 2005;**87**:737–45.
78. Boutet E, Lieberherr D, Tognolli M, Schneider M, Bairoch A. UniProtKB/Swiss-Prot: the manually annotated section of the UniProt KnowledgeBase. *Methods Mol Biol* 2007;**406**:89–112.

79. Kelly KO, Deutscher MP. Characterization of *Escherichia coli* RNase PH. *J Biol Chem* 1992;**267**:17153–8.
80. Li Z, Deutscher MP. Maturation pathways for *E. coli* tRNA precursors: A random multienzyme process *in vivo*. *Cell* 1996;**86**:503–12.
81. Ishii R, Nureki O, Yokoyama S. Crystal structure of the tRNA processing enzyme RNase PH from *Aquifex aeolicus*. *J Biol Chem* 2003;**278**:32397–404.
82. Harlow LS, Kadziola A, Jensen KF, Larsen S. Crystal structure of the phosphorolytic exoribonuclease RNase PH from *Bacillus subtilis* and implications for its quaternary structure and tRNA binding. *Protein Sci* 2004;**13**:668–77.
83. Wen T, Oussenko IA, Pellegrini O, Bechhofer DH, Condon C. Ribonuclease PH plays a major role in the exonucleolytic maturation of CCA-containing tRNA precursors in *Bacillus subtilis*. *Nucleic Acids Res* 2005;**33**:3636–43.
84. Li Z, Pandit S, Deutscher MP. 3′ Exoribonucleolytic trimming is a common feature of the maturation of small, stable RNAs in *Escherichia coli*. *Proc Natl Acad Sci USA* 1998;**95**:2856–61.
85. Lizano E, Scheibe M, Rammelt C, Betat H, Morl M. A comparative analysis of CCA-adding enzymes from human and *E. coli*: Differences in CCA addition and tRNA 3′-end repair. *Biochimie* 2008;**90**(5), 762–72.
86. Lin-Chao S, Chiou NT, Schuster G. The PNPase, exosome and RNA helicases as the building components of evolutionarily-conserved RNA degradation machines. *J Biomed Sci* 2007;**14**:523–32.
87. Jones GH, Bibb MJ. Guanosine pentaphosphate synthetase from *Streptomyces antibioticus* is also a polynucleotide phosphorylase. *J Bacteriol* 1996;**178**:4281–8.
88. Kuroda A, Murphy H, Cashel M, Kornberg A. Guanosine tetra- and pentaphosphate promote accumulation of inorganic polyphosphate in *Escherichia coli*. *J Biol Chem* 1997;**272**:21240–3.
89. Sengupta J, Agrawal RK, Frank J. Visualization of protein S1 within the 30S ribosomal subunit and its interaction with messenger RNA. *Proc Natl Acad Sci USA* 2001;**98**:11991–6.
90. Frangeul L, Nelson KE, Buchrieser C, Danchin A, Glaser P, Kunst F. Cloning and assembly strategies in microbial genome projects. *Microbiology* 1999;**145**(Pt 10)), 2625–34.
91. Bordeau V, Felden B. Ribosomal protein S1 induces a conformational change of tmRNA; more than one protein S1 per molecule of tmRNA. *Biochimie* 2002;**84**:723–9.
92. Saguy M, Gillet R, Skorski P, Hermann-Le Denmat S, Felden B. Ribosomal protein S1 influences *trans*-translation *in vitro* and *in vivo*. *Nucleic Acids Res* 2007;**35**:2368–76.
93. Amblar M, Barbas A, Fialho AM, Arraiano CM. Characterization of the functional domains of *Escherichia coli* RNase II. *J Mol Biol* 2006;**360**:921–33.
94. Schubert M, Edge RE, Lario P, Cook MA, Strynadka NC, Mackie GA, et al. Structural characterization of the RNase E S1 domain and identification of its oligonucleotide-binding and dimerization interfaces. *J Mol Biol* 2004;**341**:37–54.
95. Durand S, Richard G, Bisaglia M, Laalami S, Bontems F, Uzan M. Activation of RegB endoribonuclease by S1 ribosomal protein requires an 11 nt conserved sequence. *Nucleic Acids Res* 2006;**34**:6549–60.
96. Odaert B, Saida F, Aliprandi P, Durand S, Crechet JB, Guerois R, et al. Structural and functional studies of RegB, a new member of a family of sequence-specific ribonucleases involved in mRNA inactivation on the ribosome. *J Biol Chem* 2007;**282**:2019–28.
97. Karlin S, Brocchieri L, Mrazek J, Kaiser D. Distinguishing features of delta-proteobacterial genomes. *Proc Natl Acad Sci USA* 2006;**103**:11352–7.
98. Grishin NV. KH domain: One motif, two folds. *Nucleic Acids Res* 2001;**29**:638–43.
99. Inoue K, Chen J, Tan Q, Inouye M. Era and RbfA have overlapping function in ribosome biogenesis in *Escherichia coli*. *J Mol Microbiol Biotechnol* 2006;**11**:41–52.

100. Heeb S, Kuehne SA, Bycroft M, Crivii S, Allen MD, Haas D, et al. Functional analysis of the post-transcriptional regulator RsmA reveals a novel RNA-binding site. *J Mol Biol* 2006;**355**:1026–36.
101. Stickney LM, Hankins JS, Miao X, Mackie GA. Function of the conserved S1 and KH domains in polynucleotide phosphorylase. *J Bacteriol* 2005;**187**:7214–21.
102. Andrade JM, Arraiano CM. PNPase is a key player in the regulation of small RNAs that control the expression of outer membrane proteins. *RNA* 2008;**14**:543–51.
103. Lin PH, Lin-Chao S. RhlB helicase rather than enolase is the beta-subunit of the *Escherichia coli* polynucleotide phosphorylase (PNPase)-exoribonucleolytic complex. *Proc Natl Acad Sci USA* 2005;**102**:16590–5.
104. Khemici V, Toesca I, Poljak L, Vanzo NF, Carpousis AJ. The RNase E of *Escherichia coli* has at least two binding sites for DEAD-box RNA helicases: Functional replacement of RhlB by RhlE. *Mol Microbiol* 2004;**54**:1422–30.
105. Hilleren P, Parker R. Mechanisms of mRNA surveillance in eukaryotes. *Annu Rev Genet* 1999;**33**:229–60.
106. Liou GG, Chang HY, Lin CS, Lin-Chao S. DEAD box RhlB RNA helicase physically associates with exoribonuclease PNPase to degrade double-stranded RNA independent of the degradosome-assembling region of RNase E. *J Biol Chem* 2002;**277**:41157–62.
107. Koo JT, Choe J, Moseley SL. HrpA, a DEAH-box RNA helicase, is involved in mRNA processing of a fimbrial operon in *Escherichia coli*. *Mol Microbiol* 2004;**52**:1813–26.
108. Burgess SM, Guthrie C. Beat the clock: Paradigms for NTPases in the maintenance of biological fidelity. *Trends Biochem Sci* 1993;**18**:381–4.
109. Worrall JA, Howe FS, McKay AR, Robinson CV, Luisi BF. Allosteric activation of the ATPase activity of the *Escherichia coli* RhlB RNA helicase. *J Biol Chem* 2008;**283**(9), 5567–76.
110. Russell R. RNA misfolding and the action of chaperones. *Front Biosci* 2008;**13**:1–20.
111. Medigue C, Krin E, Pascal G, Barbe V, Bernsel A, Bertin PN, et al. Coping with cold: The genome of the versatile marine Antarctica bacterium *Pseudoalteromonas haloplanktis* TAC125. *Genome Res* 2005;**15**:1325–35.
112. Brown MR, Kornberg A. Inorganic polyphosphate in the origin and survival of species. *Proc Natl Acad Sci USA* 2004;**101**:16085–7.
113. Fraley CD, Rashid MH, Lee SS, Gottschalk B, Harrison J, Wood PJ, et al. A polyphosphate kinase 1 (*ppk1*) mutant of *Pseudomonas aeruginosa* exhibits multiple ultrastructural and functional defects. *Proc Natl Acad Sci USA* 2007;**104**:3526–31.
114. McMahon KD, Yilmaz S, He S, Gall DL, Jenkins D, Keasling JD. Polyphosphate kinase genes from full-scale activated sludge plants. *Appl Microbiol Biotechnol* 2007;**77**:167–73.
115. Zhang H, Ishige K, Kornberg A. A polyphosphate kinase (PPK2) widely conserved in bacteria. *Proc Natl Acad Sci USA* 2002;**99**:16678–83.
116. Raffaelli N, Finaurini L, Mazzola F, Pucci L, Sorci L, Amici A, et al. Characterization of *Mycobacterium tuberculosis* NAD kinase: Functional analysis of the full-length enzyme by site-directed mutagenesis. *Biochemistry* 2004;**43**:7610–7.
117. Taghbalout A, Rothfield L. RNaseE and the other constituents of the RNA degradosome are components of the bacterial cytoskeleton. *Proc Natl Acad Sci USA* 2007;**104**:1667–72.
118. Noria S, Danchin A. *Uehara memorial foundation symposium genome science: Towards a new paradigm?*. 2002; *International Congress Series* **1246**:3–13, Tokyo.
119. Chakrabarty AM. Nucleoside diphosphate kinase: Role in bacterial growth, virulence, cell signalling and polysaccharide synthesis. *Mol Microbiol* 1998;**28**:875–82.
120. Oslancova A, Janecek S. Evolutionary relatedness between glycolytic enzymes most frequently occurring in genomes. *Folia Microbiol (Praha)* 2004;**49**:247–58.

121. Scheibe M, Bonin S, Hajnsdorf E, Betat H, Morl M. Hfq stimulates the activity of the CCA-adding enzyme. *BMC Mol Biol* 2007;**8**:92.
122. Kushner SR. mRNA decay in prokaryotes and eukaryotes: Different approaches to a similar problem. *IUBMB Life* 2004;**56**:585–94.
123. Campos-Guillen J, Bralley P, Jones GH, Bechhofer DH, Olmedo-Alvarez G. Addition of poly (A) and heteropolymeric 3' ends in *Bacillus subtilis* wild-type and polynucleotide phosphorylase-deficient strains. *J Bacteriol* 2005;**187**:4698–706.
124. Jourdan SS, McDowall KJ. Sensing of 5' monophosphate by *Escherichia coli* RNase G can significantly enhance association with RNA and stimulate the decay of functional mRNA transcripts *in vivo*. *Mol Microbiol* 2008;**67**:102–15.
125. Kaga N, Umitsuki G, Nagai K, Wachi M. RNase G-dependent degradation of the eno mRNA encoding a glycolysis enzyme enolase in *Escherichia coli*. *Biosci Biotechnol Biochem* 2002;**66**:2216–20.
126. Okada Y, Wachi M, Hirata A, Suzuki K, Nagai K, Matsuhashi M. Cytoplasmic axial filaments in *Escherichia coli* cells: Possible function in the mechanism of chromosome segregation and cell division. *J Bacteriol* 1994;**176**:917–22.
127. Wachi M, Umitsuki G, Shimizu M, Takada A, Nagai K. *Escherichia coli cafA* gene encodes a novel RNase, designated as RNase G, involved in processing of the 5' end of 16S rRNA. *Biochem Biophys Res Commun* 1999;**259**:483–8.
128. Kaberdin VR, Bizebard T. Characterization of *Aquifex aeolicus* RNase E/G. *Biochem Biophys Res Commun* 2005;**327**:382–92.
129. Deana A, Belasco JG. The function of RNase G in *Escherichia coli* is constrained by its amino and carboxyl termini. *Mol Microbiol* 2004;**51**:1205–17.
130. Zeller ME, Csanadi A, Miczak A, Rose T, Bizebard T, Kaberdin VR. Quaternary structure and biochemical properties of mycobacterial RNase E/G. *Biochem J* 2007;**403**:207–15.
131. Britton RA, Wen T, Schaefer L, Pellegrini O, Uicker WC, Mathy N, et al. Maturation of the 5' end of *Bacillus subtilis* 16S rRNA by the essential ribonuclease YkqC/RNase J1. *Mol Microbiol* 2007;**63**:127–38.
132. Suzuki K, Babitzke P, Kushner SR, Romeo T. Identification of a novel regulatory protein (CsrD) that targets the global regulatory RNAs CsrB and CsrC for degradation by RNase E. *Genes Dev* 2006;**20**:2605–17.
133. Li H. Complexes of tRNA and maturation enzymes: Shaping up for translation. *Curr Opin Struct Biol* 2007;**17**:293–301.
134. Callaghan AJ, Marcaida MJ, Stead JA, McDowall KJ, Scott WG, Luisi BF. Structure of *Escherichia coli* RNase E catalytic domain and implications for RNA turnover. *Nature* 2005;**437**:1187–91.
135. Kaberdin VR, Miczak A, Jakobsen JS, Lin-Chao S, McDowall KJ, von Gabain A. The endoribonucleolytic N-terminal half of *Escherichia coli* RNase E is evolutionarily conserved in *Synechocystis* sp. and other bacteria but not the C-terminal half, which is sufficient for degradosome assembly. *Proc Natl Acad Sci USA* 1998;**95**:11637–42.
136. Lee K, Zhan X, Gao J, Qiu J, Feng Y, Meganathan R, et al. RraA. a protein inhibitor of RNase E activity that globally modulates RNA abundance in *E. coli*. *Cell* 2003;**114**:623–34.
137. Gao J, Lee K, Zhao M, Qiu J, Zhan X, Saxena A, et al. Differential modulation of *E. coli* mRNA abundance by inhibitory proteins that alter the composition of the degradosome. *Mol Microbiol* 2006;**61**:394–406.
138. Bebrone C. Metallo-beta-lactamases (classification, activity, genetic organization, structure, zinc coordination) and their superfamily. *Biochem Pharmacol* 2007;**74**:1686–701.
139. Dominski Z. Nucleases of the metallo-beta-lactamase family and their role in DNA and RNA metabolism. *Crit Rev Biochem Mol Biol* 2007;**42**:67–93.

140. Redko Y, Li de Lasierra-Gallay I, Condon C. When all's zed and done: The structure and function of RNase Z in prokaryotes. *Nat Rev Microbiol* 2007;**5**:278–86.
141. Pellegrini O, Nezzar J, Marchfelder A, Putzer H, Condon C. Endonucleolytic processing of CCA-less tRNA precursors by RNase Z in *Bacillus subtilis*. *EMBO J* 2003;**22**:4534–43.
142. Ezraty B, Dahlgren B, Deutscher MP. The RNase Z homologue encoded by *Escherichia coli elaC* gene is RNase BN. *J Biol Chem* 2005;**280**:16542–5.
143. Perwez T, Kushner SR. RNase Z in *Escherichia coli* plays a significant role in mRNA decay. *Mol Microbiol* 2006;**60**:723–37.
144. Vogel A, Schilling O, Spath B, Marchfelder A. The tRNase Z family of proteins: Physiological functions, substrate specificity and structural properties. *Biol Chem* 2005;**386**:1253–64.
145. Even S, Pellegrini O, Zig L, Labas V, Vinh J, Brechemmier-Baey D, et al. Ribonucleases J1 and J2: Two novel endoribonucleases in *B. subtilis* with functional homology to *E. coli* RNase E. *Nucleic Acids Res* 2005;**33**:2141–52.
146. Mathy N, Benard L, Pellegrini O, Daou R, Wen T, Condon C. 5′-to-3′ exoribonuclease activity in bacteria: Role of RNase J1 in rRNA maturation and 5′ stability of mRNA. *Cell* 2007;**129**:681–92.
147. de la Sierra-Gallay IL, Zig L, Jamalli A, Putzer H. Structural insights into the dual activity of RNase J. *Nat Struct Mol Biol* 2008;**15**:206–12.
148. Evguenieva-Hackenberg E, Klug G. RNase III processing of intervening sequences found in helix 9 of 23S rRNA in the alpha subclass of Proteobacteria. *J Bacteriol* 2000;**182**:4719–29.
149. Allsopp MT, Van Heerden H, Steyn HC, Allsopp BA. Phylogenetic relationships among *Ehrlichia ruminantium* isolates. *Ann N Y Acad Sci* 2003;**990**:685–91.
150. Yamamoto Y, Sunohara T, Jojima K, Inada T, Aiba H. SsrA-mediated *trans*-translation plays a role in mRNA quality control by facilitating degradation of truncated mRNAs. *RNA* 2003;**9**:408–18.
151. Dulebohn D, Choy J, Sundermeier T, Okan N, Karzai AW. *Trans*-translation: The tmRNA-mediated surveillance mechanism for ribosome rescue, directed protein degradation, and nonstop mRNA decay. *Biochemistry* 2007;**46**:4681–93.
152. Richards J, Mehta P, Karzai AW. RNase R degrades non-stop mRNAs selectively in an SmpB-tmRNA-dependent manner. *Mol Microbiol* 2006;**62**:1700–12.
153. Li X, Yagi M, Morita T, Aiba H. Cleavage of mRNAs and role of tmRNA system under amino acid starvation in *Escherichia coli*. *Mol Microbiol* 2008;**68**(2), 462–73.
154. Redko Y, Bechhofer DH, Condon C. Mini-III, an unusual member of the RNase III family of enzymes, catalyzes 23S ribosomal RNA maturation in *B. subtilis*. *Mol Microbiol* 2008;**68**:1096–106.
155. Morishita R, Kawagoshi A, Sawasaki T, Madin K, Ogasawara T, Oka T, et al. Ribonuclease activity of rat liver perchloric acid-soluble protein, a potent inhibitor of protein synthesis. *J Biol Chem* 1999;**274**:20688–92.
156. Hunt A, Rawlins JP, Thomaides HB, Errington J. Functional analysis of 11 putative essential genes in Bacillus subtilis. *Microbiology* 2006;**152**:2895–907.
157. Condon C. RNA processing and degradation in *Bacillus subtilis*. *Microbiol Mol Biol Rev* 2003;**67**:157–74.
158. Gattiker A, Michoud K, Rivoire C, Auchincloss AH, Coudert E, Lima T, et al. Automated annotation of microbial proteomes in SWISS-PROT. *Comput Biol Chem* 2003;**27**:49–58.
159. Callahan C, Deutscher MP. Identification and characterization of the Escherichia coli rbn gene encoding the tRNA processing enzyme RNase BN. *J Bacteriol* 1996;**178**:7329–32.
160. Pascal G, Medigue C, Danchin A. Universal biases in protein composition of model prokaryotes. *Proteins* 2005;**60**:27–35.
161. Kochiwa H, Tomita M, Kanai A. Evolution of ribonuclease H genes in prokaryotes to avoid inheritance of redundant genes. *BMC Evol Biol* 2007;**7**:128.

162. Ohtani N, Yanagawa H, Tomita M, Itaya M. Identification of the first archaeal Type 1 RNase H gene from *Halobacterium* sp. NRC-1: Archaeal RNase HI can cleave an RNA-DNA junction. *Biochem J* 2004;**381**:795–802.
163. Knizewski L, Ginalski K. *Bacillus subtilis* YkuK protein is distantly related to RNase H. *FEMS Microbiol Lett* 2005;**251**:341–6.
164. Vincent HA, Deutscher MP. Substrate recognition and catalysis by the exoribonuclease RNase R. *J Biol Chem* 2006;**281**:29769–75.
165. Cheng ZF, Deutscher MP. Purification and characterization of the *Escherichia coli* exoribonuclease RNase R. Comparison with RNase II. *J Biol Chem* 2002;**277**:21624–9.
166. Purusharth RI, Klein F, Sulthana S, Jager S, Jagannadham MV, Evguenieva-Hackenberg E, et al. Exoribonuclease R interacts with endoribonuclease E and an RNA helicase in the psychrotrophic bacterium *Pseudomonas syringae* Lz4W. *J Biol Chem* 2005;**280**:14572–8.
167. Cairrao F, Chora A, Zilhao R, Carpousis AJ, Arraiano CM. RNase II levels change according to the growth conditions: Characterization of *gmr*, a new *Escherichia coli* gene involved in the modulation of RNase II. *Mol Microbiol* 2001;**39**:1550–61.
168. Cudny H, Deutscher MP. Apparent involvement of ribonuclease D in the 3′ processing of tRNA precursors. *Proc Natl Acad Sci USA* 1980;**77**:837–41.
169. Deutscher MP, Marlor CW, Zaniewski R. Ribonuclease T: New exoribonuclease possibly involved in end-turnover of tRNA. *Proc Natl Acad Sci USA* 1984;**81**:4290–3.
170. Fukushima S, Itaya M, Kato H, Ogasawara N, Yoshikawa H. Reassessment of the *in vivo* functions of DNA polymerase I and RNase H in bacterial cell growth. *J Bacteriol* 2007;**189**:8575–83.
171. Deana A, Celesnik H, Belasco JG. The bacterial enzyme RppH triggers messenger RNA degradation by 5′ pyrophosphate removal. *Nature* 2008;**451**:355–8.
172. Danchin A. Phylogeny of adenylyl cyclases. *Adv Second Messenger Phosphoprotein Res* 1993;**27**:109–62.
173. Sismeiro O, Trotot P, Biville F, Vivares C, Danchin A. *Aeromonas hydrophila* adenylyl cyclase 2: A new class of adenylyl cyclases with thermophilic properties and sequence similarities to proteins from hyperthermophilic archaebacteria. *J Bacteriol* 1998;**180**:3339–44.
174. Anantharaman V, Aravind L. The NYN domains: Novel predicted RNAses with a PIN domain-like fold. *RNA Biol* 2006;**3**:18–27.
175. Milne L, Perrin DM, Sigman DS. Oligoribonucleotide-based gene-specific transcription inhibitors that target the open complex. *Methods* 2001;**23**:160–8.
176. Datta AK, Niyogi K. A novel oligoribonuclease of Escherichia coli. II. Mechanism of action. *J Biol Chem* 1975;**250**:7313–9.
177. Sekowska A, Kung HF, Danchin A. Sulfur metabolism in *Escherichia coli* and related bacteria: Facts and fiction. *J Mol Microbiol Biotechnol* 2000;**2**:145–77.
178. Mechold U, Ogryzko V, Ngo S, Danchin A. Oligoribonuclease is a common downstream target of lithium-induced pAp accumulation in *Escherichia coli* and human cells. *Nucleic Acids Res* 2006;**34**:2364–73.
179. Zhang XX, Lilley AK, Bailey MJ, Rainey PB. Functional and phylogenetic analysis of a plant-inducible oligoribonuclease (*orn*) gene from an indigenous *Pseudomonas* plasmid. *Microbiology* 2004;**150**:2889–98.
180. Mechold U, Fang G, Ngo S, Ogryzko V, Danchin A. YtqI from *Bacillus subtilis* has both oligoribonuclease and pAp-phosphatase activity. *Nucleic Acids Res* 2007;**35**:4552–61.
181. Oussenko IA, Sanchez R, Bechhofer DH. *Bacillus subtilis* YhaM, a member of a new family of 3′-to-5′ exonucleases in gram-positive bacteria. *J Bacteriol* 2002;**184**:6250–9.
182. Romling U, Amikam D. Cyclic di-GMP as a second messenger. *Curr Opin Microbiol* 2006;**9**:218–28.

183. Christen M, Christen B, Folcher M, Schauerte A, Jenal U. Identification and characterization of a cyclic di-GMP-specific phosphodiesterase and its allosteric control by GTP. *J Biol Chem* 2005;**280**:30829–37.
184. Gupta RS. The natural evolutionary relationships among prokaryotes. *Crit Rev Microbiol* 2000;**26**:111–31.
185. Yamagata A, Kakuta Y, Masui R, Fukuyama K. The crystal structure of exonuclease RecJ bound to Mn2+ ion suggests how its characteristic motifs are involved in exonuclease activity. *Proc Natl Acad Sci USA* 2002;**99**:5908–12.
186. Padmanabhan S, Zhou K, Chu CY, Lim RW, Lim LW. Overexpression, biophysical characterization, and crystallization of ribonuclease I from *Escherichia coli*, a broad-specificity enzyme in the RNase T2 family. *Arch Biochem Biophys* 2001;**390**:42–50.
187. Brezellec P, Hoebeke M, Hiet MS, Pasek S, Ferat JL. DomainSieve: a protein domain-based screen that led to the identification of dam-associated genes with potential link to DNA maintenance. *Bioinformatics* 2006;**22**:1935–41.
188. Oussenko IA, Sanchez R, Bechhofer DH. *Bacillus subtilis* YhcR, a high-molecular-weight, nonspecific endonuclease with a unique domain structure. *J Bacteriol* 2004;**186**:5376–83.
189. Condon C. Shutdown decay of mRNA. *Mol Microbiol* 2006;**61**:573–83.
190. Masuda Y, Miyakawa K, Nishimura Y, Ohtsubo E. *chpA* and *chpB*, *Escherichia coli* chromosomal homologs of the *pem* locus responsible for stable maintenance of plasmid R100. *J Bacteriol* 1993;**175**:6850–6.
191. Munoz-Gomez AJ, Santos-Sierra S, Berzal-Herranz A, Lemonnier M, Diaz-Orejas R. Insights into the specificity of RNA cleavage by the *Escherichia coli* MazF toxin. *FEBS Lett* 2004;**567**:316–20.
192. Tsuchimoto S, Ohtsubo E. Autoregulation by cooperative binding of the PemI and PemK proteins to the promoter region of the *pem* operon. *Mol Gen Genet* 1993;**237**:81–8.
193. Pellegrini O, Mathy N, Gogos A, Shapiro L, Condon C. The *Bacillus subtilis ydcDE* operon encodes an endoribonuclease of the MazF/PemK family and its inhibitor. *Mol Microbiol* 2005;**56**:1139–48.
194. Pandey DP, Gerdes K. Toxin-antitoxin loci are highly abundant in free-living but lost from host-associated prokaryotes. *Nucleic Acids Res* 2005;**33**:966–76.
195. Otsuka Y, Yonesaki T. A novel endoribonuclease, RNase LS, in *Escherichia coli*. *Genetics* 2005;**169**:13–20.
196. Hecker M, Volker U. Towards a comprehensive understanding of *Bacillus subtilis* cell physiology by physiological proteomics. *Proteomics* 2004;**4**:3727–50.
197. Danchin A. Archives or palimpsests? Bacterial genomes unveil a scenario for the origin of fife *Biol Theory* 2007;**2**:52–61.

RNA Processing and Decay in Bacteriophage T4

Marc Uzan

Institut Jacques Monod, UMR 7592, CNRS-Universités Paris 6 & Paris 7, 2 Place Jussieu, 75251 Paris, Cedex 05, France

I. Introduction 44
II. Part 1: Mechanisms and Regulation of mRNA Degradation During T4
   Development 45
   A. The Host RNase E Plays an Important Role in Bacteriophage T4 mRNA
      Degradation 47
   B. mRNA Degradation and Regulation at Early Stages: T4 Endoribonuclease
      RegB and its Partners 50
   C. mRNA Degradation During the Middle and Late Periods 59
   D. Other Factors 66
III. Part 2: Transfer RNA Processing 66
   A. Physiology and Distribution Among Bacteriophages 66
   B. Genetic Organization and Transcription of the T4 tRNA Genes 69
   C. Processing of the tRNA Precursors 71
   References 78

Bacteriophage T4 is the archetype of virulent phage. It has evolved very efficient strategies to subvert host functions to its benefit and to impose the expression of its genome. T4 utilizes a combination of host and phage-encoded RNases and factors to degrade its mRNAs in a stage-dependent manner. The host endonuclease RNase E is used throughout the phage development. The sequence-specific, T4-encoded RegB endoribonuclease functions in association with the ribosomal protein S1 to functionally inactivate early transcripts and expedite their degradation. T4 polynucleotide kinase plays a role in this process. Later, the viral factor Dmd protects middle and late mRNAs from degradation by the host RNase LS. T4 codes for a set of eight tRNAs and two small, stable RNA of unknown function that may contribute to phage virulence. Their maturation is assured by host enzymes, but one phage factor, Cef, is required for the biogenesis of some of them. The tRNA gene cluster also codes for a homing DNA endonuclease, SegB, responsible for spreading the tRNA genes to other T4-related phage.

I. Introduction

Bacteriophages, and in particular, T2 and T4, were instrumental in the discovery of mRNA. Indeed, it was observed early on that infection of *Escherichia coli* by the virulent phage T2 or T4 elicited the synthesis of phage-specific proteins, whereas the synthesis of host RNAs, whether stable or unstable, was abruptly arrested. This provided a suitable system to address the then primordial question of the way the information was transferred from DNA to proteins. Early studies were able to show that a small fraction of RNA with a high rate of turnover appeared soon after phage infection. Of major interest was the finding that this RNA fraction had a base composition identical to that of the phage DNA instead of the host DNA (*1–3*). Direct DNA/RNA hybridization experiments further showed that this RNA was a strict copy of the T4 DNA template (*4*). Furthermore, the bulk of this RNA was found to be associated with the ribosomes of the infected cells (*5*). The formal demonstration that this phage RNA fraction had the characteristics of mRNA, namely that it was unstable and directed the synthesis of the phage proteins on unspecialized pre-existing host ribosomes, came soon after, and was extended to bacterial mRNAs (*6, 7*).

The 168.9 kb T4 genome contains 300 genes and ORFs. T4's development is largely determined by a precisely coordinated subversion of the host RNA polymerase to permit initiation from different classes of T4-specific promoters. The synthesis of regulatory elements that sequentially modify the transcription apparatus in a cascade of events ensures the temporal ordering of the developmental program. Early, middle, and late promoters can be distinguished on the basis of their characteristic sequences and the *trans*-acting factors controlling their activation. Control of phage development is also exercised at the posttranscriptional level, in particular when inhibition of the genetic activity is required. Extensive information about the biology of bacteriophage T4 can be found in two monographs devoted to this phage (*8, 9*) and in a more recent review on the T4 genome (*10*). A DNA microarray study has assigned a temporal mode of transcription to every gene and ORF of T4 (*11*).

Decades of studies on bacteriophage T4 genome expression have provided a wealth of examples of gene expression regulation that involve mRNA processing and degradation. In addition, T4 codes for tRNAs and other small, stable RNA species, all of which are processed through either enzymatic or self-cleavage reactions; three T4 genes are interrupted by group I introns that are self-spliced. In this review, emphasis will be put on recent results revealing new pathways of mRNA degradation that take place in a stage-dependent manner during T4 development. The pathways of tRNA maturation and some aspects of their biology will also be presented. Earlier reviews on the processing and degradation of T4 mRNA and on some aspects of the posttranscriptional regulation can be found in refs. (*12, 13*). The pathways of T4 tRNA maturation have been reviewed in refs. (*14–16*). The questions that relate to the biology of group I introns in T4 and related phages will not be reviewed here; reviews can be found in refs. (*17, 18*).

II. Part 1: Mechanisms and Regulation of mRNA Degradation During T4 Development

Immediately after the phage has injected its DNA into the bacterial cell, early promoters are recognized by the host RNA polymerase and viral mRNAs are synthesized. This occurs in a soup of cellular RNases, some of which are known to attack and degrade mRNA. The cellular RNases involved in mRNA turnover are the 3′–5′ exoribonucleases, polynucleotide phosphorylase (PNPase), RNase II, RNase R, and oligoribonuclease; and the endoribonucleases, RNase E, G, Z, and III. Of the four endonucleases, RNase E plays a major role in *E. coli* mRNA degradation. It has target sites in almost every transcript and it accelerates mRNA degradation by creating new entry sites for the 3′–5′ exoribonucleases. Like its paralogue RNase G, RNase E is activated by 5′-monophosphate RNA termini so that it can trigger degradation, often in a wave of 5′–3′ polarity. This pathway depends on the recently identified RNA pyrophosphohydrolase, RppH, which removes a pyrophosphate moiety from the 5′-triphosphate termini of primary transcripts (*19, 20*). The C-terminal domain of RNase E is the scaffold for a dedicated RNA degradation complex, called the degradosome, which, in addition to RNase E, contains the exoribonuclease PNPase, the RNA helicase RhlB, and the glycolytic enzyme enolase (see chapter in this volume by Carpousis and colleagues). Polyadenylation of mRNAs by poly(A) polymerase (PAP) also facilitates mRNA degradation from the 3′ end. Extensive information on the mechanisms and regulation of mRNA degradation in *E. coli* can be found in refs. (*21–28*) and in several other chapters of this volume (Hajnsdorf and Regnier; Arraiano and colleagues; Dreyfus).

Among the reactions that take place very soon after T4 infection, and that accompany the initial synthesis (and processing) of the viral transcripts, is the modification of a number of host proteins. Covalent modifications are carried out by T4-encoded ADP-ribosyltransferase isoenzymes, one of which is injected along with the phage DNA, such that it functions immediately after infection. A well-known target of this modification is the α-subunit of the host RNA polymerase. But other as yet unidentified *E. coli* proteins are ADP-ribosylated (*29–31*). Noncovalent modifications, such as the binding of T4 polypeptides to host proteins, include the well-described alteration of host RNA polymerase by several T4 proteins that lead to a modification of its specificity throughout phage development (*10*). RNases and other factors involved in RNA degradation and processing may be the target of these modifications. For example, the host poly(A) polymerase, involved in mRNA turnover, is inhibited by an unknown mechanism after infection (*32, 33*), and the 3′ exoribonuclease RNase D, involved in tRNA maturation, is modified (*34*). Therefore, unless it has been confirmed that a given cellular RNase or modifying enzyme remains unaltered after T4 infection, it should not be taken for granted that all the actors involved in host RNA processing and degradation remain functional and are used in T4 mRNA catabolism. Another point deserves mentioning here. Bacteriophage T4 is a fast-developing virulent phage; a complete cycle takes 30 min at 37 °C, resulting in the production of 200 or more virions per bacterium. As mentioned above, infection with this phage triggers a program of development that includes the activation and the inhibition of the different classes of genes in a cascade. Therefore, it is unlikely that a steady state regime between the synthesis and the degradation of mRNAs is achieved at any moment in the infection cycle.

Infection with bacteriophage T4 expedites host mRNA degradation. The two long-lived *E. coli* mRNAs, *lpp* and *ompA*, are dramatically destabilized after infection with T4, for example. The host endonucleases, RNases E and G, are responsible for this increased rate of degradation. Phage-induced host mRNA destabilization requires the degradosome. Indeed, the *lpp* mRNA is not destabilized after infection of a strain that carries a nonsense mutation in the middle of the *rne* gene (encoding RNase E), leading to a protein unable to assemble the degradosome. A viral factor is also involved, since a phage carrying the Δ*tk2* deletion that removes an 11.3 kb region of the T4 genome, from the *tk* gene to ORF *nrdC*.2, loses the ability to destabilize host transcripts. The gene implicated has not yet been identified however (*35*). There is certainly an advantage for a virulent phage to accelerate host mRNA degradation immediately after infection, as this contributes to making the translation apparatus more rapidly available for viral mRNAs and facilitates the transition from the host to phage gene expression.

The same strategy seems to be widely used by other phages and viruses. Bacteriophage T2, which is closely related to T4, and T7, also induce host mRNA degradation (*35*). Infection by several animal viruses is also accompanied by an acceleration of cellular mRNA degradation and protein synthesis inhibition (*36–39*). Although the mechanisms involved are undoubtedly different, type I Herpes simplex, like bacteriophage T4, codes for a viral function necessary for this process (*40*).

T4 mRNA is slightly more stable on average than *E. coli* mRNA. The first measurements of T4 mRNA half-life were undertaken to understand whether the large amounts of RNA that accumulate at late stages of phage development were a result of stabilization of RNA synthesized earlier or a product of temporally regulated transcription activation of a portion of the phage genome.

To discriminate between the two hypotheses, the half-life of the bulk RNA was measured at different times after infection and was found to be roughly constant (3.5 min at 37 °C) (*41*). Bolle et al. (*42*) reached a similar conclusion, showing that early and late T4 RNAs decay at the same rate. A constant half-life of about 8 min was also found throughout phage infection carried out at 30 °C (*43*). At 43 °C, the half-life of the bulk T4 mRNA synthesized during the late period was found to be 4.5 min (*44*). All these values are in relatively good agreement and show that the T4 transcripts are on average slightly more stable than *E. coli* mRNAs (average half-life 4 min at 37 °C). As described below, the constant rate of bulk mRNA degradation throughout the T4 life cycle is the result of various and elaborate patterns of mRNA degradation that take place in a stage-dependent manner.

A. The Host RNase E Plays an Important Role in Bacteriophage T4 mRNA Degradation

The finding that *E. coli* RNase E has a role in T4 mRNA degradation arose from studies on the expression of the T4 gene 32. Gene 32 encodes a single-stranded DNA binding protein (gp32) essential for replication, recombination, and repair of T4 DNA (*45–47*). It appears after a few minutes of infection, reaches a maximum around the 12–14th min and declines thereafter (*48, 49*). The expression of gene 32 is submitted to multiple transcriptional and posttranscriptional regulatory events. It is transcribed initially both from a proximal middle promoter and by read-through from distant early and middle promoters, resulting in the synthesis of monocistronic and polycistronic transcripts, respectively. Later in the phage cycle, a monocistronic mRNA, initiated at a proximal late promoter, predominates (*50, 51*). All transcripts end at a strong Rho-independent terminator immediately downstream of gene 32. In addition to being temporally regulated at the transcriptional level, the gene 32 protein inhibits its own translation when in excess over its primary ligand, single-stranded DNA (*48, 49, 52, 53*). This regulation is achieved through binding of gp32 to a pseudo-knotted RNA structure located some 40 nucleotides (nt) upstream of the gene 32 translation initiation codon. This binding is thought to nucleate cooperative binding to an unstructured A+U-rich sequence that overlaps the ribosome binding site (*54–56*).

The gene 32 transcripts are more stable than any other T4 mRNAs. A half-life of 15 min was measured at 30 °C and, under derepression conditions (in a T4 gene 32 mutant infection), the half-life can reach 30 min (*49, 52*), indicating that translation of the gene 32 mRNA positively affects its stability. All the gene 32 mRNA species are processed by RNase E, 71 nt upstream of the translation initiation codon of the gene (*50, 57*). This was the first mRNA shown to be processed by RNase E. In addition to the cleavage at −71, two other major cleavages were identified, one far upstream in the polycistronic transcripts (−1340) and the other at the end of the coding sequence of gene 32 (+831) (51, 58). The conservation of all three RNase E processing sites in five different T4-related phages, in spite of significant changes in the organization of the upstream regions, suggests that these cleavages play an important role in the control of expression of gene 32 and/or its upstream genes (58). The new 3' ends created by RNase E processing are potential entry sites for the host 3'–5' exoribonucleases. In fact, the portions of the transcript upstream of the −71 and −1340 cleavage sites were shown to be rapidly degraded (51, 57). Apparently, RNase E cleavage at +831 has no consequences on the functional decay of the gene 32 mRNA, while it affects the chemical decay (12). This may be interpreted as meaning that mRNAs that undergo RNase E processing at this site are only those that have been already translationally inactivated. Under normal condition of translation, the ribosome likely hinders the accessibility of RNase E to this site because of its close proximity to the gene 32 translation termination codon. The RNase E cleavage at +831 is likely to be more important under repression conditions (excess of gp32 over single stranded DNA), to promote rapid elimination of the untranslated gene 32 transcripts. The pattern of gene 32 regulation shows striking similarities to that of the *E. coli rpsO* gene that codes for the ribosomal protein S15. Both genes autogenously regulate their own translation and both carry a pseudoknot in their translational operator (55, 59). Both mRNAs are also cleaved by RNase E at the end of the structural gene. Interestingly, ribosomes were shown to inhibit the distal RNase E cleavage that induces the decay of the *rpsO* mRNA (60).

The results obtained with the gene 32 transcripts prompted Krisch and collaborators to look for other transcripts possibly affected by RNase E. Infection of a temperature-sensitive *rne* mutant with T4 resulted in a 4-fold increase in the chemical stability of total phage RNA isolated at the onset of the late period, at the nonpermissive temperature. The degree of stabilization of several individual T4 mRNAs examined at middle stages of development varied from 28-fold (for the gene 32 mRNAs) to 1.5-fold (44) and the functional stability of total T4 mRNAs was increased 6-fold. Thus, the mutational inactivation of RNase E has a marked effect on T4 development. Indeed, T4 growth is slower, resulting in a 50% reduction in phage yield compared to infection of the wild-type host (44). The analysis of the rate of individual T4 protein synthesis during the course of infection shows that the synthesis of many proteins is decreased and delayed (57). These studies brought a decisive contribution to establishing the notion that RNase E is a major actor in mRNA decay. Krisch and colleagues soon demonstrated that RNase E plays a major role in *E. coli* mRNA degradation as well (61).

RNase E also participates in the degradation of early transcripts. Total T4 RNA synthesized during the first 2 min of infection of the temperature-sensitive *rne* host mutant is stabilized 3-fold at nonpermissive temperatures

(Sanson and Uzan, unpublished results). Therefore, RNase E plays a role in T4 mRNA processing and degradation until the onset of the late period at least. Past this time, its role has not been investigated, although some data suggest that its activity could be modulated later in the T4 cycle (see paragraph 3 of Part I).

The findings by Krisch and collaborators naturally led to the study of RNase E specificity. It is now thought that RNase E recognizes a sequence characterized by a high A+U content rather than a specific ordered sequence (62–66). The bacteriophage T4 genome is 65.5% A+T-rich (10), while that of the host is 50%. It may be surmised therefore that T4 transcripts probably contain a higher density of RNase E targets sites than E. coli transcripts.

One of the most fascinating discoveries in recent years in the field of mRNA degradation is that of machineries dedicated to the degradation of mRNAs and stable RNAs, such as the degradosome in E. coli or the exosome in Eucaryotes (67, 68). The composition of the degradosome is not constant and seems to respond to changes in bacterial growth conditions. Also, some variability in composition has been described throughout the Proteobacteria (21); see chapter by E. Evgueniva-Hackenberg and G. Klug in this volume. Phage infection may alter the composition or activity of the degradosome. Expression of the T7 gene 0.7 protein kinase from a plasmid in uninfected cells results in heavy phosphorylation of the C-terminal part of RNase E and of the RhlB helicase. Under these conditions, some RNase E processed E. coli transcripts synthesized by T7 RNA polymerase are more stable (69), suggesting an explanation as to why T7 transcripts are more stable than E. coli mRNAs (70). Growth of bacteriophage T4 on an E. coli strain carrying the $rne\Delta131$ mutation, unable to assemble the degradosome, is unchanged relative to infection of a wild-type strain (71) (Durand and Uzan, unpublished data). However, the $rne\Delta131$ mutation has no effect on the growth of E. coli either, despite affecting the stability of several individual transcripts (72–75). Therefore, the question of whether the degradosome plays a role in the turnover of some T4 mRNAs or is modified after infection remains open.

RNase E is involved in the mRNA processing and degradation of several T4-related phages (58, 76, 77). Careful studies of the filamentous phage f1 RNA processing and decay also showed that one of the major pathways of degradation involves a cascade of RNase E cleavages in the 5′–3′ orientation (78, 79).

RNase G is a paralogue of RNase E in E. coli. It is involved in 16S rRNA maturation (80). It has a strong homology to the N-terminal domain of RNase E but is unable to assemble a degradosome, as it lacks an equivalent of the C-terminal domain of RNase E. The target sites for RNase G share strong similarities with those of RNase E (A+U-rich sequences in single-stranded RNAs) (81, 82). RNase G has been shown to contribute to the destabilization of a few *E. coli* transcripts (83–85). The plating efficiency of bacteriophage T4 is reduced by 30% on a strain deficient in RNase G (*rng*::Tn5) relative to a wild-type strain (Durand *et al.*, 2008, submitted for publication). As will be shown below, this endoribonuclease indeed participates in the processing and decay of several phage transcripts (86) (Durand *et al.*, 2008, submitted for publication). Nevertheless, it seems clear that it does not have the same general effect on phage mRNA as RNase E.

B. mRNA Degradation and Regulation at Early Stages: T4 Endoribonuclease RegB and its Partners

Soon after infection of *E. coli* by bacteriophage T4, several enzymes involved in pyrimidine metabolism are induced. Four of them are encoded by genes that are linked on the T4 chromosome, among them dihydrofolate reductase. With the hope of identifying new genes involved in the coordinated regulation of these clustered genes and possibly other prereplicative (early and middle) genes, Hall and colleagues (87) isolated T4 mutants able to overproduce dihydrofolate reductase by selecting phages able to grow on medium supplemented with folic acid analogs. Several independently isolated T4 mutant phages were found to contain large deletions in the region of the *tk* and *denV* genes on the T4 chromosome. The deletions led to the overproduction of several enzymes involved in nucleotide metabolism, including dihydrofolate reductase. The effect occurred at the posttranscriptional level. Chace and Hall (87) postulated that the loss of at least one gene, called *regB* (for *reg*ulation), eliminated in all these overlapping deletions, could be responsible for the observed phenotype. It was later found that T4-infected *E. coli* cells contain an endoribonuclease that shows unique properties. Efficient cleavages occur within the Shine–Dalgarno (SD) sequence of some T4 early genes, systematically in the middle of the GGAG sequence. It was further shown that the nucleolytic cleavages, detected very soon after infection, depended on the synthesis of phage proteins, since addition of chloramphenicol, a potent protein synthesis inhibitor to the growth medium prior to infection prevented the processing (88). Shortly afterwards, the gene for this phage-encoded endonuclease was formally identified as the *regB* gene, postulated earlier to down-regulate a number of early phage proteins (89).

The bacteriophage T4 RegB endoribonuclease is a 153 residue protein, synthesized shortly after infection. Its gene is transcribed from a typical early promoter that is turned off 2–3 min after infection. The *regB* gene is also regulated at the posttranscriptional level, suggesting that the production of this nuclease must be tightly regulated. Indeed, RegB efficiently cleaves its own transcript in the SD sequence, indicating that RegB auto-controls its own synthesis. Three other cleavages of weaker efficiency occur in the *regB* coding sequence that probably contributes to *regB* mRNA breakdown (89). Mutants of *regB* are viable on laboratory *E. coli* strains. Their plaques are slightly smaller in minimal medium than those of the wild-type phage. Despite the fact that the RegB nuclease seems dispensable for T4 growth, the *regB* gene is widely distributed among T4-related phages. The *regB* sequence was determined from 35 different T4-related phages. Thirty-two of these showed striking sequence conservation, while three other sequences diverged significantly. The ability of the RegB nucleases from these different phages to cut their own SD sequences seems to have been conserved (76). The RegB primary sequence shows no obvious homology with other proteins from various databases. However, its structure shows strong homology to other *E. coli* ribonucleases of the toxin/antitoxin systems (90) (see below and see also chapter by Yamaguchi and Inouye in this volume).

The most efficient RegB cleavages are located in the intergenic regions of early genes, mostly in translation initiation regions. Indeed, the GGAG motif is one of the most frequent SD sequences encountered in T4 (88, 91, 92). However, some rare but efficient RegB cuts have also been detected within coding sequences. The RNA must be single stranded (86, 88, 89, 91–93).

The consequence of RegB cleavage within translation initiation regions is the functional inactivation of the transcripts. The *motA* gene, which codes for a middle mode transcriptional activator, provides a typical example of an early gene whose mRNA is cleaved efficiently by RegB in its SD sequence The rate of MotA protein synthesis increases from the very beginning of infection to reach a maximum at 4 min and declines abruptly thereafter (94). In a *regB* mutant infection, the rate of MotA protein synthesis continues for a longer time, resulting in twice the accumulation of MotA compared to when RegB is functional (91). The abrupt arrest of MotA protein synthesis after 4 min results both from the sudden inhibition of *motA* transcription and the functional inactivation of its mRNA by RegB. The synthesis of a number of other early proteins follows an identical pattern of synthesis in the absence of RegB (43). Thus, RegB contributes to confine the synthesis of many early proteins to a very short period of phage development. However, it also stimulates the synthesis of a few middle proteins, such as T4 DNA polymerase, encoded by gene 43 (43, 91).

RegB accelerates the degradation of most early, but not middle or late mRNAs. Indeed, early mRNAs are stabilized 3-fold on average in a *regB* mutant infection compared to wild-type infection. After the third minute of infection, mRNAs decay with a constant half-life of about 8 min throughout T4 development at 30 °C, irrespective of the presence or the absence of a functional RegB nuclease (43). Thus, the RegB nuclease down-regulates the translation of many early T4 genes. It is thought that, by providing a mechanism that frees the translation apparatus from abundant early mRNAs, RegB facilitates the transition between early and subsequent phases of T4 gene expression (reviewed in ref. (93)). As mentioned above, T4 early mRNA is destabilized by the host RNase E. In infection of a temperature-sensitive *rne* host with a T4 *regB* mutant, the half-life of the early RNA is 50 min at the nonpermissive temperature (Sanson and Uzan, unpublished result). Therefore, the T4 RegB and host RNase E endonucleases are major actors in early mRNA turnover.

The RegB endonuclease requires a cofactor to act efficiently. When assayed *in vitro*, RegB activity is extremely low but can be stimulated up to 100-fold by the ribosomal protein S1, depending on the RNA substrate (92, 95, 96). The *E. coli* ribosomal protein S1 is an RNA-binding protein that plays an essential role in translation, clamping the ribosomes to mRNAs (97, 98). It contains six homologous RNA binding domains, each of about 70 amino acids, called S1 domains. The two N-terminal domains are involved in ribosome binding while the four carboxy-terminal domains are devoted to mRNA interactions. Similar domains can be found in a large number of RNA-associated proteins from bacteria to humans, in particular among the RNases (e.g., RNase E, PNPase, and RNase II all have S1 domains) and RNA-binding proteins of the yeast and archaeal exosomes (68, 99–101). S1 seems to be a multifunctional protein involved in several unrelated processes that employ its RNA-binding and, possibly, single-stranded DNA-binding properties: it is a subunit of the Qβ phage RNA replicase; it is found associated with poly(A) tails *in vivo*; it interacts directly with RNA polymerase to stimulate its cycling *in vitro* (97, 102–105).

1. THE RegB/S1 TARGET SITE

Not all intergenic GGAG sequences are processed by RegB (88, 106), suggesting that this motif is necessary but not sufficient for cleavage. Intergenic GGAG/U motifs carried by a few early, most middle, and all late transcripts are resistant to RegB during phage infection. Thus, although the RegB nuclease is synthesized shortly after infection, most of the transcripts synthesized after a few minutes of phage development escape its action. The GGAG-containing mRNAs that escape processing by RegB during T4 infection are not substrates of this enzyme, either *in vitro*, or in uninfected cells producing both RegB and the target T4 mRNA *in vivo* (43). The possibility that RegB is inactivated late in the phage cycle has not been ruled out, but this hypothesis is not necessary to account for the RegB resistance of late transcripts.

In an attempt to identify the RNA sequence and/or structure required for RegB cleavage, a SELEX (systematic evolution of ligands by exponential enrichment; see ref. (107)) experiment was carried out, based on the selection of RNA molecules cleaved by RegB in the presence of the ribosomal protein S1 (108). The selected RNAs all contained the GGAG tetranucleotide, with one exception (GGAC). There was a preference for the sequence GGAGG, a bias that is not found among the natural T4 mRNA sequences (see ref. (92)).

Unfortunately, no other conserved sequence or structure was found among the different selected RNA molecules. However, in most cases, the GGAG tetranucleotide was found in the 5′ portion of the randomized region, suggesting that the nucleotides 3′ to the conserved GGAG(G) motif play a role.

A decisive answer to the question of RegB specificity was recently provided by Durand *et al.* (92). Progressive deletion analysis, alignment of well-cleaved *versus* resistant T4 RNA sequences and site-directed mutagenesis showed that the strong intergenic RegB cleavage sites share the following consensus: GGA-GRAYARAA, where R is a purine and Y a pyrimidine. The portion of the cleavage site 3′ to the highly conserved GGAG motif, although degenerate, exhibits strong constraints: an extreme rarity of C-residues in positions 5 and 6; a total absence of C in position 9; a strong bias in favor of A-residues in positions 6, 8, 10, and 11; and a strong bias in favor of a pyrimidine in position 7 (92). This unusually long nuclease recognition motif is reminiscent of cleavage sites for some mammalian endoribonucleases that function with auxiliary factors. One possible model assumes that the auxiliary factor(s) binds the long nucleotide sequence and recruits the endonuclease (109).

The data provided by Durand *et al.* (92) also shed light on the mechanism of S1 protein enhancement of cleavage by RegB. Clearly, RegB alone is not able to discriminate among the various GGAG-containing RNAs. It cleaves the GGA triplet at a very low rate, irrespective of the nucleotide context. The stimulation of RegB cleavage by S1 depends on the nucleotides immediately 3′ to GGA. The closer the sequence is to the consensus shown above, the greater the stimulation by S1. Therefore, the S1 protein plays a prominent role in the site selection process, presumably through specific interactions with the 11 nucleotide conserved sequence, or a part of it (92).

2. RegB Triggers a Pathway of mRNA Degradation that Involves T4 Polynucleotide Kinase and Host RNase G and RNase E

RegB could accelerate mRNA decay by increasing the number of entry sites for 3′–5′ exoribonucleases within long polycistronic early mRNAs. Although plausible, the existence of such a pathway of degradation is not supported by any evidence yet. An alternative pathway was suggested by the finding that some cleavages by RegB, at sites upstream of translation initiation regions, are followed by cuts within an A-rich sequence located a short distance downstream. These so-called secondary cleavages occur within 2 min of the RegB primary cuts at 30 °C. The fact that they do not occur in a *regB* mutant infection was taken as the evidence that RegB triggers a degradation pathway that involves a cascade of endonucleolytic cuts in the 5′–3′ orientation (91). Secondary cuts have been detected in seven different early transcripts. RNase G and RNase E were shown to be responsible for cutting at secondary sites, with RNase G playing a major role (Durand *et al.*, 2008, submitted for publication) (86). This result may seem paradoxical, since RegB produces 5'-hydroxyl RNA termini (88, 110), while both RNases G and E have a marked preference for RNA substrates bearing a monophosphate at their 5' extremities (81, 111, 112). In fact, Durand et al. (2008, submitted for publication) showed that the secondary cleavages are abolished in an infection with a phage that carries a deletion in the *pseT* gene, encoding the 5' polynucleotide kinase/3' phosphatase (PNK). This enzyme catalyzes both the phosphorylation of 5'-hydroxyl polynucleotide termini and the hydrolysis of 3'-phosphomonoesters and 2':3'-cyclic phosphodiesters. In addition, many cleavages detected in one early transcript (*cef* mRNA) over a distance of 200 nt downstream of the initial RegB cut, that are mostly generated by RNase E and a few by RNase G, disappear or are strongly weakened in the PNK mutant infection. The inactivation of PNK increases the stability of the RegB-processed transcripts to various degrees depending on the mRNA. The 2-min delay that separates RegB primary cuts from the secondary cleavages most probably reflects the very progressive rate of accumulation of PNK after T4 infection. The availability of a mutant affected only in the phosphatase activity (*pseT1*) made it possible to show that the phosphatase activity of PNK also contributes to mRNA destabilization from the 3' terminus, although marginally. This presumably occurs through the conversion of 3'-phosphate into 3'-hydroxyl termini, making RNAs better substrates for PNPase, the only host 3' exoribonuclease that requires a 3'-hydroxyl terminus to act efficiently (Durand et al., 2008, submitted for publication). Thus, RegB triggers a pathway of degradation from the 5' RNA end (schematically represented in Fig. 1), which involves the conversion of the RegB-generated 5'-hydroxyl termini into monophosphate termini by the kinase activity of PNK and a conversion of the 3'-phosphate termini to 3'-hydroxyl by its phosphatase activity. This provides access to RNase G and E, leading to 5'–3' degradation, and to PNPase leading to degradation in the opposite orientation, respectively. This shows that the status of the 5' and 3' RNA extremities plays a major role in mRNA degradation (see also (20)).

Until now, the only role attributed to T4 PNK was to counteract a defense mechanism developed by some bacterial hosts. The Snyder and Kaufman's groups showed that T4 infection of these *E. coli* strains activates a host RNase, PrrC that cleaves within the tRNA^{Lys} anticodon loop, leading to protein synthesis arrest and phage cycle abortion. PrrC activation is mediated by the T4-encoded peptide Stp. T4 PNK overcomes this host defense by converting the 5'-hydroxyl and 2' 3'-cyclic phosphate termini left by the PrrC nuclease into 5'-phosphate and 3'-hydroxyl ends that can then be joined by T4 RNA ligase (113–116). For the first time, the data obtained by Durand et al. assign a role for the T4 PNK in the development of the phage, not as an RNA repair enzyme in response to damage caused by the host, but to facilitate endonucleolytic processing of some of its own mRNAs. Therefore, the *regB* and *pseT* genes are

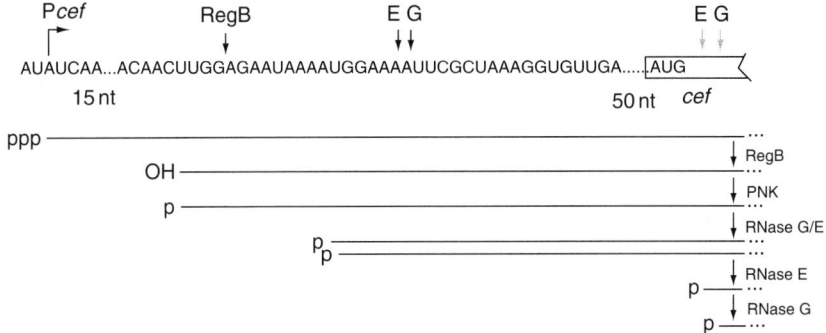

FIG. 1. Initial steps in the cascade of reactions that leads to degradation of the T4 *cef* early mRNA with 5′–3′ polarity. The sequence of the RNA of the *cef* upstream region is shown above the figure. The number of nucleotides not shown are indicated below the sequence. The broken arrow shows the transcriptional start. Only a portion of the *cef* gene is shown (broken rectangle). The vertical arrows above the sequence indicate the sites of cleavage by RegB, RNase E (E), and RNase G (G). A black arrow means that the cleavage is efficient whereas a grey arrow means a weaker cleavage. The horizontal lines below the sequence symbolize the different mRNA species that appear successively after processing of the primary transcript by RegB. In each case, the nature of the 5′ terminus is indicated: ppp, triphosphate; p, monophosphate; OH, hydroxyl. The cleavages by RNases G and E depend on prior conversion of the 5′-OH terminus generated by RegB into a 5′–monophosphate by the T4 PNK.

expected to be found associated in the same phage genomes among the T4-related bacteriophages. PNKs found in phages unrelated to T4 and in some bacteria (*117*, *118*) (but not *E. coli*) were postulated to be involved in RNA repair mechanisms because of the systematic presence of a gene for RNA ligase in the same genome. The results of Durand *et al.* (2008, submitted for publication) should prompt one to consider that, in addition to a role in RNA repair, the prokaryotic PNKs might participate in the regulation of mRNA degradation. The efficiency of degradation will depend on the density of RNase E and RNase G target sites in the RegB-processed transcripts.

The physiological justification of the RegB-mediated degradation pathway described above is not immediately obvious. Indeed, why generate 5′-hydroxyl RNA termini if it is to convert them into monophosphate with another enzyme? Durand *et al.* (2008, submitted for publication) suggested that the main role of RegB is functional inactivation of the mRNA, not degradation. Immediately after infection, phage gene expression is in competition with host gene expression. To win this competition, T4 very quickly produces a whole arsenal of molecular weapons directed against practically every stage of host gene expression, from the degradation of bacterial DNA, to the modification of the translation apparatus and the covalent modification (ADP-ribosylation) of many host proteins. We proposed that this arsenal includes the presence of translation enhancers on some T4 early mRNAs and these enhancers are the A-rich sequences that flank the strong GGAG RegB cleavage sites, recognized by the S1 ribosomal protein. At very early times of infection, S1 would strongly activate the translation of these phage transcripts, leading to the synthesis of enough viral proteins to assure further development of the phage. The synthesis of RegB would then trigger either a complete mRNA inactivation, by cleavage within the SD sequence, or a decreased rate of translation by endonucleolytic destruction of the translation enhancers. In fact, the RNase G and E cleavages within the A-rich sequences are the strongest after the upstream RegB cleavages. Maintaining these translational enhancers would preclude efficient middle and late mRNAs translation.

3. RegB Catalysis and Structure

Although the RegB primary sequence is unrelated to the RNases of the Barnase family, it shares their characteristics. (1) RegB produces 5'-hydroxyl and 3'-phosphate termini (88). Using ^{31}P nuclear magnetic resonance, it was shown that the first product after RegB cleavage is in fact a cyclic 2',3'-phosphodiester (110). (2) A search for the residues critical to RegB activity, via systematic histidine-to-alanine substitutions, reveals that only the H48A and the H68A substitutions significantly reduce RegB activity, without changing its ability to bind the substrate or affecting its overall structure. Therefore, RegB is a cyclizing ribonuclease, with two histidines (in positions 48 and 68) as potent catalytic residues (110).

NMR was used to solve the structure of RegB and to map its interactions with two RNA substrates. Despite the absence of any sequence homology and a different organization of the active site residues, RegB shares structural similarities with two *E. coli* ribonucleases of toxin/antitoxin family: YoeB and RelE (90) (see Fig. 2). YoeB and RelE are involved in mRNA inactivation under nutritional stress conditions. They are active on mRNAs being translated (119, 120). Interestingly, RelE, and in some cases, YoeB, recognize triplets on mRNAs, which they cleave between the second and third nucleotide, as does RegB (see above). It has been proposed that RegB, RelE, and YoeB form a new structural and functional family of ribonucleases specialized in mRNA inactivation within the ribosome (90).

The finding that T4 codes for an RNase that shares homology with other toxin RNases raises the question of whether RegB has an antitoxin partner that would be encoded by the T4 genome. The answer is unknown at the moment. However, it could be argued that RegB does not require an antitoxin to block its activity after the early stages, since there are no targets for RegB in the transcripts synthesized after this time (43, 92). The early promoters are shut down after 3 min of infection at 30 °C, so that no early transcript is made *de novo* after this time. In addition, a cohort of RNases and other factors of

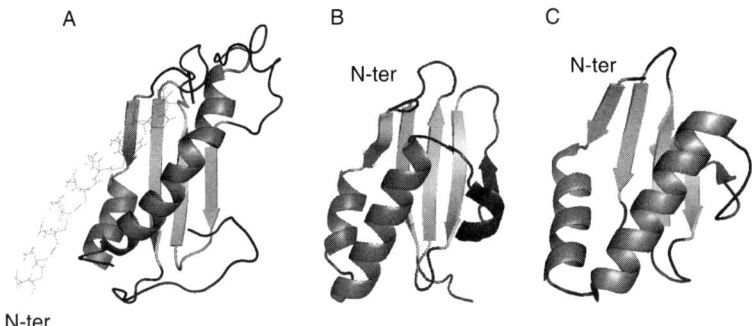

FIG. 2. Comparison of RegB, RelE, and YoeB structures. The structures of RegB (A) (*90*), RelE (B) (*244*), and YoeB (C) (*245*) are shown. The first α-helix of RegB is absent in the two other endonucleases. For this reason, it has been drawn in pale orange to better emphasize the similarities of the rest of the structure with the two other nucleases. The two conserved α-helices are in red and orange, and the conserved four-stranded β-sheet is in cyan. The N-terminus is indicated in each case.

both phage and cellular origin contribute to the swift elimination of these transcripts from the infected cells (see above). We suggest that because RegB acts within a narrow window of time during the early stages of development, the phage does not necessarily code for an anti-RegB.

4. MECHANISM OF ACTION OF S1 IN THE REGB CLEAVAGE REACTION

The mechanism of S1 activation of RegB is not understood at the moment. Lebars *et al.* (*96*) described a small structured 30-mer RNA that was efficiently and specifically cleaved by RegB in the middle of the CCAC sequence in the absence of S1. Stimulation by S1 was only about 20%. In this RNA, the G in the 3' position of the GGAG motif is base-paired while the rest of the sequence is unpaired (in a loop). Any modification that perturbs this structural arrangement leads to a decrease in cleavage efficiency. Thus, RegB is able to perform efficient and specific cleavages alone, provided the GGA sequence is unpaired and the fourth nucleotide of this motif is partly constrained. This suggests that, when S1 is required for efficient cleavage of natural T4 transcripts, the role of this protein is to promote a similar constraint on the RNA, presumably *via* its interaction with the A-rich sequence 3' to the cleavage site (*92*).

The affinity of S1 for the A-rich sequence that characterizes strong RegB sites is not better than for other RNA sequences (Durand and Uzan, unpublished data). Therefore, the function of the A-rich sequence is not simply to recruit S1 locally. Rather, it could be to establish some crucial interactions with particular residues of S1 that maintain the G–A covalent bond in a conformation that RegB will cut easily. The RegB and S1 proteins do not seem to interact directly. Indeed, the S1 protein is not retained on a column of RegB-bound Ni-agarose, either in the presence or the absence of a RegB substrate RNA (Hu and Uzan, unpublished data). In addition, RegB has a very low affinity for its RNA substrates (*108, 110*; Durand and Uzan, unpublished observations). Therefore, it is likely that the first step in the S1 activation pathway involves S1 interaction with the RNA. As RegB is easily inhibited by RNA secondary structures, one possibility was that S1 stimulates RegB through its RNA unwinding ability (*121, 122*). However, Lebars et al. (*96*) presented evidence that invalidated this hypothesis.

The entire S1 protein is not necessary to achieve efficient stimulation of the RegB reaction. The portion of the protein containing the four C-terminal S1 modules (*3–4–5–6*) perfectly mimics the effect of the entire protein. Depending on the substrate, domain 6 can be removed without affecting the extent of stimulation. The smallest domain combination able to stimulate the cleavage reaction significantly is the bimodule 4–5 (*92, 123*). This suggests that the S1 domains 4 and 5 cooperate to form an RNA binding surface able to interact with the nucleotides of RegB target sites. Module 3 could help stabilize the interaction with the RNA. The determination of the global organization of the C-terminal domains of S1, by small angle X-ray scattering (SAXS), supports this view. Indeed, this technique showed that the two adjacent domains, 4 and 5, are tightly associated, forming a rigid rod, while domain 3 has no or only a weak interaction with the others. Interestingly, domain 6 was also shown to be dispensable for other functions of S1, such as translation (*124*), transcription cycling (*105*), autoregulation of the S1 gene (*125*), and replication of the Qβ phage RNA (*126*).

The residues of the S1 C-terminal domain that participate in the interaction with three RNAs that differed in sequence and structure (two RegB substrates and one RNA carrying a translation initiation region) were identified by NMR spectroscopy. The results indicate that these three RNAs bind the same area of the protein through a set of nonspecific and specific (RNA-dependent) interactions. This result strongly supports the hypothesis that, despite the variety of S1 targets, they all interact with the same sites on the S1 protein (*127*).

In vivo, the S1 protein is found associated with the ribosome, as a genuine ribosomal protein, but it is also found bound to RNA polymerase (*104*). A recent estimate of the proportion of bound and free S1 protein in *E. coli* indicates that the concentration of S1 would be in large excess over ribosomes, such that about 50% of the protein would be in a free (unbound) form (*105*). Under which form the S1 protein participates in the RegB cleavage reaction is not yet known. However, the structural and mechanistic analogy of RegB to the two *E. coli* toxins, YoeB and RelE (*90*), which depend on translating ribosomes for activity (*119*), militates in favor of the participation of the ribosome in RegB processing *in vivo*.

In conclusion, although interesting information has been obtained on the S1/RNA interaction and on the global structure of the S1 C-terminal domain, the mechanism of activation of RegB by S1 is still mysterious. Whether S1 participates in the RegB reaction as a free protein or associated with the ribosome or other partners *in vivo* remains to be determined. These two questions are clearly major challenges for the field in the near future.

5. EFFECT OF RegB ON HOST MRNAS

E. coli translation is dramatically inhibited following T4 infection. The mechanism involved is still unknown (reviewed in ref. (*128*)). The early synthesis of RegB following T4 infection, and its ability to inactivate mRNAs functionally and to stimulate their degradation, suggested that it might take part in host translational shut off. Upon induction of RegB expression from a low copy number plasmid, the colony forming capacity of the cells drops 10^5-fold. This extreme toxicity suggested that at least some bacterial mRNAs, coding for essential functions, might be targets for RegB. The induction of *regB* gene expression causes the *E. coli* K12 cells to filament, suggesting that at least one protein involved in septation is underproduced. Consistent with this idea, we have found that the *fts*Z transcript, coding for a tubulin-like protein involved in the early steps of septation (*129*), is efficiently cleaved by RegB in its SD sequence (Hu and Uzan, unpublished data). No other individual bacterial mRNAs were examined and we found that the half-life of the bulk host mRNA before and after RegB induction from a plasmid is unchanged. This may mean that RegB targets are rare in *E. coli* cells, in agreement with the narrow specificity of this RNase (see above).

The very high toxicity caused by RegB in *E. coli* cells has been exploited to derive a strategy of cloning of DNA fragments within the *regB* gene. This provides a positive selection since insertion of the new DNA fragment disrupts the nuclease gene and abolishes the toxic effect upon induction conditions (*130*).

C. mRNA Degradation During the Middle and Late Periods

After 5 min of infection at 30 °C, the production of a number of early proteins declines strongly. This is because the early promoters are shut-down abruptly after the third minute of infection, by a mechanism that is not yet understood (*131, 132*). In addition, as described above, the host and phage endoribonucleases, RNase E, G, and RegB, and the T4 PNK combine to inactivate and degrade early mRNAs. In the meantime, the middle promoters are activated, a process that requires the two T4-encoded transcriptional activators, MotA and AsiA (*133–135*). The viral genes that are expressed at this stage of the phage cycle are transcribed from a proximal middle promoter and/or by read through from distant upstream promoters (early and middle), provided no strong transcription terminator lies on the way.

At the end of the middle period, middle transcription is inhibited and late transcription becomes predominant (136). This is the consequence of the changed specificity of the modified RNA polymerase. Indeed, transcription initiation at the late promoters requires the T4-encoded late σ-factor, gp55, which replaces the major host σ^{70}, and the T4-encoded gp33, which ensures the coupling of late transcription with ongoing viral DNA replication (137–139). In addition to this transcriptional regulation, the translation of a number of transcripts is inhibited by the RegA translational repressor. This small protein was shown to compete with the ribosome for binding to the translation initiation regions of mRNAs (140, 141). As a consequence of these negative regulatory circuits, the levels of most middle mRNAs decrease markedly when late transcription becomes very active, some 12–15 min postinfection at 30 °C (11, 50, 136, 142, 143).

1. THE T4 DMD GENE CONTROLS THE STABILITY OF MIDDLE AND LATE MRNAS IN A STAGE-DEPENDENT MANNER

An amber mutation in ORF 61.5, located between the genes 41 (DNA helicase) and 61 (DNA primase) in the T4 genome, leads to strong inhibition of phage development (144). This ORF was subsequently renamed *dmd* for *d*iscrimination of *m*essages for *d*egradation. Pulse-labeling of T4 proteins at time intervals after infection with the *dmd* mutant showed that protein synthesis is normal until the beginning of the middle period and collapses thereafter. Northern blot analysis revealed a dramatic decrease in the accumulation of three late transcripts (those of genes 23, 51, and 37). Furthermore, a number of endonucleolytic cleavages could be detected by primer extension in middle (*uvsY*) and late (gene 23 and *soc*) transcripts, which were not present in wild-type phage infection (144–146). The deleterious effect of the lack of a functional Dmd on middle and late mRNAs seems to be facilitated by the RNA chaperone Hfq (147), which has been shown to play a role in the degradation of several E. coli mRNAs (148–150)

Infection of E. coli cells with a *dmd* mutant phage that carries an additional mutation in *motA* leading to partial reversion of the *dmd* phenotype (see below), results in a shortening of the chemical and functional half-lives of several middle and late transcripts (151). These data strongly suggested that the arrest of protein synthesis in the *dmd* mutant is due to mRNA destabilization and that the function of the Dmd protein is to inhibit an endoribonuclease that targets middle and late transcripts.

2. THE HOST RNLA GENE CONTROLS THE ACTIVITY OF A NEW RNASE, RNASE LS

With the aim of identifying the activity responsible for the destabilization of late transcripts in the absence of Dmd, extragenic suppressors (*ssf1–7*) able to restore the growth of *dmd* mutant phages were isolated. One of them (*ssf5*) is in the gene for the middle transcription activator, *motA* (*151, 152*). In the *motA dmd* double mutant infection, late protein synthesis is partially restored and phage production goes from 0.1 to 3 phages per bacterium, which is still at least one order of magnitude below the level obtained with wild-type T4. Ueno and Yonesaki (*151*) showed evidence supporting the notion that the partial reversion of the *dmd* phenotype by the *motA ssf5* mutation results from a delay in the activation of the degradation of late transcripts. The other T4 suppressors also had a very moderate effect on the *dmd* phenotype, which suggested that the endoribonuclease responsible for middle and late mRNA destabilization in the *dmd* mutant is encoded by the host. In support of this view, they found that the *soc* mRNA produced from a plasmid in uninfected bacteria is cleaved at the same sites as those observed after infection by a *dmd* mutant phage (*71, 152*).

The mutational inactivation of the host endoribonucleases G, III, P or I, or the *rne*Δ131 mutation, which produces a truncated form of RNase E (lacking the C-terminal domain) unable to assemble the degradosome (*72, 153*), could not restore the growth of the *dmd* mutant phage. Although the degradation of some *E. coli* mRNAs requires the degradosome, many others can be processed by the truncated form of RNase E synthesized in the *rne*Δ131 mutant. Since none of the known host endonucleases was responsible, it was assumed that the Dmd protein would target a new endoribonuclease that was called RNase LS, for *l*ate gene *s*ilencing in T4. Several *E. coli* mutants able to support the growth of a *dmd* mutant phage were isolated, among which two were able to very efficiently reverse the *dmd* phenotype. Both mutations were mapped within the ORF *yfjN*, which was renamed *rnlA* (*71, 145*). In the infection of *E. coli* carrying the *rnlA2* allele with T4 *dmd*, the half life of the late *soc* transcript was indistinguishable from that measured in wild-type phage infection. In addition, the *dmd* mutant-specific cleavages in the *soc* mRNA were suppressed (*71*). Similar effects of *rnlA* were described for the *uvsY* middle transcript (*146*). These results showed that *rnlA* either coded for the sought-after endoribonuclease or controlled its activity (as we will see below, the product of the *rnlA* gene indeed has ribonucleolytic activity). They also comfort the notion that this RNase is inhibited by the T4-encoded Dmd protein.

3. A ROLE OF DMD IN MIDDLE mRNA DECAY AT LATE STAGES

During T4 infection, middle transcripts are destabilized when late transcription becomes very active. In a *dmd* mutant infection, however, a small fraction of middle transcripts persist at late stages and are degraded at a decreased rate compared to the wild-type infection (146, 151). Thus, it seems that the Dmd protein plays a positive role in middle mRNA decay at late stages. Dmd therefore has apparently two opposite roles: it protects middle and late transcripts from degradation when these RNA species are synthesized at the right time, while it expedites the degradation of the middle transcripts when these are no longer required (during the late period). However, this latter effect seems marginal and could be indirect (see below).

Kanesaki et al. (146) proposed that Dmd inhibits RNase LS, but activates RNase E, so that in the *dmd* mutant infection, the efficiency of RNase E would decrease at late stages. The slight weakening of the RNase E processing 71 nt upstream of the gene 32 ORF (see above) in the *dmd* mutant infection was taken as an evidence supporting this view (146). In a *dmd*-proficient infection, the late transcripts are thought to be protected from the attack by the Dmd-activated RNase E because they are actively translated. However, it is not clear why the middle transcripts would require the Dmd protein at late stages, when most of them have ceased to be translated. In addition, the Dmd protein, a polypeptide of 60 residues, does not interact with RNase E *in vitro* (146).

4. BIOCHEMICAL CHARACTERIZATION OF THE RNASE LS

Purified his-tagged RnlA protein is able to cleave the *soc* transcript *in vitro* at only one site among the three usually observed *in vivo* after infection with the *dmd* mutant phage. This cleavage is inhibited by the purified Dmd protein (154). This is an important result as it shows that RnlA is itself an RNase that responds directly to Dmd. However, whether RnlA has targets in other T4 mRNAs is unknown.

An S30 extract from *rnlA*-proficient bacteria can catalyse all of the cleavages observed in the *soc* RNA *in vivo*, whereas none of these cleavages could be observed with an S30 extract derived from a *rnlA* mutant (145). Further fractionation by ultra-centrifugation showed that the RNase activity is associated with the fast sedimenting material (100,000g pellet) and that it is inhibited by Dmd. When this preparation was treated with high-salt solutions and recentrifuged, the RNase activity was lost in both the light and the heavy fractions. However, it could be recovered after combination of these fractions, consistent with the idea that multiple factors must associate to produce fully active RNase LS. After sedimentation through a sucrose gradient, the RNase LS activity was again lost, but could be recovered in two broad peaks. One peak, consisting of low molecular weight material, showed activity when complemented with the heavy fraction of the high-salt wash, and the other, consisting of very high molecular weight material (>1000 kDa), was active when complemented with the light fraction of the high-salt wash. The heavy material was shown to include the RnlA protein (40 kDa) (154). Somewhat puzzling is the fact that only one site could be cleaved in the *soc* RNA after sucrose gradient fractionation and complementation, and that this site was different from the site cleaved by purified RnlA protein.

The molecular weight of the degradosome is estimated to be about 1500 kDa (*21*). However, as mentioned above, the *dmd* phenotype is not reversed in infection of a RNase E host mutant (*rne*Δ131) unable to assemble the degradosome (*71*), suggesting that the multiprotein complex that constitutes RNase LS is not simply a modification of the host degradosome to contain the RnlA protein as a consequence of T4 infection.

More than 10 proteins cosediment with RnlA in sucrose gradients (*154*). One of these was identified as triose phosphate isomerase. It is present in stoichiometric amounts relative to RnlA and binds to it very tightly. The meaning of this interaction is unclear at the moment. Otsuka *et al.* (*154*) pointed out the fact that triose phosphate isomerase functions within a large complex containing proteins involved in carbohydrate metabolism. They found that a mutation in the gene for triose phosphate isomerase is able to partially allow the growth of T4 *dmd* mutant, suggesting that RnlA and triose phosphate isomerase functionally interact.

The nature and function of the other partners of RnlA besides triose phosphate isomerase remain to be discovered. In particular, it is unclear at the moment whether the RNase LS entity carries only one RNase activity, presumably that of the RnlA protein, or if other RNases participate in the complex, and how the activity of RnlA is modulated by the other components of the complex.

5. Specificity and Mode of Action of RNase LS. Coupling with Translation

The way RNase LS selects its cleavage site is not yet understood. No sequence or structural motif seems to be shared by the RNase LS target sites, apart from the fact that most of the ~30 cleavages analyzed so far in various middle and late transcripts occur 3′ to a pyrimidine in single-stranded RNA. The nucleotides 3′ to the cleavage site might play a role (*144, 146, 155, 156*).

Interestingly, RNase LS is stimulated by ongoing translation of the mRNAs. Indeed, the elimination of the SD sequence, or the presence of a stop codon upstream of RNase LS cleavage sites, abolishes some of the *rnlA*-dependent cleavages in the *soc* and *uvsY* transcripts. Accordingly, disruption of either the SD sequence or the translation initiation codon of the late *soc* gene or of the *uvsY* middle transcript, results in a marked stabilization of the transcripts in a *dmd* mutant background (*146, 155*). Some other cleavages, however, are independent on translation. Furthermore, premature termination of translation by nonsense codons triggers cleavages by RNase LS at some distance (20–25 nt) downstream of the stop codon. The cleavage always occurs 3' to a pyrimidine (71, 146, 156). Yamanishi and Yonesaki (156) pointed out that what may be common to translation elongation and termination is the pausing of the ribosomes at rare codons and stop codons, respectively. They proposed that RNase LS is stimulated when the ribosome slows down or pauses on the mRNA, possibly due to the unwinding property of the ribosomes that would maintain the RNA in a locally unstructured conformation. In the absence of translation, a number of potential RNase LS sites would be masked by secondary structure. In this respect, it would be interesting to determine whether the translation-independent sites are located systematically in single-stranded regions. Whether this is the only role of the ribosome in RNase LS activation is an open question.

It has not been determined whether cleavage by RNase LS produces 5'-hydroxyl or 5'-phosphate termini. Therefore, the consequences of RNase LS cleavages on the activity of other RNases that are sensitive to the status of 5' and 3' RNA termini (RNases E and G, PNPase) cannot yet be anticipated.

6. THE ROLE OF RNASE LS IN *E. COLI*

RNase LS seems to play a minor role in noninfected bacteria, since a mutation in the *rnlA* gene, whether a point mutation or an insertion, only reduces the size of the colonies on minimum medium and has no effect on growth in rich medium. RNA is stabilized by 30% on average in an *rnlA* mutant. However, RNase LS might participate in the degradation of specific mRNAs as reflected by the prolonged functional lifetime of several mRNAs in the *rnlA* mutant. The *rpsO* and *bla* mRNAs are stabilized 2 and 3-fold, respectively in the *rnlA* mutant, while the *rpsA* and *ompA* transcripts are unaffected. A 307 nucleotide fragment internal to the 23S rRNA, already present in *rnlA*-proficient bacteria, accumulates in the absence of a functional *rnlA* gene, suggesting that the RNase LS plays a role in the degradation of this RNA species. The role of this RNA is unclear. It could function as an mRNA as it has the potential to code for a small protein of 27 amino acids. However, the synthesis of this protein *in vivo* has not been demonstrated (145).

7. RNASE LS: A DEFENSE MECHANISM AGAINST VIRULENT BACTERIOPHAGES?

Although the RnlA protein seems to play a role in determining the rate of degradation of some *E. coli* transcripts, it may be regarded as a defense mechanism evolved to fight against invasion by some virulent bacteriophages, phage T4 included. Even if infection by a phage devoid of Dmd kills the cell, its development is aborted by RNase LS-mediated degradation of its mRNAs. This precludes its propagation throughout the bacterial population. In this context, Dmd seems to be the phage's response to overcome the host's defense mechanism. In this respect, it would be interesting to look at the distribution of the *dmd* gene in the many T4-related phages that have been isolated so far. If phages devoid of *dmd* exist in nature, there must be *E. coli* strains on which they can propagate. Do these bacteria lack the *rnlA* gene?

8. CONTROL OF GENE 32 mRNA DEGRADATION

As mentioned earlier in this chapter, gene 32 transcripts are unusually stable. Two features of the 5' leader sequence of gp32 mRNA contribute to this exceptional stability: the pseudo-knot structure near the 5' end and the adjacent A+U-rich sequences. The fusion of this leader sequence upstream of any RNA leads to the stabilization of the chimeric transcript, but only after T4 infection, indicating that a phage encoded factor is required for this stability in addition to the *cis*-determinants (157). The nature of this factor has not been determined but it might well be the Dmd protein, assuming that gene 32 mRNA synthesized from a plasmid in noninfected cells is attacked by RNase LS. It is also possible that the Rho-independent transcription terminator at the 3' end of the molecule has a role in stabilizing gene 32 transcripts, by limiting the extent of 3' exonucleolytic digestion. The gene 32 mRNA stability seems to depend on the temperature of infection. At 30 °C and 37 °C, all mRNA species synthesized during infection are rather stable, with the fragment processed at −71 and the late transcript being the most stable species (50, 52). At 43 °C, however, the half-life of the gene 32 mRNA is only 4.5 min, which is comparable to the average half-life of T4 transcripts (44). It is possible that the pseudo-knot structure is partly denatured at this temperature, allowing RNase E and G to act from the 5' extremities of the processed RNAs. Alternatively, a *trans*-acting factor necessary for gene 32 mRNA stabilization may not be produced in sufficient quantity, or in the correct conformation, at high temperature.

The degradation of gene 32 mRNA is also regulated in a stage-dependent manner. The accumulation of gene 32 poly- and monocistronic middle transcripts collapses after 10 min of infection at 37 °C, while the late transcript accumulates abundantly to become practically the sole gene 32 RNA species at 15 min postinfection. It declines afterwards. The accumulation of the very stable −71 processed species (see above) parallels that of the middle transcripts, with a slight delay; it almost disappears from the infected cells after 10 min. However, when late transcription cannot take place (e.g., in an infection with a mutant of the late sigma factor, gp55), this processed RNA accumulates to huge amounts, with a half-life of more than 20 min (50). Thus, either a late viral function is required to actively destabilize this RNase E-processed species when late transcription is elicited, or the conversion by RNase E of the late transcript into the −71 species is inhibited at late stages. No data is available yet, to decide between these two possibilities. Competition between middle and late gene 32 transcripts for translation should not be involved; however, since the −71 species leader sequence carries all the signals required for translation and autogenous regulation and is included in the sequence of the late transcript (which starts at -150). This was the first description of a temporally controlled mRNA degradation pathway during the development of bacteriophage T4 (50). The factors responsible for the late destabilization of gp32 transcript are presently unknown.

D. Other Factors

RNase III plays no general role in the decay of T4 mRNAs as the functional lifetime of total phage mRNA is unchanged in infection of an RNase III-deficient host relative to an RNase III-proficient one (44). However, this endonuclease might play a role in the degradation of a few specific transcripts, as is the case in *E. coli* (158, 159) and in phage λ (160, 161). In fact, two RNase III-dependent cleavages have been detected in the T4 gene 49 mRNA (162). Although the consequences of this processing have not been analyzed, it seems likely that it contributes to the destabilization of this transcript at the early stages of T4 development. RNase III plays a decisive role in the biosynthesis of one T4 tRNA (see below). This is in contrast with the situation that prevails during infection of *E. coli* with phage T7 (163, 164) or the infection of *B. subtilis* with phage SP82 (165). In both cases, the phage mRNAs are abundantly processed by RNase III.

Whether the host 3′–5′ exoribonucleases, RNase II, PNPase, RNase R, or the RNA pyrophosphohydrolase, RppH (19, 20) are implicated in T4 mRNA turnover has not yet been determined. RNase II is likely to remain active after T4 infection as it plays a role in T4 tRNA maturation (166) (see below). The persistence of the activity of the other enzymes after T4 infection remains to be demonstrated.

The *E. coli* poly(A) polymerase (PAP), encoded by the *pcnB* gene, adds poly (A) tails to the 3′ ends of *E. coli* mRNAs (23) and contributes to the destabilization of transcripts; see chapter in this volume by Hajnsdorf and Regnier. T4 mRNAs are probably not polyadenylated, however. Indeed, it has been found that after infection with the closely related bacteriophage T2, host poly(A) polymerase activity is inhibited (32). Also, no poly(A) extension could be detected at the 3′ end of the *soc* and *uvsY* transcripts after infection with T4 (33), suggesting that bacteriophage T4 infection also leads to PAP inhibition. This could occur through ADP-ribosylation of the protein, for example.

III. Part 2: Transfer RNA Processing

A. Physiology and Distribution Among Bacteriophages

The bacteriophage T4 genome codes for 8 tRNAs specific for the amino acids Gln, Leu, Gly, Pro, Ser, Thr, Ile, and Arg, and two slightly larger, stable RNAs, called species 1 (140 nt; also called C) and species 2 (120 nt; also called D), whose function is still unknown. Unlike the tRNAs, these two RNA species have few if any modified nucleotides (species 1 from the closely related phage T2 has one pseudouridine) (*167–171*). The sequence of the T4 tRNAs are different from those of the corresponding *E. coli* isoacceptors. Evidence that these tRNAs are used in translation is based on the ability to isolate nonsense suppressor derivatives of tRNA^{Ser}, tRNA^{Gln}, tRNA^{Leu}, tRNA^{Ile}, tRNA^{Gly}, and tRNA^{Arg} (*172–178*). Also, some of the T4 tRNAs have been shown to participate in protein synthesis *in vitro* (*179*). None of the T4 encoded tRNAs are essential for T4 growth on laboratory strains of *E. coli*, since a phage mutant completely lacking the tRNA genes is viable (*168, 169*). However, some "natural" *E. coli* strains, like CT439, restrict the growth of tRNA-defective T4 mutants (*180*).

The persistence of the tRNA genes in T4, as well as in other T4-related phages (see below), suggests the existence of a selection pressure to keep these genes. As mentioned above, the bacteriophage T4 genome is 65.5% A+T-rich (*10*) while that of the host is 50% A+T-rich. Therefore, the codon usage frequency likely differs between the parasite and its host. The raison d'être of the phage-encoded tRNAs is likely to compensate for the low level of some host tRNA isoacceptors to make viral mRNA translation more efficient, thus, contributing to phage virulence. In fact, the T4 tRNAs read codons that are systematically more frequent in T4 genes than in *E. coli* genes. However, a calculation of the occurrence of the optimal *E. coli* and T4 codons (those read by the phage tRNAs) in T4 genes led to the suggestion that tRNAs from the host are predominantly used in translation of highly expressed T4 genes, while T4 tRNAs are used more frequently for weakly expressed genes (*10, 181*). The reason for this differential use of tRNAs from T4 and its host is not clear at the moment. It is likely that the tRNA genes confer a selective advantage on phage growth, but only detectable over many generations.

T4 mutants have been isolated that were unable to grow on CT439, but able to grow on a normal *E. coli* laboratory strain (*173*). Of the five complementation groups defined by the mutations, two altered the production of T4 tRNAs. One of the mutations led to the lack of tRNA^{Ile} and mapped within the structural gene for this tRNA. Revertants of this mutation showing varying abilities to grow on CT439 were then isolated. A positive correlation was established between the ability of the revertants to grow on CT439 and the amount of tRNA^{Ile} produced (*182*). This suggested that the inability of CT439 to grow T4 mutants lacking the tRNA genes might result from the lack of an appropriate tRNA^{Ile} isoacceptor in this natural isolate. Indeed, CT439 was found to be lacking a tRNA^{Ile} gene (Plunkett, 1988, quoted in ref. (*14*)).

Genes coding for tRNAs and other small stable RNAs are found in other T4-related phages. Early heteroduplex analysis of DNA from the two closely related phages, T2 and T4, showed that the regions of the genome containing the T4 tRNA genes present a strong homology (*169*). The sequences of stable RNA species 1 and species 2 from T2, T4, and T6 are strikingly conserved among the three phages (*170, 171*). A search (based on chain length and the presence of modified nucleotides) for tRNAs synthesized after infection by the T4-related phages T2, T6, and RB69 revealed that all three phages code for a set of tRNAs. Phage T2 codes for 8 tRNAs, while T6 and RB69, code for 6 and 2 tRNAs, respectively. Also, the tRNA^{Gln} from two different isolates of phage T2 (T2L and T2H) differ by one base substitution in the anticodon stem (*183*). Sequence analysis revealed that the RB1 and RB2 phages code for 10 tRNAs and two small RNA species. The genomes of five other T4-related phage (three coliphages, RB43, RB49, RB69, the *Aeromonas salmonicida* phage 44RR and the *A. hydrophila* phage Aeh1) were scanned for the presence of putative tRNA genes. A great diversity in the number of tRNA genes was found, ranging from no gene in RB49, to 24 in Aeh1. The only putative tRNA gene common to all phages is a tRNA with a CAU anticodon, which should read Met, but with the signature of an Ile tRNA (*184*). Some other phages, unrelated to T4, also code for tRNA. This is the case of the two closely related phages, T5 and BF23 (*185–187*).

The T4 tRNA^{Ile} with the CAU (Met) anticodon was found to read the AUA (Ile) codon, rather than AUG (*179, 188*). The sequence of the tRNA (*188*) initially showed that the anticodon is NAU, where, at the time, N was a unknown modified nucleotide. However, the sequence of the tRNA^{Ile} gene (*189*) revealed that the anticodon is CAU, which should read the AUG Met codon. Identical features are shared by the minor *E. coli* tRNA^{Ile}, which reads AUA. The C of its CAU anticodon is modified by the addition of a lysine moiety, and the modified nucleotide is hence called lysidine (*190, 191*). It is likely that the T4 tRNA^{Ile} anticodon carries this same modification. Thus, the ability of the T4 and *E. coli* tRNA^{Ile} species to be recognized by the isoleucyl-tRNA synthetase (IleRS) and to read the AUA codon depends only on a modification on the wobble position of the anticodon. Remarkably, no tRNA^{Ile} with unmodified C can be found after T4 infection (*188*). On this basis, it has been hypothesized that in T4 infection, the production of mature tRNA^{Ile} depends on this modification by the host (*192*). An undermodified tRNA^{Ile} would be abnormally processed and degraded swiftly. The reason why this particular tRNA is found systematically in the T4-related phages analyzed so far is an open question. The AUA codon is rare in host genes. Accordingly, the intracellular levels of the corresponding host isoacceptor are only a small fraction of the major host tRNA^{Ile} isoacceptor, which reads AU(C/U) (*193*). In the T4 genes however, the AUA codons are three times more frequent than in *E. coli* (*10*). Therefore, it would seem advantageous to T4 and related phages to produce an AUA-reading tRNA^{Ile} to sustain efficient translation of their own genes. It remains to be seen if this surmise is compatible with the fact that, after T4 infection, the tRNA^{Ile} is underproduced relative to the others (see below).

B. Genetic Organization and Transcription of the T4 tRNA Genes

The genes for tRNAs and the two stable RNAs are clustered in one place of the T4 genome. Before the DNA sequence of the region was known, all the tRNAs, RNA species 1 and 2, and three dimeric tRNA precursors had been sequenced. Moreover, the general ordering of the tRNA genes had been determined using overlapping deletions that cover the region. The determination of the DNA sequence filled the gaps between the structural genes (*143, 189, 194*). The genes for tRNAs Gln, Leu, Gly, Pro, Ser, Thr, and Ile, are tightly packaged in that order. Spacers of 1–10 nt in length separate the tRNA genes. This group of genes is called subcluster I. The genes for tRNA^{Arg} and for RNA species 2 and 1, in this order, define subcluster II. They are located about 500 nt downstream of the first cluster. In between lies the *segB* gene, which codes for a homing endonuclease (*195*) (see Fig. 3). This is a site-specific DNA endonuclease which mediates the nonreciprocal transfer of its own gene and flanking sequences, in this case, the tRNA gene region, to other phage genomes during coinfection. The function of SegB might therefore be to assure the spreading of the tRNA genes among T4-related phages (*196*).

The transcription of the tRNA region starts shortly after infection and persists until late times (*136, 142, 143*). The genes of the two subclusters are cotranscribed from upstream promoters, giving rise to polycistronic primary transcripts that all end at a strong *rho*-independent transcription terminator, downstream of the gene for stable RNA species 1 (*143, 189, 197, 198*). Despite the fact that the tRNA genes are cotranscribed, tRNA^{Ile} accumulates to a 3–4-fold lower level relative to the other T4 tRNAs. The reason for this peculiarity is not completely understood. However, the fact that this feature can be reproduced *in vitro*, with crude extracts as sources of processing enzymes and a purified transcript as precursor, indicates that it is at least in part the consequence of abnormal processing or instability (*192*).

During the prereplicative period, subcluster I is transcribed from both distal early and one proximal middle promoter located some 60 nt upstream of the first tRNA gene. Transcription of subcluster II starts with some delay relative to subcluster I, consistent with the discovery of a terminator, partially dependent on Rho, somewhere between the two subclusters (*142, 143, 197, 199–204*). The exact location of the terminator(s) has not been determined. In their 1983 review, Schmidt and Apirion (*15*) speculated that it might lie within the tRNA^{Ile} gene because part of tRNA^{Ile} can be folded into an alternative secondary structure resembling a transcription terminator. If so, this may also contribute to lowering the level of production of this tRNA species. The origin of the abundant transcription of the tRNA genes at late times is not clear. Sequence inspection reveals the existence of a typical late promoter some

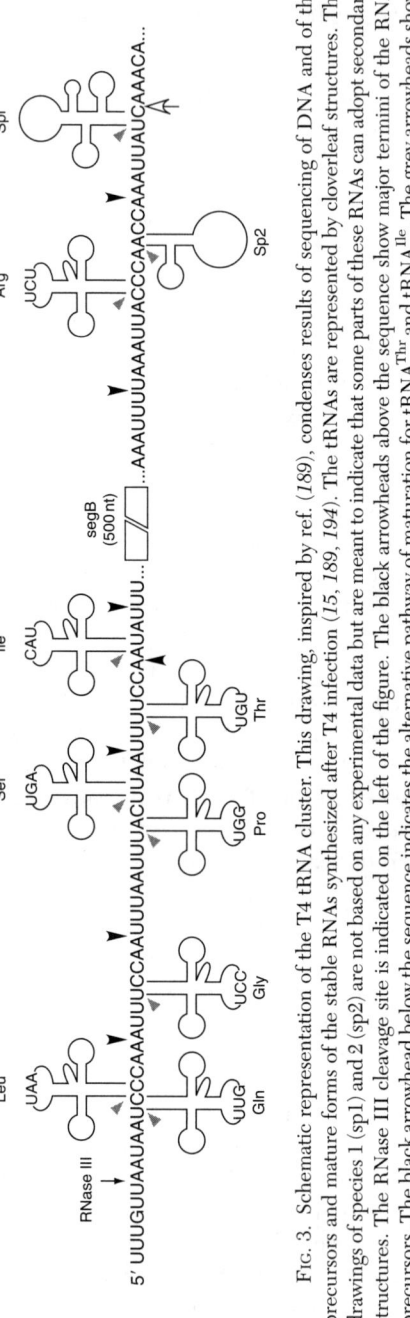

FIG. 3. Schematic representation of the T4 tRNA cluster. This drawing, inspired by ref. (189), condenses results of sequencing of DNA and of the precursors and mature forms of the stable RNAs synthesized after T4 infection (15, 189, 194). The tRNAs are represented by cloverleaf structures. The drawings of species 1 (sp1) and 2 (sp2) are not based on any experimental data but are meant to indicate that some parts of these RNAs can adopt secondary structures. The RNase III cleavage site is indicated on the left of the figure. The black arrowheads above the sequence show major termini of the RNA precursors. The black arrowhead below the sequence indicates the alternative pathway of maturation for tRNA^{Thr} and tRNA^{Ile}. The grey arrowheads show the sites of cleavage by RNase P. The empty arrowhead (on the right) marks the position of the self-cleavage identified by Apirion and collaborators (206, 238). The segB gene is symbolized by a broken rectangle.

330 nt upstream of the tRNA cluster, which was shown to be active *in vitro* and to depend on the late sigma factor, gp55 (*203*). Yet, *in vivo*, the mutational inactivation of gp55 has no effect (or rather, a positive effect) on the rate of transcription of the tRNA region (*143*).

As described above, the transcription of the tRNA clusters is temporally regulated at the levels of both initiation and termination of transcription. From a functional point of view, the transcripts from the tRNA region are mosaic RNAs, since portions of them are precursors to the stable RNAs while other portions function as mRNAs (of *segB* and upstream genes). The 5' ends of the diverse tRNA primary transcripts are variable since they originate at different promoters. However, a 5' end common to all of them is generated posttranscriptionally *via* cleavage by RNase III upstream of the first tRNA gene (*202, 205, 206*) (Fig. 3).

C. Processing of the tRNA Precursors

Host RNases are responsible for the processing of the primary transcript of the tRNA region into the mature stable RNAs. A T4 factor is also required for the maturation of some tRNAs and species 2 (see below). We do not know whether processing occurs cotranscriptionally, which is likely, or after the transcript has been fully synthesized. Long primary transcripts (about 2.8 kb) synthesized *in vitro* can be processed to mature tRNAs by cellular extracts (*197, 200, 201, 207*), indicating that coupling, should it occur *in vivo*, is not an absolute condition for processing. The earliest steps in T4 tRNA processing are the least understood. They will be presented last. As in all organisms examined, the mature forms of T4 tRNAs bear the CCA_{OH} sequence at the end of the acceptor stem. This motif is essential for aminoacylation and interaction with the ribosome. Intriguingly, the mature forms of RNA species 1 and 2 also end with CCA. Several bases of the final tRNA are modified. These modifications contribute to the L-shape folding of the molecule and in some cases, when it affects a nucleotide located in or near the anticodon sequence, contribute to codon recognition specificity (see above).

1. RNase P, Nucleotidyl Transferase, 3' Exonucleases, and RNase BN or Z

Six tRNAs derive from three dimeric tRNA precursors. These are the $tRNA^{Gln}$-$tRNA^{Leu}$, $tRNA^{Pro}$-$tRNA^{Ser}$, and $tRNAs^{Thr}$-$tRNA^{Ile}$ precursors, which occur in that order in the genome. All three precursors have been sequenced (*192, 208–210*). They all contain extra 5' and 3' residues and a few nucleotides in between. Precursors for $tRNA^{Gly}$, $tRNA^{Arg}$, and RNA species 1 and 2 are monomeric.

Although not all the tRNA precursors accumulate during infection of a RNase P-deficient host, the 5′ leaders from all tRNA precursors are removed by RNase P cleavage; see chapter in this volume by Hartmann and colleagues. This step generates the 5′ termini of mature tRNAs and species 1 and 2 (*211–213*). Apart from this common feature, the maturation pathway for each stable RNA depends on whether the precursor is monomeric or dimeric and also on whether the CCA is genomically encoded or not. In bacteriophage T4, two situations occur. The CCA sequence may be genomically encoded, like all of the tRNAs of its host *E. coli*. This is the case for the tRNA Leu, Gly, Thr, Arg, and species 2 of T4 (Fig. 3). The additional 3′ residues must be trimmed back to the CCA sequence by any of the four host 3′ exonucleases, RNases T, PH, D, and II, with RNase T and PH being the most efficient (*34, 166, 214–216*). The T4 genes for tRNA Gln, Pro, Ser, Ile, and species 1 do not code for the CCA terminal sequence (Fig. 3), a situation that is also encountered in other bacteria, in Eukarya, and in the Archaea. In these organisms, the CCA sequence is added posttranscriptionally in two steps. The 3′ extra residues are removed by an endonucleolytic cleavage 3′ to the discriminator base, followed by CCA addition by nucleotidyl transferase. The endonuclease responsible is called RNase Z, or tRNase Z (*217, 218*). The *E. coli* enzyme responsible for removing the 3′ trailers of the CCA-less T4 tRNA species was identified a number of years ago as RNase BN, initially thought to be a 3′ exoribonuclease (*219, 220*). However, the gene encoding RNase BN, which was initially misidentified, was recently shown to be *elaC*, known to encode RNase Z (*221*). Thus, RNase BN is probably not an exonuclease but an endonuclease, leading to even more conservation in the pathways of tRNA maturation in eukaryotic and prokaryotic organisms. In the following, I shall use the term RNase BN instead of RNase Z to follow the recommendation of Ezraty *et al.* (*221*).

The sequence of tRNA processing steps has been established by analyzing the precursors that accumulate in various host mutants. The maturation pathways for each stable T4 RNA have been reviewed in detail in ref. (*15*). They will be briefly presented in the following.

The dimeric precursor of $tRNA^{Pro}$ and $tRNA^{Ser}$ (both lacking an encoded CCA) accumulates in mutants deficient for either RNase BN, nucleotidyl transferase or RNase P. Some dimeric precursors that accumulate during T4 infection of an *rnpA* mutant end with the CCA sequence (*219, 222*). This shows that RNase P depends on prior addition of the CCA sequence to the 3′ end of the dimeric molecule, one of the first observations that showed that RNase P, which acts at the 5′ end of tRNA molecules, is sensitive to the presence of the 3′ CCA sequence (*223*). This suggested that RNase P recognizes tRNA structure rather than sequence. Therefore, the sequence of events is, first, the elimination of the extra residues from the 3′ end of the dimeric precursor by RNase BN, followed by CCA addition by nucleotidyl transferase and finally, cleavage by RNase P to generate the mature 5′ termini. This step separates two, still immature, tRNAs. Indeed, the distal tRNA^{Ser} undergoes a last nucleotide modification that results in a 2′-O-methylguanosine in the D-arm. The proximal tRNA^{Pro} still contains extra residues at its 3′ end. The mature form of tRNA^{Pro} is obtained after the action of RNase BN and nucleotidyl transferase.

The tRNA^{Gln}-tRNA^{Leu} dimeric precursor also accumulates in infection of *rnpA* mutant cells (*213*), but not in RNase BN or nucleotidyl transferase deficient strains. In agreement, the tRNA^{Leu} su$^+$ requires neither RNase BN nor the CCA adding enzyme to manifest its suppressor activity (*177*). This is because the CCA sequence of tRNA^{Leu} is genomically encoded. The activity of tRNA^{Gln} su$^+$, however, does require a functional nucleotidyl transferase but not RNase BN (*224*). This is because after cleavage of the dimeric precursor by RNase P, the tRNA^{Gln} has no extra 3′ nucleotides (see Fig. 3). A CCA sequence must then be added to complete the maturation of the tRNA^{Gln} (*210, 225*).

The third dimeric precursor that includes tRNAs^{Thr} and tRNA^{Ile}, is the only one with an interstitial CCA sequence. This precursor does not accumulate in an RNase P mutant host. It does accumulate, however, in mutants of RNase BN or nucleotidyl transferase (*192, 219*), consistent with lack of CCA in the 3′ trailer sequence of the distal tRNA in the dimeric precursor. In the absence of RNase P, an immature monomeric form of tRNA^{Thr} appears that possesses an unprocessed 5′ leader but a 3′ matured terminus. It has been postulated that the Thr-Ile dimer is processed *via* two pathways. One is identical to that of the tRNA^{Pro}-tRNA^{Ser} precursor, where maturation of the distal tRNA by RNase P is dependent on the proper maturation at the 3′ end of the dimeric precursor, *via* RNase BN and addition of the CCA sequence. The second pathway was proposed to involve another endonuclease activity, different from RNase P, that cleaves the dimeric precursor in several locations downstream of the interstitial CCA sequence, generating tRNA^{Ile} (even in RNase BN or nucleotidyl transferase mutants). Interestingly, the pre-tRNA^{Gly} and dimeric tRNA^{Gln}-tRNA^{Leu} precursor sequences end with the CCA sequence. This could arise from trimming but it could also arise from an endonucleolytic cleavage of the primary transcript immediately 3′ to the CCA sequence (*192*). The identity of the endonuclease involved remains to be discovered, but possible candidates are RNase E or RNase G (see below).

As mentioned above, the monomeric tRNA^{Gly} precursor ends with the encoded 3′ CCA, which may arise from direct endonucleolytic cleavage of the primary transcript after the CCA motif. Presumably, the only processing step then required is the removal of its 5′ leader by RNase P. The tRNA^{Arg} and species 2 RNA precursors also carry a genomically encoded 3′ CCA. Accordingly, neither requires RNase BN or nucleotidyl transferase for their maturation. Although the 5′ terminus of tRNA^{Arg} is matured by RNase P, this tRNA does not accumulate in T4 infection of a temperature-sensitive RNase P mutant at the nonpermissive temperature. Its pathway of maturation is not yet completely understood. The 3′ end of the species 1 RNA precursor is generated by self-cleavage, helped by a protein initially called RNase F (see below). Its CCA must be added posttranscriptionally.

Aside from specificities in the pathway of maturation for each tRNA, the following general principles have emerged from these pioneer studies of the T4 tRNA maturation pathways that have greatly contributed to establishing our present knowledge of the maturation of stable RNAs in many biological systems: (1) Pre-tRNAs contain extra 5′ and 3′ nucleotides that must be removed to generate the functional, mature forms of tRNA; (2) RNase P generates the mature 5′ end of all tRNA in practically every organism examined; for discussion on recent exceptions, see chapter by Hartmann and colleagues in this volume; (3) The presence of the CCA sequence at the 3′ terminus of the acceptor stem is a universal feature and may affect the activity of RNase P at the 5′ end of the tRNA precursor; and (4) The maturation enzymes, whether RNase P, the CCA-adding enzyme, the enzymes of tRNA modification and possibly, the 3′ exoribonuclease RNase D, recognize structures rather than sequences in the precursor, a result that also implies that the tRNAs are folded in their precursor forms (*34, 178, 210, 226–228*). In full agreement with these functional studies, recent structural analyses reveal interactions of some of these enzymes with several regions of the folded tRNAs (*229*).

2. INITIAL PROCESSING EVENTS

Inspection of the DNA sequence of the tRNA cluster (*189, 194*) indicates that the monomeric and dimeric precursors must be generated from the primary transcripts by endonucleolytic cleavages. In fact, apart from the RNase III cleavage upstream of the first gene of the cluster, responsible for generating the 5′ terminus of the dimeric tRNA^{Gln}-tRNA^{Leu} precursor, and a self-cleavage event that gives rise to the 3′ end of species 1, the enzymes repsonsible for the other processing events are still unknown. Some activities have nevertheless been implicated. These processing events are reviewed here.

RNase III. As mentioned above, one of the first efficient processing events of the tRNA primary transcripts is performed by the host endonuclease RNase III. It was initially observed that expression of the suppressor phenotype associated with the T4 $sup_2{}^+$ mutation in the tRNA^{Gln} gene, is abolished in infection of an RNaseIII-deficient (*rnc*) host. Concomitantly, the production of the tRNA^{Gln} is dramatically reduced, while that of the remaining seven tRNAs is normal. This includes tRNA^{Leu}, which is linked to tRNA^{Gln} in the dimeric precursor (*230*). A strong RNase III cleavage occurs 6 nt upstream of the gene for tRNA^{Gln} (*202, 205, 206*). When *rnc* bacteria are infected with T4Δ27, a deletion mutant phage in which only the genes for tRNA^{Gln}, tRNA^{Leu}, and species 1 are left, a transient precursor RNA with an additional 50 nt at its 5′ end replaces the precursor found in infection of an RNase III-proficient strain. *In vitro*, the longest precursor can be converted into the shorter one and eventually into the mature tRNA^{Gln} by crude extracts from RNase III-proficient cells but not from *rnc* mutant cells. Partially purified RNase III was shown to carry out the same conversion (*205*). Thus, preventing the RNase III cleavage upstream of the tRNA^{Gln} structural gene apparently leads to a rapid degradation of tRNA^{Gln}, but has no effect on the adjacent tRNA^{Leu}. In fact, in a T4 infection of a mutant host lacking both RNase P and RNase III, a dimer of the two tRNAs is found, but whose 5′ leader has only two extra nucleotides instead of six (*231*). In infection of an RNase III mutant host (RNase P-proficient), the cleavage by RNase P at the internal site of the dimer remains very efficient, while the rate of cleavage at the 5′ site of the shorter dimer becomes sufficiently low to enable some nonspecific RNase to degrade the tRNA molecule before the cleavage by RNase P occurs (*231*). This finding underlines the fact that each tRNA maturation step is probably in competition with the degradation process. The rate of the maturation reactions must be higher than the rate of the degradation reactions to produce functional tRNAs.

RNase E. In *E. coli*, RNase E is involved in the processing of precursors of the 5S and 16S rRNAs and of many tRNAs. The primary polycistronic tRNA transcripts are cleaved by RNase E to produce precursors that are subsequently matured by RNase P (*11, 80, 232*). In an infection of RNase E-deficient *E. coli* cells with phage T4Δ27, a large precursor of the three encoded stable RNA species (tRNA^{Gln}, tRNA^{Leu}, and species 1) and a smaller precursor to species 1, called p2sp1, accumulate with a concomitant dramatic reduction of the mature form of species 1. Mature tRNA Gln and Leu are also reduced, but to a lesser extent. The large polycistronic precursor is processed at its 5′ end by RNase III, as it is in the presence of RNase E, and terminates at the strong 3′ terminator. The absence of RNase E does not prevent the maturation of the large precursors *in vitro*. Indeed, a crude extract from *rne* mutant cells is able to carry out total maturation of the three stable RNA species, albeit at a reduced rate when compared to an extract from RNase E-proficient cells. RNase E cleavage sites within the tRNA precursors have not been found (*198, 233*). It has been suggested that RNase E is not involved directly in the processing of the T4 tRNA precursors but helps other processing enzymes to accomplish their reactions in some way (e.g., removal of inhibitory secondary structures that block the progression of 3′ exoribonucleases). Also, *rne* mutations have very pleïotropic effects in cells. In particular, they could affect the level of other RNases and factors involved in T4 tRNA maturation. For example, RNase E is responsible for the maturation of the M1 RNA responsible for the catalytic activity of RNase P (234).

RNase PC, RNase F and self-cleavage. An *E. coli* 100,000 × g (S-100) supernatant fraction is able to carry out the complete processing of an *in vitro* synthesized primary transcript from the tRNA region, indicating that all the enzymes necessary for maturation of the T4 stable RNAs are present in this crude cellular extract. Further fractionation allowed Goldfarb and Daniel (235) to isolate an endonuclease activity that cleaves the polycistronic RNA precursor into RNA fragments whose size is compatible with the monomeric and dimeric stable RNAs. This nuclease exhibits a very strong affinity to phosphocellulose, hence its name, RNase PC. RNase PC products obtained *in vitro* can be further processed to the mature forms of tRNA^{Arg} and species 2 RNA by S-100 extracts. Thus, this purified fraction seems to contain an activity involved in the initial steps of tRNA processing. However, an excess of cleavages was observed, indicating that this preparation is contaminated by other endonucleases.

Apirion and collaborators partially purified a protein, called RNase F, which generates the 3′ terminus of species 1 RNA from its precursor, p2Sp1, *in vitro* (236, 237). This protein is different from the RNases III, E, and P. Later on, convincing evidence was reported to show that the large precursor, p2Sp1, is capable of self-cleavage at the same site as that stimulated by RNase F (238). The spontaneous cleavage reaction is slower than in the presence of RNase F. It depends on the presence of both a nonionic detergent and ammonium ions (100–200 mM). The protein described initially as RNase F is now thought to assist RNA autocatalytic cleavages, without bearing the catalytic activity itself (206). Since this time, several proteins have been described that assist group I intron self-splicing in T4 and other organisms (239, 240), making the above-mentioned hypothesis quite plausible. Whether RNase F is the same protein as RNase PC is not known; no other self-cleavages were reported to occur elsewhere in the primary transcripts of the tRNA cluster. The isolation of a mutant of the "RNase F" gene seems mandatory to understand the mechanism exactly. The fact that self-cleavage occurs in a deletion of the *segB* gene (T4Δ27) indicates that the SegB protein is not involved in this processing event.

3. Cef, a T4 Factor Involved in the Biosynthesis of Several tRNAs and RNA Species 2

Genetical approaches were undertaken to better understand the functions of the T4 tRNAs and also to look for possible factors that would affect tRNA biosynthesis. In addition to the search for T4 mutants unable to grow on the natural host isolate CT439 (see above (182)), two groups exploited the availability of a positive selection for suppressor-negative derivatives of T4 carrying a nonsense suppressor mutation in the gene for tRNA^{Ser} (*173, 241*). Apart from many mutations isolated within the gene for tRNA^{Ser} itself, these selections led to the isolation of mutants, called *M1* (*173*) or *mb* (*241*) or "gene II" (*182*), which map far from the tRNA cluster, upstream of gene 39 on the T4 genome. All these independently isolated mutants show strictly the same phenotype: the absence of tRNA^{Pro}, tRNA^{Ser}, tRNA^{Ile}, and species 2 and reduced levels of the other tRNAs. The M1 mutation fails to complement HA4, a mutation in "gene II," thus showing that these mutations are in the same gene. We have sequenced several of these T4 mutants (mb-3, -5, and -6 (*180*), M101 (*173*) and AA2X2, and HA4X2 (*182*)). The sequence alterations are all located in the *cef* gene and lead to inactivation of the gene by nonsense, deletion, or frameshift mutations (Mathy and Uzan, unpublished results). The T4 *cef* gene is expressed shortly after infection. It codes for a protein of 71 residues (*91*) that shows no homology to any other protein in the data bases. T4 *cef* deletion mutants grow normally on laboratory strains.

Nierlich *et al.* (*207*) found that *in vitro* processing of tRNA primary transcripts by extracts from uninfected cells fail to produce tRNA^{Pro}, tRNA^{Ser}, tRNA^{Ile}, and species 2, thus supporting the notion that a T4-encoded factor promotes the biosynthesis of these RNA species. However, other researchers were able to obtain the complete maturation of nine out of the ten T4 stable RNAs (tRNA^{Ile} processing was abnormal (*192*)) using extracts from uninfected cells (*197*). Nevertheless, they observed that the tRNA^{Pro}, tRNA^{Ser}, and the species 2 were less stable than the other products upon prolonged incubation.

A common feature of the tRNA^{Pro}, tRNA^{Ser}, and tRNA^{Ile} precursors is that their extra 3′ residues do not include the CCA sequence. These trailers are removed by RNase BN, before the addition of the CCA sequence (*209, 220*). In the absence of Cef, the tRNA^{Pro}-tRNA^{Ser} dimeric precursor does not accumulate as it does in the absence of RNase BN, suggesting that the role of Cef is not simply to promote RNase BN action. Taken together, these data suggest that the Cef protein protects the tRNAs and species 2 from 3′ exonucleolytic degradation and that RNase BN processing and 3′ exonucleolytic degradation are in competition. In the absence of Cef, the 3′ exonucleolytic pathway would be predominant for tRNAs that have no encoded CCA residues in their 3′ trailers. However, the presence of the 3′ CCA sequence is not sufficient to confer protection as shown by the fact that maturation of species 2, which carries an encoded 3′ CCA sequence, also requires the Cef factor. Species 2 cannot adopt a tRNA-like structure, which suggests that both the CCA sequence and a genuine tRNA structure combine to antagonize the action of the 3′exonuclease(s).

The above model is reminiscent of a pathway of tRNA maturation that has been described in Eukaryotic cells. In yeast, the 3′ sequences of many pre-tRNAs are removed by RNase Z–mediated endonucleolytic cleavage.

Interestingly, an additional factor is required in this processing step. This is the protein Lhp1p, the yeast homolog of the human La protein, thought to have a role of an RNA chaperone. In the absence of Lhp1p, the extra 3' residues are digested by an exonuclease (242). Consistent with this result, the Xenopus La protein was found to prevent tRNA precursors from degradation by a 3' exonuclease in cell extracts (243). Based on these data, it is tempting to propose that the T4 Cef protein is a functional homologue of the eucaryotic La protein.

The function of RNase BN (RNase Z) in the host is not understood since the 3' trailers of all *E. coli* tRNAs include the CCA sequence that inhibit the activity of this RNase. However, RNase Z was recently found to play a role in mRNA degradation (28). If the T4 mRNAs were also substrates for RNase BN/Z, one might also expect the Cef protein to play a role in phage mRNA turnover.

Acknowledgments

I am indebted to C. Condon for asking me to write this review and for his critical reviewing of the manuscript, and to F. Bontems for fruitful discussions and for providing Fig. 2 of this review. I warmly thank my direct collaborators, S. Durand and G. Richard, for stimulating discussions.

References

1. Volkin E, Astrachan L. Phosphorus incorporation in *Escherichia coli* ribo-nucleic acid after infection with bacteriophage T2. *Virology* 1956;**2**:149–61.
2. Astrachan L, Volkin E. Properties of ribonucleic acid turnover in T2-infected *Escherichia coli*. *Biochim Biophys Acta* 1958;**29**:536–44.
3. Volkin E, Astrachan L. Metabolism of RNA phosphorus in *Escherichia coli* infected with bacteriophage T7. *Virology* 1958;**6**:545–55.
4. Geiduschek EP, Nakamoto T, Weiss SB. The enzymatic synthesis of RNA: Complementary interaction with DNA. *Proc Natl Acad Sci USA* 1961;**47**:1405–15.
5. Nomura M, Hall BD, Spiegelman S. Characterization of RNA synthesis in *Escherichia coli* after bacteriophage T2 infection. *J Mol Biol* 1960;**2**:306–26.
6. Brenner S, Jacob F, Meselson M. An unstable intermediate carrying information from genes to ribosomes for protein synthesis. *Nature* 1961;**190**:576–81.
7. Gros F, Hiatt H, Gilbert W, Kurland CG, Risebrough RW, Watson JD. Unstable ribonucleic acid revealed by pulse labelling of *Escherichia coli*. *Nature* 1961;**190**:581–5.
8. Mathews CK, Kutter EM, Mosig G, Berget PB. *Bacteriophage T4*. American Society for Microbiology, Washington, DC.
9. Karam JD. *Molecular biology of bacteriophage T4*. American Society for Microbiology, Washington, DC.
10. Miller ES, Kutter E, Mosig G, Arisaka F, Kunisawa T, Ruger W. Bacteriopage T4 genome. *Microbiol Mol Biol Rev* 2003;**67**:86–156.
11. Luke K, Radek A, Liu X, Campbell J, Uzan M, Haselkorn R, *et al.* Microarray analysis of gene expression during bacteriophage T4 infection. *Virology* 2002;**299**:182–91.

12. Carpousis AJ, Krisch H. mRNA processing and degradation. In: Karam JD, editor. *Molecular biology of bacteriophage T4*. Washington, DC: American Society for Microbiology; 1994.
13. Sanson B, Uzan M. Posttranscriptional controls in bacteriophage T4: Roles of the sequence-specific endoribonuclease RegB. *FEMS Microbiol Rev* 1995;17:141–50.
14. Mosig G. Synthesis and maturation of T4-encoded tRNAs. In: Karam JD, editor. *Molecular bioloby of bacteriophage T4*. Washington, DC: American Society for Microbiology; 1994.
15. Schmidt FJ, Apirion D. T4 transfer RNAs: Paradigmatic system for the study of RNA processing. In: Mathews CK, Kutter EM, Mosig G, Berget PB, editors. *Bacteriophage T4*. Washington, DC: American Society for Microbiology; 1983. p. 208–17.
16. Mazzara GP, Plunkett I,G, McClain WH. Maturation events leading to transfer RNA and riibosomal RNA. In: Goldstein L, Prescott DM, editors. *Cell biology: A comprehensive treatise*. 3. New York: Academic Press; 1980.
17. Shub DA, Coetzee T, Hall DH, Belfort M. The self-splicing introns of bacteriophage T4. In: Karam JD, editor. *Molecular biology of bacteriophage T4*. vols. 186–192. Washington, DC: Amercican Society for Microbiology; 1994.
18. Belfort M. Two for the price of one: A bifunctional intron-encoded DNA endonuclease-RNA maturase. *Genes Dev* 2003;17:2860–3.
19. Celesnik H, Deana A, Belasco JG. Initiation of RNA decay in *Escherichia coli* by 5′ pyrophosphate removal. *Mol Cell* 2007;27:79–90.
20. Deana A, Celesnik H, Belasco JG. The bacterial enzyme RppH triggers messenger RNA degradation by 5′ pyrophosphate removal. *Nature* 2008;451:355–8.
21. Carpousis AJ. The RNA degradosome of *Escherichia coli*: An mRNA-degrading machine assembled on RNase E. *Annu Rev Microbiol* 2007;61:71–87.
22. Deutscher MP. Degradation of RNA in bacteria: Comparison of mRNA and stable RNA. *Nucleic Acids Res* 2006;34:659–66.
23. Dreyfus M, Regnier P. The poly(A) tail of mRNAs: Bodyguard in eukaryotes, scavenger in bacteria. *Cell* 2002;111:611–3.
24. Steege DA. Emerging features of mRNA decay in bacteria. *RNA* 2000;6:1079–90.
25. Coburn GA, Mackie GA. Degradation of mRNA in *Escherichia coli*: An old problem with some new twists. *Prog Nucleic Acid Res Mol Biol* 1999;62:55–108.
26. Grunberg-Manago M. Messenger RNA stability and its role in control of gene expression in bacteria and phages. *Annu Rev Genet* 1999;33:193–227.
27. Cheng ZF, Deutscher MP. An important role for RNase R in mRNA decay. *Mol Cell* 2005;17:313–8.
28. Perwez T, Kushner SR. RNase Z in *Escherichia coli* plays a significant role in mRNA decay. *Mol Microbiol* 2006;60:723–37.
29. Horvitz HR. Control by bacteriophage T4 of two sequential phosphorylations of the alpha subunit of *Escherichia coli* RNA polymerase. *J Mol Biol* 1974;90:727–38.
30. Koch T, Ruger W. The ADP-ribosyltransferases (gpalt) of bacteriophages T2, T4, and T6: Sequencing of the genes and comparison of their products. *Virology* 1994;203:294–8.
31. Koch T, Raudonikiene A, Wilkens K, Rüger W. Overexpression, purification and characterization of the ADP-ribosyltranscferase (gpalt) of bacteriophage T4: ADP-ribosylation of *E. coli* RNA polymerase modulates T4 "early" transcription. *Gene Expr* 1995;4:1–12.
32. Hurwitz J, Furth JJ, Anders M, Ortiz PJ, August JT. The enzymatic incorporation of ribonucleotides into RNA and the role of DNA. *Cold Spring Harb Symp Quant Biol* 1961;26:91–100.
33. Yonesaki T. Scarce adenylation in bacteriophage T4 mRNAs. *Genes Genet Syst* 2002;77:219–25.
34. Cudny H, Deutscher MP. Apparent involvement of ribonuclease D in the 3′ processing of tRNA precursors. *Proc Natl Acad Sci USA* 1980;77:837–41.

35. Ueno H, Yonesaki T. Phage-induced change in the stability of mRNAs. *Virology* 2004;**329**:134–41.
36. Rice AP, Roberts BE. Vaccinia virus induces cellular mRNA degradation. *J Virol* 1983;**47**:529–39.
37. Ooi BG, Miller LK. Regulation of host RNA levels during baculovirus infection. *Virology* 1988;**166**:515–23.
38. Agy MB, Wambach M, Foy K, Katze MG. Expression of cellular genes in CD4 positive lymphoid cells infected by the human immunodeficiency virus, HIV-1: Evidence for a host protein synthesis shut-off induced by cellular mRNA degradation. *Virology* 1990;**177**:251–8.
39. Sorenson CM, Hart PA, Ross J. Analysis of herpes simplex virus-induced mRNA destabilizing activity using an *in vitro* mRNA decay system. *Nucleic Acids Res* 1991;**19**:4459–65.
40. Everly DN, Jr, Feng P, Mian IS, Read GS. mRNA degradation by the virion host shutoff (Vhs) protein of herpes simplex virus: Genetic and biochemical evidence that Vhs is a nuclease. *J Virol* 2002;**76**:8560–71.
41. Greene R, Korn D. The instability of T4 messenger RNA. *J Mol Biol* 1967;**28**:435–43.
42. Bolle A, Epstein RH, Salser W, Geiduschek EP. Transcription during bacteriophage T4 development: Synthesis and relative stability of early and late RNA. *J Mol Biol* 1968;**31**:325–48.
43. Sanson B, Hu R-M, Troitskaya E, Mathy N, Uzan M. Endoribonuclease RegB from bacteriophage T4 is necessary for the degradation of early but not middle or late mRNAs. *J Mol Biol* 2000;**297**:1063–74.
44. Mudd EA, Carpoussis AJ, Krisch HM. E. coli RNase E has a role in the decay of bacteriophage t4 mRNA. *Genes Dev* 1990;**4**:873–81.
45. Alberts BM, Frey L. T4 bacteriophage gene 32: A structural protein in the replication and recombination of DNA. *Nature* 1970;**227**:1313–8.
46. Nossal NG. The bacteriophage T4 DNA replication fork. In: Karam JD, editors. *Molecular biology of bacteriophage T4*.Washington, DC: American Society for Microbiology; 1994.
47. Mosig G. Homologous recombination. In: Karam JD, editor. *Molecular biology of bacteriophage T4*. Washington, DC: American Society for Microbiology; 1994.
48. Krisch HM, Bolle A, Epstein RH. Regulation of the synthesis of bacteriophage T4 gene 32 protein. *J Mol Biol* 1974;**88**:89–104.
49. Gold L, O'Farrell PZ, Russel M. Regulation of gene 32 expression during bacteriophage T4 infection of *Escherichia coli*. *J Biol Chem* 1976;**251**:7251–62.
50. Belin D, Mudd EA, Prentki P, Yi YY, Krisch HM. Sense and antisense transcription of bacteriophage T4 gene 32. Processing and stability of the mRNAs. *J Mol Biol* 1987;**194**:231–43.
51. Carpousis A, Mudd EA, Krisch HM. Transcription and messenger RNA processing upstream of bacteriophage T4 gene 32. *Mol Gen Genet* 1989;**219**:39–48.
52. Russel M, Gold L, Morrissett H, O'Farrell PZ. Translational, autogenous regulation of gene 32 expression during bacteriophage T4 infection. *J Biol Chem* 1976;**251**:7263–70.
53. Lemaire G, Gold L, Yarus M. Autogenous translational repression of bacteriophage T4 gene 32 expression *in vitro*. *J Mol Biol* 1978;**126**:73–90.
54. Krisch HM, Allet B. Nucleotide sequences involved in bacteriophage T4 gene 32 translational self-regulation. *Proc Natl Acad Sci USA* 1982;**79**:4937–41.
55. McPheeters DS, Stormo GD, Gold L. Autogenous regulatory site on the bacteriophage T4 gene 32 messenger RNA. *J Mol Biol* 1988;**201**:517–35.
56. Shamoo Y, Webster KR, Williams KR, Konigsberg WH. A retrovirus-like zinc domain is essential for translational repression of bacteriophage T4 gene 32. *J Biol Chem* 1991;**266**:7967–70.

57. Mudd EA, Prentki P, Belin D, Krish HM. Processing of unstable bacteriophage T4 gene 32 mRNAs into a stable species requires *Escherichia coli* ribonuclease E. *EMBO J* 1988;**7**:3601–7.
58. Loayza D, Carpousis AJ, Krisch HM. Gene 32 transcription and mRNA processing in T4-related bacteriophages. *Mol Microbiol* 1991;**5**:715–25.
59. Philippe C, Eyermann F, Benard L, Portier C, Ehresmann B, Ehresmann C. Ribosomal protein S15 from *Escherichia coli* modulates its own translation by trapping the ribosome on the mRNA initiation loading site. *Proc Natl Acad Sci USA* 1993;**90**:4394–8.
60. Braun F, Le Derout J, Régnier P. Ribosomes inhibit an RNase E cleavage which induces the decay of the *rpso* mRNA of *Escherichia coli*. *EMBO J* 1998;**17**:4790–7.
61. Mudd EA, Krisch HM, Higgins CF. RNase E, an endoribonuclease, has a general role in the chemical decay of *E. coli* mRNA: Evidence that *rne* and *ams* are the same genetic locus. *Mol Microbiol* 1990;**4**:2127–35.
62. Ehretsmann CP, Carpousis AJ, Krisch HM. Specificity of *Escherichia coli* endoribonuclease RNase E: *In vivo* and *in vitro* analysis of mutants in a bacteriophage T4 mRNA processing site. *Genes Dev* 1992;**6**:149–59.
63. Lin-Chao S, Wong T-T, McDowall KJ, Cohen SN. Effects of nucleotide sequence on the specificity of *rne*-dependent and RNase E-mediated cleavages of Rna I encoded by pbr322 plasmid. *J Biol Chem* 1994;**269**:10797–803.
64. McDowall KJ, Lin-Chao S, Cohen SN. A+U content rather than a particular nucleotide order determines the specificity of RNase E cleavage. *J Biol Chem* 1994;**269**:10790–6.
65. Kaberdin VR, Walsh AP, Jakobsen T, McDowall KJ, von Gabain A. Enhanced cleavage of RNA mediated by an interaction between substrates and the arginine-rich domain of *E. coli* ribonuclease E. *J Mol Biol* 2000;**301**:257–64.
66. Kaberdin VR. Probing the substrate specificity of *Escherichia coli* RNase E using a novel oligonucleotide-based assay. *Nucleic Acids Res* 2003;**31**:4710–6.
67. Carpousis AJ, Van Houwe G, Ehretsmann C, Krisch HM. Copurification of *E. coli* RNase E and PNPase: Evidence for a specific association between two enzymes important in RNA processing and degradation. *Cell* 1994;**76**:889–900.
68. Mitchell P, Tollervey D. Musing on the structural organization of the exosome complex. *Nat Struct Biol* 2000;**7**:843–6.
69. Marchand I, Nicholson AW, Dreyfus M. Bacteriophage T7 protein kinase phosphorylates RNase E and stabilizes mRNAs synthesized by T7 RNA polymerase. *Mol Microbiol* 2001;**42**:767–76.
70. Summers WC. The process of infection with coliphage T7. IV. Stability of RNA in bacteriophage-infected cells. *J Mol Biol* 1970;**51**:671–8.
71. Otsuka Y, Ueno H, Yonesaki T. *Escherichia coli* endoribonucleases involved in cleavage of bacteriophage T4 mRNAs. *J Bacteriol* 2003;**185**:983–90.
72. Kido M, Yamanaka K, Mitani T, Niki H, Ogura T, Hiraga S. RNase E polypeptides lacking a carboxyl-terminal half suppress a *mukb* mutation in *Escherichia coli*. *J Bacteriol* 1996;**178**:3917–25.
73. Lopez PJ, Marchand I, Joyce SA, Dreyfus M. The C-terminal half of RNase E, which organizes the *Escherichia coli* degradosome, participates in mRNA degradation but not rRNA processing *in vitro*. *Mol Microbiol* 1999;**33**:188–99.
74. Leroy A, Vanzo NF, Sousa S, Dreyfus M, Carpousis AJ. Function in *Escherichia coli* of the non-catalytic part of RNase E: role in the degradation of ribosome-free mRNA. *Mol Microbiol* 2002;**45**:1231–43.
75. Khemici V, Carpousis AJ. The RNA degradosome and poly(A) polymerase of *Escherichia coli* are required *in vivo* for the degradation of small mRNA decay intermediates containing REP-stabilizers. *Mol Microbiol* 2004;**51**:777–90.

76. Piesiniene L, Truncaite L, Zajanckauskaite A, Nivinskas R. The sequences and activities of RegB endoribonucleases of T4-related bacteriophages. *Nucleic Acids Res* 2004;**32**:5582–95.
77. Truncaite L, Zajanckauskaite A, Arlauskas A, Nivinskas R. Transcription and RNA processing during expression of genes preceding DNA ligase gene 30 in T4-related bacteriophages. *Virology* 2006;**344**:378–90.
78. Kokoska RJ, Steege DA. Appropriate expression of filamentous phage f1 DNA replication genes II and X requires RNase E-dependent processing and separate mRNAs. *J Bacteriol* 1998;**180**:3245–9.
79. Goodrich AF, Steege DA. Roles of polyadenylation and nucleolytic cleavage in the filamentous phage mRNA processing and decay pathways in *Escherichia coli*. *RNA* 1999;**5**:972–85.
80. Li Z, Pandit S, Deutscher MP. RNase G (cafa protein) and RNase E are both required for the 5′ maturation of 16S ribosomal RNA. *EMBO J* 1999;**18**:2878–85.
81. Tock MR, Walsh AP, Carroll G, McDowall KJ. The cafa protein required for the 5′-maturation of 16 S rRNA is a 5′-end-dependent ribonuclease that has context-dependent broad sequence specificity. *J Biol Chem* 2000;**275**:8726–32.
82. Hambreus G, Rutberg T. *Escherichia coli* RNase E and RNase G cleave a *Bacilllus subtilis* transcript at the same site in a structure-dependent manner. *Arch Microbiol* 2004;**181**:137–43.
83. Umitsuki G, Wachi M, Takada A, Hikichi T, Nagai K. Involvement of RNase G in *in vivo* mRNA metabolism in *Escherichia coli*. *Genes Cells* 2001;**6**:403–10.
84. Kaga N, Umitsuki G, Nagai K, Wachi M. RNase G-dependent degradation of the eno mRNA encoding a glycolysis enzyme enolase in *Escherichia coli*. *Biosci Biotechnol Biochem* 2002;**66**:2216–20.
85. Lee K, Bernstein JA, Cohen SN. RNase G complementation of rne null mutation identifies functional interrelationships with RNase E in *Escherichia coli*. *Mol Microbiol* 2002;**43**:1445–56.
86. Zajanckauskaite A, Truncaite L, Strazdaite-Zieliene Z, Nivinskas R. Involvement of the *Escherichia coli* endoribonucleases G and E in the secondary processing of RegB-cleaved transcripts of bacteriophage T4. *Virology* 2008;**375**:342–53.
87. Chace KV, Hall DH. Characterization of new regulatory mutants of bacteriophage T4. II. New class of mutants. *J Virol* 1975;**15**:929–45.
88. Uzan M, Favre R, Brody E. A nuclease that cuts specifically in the ribosome binding site of some T4 mRNAs. *Proc Natl Acad Sci USA* 1988;**85**:8895–9.
89. Ruckman J, Parma D, Tuerk C, Hall DH, Gold L. Identification of a T4 gene required for bacteriophage mRNA processing. *New Biol* 1989;**1**:54–65.
90. Odaert B, Saida F, Aliprandi P, Durand S, Crechet JB, Guerois R, et al. Structural and functional studies of RegB, a new member of a family of sequence-specific ribonucleases involved in mRNA inactivation on the ribosome. *J Biol Chem* 2007;**282**:2019–28.
91. Sanson B, Uzan M. Dual role of the sequence-specific bacteriophage T4 endoribonuclease RegB: mRNA inactivation and mRNA destabilization. *J Mol Biol* 1993;**233**:429–46.
92. Durand S, Richard G, Bisaglia M, Laalami S, Bontems F, Uzan M. Activation of RegB endoribonuclease by S1 ribosomal protein requires an 11 nt conserved sequence. *Nucleic Acids Res* 2006;**34**:6549–60.
93. Uzan M. Bacteriophage T4 RegB endoribonuclease. *Methods Enzymol* 2001;**342**:467–80.
94. Uzan M, Leautey J, d'Aubenton-Carafa Y, Brody E. Identification and biosynthesis of the bacteriophage T4 mot regulatory protein. *EMBO J* 1983;**2**:1207–12.
95. Ruckman J, Ringquist S, Brody E, Gold L. The bacteriophage T4 RegB ribonuclease. Stimulation of the purified enzyme by ribosomal protein S1. *J Biol Chem* 1994;**269**:26655–62.
96. Lebars I, Hu RM, Lallemand JY, Uzan M, Bontems F. Role of the substrate conformation and of the S1 protein in the cleavage efficiency of the T4 endoribonuclease RegB. *J Biol Chem* 2001;**276**:13264–72.

97. Subramanian A-R. Structure and functions of ribosomal protein S1. *Prog Nucleic Acid Res Mol Biol* 1983;**28**:101–42.
98. Sorensen MA, Fricke J, Pedersen S. Ribosomal protein S1 is required for translation of most, if not all, natural mRNAs in *Escherichia coli in vivo*. *J Mol Biol* 1998;**280**:561–9.
99. Bycroft M, Grunert S, Murzin AG, Proctor M, St Johnston D. NMR solution structure of a dsRNA binding domain from *Drosophila* staufen protein reveals homology to the N-terminal domain of ribosomal protein S5 [published erratum appears in EMBO J 1995 Sep 1; 14(17), 4385]. *EMBO J* 1995;**14**:3563–71.
100. Schubert M, Edge RE, Lario P, Cook MA, Strynadka NC, Mackie GA, et al. Structural characterization of the RNase E S1 domain and identification of its oligonucleotide-binding and dimerization interfaces. *J Mol Biol* 2004;**341**:37–54.
101. Buttner K, Wenig K, Hopfner KP. Structural framework for the mechanism of archaeal exosomes in RNA processing. *Mol Cell* 2005;**20**:461–71.
102. Miranda G, Schuppli D, Barrera I, Hausherr C, Sogo JM, Weber H. Recognition of bacteriophage Qb plus strand RNA as a template by Qb replicase: Role of RNA interactions mediated by ribosomal proteins S1 and host factor. *J Mol Biol* 1997;**267**:1089–103.
103. Kalapos MP, Paulus H, Sarkar N. Identification of ribosomal protein S1 as a poly(A) binding protein in *Escherichia coli*. *Biochimie* 1997;**79**:493–502.
104. Sukhodolets MV, Garges S. Interaction of *Escherichia coli* RNA polymerase with the ribosomal protein S1 and the Sm-like atpase Hfq. *Biochemistry* 2003;**42**:8022–34.
105. Sukhodolets MV, Garges S, Adhya S. Ribosomal protein S1 promotes transcriptional cycling. *RNA* 2006;**12**:1505–13.
106. Uzan M, Brody E, Favre R. Nucleotide sequence and control of transcription of the bacteriophage T4 *mota* regulatory gene. *Mol Microbiol* 1990;**4**:1487–96.
107. Tuerk C, Gold L. Systematic evolution of ligands by expontial enrichement: RNA ligands to bacteriophage T4 DNA polymerase. *Science* 1990;**249**:505–10.
108. Jayasena VK, Brown D, Shtatland T, Gold L. *In vitro* selection of RNA specifically cleaved by bacteriophage T4 RegB endonuclease. *Biochemistry* 1996;**35**:2349–56.
109. Dodson RE, Shapiro DJ. Regulation of pathways of mRNA destabilization and stabilization. *Prog Nucleic Acid Res Mol Biol* 2002;**72**:129–64.
110. Saida F, Uzan M, Bontems F. The phage T4 restriction endoribonuclease RegB: A cyclizing enzyme that requires two histidines to be fully active. *Nucleic Acids Res* 2003;**31**:2751–8.
111. Mackie GA. Ribonuclease E is a 5′-end-dependent endonuclease. *Nature* 1998;**395**:720–3.
112. Mackie GA. Stabilization of circular rpsT mRNA demonstrates the 5′-end dependence of RNase E action *in vivo*. *J Biol Chem* 2000;**275**:25069–72.
113. Severinov K, Kashlev M, Severinova E, Bass L, McWilliams K, Kutter E, et al. A non-essential domain of *Escherichia coli* RNA polymerase required for the action of the termination factor Alc. *J Biol Chem* 1994;**269**:14254–9.
114. Kaufmann G. Anticodon nucleases. *Trends Biochem Sci* 2000;**25**:70–4.
115. Amitsur M, Benjamin S, Rosner R, Chapman-Shimshoni D, Meidler R, Blanga S, et al. Bacteriophage T4-encoded Stp can be replaced as activator of anticodon nuclease by a normal host cell metabolite. *Mol Microbiol* 2003;**50**:129–43.
116. Blanga-Kanfi S, Amitsur M, Azem A, Kaufmann G. PrrC-anticodon nuclease: Functional organization of a prototypical bacterial restriction RNase. *Nucleic Acids Res* 2006;**34**:3209–19.
117. Zhu H, Yin S, Shuman S. Characterization of polynucleotide kinase/phosphatase enzymes from Mycobacteriophages omega and Cjw1 and vibriophage KVP40. *J Biol Chem* 2004;**279**:26358–69.
118. Blondal T, Hjorleifsdottir S, Aevarsson A, Fridjonsson OH, Skirnisdottir S, Wheat JO, et al. Characterization of a 5′-polynucleotide kinase/3′-phosphatase from bacteriophage RM378. *J Biol Chem* 2005;**280**:5188–94.

119. Gerdes K, Christensen SK, Lobner-Olesen A. Prokaryotic toxin-antitoxin stress response loci. *Nat Rev Microbiol* 2005;**3**:371–82.
120. Condon C. Shutdown decay of mRNA. *Mol Microbiol* 2006;**61**:573–83.
121. Rajkowitsch L, Schroeder R. Dissecting RNA chaperone activity. *RNA* 2007;**13**:2053–60.
122. Thomas JO, Szer W. RNA-helix-destabilizing proteins. *Prog Nucleic Acid Res Mol Biol* 1982;**27**:157–87.
123. Bisaglia M, Laalami S, Uzan M, Bontems F. Activation of the RegB endoribonuclease by the S1 ribosomal protein is due to cooperation between the S1 four C-terminal modules in a substrate-dependant manner. *J Biol Chem* 2003;**278**:15261–71.
124. Boni IV, Artamonova VS, Dreyfus M. The last RNA-binding repeat of the *Escherichia coli* ribosomal protein S1 is specifically involved in autogenous control. *J Bacteriol* 2000;**182**:5872–9.
125. Skorski P, Proux F, Cheraiti C, Dreyfus M, Hermann-Le Denmat S. The deleterious effect of an insertion sequence removing the last twenty percent of the essential *Escherichia coli* rpsA gene is due to mRNA destabilization, not protein truncation. *J Bacteriol* 2007;**189**:6205–12.
126. Guerrier-Takada C, Subramanian AR, Cole PE. The activity of discrete fragments of ribosomal protein S1 in Q beta replicase function. *J Biol Chem* 1983;**258**:13649–52.
127. Aliprandi P, Sizun C, Perez J, Mareuil F, Caputo S, Leroy JL, et al. S1 ribosomal protein functions in translation initiation and ribonuclease RegB activation are mediated by similar RNA–protein interactions: An NMR and SAXS analysis. *J Biol Chem* 2008;**283**:13289–301.
128. Wiberg JS, Karam JD. Translational regulation in T4 phage development. In: CK Mathews, EM Kutter, G Mosig, PB Berget editors. *Bacteriophage T4*. Washington, DC: American Society for Microbiology; 1983.
129. Erickson HP. Ftsz, a tubulin homologue in prokaryote cell division. *Trends Cell Biol* 1997;**7**:362–7.
130. Saida F, Uzan M, Lallemand JY, Bontems F. New system for positive selection of recombinant plasmids and dual expression in yeast and bacteria based on the restriction ribonuclease RegB. *Biotechnol Prog* 2003;**19**:727–33.
131. Orsini G, Igonet S, Pene C, Sclavi B, Buckle M, Uzan M, et al. Phage T4 early promoters are resistant to inhibition by the anti-sigma factor AsiA. *Mol Microbiol* 2004;**52**:1013–28.
132. Pène C, Uzan M. The bacteriophage T4 anti-sigma factor, AsiA, is not necessary for the inhibition of early promoters *in vivo*. *Mol Microbiol* 2000;**35**:1180–91.
133. Mattson T, Richardson J, Goodin D. Mutant of bacteriophage T4D affecting expression of many early genes. *Nature* 1974;**250**:48–50.
134. Ouhammouch M, Adelman K, Harvey SR, Orsini G, Brody EN. Bacteriophage T4 mota and AsiA proteins suffice to direct *Escherichia coli* RNA polymerase to initiate transcription at T4 middle promoters. *Proc Natl Acad Sci USA* 1995;**92**:1451–5.
135. Stitt B, Hinton D. Regulation of middle-mode transcription. In: JD Karam editor. *Molecular biology of bacteriophage T4*. vols. 142–160. Washington, DC: American Society for Microbiology; 1994. p. 142–60.
136. Young ET, Mattson T, Seizer G, Houwe GV, Bolle A, Epstein R. Bacteriophage T4 gene transcription studied by hybridization to cloned restriction fragments. *J Mol Biol* 1980;**138**:423–45.
137. Williams KP, Kassavetis GA, Herendeen DR, Geiduschek EP. Regulation of late gene expression. In: JD Karam editor. *Molecular biology of bacteriophage T4*. Washington, DC: American Society for Microbiology; 1994. p. 161–75.
138. Kolesky S, Ouhammouch M, Brody EN, Geiduschek EP. Sigma competition: The contest between bacteriophage T4 middle and late transcription. *J Mol Biol* 1999;**291**:267–81.

139. Nechaev S, Kamali-Moghaddam M, Andre E, Leonetti JP, Geiduschek EP. The bacteriophage T4 late-transcription coactivator gp33 binds the flap domain of *Escherichia coli* RNA polymerase. *Proc Natl Acad Sci USA* 2004;**101**:17365–70.
140. Winter RB, Morrissey L, Gauss P, Gold L, Hsu T, Karam J. Bacteriophage T4 *regA* protein binds to mRNAs and prevents translation initiation. *Proc Natl Acad Sci USA* 1987;**84**:7822–6.
141. Gordon J, Sengupta TK, Phillips CA, O'Malley SM, Williams KR, Spicer EK. Identification of the RNA binding domain of T4 regA protein by structure-based mutagenesis. *J Biol Chem* 1999;**274**:32265–73.
142. Pulitzer JF, Colombo M, Ciaramella M. New control elements of bacteriophage T4 pre-replicative transcription. *J Mol Biol* 1985;**182**:249–63.
143. Broida J, Abelson J. Sequence organization and control of transcription in the bacteriophage T4 tRNA region. *J Mol Biol* 1985;**185**:545–63.
144. Kai T, Selick HE, Yonesaki T. Destabilization of bacteriophage T4 mRNAs by a mutation of gene 61.5. *Genetics* 1996;**144**:7–14.
145. Otsuka Y, Yonesaki T. A novel endoribonuclease, RNase LS, in *Escherichia coli*. *Genetics* 2005;**169**:13–20.
146. Kanesaki T, Hamada T, Yonesaki T. Opposite roles of the dmd gene in the control of RNase E and RNase LS activities. *Genes Genet Syst* 2005;**80**:241–9.
147. Ueno H, Yonesaki T. Role of *Escherichia coli* Hfq in late-gene silencing of bacteriophage T4 dmd mutant. *Genes Genet Syst* 2002;**77**:301–8.
148. Tsui H-CT, Feng G, Winkler ME. Negative regulation of *mutS* and *mutH* repair gene expression by the Hfq and RpoS global regulators of *Escherichia coli* K-12. *J Bacteriol* 1997;**179**:7476–87.
149. Vytvytska O, Moll I, Kaberdin VR, von Gabain A, Blasi U. Hfq (HF1) stimulates ompA mRNA decay by interfering with ribosome binding. *Genes Dev* 2000;**14**:1109–18.
150. Folichon M, Arluison V, Pellegrini O, Huntzinger E, Regnier P, Hajnsdorf E. The poly(A) binding protein Hfq protects RNA from RNase E and exoribonucleolytic degradation. *Nucleic Acids Res* 2003;**31**:7302–10.
151. Ueno H, Yonesaki T. Recognition and specific degradation of bacteriophage T4 mRNAs. *Genetics* 2001;**158**:7–17.
152. Kai T, Ueno H, Yonesaki T. Involvement of other bacteriophage T4 genes in the blockade of protein synthesis and mRNA destabilization by a mutation of gene 61.5. *Virology* 1998;**248**:148–55.
153. Vanzo NF, Li YS, Py B, Blum E, Higgins CF, Raynal LC, et al. Ribonuclease E organizes the protein interactions in the *Escherichia coli* RNA degradosome. *Genes Dev* 1998;**12**:2770–81.
154. Otsuka Y, Koga M, Iwamoto A, Yonesaki T. A role of RnlA in the RNase LS activity from *Escherichia coli*. *Genes Genet Syst* 2007;**82**:291–9.
155. Kai T, Yonesaki T. Multiple mechanisms for degradation of bacteriophage T4 soc mRNA. *Genetics* 2002;**160**:5–12.
156. Yamanishi H, Yonesaki T. RNA cleavage linked with ribosomal action. *Genetics* 2005;**171**:419–25.
157. Gorski K, Roch JM, Prentki P, Krisch HM. The stability of bacteriophage T4 gene 32 mRNA: A 5' leader sequence that can stabilize mRNA transcripts. *Cell* 1985;**43**:461–9.
158. Portier C, Dondon L, Grunberg-Manago M, Régnier P. The first step in the functional inactivation of the *Escherichia coli* polynucleotide phosphorylase messenger is a ribonuclease III processing at the 5' end. *EMBO J* 1987;**6**:2165–70.
159. Bardwell JCA, Régnier PC, Chen P, Nakamura Y, Grunberg-Manago M. Autoregulation of RNase III operon by mRNA Processing. *EMBO J* 1989;**8**:3401–7.
160. Schmeissner U, McKenney K, Rosenberg M, Court D. Removal of a terminator structure by RNA processing regulates *int* gene expression. *J Mol Biol* 1984;**176**:39–53.

161. Oppenheim AB, Kobiler O, Stavans J, Court DL, Adhya S. Switches in bacteriophage lambda development. *Annu Rev Genet* 2005;**39**:409–29.
162. Barth KA, Powell D, Trupin M, Mosig G. Regulation of two nested proteins from gene 49 (recombination endonuclease VII) and of a lambda RexA-like protein of bacteriophage T4. *Genetics* 1988;**120**:329–43.
163. Dunn J, Studier FW. T7 early RNAs and *Escherichia coli* ribosomal RNAs are cut from large precursor RNAs *in vivo* by ribonuclease III. *Proc Natl Acad Sci USA* 1973;**70**:3296–300.
164. Dunn JJ, Studier FW. Complete nucleotide sequence of bacteriophage T7 DNA and the locations of T7 genetic elements. *J Mol Biol* 1983;**166**:477–535.
165. Panganiban AT, Whiteley HR. *Bacillus subtilis* RNAase III cleavage sites in phage SP82 early mRNA. *Cell* 1983;**33**:907–13.
166. Birenbaum M, Schlessinger D, Ohnishi Y. Altered bacteriophage T4 ribonucleic acid metabolism in a ribonuclease II-deficient mutant of *Escherichia coli*. *J Bacteriol* 1980;**142**:327–30.
167. Daniel V, Sarid S, Littauer UZ. Bacteriophage induced transfer RNA in *Escherichia coli*. New transfer RNA molecules are synthesized on the bacteriophage genome. *Science* 1970;**167**:1682–8.
168. McClain WH, Guthrie C, Barrell BG. Eight transfer RNAs induced by infection of *Escherichia coli* with bacteriophage T4. *Proc Natl Acad Sci USA* 1972;**69**:3703–7.
169. Wilson JH, Kim JS, Abelson JN. Bacteriophage T4 transfer RNA. 3. Clustering of the genes for the T4 transfer RNAs. *J Mol Biol* 1972;**71**:547–56.
170. Pinkerton TC, Paddock G, Abelson J. Bacteriophage T4 tRNA Leu. *Nat New Biol* 1972;**240**:88–90.
171. Plunkett G, Mazzara GP, McClain WH. Characterization of bacteriophage T4 and D RNA, alow-molecular-weight RNA of unknown function. *Arch Biochem Biophys* 1981;**210**:298–306.
172. McClain WH. UAG suppressor coded by bacteriophage T4. *FEBS Lett* 1970;**6**:99–101.
173. McClain WH, Guthrie C, Barrell BG. The psu1+ amber suppressor gene of bacteriophage T4: Identification of its amino acid and transfer RNA. *J Mol Biol* 1973;**81**:157–71.
174. Wilson JH, Kells S. Bacteriophage T4 transfer RNA. I. Isolation and characterization of two-phage-coded nonsense suppressors. *J Mol Biol* 1972;**69**:39–56.
175. Comer MM, Guthrie C, McClain WH. An ochre suppressor of bacteriophage T4 that is associated with a transfer RNA. *J Mol Biol* 1974;**90**:665–76.
176. Kao SH, McClain WH. U-G-A suppressor of bacteriophage T4 associated with arginine transfer RNA. *J Biol Chem* 1977;**252**:8254–7.
177. Foss K, Kao S, McClain WH. Three suppressor forms of bacteriophage T4 leucine transfer RNA. *J Mol Biol* 1979;**135**:1013–21.
178. McClain WH, Foss K, Schneider J, Guerrier-Takada C, Altman S. Suppressor and novel mutants of bacteriophage T4 tRNA(Gly). *J Mol Biol* 1987;**193**:223–6.
179. Scherberg NH, Weiss SB. T4 transfer RNAs: Codon recognition and translational properties. *Proc Natl Acad Sci USA* 1972;**69**:1114–8.
180. Wilson JH. Function of the bacteriophage T4 transfer RNAs. *J Mol Biol* 1973;**74**:753–7.
181. Kunisawa T. Synonymouscondon preferences in bacteriophage T4: A distinctive use of transfer RNA from T4 and its host *Escherichia coli*. *J Theor Biol* 1992;**159**:287–98.
182. Guthrie C, McClain WH. Conditionally lethal mutants of bacteriophage T4 defective in production of a transfer RNA. *J Mol Biol* 1973;**81**:137–55.
183. Likover Moen TL, Seidman JG, McClain WH. A catalogue of transfer RNA-like molecules synthesized following infection of *Escherichia coli* by T-even bacteriophages. *J Biol Chem* 1978;**253**:7910–7.

184. Nolan JM, Petrov V, Bertrand C, Krisch HM, Karam JD. Genetic diversity among five T4-like bacteriophages. *Virol J* 2006;**3**:30.
185. Hunt C, Desai SM, Vaughan J, Weiss SB. Bacteriophage T5 transfer RNA. Isolation and characterization of tRNA species and refinement of the tRNA gene map. *J Biol Chem* 1980;**255**:3164–73.
186. Ksenzenko VN, Shlyapnikov MG, Azbarov VG, Garcia O, Kryukov VM, Bayev AA. Nucleotide sequence of the bacteriophage T5 DNA fragment containing a distal part of tRNA gene region. *Nucleic Acids Res* 1987;**15**:5480–1.
187. Ikemura T, Okada K, Ozeki H. Clustering of transfer RNA genes in bacteriophage BF23. *Virology* 1978;**90**:142–6.
188. Guthrie C, McClain WH. Rare transfer ribonucleic acid essential for phage growth. Nucleotide sequence comparison of normal and mutant T4 isoleucine-accepting transfer ribonucleic acid. *Biochemistry* 1979;**18**:3786–95.
189. Fukada K, Abelson J. DNA sequence of a T4 transfer RNA gene cluster. *J Mol Biol* 1980;**139**:377–91.
190. Muramatsu T, Nishikawa K, Nemoto F, Kuchino Y, Nishimura S, Miyazawa T, *et al*. Codon and amino-acid specificities of a transfer RNA are both converted by a single posttranscriptional modification. *Nature* 1988;**336**:179–81.
191. Muramatsu T, Yokoyama S, Horie N, Matsuda A, Ueda T, Yamaizumi Z, *et al*. A novel lysine-substituted nucleoside in the first position of the anticodon of minor isoleucine tRNA from *Escherichia coli*. *J Biol Chem* 1988;**263**:9261–7.
192. Guthrie C, Scholla CA. Asymmetric maturation of a dimeric transfer RNA precursor. *J Mol Biol* 1980;**139**:349–75.
193. Harada F, Nishimura S. Purification and characterization of AUA specific isoleucine transfer ribonucleic acid from *Escherichia coli* B. *Biochemistry* 1974;**13**:300–7.
194. Mazzara GP, Plunkett G, III, McClain WH. DNA sequence of the transfer RNA region of bacteriophage T4: Implications for transfer RNA synthesis. *Proc Natl Acad Sci USA* 1981;**78**:889–92.
195. Sharma M, Ellis RL, Hinton DM. Identification of a family of bacteriophage T4 genes encoding proteins similar to those present in group I introns of fungi and phage. *Proc Natl Acad Sci USA* 1992;**89**:6658–62.
196. Brok-Volchanskaya VS, Kadyrov FA, Sivogrivov DE, Kolosov PM, Sokolov AS, Shlyapnikov MG, *et al*. Phage T4 SegB protein is a homing endonuclease required for the preferred inheritance of T4 tRNA gene region occurring in co-infection with a related phage. *Nucleic Acids Res* 2008;**36**:2094–105.
197. Goldfarb A, Broida J, Abelson J. Transcription *in vitro* of an isolated fragment of bacteriophage T4 genome. *J Mol Biol* 1982;**160**:579–91.
198. Pragai B, Apirion D. Processing of bacteriophage T4 transfer RNAs: Structural analysis and *in vitro* processing of precursors that accumulate in RNase E$^-$ strains. *J Mol Biol* 1982;**154**:465–84.
199. Mattson T, van Houwe G, Epstein RH. Isolation and characterization of conditional lethal mutations in the *mot* gene of bacteriophage T4. *J Mol Biol* 1978;**126**:551–70.
200. Goldfarb A, Daniel V. Transcriptional control of two gene subclusters in the tRNA operon of bacteriophage T4. *Nature* 1980;**286**:418–20.
201. Goldfarb A, Daniel V. Mapping of transcription units in the bacteriophage T4 tRNA gene cluster. *J Mol Biol* 1981;**146**:393–412.
202. Barkay T, Goldfarb A. Processing of bacteriophage T4 primary transcripts with ribonuclease III. *J Mol Biol* 1982;**162**:299–315.
203. Goldfarb A, Malik S. Changed promoter specificity and antitermination properties displayed *in vitro* by bacteriophage T4-modified RNA polymerase. *J Mol Biol* 1984;**177**:87–105.

204. Guild N, Gayle M, Sweeney R, Hollingsworth TM, Modeer T, Gold L. Transcriptional activation of bacteriophage T4 middle promoters by the MotA protein. *J Mol Biol* 1988;**199**:241–58.
205. Pragai B, Apirion D. Processing of bacteriophage T4 tRNAs. The role of RNAase III. *J Mol Biol* 1981;**153**:619–30.
206. Gurevitz M, Apirion D. The ribonuclease-III-processing site near the 5′ end of an RNA precursor of bacteriophage T4 and its effect on termination. *Eur J Biochem* 1985;**147**:581–6.
207. Nierlich DP, Lamfrom H, Sarabhai A, Abelson J. Transfer RNA synthesis *in vitro*. *Proc Natl Acad Sci USA* 1973;**70**:179–82.
208. Barrell BG, Seidman JG, Guthrie C, McClain WH. Transfer RNA biosynthesis: The nucleotide sequence of a precursor to serine and proline transfer RNAs. *Proc Natl Acad Sci USA* 1974;**71**:413–6.
209. Seidman JG, Barrell BG, McClain WH. Five steps in the conversion of a large precursor RNA into bacteriophage proline and serine transfer RNAs. *J Mol Biol* 1975;**99**:733–60.
210. Guthrie C. The nucleotide sequence of the dimeric precursor to glutamine and leucine transfer RNAs coded by bacteriophage T4. *J Mol Biol* 1975;**95**:529–47.
211. Sakano H, Shimura Y, Ozeki H. Studies on T4-tRNA biosynthesis: Accumulation of precursor tRNA molecules in a temperature sensitive mutant of *Escherichia coli*. *FEBS Lett* 1974;**40**:312–6.
212. Abelson J, Fukada K, Johnson P, Lamfrom H, Nierlich DP, Otsuka A, et al. Bacteriophage T4 tRNAs: Structure, genetics, and biosynthesis. *Brookhaven Symp Biol* 1975;**26**:77–88.
213. Guthrie C, Seidman JG, Comer MM, Bock RM, Schmidt FJ, Barrell BG, et al. The biology of bacteriophage T4 transfer RNAs. *Brookhaven Symp Biol* 1975;**26**:106–23.
214. Zhang JR, Deutscher MP. Transfer RNA is a substrate for RNase D *in vivo*. *J Biol Chem* 1988;**263**:17909–12.
215. Kelly KO, Deutscher MP. The presence of only one of five exoribonuclease is sufficient to support the growth of *Escherichia coli*. *J Bacteriol* 1992;**174**:6682–4.
216. Reuven NB, Deutscher MP. Multiple exoribonucleases are required for the 3′ processing of *Escherichia coli* tRNA precursors *in vivo*. *FASEB J* 1993;**7**:143–8.
217. Vogel A, Schilling O, Spath B, Marchfelder A. The tRNase Z family of proteins: Physiological functions, substrate specificity and structural properties. *Biol Chem* 2005;**386**:1253–64.
218. Redko Y, Li de Lasierra-Gallay I, Condon C. When all's zed and done:The structure and function of RNase Z in prokaryotes. *Nat Rev Microbiol* 2007;**5**:278–86.
219. Seidman JG, Schmidt FJ, Foss K, McClain WH. A mutant of *Escherichia coli* defective in removing 3′ terminal nucleotides from some transfer RNA precursor molecules. *Cell* 1975;**5**:389–400.
220. Schmidt FJ, McClain WH. An *Escherichia coli* ribonuclease which removes an extra nucleotide from a biosynthetic intermediate of bacteriophage T4 proline transfer RNA. *Nucleic Acids Res* 1978;**5**:4129–39.
221. Ezraty B, Dahlgren B, Deutscher MP. The RNase Z homologue encoded by *Escherichia coli* elaC gene is RNase BN. *J Biol Chem* 2005;**280**:16542–5.
222. Seidman JG, McClain WH. Three steps in conversion of large precursor RNA into serine and proline transfer RNAs. *Proc Natl Acad Sci USA* 1975;**72**:1491–5.
223. Guerrier-Takada C, McClain WH, Altman S. Cleavage of tRNA precursors by the RNA subunit of *E. coli* ribonuclease P (M1 RNA) is influenced by 3′-proximal CCA in the substrates. *Cell* 1984;**38**:219–24.

224. Deutscher MP, Foulds J, McClain WH. Transfer ribonucleic acid nucleotidyl-transferase plays an essential role in the normal growth of *Escherichia coli* and in the biosynthesis of some bacteriophage T4 transfer ribonucleic acids. *J Biol Chem* 1974;**249**:6696–9.
225. McClain WH, Seidman JG, Schmidt FJ. Evolution of the biosynthesis of 3'-terminal C-C-A residues in T-even bacteriophage transfer RNAs. *J Mol Biol* 1978;**119**:519–36.
226. McClain WH, Barrell BG, Seidman JG. Nucleotide alterations in bacteriophage T4 serine transfer RNA that affect the conversion of precursor RNA into transfer RNA. *J Mol Biol* 1975;**99**:717–32.
227. McClain WH, Seidman JG. Genetic perturbations that reveal tertiary conformation of tRNA precursor molecules. *Nature* 1975;**257**:106–10.
228. Manale A, Guthrie C, Colby D. S1 nuclease as a probe for the conformation of a dimeric tRNA precursor. *Biochemistry* 1979;**18**:77–83.
229. Nakanishi K, Nureki O. Recent progress of structural biology of tRNA processing and modification. *Mol Cells* 2005;**19**:157–66.
230. McClain WH. A role for ribonuclease III in synthesis of bacteriophage T4 transfer RNAs. *Biochem Biophys Res Commun* 1979;**86**:718–24.
231. Gurevitz M, Apirion D. Interplay among processing and degradative enzymes and a precursor ribonucleic acid in the selective maturation and maintenance of ribonucleic acid molecules. *Biochemistry* 1983;**22**:4000–5.
232. Ghora BK, Apirion D. Structural analysis and *in vitro* processing to p5 rRNA of a 9S RNA molecule isolated from an *rne* mutant of *E. coli*. *Cell* 1978;**15**:1055–66.
233. Gurevitz M, Apirion D. Processing of bacteriophage T4 tRNAs: A precursor of species 1 RNA. *FEBS Lett* 1983;**159**:180–4.
234. Lundberg U, Altman S. Processing of the precursor to the catalytic RNA subunit of RNase P from *Escherichia coli*. *RNA* 1995;**1**:327–34.
235. Goldfarb A, Daniel V. An *Escherichia coli* endonuclease responsible for primary cleavage of *in vitro* transcripts of bacteriophage T4 tRNA gene cluster. *Nucleic Acids Res* 1980;**8**:4501–16.
236. Watson N, Apirion D. Ribonuclease F, a putative processing endoribonuclease from *Escherichia coli*. *Biochem Biophys Res Commun* 1981;**103**:543–51.
237. Gurevitz M, Watson N, Apirion D. A cleavage site of ribonuclease F. *Eur J Biochem* 1982;**124**:553–9.
238. Watson N, Gurevitz M, Ford J, Apirion D. Self cleavage of a precursor RNA from bacteriophage T4. *J Mol Biol* 1984;**172**:301–23.
239. Lambowitz AM, Perlman PS. Involvement of aminoacyl-tRNA synthetases and other proteins in group I and group II intron splicing. *Trends Biochem Sci* 1990;**15**:440–4.
240. Coetzee T, Herschlag D, Belfort M. *Escherichia coli* proteins, including ribosomal protein S12, facilitate *in vitro* splicing of phage T4 introns by acting as RNA chaperones. *Genes Dev* 1994;**8**:1575–88.
241. Wilson JH, Abelson JN. Bacteriophage T4 transfer RNA. II. Mutants of T4 defective in the formation of functional suppressor transfer RNA. *J Mol Biol* 1972;**69**:57–73.
242. Yoo CJ, Wolin SL. The yeast La protein is required for the 3' endonucleolytic cleavage that matures tRNA precursors. *Cell* 1997;**89**:393–402.
243. Lin-Marq N, Clarkson SG. Efficient synthesis, termination and release of RNA polymerase III transcripts in *Xenopus* extracts depleted of La protein. *EMBO J* 1998;**7**:2033–41.
244. Takagi H, Kakuta Y, Okada T, Yao M, Tanaka I, Kimura M. Crystal structure of archaeal toxin-antitoxin RelE–RelB complex with implications for toxin activity and antitoxin effects. *Nat Struct Mol Biol* 2005;**12**:327–31.
245. Kamada K, Hanaoka F. Conformational change in the catalytic site of the ribonuclease YoeB toxin by YefM antitoxin. *Mol Cell* 2005;**19**:497–509.

Endonucleolytic Initiation of mRNA Decay in *Escherichia coli*

AGAMEMNON J. CARPOUSIS,[*]
BEN F. LUISI,[†] AND
KENNETH J. MCDOWALL[‡]

[*]*Laboratoire de Microbiologie et Génétique Moléculaires, CNRS et Université Paul Sabatier, 118 route de Narbonne, 31062 Toulouse, France*

[†]*Department of Biochemistry, University of Cambridge, Cambridge, United Kingdom*

[‡]*Astbury Centre for Structural Molecular Biology, Faculty of Biological Sciences, University of Leeds, England, United Kingdom*

I. Background	92
A. Stable Versus Unstable RNA	92
B. Enzymes Involved in RNA Processing and Degradation	93
C. mRNA Structure in *E. coli*	95
D. Chemical Versus Functional Inactivation of mRNA	96
II. General Features of mRNA Degradation in *E. coli*	97
A. Rapid Turnover of mRNA	97
B. RNase E	98
C. Endonucleolytic Initiation of mRNA Degradation	100
D. The RNA Degradosome	101
E. The Influence of the RNA 5′ End on RNase E Activity	101
III. Evidence for Diverse Pathways of mRNA Degradation in *E. coli*	102
A. Limitations of Endonucleolytic Initiation of mRNA Degradation	102
B. Degradation of the *rpsO* mRNA	103
C. Degradation of the *rpsT* mRNA and Internal Entry	104
D. Translation and Cellular Localization of the mRNA Degradation Machinery	105
IV. The Structure of RNase E	106
A. The Organization of RNase E Functional Domains	106
B. The RNase E Catalytic Domain is a Composite of Recurrent Structural Subdomains	110
C. RNase E Catalytic Domain Quaternary Structure	110
V. Substrate Specificity and Catalytic Mechanism of RNase E	112
A. The Active Site and Cleavage Mechanism	112
B. Catalytic Activation and a Model of Allosteric Switching	114
C. The Puzzle of 5′-End Bypass	116
D. The Functional Role of the Quaternary Structure	117

VI. Modulators and Adaptors of RNase E... 118
   A. RNase E Inhibitors 118
   B. Recruitment of RNase E by Hfq and sRNA 118
VII. Other Endoribonucleases in *E. coli*... 119
   A. RNase G... 119
   B. Overlapping and Essential Functions of RNase E and RNase G......... 120
   C. RNase III.. 121
   D. RNase P, RNase Z, and RNase LS 122
   E. Conclusion 123
VIII. Phylogenetic Distribution of RNase E and The RNA Degradosome......... 123
   A. RNase E/G in the *Proteobacteria*... 123
   B. A Wider Evolutionary Perspective on RNase E/G......................... 124
   References... 125

Instability is a fundamental property of mRNA that is necessary for the regulation of gene expression. In *E. coli*, the turnover of mRNA involves multiple, redundant pathways involving 3′-exoribonucleases, endoribonucleases, and a variety of other enzymes that modify RNA covalently or affect its conformation. Endoribonucleases are thought to initiate or accelerate the process of mRNA degradation. A major endoribonuclease in this process is RNase E, which is a key component of the degradative machinery amongst the *Proteobacteria*. RNase E is the central element in a multienzyme complex known as the RNA degradosome. Structural and functional data are converging on models for the mechanism of activation and regulation of RNase E and its paralog, RNase G. Here, we discuss current models for mRNA degradation in *E. coli* and we present current thinking on the structure and function of RNase E based on recent crystal structures of its catalytic core.

I. Background

In this section, we will briefly introduce a number of concepts useful for discussing mechanisms of RNA processing and degradation.

A. Stable Versus Unstable RNA

The RNA of all living cells can be crudely divided between those that are stable and those that are not. "Stable" RNA includes the structured species, such as transfer RNA (tRNA) and ribosomal RNA (rRNA), which are key components of the translational machinery. "Unstable" RNA includes messenger RNA (mRNA). One obvious and important consequence of mRNA instability is that it permits the protein synthesizing machinery to adjust rapidly to changes in the program of transcription. In this sense, it can be argued that mRNA instability is a fundamentally important property in the regulation of gene expression. Furthermore, most stable RNAs are processed from larger primary transcripts into their mature forms. The portions of the primary transcript that are not part of the mature RNA are degraded in a process that closely resembles the degradation of mRNA. The nucleotides produced by RNA degradation are recycled to replenish the pool of precursors for transcription. This cycle of RNA synthesis and degradation represents an important flux in the nucleotide metabolism of the cell. The processing and degradation of RNA is catalyzed by enzymes that operate at distinct steps in these processes.

B. Enzymes Involved in RNA Processing and Degradation

Ribonucleases are enzymes that break the phosphodiester linkage of the RNA chain. The ribonucleases can be divided into two categories based on their mode of action: exoribonucleases and endoribonucleases. The "exos" degrade the RNA chain from either the 3′ end or the 5′ end, removing one nucleotide at a time. In this review, we will refer to enzymes with a 3′–5′ directionality as 3′-exoribonucleases and the enzymes with a 5′–3′ directionality as 5′-exoribonucleases. The exos are further classified as either "distributive" or "processive." Distributive ribonucleases release the RNA substrate after each catalytic step and need to rebind to the substrate for the subsequent step. Processive enzymes can perform the catalytic cycle repeatedly without release of the RNA substrate. In the extreme case, a highly processive exoribonuclease can completely degrade an RNA molecule in a single continuous process, whereas a distributive exoribonuclease "nibbles" at its substrate discontinuously. The principal exos in *E. coli* RNA processing and degradation are RNase II, RNase R, and PNPase (see Chapters by Arraiano, Deutscher, and Hajnsdorf, and Régnier in this volume). These ribonucleases are 3′-exonucleases that are processive for longer RNA species, but become distributive for RNA species shorter than 10 nt.

The breakage of the phosphodiester linkage of the RNA chain can be performed by water (hydrolysis) or orthophosphate (phosphorlysis). To the best of our knowledge, all of the known phosphorlysis ribonucleases are part of a family of 3′-exoribonucleases related to PNPase (polynucleotide phosphorylase) (*1*). There are important differences between hydrolytic ribonucleases such as RNase II and RNase R, and phosphorlytic ribonucleases such as PNPase. The hydrolytic enzymes produce a nucleotide monophosphate (NMP) in a reaction that is for all practical purposes irreversible. In contrast, two phosphates rather than one are in the nucleotide diphosphate (NDP) product of PNPase. In the presence of high NDP and low phosphate concentrations, PNPase can catalyze a nontemplated synthesis reaction to produce polynucleotides. However, the concentration of NDP and phosphate *in vivo* should favor net degradation. Note also that the NDP product of PNPase can be recycled to the triphosphate form (NTP) in a reaction requiring a single high-energy phosphate donor, whereas the NMP of the hydrolytic reaction requires two high-energy phosphate donors. Most organisms that have been studied, eukaryotic as well as prokaryotic, have both hydrolytic and phosphorlytic ribonucleases. The coexistence of hydrolytic and phosphorlytic enzymes in many different types of living organisms suggests that their distinct characteristics might enable the cell to meet the need for rapid RNA turnover while at the same time maintaining energy efficiency. The chapter by Arraiano in this volume details the role of the 3'-exoribonucleases in *E. coli* mRNA degradation.

Endoribonucleases break the RNA chain by hydrolysis of a phosphodiester bond so that the products are two fragments of RNA. In this sense, the activity of an endoribonuclease corresponds to the activity of a DNA restriction enzyme. However, an important difference is that endoribonucleases are not necessarily sequence-specific. Instead, they can be distinguished by the effect of the substrate's structure on enzymatic activity. Classic examples of single-strand-specific and double-strand-specific endoribonucleases in *E. coli* are RNase E and RNase III, respectively. Some endoribonucleases recognize a more elaborate structure, as is the case for the tRNA processing enzyme, RNase P. Enzymes with sequence specificity include the RegB endoribonuclease encoded by bacteriophage T4, which has a preference for GAGG sequences in the translation initiation region (TIR) of mRNAs, and the sequence-specific endoribonucleases encoded by certain toxin–antitoxin systems. The Chapters by Uzan and Inouyé in this volume describe, respectively, RegB and sequence-specific endoribonucleases encoded by toxin–antitoxin systems.

We briefly turn now to several other types of enzymes involved in RNA processing and degradation. *E. coli* has two enzymes that covalently modify the ends of the RNA: poly(A) polymerase (PAP) and the recently discovered RNA 5'-pyrophosphohydrolase, RppH. PAP of *E. coli* adds short oligo(A) extensions to the 3' end of RNA molecules. These 3' extensions promote RNA degradation by serving as platforms that facilitate their recruitment of 3' exoribonucleases. For more detail about this pathway, we direct the interested reader to a selection of reviews (2–4) as well as to the chapter by Hajnsdorf and Régnier in this volume. RppH converts the 5' triphosphate end of a primary transcript to a 5' monophosphate by pyrophosphate removal (5, 6). This enzyme appears to have a function analogous to the 5' decapping enzymes in eukaryotes, with interesting consequences for RNA turnover that we will return to in greater detail later. The DEAD-box RNA helicase, RhlB, is an RNA unwinding enzyme that facilitates the degradation of structured RNA. Double-stranded regions of RNA inhibit degradation by PNPase, but RhlB can unwind or remodel these regions using ATP hydrolysis to transform the RNA into a better, more single-stranded substrate (7–9).

Finally, oligoribonuclease is the "mopping up" enzyme involved in the finishing stages of *E. coli* RNA degradation (*10*). The exoribonucleases described earlier cannot efficiently degrade the last few nucleotides in an RNA chain. The residual product is degraded by oligoribonuclease which specifically degrades oligonucleotides of 2–5 residues in length. The observations that oligoribonuclease is an essential enzyme in *E. coli* and that related enzymes are found in many prokaryotic and eukaryotic organisms suggests an important role for this enzyme family in RNA metabolism. It has been suggested that the accumulation of short oligonucleotides is deleterious in *E. coli*, either because the pool of free nucleotides is depleted by oligonucleotide accumulation or because the oligonucleotides at high concentrations inhibit key enzymes (*10, 11*, and see the Chapter by Danchin in this volume).

C. mRNA Structure in *E. coli*

mRNA in *E. coli* can be either monocistronic or polycistronic. The nascent, primary transcript contains a 5′ triphosphate end that originates from the initiating nucleotide in transcription. Unlike eukaryotic organisms, *E. coli* messages do not have a 5′ methyl-guanosine (cap) modification. However, like the eukaryotic cap, the bacterial triphosphate end offers some protection against mRNA degradation. As mentioned earlier, an mRNA-acting enzyme that converts the 5′ end from triphosphate to monophosphate has recently been described (*5, 6*). The presence of RNA stem-loop structures at the 5′ end can also protect against mRNA degradation. The influence of the RNA 5′ end on mRNA degradation is now thought to be primarily due to an effect on the activities of the endoribonuclease, RNase E, and of its paralog, RNase G. This 5′-end effect, which is one of the principal subjects of this review, will be discussed in the sections that follow.

Other important elements in the 5′ untranslated region (5′ UTR) of the mRNA are the AUG translation start codon and the "Shine–Dalgarno" (SD) element, or ribosome binding site (rbs), which is located just upstream of the AUG. The SD element and the AUG encompass the translation initiation region (TIR). The SD element is involved in a base-pairing interaction with the 3′ end of the 16S rRNA that is important for translation initiation. The "strength" of the TIR depends on a number of features, including the extent of base-pairing between the SD and its complementary sequence at the 3′ end of 16S rRNA. RNA secondary structure, RNA binding proteins, or regulatory RNAs that block access to the TIR can each inhibit translation. Thus, it seems that a polycistronic transcript might pose a problem for translation, since there is often little if any spacing between cistrons. This is often overcome by a mechanism of translation coupling involving the overlap of the translation stop codon of the upstream reading frame and the AUG initiation codon of the downstream reading frame.

An important feature of mRNA structure in *E. coli* is that translation is coupled to transcription. That is, ribosomes begin to translate the message while it is still being synthesized, often forming polyribosomes, a chain of ribosomes attached to a single elongating mRNA. Therefore, the 3' end of an actively translated message is often still engaged in the catalytic site of the transcribing RNA polymerase while the 5' end is being translated into polypeptide. Although we know of no experimental work directly addressing the issue, we speculate that the 3' end of the nascent transcript is protected from 3'-exonucleases by the RNA polymerase elongation complex. Free mRNA 3' ends, the targets for exonuclease-mediated destruction, are formed by two different mechanisms of transcription termination in *E. coli*. In Rho-independent or intrinsic termination, transcription is arrested at a site that results in the formation of an RNA stem-loop followed by a short U-rich segment. Intrinsic termination is a function of the core transcription machinery and does not require accessory factors. Note that intrinsic termination produces a transcript with a 3' stem-loop structure that protects the message from degradation by 3'-exoribonucleases (*12–14*).

A different type of transcription termination is mediated by Rho, a helicase that loads onto the nascent transcript and terminates transcription in a mechanism requiring ATP hydrolysis (*15, 16*). Rho-mediated termination does not occur at a precise position. Instead, termination is believed to occur at heterogeneous positions that can be hundreds of base pairs downstream of the last cistron in the message. The available evidence suggests that Rho translocates along the nascent RNA with a 5'–3' directionality and that transcription termination is achieved when Rho catches up with the RNA polymerase and disrupts the elongation complex. The intricate relationship between translation, Rho-dependent termination and mRNA degradation has not yet been fully elucidated. A recent report suggests that, at the genomic level, Rho-mediated transcription termination appears to be involved in silencing the expression of foreign genes as a potentially protective mechanism to counter phage infection (*17*).

D. Chemical Versus Functional Inactivation of mRNA

In this chapter, when we refer to RNA degradation, we are generally referring to the enzymatic processes that lead to a decrease in the abundance of primary transcripts or stable intermediates. This is often called chemical degradation, as it does not take into account whether the RNAs are functional. Chemical degradation in this context is enzymatic and should not be confused with the hydrolysis that occurs spontaneously and that can be enhanced for example by alkaline conditions. It is obvious that the chemical degradation of an mRNA also results in functional inactivation since the degraded transcript can no longer support translation. However, it is possible that the functional inactivation of an mRNA occurs before transcripts are no longer detectable. For instance, the inhibition of translation can trigger mRNA degradation by a mechanism in which the untranslated mRNA is the target of the degradation machinery. In this scenario, the functional inactivation is due to the inhibition of translation. Further information on the relationship between translation and mRNA degradation can be found in a recent review (18) and in the chapter by Dreyfus in this volume.

II. General Features of mRNA Degradation in E. coli

In this section, where possible, we will only cite a selection of research articles, since much of the material introduced here has been covered in earlier reviews. For a detailed bibliography, we direct the reader to a selection of reviews including two chapters that appeared in earlier volumes of *Progress in Nucleic Acids Research and Molecular Biology* (2, 12–14, 19–25).

A. Rapid Turnover of mRNA

Instability has been recognized as a characteristic property of mRNA since its discovery nearly half a century ago. Indeed, one of the earliest studies characterizing mRNA in *E. coli* estimated that the chemical half-life of bulk mRNA could be as short as 10 s (26, 27). Based on this measurement, it was proposed that each message might only be translated once to produce a single protein molecule. In practical terms, this reduces the control of gene expression to the control of transcription initiation. We now know that the situation is more complex. Although short lived, messages can support repeated rounds of translation initiation to form polyribosomes, and there are many examples where posttranscriptional control has a major role in the regulation of gene expression.

Seminal work on the *trp* operon showed that six genes are expressed as part of a large polycistronic message. Although an analysis of the 3' end of the *trp* message suggested that mRNA degradation proceeds in the 3'–5' direction, it was subsequently realized that the half-life of *trp* mRNA is actually shorter than the time required to transcribe the entire operon (28, 29). That is, cistrons in the 5' end of the mRNA are translated and degraded before the mRNA corresponding to the cistrons in the 3' end are synthesized. This observation, taken together with subsequent studies of other operons, gave rise to the notion that mRNA degradation in *E. coli* proceeds in a 5′–3′ wave that follows transcription and translation of the mRNA. The simplest hypothesis to explain this directionality would be degradation by a 5′-exoribonuclease. However, after several decades of intensive experimental work, there has been no definitive identification of a 5′-exoribonuclease in *E. coli*.

In 1973, an alternative model was proposed suggesting that mRNA degradation in *E. coli* is initiated by endoribonucleolytic cleavage (*30*). After fragmentation of the mRNA by an endoribonuclease, 3′-exoribonucleases such as RNase II or PNPase could act to degrade the mRNA fragments into its constituent nucleotides. The principal feature of this model was that it reconciled the apparent contradiction between chemical degradation by 3′-exoribonucleases with an overall 5′–3′ directionality. At the time, experimental evidence that either RNase II or PNPase were involved in mRNA degradation was tenuous and there was no obvious candidate for the hypothetical endoribonuclease involved in mRNA turnover. The field underwent a hiatus of nearly fifteen years before enzymes involved in mRNA degradation were clearly identified and characterized.

B. RNase E

Interest in the mechanism of mRNA degradation was revived in the 1980s due in part to work using *E. coli* as an expression host for recombinant proteins. It became apparent that driving expression with strong transcription and translation signals was not always enough for high yields of recombinant protein. In some cases, mRNA instability appeared to be a factor contributing to poor expression. In 1986, an important article that marks the revival of the study of mRNA degradation appeared (*31*). The fate of mRNA was analyzed in mutant strains of *E. coli* in which the genes encoding RNase II (*rnb*) and PNPase (*pnp*) were inactivated. The inactivation of both enzymes together is lethal and mRNA fragments 100–1500 nt in length accumulate in an rnb^{ts} *pnp* mutant at the nonpermissive temperature. These results validated the importance of RNase II and PNPase in mRNA degradation and renewed interest in identifying endoribonucleases involved in mRNA degradation.

RNase E was originally discovered as an *E. coli* rRNA-processing enzyme (*32, 33*). RNase E is a single-strand-specific endoribonuclease with a preference for AU-rich sequences (*34–37*). The first clue that RNase E might be involved in mRNA degradation came from the study of bacteriophage T4. Using a temperature-sensitive *rne* mutant *E. coli* strain, it was demonstrated that RNase E had an important role in the processing and degradation of the bacteriophage's mRNA (*38, 39*). At the nonpermissive temperature, there was significant stabilization of many bacteriophage T4 messages. Moreover, the chemically stabilized T4 messages were found to be capable of supporting protein synthesis for extended periods after the inhibition of transcription by treatment with rifampicin. In the wild-type background, the arrest of transcription by rifampicin resulted in the rapid turnover of the existing mRNA and in the concurrent loss of capacity to synthesize bacteriophage T4 proteins. These studies were corroborated by *in vitro* work showing that the stabilization of the bacteriophage T4 messages correlated with the inactivation of the endoribonuclease activity of RNase E at the nonpermissive temperature.

It was initially unclear if the role of RNase E in mRNA degradation was a peculiarity of bacteriophage T4 or a general feature of mRNA degradation in *E. coli*. At the time, there was a bias against the idea that an enzyme such as RNase E might be involved in mRNA degradation because it already had a clearly established role in rRNA processing. The thinking was that enzymes involved in stable RNA processing would be distinct from enzymes involved in unstable RNA degradation. Nevertheless, in a relatively short period, several different groups presented evidence that RNase E was an important enzyme in the degradation of *E. coli* mRNA (*40–43*). Amongst the evidence that RNase E was involved in mRNA degradation was the finding that two previously identified genes, *rne* (RNase E) and *ams* (affects mRNA stability) were actually different mutant alleles of the same gene, encoding RNase E (*44, 45*). The *ams* allele was originally identified and characterized as a temperature-sensitive mutation that slowed the chemical degradation of mRNA at the nonpermissive temperature (*46*). Subsequent work revealed that the mutant alleles encoded RNase E variants with single amino acid substitutions that led to inactivation of the enzyme at high temperature (*44*).

It should be noted here that the effects of the *rne* and *ams* alleles on *E. coli* mRNA degradation were not nearly as dramatic as their effect on bacteriophage T4 mRNA degradation. At the nonpermissive temperature, *E. coli* mRNA continues to turnover albeit more slowly. Nevertheless, subsequent studies of specific mRNAs confirmed the importance of RNase E in *E. coli* mRNA turnover and suggested that an endoribonucleolytic cleavage by RNase E is often the initiating event in *E. coli* mRNA degradation. We will return to this issue later when we consider the possibility that *E. coli* has parallel or alternative pathways of mRNA degradation independent of RNase E.

The work from this period was a landmark in the field of mRNA degradation. It firmly established, for the first time in any organism, a working model for the mechanism of mRNA degradation. It also dispelled the notion that enzymes involved in the maturation of stable RNA are distinct from the enzymes involved in the degradation of mRNA. We know today that in *E. coli* there is a significant overlap between the machineries involved in RNA processing and degradation, and that this is also the case in eukaryotic organisms (see (*11*) and the chapter by Deutscher in this volume covering a related topic: maturation and quality control of rRNA).

C. Endonucleolytic Initiation of mRNA Degradation

The developments in the field of RNA decay have cumulatively provided a picture of the generic process of mRNA turnover in *E. coli*. Under normal laboratory conditions of balanced growth, the principal pathway for mRNA degradation in *E. coli* is believed to be mediated by RNase E (Fig. 1). In the model, RNase E inactivates mRNA by endoribonucleolytic cleavage to produce mRNA fragments, which are then digested to nucleotides by 3'-exoribonucleases, with help from RhlB, PAP, and oligoribonuclease. Here, the mRNA is depicted as polyribosomal in keeping with the notion that mRNA degradation in *E. coli* proceeds in a 5'–3' wave that follows transcription and translation of the mRNA. However, we are not aware of any direct evidence that polyribosomal mRNA is the actual target of RNase E. The complex relationship between translation and mRNA degradation is the subject of the chapter by Dreyfus in this volume. In wild type cells, it is difficult to detect the mRNA fragments depicted in Fig. 1 since they are rapidly digested by the 3'-exoribonucleases. That is, under normal circumstances, the initial cleavage by RNase E appears to be the slowest of the major steps that occur in the chemical degradation of mRNA. The elucidation of this pathway is based on the biochemical characterization of the enzymes in Fig. 1 and physiological studies employing *E. coli* strains with mutations in the genes encoding these enzymes. RNase E and oligoribonuclease (oligo-RNase) are the only essential enzymes in

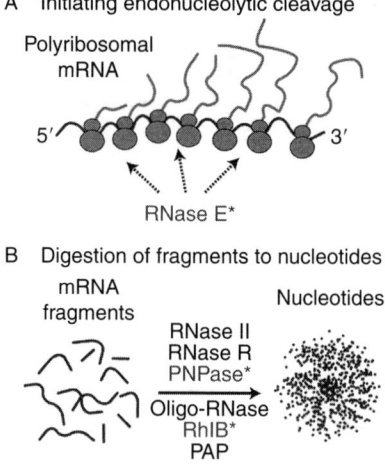

FIG. 1. Endonucleolytic initiation of mRNA degradation in *E. coli*. (A) Attack of polyribosomal mRNA by RNase E. (B) Digestion of mRNA fragments to nucleotides by a combination of exoribonucleases and accessory factors that facilitate this process. Three of the enzymes in the figure (asterisks) are components of the multienzyme RNA degradosome.

this pathway. Strains in which both RNase II and PNPase, or both RNase R and PNPase have been inactivated are also not viable (*11*). Interestingly, strains in which both RNase II and RNase R have been inactivated are viable, suggesting that PNPase by itself can perform all of the essential functions of the exoribonucleases.

D. The RNA Degradosome

Three of the enzymes in Fig. 1, RNase E, PNPase, and RhlB, are components of a multienzyme complex known as the RNA degradosome. The RNA degradosome was discovered during work in two different laboratories on the purification and characterization of RNase E, and the identification of factors affecting the activity of PNPase (*34, 47, 48*). The major components of this multienzyme complex are RNase E, PNPase, RhlB, and enolase (*7, 49*). More recent work suggests that the composition of the RNA degradosome might vary depending on conditions of growth or stress (*50–53*). As a general comment, the available evidence indicates that most if not all of the RNase E in the cell is part of the multienzyme RNA degradosome. For this reason, in the section below on the structure of RNase E, we detail the protein–protein interactions between RNase E and the other components of the RNA degradosome. A comprehensive review on the RNA degradosome of *E. coli* has appeared recently (*54*) and we direct the interested reader to this article for further background information.

E. The Influence of the RNA 5′ End on RNase E Activity

Clues that the 5′ end of an RNA substrate can influence the endoribonucleolytic activity of RNase E came from the study of the *ompA* mRNA, which encodes an outer membrane porin protein, and RNAI, the antisense RNA regulator of ColE1-type plasmid replication. The stability of the *ompA* message varies with growth rate, and we now know that this is due to the translational control by a small regulatory RNA (sRNA) (*55, 56*). Under growth conditions where the *ompA* mRNA is efficiently translated, it has a half-life of 15–20 min, which is one of the longest in *E. coli* (*57*). When the 5′ UTR of the relatively labile *bla* mRNA (half-life ∼3 min) was replaced with the corresponding segment of *ompA* mRNA, its stability increased to a level similar to that of the donor transcript (*58*). Subsequent characterization revealed that the salient feature required for increased stability was a stem-loop no more than 2–4 nt from the extreme 5′ end (*59*). Such a structure was then shown to stabilize RNAI (*60*), which importantly was known to be a substrate of RNase E (*37, 61*). This implied that RNase E has exceptional substrate specificity for an endoribonuclease; that is to say, it preferentially cleaves RNAs that have several unpaired nucleotides at the 5′ end.

In a separate line of investigation, it was found that RNase E cleavage of natural substrates such as *rpsT* mRNA can be enhanced *in vitro* by the presence of a 5′ monophosphate on the RNA. Substrates of the same sequence, but with a hydroxyl or triphosphate group at the 5′ end were cleaved less efficiently (62–65). This led to the suggestion that RNase E contains a binding pocket for a 5′ monophosphate group (64). Moreover, the hybridization of a complementary DNA oligonucleotide to the 5′ end of *rpsT* mRNA was shown to inhibit RNase E, even though the 5′ end was monophosphorylated. This indicated that a single-strandedness at the 5′ end is a prerequisite for recognition of the 5′ monophosphate group. The importance of access to the 5′ end *in vivo* was demonstrated by the substantial increase in stability conferred on *rpsT* mRNA when it was circularized (64, 66). In these experiments, the only chemical difference between the linear and circular substrates was the 5′ triphosphate and 3′ hydroxyl versus the covalent linkage of the 5′ end to the 3′ end by a phosphodiester bond.

Recently the X-ray crystal structures of the catalytic domain of *E. coli* RNase E in complexes with oligonucleotide substrates (see later text) confirmed the existence of a pocket that binds a 5′-monophosphate group. The residues that contact the 5′ monophosphate are conserved in *E. coli* RNase G and other homologs (67). To our knowledge, all *E. coli* RNase E homologs that have been tested have been found to be capable of stimulation by the presence of a 5′ monophosphate on substrate RNA (63, 68–70). Moreover, the recent discovery of the RppH 5′-pyrophosphatase (5, 6) explains why RNAs such as *ompA* mRNA with a 5′ terminal stem-loop can be more stable *in vivo* (59, 60), even though a triphosphate is sufficient *in vitro* to block 5′ end-stimulated cleavage by RNase E (64). A 5′ stem-loop reduces the efficient with which RppH converts a triphosphate to a monophosphate group (5, 6) and sterically blocks the interaction with the 5′ monophosphate binding pocket of RNase E (67). The work described in this section has led to the notion that "the end defines the means in bacterial mRNA decay" (71).

III. Evidence for Diverse Pathways of mRNA Degradation in *E. coli*

A. Limitations of Endonucleolytic Initiation of mRNA Degradation

The initiation of transcription and translation can be described as the binding of RNA polymerase to promoter DNA and the assembly of the ribosomal subunits and initiator tRNA at the start codon of mRNA, respectively. These interactions are essential for transcription and translation. In contrast, the initiation of mRNA decay cannot be categorized by a specific interaction. No single gene has been identified that when inactivated blocks *E. coli* mRNA degradation completely. Although RNase E is essential for cell viability (46, 72), in general mRNA decay is only slowed in strains defective for this enzyme (46, 73). Thus, it appears that there is redundancy in the enzymes that undertake the initial nucleolytic step or that multiple pathways exist by which mRNA can be degraded. The degradation of mRNA in eukaryotes by either a 5'- or 3'-exonucleolytic pathway provides a precedent for the coexistence of distinct pathways (for recent reviews see (74–77)).

The general model depicted in Fig. 1 is useful for thinking about mRNA decay in *E. coli*; however, alternate pathways should also be considered. Transcriptome analysis suggests that a significant proportion of transcripts are not stabilized by inactivation of RNase E (73). Moreover, endoribonucleases other than RNase E, including RNase G, RNase III, RNase P, and RNase Z have been implicated in mRNA decay (11, 19, 21, 73, 78, 79). However, strains in which RNase E and each of the endoribonucleases mentioned above have been inactivated exhibit mRNA degradation rates that are only slightly slower than when RNase E alone has been inactivated. Thus, compared to RNase E, the other endoribonucleases appear to have a more limited role in mRNA turnover. In a strain in which RNase E, PNPase, RNase II, and PAP have been inactivated, mRNA degradation is strongly impeded (80). Taken together, these results suggest that, in the absence of RNase E activity, *E. coli* has an alternative mRNA degradation route provided by PAP and the exoribonucleases. DNA microarray studies suggest that mRNA polyadenylation in *E. coli* is more widespread than previously thought and that the steady-state level of a significant cross-section of transcripts increases when the gene encoding PNPase is disrupted (81, 82). One important unanswered question is what proportion of mRNA degradation is initiated by 3' polyadenylation and whether this proportion changes in response to growth conditions. Other key questions are what proportion of the mRNA is functionally stabilized by inactivation of PAP and the exoribonucleases, and to what extent RNase E and other endoribonucleases are capable of functionally inactivating mRNA independent of 3' exonucleolytic attack.

B. Degradation of the *rpsO* mRNA

The degradation of certain transcripts has been studied in detail. For example, it has been shown that *rpsO* mRNA is cleaved by RNase E in its 3' UTR; moreover, inactivation of RNase E blocks cleavage at this site and results in the mRNA being degraded at a slower rate (40). Cleavage by RNase E removes the stem-loop of the transcriptional terminator from the 3' end. Such structures are known to impede 3' exonucleolytic attack (for reviews see (12–14)), and the upstream product of RNase E cleavage is more stable in the absence of the 3′ exonuclease PNPase (*83*). Thus, it can be concluded that RNase E cleavage functions to limit the longevity of *rpsO* mRNA. Although RNase E cleavage of *rpsO* mRNA can be referred to as the initial endonucleolytic step in its degradation, it should be understood that it is not required for mRNA degradation. In the absence of functional RNase E, the *rpsO* mRNA is still degraded, albeit at a reduced rate (*40*). This is explained by the finding that 3′ stem-loops restrict rather than abolish 3′-exonucleolytic attack.

The degradation of the *rpsO* mRNA is virtually blocked however, when cells are defective not only for RNase E, but also PNPase, RNase II, and poly(A) polymerase (*80*). As described earlier in this chapter, the latter enhances 3′ exonucleolytic attack (*4, 80, 84–87*). Thus, it appears that *rpsO* mRNA is degraded primarily by a 3′–5′ pathway that is accelerated by RNase E-mediated removal of the 3′ terminal stem-loop. The 3′ exonuclease RNase II, has been shown to antagonize mRNA degradation by trimming the 3′ oligo(A) tails that are used by PNPase to access its substrates (*88*). The degradation of *rpsO* mRNA is fully consistent with and exemplifies the general model (Fig. 1). Other examples of this mode of mRNA degradation include RNase E cleavage in the 3′ UTR of *cspA* mRNA and the intergenic regions of the polycistronic *papAB* (*89*) and *ftzQAZ* mRNAs (*90*). Thus, cleavage of such regions may be a common mechanism for accelerating the decay of segments that lie upstream.

C. Degradation of the *rpsT* mRNA and Internal Entry

The best studied mRNA that is subject to 5′ end-stimulated decay is the *rpsT* mRNA, a known substrate of RNase E (*91*). Its longevity can be increased via the addition of a 5′ terminal stem-loop (*5, 92*) or disruption of the *rppH* gene encoding the RppH 5′-pyrophosphohydrolase (*5*), which is related to eukaryotic decapping enzyme Dcp2. Whether conversion of the triphosphate to a monophosphate group at the 5′ end places an upper limit on the rate of mRNA decay remains to be determined. If it does, then increasing the level of RNase E activity in the cell would not be expected to increase the overall rate of degradation. The major site of RNase E cleavage in *rpsT* is within the coding region, ∼300 nt from the 5′ end. Thus, it appears that a 5′ monophosphate can stimulate cleavage at a distance and in competition with translating ribosomes *in vivo*. 3′ UTRs and intergenic regions might also be targets for 5′ end-stimulated cleavage, although not in the case of *rpsO* mRNA (discussed further later). The downstream product of RNase E cleavage of *rpsT* is not readily detectable, suggesting that its 3′ exonucleolytic decay is enhanced by RNase E cleavage (*93*).

Interestingly, disruption of *rppH* was found to cause a greater stabilization of *rpsT* and other mRNAs including *yeiP* than disruption of *rne*, suggesting that these transcripts may be degraded by multiple, 5′ monophosphate-stimulated pathways (5). The *yeiP* mRNA is known to be cleaved by RNase G (73), an RNase E paralog that is also 5′-end-dependent (70, 94). It has also been reported for RNAI that 3′ oligoadenylation by PAP is affected by its 5′ phosphorylation status (95). The possibility that pyrophosphohydrolysis also functions to stimulate 3′ exoribonucleolytic attack of mRNA merits further investigation. Although RppH appears to function as a key regulator of 5′-end-dependent mRNA decay, the enzyme is nonessential and disruption appears to enhance the stability of only a proportion of the *E. coli* transcripts (5), including those known to be stabilized by disruption of RNase E (73). This can be interpreted as support for the existence of alternative pathway(s) by which RNase E can initiate mRNA decay independent of 5′-end stimulation. It has already been reported that the addition of a 5′ stem-loop to the *rpsO* transcript does not affect the rate of its RNase E-mediated decay *in vivo* (96), and the insertion of a segment cleaved by RNase E into a similarly protected transcript resulted in its destabilization *in vivo* (92). This form of RNase E cleavage has been termed "internal entry" (92). However, to date, there is no report of the importance of internal entry relative to 5′ monophosphate-stimulated cleavage with regard to the degradation of the mRNA pool, its requirements in terms of the substrate and enzyme, and the general significance for gene regulation. Nevertheless, a larger number of transcripts appear to be stabilized by inactivation of RNase E versus RppH (see later text). This is consistent with the notion that the decay of many transcript could be initiated by RNase E cleavage independent of an interaction with a 5′ monophosphate generated by RppH.

D. Translation and Cellular Localization of the mRNA Degradation Machinery

The 3′–5′ directionality of mRNA decay raises the possibility that transcripts will become truncated at their 3′ ends before ribosomes have completed their translation into polypeptide. In such situations, we speculate that tmRNA with its dual tRNA- and mRNA-like nature may facilitate 3′ exonucleolytic attack via its ability to release stalled ribosomes (97–100). In addition, it is conceivable that the translation and degradation of a bacterial mRNA may not be entirely concurrent. RNase E and the other components of the RNA degradosome have been shown to associate with the inner surface of cells (101) and recently interactions with the bacterial cytoskeleton (102, 103) and the inner cytoplasmic membrane have been described (104). Thus, when engaged with the degradation machinery, an mRNA might be isolated from the translational machinery. We speculate that the converse may also be true; for example, it has been shown for two mRNAs that increased ribosome binding is sufficient to enhance their stability (*105*, *106*). The significance of the subcellular localization of the degradosome and the possible coordination of translation and RNA degradation are both areas that merit investigation.

IV. The Structure of RNase E

A. The Organization of RNase E Functional Domains

In comparison with the other proteins encoded in the *E. coli* genome, RNase E (accession number P21513) is comparatively large, comprising 1061 amino acids in strain K12. The N-terminal half of *E. coli* RNase E, which encompasses the ribonucleolytic function, shares a high degree of sequence similarity with its paralog, RNase G (corresponding to roughly 50% similarity), and is well conserved among the many homologs identified in bacteria (*107–109*). Homologs have also been identified in the nuclear or chloroplast genomes of certain plants and algae. In contrast to the conservation of the N-terminal half, the C-terminal half of *E. coli* RNase E is much more divergent even amongst the closest homologs in other γ-*Proteobacteria* (*110*). This portion varies widely in length and sequence composition amongst homologous enzymes, and in some species, RNase E resembles a circularly permuted version of the *E. coli* enzyme, in which the variable portion proceeds rather than follows the conserved catalytic region. A notable example of this segment shuffling is RNase E from *Streptomyces coelicolor* (*109*). In view of this segment shuffling, it might be better to refer to the conserved N-terminal half as the "catalytic core" of RNase E.

For *E. coli* RNase E, the C-terminal half, which we will refer to as the noncatalytic region, provides the principal protein–protein interactions of the RNA degradosome (*111*, *112*). This portion of RNase E has low sequence complexity, but of a type that characterizes proteins predicted to have little intrinsic structure in the absence of binding partners. Corroborating this, the isolated recombinant noncatalytic region has little secondary structure as determined by circular dichroism, and its shape in solution is extended rather than compact as judged by X-ray solution scattering (*113*). It is possible that the noncatalytic region gains secondary structure or compacts upon association with the other degradosome components, and this awaits experimental evaluation. The recombinant noncatalytic region can be used to reconstitute a minidegradosome incorporating highly purified RhlB and PNPase (*50*), suggesting that these protein–protein interactions do not require cofactors. The noncatalytic region of RNase E corresponds to nearly 550 amino acids and is one of the longest regions of natively unstructured protein predicted for the *E. coli* genome. Comparably, large noncatalytic regions are conserved throughout homologs in the *Proteobacteria* (*110*). Conservation of the coding sequence, and transcription and translation of such an extensive region must represent a burden on organism fitness, so it is likely to have an important beneficial role to have been maintained in evolution.

The noncatalytic region of RNase E is punctuated by small segments of comparative sequence conservation amongst the *Enterobacteria* although these segments are not as well conserved as sequences in the catalytic core (*110*). Four segments were identified that are predicted to have propensity for forming secondary structure (*113*). These segments are depicted in Fig. 2 and are annotated as A, B, C, and D. Segments B, C, and D overlap with regions identified as protein–protein interaction sites in yeast two-hybrid studies (*111*). Segment A encompasses an amphipathic α-helix that can associate with membranes and localizes the degradosome to the inner cytoplasmic membrane (*104*). Segment B has characteristics of a coiled coil and may engage RNA, but is not stably folded in isolation (*113*). A 69-residue conserved segment downstream of segment B is the interaction site with the C-terminal domain of DEAD-box RNA helicase, RhlB (*114, 115*). Segment C is the enolase binding site (*116*), and segment D interacts with PNPase (*113*).

Insight into the recognition of segments C and D have been provided by crystal structures in complex with enolase and PNPase, respectively (*116*) (Nurmohamed *et al.*, submitted) (Fig. 3). Enolase is a homodimer with an extensive protein–protein interface formed between the β-barrel (belonging to the trioseisomerase (TIM) barrel structural family) structural domain. A helical subdomain lies on one surface of the enolase β-barrel domain and the symmetry-related pair

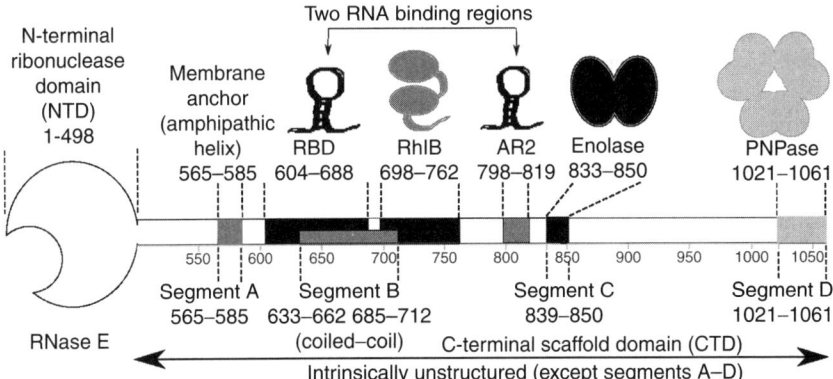

FIG. 2. Schematic of the protein–protein and protein–RNA recognition sites in the noncatalytic region of RNase E.

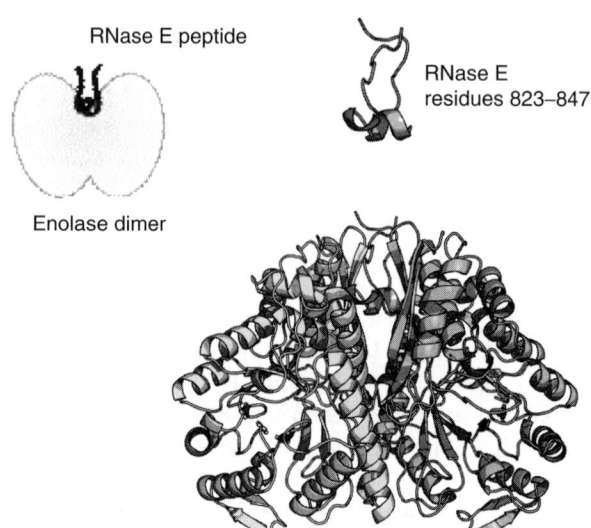

Fig. 3. Crystal structure of *E. coli* enolase in complex with the C-peptide. A schematic of the enolase dimer complex with the RNase E peptide is shown in the upper left panel. The right panel shows a peptide backbone trace of the complex, with side chains and water molecules not shown for clarity. The RNase E peptide encompassing residues 823–847 is engaged in a groove between the protomers of the enolase dimer. The isolated peptide is shown in the upper panel. The model was provided courtesy of Salima Nurmohamed.

form an inter-protomer groove that is well conserved even in eukaryotic enolase (*116, 117*). One molecule of segment C is engaged to the enolase dimer. There is a mismatch of molecular symmetry of the C peptide and the dimeric enolase, so the complex is asymmetric with two equivalent ways for the C peptide to engage. It seems likely that in the degradosome, one enolase dimer might bind each RNase E protomer, but the available data from cell-extracted degradosomes is not clear on the stoichiometry. The stoichiometry of enolase and perhaps the other degradosome components may be modulated by regulatory binding partners, such as RraA and RraB (*53, 118, 119*).

The C peptide is tightly nestled into the enolase inter-protomer groove, and the peptide forms a short helix and extended strand. The binding site is at a distance from the active site and peptide binding does not induce any appreciable conformational changes, accounting for the observation that enolase enzymatic activity is not affected by recruitment into the RNA degradosome (*7, 113*). The sequence of segment C is conserved in RNase E from other Gram-negative bacteria, including pathogens such as *Salmonella typhimurium, Yersinia pestis, Y. pseudotuberculosis, Shigella flexneri*, and *Erwinia cartovora*, and the constellation of contacts between enolase and RNase E peptide may account for the conservation of the key recognition determinants that support the internal contacts within the folded peptide and the extrinsic contacts to the enolase (*116*). The termini of the C-peptide are surface-exposed in the complex with enolase, so there is a clear entry and exit paths for the peptide in the groove between the subunits.

The high-resolution crystal structure of the catalytic core of *E. coli* PNPase in complex with peptide D has been solved at 2.5 Å resolution (Nurmohamed *et al.*, submitted). This structure reveals that a segment of about 20 residues from the D-peptide engages an exposed groove between protomers on the surface of the homo-trimeric PNPase (Fig. 4). The D-peptide assumes a β-strand conformation with its peptide backbone forming β-sheet hydrogen bonding interactions with the surface-exposed terminal ridge of a β-sheet from PNPase. This type of interaction generates a pseudo-continuous antiparallel β sheet that extends from deep within the PNPase core and terminates with the RNase E D-peptide. The side chains of the D-peptide form van der Waals interactions with residues on either side of the intra-protomer groove in PNPase. One molecule of D-peptide is engaged to each protomer of the PNPase homo-trimer, consistent with solution studies by nondissociating mass spectrometry (*113*). The mode of D-peptide binding indicates an optimal binding stoichiometry of one PNPase protomer engaging each RNase E protomer, with implications for the organization of the interactions in the RNA

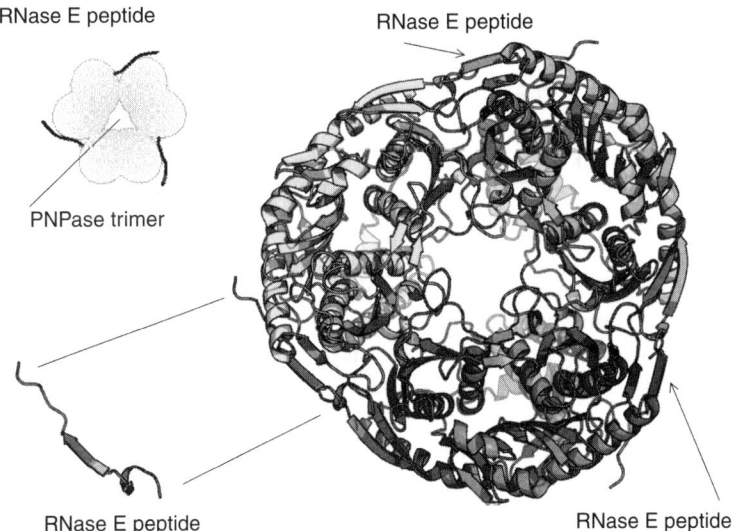

Fig. 4. Crystal structure of *E. coli* polynucleotide phosphorylase (PNPase) complexed with its recognition site from RNase E. The isolated peptide is shown in the lower left panel. A schematic of the complex of the three copies of the RNase E peptide with the PNPase is shown in the upper left panel. The model was provided courtesy of Salima Nurmohamed.

degradosome (see (54, 110)). The dissociation constant of the D-peptide is estimated to be roughly 1 μM by isothermal titration calorimetry (Nurmohamed et al., submitted). Although this binding represents a weak molecular affinity, binding of the trimeric PNPase to the full-length tetrameric RNase E may be much greater due to the localization of the binding partners. Curiously, the shuffled section of S. coelicolor RNase E mentioned earlier forms a stable complex with polynucleotide phosphorylase (109), but has no clear homologous sequence to the E. coli D-peptide. Nonetheless, S. coelicolor RNase E may engage its PNPase partner with a similar extended-sheet mode as seen for the E. coli pair, since this interaction can be accommodated with a different set of amino acid pairs in RNase E and PNPase, as long as they provide complementary match of van der Waals surfaces.

B. The RNase E Catalytic Domain is a Composite of Recurrent Structural Subdomains

The crystal structure of the catalytic domain of RNase E (67), corresponding to residues 1–529, is shown in Fig. 5. The structure can be decomposed into individual subdomains that closely resemble the globular folds of RNase H, DNase I and the ancient and ubiquitous S1 RNA-binding domain (120). The S1 domain was identified from sequence similarity with other family members, and it is interesting to note that a similar S1 fold is found in the 3′-exonucleolytic partner of RNase E in the degradosome, polynucleotide phosphorylase (121). In contrast, the RNase H and DNase I subdomains have highly divergent sequences from other members of their respective structural families, and their folds were not predicted in advance of the crystal structure. Another domain, referred to as the 5′-sensor for its role in recognition of the 5′-end of the RNA (see later text) has a mixed α/β fold, which forms a unitary structure with the S1 domain. The S1 domain and 5′-sensor move together as one unit with accommodation of substrate, and we will return to this point later. In the RNase E homologs of plants, algae, and certain bacteria, such as the α-proteobacterium *Caulobacter crescentus*, there are insertions into a loop region within the S1 domain (109, 122). The function of these insertions is unknown currently. The last domain, referred to as the small domain, forms protomer–protomer interfaces.

C. RNase E Catalytic Domain Quaternary Structure

Solution studies indicate that the isolated catalytic domain of *E. coli* RNase E forms a stable homotetramer (123). RNase G shares high level of sequence identity (35%) and similarity (50%) with RNase E catalytic domain, but is predominantly a homodimer (124). The Mycobacterial RNase E/G homolog is in equilibrium between dimeric and tetrameric forms (68). The crystal

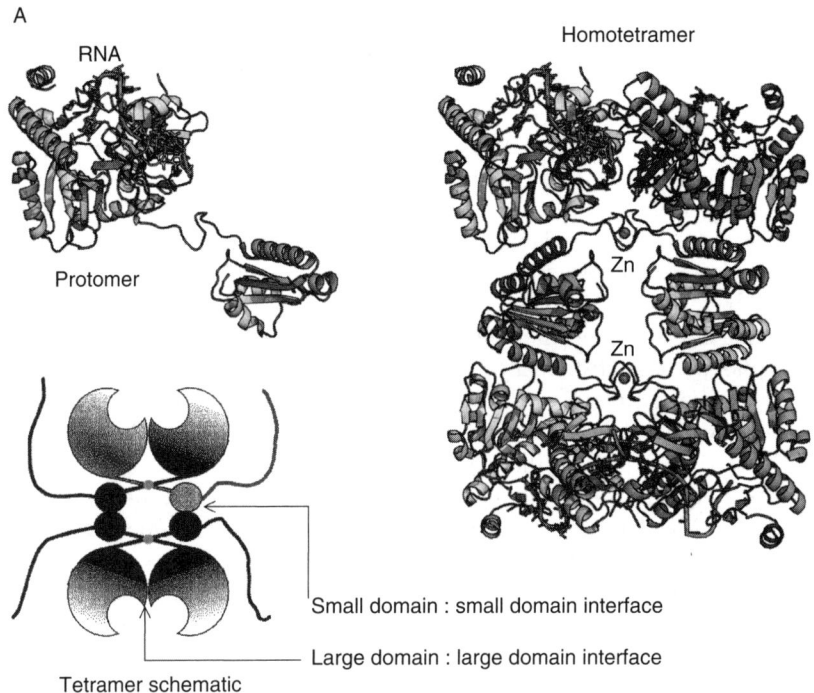

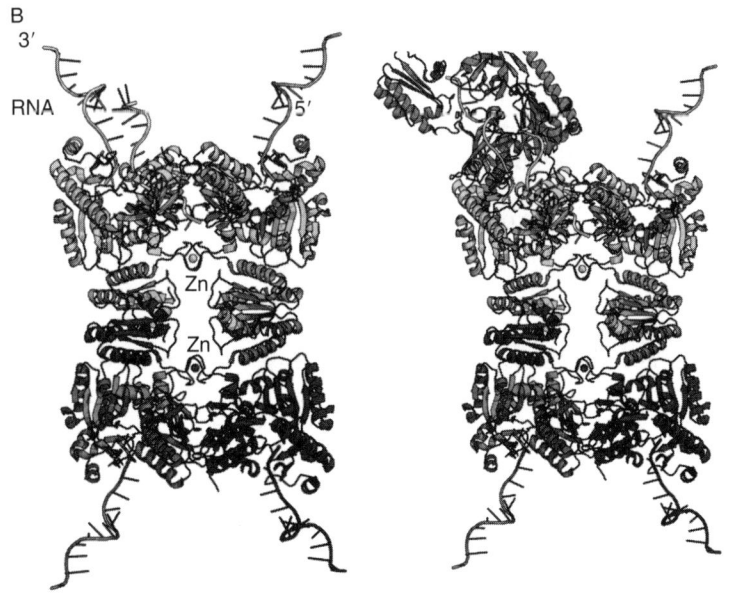

structure of *E. coli* RNase E reveals that the tetramer is formed as a dimer-of-dimers through two nonequivalent interfaces. One is made by self-complementary face of the DNase I subdomain; the second is made by small domain–small domain interactions.

In isolation, each protomer appears elongated, with a large domain comprising the cluster of the main domains (S1, 5′-sensor, RNase H, and DNase I) followed by an elongated linker region, and then the small domain. Each dimer is made by the cross-like packing of these elongated protomers, so that the subunit organization resembles a pair of scissors (Fig. 5A). The main dimer contact is formed by the DNase I domain, and this is an extensive hydrophobic interface. The small domain forms the dimer–dimer interface; this is also predominantly hydrophobic, but substantially smaller than the DNase I–DNase I interface. At the junction point where the two subunits of the dimer cross, there is a zinc binding site (67). Four cysteines coordinate the zinc, with two of these cysteines contributed from each of the paired protomers. This gives an organometallic bond that links the two protomers physically through a shared metal coordination scheme that has been termed a "zinc-link" (125). The coordinating cysteine motif consensus, CxxCxGxG is well conserved in all known RNase E and RNase G homologs to date. The plastid RNase E/G homolog from *Arabidopsis thaliana* and other plants and algae have a zinc-link motif (122). Disruption of the zinc link by mutations causes destabilization of the tetramer and these variants exist predominantly as a dimer. Considering the extensive hydrophobic interface of the DNase I domain seen in the native structure, the lost interaction is likely to be between the small domain and small domain interface, suggesting that the zinc-link is important for correctly positioning the small domains to make the dimer–dimer interaction.

V. Substrate Specificity and Catalytic Mechanism of RNase E

A. The Active Site and Cleavage Mechanism

The active site residues D346 and D303 coordinate a single magnesium ion (Fig. 6). The role of these coordinating residues has been corroborated by mutation of the aspartates to asparagines, which involves single atom substitutions. The individual substitutions each decrease the activity on 13-mer RNA

FIG. 5. Subunit organization of the catalytic core of RNase E. (A) The lower left panel shows a schematic of the tetramer, showing the arrangement of subunits. The top right panel presents a view of the tetramer with 13-mer RNA and the top left panel shows an isolated protomer. (B) View of the tetramer with 15-mer RNA which mimics a structured RNA loop. In the crystal structure with this RNA, the 5′ end of the 15-mer RNA is engaged in the 5′ sensing pocket of one protomer and the 3′ end extends to the active site of a neighboring subunit in the adjacent crystal cell, shown in the panel on the right. This structure was used to generate speculative models of RNA engaged with an RNA hairpin (such as the model shown in Fig. 7B).

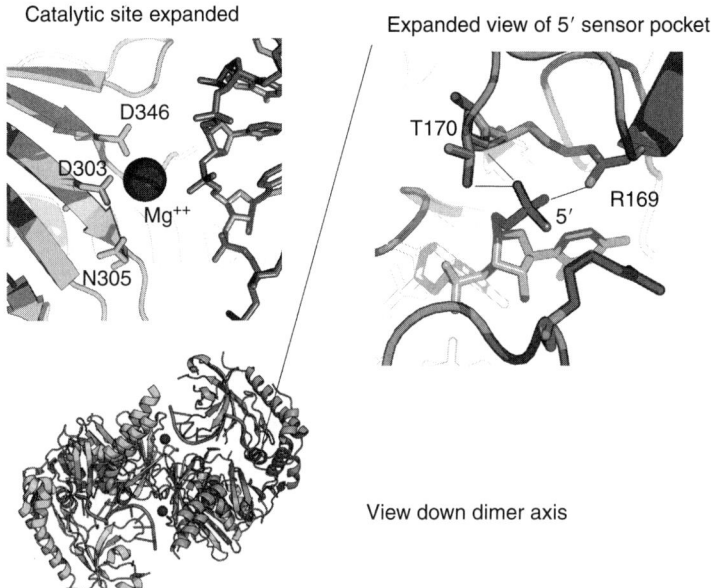

FIG. 6. The active site and 5′-monophosphate-binding pocket of RNase E. The RNA and the side chains of N305, D303, and D346 are shown in stick representation. The S1 domain, which helps to engage and orient the RNA, is not shown for clarity. The 5′ terminal phosphate forms hydrogen bonds (indicated by lines) with the OH group and amide NH of T170 and the guanidine group of R169.

substrate below the limit of detection (67). N305 forms a hydrogen bond with D303 to support its metal-coordinating geometry, and making the conserved substitution of N to D at this position is also strongly inhibiting. The importance of the metal is also shown by the sensitivity of catalytic activity to the chelator EDTA. The resolution of the crystal structure is not sufficient to visualize the hydration pattern, but it is anticipated that the metal will be coordinated by water as well as the carboxylate oxygens of the D303 and D346. The hydration shell around a magnesium ion is used frequently in a variety of ribonucleases to activate hydrolytic attack on the RNA backbone (126). High resolution structures reveal that often a pair of magnesium ions is required for forming the attacking hydroxyl group (127), as seen for instance in the DEDDh structural family of nucleases (128). However, the RNase E crystal structure reveals only one magnesium at the active site. The metal interacts with the phosphate backbone of the RNA, so it seems likely that the metal is positioned and held in orientation by the substrate itself. The OH of a water molecule makes a nucleophilic attack on the phosphate, presumably by a nucleophilic substitution (S_N2) mechanism that proceeds through a pentavalent intermediate with inversion of the prochiral center. The 2′-OH group at the scissile position does not appear to be involved, since substitution to fluorine has little effect on catalytic rates (129). The crystal structure used RNA substrates with 2′-O-methyl modifications, which may place steric restrictions on the approach to the transition state. This may explain why the reaction did not proceed in the crystal structure, even in the presence of Mg^{++}.

B. Catalytic Activation and a Model of Allosteric Switching

As described earlier, biochemical studies have shown that the 5′ end of the RNA substrate has a strong effect on the nuclease activity of RNase E and RNase G (64, 70, 94, 130). A substrate with a 5′-monophosphate is greatly preferred over a 5′-triphosphate or 5′-OH group, with cleavage rates at least an order of magnitude greater depending on the RNA. The effects of a 5′-diphosphate have not been explored. Additionally, the terminal base of the substrate must be accessible and not sequestered in a secondary structure, such as a hairpin. The question arises as to how the 5′ end of the substrate is recognized, and how this recognition boosts the catalytic power of the enzyme.

The crystal structure of the RNase E catalytic domain with bound 5′-monophosphate RNA reveals that the 5′ end of the substrate is bound in a pocket within a subdomain encompassing residues 118–215. The 5′-monophosphate group is engaged in a semicircular ring of hydrogen bonding interactions with the side chains of T170 and R169 and the main chain amide of T170 (Fig. 6). It is clear that a substrate with a 5′-OH group could not form this interaction network. A triphosphate group may be sterically occluded from the site. Substitution of the residues that contact the phosphate results in reduced cleavage rates for substrate with a 5′-monophosphate to roughly the same rate observed for the 5′-OH substrate (67). While the crystallographic data explains the recognition of the terminus, they do not explain how catalytic power is boosted by the recognition of the terminal phosphate group. It was proposed that the engagement of the 5′-monophosphate in the binding pocket triggers a conformational change that swings the S1 domain down onto the RNA to organize the catalytic site (the "mouse trap" model for allosteric activation). In accord with this model, protease digestion patterns suggest that the RNase E structure changes according to the terminal group of the substrate (123), and neutron solution scattering profiles indicate that the 5′-monophosphate group causes greater compaction of the structure (131).

The crystal structure has now been solved for the apo-form of the RNase E catalytic domain in two different crystal lattices (132). Both structures corroborate the mousetrap conformational switching, but reveal large conformational changes compared to the RNA bound form, whereby the S1 and 5′ sensing domains move together as one body to clamp down on the substrate (Fig. 7A).

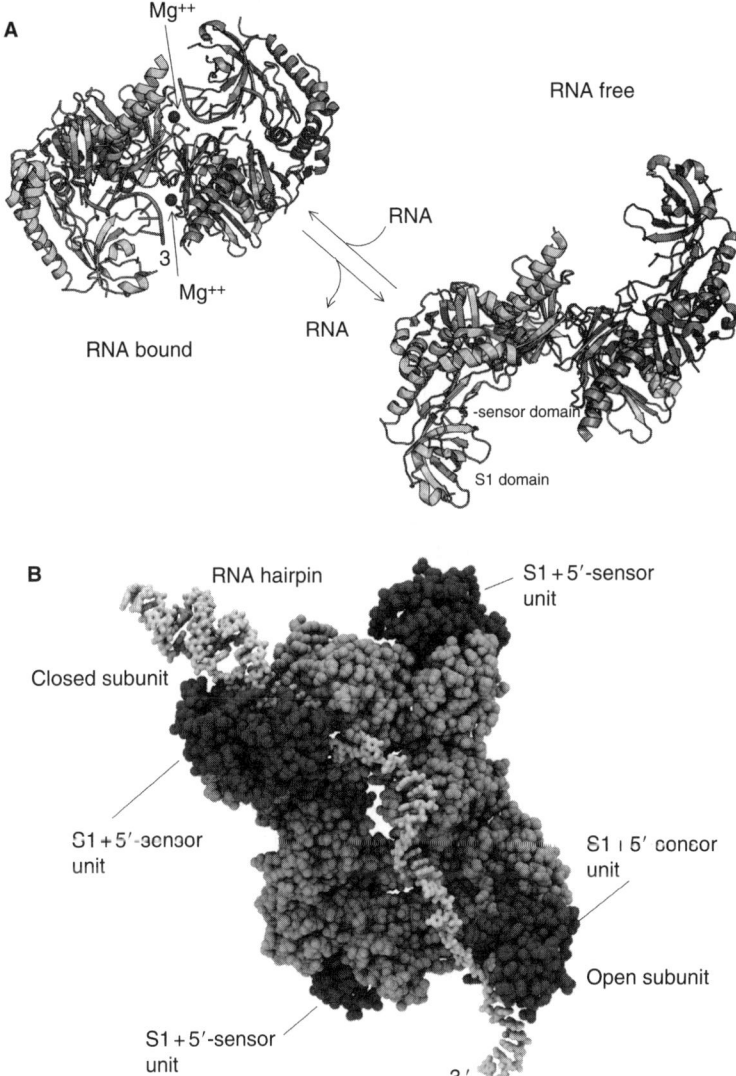

FIG. 7. Structural changes associated with RNA binding. (A) A view of the RNA bound form along the twofold axis of the principle dimer in the upper left panel, and the same view of the apo-form in the lower right panel. (B) A speculative model of the hybrid state with one subunit closed on the substrate in the upper left and the diagonally opposite subunit in the open state with the RNA engaged to the surface of the S1 domain. The model shows all the atoms of the protein and RNA in space filling representation.

This structural switch generates a groove to bind single-stranded RNA, organizes the catalytic site, and presents the phosphodiester backbone in an optimal geometry to favor cleavage. There is no clear density for magnesium at the active site in the apo-structure, even though the crystallization buffer contained magnesium at high concentration. The crystallographic resolution is limited for these structures, so caution must be used in interpreting the apparent absence of Mg^{++} in the apo-form, but it seems likely that one or two metals are corecruited with the RNA substrate and that this accompanies the allosteric switch and contributes to the binding energy of the substrate.

While the structural data corroborate the activational switch hypothesis, the question still remains how the engagement of the 5′ end of the substrate triggers the structural change. For instance, does the occupation of the phosphate-binding site in the 5′ sensing domain favor domain–domain interactions that stabilize the closed, active structural state? A sulfate ion occupies the site in the open form of the enzyme, so this seems unlikely to be sufficient to throw the conformational switch. It is possible that the 5′-terminus binding does not induce any conformational switch *per se*, but simply favors substrate binding. This is borne out by studies with RNase G, which show that the interaction with the 5′ monophosphate contributes to the overall affinity of binding (*133*). We thus envisage that the 5′-end dependency originates from the higher affinity of the 5′-monophosphate substrate. Binding the RNA increases the local concentration, and the equilibrium between open and closed forms becomes favored when the RNA is accommodated into the groove and in turn helps to stabilize the formation of the groove. Thus, a substrate with a 5′-OH group will bind and be cut, but not as frequently. A substrate with a 5′-triphosphate will sterically impede the structural transition required for cutting.

The wide sweep of the swinging arms formed by the consolidated 5′-sensing domain and the S1 domain may help to capture RNA. The structure of RNase E in complex with a small fragment of RNA shows that the RNA is bound on the surface of the S1 domain. Thus, the S1 domain in the tetramer may work to engage an mRNA at several points to keep the substrate in proximity for processive cutting (Fig. 7B).

C. The Puzzle of 5′-End Bypass

Notwithstanding the 5′-end preferences of most simple substrates, there have been some substrates identified in which the requirement seems to be bypassed, as indicated earlier. These are structured RNA substrates, and they can be cleaved with high activity by RNase E even if the 5′ terminus is sequestered in secondary structure and regardless of whether the 5′ end has a monophosphate, a triphosphate, or a hydroxyl group (*92, 134*). For these substrates, we envisage that the fold of the RNA organizes the active site of

RNase E for cleavage. A structured RNA, for instance, could be accommodated in the large cleft seen in the open state structure. All that would be required is for the RNA to present a single stranded region to the two aspartates at the active site. The S1 domain could make a stabilizing interaction with the large structured RNA, and this is supported by the crystal structure of RNase E with small fragments of RNA, which reveal an open conformation, but with RNA engaged on the surface of the S1 domain at the same site as seen in the closed state (132). It this way, the active site could be formed without requiring binding in the 5′ sensing pocket. In effect, the RNA structure plays the honorary role of protein in helping to form a channel for the single-stranded region to be presented to the active site aspartates. RNA fold rather than sequence would be the determinant for this activity.

D. The Functional Role of the Quaternary Structure

The question naturally arises as to whether the quaternary structure is required for function. The main dimer interface that is formed by the two DNase I domains is important for catalysis since the active site is a structural composite of elements from both of the subunits. Explicitly, the magnesium coordination site of subunit A is matched by the S1/5′-sensor and RNase H fold of subunit B. The same crossover, pretzel-like organization is expected for the RNase G homologs.

What about the dimer-of-dimers? Mutation of the cysteine residues that coordinate the Zn substantially decrease the activity for cleaving 10-mer substrates (125). However, E. coli mutants in which RNase E has been truncated and lacks the zinc-link region are viable, whereas RNase E null mutants are not viable, suggesting that catalytic function still operates, but the cells have a severely impaired growth (135). Thus, it is clear that the Zn-link and small domain interactions are important for organism fitness, suggesting that the quaternary structure of RNase E is required for an important aspect of the enzyme function. Other *in vivo* experiments also support this notion; a truncated version of RNase E is dominant negative for autoregulatory control of RNase E expression (136). A truncated RNase E homolog of *Y. pseudotuberculosis* is also a dominant negative for plasmid gene expression (137). The truncations are in the small domain and are likely to affect dimer–dimer interactions.

A clue as to how this might work is provided by the crystal structure of RNase E in the RNA-free apo-form, which reveals that the quaternary structure is highly flexible (132). It seems likely that the flexibility might be required to accommodate large structured RNA. The tetramer can flex at interfaces between the small domain of one subunit and the large domain of the partner subunit. The interfaces that remain invariant are ones that are made between like domains, namely the DNase I–DNase I interactions, and the small domain–small domain interactions. The quaternary flexibility results from repacking at the interfaces between different types of domain—for instance, the small domain of one subunit packing against the large domain of the partner subunit.

VI. Modulators and Adaptors of RNase E
A. RNase E Inhibitors

Recently, two protein inhibitors of RNase E named RraA and RraB were identified in genetic screens (118). These proteins inhibit RNase E activity *in vivo* and *in vitro*, and appear to work by interacting with the noncatalytic region of RNase E. RraA binding can not be localized to a single specific site in the noncatalytic region of RNase E whereas RraB has been shown to interact in the same region where RhlB binds. The failure to localize RraA binding to a single region suggests that is recognizes multiple sites in the noncatalytic region. *In vivo* expression of RraA or RraB had differential effects on the level of certain specific transcripts and on the composition of the RNA degradosome in an RNase E pull-down experiment (53). Both inhibitors significantly reduced the amount of PNPase in the RNA degradosome. RraA also significantly reduced the amount of RhlB and enolase that are bound. In these experiments, the DEAD-box RNA helicase CsdA was bound to RNase E in the cells where RraB was expressed. CsdA was previously shown to associate with RNase E in response to cold shock (50, 51). The work with RraA and RraB is the first to suggest that *E. coli* encodes protein factors that can affect RNase E activity and modulate the composition of the RNA degradosome.

B. Recruitment of RNase E by Hfq and sRNA

Hfq is an abundant *E. coli* RNA binding protein involved in a variety of processes including the posttranscriptional control of gene expression by small regulatory RNA (sRNA) (138–141). Two well-characterized sRNAs, RyhB, and SgrS, as well as many other sRNAs, require Hfq for function. The targets of RyhB are mRNAs involved in iron metabolism (142). The target of SgrS is the *ptsG* mRNA encoding a key component of the phosphosugar transport system (143). Hfq bound to the RyhB and SgrS sRNAs protects them from degradation by RNase E and mediates their interaction with target mRNA (52, 144). RyhB and SgrS block translation initiation by binding to the target mRNAs. Translational repression is coupled to the degradation of both the sRNA and the target mRNA in a coordinated reaction that is mediated by RNase E (52, 144). The degradation of the untranslated mRNA appears to be a secondary event making translational inhibition irreversible (145, 146). A complex containing RNase E, Hfq, and SgrS is formed under conditions of phosphosugar stress. The formation of the complex with Hfq and SgrS requires the same region of RNase E that is necessary for the formation of the canonical RNA degradosome and evidence suggests that the degradosome is remodeled because of the new interaction. In strains expressing variants of RNase E lacking the region required for the interaction with the Hfq–SgrS complex, the translation of *ptsG* is inhibited, but the mRNA is not degraded. These results have been interpreted as evidence that the physical interaction of RNase E with the Hfq–SgrS complex is required to recruit RNase E to the mRNA target to initiate degradation.

VII. Other Endoribonucleases in *E. coli*

A. RNase G

As described earlier, *E. coli* contains a paralog of RNase E called RNase G (*44, 147, 148*). The sequence similarity includes residues that the X-ray crystal structure of RNase E revealed as forming the 5′ monophosphate-binding pocket (*67*). Moreover, it has also been shown that RNase G preferentially cleaves substrates with a 5′ monophosphate group *in vitro* (*63, 70*). Early biochemical analysis of RNase G revealed that it has broad sequence specificity and could cut a segment of mRNA at multiple sites *in vivo* (*70*). This led to the suggestion, which has been borne out by subsequent experiments, that RNase G has a role in the degradation of mRNA (*73, 149–153*). Transcripts that are known to be stabilized in strains deficient for RNase G include *adhE* and *eno* mRNA; moreover, the protein products of these transcripts, alcohol dehydrogenase E and enolase, respectively, accumulate (*149, 150*). This indicates that RNase G limits the longevity of a proportion of translationally competent transcripts in *E. coli*.

Recently, it has been shown using strains containing variants of RNase G that the normal decay of *adhE* and *eno* mRNA by this ribonuclease requires the stimulation provided by interaction with a 5′ monophosphate group (*133*). Consistent with this, it has been confirmed that the *adhE* and *eno* transcripts that accumulated in the absence of RNase G have a 5′ monophosphorylated end (Kime and McDowall, unpublished data). The origin of the 5′ monophosphate group on *adhE* mRNA is RNase III cleavage of the 5′ UTR of the nascent transcript (*154*), whereas for *eno* it appears to be RNase E cleavage within the intergenic region shared with *pyrG* (*155*). Cleavage by RNase G of intermediates generated by RNase E has also been reported for the transcript of *rnr* (*152*), which encodes the 3′-exonuclease RNase R (*156, 157*). Indeed, it has been suggested that RNase G acts primarily on intermediates generated by other endonucleases (*73*). The degradation of *eno* and *rnr* transcripts does not appear to be dependent on the 5′-terminal pyrophosphohydrolase RppH (5), thus the RNase E cleavages that generate intermediates of these transcripts for RNase G may represent additional examples of internal entry.

About a third of all the mRNAs identified as having increased abundance as a result of disrupting RNase G (73) are also increased by disruption of the gene encoding of the RppH pyrophosphohydrolase (5). Moreover, all of these genes that have been assigned to transcriptional units are located immediately downstream of a nascent 5′ end (158). Thus, it seems likely that RppH-mediated generation of a 5′ monophosphate may stimulate RNase G to cleave at least some full-length transcripts. To the best of our knowledge, the positions at which RNase G cleaves mRNA and the subsequent steps that occur have not been investigated for any transcript whose degradation is dependent on this enzyme. It is known however that the fusion of the 5′- UTR of adhE mRNA to lacZ mRNA results in its degradation becoming dependent on RNase G (150).

Of the genes whose mRNA increase in abundance as a result of disrupting RNase G, a number encode products associated with sugar metabolism centered on glycolysis (adhE, pgi, glk, nagB, acs, eno, tpiA) (73). Moreover, 2D-PAGE analysis has revealed that the products of the epd, fba, and pgk genes increase in abundance in addition to those of eno and adhE (Jourdan and McDowall, unpublished data), and it has been shown that strains defective in RNase G produce increased levels of pyruvic acid (159). Considered together, these results suggest that RNase G is a regulator of central metabolism. The incorporation of enolase in the degradosome (7, 49) and the finding that it is required for the normal decay of the mRNA encoding the glucose transporter in response to phosphosugar stress (160) is further evidence of a link between RNA turnover and central metabolism.

B. Overlapping and Essential Functions of RNase E and RNase G

Interestingly, overproduction of RNase G can restore viability to cells defective in RNase E (73, 78). Although there is some uncertainty regarding the extent to which this can be achieved by RNase G with a wild-type sequence (161), this complementation is associated with the return to near wild-type levels of the mRNAs of many genes associated with central metabolism (73). This has led to the suggestion that the lethality normally associated with the disruption of RNase E may result at least in part from the perturbation of central metabolic processes in E. coli. There is also evidence that lethality is due to an absolute requirement for RNase E in the processing of certain precursors of tRNA (162, 163). A suggestion that a decreased level of FtsZ is responsible for the inviability of cells deficient in RNase E has not been confirmed by subsequent experiments (164).

The role of RNase G is not restricted to mRNA degradation. It also cooperates with RNase E in the maturation of 16S rRNA (165, 166), 6S RNA (167) and the secG and leuU tRNAs (163). However, the maturation of these RNAs is not completely blocked in the absence of RNase G. This suggests that the role of RNase G in processing as well as mRNA degradation is largely overlapped by RNase E. The finding that RNase E is essential for cell viability and the normal decay of many if not most mRNAs in a background that is wild-type for RNase G has been widely interpreted as evidence for RNase G being unable to cleave a substantial proportion of the sites cleaved by RNase E. Indeed, RNase G has been described as a minor nuclease (5). However, as RNase E provides the platform for the assembly of the degradosome (111, 112) and is known to associate with the cytoskeleton (102, 103) and inner cytoplasmic membrane (101, 104), it is increasingly possible that some of the effects of disrupting RNase E are secondary due to perturbation of cellular complexes or structures rather than the loss of its endonucleolytic activity. Thus, 'with regard to their endonucleolytic activities, there may be more overlap between RNase E and RNase G' than was initially appreciated (78).

C. RNase III

Another *E. coli* endoribonuclease involved in mRNA degradation is RNase III, which is specific for double-stranded RNA and has a well-documented role in the maturation of 16S and 23S rRNA from the 30S primary transcripts (for a review see (168) and the chapter by Deutscher in this volume). In strains defective for RNase III, the rRNA is still matured indicating that there is an RNase III-independent means of processing 30S RNA. However, the 30S precursor is present at a higher level indicating that processing in the absence of RNase III is less efficient. In addition to cleaving the 5' UTR of *adhE* mRNA (154) (see earlier text), RNase III cleaves the intergenic regions of several *E. coli* polycistronic mRNAs between, for example, *rplL* and *rpoB* (169), *rpsO* and *pnp* (170, 171), *metY* (encoding a tRNA) and *nusA* (172) and on both sides of *dicF* (encoding a small regulatory RNA) (173). By generating upstream and downstream products with 5' monophosphate and 3' hydroxyl groups respectively, RNase III cleavage has the potential to stimulate cleavage by RNase E and G and to provide a new end for 3' exonucleolytic attack.

Disruption of RNase III has been shown to affect the abundance of other *E. coli* mRNAs (174, 175) and the expression of several of its operons (176). RNase III is also involved in the processing and degradation of the mRNA of coliphages such as T7, T3, and lambda (for reviews see (21, 177)). Recently, it has been shown that RNase III cleavage modulates translational repression by the N protein of lambda. Moreover, this repression is growth-rate sensitive and it has been suggested that the level of RNase III activity provides a measure of the physiological state of *E. coli* that inputs into the lysis/lysogeny decision (*178*). RNase III has recently been found to be also involved in the decay of several mRNAs targeted by sRNA and in the decay of several sRNAs themselves (*179–182*).

RNase III regulates its own production by cleaving its *rnc* message within the 5' UTR (*183*). This process is independent of *rnc* translation (*184*). To the best of our knowledge, the possibility that the cleavage of the 5' UTR by RNase III stimulates 5' end-dependent cleavage by RNase E and/or RNase G remains to be investigated. As indicated earlier, RNase III cleaves upstream of the *pnp* gene encoding PNPase. This cleavage occurs within a long stem-loop and allows PNPase to autoregulate its production: PNPase removes the 5' half of the remaining stump of double-stranded RNA, which in turn decreases the stability of the *pnp* transcript (*185*). For this example, there is evidence that decreased mRNA stability is associated with increased vulnerability to RNase E (*186*). The resulting RNase E cleavage is probably stimulated by the 5' end, which through the combined actions of RNase III and PNPase becomes both monophosphorylated and single-stranded, respectively.

Several bacterial homologs of RNase III have been characterized (*187–194*) and homologs are found in eukaryotes, including animals (reviewed in (*195–197*)). Like the *E. coli* enzyme, they are double-strand-specific endoribonucleases involved in the processing of almost every class of RNA. Much interest has focused on Dicer, the component of the RNAi pathway that generates the small interfering RNAs from double-stranded RNA and matures microRNAs (for reviews see (*198–200*)).

D. RNase P, RNase Z, and RNase LS

RNase P and RNase Z, two endoribonuclease known primarily in bacteria for their roles in tRNA maturation (*201, 202*), have also been implicated in mRNA degradation in *E. coli*. RNase P cuts within intergenic regions of several polycistronic RNAs including *tnaAB*, *rbsDACBK*, *lacZYA*, *secG-leuU*, and *hisG-I* (*203–205*). For all of these examples there is evidence that cleavage by RNase E may precede cleavage by RNase P. *E. coli* RNase P also cuts CI RNA (*206*), the immunity factor of phage-plasmid P4, the adenine riboswitch (*207*) and it is involved in the maturation of 4.5S RNA (*208*), which is an essential component of the signal recognition particle of the protein secretory apparatus. *E. coli* RNase Z, unlike many of its orthologs, is nonessential and its role has been studied on a genome-wide scale using gene arrays. This revealed that five different mRNAs were significantly more stable in a mutant defective for RNase Z and RNase E versus mutants defective in RNase E or RNase Z alone (*79*). Thus, the participation of RNase Z in mRNA decay appears relatively minor. Whether *E. coli* RNase Z has another role remains to be determined. Although the essential function of RNase Z in other organisms including *B. subtilis* is the maturation of tRNA precursors lacking a CCA determinant, all *E. coli* tRNA precursors have this determinant (for a review see (209)).

The most recent endoribonuclease to be implicated in mRNA degradation, apart from the bacterial toxin–antitoxin systems that function under specific conditions (see Chapter by Inouyé in this volume), is RNase LS. This enzyme was initially described as the activity that induces late-gene silencing in bacteriophage T4, hence its LS designation (210). The identification of a gene required for RNase LS activity has allowed the impact of this gene on the degradation of *E. coli* mRNAs to be assessed (211). Inactivation of the *rnlA* gene, now thought to encode RNase LS, causes a modest increase in the functional half-life of bulk mRNA and an increase in the chemical stability of *rpsO* and *bla* mRNA, two of four transcripts that were analyzed. The stabilities of *ompA* and *rpsA* were unaffected.

E. Conclusion

From the work presented in this section, it is evident that a number of endoribonucleases other than members of the RNase E/G family participate in the degradation of *E. coli* mRNA. Of these additional enzymes, only RNase P is essential. However, the inviability of cells lacking RNase P likely reflects the requirement for this enzyme in tRNA maturation and not perturbation of mRNA decay.

VIII. Phylogenetic Distribution of RNase E and The RNA Degradosome

This section is not intended to be a comprehensive analysis of the phylogenetic relationship of the members of the RNase E/G family of endoribonucleases. The reader interested in more detail is directed to previously published work (108–110) and the chapter by Danchin in this volume.

A. RNase E/G in the *Proteobacteria*

E. coli is a member of the γ-*Proteobacteria*, a large family of Gram-negative bacteria that includes important pathogens such as *Salmonella typhi*, *Yersinia pestis*, *Vibrio cholera*, *Haemophilus influenza*, and *Legionella pneumophilia*. The coexistence of RNase E and RNase G appears to be a general feature of this family of bacteria as well as the related β-*Proteobacteria* ((108) and Carpousis, unpublished data). Outside of the β- and γ-*Proteobacteria*, many, but not all, bacteria have ribonucleases that are members of the RNase E/G family of enzymes. However, these organisms often encode a single gene and the longer RNase E-like form is not always present. For example, *Synechocystis*, a member of the *Cyanobacteria*, encodes a short form of the enzyme whereas *Streptomyces coelicolor*, a member of the *Actinobacteria*, encodes an unusual form in which the catalytic core is flanked by long N- and C-terminal extensions (*109, 112*). The α-*Proteobacteria* encode a longer RNase E-like protein that is, however, distinct from its homologs in the β- and γ-*proteobacteria* (*109*) and they lack the shorter RNase G form of the enzyme. The RNase E-based RNA degradosome has been characterized in only a few other *Proteobacteria* and the results suggest that its composition is variable. *Rhodobacter capsulatus* (α-*Proteobacteria*) has a complex containing RNase E, RNase R and two DEAD-box proteins; *Pseudomonas syringae* (γ-*Proteobacteria*) has a complex containing RNase E, a DEAD-box protein and the transcription termination factor Rho (*212, 213*). This plasticity in composition is associated with the sequence variability of the noncatalytic region of RNase E. It has been proposed that the recruitment of different partners to RNase E might confer new regulatory properties and that these changes might contribute to the adaptation of the *Proteobacteria* to the wide variety of environmental niches that they occupy (*110*).

B. A Wider Evolutionary Perspective on RNase E/G

RNase J is a ribonuclease discovered recently in the Gram-positive bacterium *Bacillus subtilis*, which notably lacks a homolog of RNase E/G. Initially characterized as a novel endoribonuclease with similar cleavage specificity to RNase E (*214*), more recent work has shown that RNase J also has a 5′-exoribonuclease activity (*215–217*). This is the first time that an enzyme with a 5′-exoribonuclease activity has been identified and characterized in a prokaryotic organism. There has been longstanding interesting in elucidating the mechanism of RNA degradation in *B. subtilis* since this model organism, representative of a large number of Gram-positive bacteria, has a distinctly different repertoire of exo and endoribonucleases. The chapter by Bechhofer in this volume is an update on our understanding on RNA degradation in *B. subtilis*. We note here that *E. coli* does not encode a homolog of RNase J and that RNase J homologs are generally absent in the β- and γ-*Proteobacteria* (*214*)(Clouet-d'Orval and Carpousis, unpublished data). This is not the case for other *Proteobacteria* where RNase J homologs coexist with RNase E/G homologs. The *Proteobacteria* can be divided into two classes based on RNase J and RNase E/G content. One class contains an RNase E homolog and an RNase G homolog, but lacks an RNase J homolog (β- and γ-*Proteobacteria*). The other class contains a single representative of the RNase E/G family and an RNase J homolog (α-, δ-, and ε-*Proteobacteria*). It will be interesting in future work to explore the interrelationship between the activity of the RNase J homologs and the RNase E/G homologs in the degradation of RNA in *Proteobacteria* that contain both types of enzyme.

ACKNOWLEDGMENTS

Research in our groups is supported as follows: Carpousis, Centre National de la Recherche Scientifique with additional funding from the Agence Nationale de la Recherche (grant NT05_1-44659); Luisi, the Wellcome Trust; McDowall, the UK Biotechnology and Biological Sciences Research Council and facilities provided by the Wellcome Trust. We thank Maja Gorna and Zbyszek Pietras for helpful discussions and help with some of the figures.

REFERENCES

1. Symmons MF, Williams MG, Luisi BF, Jones GH, Carpousis AJ. Running rings around RNA: A superfamily of phosphate-dependent RNases. *Trends Biochem Sci* 2002;**27**:11–8.
2. Carpousis AJ, Vanzo NF, Raynal LC. mRNA degradation: A tale of poly(A) and multiprotein machines. *Trends Genet* 1999;**15**:24–8.
3. Dreyfus M, Regnier P. The poly(A) tail of mRNAs. Bodyguard in eukaryotes, scavenger in bacteria. *Cell* 2002;**111**:611–3.
4. Cohen SN. Surprises at the 3′ end of prokaryotic RNA. *Cell* 1995;**80**:829–32.
5. Deana A, Celesnik H, Belasco JG. The bacterial enzyme RppH triggers messenger RNA degradation by 5′ pyrophosphate removal. *Nature* 2008;**451**:355–8.
6. Celesnik H, Deana A, Belasco JG. Initiation of RNA decay in *Escherichia coli* by 5′ pyrophosphate removal. *Mol Cell* 2007;**27**:79–90.
7. Py B, Higgins CF, Krisch HM, Carpousis AJ. A DEAD-box RNA helicase in the *Escherichia coli* RNA degradosome. *Nature* 1996;**381**:169–72.
8. Coburn GA, Miao X, Briant DJ, Mackie GA. Reconstitution of a minimal RNA degradosome demonstrates functional coordination between a 3′ exonuclease and a DEAD-box RNA helicase. *Genes Dev* 1999;**13**:2594–603.
9. Khemici V, Carpousis AJ. The RNA degradosome and poly(A) polymerase of *Escherichia coli* are required *in vivo* for the degradation of small mRNA decay intermediates containing REP-stabilizers. *Mol Microbiol* 2004;**51**:777–90.
10. Ghosh S, Deutscher MP. Oligoribonuclease is an essential component of the mRNA decay pathway. *Proc Natl Acad Sci USA* 1999;**96**:4372–7.
11. Deutscher MP. Degradation of RNA in bacteria: Comparison of mRNA and stable RNA. *Nucleic Acids Res* 2006;**34**:659–66.
12. Steege DA. Emerging features of mRNA decay in bacteria. *RNA* 2000;**6**:1079–90.
13. Belasco JG, Higgins CF. Mechanisms of mRNA decay in bacteria: A perspective. *Gene* 1988;**72**:15–23.
14. Regnier P, Arraiano CM. Degradation of mRNA in bacteria: Emergence of ubiquitous features. *Bioessays* 2000;**22**:235–44.
15. Platt T. Transcription termination and the regulation of gene expression. *Annu Rev Biochem* 1986;**55**:339–72.
16. Platt T. Rho and RNA: Models for recognition and response. *Mol Microbiol* 1994;**11**:983–90.

17. Cardinale CJ, Washburn RS, Tadigotla VR, Brown LM, Gottesman ME, Nudler E. Termination factor Rho and its cofactors NusA and NusG silence foreign DNA in *E. coli. Science* 2008;**320**:935–8.
18. Deana A, Belasco JG. Lost in translation: The influence of ribosomes on bacterial mRNA decay. *Genes Dev* 2005;**19**:2526–33.
19. Coburn GA, Mackie GA. Degradation of mRNA in *Escherichia coli*: An old problem with some new twists. *Prog Nucleic Acid Res Mol Biol* 1999;**62**:55–108.
20. Nierlich DP, Murakawa GJ. The decay of bacterial messenger RNA. *Prog Nucleic Acid Res Mol Biol* 1996;**52**:153–216.
21. Grunberg-Manago M. Messenger RNA stability and its role in control of gene expression in bacteria and phages. *Annu Rev Genet* 1999;**33**:193–227.
22. Kushner SR. mRNA decay in *Escherichia coli* comes of age. *J Bacteriol* 2002;**184**:4658–65.
23. Ehretsmann CP, Carpousis AJ, Krisch HM. mRNA degradation in procaryotes. *FASEB J* 1992;**6**:3186–92.
24. Cohen SN, McDowall KJ. RNase E: Still a wonderfully mysterious enzyme. *Mol Microbiol* 1997;**23**:1099–106.
25. Rauhut R, Klug G. mRNA degradation in bacteria. *FEMS Microbiol Rev* 1999;**23**:353–70.
26. Gros F, Hiatt H, Gilbert W, Kurland CG, Risebrough RW, Watson JD. Unstable ribonucleic acid revealed by pulse labelling of *Escherichia coli*. *Nature* 1961;**190**:581–5.
27. Gros F, Gilbert W, Hiatt HH, Attardi G, Spahr PF, Watson JD. Molecular and biological characterization of messenger RNA. *Cold Spring Harb Symp Quant Biol* 1961;**26**:111–32.
28. Morse DE, Mosteller RD, Yanofsky C. Dynamics of synthesis, translation, and degradation of trp operon messenger RNA in *E. coli*. *Cold Spring Harb Symp Quant Biol* 1969;**34**:725–40.
29. Morse DE, Yanofsky C. Polarity and the degradation of mRNA. *Nature* 1969;**224**:329–31.
30. Apirion D. Degradation of RNA in *Escherichia coli*. A hypothesis. *Mol Gen Genet* 1973;**122**:313–22.
31. Donovan WP, Kushner SR. Polynucleotide phosphorylase and ribonuclease II are required for cell viability and mRNA turnover in *Escherichia coli* K-12. *Proc Natl Acad Sci USA* 1986;**83**:120–4.
32. Ghora BK, Apirion D. Structural analysis and *in vitro* processing to p5 rRNA of a 9S RNA molecule isolated from an rne mutant of *E. coli*. *Cell* 1978;**15**:1055–66.
33. Misra TK, Apirion D. RNase E, an RNA processing enzyme from *Escherichia coli*. *J Biol Chem* 1979;**254**:11154–9.
34. Ehretsmann CP, Carpousis AJ, Krisch HM. Specificity of *Escherichia coli* endoribonuclease RNase E: *in vivo* and *in vitro* analysis of mutants in a bacteriophage T4 mRNA processing site. *Genes Dev* 1992;**6**:149–59.
35. McDowall KJ, Lin-Chao S, Cohen SN. A + U content rather than a particular nucleotide order determines the specificity of RNase E cleavage. *J Biol Chem* 1994;**269**:10790–6.
36. Lin-Chao S, Wong TT, McDowall KJ, Cohen SN. Effects of nucleotide sequence on the specificity of rne-dependent and RNase E-mediated cleavages of RNA I encoded by the pBR322 plasmid. *J Biol Chem* 1994;**269**:10797–803.
37. Tomcsanyi T, Apirion D. Processing enzyme ribonuclease E specifically cleaves RNA I. An inhibitor of primer formation in plasmid DNA synthesis. *J Mol Biol* 1985;**185**:713–20.
38. Mudd EA, Prentki P, Belin D, Krisch HM. Processing of unstable bacteriophage T4 gene 32 mRNAs into a stable species requires *Escherichia coli* ribonuclease E. *EMBO J* 1988;**7**:3601–7.
39. Mudd EA, Carpousis AJ, Krisch HM. *Escherichia coli* RNase E has a role in the decay of bacteriophage T4 mRNA. *Genes Dev* 1990;**4**:873–81.

40. Régnier P, Hajnsdorf E. Decay of mRNA encoding ribosomal protein S15 of *Escherichia coli* is initiated by an RNase E-dependent endonucleolytic cleavage that removes the 3′ stabilizing stem and loop structure. *J Mol Biol* 1991;**217**:283–92.
41. Taraseviciene L, Miczak A, Apirion D. The gene specifying RNase E (rne) and a gene affecting mRNA stability (ams) are the same gene. *Mol Microbiol* 1991;**5**:851–5.
42. Mudd EA, Krisch HM, Higgins CF. RNase E, an endoribonuclease, has a general role in the chemical decay of *Escherichia coli* mRNA: Evidence that rne and ams are the same genetic locus. *Mol Microbiol* 1990;**4**:2127–35.
43. Babitzke P, Kushner SR. The Ams (altered mRNA stability) protein and ribonuclease E are encoded by the same structural gene of *Escherichia coli*. *Proc Natl Acad Sci USA* 1991;**88**:1–5.
44. McDowall KJ, Hernandez RG, Lin-Chao S, Cohen SN. The ams-1 and rne-3071 temperature-sensitive mutations in the ams gene are in close proximity to each other and cause substitutions within a domain that resembles a product of the *Escherichia coli* mre locus. *J Bacteriol* 1993;**175**:4245–9.
45. Cormack RS, Genereaux JL, Mackie GA. RNase E activity is conferred by a single polypeptide: Overexpression, purification, and properties of the ams/rne/hmp1 gene product. *Proc Natl Acad Sci USA* 1993;**90**:9006–10.
46. Ono M, Kuwano M. A conditional lethal mutation in an *Escherichia coli* strain with a longer chemical lifetime of messenger RNA. *J Mol Biol* 1979;**129**:343–57.
47. Carpousis AJ, Van Houwe G, Ehretsmann C, Krisch HM. Copurification of *E. coli* RNAase E and PNPase: Evidence for a specific association between two enzymes important in RNA processing and degradation. *Cell* 1994;**76**:889–900.
48. Py B, Causton H, Mudd EA, Higgins CF. A protein complex mediating mRNA degradation in *Escherichia coli*. *Mol Microbiol* 1994;**14**:717–29.
49. Miczak A, Kaberdin VR, Wei CL, Lin-Chao S. Proteins associated with RNase E in a multicomponent ribonucleolytic complex. *Proc Natl Acad Sci USA* 1996;**93**:3865–9.
50. Khemici V, Toesca I, Poljak L, Vanzo NF, Carpousis AJ. The RNase E of *Escherichia coli* has at least two binding sites for DEAD-box RNA helicases: Functional replacement of RhlB by RhlE. *Mol Microbiol* 2004;**54**:1422–30.
51. Prud'homme-Genereux A, Beran RK, Iost I, Ramey CS, Mackie GA, Simons RW. Physical and functional interactions among RNase E, polynucleotide phosphorylase and the cold-shock protein, CsdA: Evidence for a "cold shock degradosome." *Mol Microbiol* 2004;**54**:1409–21.
52. Morita T, Maki K, Aiba H. RNase E-based ribonucleoprotein complexes: Mechanical basis of mRNA destabilization mediated by bacterial noncoding RNAs. *Genes Dev* 2005;**19**:2176–86.
53. Gao J, Lee K, Zhao M, Qiu J, Zhan X, Saxena A, et al. Differential modulation of *E. coli* mRNA abundance by inhibitory proteins that alter the composition of the degradosome. *Mol Microbiol* 2006;**61**:394–406.
54. Carpousis AJ. The RNA degradosome of *Escherichia coli*: An mRNA-degrading machine assembled on RNase E. *Annu Rev Microbiol* 2007;**61**:71–87.
55. Rasmussen AA, Eriksen M, Gilany K, Udesen C, Franch T, Petersen C, et al. Regulation of ompA mRNA stability: The role of a small regulatory RNA in growth phase-dependent control. *Mol Microbiol* 2005;**58**:1421–9.
56. Udekwu KI, Darfeuille F, Vogel J, Reimegard J, Holmqvist E, Wagner EG. Hfq-dependent regulation of ompA synthesis is mediated by an antisense RNA. *Genes Dev* 2005;**19**:2355–66.
57. von Gabain A, Belasco JG, Schottel JL, Chang AC, Cohen SN. Decay of mRNA in *Escherichia coli*: Iinvestigation of the fate of specific segments of transcripts. *Proc Natl Acad Sci USA* 1983;**80**:653–7.
58. Belasco JG, Nilsson G, von Gabain A, Cohen SN. The stability of *E. coli* gene transcripts is dependent on determinants localized to specific mRNA segments. *Cell* 1986;**46**:245–51.

59. Emory SA, Bouvet P, Belasco JG. A 5′-terminal stem-loop structure can stabilize mRNA in *Escherichia coli*. *Genes Dev* 1992;**6**:135–48.
60. Bouvet P, Belasco JG. Control of RNase E-mediated RNA degradation by 5′-terminal base pairing in *E. coli*. *Nature* 1992;**360**:488–91.
61. Lin-Chao S, Cohen SN. The rate of processing and degradation of antisense RNAI regulates the replication of ColE1-type plasmids *in vivo*. *Cell* 1991;**65**:1233–42.
62. Feng Y, Vickers TA, Cohen SN. The catalytic domain of RNase E shows inherent 3′ to 5′ directionality in cleavage site selection. *Proc Natl Acad Sci USA* 2002;**99**:14746–51.
63. Jiang X, Belasco JG. Catalytic activation of multimeric RNase E and RNase G by 5′-monophosphorylated RNA. *Proc Natl Acad Sci USA* 2004;**101**:9211–6.
64. Mackie GA. Ribonuclease E is a 5′-end-dependent endonuclease. *Nature* 1998;**395**:720–3.
65. Walsh AP, Tock MR, Mallen MH, Kaberdin VR, Gabain Av A, McDowall KJ. Cleavage of poly (A) tails on the 3′-end of RNA by ribonuclease E of *Escherichia coli*. *Nucleic Acids Res* 2001;**29**:1864–71.
66. Mackie GA. Stabilization of circular rpsT mRNA demonstrates the 5′-end dependence of RNase E action *in vivo*. *J Biol Chem* 2000;**275**:25069–72.
67. Callaghan AJ, Marcaida MJ, Stead JA, McDowall KJ, Scott WG, Luisi BF. Structure of *Escherichia coli* RNase E catalytic domain and implications for RNA turnover. *Nature* 2005;**437**:1187–91.
68. Zeller ME, Csanadi A, Miczak A, Rose T, Bizebard T, Kaberdin VR. Quaternary structure and biochemical properties of mycobacterial RNase E/G. *Biochem J* 2007;**403**:207–15.
69. Kaberdin VR, Bizebard T. Characterization of *Aquifex aeolicus* RNase E/G. *Biochem Biophys Res Commun* 2005;**327**:382–92.
70. Tock MR, Walsh AP, Carroll G, McDowall KJ. The CafA protein required for the 5′-maturation of 16 S rRNA is a 5′- end-dependent ribonuclease that has context-dependent broad sequence specificity. *J Biol Chem* 2000;**275**:8726–32.
71. Schoenberg DR. The end defines the means in bacterial mRNA decay. *Nat Chem Biol* 2007;**3**:535–6.
72. Apirion D, Lassar AB. A conditional lethal mutant of *Escherichia coli* which affects the processing of ribosomal RNA. *J Biol Chem* 1978;**253**:1738–42.
73. Lee K, Bernstein JA, Cohen SN. RNase G complementation of rne null mutation identifies functional interrelationships with RNase E in *Escherichia coli*. *Mol Microbiol* 2002;**43**:1445–56.
74. Garneau NL, Wilusz J, Wilusz CJ. The highways and byways of mRNA decay. *Nat Rev Mol Cell Biol* 2007;**8**:113–26.
75. Meyer S, Temme C, Wahle E. Messenger RNA turnover in eukaryotes: Pathways and enzymes. *Crit Rev Biochem Mol Biol* 2004;**39**:197–216.
76. Moore MJ. From birth to death: The complex lives of eukaryotic mRNAs. *Science* 2005;**309**:1514–8.
77. Parker R, Song H. The enzymes and control of eukaryotic mRNA turnover. *Nat Struct Mol Biol* 2004;**11**:121–7.
78. Ow MC, Perwez T, Kushner SR. RNase G of *Escherichia coli* exhibits only limited functional overlap with its essential homologue, RNase E. *Mol Microbiol* 2003;**49**:607–22.
79. Perwez T, Kushner SR. RNase Z in *Escherichia coli* plays a significant role in mRNA decay. *Mol Microbiol* 2006;**60**:723–37.
80. Hajnsdorf E, Braun F, Haugel-Nielsen J, Regnier P. Polyadenylylation destabilizes the rpsO mRNA of *Escherichia coli*. *Proc Natl Acad Sci USA* 1995;**92**:3973–7.
81. Mohanty BK, Kushner SR. The majority of *Escherichia coli* mRNAs undergo posttranscriptional modification in exponentially growing cells. *Nucleic Acids Res* 2006;**34**:5695–704.

82. Mohanty BK, Kushner SR. Genomic analysis in *Escherichia coli* demonstrates differential roles for polynucleotide phosphorylase and RNase II in mRNA abundance and decay. *Mol Microbiol* 2003;**50**:645–58.
83. Braun F, Hajnsdorf E, Regnier P. Polynucleotide phosphorylase is required for the rapid degradation of the RNase E-processed rpsO mRNA of *Escherichia coli* devoid of its 3′ hairpin. *Mol Microbiol* 1996;**19**:997–1005.
84. Blum E, Carpousis AJ, Higgins CF. Polyadenylation promotes degradation of 3′-structured RNA by the *Escherichia coli* mRNA degradosome *in vitro*. *J Biol Chem* 1999;**274**:4009–16.
85. O'Hara EB, Chekanova JA, Ingle CA, Kushner ZR, Peters E, Kushner SR. Polyadenylylation helps regulate mRNA decay in *Escherichia coli*. *Proc Natl Acad Sci USA* 1995;**92**:1807–11.
86. Xu F, Lin-Chao S, Cohen SN. The *Escherichia coli* pcnB gene promotes adenylylation of antisense RNAI of ColE1-type plasmids *in vivo* and degradation of RNAI decay intermediates. *Proc Natl Acad Sci USA* 1993;**90**:6756–60.
87. Sarkar N. Polyadenylation of mRNA in prokaryotes. *Annu Rev Biochem* 1997;**66**:173–97.
88. Hajnsdorf E, Steier O, Coscoy L, Teysset L, Regnier P. Roles of RNase E, RNase II and PNPase in the degradation of the rpsO transcripts of *Escherichia coli*: Stabilizing function of RNase II and evidence for efficient degradation in an ams pnp rnb mutant. *EMBO J* 1994;**13**:3368–77.
89. Nilsson P, Naureckiene S, Uhlin BE. Mutations affecting mRNA processing and fimbrial biogenesis in the *Escherichia coli* pap operon. *J Bacteriol* 1996;**178**:683–90.
90. Cam K, Rome G, Krisch HM, Bouche JP. RNase E processing of essential cell division genes mRNA in *Escherichia coli*. *Nucleic Acids Res* 1996;**24**:3065–70.
91. Mackie GA. Specific endonucleolytic cleavage of the mRNA for ribosomal protein S20 of *Escherichia coli* requires the product of the ams gene *in vivo* and *in vitro*. *J Bacteriol* 1991;**173**:2488–97.
92. Baker KE, Mackie GA. Ectopic RNase E sites promote bypass of 5′-end-dependent mRNA decay in *Escherichia coli*. *Mol Microbiol* 2003;**47**:75–88.
93. Mackie GA. Stabilization of the 3′ one-third of *Escherichia coli* ribosomal protein S20 mRNA in mutants lacking polynucleotide phosphorylase. *J Bacteriol* 1989;**171**:4112–20.
94. Jiang X, Diwa A, Belasco JG. Regions of RNase E important for 5′-end-dependent RNA cleavage and autoregulated synthesis. *J Bacteriol* 2000;**182**:2468–75.
95. Feng Y, Cohen SN. Unpaired terminal nucleotides and 5′ monophosphorylation govern 3′ polyadenylation by *Escherichia coli* poly(A) polymerase I. *Proc Natl Acad Sci USA* 2000;**97**:6415–20.
96. Marujo PE, Braun F, Haugel-Nielsen J, Le Derout J, Arraiano CM, Regnier P. Inactivation of the decay pathway initiated at an internal site by RNase E promotes poly(A)-dependent degradation of the rpsO mRNA in *Escherichia coli*. *Mol Microbiol* 2003;**50**:1283–94.
97. Dulebohn D, Choy J, Sundermeier T, Okan N, Karzai AW. *Trans*-translation: The tmRNA-mediated surveillance mechanism for ribosome rescue, directed protein degradation, and nonstop mRNA decay. *Biochemistry* 2007;**46**:4681–93.
98. Moore SD, Sauer RT. The tmRNA system for translational surveillance and ribosome rescue. *Annu Rev Biochem* 2007;**76**:101–24.
99. Withey JH, Friedman DI. The biological roles of *trans*-translation. *Curr Opin Microbiol* 2002;**5**:154–9.
100. Withey JH, Friedman DI. A salvage pathway for protein structures: tmRNA and *trans*-translation. *Annu Rev Microbiol* 2003;**57**:101–23.
101. Liou GG, Jane WN, Cohen SN, Lin NS, Lin-Chao S. RNA degradosomes exist *in vivo* in *Escherichia coli* as multicomponent complexes associated with the cytoplasmic membrane via the N-terminal region of ribonuclease E. *Proc Natl Acad Sci USA* 2001;**98**:63–8.

102. Taghbalout A, Rothfield L. RNaseE and the other constituents of the RNA degradosome are components of the bacterial cytoskeleton. *Proc Natl Acad Sci USA* 2007;**104**:1667–72.
103. Taghbalout A, Rothfield L. RNaseE and RNA helicase B play central roles in the cytoskeletal organization of the RNA degradosome. *J Biol Chem* 2008;**283**:13850–5.
104. Khemici V, Poljak L, Luisi BF, Carpousis AJ. The RNase E of *Escherichia coli* is a membrane binding protein. *Mol. Microbiol.* 2008;**70**:799–813.
105. Arnold TE, Yu J, Belasco JG. mRNA stabilization by the ompA 5′ untranslated region: Two protective elements hinder distinct pathways for mRNA degradation. *RNA* 1998;**4**:319–30.
106. Wagner LA, Gesteland RF, Dayhuff TJ, Weiss RB. An efficient Shine–Dalgarno sequence but not translation is necessary for lacZ mRNA stability in *Escherichia coli*. *J Bacteriol* 1994;**176**:1683–8.
107. Carpousis AJ. The *Escherichia coli* RNA degradosome: Structure, function and relationship to other ribonucleolytic multienzyme complexes. *Biochem Soc Trans* 2002;**30**:150–5.
108. Condon C, Putzer H. The phylogenetic distribution of bacterial ribonucleases. *Nucleic Acids Res* 2002;**30**:5339–46.
109. Lee K, Cohen SN. A *Streptomyces coelicolor* functional orthologue of *Escherichia coli* RNase E shows shuffling of catalytic and PNPase-binding domains. *Mol Microbiol* 2003;**48**:349–60.
110. Marcaida MJ, DePristo MA, Chandran V, Carpousis AJ, Luisi BF. The RNA degradosome: Life in the fast lane of adaptive molecular evolution. *Trends Biochem Sci* 2006;**31**:359–65.
111. Vanzo NF, Li YS, Py B, Blum E, Higgins CF, Raynal LC, et al. Ribonuclease E organizes the protein interactions in the *Escherichia coli* RNA degradosome. *Genes Dev* 1998;**12**:2770–81.
112. Kaberdin VR, Miczak A, Jakobsen JS, Lin-Chao S, McDowall KJ, von Gabain A. The endoribonucleolytic N-terminal half of *Escherichia coli* RNase E is evolutionarily conserved in *Synechocystis* sp. And other bacteria but not the C-terminal half, which is sufficient for degradosome assembly. *Proc Natl Acad Sci USA* 1998;**95**:11637–42.
113. Callaghan AJ, Aurikko JP, Ilag LL, Gunter Grossmann J, Chandran V, Kuhnel K, et al. Studies of the RNA degradosome-organizing domain of the *Escherichia coli* ribonuclease RNase E. *J Mol Biol* 2004;**340**:965–79.
114. Chandran V, Poljak L, Vanzo NF, Leroy A, Miguel RN, Fernandez-Recio J, et al. Recognition and cooperation between the ATP-dependent RNA helicase RhlB and ribonuclease RNase E. *J Mol Biol* 2007;**367**:113–32.
115. Worrall JA, Howe FS, McKay AR, Robinson CV, Luisi BF. Allosteric activation of the ATPase activity of the *Escherichia coli* RhlB RNA helicase. *J Biol Chem* 2008;**283**:5567–76.
116. Chandran V, Luisi BF. Recognition of enolase in the *Escherichia coli* RNA degradosome. *J Mol Biol* 2006;**358**:8–15.
117. Kuhnel K, Luisi BF. Crystal structure of the *Escherichia coli* RNA degradosome component enolase. *J Mol Biol* 2001;**313**:583–92.
118. Lee K, Zhan X, Gao J, Qiu J, Feng Y, Meganathan R, et al. RraA. A protein inhibitor of RNase E activity that globally modulates RNA abundance in E. coli. *Cell* 2003;**114**:623–34.
119. Regonesi ME, Del Favero M, Basilico F, Briani F, Benazzi L, Tortora P, et al. Analysis of the *Escherichia coli* RNA degradosome composition by a proteomic approach. *Biochimie* 2006;**88**:151–61.
120. Bycroft M, Hubbard TJ, Proctor M, Freund SM, Murzin AG. The solution structure of the S1 RNA binding domain: A member of an ancient nucleic acid-binding fold. *Cell* 1997;**88**:235–42.
121. Symmons MF, Jones GH, Luisi BF. A duplicated fold is the structural basis for polynucleotide phosphorylase catalytic activity, processivity, and regulation. *Structure Fold Des* 2000;**8**:1215–26.

122. Mudd EA, Sullivan S, Gisby MF, Mironov A, Kwon CS, Chung WI, et al. A 125 kDa RNase E/G-like protein is present in plastids and is essential for chloroplast development and autotrophic growth in Arabidopsis. *J Exp Bot* 2008;**59**:2597–2610.
123. Callaghan AJ, Grossmann JG, Redko YU, Ilag LL, Moncrieffe MC, Symmons MF, et al. Quaternary structure and catalytic activity of the *Escherichia coli* ribonuclease E amino-terminal catalytic domain. *Biochemistry* 2003;**42**:13848–55.
124. Briant DJ, Hankins JS, Cook MA, Mackie GA. The quaternary structure of RNase G from *Escherichia coli*. *Mol Microbiol* 2003;**50**:1381–90.
125. Callaghan AJ, Redko Y, Murphy LM, Grossmann JG, Yates D, Garman E, et al. Zn-link: A metal-sharing interface that organizes the quaternary structure and catalytic site of the endoribonuclease, RNase E. *Biochemistry* 2005;**44**:4667–75.
126. Worrall JA, Luisi BF. Information available at cut rates: structure and mechanism of ribonucleases. *Curr Opin Struct Biol* 2007;**17**:128–37.
127. Yang W, Lee JY, Nowotny M. Making and breaking nucleic acids: Two-Mg2 + -ion catalysis and substrate specificity. *Mol Cell* 2006;**22**:5–13.
128. Cheng Y, Patel DJ. Crystallographic structure of the nuclease domain of 3′ηEξo, a DEDDh family member, bound to rAMP. *J Mol Biol* 2004;**343**:305–12.
129. Redko Y, Tock MR, Adams CJ, Kaberdin VR, Grasby JA, McDowall KJ. Determination of the catalytic parameters of the N-terminal half of *Escherichia coli* ribonuclease E and the identification of critical functional groups in RNA substrates. *J Biol Chem* 2003;**278**:44001–8.
130. Spickler C, Stronge V, Mackie GA. Preferential cleavage of degradative intermediates of rpsT mRNA by the *Escherichia coli* RNA degradosome. *J Bacteriol* 2001;**183**:1106–9.
131. Grossmann JG, Callaghan AJ, Marcaida MJ, Luisi BF, Alcock FH, Tokatlidis K. Complementing structural information of modular proteins with small angle neutron scattering and contrast variation. *Eur Biophys J* 2008;**37**:603–11.
132. Koslover DJ, Callaghan AJ, Marcaida MJ, Garman EF, Martick M, Scott WG, et al. The crystal structure of the *Escherichia coli* RNase E apoprotein and a mechanism for RNA degradation. *Structure* 2008;**16**:1238–44.
133. Jourdan SS, McDowall KJ. Sensing of 5′ monophosphate by *Escherichia coli* RNase G can significantly enhance association with RNA and stimulate the decay of functional mRNA transcripts *in vivo*. *Mol Microbiol* 2008;**67**:102–15.
134. Hankins JS, Zappavigna C, Prud'homme-Genereux A, Mackie GA. Role of RNA structure and susceptibility to RNase E in regulation of a cold shock mRNA, cspA mRNA. *J Bacteriol* 2007;**189**:4353–8.
135. Caruthers JM, Feng Y, McKay DB, Cohen SN. Retention of core catalytic functions by a conserved minimal ribonuclease E peptide that lacks the domain required for tetramer formation. *J Biol Chem* 2006;**281**:27046–51.
136. Briegel KJ, Baker A, Jain C. Identification and analysis of *Escherichia coli* ribonuclease E dominant-negative mutants. *Genetics* 2006;**172**:7–15.
137. Yang J, Jain C, Schesser K. RNase E regulates the *Yersinia* type 3 secretion system. *J Bacteriol* 2008;**190**:3774–8.
138. Wagner EG, Altuvia S, Romby P. Antisense RNAs in bacteria and their genetic elements. *Adv Genet* 2002;**46**:361–98.
139. Gottesman S. The small RNA regulators of *Escherichia coli*: Roles and mechanisms. *Annu Rev Microbiol* 2004;**58**:303–28.
140. Carpousis AJ. Degradation of targeted mRNAs in *Escherichia coli*: Regulation by a small antisense RNA. *Genes Dev* 2003;**17**:2351–5.
141. Brennan RG, Link TM. Hfq structure, function and ligand binding. *Curr Opin Microbiol* 2007;**10**:125–33.

142. Masse E, Salvail H, Desnoyers G, Arguin M. Small RNAs controlling iron metabolism. *Curr Opin Microbiol* 2007;**10**:140–5.
143. Vanderpool CK, Gottesman S. Involvement of a novel transcriptional activator and small RNA in posttranscriptional regulation of the glucose phosphoenolpyruvate phosphotransferase system. *Mol Microbiol* 2004;**54**:1076–89.
144. Masse E, Escorcia FE, Gottesman S. Coupled degradation of a small regulatory RNA and its mRNA targets in *Escherichia coli*. *Genes Dev* 2003;**17**:2374–83.
145. Morita T, Mochizuki Y, Aiba H. Translational repression is sufficient for gene silencing by bacterial small noncoding RNAs in the absence of mRNA destruction. *Proc Natl Acad Sci USA* 2006;**103**:4858–63.
146. Aiba H. Mechanism of RNA silencing by Hfq-binding small RNAs. *Curr Opin Microbiol* 2007;**10**:134–9.
147. McDowall KJ, Cohen SN. The N-terminal domain of the rne gene product has RNase E activity and is nonoverlapping with the arginine-rich RNA-binding site. *J Mol Biol* 1996;**255**:349–55.
148. Taraseviciene L, Bjork GR, Uhlin BE. Evidence for an RNA binding region in the *Escherichia coli* processing endoribonuclease RNase E. *J Biol Chem* 1995;**270**:26391–8.
149. Kaga N, Umitsuki G, Nagai K, Wachi M. RNase G-dependent degradation of the eno mRNA encoding a glycolysis enzyme enolase in *Escherichia coli*. *Biosci Biotechnol Biochem* 2002;**66**:2216–20.
150. Umitsuki G, Wachi M, Takada A, Hikichi T, Nagai K. Involvement of RNase G in in vivo mRNA metabolism in *Escherichia coli*. *Genes Cells* 2001;**6**:403–10.
151. Zajanckauskaite A, Truncaite L, Strazdaite-Zieliene Z, Nivinskas R. Involvement of the *Escherichia coli* endoribonucleases G and E in the secondary processing of RegB-cleaved transcripts of bacteriophage T4. *Virology* 2008;**375**:342–53.
152. Cairrao F, Arraiano CM. The role of endoribonucleases in the regulation of RNase R. *Biochem Biophys Res Commun* 2006;**343**:731–7.
153. Ueno H, Yonesaki T. Phage-induced change in the stability of mRNAs. *Virology* 2004;**329**:134–41.
154. Aristarkhov A, Mikulskis A, Belasco JG, Lin EC. Translation of the adhE transcript to produce ethanol dehydrogenase requires RNase III cleavage in *Escherichia coli*. *J Bacteriol* 1996;**178**:4327–32.
155. Kime LS, Jourdan S, McDowall KJ. Identifying and characterizing substrates of the RNase E/G family of enzymes. *Methods Enzymol* **448** (in press).
156. Vincent HA, Deutscher MP. Substrate recognition and catalysis by the exoribonuclease RNase R. *J Biol Chem* 2006;**281**:29769–75.
157. Chen C, Deutscher MP. Elevation of RNase R in response to multiple stress conditions. *J Biol Chem* 2005;**280**:34393–6.
158. Gama-Castro S, Jimenez-Jacinto V, Peralta-Gil M, Santos-Zavaleta A, Penaloza-Spinola MI, Contreras-Moreira B, et al. RegulonDB (version 6.0): Gene regulation model of *Escherichia coli* K-12 beyond transcription, active (experimental) annotated promoters and Textpresso navigation. *Nucleic Acids Res* 2008;**36**:D120–24.
159. Sakai T, Nakamura N, Umitsuki G, Nagai K, Wachi M. Increased production of pyruvic acid by *Escherichia coli* RNase G mutants in combination with cra mutations. *Appl Microbiol Biotechnol* 2007;**76**:183–92.
160. Morita T, Kawamoto H, Mizota T, Inada T, Aiba H. Enolase in the RNA degradosome plays a crucial role in the rapid decay of glucose transporter mRNA in the response to phosphosugar stress in *Escherichia coli*. *Mol Microbiol* 2004;**54**:1063–75.
161. Deana A, Belasco JG. The function of RNase G in *Escherichia coli* is constrained by its amino and carboxyl termini. *Mol Microbiol* 2004;**51**:1205–17.

162. Li Z, Deutscher MP. RNase E plays an essential role in the maturation of *Escherichia coli* tRNA precursors. *RNA* 2002;**8**:97–109.
163. Mohanty BK, Kushner SR. Ribonuclease P processes polycistronic tRNA transcripts in *Escherichia coli* independent of ribonuclease E. *Nucleic Acids Res* 2007;**35**:7614–25.
164. Tamura M, Lee K, Miller CA, Moore CJ, Shirako Y, Kobayashi M, et al. RNase E maintenance of proper FtsZ/FtsA ratio required for nonfilamentous growth of *Escherichia coli* cells but not for colony-forming ability. *J Bacteriol* 2006;**188**:5145–52.
165. Wachi M, Umitsuki G, Shimizu M, Takada A, Nagai K. *Escherichia coli* cafa gene encodes a novel RNase, designated as RNase G, involved in processing of the 5′ end of 16S rRNA. *Biochem Biophys Res Commun* 1999;**259**:483–8.
166. Li Z, Pandit S, Deutscher MP. RNase G (cafa protein) and RNase E are both required for the 5′ maturation of 16S ribosomal RNA. *EMBO J* 1999;**18**:2878–85.
167. Kim KS, Lee Y. Regulation of 6S RNA biogenesis by switching utilization of both sigma factors and endoribonucleases. *Nucleic Acids Res* 2004;**32**:6057–68.
168. Gegenheimer P, Apirion D. Processing of procaryotic ribonucleic acid. *Microbiol Rev* 1981;**45**:502–41.
169. Barry G, Squires C, Squires CL. Attenuation and processing of RNA from the rplJL—rpoBC transcription unit of *Escherichia coli*. *Proc Natl Acad Sci USA* 1980;**77**:3331–5.
170. Portier C, Dondon L, Grunberg-Manago M, Regnier P. The first step in the functional inactivation of the *Escherichia coli* polynucleotide phosphorylase messenger is a ribonuclease III processing at the 5′ end. *EMBO J* 1987;**6**:2165–70.
171. Takata R, Mukai T, Hori K. RNA processing by RNase III is involved in the synthesis of *Escherichia coli* polynucleotide phosphorylase. *Mol Gen Genet* 1987;**209**:28–32.
172. Regnier P, Grunberg-Manago M. Cleavage by RNase III in the transcripts of the met Y-nus-A-infB operon of *Escherichia coli* releases the tRNA and initiates the decay of the downstream mRNA. *J Mol Biol* 1989;**210**:293–302.
173. Faubladier M, Cam K, Bouche JP. *Escherichia coli* cell division inhibitor DicF-RNA of the dicB operon. Evidence for its generation *in vivo* by transcription termination and by RNase III and RNase E-dependent processing. *J Mol Biol* 1990;**212**:461–71.
174. Santos JM, Drider D, Marujo PE, Lopez P, Arraiano CM. Determinant role of *E. coli* RNase III in the decay of both specific and heterologous mRNAs. *FEMS Microbiol Lett* 1997;**157**:31–8.
175. Freire P, Amaral JD, Santos JM, Arraiano CM. Adaptation to carbon starvation: RNase III ensures normal expression levels of bolA1p mRNA and sigma(S). *Biochimie* 2006;**88**:341–6.
176. Talkad V, Achord D, Kennell D. Altered mRNA metabolism in ribonuclease III-deficient strains of *Escherichia coli*. *J Bacteriol* 1978;**135**:528–41.
177. Nicholson AW. Function, mechanism and regulation of bacterial ribonucleases. *FEMS Microbiol Rev* 1999;**23**:371–90.
178. Wilson HR, Yu D, Peters HK, III, Zhou JG, Court DL. The global regulator RNase III modulates translation repression by the transcription elongation factor N. *EMBO J* 2002;**21**:4154–61.
179. Darfeuille F, Unoson C, Vogel J, Wagner EG. An antisense RNA inhibits translation by competing with standby ribosomes. *Mol Cell* 2007;**26**:381–92.
180. Vogel J, Bartels V, Tang TH, Churakov G, Slagter-Jager JG, Huttenhofer A, et al. RNomics in *Escherichia coli* detects new sRNA species and indicates parallel transcriptional output in bacteria. *Nucleic Acids Res* 2003;**31**:6435–43.
181. Viegas SC, Pfeiffer V, Sittka A, Silva IJ, Vogel J, Arraiano CM. Characterization of the role of ribonucleases in *Salmonella* small RNA decay. *Nucleic Acids Res* 2007;**35**:7651–64.

182. Afonyushkin T, Vecerek B, Moll I, Blasi U, Kaberdin VR. Both RNase E and RNase III control the stability of sodB mRNA upon translational inhibition by the small regulatory RNA ryhB. *Nucleic Acids Res* 2005;**33**:1678–89.
183. Bardwell JC, Regnier P, Chen SM, Nakamura Y, Grunberg-Manago M, Court DL. Autoregulation of RNase III operon by mRNA processing. *EMBO J* 1989;**8**:3401–7.
184. Matsunaga J, Simons EL, Simons RW. *Escherichia coli* RNase III (rnc) autoregulation occurs independently of rnc gene translation. *Mol Microbiol* 1997;**26**:1125–35.
185. Jarrige AC, Mathy N, Portier C. PNPase autocontrols its expression by degrading a double-stranded structure in the PNP mRNA leader. *EMBO J* 2001;**20**:6845–55.
186. Hajnsdorf E, Carpousis AJ, Regnier P. Nucleolytic inactivation and degradation of the RNase III processed PNP message encoding polynucleotide phosphorylase of *Escherichia coli*. *J Mol Biol* 1994;**239**:439–54.
187. Chang SA, Bralley P, Jones GH. The absB gene encodes a double strand-specific endoribonuclease that cleaves the read-through transcript of the rpsO-pnp operon in *Streptomyces coelicolor*. *J Biol Chem* 2005;**280**:33213–9.
188. Oguro A, Kakeshita H, Nakamura K, Yamane K, Wang W, Bechhofer DH. Bacillus subtilis RNase III cleaves both 5'- and 3'-sites of the small cytoplasmic RNA precursor. *J Biol Chem* 1998;**273**:19542–7.
189. Herskovitz MA, Bechhofer DH. Endoribonuclease RNase III is essential in *Bacillus subtilis*. *Mol Microbiol* 2000;**38**:1027–33.
190. Huntzinger E, Boisset S, Saveanu C, Benito Y, Geissmann T, Namane A, et al. Staphylococcus aureus RNAIII and the endoribonuclease III coordinately regulate spa gene expression. *EMBO J* 2005;**24**:824–35.
191. Olmedo G, Guzman P. Mini-III, a fourth class of RNase III catalyses maturation of the *Bacillus subtilis* 23S ribosomal RNA. *Mol Microbiol* 2008;**68**:1073–6.
192. Redko Y, Bechhofer DH, Condon C. Mini-III, an unusual member of the RNase III family of enzymes, catalyses 23S ribosomal RNA maturation in *B. subtilis*. *Mol Microbiol* 2008;**68**:1096–106.
193. Viegas SC, Schmidt D, Kasche V, Arraiano CM, Ignatova Z. Effect of the increased stability of the penicillin amidase mRNA on the protein expression levels. *FEBS Lett* 2005;**579**:5069–73.
194. Conrad C, Rauhut R, Klug G. Different cleavage specificities of RNases III from *Rhodobacter capsulatus* and *Escherichia coli*. *Nucleic Acids Res* 1998;**26**:4446–53.
195. Drider D, Condon C. The continuing story of endoribonuclease III. *J Mol Microbiol Biotechnol* 2004;**8**:195–200.
196. Conrad C, Rauhut R. Ribonuclease III: New sense from nuisance. *Int J Biochem Cell Biol* 2002;**34**:116–29.
197. MacRae IJ, Doudna JA. Ribonuclease revisited: Structural insights into ribonuclease III family enzymes. *Curr Opin Struct Biol* 2007;**17**:138–45.
198. Dykxhoorn DM, Novina CD, Sharp PA. Killing the messenger: Short RNAs that silence gene expression. *Nat Rev Mol Cell Biol* 2003;**4**:457–67.
199. Esquela-Kerscher A, Slack FJ. Oncomirs—microRNAs with a role in cancer. *Nat Rev Cancer* 2006;**6**:259–69.
200. He L, Hannon GJ. MicroRNAs: Small RNAs with a big role in gene regulation. *Nat Rev Genet* 2004;**5**:522–31.
201. Pellegrini O, Nezzar J, Marchfelder A, Putzer H, Condon C. Endonucleolytic processing of CCA-less tRNA precursors by RNase Z in *Bacillus subtilis*. *EMBO J* 2003;**22**:4534–43.
202. Schedl P, Roberts J, Primakoff P. *In vitro* processing of *E. coli* tRNA precursors. *Cell* 1976;**8**:581–94.
203. Li Y, Altman S. A specific endoribonuclease, RNase P, affects gene expression of polycistronic operon mRNAs. *Proc Natl Acad Sci USA* 2003;**100**:13213–8.

204. Li Y, Altman S. Polarity effects in the lactose operon of *Escherichia coli*. *J Mol Biol* 2004;**339**:31–9.
205. Alifano P, Rivellini F, Piscitelli C, Arraiano CM, Bruni CB, Carlomagno MS. Ribonuclease E provides substrates for ribonuclease P-dependent processing of a polycistronic mRNA. *Genes Dev* 1994;**8**:3021–31.
206. Forti F, Sabbattini P, Sironi G, Zangrossi S, Deho G, Ghisotti D. Immunity determinant of phage-plasmid P4 is a short processed RNA. *J Mol Biol* 1995;**249**:869–78.
207. Altman S, Wesolowski D, Guerrier-Takada C, Li Y. RNase P cleaves transient structures in some riboswitches. *Proc Natl Acad Sci USA* 2005;**102**:11284–9.
208. Bothwell AL, Garber RL, Altman S. Nucleotide sequence and *in vitro* processing of a precursor molecule to *Escherichia coli* 4.5 S RNA. *J Biol Chem* 1976;**251**:7709–16.
209. Redko Y, Li de Lasierra-Gallay I, Condon C. When all's zed and done: The structure and function of RNase Z in prokaryotes. *Nat Rev Microbiol* 2007;**5**:278–86.
210. Kai T, Selick HE, Yonesaki T. Destabilization of bacteriophage T4 mRNAs by a mutation of gene 61.5. *Genetics* 1996;**144**:7–14.
211. Otsuka Y, Koga M, Iwamoto A, Yonesaki T. A role of RnlA in the RNase LS activity from *Escherichia coli*. *Genes Genet Syst* 2007;**82**:291–9.
212. Jager S, Fuhrmann O, Heck C, Hebermehl M, Schiltz E, Rauhut R, *et al*. An mRNA degrading complex in *Rhodobacter capsulatus*. *Nucleic Acids Res* 2001;**29**:4581–8.
213. Purusharth RI, Klein F, Sulthana S, Jager S, Jagannadham MV, Evguenieva-Hackenberg E, *et al*. Exoribonuclease R interacts with endoribonuclease E and an RNA helicase in the psychrotrophic bacterium *Pseudomonas syringae* Lz4W. *J Biol Chem* 2005;**280**:14572–8.
214. Even S, Pellegrini O, Zig L, Labas V, Vinh J, Brechemmier-Baey D, *et al*. Ribonucleases J1 and J2: two novel endoribonucleases in *B.subtilis* with functional homology to *E.coli* RNase E. *Nucleic Acids Res* 2005;**33**:2141–52.
215. Britton RA, Wen T, Schaefer L, Pellegrini O, Uicker WC, Mathy N, *et al*. Maturation of the 5' end of *Bacillus subtilis* 16S rRNA by the essential ribonuclease YkqC/RNase J1. *Mol Microbiol* 2007;**63**:127–38.
216. Mathy N, Benard L, Pellegrini O, Daou R, Wen T, Condon C. 5'–3' exoribonuclease activity in bacteria: role of RNase J1 in rRNA maturation and 5' stability of mRNA. *Cell* 2007;**129**:681–92.
217. de la Sierra-Gallay IL, Zig L, Jamalli A, Putzer H. Structural insights into the dual activity of RNase J. *Nat Struct Mol Biol* 2008;**15**:206–12.

Poly(A)-Assisted RNA Decay and Modulators of RNA Stability

> Philippe Régnier and
> Eliane Hajnsdorf
>
> CNRS UPR9073, Affiliated with the
> University of Paris 7, Denis Diderot, Institut
> de Biologie Physico-Chimique, 13, rue
> Pierre et Marie Curie, 75005 Paris, France

I. Introduction	138
II. Regulation of Gene Expression in Response to Environmental Signals	139
A. Growth-Phase and Growth-Rate	139
B. O_2, Light, Nutrients, and Starvation	140
C. Temperature	141
III. Polyadenylation and Poly(A)-Dependent Decay	142
A. Diversity of Poly(A) Functions	142
B. Characterization of Polyadenylated RNAs	143
C. Polyadenylation Affects RNA Stability	144
D. The Enzymes of Poly(A)-Metabolism	145
E. A Mechanistic View of Poly(A)-Metabolism and Poly(A)-Assisted RNA Decay	148
F. Regulation and Cofactors of the Poly(A)-Dependent Machinery of Degradation	153
G. Roles of Poly(A)-Dependent Decay	156
H. Heterogeneous Tails and RNA Stability	157
IV. Other Modulators of RNA Stability	158
A. Regulators of RNase E	158
B. Gmr, The Regulator of RNase II	159
C. Proteins Hfq, CsrA, and ncRNAs	160
D. tmRNA and Quality Control of RNA	165
E. Destabilizing Riboswitches	166
V. Conclusions and Open Questions	167
References	169

In *Escherichia coli*, RNA degradation is orchestrated by the degradosome with the assistance of complementary pathways and regulatory cofactors described in this chapter. They control the stability of each transcript and regulate the expression of many genes involved in environmental adaptation. The poly

(A)-dependent degradation machinery has diverse functions such as the degradation of decay intermediates generated by endoribonucleases, the control of the stability of regulatory non coding RNAs (ncRNAs) and the quality control of stable RNA. The metabolism of poly(A) and mechanism of poly(A)-assisted degradation are beginning to be understood. Regulatory factors, exemplified by RraA and RraB, control the decay rates of subsets of transcripts by binding to RNase E, in contrast to regulatory ncRNAs which, assisted by Hfq, target RNase E to specific transcripts. Destabilization is often consecutive to the translational inactivation of mRNA. However, there are examples where RNA degradation is the primary regulatory step.

I. Introduction

RNA turnover has long been considered to be an ancillary scavenging process whose role is to get rid of inactivated RNAs and to recycle nucleotides. After much investigation, a different picture has now emerged, where the radically different pathways of RNA degradation characterized in several eukaryotic and prokaryotic lineages also play a role in regulating the expression of many genes. Although there were hints of a diversity of decay processes in bacteria, it was accepted until recently that bacterial mechanisms of degradation all shared large degree of similarity with the endonucleolytic process of decay mediated by RNase E in *E. coli*. This process was often contrasted to the 5′–3′ exonucleolytic pathways of RNA degradation found in eukaryotic cells (*1*). However, this boundary was recently blurred when 5′–3′ exonucleolytic degradation was also found to take place in *Bacillus subtilis* (*2*) and 5′ end dephosphorylation, reminiscent of the decapping step of eukaryotic mRNA decay, was found to be the first step in the degradation of some *E. coli* transcripts (*3, 4*).

Although important features of RNA decay in *E. coli* presumably remain to be elucidated, the major route of degradation orchestrated by the RNA degradosome is now well-characterized. This multienzyme complex is described in the chapter by Carpousis and colleagues in this volume. In addition to RNase E and other degradosome components, this pathway involves RNA modifying enzymes, exoribonucleases and regulatory factors, which will be the main focus of this chapter. A special emphasis is given to the polyadenylation or poly(A)-dependent decay pathway, dedicated to the degradation of structured RNA fragments produced by RNase E cleavage. This pathway also controls the intracellular concentrations of small regulatory RNAs and participates in the quality control of RNAs. Other cofactors and mechanisms affecting RNA stability, such as the RNA binding protein Hfq, and gene-specific processes involving small ncRNAs and riboswitches are also described. There are many reports suggesting that modulations of RNA stability participate in the adaptation of bacteria to changes in environmental conditions and physiological status. These data that were previously reviewed by Takayama and Kjelleberg (5) are summarized at the beginning of this chapter.

II. Regulation of Gene Expression in Response to Environmental Signals

A. Growth-Phase and Growth-Rate

A global analysis of mRNAs, using microarrays, indicates that the stability of *E. coli* transcripts is similar during growth in rich and defined medium (6). However, in several bacteria there are specific examples of mRNAs whose decay-rate depends on growth-rate. The *ompA* mRNA coding for a porin of *E. coli* is one of the earliest reported examples of an adaptation of mRNA stability to the rate of growth. Its half-life drops from 15 to 4 min upon entry into stationary phase (7). This was attributed to facilitation of the RNase E cleavage event that is rate-limiting for decay (8). Further investigation produced evidence that the destabilization observed when cells enter stationary phase is due to the synthesis of the MicA ncRNA, which interferes with translation initiation (9). A lack of protection by ribosomes probably explains why the *ompA* mRNA is more efficiently degraded by RNase E. The RNA binding protein Hfq, which was also demonstrated to be involved in the posttranscriptional regulation of OmpA expression (10), is essential for the recognition of the *ompA* mRNA by MicA. More generally, the MicA and RybB ncRNAs downregulate the expression of several outer membrane proteins in response to envelope stress, induced by temperature shock, ethanol, hyperosmolarity, and entry into stationary phase (11–14). The activation of σ^E by misfolded porins, which accumulate in the periplasmic space under stress conditions, triggers a cascade of events that culminates in the destabilization of transcripts encoding outer membrane proteins. As in the case of *ompA* mRNA, destabilization of at least some of these transcripts may be consecutive to the inhibition of translation caused by Hfq-mediated annealing of the small ncRNA in the vicinity of the ribosome binding site.

The modulation of mRNA decay is also a key means by which *B. subtilis* adjusts to growth phase. The half-life of the *sdh* transcript, encoding succinate dehydrogenase, decreases more than sixfold upon entry into stationary phase (15) while the *aprE* transcript, encoding subtilisin, becomes more stable during early stationary phase and is subsequently destabilized (16). Increased mRNA stability also contributes to the upregulation of the RpoS sigma factor of *Salmonella dublin* upon entry into stationary phase (17). The stability of transcripts encoding chloramphenicol acetyltransferase and the ribosome modulation factor Rmf of *E. coli* (7, 18, 19), the CryIVB toxin of Cyanobacteria (20) and some virulence genes of *Streptococcus pyogenes* (21) also depends upon the growth-phase.

RNA degradation may also be affected by changes in genetic programs, such as under conditions of phage infection or specific developmental processes. The *ompA* and *lpp* transcripts of *E. coli* are destabilized following infection by bacteriophage T4, while the *soc* mRNA is stabilized, for example (22). T7 phage infection causes phosphorylation of both the scaffolding domain of RNase E and the RNA helicase RhlB of the degradosome, a modification that may account for the specific stabilization of transcripts synthesized by the T7 RNA polymerase (23). The expression of genes involved in developmental cycles of bacteria and phages are also governed by mRNA stability. The early transcripts of bacteriophage T4 are cleaved specifically by the phage-encoded RegB nuclease (24) (See also chapter in this volume by Marc Uzan), while the hemagglutinin *mbhA* transcript becomes dramatically more stable during fruiting body formation in *Myxococcus xanthus* (25).

B. O_2, Light, Nutrients, and Starvation

Changes in environmental conditions can also have effects on the half-lives of particular RNAs. The metabolic adaptation to carbon supply, for example, is in part mediated by modifications in mRNA stability. The best characterized example is the *ptsG* mRNA, encoding the major glucose transporter IICB(Glc), whose stability decreases dramatically to prevent the intracellular accumulation of hexose phosphates (26). Glucose-phosphate stress induces the synthesis of the small regulatory SgrS ncRNA, which inhibits the translation of *ptsG* (27). The translationally inactive messenger becomes sensitive to RNase E and the IICB(Glc) transporter is no longer synthesized (see later text). Global mRNA stability increases upon adaptation of both *Lactococcus lactis* and *E. coli* to carbon starvation (28, 29). A gene-specific mRNA stabilization also accounts for the overexpresion of the GsiB stress protein that occurs when *B. subtilis* is starved for glucose, or exposed to heat, ethanol, or salt stress (30). Salt concentration affects the stability of the *Haloferax mediterranei gvpD* mRNA, involved in gas vesicule formation (31). High levels of ammonium ion (NH_4^+), O_2, and high temperature, which are not favorable to nitrogen fixation, destabilize transcripts of the nitrogen fixation genes of *Klebsiella pneumoniae* (32). In *B. subtilis*, mRNA stabilization is one of the means used by the cell to adapt the synthesis of threonyl-tRNA synthetase to amino acid availability (33). Threonyl-tRNA synthetase transcripts resulting from transcription antitermination are stabilized by a ribonucleolytic maturation mediated by ribonuclease JI (34), an enzyme which exhibits both endo and exonucleolytic activity (2). The uptake and storage of iron are controlled by complex regulatory circuits which also involve changes in mRNA stability (35, 36). The downregulation of proteins implicated in iron storage, protection against highly reactive radicals and more generally in the redistribution of intracellular iron (37) is mediated by the RyhB regulatory ncRNA (38). It is not clear whether formation of the RyhB–mRNA complex triggers RNA decay or whether the degradosome gets rid of nontranslated transcripts inactivated by the ncRNA (36, 39). An impact of iron on mRNA stability was also described in *Synechococcus* and *Anabaena*, where the ferredoxin transcript is more stable in iron supplemented media (40). In *B. subtilis*, induction of the *ermC* rRNA methylase by erythomycin results in part from a stabilization of the transcript (41). It is also worth mentioning that amino acid supply modulates an endonucleolytic cleavage which may inactivate the RNase E transcript (42).

Light has been reported to modify mRNA stability in Cyanobacteria. Dark conditions cause an induction of the expression of a histidine kinase (43) and repression of that of a reaction center ortholog (44). These events are correlated with the accumulation and the destabilization of the respective transcripts. In *Synechococcus*, light modifies the stability of *psbA* transcripts, encoding light-regulated proteins involved in photosynthesis (45).

Anaerobiosis has been reported to delay rRNA maturation and to prolong the half-lives of specific *E. coli* transcripts (46). Oxygen levels have also been suggested to control RNA stability in *Klebsiella pneumoniae* (see earlier text) and in *Rhodobacter capsulatus*, where transcripts encoding components of the light harvesting complex are degraded more rapidly under conditions of high oxygen tension, which inhibits their synthesis (47). The composition and activity of the *R. capsulatus* degradosome vary under different oxygen concentrations (48). See also the chapter in this volume by Evguenieva-Hackenberg and Klug.

C. Temperature

In some instances, such as those of the *E. coli* and *R. capsulatus* polynucleotide phosphorylase (PNPase) and CspA cold-shock proteins, overexpression at low temperature has been shown to be the consequence of mRNA stabilization (49–52). A similar observation has recently been reported for RNase R in *E. coli* (53). The *desA* and *desB* lipid desaturase transcripts of *Synechococcus* (54) and the mRNAs encoding the CrhC RNA helicase (55) and the RbpA1 binding protein (56) of the Cyanobacterium *Anabaena* are selectively stabilized at low temperature, allowing adaptation of lipid metabolism and overexpression of the Crh helicase. Cold induction of expression of enzymes of the *bkd* operon, which catalyze the degradation of branched amino acids in *B. subtilis*, is also mediated by an increase in the stability of the corresponding transcripts (57). A drop in temperature also stabilizes the DsrA ncRNA, which activates translation of the *rpoS* gene, encoding the stationary phase sigma factor σ^S (58).

A modification of the RNA degradosome at low temperature, namely the replacement of the resident RhlB RNA helicase by its cold-induced CsdA homologue, may account for some of the mRNA stabilization observed at low temperature in *E. coli* (59). Interestingly, several other putative RNA helicases can also bind to two sites in the *E. coli* degradosome (60).

High temperatures have also been observed to affect mRNA stability and processing in different bacteria. In *E. coli*, the *rpsO* transcript becomes more stable when cells grown in minimum medium are shifted to higher temperature (61). Interestingly, this is correlated with an inhibition of the RNase E cleavage which initiates its decay. High temperature also stabilizes the Hsp1 heat shock chaperonin transcript (62) and the *ompA* messenger, thus compensating for its increased sensitivity to RNase E (63). Heat-induced stabilization of mRNA presumably also accounts, at least in part, for the overproduction of certain extracellular enzymes of *Erwinia carotovora* (64), of the heat shock sigma factor of the thermophilic bacteria *Colwellia maris* (65) and of the ClpL ATPase subunit of the Clp proteolytic complex of *Oenococcus oeni* (66). In *Staphylococcus aureus*, both cold and heat were reported to have general stabilizing effects on mRNA (67).

The examples cited earlier give an idea of the kinds of environmental cues that affect mRNA stability. In the remainder of this chapter, we will look at the underlying mechanisms involved.

III. Polyadenylation and Poly(A)-Dependent Decay

A. Diversity of Poly(A) Functions

Polyadenylation is a posttranscriptional modification of RNA that occurs in prokaryotes, eukaryotes, and organelles. The function and extent of bacterial polyadenylation first appeared in marked contrast to those of eukaryotic mRNAs (1). Indeed, the scarcity of bacterial poly(A) tails explains why they were neglected for such a long time (68). In eukaryotes, where they were originally discovered, long poly(A) tails ranging from 60 to 200 A-residues contribute to the export to the cytoplasm and promote mRNA stability and translation. On the other hand, short poly(A) tails contribute to RNA degradation in prokaryotes. Tails containing only one to three A-residues have been reproducibly characterized in bacteria (18, 69–74). Following the recent discovery that poly(A)-dependent degradation of RNA also takes place in organelles and in nuclei in eukaryotes, the destabilizing function of poly(A) is now considered as a ubiquitous mechanism involved in RNA quality control and degradation (75).

B. Characterization of Polyadenylated RNAs

The observation that pulse-labeled RNAs isolated from *E. coli*, and several other eubacteria and archaea, could be retained on oligo(dT) cellulose, was the first indication that polyadenylation takes place in prokaryotes (76, 77). However, the fraction of polyadenylated molecules was low or very low: 15–25% in *B. subtilis* and less than 2% of total labeled RNA in *E. coli* (78–80), where poly (A) tails ranged in size from 1 to 50 A-residues (78, 79, 81–83). Investigations of specific transcripts confirmed that the majority of molecules are not adenylated and that polyadenylation efficiencies vary dramatically from one mRNA species to another. Only 0.011% of bacteriophage f1 mRNA is polyadenylated (84), for example, whereas 10% of *rpsO* primary transcripts harbor short oligo(A) tails of 1–5 nucleotides (71) and 40% of *lpp* and *trpA* mRNAs have longer tails of about 10–20 As (85–87). Poly(A) tails have also been identified at the 3' ends of messengers of the *hag*, *rnpB*, *rps*, *cry1Aa* genes of *B. subtilis*, the *rmf*, *rpoS*, *bolA*, *lacZ*, and *dp1* genes of *E. coli* and bacteriophage T7 transcripts (18, 81, 88–90) (Nierlich, personal communication). A recent analysis of the *E. coli* transcriptome with oligo(dT) primed cDNA probes suggested that 90% of the transcribed ORFs are adenylated (91). It is not known, however, whether these represented full-length or fragmented transcripts. The basal level of polyadenylation, which is very low under normal conditions, increases dramatically when the enzyme responsible for their synthesis, poly(A) polymerase (PAP I), is overproduced.

In bacteria, polyadenylation is not restricted to messenger RNA. It was demonstrated, unexpectedly at the time, that a few A-residues are posttranscriptionally added to the 3' end of the small antisense RNA I, which controls initiation of ColE1-type plasmid replication (74). Even more surprising was the finding that polyadenylation accelerated the rate of degradation and reduced the intracellular concentration of RNA I, thus introducing the provocative idea that bacterial poly(A) tails act in a manner opposite to the stabilizing effect of eukaryotic tails. Several other polyadenylated ncRNAs where then characterized, either under normal physiological conditions, or in strains where the overproduction of PAP I or the lack of exoribonucleases increased both the amount and lengths of poly(A) tails. This included both *cis*- and *trans*-acting small regulatory ncRNAs (namely the Oop and C1 RNAs of bacteriophages λ and P4, respectively, and the Sok, SraK, SraL, and GlmY ncRNAs of *E. coli*), stable 16S and 23S rRNAs of *E. coli*, *B. subtilis*, and *Streptomyces*, tRNA^{Tyr} of *E. coli*, tmRNA, as well as precursors of rRNAs, tRNAs, and of the M1, 4.5S, and 6S ncRNAs (69, 72, 74, 89, 92–97). Replicative forms of the genomic RNA from bacteriophage MS2 can also be adenylated by PAP I in *E. coli* (73). Not only are stable, regulatory, and messenger RNAs polyadenylated, but many of them can be adenylated at multiple sites resulting from transcription termination, endonucleolytic processing by RNase E and RNase III and digestion by exonucleases engaged in the trimming and degradation of RNAs (*18, 69, 77, 84, 89, 94, 95, 98, 99*). Finally, some regulatory ncRNAs and full-length or fragmented mRNAs (CopA, *ompA, trxA, rpsT,* repetitive extragenic palindromic (REP)) containing mRNA fragments) stabilized upon PAP I inactivation are also presumed to be polyadenylated (*83, 100–103*). It therefore appears, that PAP I can polyadenylate all classes of bacterial RNA, but that only a small fraction of molecules, which varies from one RNA species to another, is actually polyadenylated.

C. Polyadenylation Affects RNA Stability

The first indication that poly(A) tails destabilize RNA came from the discovery that PAP I controls the stability of the small antisense RNA, RNA I, which regulates the replication of ColE1 plasmids (*74*). It was originally reported that a mutation in the *pcnB* gene (for plasmid copy number) reduced the copy number of ColE1-type plasmids, whose replication is negatively controlled by RNA I (*104*). Later on, the discovery that *pcnB* actually codes for PAP I brought the first evidence of the metabolic impact of bacterial polyadenylation, and allowed an understanding of the molecular basis of RNA I-mediated regulation of plasmid replication (*105–107*). RNA I is a 108 nucleotide long antisense RNA, which inhibits replication by forming a nonproductive duplex with RNA II, the primer of DNA replication. Inhibition of replication in the absence of PAP I results from the accumulation of a processed form of the RNA I inhibitor. It was subsequently demonstrated that polyadenylation dramatically accelerates the decay of this processed form of RNA I (*74*). It was similarly shown that a few full-length transcripts (*rpsO, ompA, trxA,* and *lpp*) also became less stable when adenylated (*18, 70, 83*). However, further investigations indicated that poly(A) is primarily involved in the catabolism of small tightly folded molecules. This, for example, is the case of the CopA, Sok, and Oop small regulatory RNAs. Their stability and their intracellular concentrations, which control the replication and the maintenance of plasmid R1 and the lysogeny of bacteriophage λ, respectively (*92, 93, 100*), increase when PAP I is inactive. The decay-rate and abundance of several recently discovered *trans*-acting ncRNAs that regulate the translation and stability of bacterial mRNAs, also depends on PAP I. This has been observed for the SraL ncRNA of *S. typhimurium,* and the MicA and GlmY ncRNAs of *E. coli* (*94, 108, 109*). In the case of the small C1 RNA, involved in immunity of the P4 prophage, polyadenylation affects both the stability of precursors and degradation of decay intermediates (*69, 110*).

In all of the examples described earlier, poly(A)-dependent degradation of full-length transcripts or regulatory ncRNAs can directly affect gene expression by modifying the internal cellular concentrations of functional mRNAs encoding specific polypeptides (see later text). However, polyadenylation is also involved in a more ancilliary function: the scavenging of decay intermediates resulting from the endonucleolytic inactivation of RNAs, consistent with its implication in destabilization of small structured RNAs. PAP I inactivation was demonstrated early on to cause an accumulation of mRNA fragments resulting from the endonucleolytic processing of the *rpsO*, *trxA*, *lpp*, *ompA*, and *rpsT* mRNAs, some of which are highly structured (*83, 99, 111–113*). Moreover, the mapping of mRNA-poly(A) tail junctions at multiple locations in many transcripts confirms that PAP I can polyadenylate and destabilize many different RNA fragments with different 3′ extremities thought to be generated by endo and exoribonucleases (*18, 78, 84, 89, 98, 99*). These observations led to the proposal that polyadenylation facilitates the exonucleolytic degradation of tightly folded decay intermediates (*103, 111, 114*). This hypothesis has been verified *in vitro* by showing that poly(A) tails accelerate the degradation of both RNA I and an mRNA fragment terminated by a stable secondary structure, by free and degradosome-associated PNPase (*115–118*). More recently, polyadenylation was also reported to be required for elimination of RNA fragments containing intercistronic REP hairpins, which are believed to protect functional monocistronic units of mRNAs from ribonucleases (*101, 102, 119, 120*).

D. The Enzymes of Poly(A)-Metabolism

The appearance of poly(A) tails, which are not detected under normal conditions, in cells either lacking the two 3′–5′ exoribonucleases, PNPase, and RNase II or in cells overproducing PAP I, led to the conclusion that the length of bacterial poly(A) tails is determined by a dynamic equilibrium between the opposing activities of PAP I and exoribonucleases (*72, 82, 103, 114*). In *E. coli*, the main enzymes involved in the metabolism of polyadenylated RNAs are PAP I, which accounts for the synthesis of most, if not all, poly(A) tails, and exonucleases of 3′–5′ polarity, capable of degrading RNAs harboring single-stranded stretches of nucleotides at their 3′ end (*70, 121*).

PAP I is encoded by the *pcnB* gene and is dispensable for growth (*104, 105, 122*). It is a 53 kDa monomer (*105, 106*), with characteristic features of the nucleotidyltransferase super-family. This family includes eukaryotic PAPs, as well as eukaryotic and prokaryotic tRNA nucleotidyltransferases, which generate the CCA 3′ extremity of tRNAs (*123–125*). The PAP I polypeptide undergoes a proteolytic maturation step, whose function is not known, 10 amino acids from its N-terminus (*106, 126*). PAP I requires a divalent cation, Mg^{2+} or Mn^{2+}, to be active (*127, 128*). The enzyme uses ATP as a substrate to polymerize AMP residues at the 3′ end of RNA primers. CTP and, to a lesser extent UTP, can also be used by PAP I, but at a much slower rate than ATP (*128–130*). These properties may explain why C-residues are sometimes detected in poly(A) tails (Hajnsdorf and Régnier, unpublished results). The addition of guanosine residues by PAP I has also been reported (*69, 129*), but this reaction is probably very slow (*128, 130*). The fact that poly(A) tails are synthesized at the 3′ ends of many primary transcripts, processed RNAs and intermediary products of exonucleolytic degradation, none of which share common 3′ terminal features, suggests that PAP I does not specifically recognize particular structural motifs (*18, 84, 89, 98, 99, 131*). In contrast, the variability of tail lengths mentioned earlier may indicate that all RNA species are not adenylated with equal efficiency. This conclusion is supported by experiments showing that PAP I preferentially adenylates RNAs harboring a 5′ monophosphorylated extremity or single stranded stretches of nucleotides at the 3′ or at the 5′ end (*129, 130, 132*). Moreover, it has been reported that 3′ terminal secondary structures and transcription terminators inhibit PAP I activity (*129*). The influence of 5′ structures on polyadenylation efficiency could explain why processing at the 5′ end of RNA I increases its sensitivity to poly(A)-dependent degradation (*132*). PAP I is a distributive enzyme (it dissociates from RNAs after addition of each, or a few nucleotides) which exhibits a preference for poly(A) primers (*133, 134*). However, PAP I can also act processively in the presence of the Hfq protein, which strongly stimulates its activity (see later text).

E. coli contains six 3′–5′ exonucleases capable of degrading single-stranded RNAs or oligonucleotides. All six of these may be involved in degradation of poly(A) (*135, 136*). The fact that poly(A) is much more abundant when two of these enzymes, RNase II and PNPase, are inactive, while inactivation of only one of them does not have such an effect, suggests that both of these enzymes can degrade poly(A) tails (*82, 121, 131, 137*). RNase II belongs to the ubiquitous RNase R family of proteins, exemplified by RNase R in *E. coli* (*135, 138–140*). This processive hydrolase (it remains bound to the molecules that it degrades after each catalytic cycle) displays maximal activity in the presence of Mg^{2+} ions and releases ribonucleoside 5′ monophosphates (*139*). The abundance of this enzyme, which represents 90% of the poly(A)-degrading activity measured in *E. coli* cellular extracts, suggests that it plays a major role in poly(A) catabolism (*141*). Consistent with this idea, it has been shown that RNase II removes the poly(A) tails synthesized by PAP I at the 3′ end of *rpsO* primary transcripts almost completely, presumably explaining why 90% of these molecules are nonadenylated (*82, 142*). However, RNase II is blocked by stable secondary structures, which cause dissociation of the enzyme from the RNA. RNase II removes single-stranded stretches of nucleotides and poly(A) tails lying downstream of transcription terminators or REP sequences, but fails to degrade RNAs upstream of these hairpins (*82, 103, 120, 142–152*). The interactions between RNase II and single-stranded RNA, deduced from crystallographic three-dimensional structures, indicate that a clamp formed by two RNA binding domains and a narrow channel leading to the catalytic site accounts for the stalling effect of double-stranded RNA (*147, 152–154*). RNAs devoid of 3′ secondary structures are almost completely degraded by RNase II (*111, 113*). The minimum substrate length is 10 nucleotides and the end products of degradation are primarily 4-nucleotide oligomers (*155, 156*).

PNPase is a phosphorylase which attacks the single-stranded 3′ end of RNAs, which are degraded processively into monoribonucleotide diphosphates (NDPs) (*157*). In the reverse reaction, which takes place at low phosphate concentrations, PNPase processively incorporates NDPs into heterogeneous polynucleotides of random sequence (*157*). However, its primary role in the cell is the 3′–5′ degradation of RNA molecules (*158*). PNPase is one of the components of a multienzymatic complex, referred to as the RNA degradosome, which participates in mRNA decay (see later text) (*159–161*). Its three dimensional structure suggests that PNPase is a homo-trimer with a small central pore, through which a single RNA molecule reaches a central cavity containing three active sites (*162*). Interestingly, several PNPase orthologs in eukaryotes and archaea also form oligomers with similar spatial organization and are involved in RNA decay (*163*). See chapter by Evguenieva-Hackenberg and Klug in this volume. The threading of single-stranded RNA through a narrow pore leading to the catalytic site may explain why PNPase mediated phosphorolysis of RNA is slowed or blocked by secondary structures (*152, 162*). However, PNPase is capable of degrading secondary structures of moderate stability that are resistant to RNase II (*103, 120, 144, 145, 164*). PNPase, free or in the degradosome, fails to degrade RNAs whose 3′ terminal nucleotides are sequestered in a secondary structure. Both forms of the enzyme only attack RNAs harboring a 3′ terminal single-stranded stretch of nucleotides, which is used as a toehold to begin the processive degradation of structured molecules. Its association with another degradosome component, the RhlB helicase, allows PNPase to disrupt very stable secondary structures that cannot be degraded by the free enzyme (*117, 160, 161*).

RNase R, the archetype of the ubiquitous RNase R family of 3′–5′ exonucleases (*135*), accounts for less than 5% of the hydrolytic activity of *E. coli* measured on poly(A), which is due primarily to RNase II (*145*). In contrast, RNase R is very efficient on structured RNAs such as rRNAs that resist the attacks of PNPase and RNase II (*147, 165*). The reaction is processive and generates end-products of 2 or 3 nucleotides (*147, 166*). Intermediate degradation products were not detected for structured substrates, indicating that the rate of the reaction is not reduced by secondary structures (*147*). Consistent with these conclusions, based on *in vitro* experiments, it has been demonstrated recently that RNase R is also involved in the degradation of mRNAs with extensive secondary structures, such as those containing REP sequences, and in the quality control of ribosomal rRNA (*102, 167*). The finding that RNase R can only attack RNAs harboring a single-stranded stretch of seven nucleotides or more at their 3' end suggested that it might be involved in the oligo(A) dependent degradation of structured RNAs (*166*). Moreover, a report showing that degradation of REP sequences is mediated by a poly(A)-dependent enzyme other than PNPase also supported this idea (*101*). This hypothesis was recently verified, when we demonstrated that poly(A)-dependent degradation of the *rpsO* mRNA is mediated by RNase R (*70, 168*) (Andrade et al., in press). These data imply that the accepted idea that PNPase is responsible for the poly(A)-dependent degradation of structured RNAs probably has to be revised (*107, 116, 160, 169*). The fact that a mutant strain lacking both PNPase and RNase R is inviable suggests that these enzymes carry out an essential function that may well be the poly(A)-assisted degradation of structured molecules (*170*). An example of poly(A)-dependent degradation mediated primarily by RNase R is shown in Fig. 1, which summarizes the actions of PAP I and exoribonucleases on the 3' extremity of the *rpsO* primary transcript.

RNase E is considered the main player of mRNA decay, in addition to being involved in the maturation of rRNAs and tRNAs (*169, 171–174*). It is the scaffold for the association of PNPase (see earlier text), the RNA helicase RhlB and enolase, a glycolytic enzyme, in the complex known as the RNA degradosome (*117, 159–161, 175*) (see chapter by Carpousis and colleagues). RNase E is a 5' end-dependent endoribonuclease which preferentially cleaves RNAs harboring single-stranded mono-phosphorylated 5' extremities (*3, 4, 176*). This enzyme, which cleaves molecules in single-stranded A–U rich sequences (*169, 171*) may remove tails of polyadenylated molecules *in vitro* (*177, 178*).

In addition to PNPase, RNase II, and RNase R, *E. coli* contains three other 3'–5' exoribonucleases, RNase T, RNase PH, and RNase D, capable of removing single-stranded nucleotides from the 3' extremities of RNAs. These enzymes have been implicated in the maturation of many stable RNAs (*179–183*). They are also blocked by secondary structures, and could potentially contribute to the degradation of oligo(A) tails added at the 3' ends of mRNAs, regulatory RNAs and precursors of stable RNAs. It appears, however, that they do not counteract the synthesis of oligo(A) tails by PAP I (*70, 82, 168*).

E. A Mechanistic View of Poly(A)-Metabolism and Poly(A)-Assisted RNA Decay

It is currently believed that most, if not all, accessible 3' extremities of RNA can be adenylated by PAP I or nibbled by exoribonucleases. Stable RNAs, such as mature tRNAs and rRNAs, are almost never subject to polyadenylation, probably because their 3' termini are covalently bound to amino acids in the first case, or buried within the ribonucleoparticles in the second (*181*). In contrast, PAP I polyadenylates the 3' extremities of abnormal tRNA precursors that cannot be aminoacylated and those of precursors of the 16S and 23S

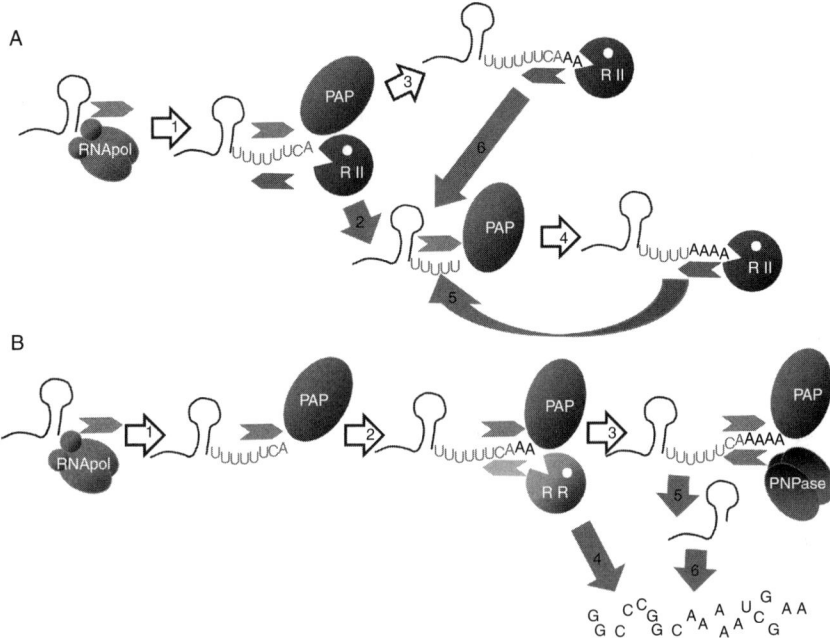

FIG. 1. Model of poly(A) metabolism at the 3' end of a primary transcript terminated at a Rho-independent terminator. This model derives from experiments performed on the *rpsO* transcript (*82, 114, 142*). Each step of this model is shown by a numbered arrow; white arrows correspond to elongation steps while shortening reactions are shown by grey arrows. Moreover, each enzyme is associated with a head-less arrow indicating the direction (elongation to the right or shortening to the left) of the reaction catalyzed. Transcription terminates at the C or A-residue following the stretch of 6 U-residues (Step 1 of panels A and B). Panel A shows how adenosine residues added by PAP I are removed by RNase II. Primary transcripts can be nibbled by 1–3 nucleotides by RNase II (Step 2). Primary transcripts and nibbled molecules are both adenylated by PAP I (PAP) (Steps 3 and 4) thus acquiring short oligo(A) extensions that are removed by RNase II (R II) (Steps 5 and 6). Both reactions are distributive and very slow. The rates of nibbling and addition of the first few A-residues range between 0.5 and 7 nucleotides per minute (*142*). The equilibrium between the opposing actions of PAP I and exoribonucleases presumably accounts for the occurrence of mRNAs harboring 1–5 A-residues in wild-type cells (*71*). Panel B: Oligo(A) tails synthesized by PAP I (Steps 2 and 3) can initiate decay of the transcript provided they are long enough to be recognized by RNase R (RR) (Step 4) or PNPase (Steps 5 and 6) (*70, 83, 98*). Oligo(A) tails containing 3 A-residues downstream of the transcription termination site (C residue) can be attacked by RNase R which completely and processively degrades the RNA. Further elongation generates tails of 5 A-residues that can be recognized by PNPase. Degradation by PNPase depends on secondary structures, which can either block the enzyme or eventually succumb to reiterative attacks (see Fig. 2 and text). All elongated transcripts are efficiently shortened by RNase II. Those harboring more than 9 A-residues are rapidly and processively degraded by RNase II (*142*) (not shown). This hierarchy of recognition of oligoadenylated RNAs by 3'–5' exoribonucleases may explain why RNase II protects the RNA from being degraded by RNase R and PNPase and also why RNase R is the major actor of poly(A)-dependent degradation of the *rpsO* transcript.

rRNAs that might emerge from the surface of the ribosome (98, 184). The 3′ extremities of adenylated RNAs and primary transcripts are also substrates of 3′–5′ exoribonucleases provided they harbor a single stranded 3′ terminus (72, 82, 142, 182) (Fig. 1). The fact that PAP I is an intrinsically distributive enzyme that dissociates readily from the RNA primer that it has elongated, means that 3′ ends of RNA are accessible to exoribonucleases after each (or a few) step(s) of adenylation. This implies that the addition of only few A-residues is sufficient to generate a toehold that can be recognized by exoribonucleases.

The actions of PAP I and exoribonucleases on typical bacterial RNAs terminated by the secondary structure of a Rho-independent terminator have been investigated with the *rpsO* transcript (Figs. 1 and 2) (82, 142). Primary transcripts can be nibbled slowly and distributively by RNase II, but not by other ribonucleases such as PNPase and RNase R, presumably because the 3′ single-stranded stretch of seven or eight nucleotides is too short (120, 166). Both nibbled RNAs and primary transcripts can be adenylated distributively by PAP I (Fig. 1A). The balance between the opposing actions of PAP I and exoribonucleases presumably accounts for the occurrence of mRNAs harboring one to five A-residues in wild-type cells (71). The synthesis of the tails is very slow; the rates of nibbling and addition of these few A-residues by RNase II and PAP I, respectively, range between 0.5 and 7 nucleotides per minute (142). Incidentally, longer tails are sometimes synthesized that are long enough to be recognized by PNPase and RNase R, in addition to RNase II (70, 83, 98). These slightly longer RNAs can then either be trimmed back by RNase II to the 3′ proximal stable secondary structure, thus preserving a functional coding sequence, or be completely degraded by PNPase or RNase R (142) (Fig. 1B).

Long tails are very rapidly and processively removed by RNase II, PNPase, and presumably RNase R. Tails longer than eight nucleotides are degraded at least 20 times faster than the short tails of one to five A-residues (142). The rapid and processive step of exonucleolytic decay produces RNAs with single-stranded stretches of 12–16 nucleotides downstream of the hairpin that are then either degraded distributively by RNase II and PNPase, or readenylated by PAP I (see earlier text) (18, 71, 74, 82, 142, 143). In the case of the *rpsO* mRNA, RNase II can remove the tail completely (Fig. 1A), while PNPase releases slightly longer RNAs harboring tails of two to four A-residues (Fig. 2A) (82, 120, 142). More generally, 3′ terminal hairpins of different structures and stabilities may block 3′–5′ exoribonucleases at different positions and thus determine the number of A-residues left at the 3′ end of the RNA (82, 144). Moreover, the other 3′–5′ exonucleases that are involved in 3′ trimming of stable RNA precursors may also contribute to the nibbling of messengers and generate tails of different lengths depending on their ability to approach secondary structures (181). On the other hand, it has been suggested that the poly(A) tails of some RNAs may be specifically removed by PNPase or RNase II (121).

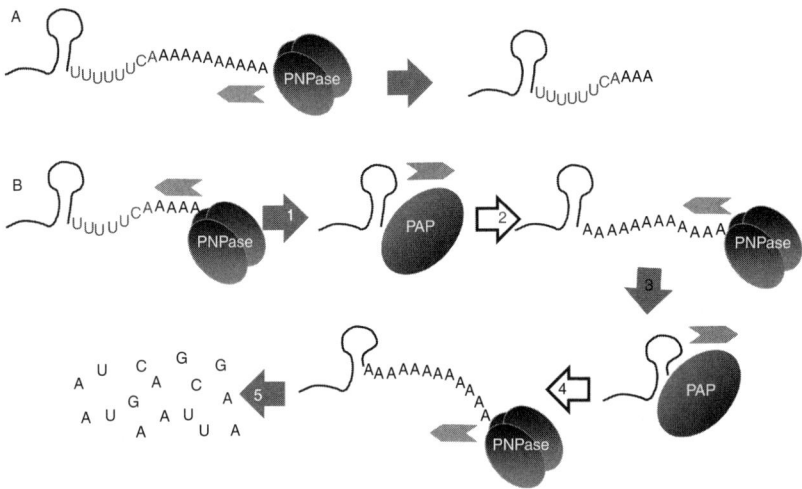

FIG. 2. Poly(A)-dependent degradation of structured RNAs by PNPase. PNPase is arrested by secondary structures, which can either slow down the degradation or completely arrest the enzyme. Panel A shows the RNA released when PNPase is impeded by a stable secondary structure. Panel B shows the current model that has been proposed to explain how secondary structures are progressively degraded by PNPase. Steps and enzymes are indicated as in Fig. 1. 3′ extensions of about 12 nucleotides allow PNPase to initiate decay. When it encounters a secondary structure, PNPase can continue to degrade the RNA of the tail (Step 1) and even remove few nucleotides from the bottom of the hairpin (Step 3) before dissociating from the RNA. These decay intermediates are then readenylated by PAP I (Steps 2 and 4) so that PNPase can initiate additional rounds of degradation that will progressively shorten the RNA (Steps 1, 3, and 5). Removal of successive RNA segments of the descending strand of the secondary structure by PNPase progressively disrupts the 3′ protecting hairpin, thus leading to the complete degradation of the RNA. One can imagine that all PNPase attempts are not efficient and that this enzyme reiterates its attack until the structure breathes and can be invaded. The down-side of this approach is that RNA presumably undergoes many unproductive cycles of polyadenylation and deadenylation before being completely degraded.

It first appeared paradoxical that inactivation of RNase II destabilizes both the *rpsO* transcript and the RNA-OUT, which represses Tn10 transposition (*82, 103, 151, 168*). This was explained, however, when it became clear that the tails removed by RNase II are destabilizing elements recognized by other exoribonucleases able to completely degrade the RNA (*82, 142, 185*) (Andrade et al., in press). *In vitro* experiments confirmed that RNase II protects the polyadenylated *rpsT* transcript from PNPase-mediated degradation (*103*).

In spite of the recent discovery that RNase R participates in the degradation of structured RNAs (*102, 166*) (Andrade et al., in press), it is still believed that PNPase is the main enzyme involved in the poly(A)-dependent degradation of RNA. Facilitation of PNPase-mediated decay by poly(A) tails was first demonstrated in the case of RNA I of ColE1 plasmids (see earlier text) (*74*);

the conclusion was that poly(A) tails provide sites where PNPase can bind and initiate the exonucleolytic degradation of RNA I (*107, 116*). Polyadenylation has since been shown to be involved in the degradation of other RNA species and of RNA in general (*83*) and it is admitted that the mechanism proposed for degradation of RNA I can be extended to the poly(A)-dependent degradation of any RNA with a 3' secondary structure (*70, 83, 84, 93, 100, 103*). The current idea is that free PNPase can carry out the complete processive degradation of RNAs containing weak secondary structures, but is blocked when it encounters stable hairpins, which cause dissociation of the ribonuclease from its substrate (Fig. 2) (*103, 107, 161, 169*). When it encounters a secondary structure, PNPase can either release RNAs devoid of a 3' toe-hold (PNPase is arrested at about 9–11 nucleotides from the stable hairpin (Fig. 2A)) or continue to degrade the RNA, removing a few nucleotides at the bottom of the hairpin before dissociating from the RNA (Fig. 2B). Localized melting of tightly folded secondary structures probably allows PNPase to invade and nibble the base of secondary structures. Several cycles of exoribonucleolytic degradation and polyadenylation would allow PNPase to reiterate its attack until the structure breathes and can be invaded. Successive removal of RNA segments from the descending strand of the secondary structure by PNPase progressively disrupts the 3' protecting hairpin, and eventually leads to the complete degradation of the RNA (*1*). Consistent with this hypothesis, it has been shown that complete degradation of RNAs polyadenylated downstream of a stable hairpin can only be achieved by PNPase *in vitro* if the RNA is continuously adenylated by PAP I (*103*). It has been proposed that the most stable of 3' protecting hairpins, which cannot be traversed by free PNPase, are degraded by PNPase associated with the degradosome, where its activity is facilitated by the RNA helicase RhlB (*160*). This helicase, activated through its association with RNase E, is thought to disrupt the hydrogen bonds between nucleotides of the hairpin, which are then removed by PNPase in the degradosome (*117, 186*). An incremental mechanism of degradation similar to that described earlier for free PNPase, probably also accounts for the degradation of stably folded RNAs by the degradosome (*103*). The clustering of poly(A) sites observed *in vivo* for several transcripts is consistent with the idea that successive adenylation events are required for progression of PNPase (*84, 98, 99*).

Recent data from our laboratory has shown that RNase R also uses poly(A) extensions to initiate degradation of mRNAs terminated by stable secondary structures of Rho-independent terminators (Andrade *et al.*, in press). Although we do not yet know the contribution of RNase R to overall poly(A)-dependent degradation of RNA, the fact that oligoadenylated *rpsO* mRNA is primarily degraded by RNase R, while PNPase plays only a minor role, opens the possibility that other primary transcripts are degraded by RNase R. In this case, and in contrast to the reiterative model described earlier for PNPase (Fig. 2), oligo(A) extensions which have escaped RNase II surveillance are likely to permit the immediate degradation of the entire RNA by RNase R (Fig. 1).

F. Regulation and Cofactors of the Poly(A)-Dependent Machinery of Degradation

The expression of enzymes involved in poly(A)-metabolism and poly(A)-assisted decay change according to growth-rate and environmental conditions. The amount of PAP I is higher in slow-growing cells, because of a more efficient transcription of the *pcnB* gene (*187*). This causes an increase in polyadenylation efficiency, which destabilizes RNA I and the Oop small ncRNA. In contrast, the stability of the *lpp* mRNA is not modified. The scarcity of PAP I, whose translation begins at a very inefficient noncanonical AUU initiation codon (*126, 188*), in addition to the competition with the poly(A) tail degrading activity of exonucleases, may explain why the long tails of 500–1000 A-residues that can be synthesized by PAP I *in vitro* are not detected in the cell (*133, 142, 189*).

Both PNPase and RNase R are transiently overexpressed after a shift to low temperature (*50, 53, 190*). In both cases, the overexpression results from stabilization of their mRNAs. RNase R is also overproduced under starvation conditions, such as slow growth in minimum medium, entry into stationary phase and nitrogen, carbon, or phosphorus starvation (*191*). PNPase is essential for growth at low temperature (*192, 193*). Moreover, a posttranscriptional mechanism of autoregulation allows adjustment of the intracellular level of PNPase, presumably to the amount of substrate available in the cell (*194–196*). RNase II levels also change according to growth medium (*197*) and intracellular levels of PNPase and RNase II have been shown to be cross-regulated (*198*). This series of observations indicates that PAP I and the exoribonucleases responsible for poly(A)-dependent degradation of structured RNA are controlled by a complex regulatory network that responds to many different signals and which may affect RNA turnover rates. This is probably the case at low temperature. Indeed, under these conditions, the increased stability of secondary structures is overcome by an overproduction of PNPase and RNase R permitting the removal of nonfunctional structured RNAs (*53*).

There is a set of data indicating that poly(A)-dependent degradation is triggered by RNase E cleavage (*1, 74, 93, 103, 107, 111, 116, 199*). It was demonstrated that the CopA RNA must be cleaved by RNase E before becoming a target for poly(A)-dependent degradation (*100*) and it was proposed that RNase E and PNPase interact functionally during the degradation of RNA I (*107*). Similarly, RNase E cleavage stimulates the poly(A)-dependent degradation of the *rpsO* mRNA (*111*). *In vitro* results with RNA I suggest that it could be PAP I itself that discriminates between the full-length message and 5′

monophosphorylated mRNA decay fragments produced by RNase E (*132*), although a physical interaction has been detected between RNase E and PAP I *in vitro* (*200*). All of these observations support the idea that coordination of the endonucleolytic and exonucleolytic pathways of decay channels decay intermediates to the poly(A)-dependent degradation machinery.

The model depicted in Fig. 1, in which 3′–5′ processive exonucleases prevent elongation of the tails, led to the idea that cofactors similar to mammalian CPSF (Cleavage and Polyadenylation Specificity Factor) and PABP II (a poly(A) binding protein (*201*) may account for the synthesis and the protection of the long poly(A) tails of up to 50 nucleotides in length, that have been detected in *E. coli* (*83*). One possible candidate was the host-factor, Hfq, which interacts with A-rich regions of RNAs (*202–205*) and intervenes in the interactions between mRNAs and small regulatory ncRNAs (*206–209*). Hfq stimulates poly(A) synthesis *in vivo* and *in vitro* by converting PAP I into a processive enzyme that remains associated with its elongating substrate (Fig. 3) (*71, 133, 134*). This can only occur if RNA already harbors an oligo(A) tail of about 20 nucleotides (*133*). Oligo(A) tails become longer and more abundant when Hfq is active (*71*). A change in substrate conformation resulting from the very strong interaction between Hfq and poly(A) tails was proposed to account for this dramatic modification of PAP I behavior (*130, 205*). Alternatively, a direct Hfq–PAP I interaction was suspected in the modification of polyadenylation efficiency at Rho-independent terminators (*188*). However, if this is so, it remains to be clarified how Hfq can have opposed positive and negative effects on the polyadenylation of Rho-independent terminators, as observed in the case of the *rpsO* transcript (*71, 188*). If functionally significant, the physical interaction that has been reported between PAP I and Hfq raises the possibility that PAP I could also be modulated by the many protein and RNA partners of Hfq, such as regulatory ncRNAs, RNase E and components of the RNA degradosome (*161, 188, 210*). Several other interactions involving PAP I have been reported, and these may reflect functional relationships which have not yet been characterized. PAP I has been reported to interact directly with PNPase, RNase E, and DEAD box RNA helicases, which may cooperate with PAP I for poly(A) synthesis and RNA decay (*188, 200*). PAP I is also thought to have some degree of association with the membrane (*211*). Compartmentalization to the membrane, in the vicinity of degradosome components, which have also been shown to colocalize with the bacterial cytoskeleton (*212, 213*) may help coordinate the different steps of poly(A)-dependent RNA degradation. A direct interaction between PNPase and the RhlB RNA helicase, distinct from the functional relationship that occurs in the degradosome (*214*), may also facilitate poly(A)-assisted degradation of double stranded RNA by PNPase. It is also worth mentioning that phosphorylation impairs PAP I activity *in vitro* and that this modification was proposed to regulate polyadenylation *in vivo* (*215*).

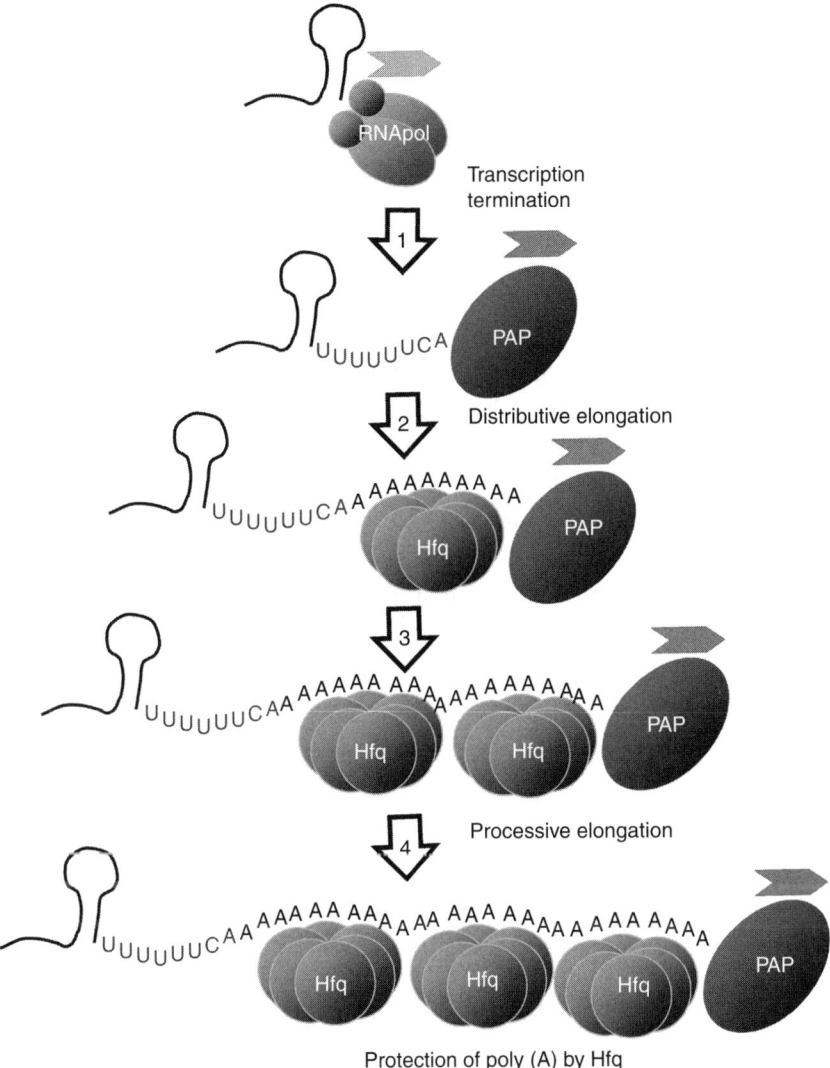

FIG. 3. Role of Hfq in poly(A) metabolism and RNA stabilization. This model derives from experiments performed on the *rpsO* transcript (71, 130, 133, 205). Steps and enzymes are indicated as in Fig. 1. Transcripts released by RNA polymerase (Step 1) are first slowly and distributively polyadenylated by PAP I (PAP) (Steps 2 and 3). Once 20–30 A-residues have been added, Hfq binds tightly to tails, and the reaction becomes processive and accelerates (Step 4). Hfq bound to poly(A) protects it from exoribonucleases. The properties of Hfq may explain why long tails are sometimes detected in bacteria despite the fact that they can be rapidly and processively degraded by exoribonucleases.

Two PABPs besides Hfq may affect poly(A) metabolism. CspE prevents poly(A) removal *in vitro* and ribosomal protein S1 can interact physically with two enzymes of poly(A) metabolism, PNPase and RNase E (*216*). It has been proposed that an endonucleolytic cleavage by RNase E, close to the RNA-poly (A) junction may remove poly(A) tails protected by poly(A)-binding proteins (*178*). However, there is no evidence that cofactors such as CspE and ribosomal protein S1 affect poly(A) metabolism *in vivo*. The fact that tails longer than 20 A-residues are infrequently detected in *E. coli* suggests that the probability that RNAs undergo many successive steps of elongation by PAP I assisted by Hfq or other cofactors without being attacked by exoribonucleases is very low and/or that tails are not efficiently protected by PABPs. The scarcity of long tails can also be explained by the rapidity of their degradation by RNase II (*142*). One could also imagine that only a few preferentially polyadenylated RNA species acquire long poly(A) tails.

G. Roles of Poly(A)-Dependent Decay

We have mentioned earlier that polyadenylation controls the stability and the activity of a few small antisense ncRNAs that control plasmid replication and maintenance, and viral lysogeny. We have also seen that poly(A) tails accelerate the scavenging of decay intermediates generated by endonucleases. Most importantly, there are indications that bacterial polyadenylation is also involved in the quality control of RNA, a function reminiscent of that of the TRAMP polyadenylation complex in the nucleus of eukaryotic cells (*217, 218*). Precursors of stable RNAs, which accumulate in strains deficient for 3′ trimming exoribonucleases, are polyadenylated (*72*). In the case of the thermosensitive mutant of tRNA^{Trp}, polyadenylation accelerates the degradation of the precursor by PNPase. This was a hint that *E. coli* has developed a poly(A)-dependent quality control system whose role is to eliminate abnormally folded RNAs that cannot be rapidly processed into active mature molecules.

It has been observed repeatedly that stabilization of mRNA consecutive to PAP I inactivation is much more dramatic when the RNA cannot be degraded by RNase E, either following RNase E inactivation or because the cleavage sites have been mutated (*70, 83, 219*). This also suggests that polyadenylation acts as a surveillance mechanism whose role is to get rid of mRNAs that accumulate abnormally in the cell when the prominent decay pathway is not operational (*83*). The slow rate of A-residue addition and the rapidity of poly(A) tail removal by RNase II may in part explain why full-length transcripts are primarily degraded by RNase E (*142*). However, it is also possible that a degradosome bound simultaneously at an internal RNase E site and at the 3′ end of the polyadenylated RNA, through PNPase and RhlB, could impair poly(A)-dependent degradation of full-length transcripts (*219*).

Finally, there are few genes whose expression has been reported to be affected by polyadenylation. Increasing polyadenylation stabilizes the RNase E and PNPase transcripts, leading to increased PNPase and RNase E levels (*220*). It has been proposed that this unexpected stabilization, in conditions which accelerate RNA decay, might be the consequence of the autoregulation of the *pnp* and *rne* genes in response to increased poly(A) synthesis (*220*). Another example is the dramatic increase of glucosamine-6-phosphate synthase (GlmS) expression in PAP I deficient cells. In this case, PAP I acts by polyadenylating and destabilizing the small *trans*-acting GlmY ncRNA which indirectly stabilizes the GlmS mRNA (*94, 221, 222*). PAP I was also reported to affect the induction of the expression of the ribosome modulating factor Rmf that occurs in the stationary phase of growth (*18*). Lastly, one can also speculate that the implication of PAP I in the stability of the MicA and SrlA regulatory ncRNAs suggests that polyadenylation also affects the translation of their mRNA targets (*109*).

H. Heterogeneous Tails and RNA Stability

Long nontemplated heterogeneous tails, containing all four ribonucleotides, were first detected at the 3′ end of spinach chloroplast RNAs, where they were demonstrated to accelerate RNA decay (*223*). Similar 3′ extensions were then identified in the cytoplasm and organelles of nearly all eukaryotic cells and in many bacteria (*224*). Long heterogeneous tails were characterized in Cyanobacteria, *Streptomyces* and *B. subtilis*, but not in Mycobacteria (*89, 225–227*). The fact that these tails were still present in *E. coli* cells lacking PAP I, but not in a *pnp* mutant led to the conclusion that they are synthesized by PNPase, functioning in polymerization rather than degradation mode. It has been suggested that PNPase synthesizes heterogeneous tails *in vivo* in the absence of PAP I and occasionally incorporates U and C residues in poly(A) tails, even when PAP I is active (*131*). PNPase is also believed to synthesize tails in chloroplasts (*228*), but not in *B. subtilis*, where heterogeneous tails were detected in the absence of PNPase (*89*). The idea is that the conversion of PNPase activity from phosphorolysis (degradation) to polymerization and the reverse might be governed by transient modifications of local phosphate concentration near the 3′ end of RNAs (*131*). However, the physiological significance of these alternate phases of RNA elongation and shortening remains mysterious. Indeed, there is no evidence that long heterogeneous tails affect RNA stability in bacteria. Moreover, the fact that the addition of five A-residues at the 3′ end of RNAs are sufficient to promote exonucleolytic degradation (*115*) suggests that long hetero- or homo-polymeric tails may have a different function in RNA metabolism. It will also be interesting to know whether the destabilizing function of posttranscriptionally added U-residues, recently characterized in eukaryotic cells, also takes place in bacteria (*229*).

IV. Other Modulators of RNA Stability

A. Regulators of RNase E

RNase E and its associated proteins constitute the major components of the large protein complex termed the degradosome. Recent studies indicate that the functions of all four major degradosome proteins are required for normal RNA decay and further suggest that components of the assembled degradosome complex work in concert to regulate the degradation of most mRNAs in *E. coli*. The cellular level and activity of RNase E in *E. coli* are subject to complex control including autoregulation, allosteric regulation by the 5′ monophosphorylated extremities of RNA and inhibition of activity by RraA and RraB. These regulators of RNase E activity were identified during experiments aimed at identifying genes whose overexpression enhances disulfide isomerase activity in *E. coli* (230, 231). The authors demonstrated that the increase in DsbC protein levels resulted from the stabilization of the *dsbC* transcripts when the *menG* or the *yjgD* genes were overproduced. These genes were renamed *rraA* and *rraB*, respectively. The observed similarity between the phenotypes of the chromosomal *rne* mutant (232) and those conferred by multicopy expression of *rraA*, led the authors to speculate that the latter might also exert its effect through modulating RNA processing by RNase E. Both the RraA and RraB proteins inhibited the endoribonucleolytic cleavage by RNase E of the 5′ untranslated region (UTR) of its own transcript strongly enough to override the autoregulation of *rne* that normally maintains a constant level of this ribonuclease. A genome-wide analysis of mRNA abundance using DNA microarrays revealed that the abundance of 18% of all *E. coli* transcripts was affected by at least twofold, when either RraA or RraB was overexpressed. Strikingly, however, not all RNAs were affected equally by these inhibitors: 7.6% of the transcriptome was stabilized equally by either protein inhibitor, 8.4% of RNAs were affected uniquely by the overexpression of RraA and 1.9% of transcripts were stabilized only by RraB (231).

The inhibition of RNA cleavage by RraA and RraB results from the direct interaction of these two proteins with RNase E. However, they interact in distinct ways with RNase E. Amino acids 694–727 of the scaffolding domain are required for the binding of RraB to RNase E (231). In contrast, RraA can inhibit the endonuclease activity of the N-terminal catalytic domain *in vitro* (230). Both proteins also affect degradosome composition. In control cells, the molar ratios of the major degradosome protein components to RNase E were: PNPase, 5.4; RhlB, 2; enolase, 1.3, while in degradosomes isolated from cells overexpressing RraA, the ratios of PNPase, RhlB, and enolase to RNase E were 0.7, 0.3, and 0.1 respectively. In cells overexpressing RraB the ratios of RNase E-bound PNPase, RhlB, and enolase dropped to 1.1, 0.9 ,and 0.6, respectively.

On the basis of these results, the authors proposed that these changes in degradosome composition may account for the selective stabilization of different sets of transcripts by RraR and RraB (*231*).

It is important to note that the transcription of *rraA* is elevated upon entry into stationary phase in a σ^s-dependent manner. In addition, the stability of the *rraA* transcript is itself dependent on RNase E activity, suggesting the involvement of a feedback circuit in the regulation of RraA levels in *E. coli* (*233*). Moreover, the absence of *rraA* has no strong phenotype. The primary and ternary structures of RraA suggest that it is related to an aldolase family and derives from an ancient fold that has been adapted to a wide-range of functions (*234*).

It is also worth noting that bacteriophage T4 has been proposed to synthesize RNase E regulators (*22, 235*); see chapter in this volume by M. Uzan. Indeed, gene *61,5* of bacteriophage T4, also designated *dmd* (discrimination of message for degradation), has opposite stabilizing and destabilizing effects on middle and late gene mRNAs, depending upon the stage of development (*236*). The *usvY* middle gene transcript is normally rapidly degraded in the middle and late stages of development. Interestingly, the stabilization of the *usvY* middle gene transcript upon inactivation of *dmd* or upon reduction of RNase E activity led to the proposal that Dmd might preserve the RNase E activity of the host cell (*235*). Interestingly Dmd has been proposed to have an opposite inhibitory activity on RNase LS.

B. Gmr, The Regulator of RNase II

We have already mentioned that RNase II is one of the major exoribonucleases of *E. coli* whose expression depends on a complex regulatory network involving the endoribonucleases RNase III and RNase E, and the exoribonuclease PNPase, all of which affect levels of the *rnb* mRNA (*198, 237*). Furthermore, deletion of the *gmr* gene (gene modulating RNase II), later designated *yciR*, lying just downstream of *rnb* causes a dramatic accumulation of RNase II (*197*). Moreover, RNase II levels depend on growth conditions; it is more abundant in rich media compared to minimum media and is sensitive to nitrogen content in the medium. Interestingly, *gmr* regulation does not affect *rnb* mRNA levels but modifies the stability of the RNase II polypeptide, whose half-life is less than 60 min in wild-type cells and more than 2 h in a *gmr* mutant. Moreover, changes in RNase II levels in response to changing growth conditions are abolished in the *gmr* mutant. Gmr is thought to act somewhere on the transduction pathway, carrying the signal from a sensor monitoring nutrients in the growth medium to the proteolytic enzymes responsible for RNase II degradation.

YciR (Gmr) belongs to a large family of proteins harboring GGDEF and EAL domains thought to be involved in metabolism of the bis-(3′–5′)-cyclic-di-guanosine monophosphate (c-di-GMP) signaling molecule (*238*). c-di-GMP is involved in cellulose metabolism, biofilm formation, mobility, and virulence. Interestingly, YciR exhibits σ^s-dependent stationary phase induction, modulated by temperature and medium composition.

C. Proteins Hfq, CsrA, and ncRNAs

Following the discovery of siRNAs and miRNAs, which regulate development, cell death, and gene silencing in eukaryotic cells, bacterial genomes such as that of *E. coli*, were found to express many small ncRNAs that are crucial regulators of stress responses and bacterial virulence. The most exhaustive searches for ncRNAs have taken place in *E. coli*, resulting in the identification of about 100 ncRNAs, some of which affect mRNA stability (95, 239–242). One large class of these ncRNAs acts by base-pairing to target mRNAs, and the specificity of action depends on that pairing. Base-pairing occurs most frequently at the 5′ end of the message and occludes the ribosome binding site and/or the start codon to inhibit translation. This is the case for the MicF (243, 244), OxyS (203, 245, 246), and Spot42 (247) ncRNAs. In contrast, the base-pairing of DsrA to one strand of an mRNA structure that occludes the translation initiation site of the *rpoS* mRNA facilitates ribosome binding and thus positively regulates translation (248, 249). Interestingly, RyhB and OxyS can act both negatively and positively depending on the targets. Base-pairing with ncRNAs also frequently affects the longevity of mRNA targets, which can either be stabilized or become more prone to degradation. For example, base-pairing of RyhB to the *sdh* and *sodB* mRNAs is associated with their degradation, while base-pairing of GadY RNA with *gadX* mRNA allows accumulation of this transcript. Because translating ribosomes are known to impede the access of endoribonucleases to cleavage sites, one early prediction was that blocking translation ultimately leads to the decay of the mRNA (250). It is also conceivable; however, that ncRNA base-pairing with its target could block the access of a ribonuclease and thus stabilize the mRNA.

As has been shown to occur in eukaryotes, base-pairing between small ncRNA and mRNA takes place in the presence of particular proteins. In bacteria, this function is achieved by the protein Hfq, an interesting RNA binding protein related to the Sm and Sm-like proteins involved in splicing. In *E. coli*, Hfq acts as a pleiotropic regulator, controlling the expression of many genes both positively and negatively, directly and indirectly (251–254) (for review (206, 207, 209, 255, 256)). Hfq binds to the majority of the small regulatory ncRNAs identified in *E. coli* (239, 257) and it is essential for the function exerted by the OxyS, DsrA, RyhB, Spot 42, MicA, and SgrS ncRNAs on their mRNA targets *rpoS*, *fhlA*, *sodB*, *sdhCDAB*, *galK*, *ompA*, and *ptsG*, respectively (9, 38, 247, 258–260). Hfq binds AU-rich single-stranded RNA, with a preference for binding next to stem-loops. The presence of several independent RNA binding surfaces on the homo-hexameric ring-shaped structure of Hfq is predicted to allow Hfq to bind to a ncRNA and its mRNA target simultaneously (261, 262). It is also possible that dimerization of the Hfq hexamer facilitates base-pairing by increasing the local concentrations of the RNAs involved in pairing (263). Duplex formation between ncRNAs and target mRNAs is believed to be essential for the regulatory action of ncRNAs. It was shown that mutations that disrupt base-pairing reduced the regulatory function of the ncRNA while compensating mutations restored it (245, 264). In several cases, the function of Hfq is to enhance base-pairing between ncRNAs and target mRNAs. It is not clear whether Hfq affects the equilibrium constant and/or the binding rate of the two RNAs. In the case of SgrS and *ptsG*, Hfq enhances the mRNA–ncRNA duplex formation by increasing its rate of binding (264). However, *cis*-acting ncRNAs that match their mRNA targets perfectly and the transacting RNA III of *Staphylococcus aureus* can act in the absence of Hfq, indicating that an extended complementarity and/or high RNA concentration may overcome the need for Hfq (Le Derout et al., unpublished results) (265, 266). In addition to facilitating RNA–RNA annealing, Hfq has been shown to regulate poly(A)-dependent mRNA decay (71, 133, 188), to affect RNA stability (258, 267–269), to modulate RNA processing by RNase E (205, 210, 257, 268, 270), and to modulate translation (10, 271). Although it is clear that Hfq binds to ncRNAs and to some mRNA targets, it is not known whether binding to both the ncRNA and the mRNA target is required for all Hfq functions. Moreover, although the multiple roles of the protein seem to be correlated with its capacity to bind RNA, large-scale analysis of protein complexes has revealed that Hfq interacts with numerous proteins and therefore could affect their function directly. This includes ribosomal proteins, RNases (RNase E, RNase G, RNase R), helicase (DeaD), and other factors involved in RNA decay (210, 272), RNA polymerase in the presence of S1 protein, PAP, and the protein H-NS (188, 273, 274).

The modification of RNA stability can be mediated synergistically by Hfq and ncRNAs, or by only one of these stability modulators acting on its own. The propensity of Hfq to bind single-stranded AU-rich sequences (202), exemplified by the direct interactions between Hfq and the OxyS, DsrA, and Spot42 ncRNAs (203, 247, 258), but also with mRNAs (205), may result in the protection of these molecules against RNase E cleavage, since Hfq binding sites and sites of RNase E cleavage share sequence similarity (205, 270). In agreement with this idea, the stability of many ncRNAs is clearly increased by Hfq (257) and the *sodB* mRNA was proposed to be directly repressed and possibly stabilized by Hfq (275). Exceptions do exist, such as the MicX ncRNA of *Vibrio cholerae*, which is processed in an RNase E and Hfq-dependent fashion into a shorter form that is more stable than the precursor transcript (276). Both full-length and processed forms have regulatory activity; they cause a similar reduction in the level of *vc0972* and *vc0620* target mRNAs likely resulting from direct interactions between MicX and sequences near the translational start site. In addition, Hfq decreases the amounts and the stabilities of the *mutS*, *miaA*, and *hfq E. coli* transcripts (*269*) and of T4 mRNAs (*277*) and has different effects on the expression of mRNA targets not known to be regulated by ncRNAs (*275*). These observations indicate that Hfq regulates the expression of many genes, either directly or indirectly, by mechanisms that modify RNA stability.

There are several examples where the synergistic action of a small ncRNA and Hfq simultaneously represses translation and destabilizes both RNAs. It is not always clear, however, whether mRNA destabilization and the arrest of translation are coupled. Messenger RNA destabilization could be the consequence of the translation block mediated by the ncRNA and Hfq. Alternatively, it is also possible that Hfq-facilitated mRNA–ncRNA annealing leads to the appearance of ribonuclease-sensitive conformations or activates the degradation machinery (see later text). Several examples illustrating such destabilizing mechanisms are described later.

The RyhB ncRNA modulates mRNAs involved in iron metabolism. Its expression is repressed by the Fur protein and it negatively regulates at least six mRNAs encoding Fe-binding or Fe-storage proteins in *E. coli*. RyhB acts on the *sodB* mRNA, encoding superoxide dismutase, and on the *sdhCDAB* mRNA as an antisense RNA (*38, 95, 239, 278*). When iron is limiting, RyhB levels rise. Hfq protein is required for the annealing of RyhB with its short discontinuous complementary target sequence, located in the vicinity of the ribosome binding site on the *sodB* mRNA. (*268, 271*). Hfq binds to the RyhB ncRNA efficiently, protecting it from degradation by RNase E and presumably presenting it to its mRNA targets. As a consequence, RyhB blocks the translation initiation site of *sodB* and triggers the rapid and concomitant degradation of RyhB and *sodB* mRNA *via* RNase E (*268*). The two annealed RNA are degraded simultaneously suggesting that the ternary Hfq–ncRNA–mRNA complex undergoes conformational modifications that facilitate degradation of the small ncRNA (*279*). Although turnover of *sodB* mRNA is RNase E-dependent, RyhB decay *in vivo* is mainly dependent on RNase III, and its cleavage by RNase III *in vitro* is facilitated upon base-pairing with the *sodB* 5' UTR (*280*).

RyhB also activates several genes by increasing their mRNA levels. Among these activated genes is *shiA*, which encodes a permease of shikimate, an aromatic compound participating in the biosynthesis of siderophores. RyhB directly pairs with the 5'-UTR of the *shiA* mRNA to disrupt an intrinsically inhibitory structure that sequesters the ribosome-binding site and the translation initiation codon. The formation of this inhibitory structure in the *shiA* mRNA correlates with instability of the mRNA. In this case, activation of mRNA expression by RyhB is correlated with the stabilization of mRNA (*281*). RyhB-dependent *shiA* activation takes place in the absence of Hfq. In contrast, Hfq was found to exert a strong antagonistic posttranscriptional inhibitory effect on *shiA* expression.

The SgrS ncRNA was identified by its ability to bind Hfq (257). SgrS is synthesized when glucose-phosphate accumulates in the cell, or upon addition of the nonmetabolizable sugar α-methyl-glucoside (260). At the same time, the *ptsG* mRNA, encoding the major glucose transporter IICB(Glc), is rapidly degraded in an RNase E/degradosome-dependent process (282). Partial base-pairing between SgrS and the translation initiation region is responsible for silencing of the target *ptsG* mRNA (264) and Hfq is required for ncRNA–mRNA annealing. Experiments performed to better understand the coupling between translation inhibition and mRNA destabilization demonstrated that physical interactions between the RNAs, Hfq and the degradation machinery specifically accelerate the degradation of translationally silenced mRNAs. Data showing that translation repression occurs in the absence of ncRNA-mediated mRNA destabilization, imply that a translational block is sufficient for gene silencing by SgrS, suggesting that the role of RNA decay is to render the translational inactivation irreversible and to ensure an efficient turnover of both mRNA and ncRNA (283). Similarly, the simultaneous degradation of the RyhB ncRNA annealed to its *sodB* mRNA target led to the conclusion that ncRNA acts stoichiometrically instead of catalytically (268). Importantly, elements involved in degradosome assembly, such as the C-terminal scaffold region of RNase E and enolase, and the RNA binding protein Hfq are required for the rapid RNase E-dependent degradation of *ptsG* mRNA in response to phosphosugar stress. In addition, Hfq was demonstrated to associate with RNase E through the C-terminal domain of the latter (27, 284). A model derived from these data postulates that Hfq bridges the SgrS ncRNA with the C-terminal scaffold region of RNase E, thus forming a nucleoprotein complex able to bring RNase E into the vicinity of the *ptsG* mRNA destined for decay (210). Finally, it was also demonstrated that SgrS-mediated destabilization only takes place if the *ptsG* mRNA is localized to the inner membrane and if the nascent peptide is inserted in the membrane (27). This observation is reminiscent of previous data showing that RNase E is predominantly localized at the membrane, presumably due to its association with the bacterial cytoskeleton (212, 213).

The GadY ncRNA was first identified as IS183 by bioinformatics-based search for small ncRNAs in *E. coli* (241). GadY coimmunoprecipitates with Hfq (257) and is stabilized upon interaction with Hfq. Overproduction of GadY results in an increase in the mRNA encoding the GadX transcriptional activator, which in turn causes increased levels of the GadA and GadB glutamate decarboxylases involved in glutamic acid resistance. The *gadY* gene is *cis*-encoded on the opposite strand to the *gadX* gene. It overlaps the 3′ end of the *gadX* gene, which does not have a Rho-independent terminator (285). Storz and collaborators have shown that the positive regulatory control of GadY exerted on *gadX* expression requires base-pairing of GadY to the 3′ UTR of the *gadX* mRNA to increase its stability. This makes GadY unique, in that it can hybridize to the 3' end of its target, presumably protecting this transcript from degradation by 3'–5' exoribonucleases. Whether Hfq is required to promote the interaction between *gadX* mRNA and GadY is not known. However, both Hfq and RNase E are required for the expression of GadA and GadB and survival under acidic conditions (286).

It is also worth mentioning other *cis*-acting ncRNAs that, in contrast to GadY, destabilize their targets. Many of them have been found in plasmids, phages, and transposons (reviewed in (265)). They mostly affect translation or mRNA stability. The Oop mRNA of bacteriophage λ forms a perfect duplex with the 3' part of the *CII* mRNA, similar to the GadY–*gadX* hybrid. However, in this case, the double-stranded structure is recognized by RNase III, which initiates its degradation (287, 288). Similarly, RNAα of the pJM1 plasmid of *Vibrio anguillarum* and the IsrR ncRNA of *Synecchocystis* were proposed to reduce the life-times of the complementary *fatA* and *fatB* transcripts, and the *isiA* mRNA, respectively (289, 290). Inhibition of translation, mediated by antisense ncRNA, often causes degradation of both mRNA and regulatory RNA by RNase III. This was shown for the CopA/CopT (291) and Sok/hok (292) ncRNA–mRNA pairs of plasmid R1 and RNA-OUT/RNA-IN of the IS10 transposon, respectively (293). In contrast to the Oop-mediated regulation mentioned earlier, the RNase III processing step is not required for control, presumably because its role is only to make an early inhibitory step irreversible. This is reminiscent of the "secondary role" attributed to RNA degradation in the case of SgrS-mediated regulation of the *ptsG* mRNA (see earlier text).

The CsrA protein (carbon storage regulator) was first characterized by Romeo and coworkers as a negative regulator of the expression of two glycogen biosynthetic operons, by controlling the level of expression of the *glgC* (ADP-glucose pyrophosphorylase) and *glgB* (glycogen branching enzyme) genes (294). A more extensive role for CsrA in directing intracellular carbon flux was then established through an examination of the effects of CsrA on several enzymes, genes, and metabolites of central carbohydrate metabolism. While the levels of these enzymes do not change dramatically in response to various physiological requirements (partial amino acid starvation, growth at temperatures between 25 and 37 °C), the levels of many, if not all, of these enzymes responds to conditions such as oxygen availability and growth rate (295).

A deletion of the *csrA* gene was used to identify a posttranscriptional regulation of the two structural genes of the glycogen biosynthetic pathway by accelerating the decay of the *glg* transcripts (296). Additional studies established that this 61 amino acid protein is an RNA binding protein (297), capable of recognizing a degenerate sequence element of the CsrB ncRNA, CAGGA (U/A/C)G, which is similar to the Shine–Dalgarno sequence (298). Romeo proposed that CsrA-mediated repression in *E. coli* involves the binding of

CsrA to the ribosome binding site of target transcripts, thereby blocking ribosome access to the mRNA. Reduced translation probably contributes to the rapid degradation of target mRNAs observed. CsrB acts as an antagonist of CsrA by sequestering this protein. A second redundant ncRNA, CsrC, functions in an analogous manner to CsrB. Both ncRNAs possess similar imperfect repeat sequences (18 in CsrB, 9 in CsrC), primarily localized in the loops of predicted hairpins, which may serve as CsrA binding elements (299). The current understanding of the complex Csr regulatory circuitry is presented in more detail elsewhere (300). CsrA can also activate gene expression by stabilizing target mRNAs. Romeo and coworkers showed that the *csrA* gene is required for motility and flagellum biosynthesis through the posttranscriptional activation of *flhDC* expression. The precise mechanism of the CsrA-mediated transcript stabilization remains to be determined (301).

The modulation of RNA decay by *csr* genes does not stop here. Inactivation of the *csrD* gene was recently shown to dramatically stabilize the CsrB/C ncRNAs, and as a consequence alter the expression of CsrA-controlled genes (302). CsrD is not a nuclease, but acts selectively on CsrB and CsrC ncRNAs together with RNase E. Strong interactions between CsrD and these ncRNAs is proposed, in both cases, to convert them to RNase E substrates. However, the sensitivity of CsrB and CsrC to PNPase, degradosome assembly, and CsrA is different.

The CsrA protein has homologues in other bacteria, generally named RsmA (repressor of secondary metabolites), and the members of this family have been shown to play important roles in global posttranscriptional regulation in *Pseudomonas aeruginosa*. In this organism, and in *P. fluorescens*, *Salmonella typhimurium*, *Legionella pneumophila*, *Serratia marcescens*, *Proteus mirabilis*, *Haemophilus influenzae*, *Helicobacter pylori* (reviewed in (303), RsmA exerts both positive and negative regulation on different genes and is antagonized by the RsmZ ncRNA (304).

D. tmRNA and Quality Control of RNA

Ribosome stalling at inefficient termination codons, rare codons recognized by nonabundant tRNAs or at the 3′ end of mRNAs lacking termination codons leads to loss of translation potential that is detrimental to cell growth. Moreover, the production of truncated polypeptides from incomplete mRNAs diverts the translation machinery away from the synthesis of functional proteins. An elegant *trans*-translation mechanism has been identified that alleviates the deleterious consequences of ribosome stalling (for a review (305)). A bifunctional RNA molecule, the tmRNA, associated with a small protein partner SmpB, first exerts a tRNA function, allowing dissociation of the mRNA from the ribosome–polypeptidyl–tmRNA complex. The latter then uses the tmRNA as a messenger to continue translating up to the translation termination codon of the small tmRNA-encoded ORF. The 11 amino acid tmRNA-encoded C-terminal tag of the resulting chimeric polypeptide is then recognized as a selective destabilizing signal by the cellular proteolytic machinery. Interestingly, in addition to ribosome rescue and degradation of nonfunctional proteins, the tmRNA also induces degradation of transcripts. Defective transcripts that promote ribosome stalling are degraded preferentially in a tmRNA-dependent manner (*306–309*). This was first demonstrated for truncated transcripts lacking a stop codon, before the analysis was extended to full-length transcripts harboring ribosome stalling sequences. Ribosome stalling at internal sites provokes transcript cleavage and the cleaved mRNA becomes a substrate for the SmpB–tmRNA complex (*310–314*). Cleavage occurs in the A-site of stalled ribosomes. However, the ribonuclease responsible for this cleavage has not been identified. RNase R is the most likely candidate for decay of the truncated mRNA, as this enzyme was previously shown to be associated with a multicomponent RNA–protein complex that includes SmpB and the tmRNA (*315*). A similar functional association between RNase R and tmRNA was also observed in *Caulobacter crescentus* (*316*). The implication of RNase R in the degradation of the *trans*-translated mRNA was recently demonstrated in *E. coli* (*317*). This RNase R activity requires both SmpB and the tmRNA. RNase R therefore appears to be an important component of the tmRNA-mediated mRNA surveillance system.

E. Destabilizing Riboswitches

Riboswitches are RNA sensors that specifically recognize metabolites and modulate gene expression in ways previously thought to be confined to proteins in both Gram-positive and Gram-negative bacteria. Riboswitches commonly occur in the 5′ UTRs of bacterial mRNAs, where they regulate the expression of the downstream coding regions by directly binding small compounds. These genes are usually responsible for the biosynthesis or transport of the metabolite being sensed. The ligands that have been found to bind specific riboswitches thus far are lysine, glycine, guanine, adenine, flavin mononucleotide (FMN), thiamine pyrophosphate (TPP), adenosylcobalamin (AdoCbl), 7-aminoethyl 7-deazaguanine (preQ1), S-adenosylmethionine (SAM), and glucosamine-6-phosphate (GlcN6P) (*318–320*). A typical riboswitch can be divided into two functional domains, an aptamer which binds specific metabolites and an expression platform that converts the metabolite binding event into changes in gene expression. The metabolite binding domain changes structure upon metabolite binding and triggers structural rearrangements in the expression platform. The aptamer exhibits extensive sequence and secondary structure conservation among many bacterial species (*321*). Most bacterial riboswitches exert their regulatory effects on transcription, by forming mutually exclusive terminator or antiterminator hairpins, or on translation, by occluding or exposing the ribosome binding site in response to metabolite binding (319). Here, we will focus on riboswitches that induce cleavage of the downstream mRNA.

The best characterized riboswitch acting on RNA stability was identified upstream of the *glmS* gene, encoding GlcN6P aminotransferase, the enzyme that produces GlcN6P, in *B. subtilis*. It is the only riboswitch described to date in which binding of the effector, GlcN6P in this case, stimulates a self-cleaving ribozyme activity targeting its own mRNA (318, 322). As predicted, the *glmS* ribozyme responds to fluctuations in the intracellular GlcN6P pool to control *glmS* mRNA stability: *glmS* mRNA levels decrease when GlcN6P concentration increases. GlcN6P-stimulated self-cleavage of the *glmS* 5′ UTR has been shown to produce RNA products with a 2′,3′ cyclic phosphate and a 5′ hydroxyl (318). RNase J1, an enzyme that exhibits both endoribonuclease (34) and 5′–3′ exoribonuclease activities (2), then destabilizes the 5′-hydroxylated 3′-cleavage product containing the *glmS* ORF (323). This makes the *glmS* riboswitch a unique case that wields its regulatory influence directly over mRNA stability, rather than affecting transcriptional elongation or translation initiation, and raises the interesting possibility that differential RNA processing or degradation caused by ligand-induced conformational changes might be the primary mechanism by which other riboswitches regulate gene expression. The demonstration that the coenzyme B12 riboswitch (AdoCbl) of *E. coli* and a similar riboswitch from *B. subtilis* are cleaved by RNase P *in vitro* suggests that these may also be examples of RNA destabilizing riboswitches. The small inhibitory effect of coenzyme B12 on RNase P cleavage of the *E. coli* riboswitch *in vitro* supports this view (324).

V. Conclusions and Open Questions

Both the dominant RNase E-dependent endonucleolytic pathway characterized in *E. coli* and the exonucleolytic poly(A)-assisted ancestral mechanism of decay, which has been conserved in most branches of the phylogenic tree, are general decay pathways involved in the degradation of the vast majority of labile mRNAs, stable rRNAs, and tRNAs under a variety of physiological conditions (155, 325). Interestingly, many of the actors of these pathways, namely PNPase, RNase II and RNase R, PAP I, protein Hfq, and RNA helicases are close relatives of components of the exosome and its cofactors, involved in the exonucleolytic degradation of eukaryotic RNAs. On the other hand, the endonucleolytic enzyme, RNase E, is specific to bacteria and organelles (135, 207, 326). Bacterial cells facing environmental changes and stress conditions activate mechanisms of RNA degradation which can either target specific mRNAs or nonspecifically destroy all kinds of RNA. Small ncRNAs often orchestrate the extinction and degradation of specific transcripts by the degradosome, while toxin–antitoxin mediated shut-down decay is believed to allow a complete reprogramming of gene expression in response to severe stress conditions (327). Although some important features of the degradation machinery undoubtedly still remain to be discovered, a major remaining question is how the cell controls the rate of degradation of each individual mRNA species by the degradosome. There are indications that the activity of the degradosome-mediated pathway of decay remains approximately constant in the cell during balanced growth conditions. Indeed, the fact that RNase E, PNPase, and other enzymes such as RNase III, autoregulate and/or cross-regulate their synthesis suggests that the intracellular levels of house-keeping ribonucleases are adjusted to the amount of substrate available in the cell (194, 195, 328, 329). However, the induction of PNPase synthesis at low temperature and the regulation of RNase E by RraA and RraB are ways in which the cells degradation capacity can be modified. Interestingly, both RraA and RraB affect degradosome composition differently and stabilize different sets of transcripts. In addition, the natively unstructured microdomains of the RNase E scaffolding C-terminal region have been proposed to allow the transient assembly of many different cofactors able to regulate degradosome activity (326). Another way to channel RNase E-mediated degradation is exemplified by the protein Hfq which brings the degradosome to mRNAs specifically inactivated by a ncRNA, effectively increasing local concentration. Such a mechanism of destabilization could be extended to the many other mRNAs silenced by ncRNAs. Moreover, further studies of the many small *E. coli* ncRNAs whose action is not yet known may identify other ncRNAs that act primarily on RNA stability, as reported for GadY. Finally, other mRNAs may prove to be regulated by riboswitches affecting mRNA stability, as described for *GlmS* transcript of *B. subtilis*, rather than at the level of transcription or translation. Interestingly enough, the *glmS* mRNA of *E. coli* is also controlled by riboregulation at the level of mRNA stability. However, in this case, the decay-rate of the transcript depends on a ncRNA whose internal concentration is controlled by PAP I. Riboregulation permits physiological responses whose rapidity, specificity, low energetic need, and/or reversibility often cannot be matched by protein effectors.

Poly(A)-dependent decay is now considered a ubiquitous mechanism responsible for the degradation of small RNAs or RNA fragments resulting from transcription of non coding regions and processing of mRNA (75). It appears, however, that some RNA species are preferentially targeted by the poly(A)-dependent degradation machinery. The structure of RNA fragments, the accessibility of their 3′ extremities, the rate of poly(A) synthesis, and the shortening of tails by RNase II, all modulate the efficiency of poly(A)-dependent degradation. Moreover, the coordination of the endonucleolytic and exonucleolytic degradation machineries may explain why RNA fragments generated by RNase E are very efficiently degraded and also why full-length primary transcripts are primarily sensitive to the RNase E endonucleolytic pathway. The many physical interactions between PAP I, Hfq, ncRNAs, and partners of the degradosome, in addition to their similar localization close to the membrane may reflect such functional relationships. The quality control of RNA is also an essential aspect of RNA metabolism, carried out, at least in part, by poly(A)-dependent enzymes and which remains to be further investigated in bacteria.

References

1. Dreyfus M, Régnier P. The poly(A) tail of mRNAs: Bodyguard in eukaryotes, scavenger in bacteria. *Cell* 2002;**111**:611–3.
2. Mathy N, Benard L, Pellegrini O, Daou R, Wen T, Condon C. 5′–3′ exoribonuclease activity in bacteria: Role of RNase J1 in rRNA maturation and 5′ stability of mRNA. *Cell* 2007;**129**:681–92.
3. Deana A, Celesnik H, Belasco JG. The bacterial enzyme RppH triggers messenger RNA degradation by 5′ pyrophosphate removal. *Nature* 2008;**451**:355–8.
4. Celesnik H, Deana A, Belasco JG. Initiation of RNA decay in *Escherichia coli* by 5′ pyrophosphate removal. *Mol Cell* 2007;**27**:79–90.
5. Takayama K, Kjelleberg S. The role of RNA stability during bacterial stress responses and starvation. *Environ Microbiol* 2000;**2**:355–65.
6. Bernstein JA, Khodursky AB, Lin P-H, Lin-Chao S, Cohen SN. Global analysis of mRNA decay and abundance in *Escherichia coli* at single-gene resolution using two-color fluorescent DNA microarrays. *Proc Natl Acad Sci USA* 2002;**99**:9697–702.
7. Nilsson G, Belasco JG, Cohen SN, von Gabain A. Growth-rate dependent regulation of mRNA stability in *Escherichia coli*. *Nature* 1984;**312**:75–7.
8. Melefors O, von Gabain A. Site specific endonucleolytic cleavages and the regulation of stability of *E. coli* ompA mRNA. *Cell* 1988;**52**:893–901.
9. Udekwu KI, Darfeuille F, Vogel J, Reimegard J, Holmqvist E, Wagner EGH. Hfq-dependent regulation of OmpA synthesis is mediated by an antisense RNA. *Genes Dev* 2005;**19**:2355–66.
10. Vytvytska O, Moll I, Kaberdin VR, von Gabain A, Bläsi U. Hfq (HFI) stimulates *ompA* mRNA decay by interfering with ribosomes binding. *Genes Dev* 2000;**14**:1109–18.
11. Papenfort K, Pfeiffer V, Mika F, Lucchini S, Hinton JC, Vogel J. SigmaE-dependent small RNAs of *Salmonella* respond to membrane stress by accelerating global omp mRNA decay. *Mol Microbiol* 2006;**62**:1674–88.
12. Johansen J, Rasmussen AA, Overgaard M, Valentin-Hansen P. Conserved small noncoding RNAs that belong to the sigmaE regulon: Role in down-regulation of outer membrane proteins. *J Mol Biol* 2006;**364**:1–8.
13. Figueroa-Bossi N, Lemire S, Maloriol D, Balbontin R, Casadesus J, Bossi L. Loss of Hfq activates the sigmaE-dependent envelope stress response in *Salmonella enterica*. *Mol Microbiol* 2006;**62**:838–52.
14. Guisbert E, Rhodius VA, Ahuja N, Witkin E, Gross CA. Hfq modulates the sigmaE-mediated envelope stress response and the sigma32-mediated cytoplasmic stress response in *Escherichia coli*. *J Bacteriol* 2007;**189**:1963–73.
15. Melin L, Rutberg L, von Gabain A. Transcriptional and posttranscriptional control of the *Bacillus subtilis* succinate dehydrogenase operon. *J Bacteriol* 1989;**171**:2110–5.

16. Resnekov O, Rutberg L, von Gabain A. Changes in the stability of specific mRNA species in response to growth stage in *Bacillus subtilis*. *Proc Natl Acad Sci USA* 1990;**87**:8355–9.
17. Paesold G, Krause M. Analysis of rpoS mRNA in *Salmonella dublin*: Identification of multiple transcripts with growth-phase-dependent variation in transcript stability. *J Bacteriol* 1999;**181**:1264–8.
18. Aiso T, Yoshida H, Wada A, Ohki R. Modulation of mRNA stability participates in stationary-phase-specific expression of ribosome modulation factor. *J Bacteriol* 2005;**187**:1951–8.
19. Kusz AE, Medberry PS, Schottel JL. Stationary phase, amino acid limitation and recovery from stationary phase modulate the stability and translation of chloramphenicol acetyltransferase mRNA and total mRNA in *Escherichia coli*. *Microbiology* 1998;**144**:739–50.
20. Soltes-Rak E, Kushner DJ, Williams DD, Coleman JR. Factors regulating cryIVB expression in the cyanobacterium—*Synechococcus* PCC 7942. *Mol Gen Genet* 1995;**246**:301–8.
21. Barnett TC, Bugryesheva JV, Scott JR. Role of mRNA stability in growth phase regulation of gene expression in the group A streptococcus. *J Bacteriol* 2007;**189**:1866–73.
22. Ueno H, Yonesaki T. Phage-induced change in the stability of mRNAs. *Virology* 2004;**329**:134–41.
23. Marchand I, Nicholson AW, Dreyfus M. Bacteriophage T7 protein kinase phosphorylates RNase E and stabilizes mRNAs synthesized by T7 RNA polymerase. *Mol Microbiol* 2001;**42**:767–76.
24. Sanson B, Hu RM, Troitskayadagger E, Mathy N, Uzan M. Endoribonuclease RegB from bacteriophage T4 is necessary for the degradation of early but not middle or late mRNAs. *J Mol Biol* 2000;**297**:1063–74.
25. Romeo JM, Zusman DR. Determinants of an unusually stable mRNA in the bacterium *Myxococcus xanthus*. *Mol Microbiol* 1992;**6**:2975–88.
26. Morita T, El-Kazzaz W, Tanaka Y, Inada T, Aiba H. Accumulation of glucose 6-phosphate or fructose 6-phosphate is responsible for destabilization of glucose transporter mRNA in *Escherichia coli*. *J Biol Chem* 2003;**278**:15608–14.
27. Kawamoto H, Morita T, Shimizu A, Inada T, Aiba H. Implication of membrane localization of target mRNA in the action of a small RNA: Mechanism of posttranscriptional regulation of glucose transporter in *Escherichia coli*. *Genes Dev* 2005;**19**:328–38.
28. Albertson NH, Nystrom T. Effects of starvation for exogenous carbon on functional mRNA stability and rate of peptide chain elongation in *Escherichia coli*. *FEMS Microbiol Lett* 1994;**117**:181–7.
29. Redon E, Loubiere P, Cocaign-Bousquet M. Role of mRNA stability during genome-wide adaptation of *Lactococcus lactis* to carbon starvation. *J Biol Chem* 2005;**280**:36380–5.
30. Jurgen B, Schweder T, Hecker M. The stability of mRNA *gsiB* of *Bacillus subtilis* is dependent on the presence of a strong ribosome binding site. *Mol Gen Genet* 1998;**258**:538–45.
31. Jager A, Samorski R, Pfeifer F, Klug G. Individual gvp transcript segments in *Haloferax mediterranei* exhibit varying half-lives, which are differentially affected by salt concentration and growth phase. *Nucleic Acids Res* 2002;**30**:5436–43.
32. Collins JJ, Roberts GP, Brill WJ. Posttranscriptional control of *Klebsiella pneumoniae* nif mRNA stability by the nifL product. *J Bacteriol* 1986;**168**:173–8.
33. Condon C, Putzer H, Grunberg-Manago M. Processing of the leader mRNA plays a major role in the induction of the *thrS* espression following threonine starvation in *Bacillus subtilis*. *Proc Natl Acad Sci USA* 1996;**93**:6992–7.
34. Even S, Pellegrini O, Zig L, Labas V, Vinh J, Brechemmier-Baey D, et al. Ribonucleases J1 and J2: two novel endoribonucleases in *B. subtilis* with functional homology to *E. coli* RNase E. *Nucleic Acids Res* 2005;**33**:2141–52.
35. Dubrac S, Touati D. Fur positive regulation of iron superoxide dismutase in *Escherichia coli*: Functional analysis of the sodB promoter. *J Bacteriol* 2000;**182**:3802–8.

36. Masse E, Salvail H, Desnoyers G, Arguin M. Small RNAs controlling iron metabolism. *Curr Opin Microbiol* 2007;**10**:140–5.
37. Masse E, Vanderpool CK, Gottesman S. Effect of RyhB small RNA on global iron use in *Escherichia coli*. *J Bacteriol* 2005;**187**:6962–71.
38. Massé E, Gottesman S. A small RNA regulates the expression of genes involved in iron metabolism in *Escherichia coli*. *Proc Natl Acad Sci USA* 2002;**99**:4620–5.
39. Aiba H. Mechanism of RNA silencing by Hfq-binding small RNAs. *Curr Opin Microbiol* 2007;**10**:134–9.
40. Bovy A, de Vrieze G, Lugones L, van Horssen P, van den Berg C, Borrias M, et al. Iron-dependent stability of the ferredoxin I transcripts from the cyanobacterial strains *Synechococcus* species PCC 7942 and *Anabaena* species PCC 7937. *Mol Microbiol* 1993;**7**:429–39.
41. Bechhofer DH, Dubnau D. Induced mRNA stability in *Bacillus subtilis*. *Proc Natl Acad Sci USA* 1987;**84**:498–502.
42. Woo W-M, Lin-Chao S. Processing of the *rne* transcript by an RNase E-independent amino acid-dependent mechanism. *J Biol Chem* 1997;**272**:15516–20.
43. Garcia-Dominguez M, Muro-Pastor MI, Reyes JC, Florencio FJ. Light-dependent regulation of cyanobacterial phytochrome expression. *J Bacteriol* 2000;**182**:38–44.
44. Agrawal GK, Kato H, Asayama M, Shirai M. An AU-box motif upstream of the SD sequence of light-dependent psbA transcripts confers mRNA instability in darkness in cyanobacteria. *Nucleic Acids Res* 2001;**29**:1835–43.
45. Kulkarni RD, Golden SS. mRNA stability is regulated by a coding-region element and the unique 5′ untranslated leader sequences of the three *Synechococcus* psbA transcripts. *Mol Microbiol* 1997;**24**:1131–42.
46. Georgellis D, Barlow T, Arvidson S, von Gabain A. Retarded RNA turnover in *Escherichia coli*: A means of maintaining gene expression during anaerobiosis. *Mol Microbiol* 1993;**9**:375–81.
47. Klug G. Endonucleolytic degradation of puf mRNA in *Rhodobacter capsulatus* is influenced by oxygen. *Proc Natl Acad Sci USA* 1991;**88**:1765–9.
48. Jager S, Hebermehl M, Schiltz E, Klug G. Composition and activity of the *Rhodobacter capsulatus* degradosome vary under different oxygen concentrations. *J Mol Microbiol Biotechnol* 2004;**7**:148–54.
49. Mathy N, Jarrige A-C, Robert-Le Meur M, Portier C. Increased expression of *Escherichia coli* polynucleotide phosphorylase at low temperatures is linked to a decrease in the efficiency of autocontrol. *J Bacteriol* 2001;**183**:3848–54.
50. Zangrossi S, Briani F, Ghisotti D, Regonesi ME, Tortora P, Dehò G. Transcriptional and posttranscriptional control of polynucleotide phosphorylase during cold acclimation in *Escherichia coli*. *Mol Microbiol* 2000;**36**:1470–80.
51. Goldenberg D, Azar I, Oppenheim AB. Differential mRNA stability of the *cspA* gene in the cold-shock response of *Escherichia coli*. *Mol Microbiol* 1996;**19**:241–8.
52. Brandi A, Pietroni P, Gualerzi CO, Pon CL. Posttranscriptional regulation of CspA expression in *Escherichia coli*. *Mol Microbiol* 1996;**19**:231–40.
53. Cairrao F, Cruz A, Mori H, Arraiano CM. Cold shock induction of RNase R and its role in the maturation of the quality control mediator SsrA/tmRNA. *Mol Microbiol* 2003;**50**:1349–60.
54. Sakamoto T, Bryant DA. Temperature-regulated mRNA accumulation and stabilization for fatty acid desaturase genes in the cyanobacterium *Synechococcus* sp. strain PCC 7002. *Mol Microbiol* 1997;**23**:1281–92.
55. Chamot D, Owttrim GW. Regulation of cold shock-induced RNA helicase gene expression in the Cyanobacterium *Anabaena* sp. strain PCC 7120. *J Bacteriol* 2000;**182**:1251–6.

56. Sato N, Nakamura A. Involvement of the 5'-untranslated region in cold-regulated expression of the rbpA1 gene in the cyanobacterium *Anabaena variabilis* M3. *Nucleic Acids Res* 1998;**26**:2192–9.
57. Nickel M, Homuth G, Bohnisch C, Mader U, Schweder T. Cold induction of the *Bacillus subtilis* bkd operon is mediated by increased mRNA stability. *Mol Genet Genomics* 2004;**272**:98–107.
58. Repoila F, Gottesman S. Signal transduction cascade for regulation of RpoS: Temperature regulation of DsrA. *J Bacteriol* 2001;**183**:4012–23.
59. Prud'homme-Genereux A, Beran RK, Iost I, Ramey CS, Mackie GA, Simons RW. Physical and functional interactions among RNase E, polynucleotide phosphorylase and the cold-shock protein, CsdA: Evidence for a "cold shock degradosome." *Mol Microbiol* 2004;**54**:1409–21.
60. Khemici V, Toesca I, Poljak L, Vanzo NF, Carpousis AJ. The RNase E of *Escherichia coli* has at least two binding sites for DEAD-box RNA helicases: Functional replacement of RhlB by RhlE. *Mol Microbiol* 2004;**54**:1422–30.
61. Le Derout J, Regnier P, Hajnsdorf E. Both temperature and medium composition regulate RNase E processing efficiency of the rpsO mRNA coding for ribosomal protein S15 of *Escherichia coli*. *J Mol Biol* 2002;**319**:341–9.
62. Rasouly A, Shenhar Y, Ron EZ. Thermoregulation of *Escherichia coli* hchA transcript stability. *J Bacteriol* 2007;**189**:5779–81.
63. Afonyushkin T, Moll I, Blasi U, Kaberdin VR. Temperature-dependent stability and translation of *Escherichia coli* ompA mRNA. *Biochem Biophys Res Commun* 2003;**311**:604–9.
64. Hasegawa H, Chatterjee A, Cui Y, Chatterjee AK. Elevated temperature enhances virulence of *Erwinia carotovora* subsp. *carotovora* strain EC153 to plants and stimulates production of the quorum sensing signal, N-acyl homoserine lactone, and extracellular proteins. *Appl Environ Microbiol* 2005;**71**:4655–63.
65. Yamauchi S, Okuyama H, Nishiyama Y, Hayashi H. The rpoH gene encoding heat shock sigma factor sigma32 of psychrophilic bacterium *Colwellia maris*. *Extremophiles* 2006;**10**:149–58.
66. Beltramo C, Grandvalet C, Pierre F, Guzzo J. Evidence for multiple levels of regulation of *Oenococcus oeni* clpP-clpL locus expression in response to stress. *J Bacteriol* 2004;**186**:2200–5.
67. Anderson KL, Roberts C, Disz T, Vonstein V, Hwang K, Overbeek R, et al. Characterization of the *Staphylococcus aureus* heat shock, cold shock, stringent, and SOS responses and their effects on log-phase mRNA turnover. *J Bacteriol* 2006;**188**:6739–56.
68. Edmonds M. A history of polyA sequences: From formation to factors to function. *Prog Nucleic Acid Res Mol Biol* 2002;**71**:285–389.
69. Briani F, Del Vecchio E, Migliorini D, Hajnsdorf E, Regnier P, Ghisotti D, et al. RNase E and polyadenyl polymerase I are involved in maturation of CI RNA, the P4 phage immunity factor. *J Mol Biol* 2002;**318**:321–31.
70. Hajnsdorf E, Braun F, Haugel-Nielsen J, Regnier P. Polyadenylylation destabilizes the *rpsO* mRNA of *Escherichia coli*. *Proc Natl Acad Sci USA* 1995;**92**:3973–7.
71. Le Derout J, Folichon M, Briani F, Dehò G, Régnier P, Hajnsdorf E. Hfq affects the length and the frequency of short oligo(A) tails at the 3' end of *Escherichia coli* rpsO mRNAs. *Nucleic Acids Res* 2003;**31**:4017–23.
72. Li Z, Pandit S, Deutscher MP. Polyadenylation of stable RNA precursors *in vivo*. *Proc Natl Acad Sci USA* 1998;**95**:12158–62.
73. van Meerten D, Zelwer M, Regnier P, Duin J. *In vivo* oligo(A) insertions in phage MS2: Role of *Escherichia coli* poly(A) polymerase. *Nucleic Acids Res* 1999;**27**:3891–8.

74. Xu F, Lin-Chao S, Cohen SN. The *Escherichia coli pcnB* gene promotes adenylylation of antisense RNAI of ColE1-type plasmids *in vivo* and degradation of RNAI decay intermediates. *Proc Natl Acad Sci USA* 1993;**90**:6756–60.
75. Anderson JT. RNA turnover: Unexpected consequences of being tailed. *Curr Biol* 2005;**15**: R635–8.
76. Sarkar N. Polyadenylation of mRNA in bacteria. *Microbiology* 1996;**142**:3125–33.
77. Sarkar N. Polyadenylation of mRNA in prokaryotes. *Annu Rev Biochem* 1997;**66**:173–97.
78. Nakazato H, Venkatesan S, Edmonds M. Polyadenylic acid sequences in *E. coli* messenger RNA. *Nature* 1975;**256**:144–6.
79. Srinivasan PR, Ramanarayanan M, Rabbani E. Presence of polyriboadenylate sequences in pulse-labeled RNA of *Escherichia coli*. *Proc Natl Acad Sci USA* 1975;**72**:2910–4.
80. Sarkar N, Langley D, Paulus H. Isolation and characterization of polyadenylate-containing RNA from *Bacillus brevis*. *Biochemistry* 1978;**17**:3468–74.
81. Kramer RA, Rosenberg M, Steitz JA. Nucleotide sequences of the 5′ and 3′ termini of bacteriophage T7 early messenger RNAs synthesized *in vivo*: Evidence for sequence specificity in RNA processing. *J Mol Biol* 1974;**89**:767–76.
82. Marujo PE, Hajnsdorf E, Le Derout J, Andrade R, Arraiano CM, Régnier P. RNase II removes the oligo(A) tails that destabilize the *rpsO* mRNA of *Escherichia coli*. *RNA* 2000;**6**:1185–93.
83. O'Hara EB, Chekanova JA, Ingle CA, Kushner ZR, Peters E, Kushner SR. Polyadenylylation helps regulate mRNA decay in *Escherichia coli*. *Proc Natl Acad Sci USA* 1995;**92**:1807–11.
84. Goodrich AF, Steege DA. Roles of polyadenylation and nucleolytic cleavage in the filamentous phage mRNA processing and decay pathways in *Escherichia coli*. *RNA* 1999;**5**:972–85.
85. Karnik P, Taljanidisz J, Sasvari-Szekely M, Sarkar N. 3′-terminal polyadenylate sequenses of *Escherichia coli* tryptophan synthetase α-subunit messenger RNA. *J Mol Biol* 1987;**196**:347–54.
86. Taljanidisz J, Karnik P, Sarkar N. Messenger ribonucleic acid for lipoprotein of the *Escherichia coli* outer membrane is polyadenylated. *J Mol Biol* 1987;**193**:507–15.
87. Mohanty BK, Kushner SR. Residual polyadenylation in poly(A) polymerase I (*pcnB*) mutants of *Escherichia coli* does not result from the activity encoded by the *f310* gene. *Mol Microbiol* 1999;**34**:1109–19.
88. Johnson MD, Popowski J, Cao G-J, Shen P, Sarkar N. Bacteriophage T7 mRNA is polyadenylated. *Mol Microbiol* 1998;**27**:23–30.
89. Campos-Guillén J, Bralley P, Jones GH, Bechhofer DH, Olmedo-Alvarez G. Addition of poly (A) and heteropoymeric 3′ ends in *Bacillus subtilis* wild-type and polynucleotide phosphorylase-deficient strains. *J Bacteriol* 2005;**187**:4698–706.
90. Cao G-J, Sarkar N. Stationary phase-specific mRNAs in *Escherichia coli* are polyadenylated. *Biochem Biophys Res Commun* 1997;**239**:46–50.
91. Mohanty BK, Kushner SR. The majority of *Escherichia coli* mRNAs undergo posttranscriptional modification in exponentially growing cells. *Nucleic Acids Res* 2006;**34**:5695–704.
92. Szalewska-Palasz A, Wrobel B, Wegrzyn G. Rapid degradation of polyadenylated oop RNA. *FEBS Lett* 1998;**432**:70–2.
93. Mikkelsen ND, Gerdes K. Sok antisense RNA from plasmid R1 is functionally inactivated by RNase E and polyadenylated by poly(A)polymerase I. *Mol Microbiol* 1997;**26**:311–20.
94. Reichenbach B, Maes A, Kalamorz F, Hajnsdorf E, Görke B. The small RNA GlmY acts upstream of the sRNA GlmZ in the activation of glmS expression and is subject to regulation by polyadenylation in *Escherichia coli*. *Nucleic Acids Res* 2008;**36**:2570–80.
95. Argaman L, Hershberg R, Vogel J, Bejerano G, Wagner EGH, Margalit H, et al. Novel small RNA-encoding genes in the intergenic regions of *Escherichia coli*. *Current Biol* 2001;**11**:941–50.

96. Bralley PJ, GH. Overexpresion of the polynucleotide phosphorylase gene (*pnp*) of *Streptomyces antibioticus* affects mRNA stability and poly(A) tail length but not ppGpp levels. *Microbiology* 2003;**149**:2173–82.
97. Bralley P, Gust B, Chang S, Chater KF, Jones GH. RNA 3'-tail synthesis in *Streptomyces*: In vitro and in vivo activities of RNase PH, the SCO3896 gene product and polynucleotide phosphorylase. *Microbiology* 2006;**152**:627–36.
98. Mohanty BK, Kushner SR. Analysis of the function of *Escherichia coli* poly(A) polymerase I in RNA metabolism. *Mol Microbiol* 1999;**34**:1094–108.
99. Haugel-Nielsen J, Hajnsdorf E, Regnier P. The *rpsO* mRNA of *Escherichia coli* is polyadenylated at multiple sites resulting from endonucleolytic processing and exonucleolytic degradation. *EMBO J* 1996;**15**:3144–52.
100. Söderbom F, Binnie U, Masters M, Wagner EGH. Regulation of plasmid R1 replication: PcnB and RNase E expedite the decay of the antisense RNA, copA. *Mol Microbiol* 1997;**26**:493–504.
101. Khemici V, Carpousis AJ. The RNA degradosome and poly(A) polymerase of *Escherichia coli* are required *in vivo* for the degradation of small mRNA decay intermediates containing REP-stabilizers. *Mol Microbiol* 2004;**51**:777–90.
102. Cheng Z-F, Deutscher MP. An important role for RNase R in mRNA decay. *Mol Cell* 2005;**17**:313–8.
103. Coburn GA, Mackie GA. Reconstitution of the degradation of the mRNA for ribosomal protein S20 with purified enzymes. *J Mol Biol* 1998;**279**:1061–74.
104. Lopilato J, Bortner S, Beckwith J. Mutations in a new chromosomal gene of *Escherichia coli* K-12, *pcnB*, reduce plasmid copy number of pBR322 and its derivatives. *Mol Gen Genet* 1986;**205**:285–90.
105. Liu J, Parkinson JS. Genetics and sequence analysis of the *pcnB* locus, an *Escherichia coli* gene involved in plasmid copy number control. *J Bacteriol* 1989;**171**:1254–61.
106. Cao G-J, Sarkar N. Identification of the gene for an *Escherichia coli* poly(A) polymerase. *Proc Natl Acad Sci USA* 1992;**89**:10380–4.
107. Cohen SN. Surprises at the 3' end of prokaryotic RNA. *Cell* 1995;**80**:829–32.
108. Andrade JM, Arraiano CM. PNPase is a key player in the regulation of small RNAs that control the expression of outer membrane proteins. *RNA* 2008;**14**:543–551.
109. Viegas SC, Pfeiffer V, Sittka A, Silva IJ, Vogel J, Arraiano CM. Characterization of the role of ribonucleases in *Salmonella* small RNA decay. *Nucleic Acids Res* 2007;**15**:7651–64.
110. Ghisotti D, Chiaramonte R, Forti F, Zangrossi S, Sironi G, Deho G. Genetic analysis of the immunity region of phage-plasmid P4. *Mol Microbiol* 1992;**6**:3405–13.
111. Hajnsdorf E, Régnier P. E. coli *rpsO* mRNA decay: RNase E processing at the beginning of the coding sequence stimulates poly(A)-dependent degradation of the mRNA. *J Mol Biol* 1999;**286**:1033–43.
112. Mackie GA. Stabilization of the 3' one third of *Escherichia coli* ribosomal protein S20 mRNA in mutants lacking polynucleotide phosphorylase. *J Bacteriol* 1989;**171**:4112–20.
113. Coburn GA, Mackie GA. Differential sensitivities of portions of the mRNA for ribosomal protein S20 to 3' exonucleases dependent on oligoadenylation and RNA secondary structure. *J Biol Chem* 1996;**271**:15776–81.
114. Hajnsdorf E, Braun F, Haugel-Nielsen J, Le Derout J, Regnier P. Multiple degradation pathways of the *rpsO* mRNA of *E. coli*. RNase E interacts with the 5' and 3' extremities of the primary transcript. *Biochimie* 1996;**78**:416–24.
115. Blum E, Carpousis AJ, Higgins CF. Polyadenylation promotes degradation of 3'-structured RNA by the *Escherichia coli* mRNA degradosome *in vitro*. *J Biol Chem* 1999;**274**:4009–16.
116. Xu F, Cohen SN. RNA degradation in *Escherichia coli* regulated by 3' adenylation and 5' phosphorylation. *Nature* 1995;**374**:180–3.

117. Coburn GA, Miao X, Briant DJ, Mackie GA. Reconstitution of a minimal RNA degradosome demonstrates functional coordination between a 3′ exonuclease and a DEAD-box RNA helicase. *Genes Dev* 1999;**13**:2594–603.
118. Mackie GA, Genereaux JL. The role of RNA structure in determining RNase E dependent cleavage sites in the mRNA for ribosomal protein S20 *in vitro*. *J Mol Biol* 1993;**234**:998–1012.
119. Plamann MD, Stauffer GV. *E.coli glyA* mRNA decay: The role of 3′ secondary structure and the effects of the *pnp* and *rnb* mutations. *Mol Gen Genet* 1990;**220**:301–6.
120. McLaren RS, Newbury SF, Dance GSC, Causton HC, Higgins CF. mRNA degradation by processive 3′–5′ exoribonucleases *in vitro* and the implications for prokaryotic mRNA decay *in vivo*. *J Mol Biol* 1991;**221**:81–95.
121. Mohanty BK, Kushner SR. Polynucleotide phosphorylase, RNase II and RNase E play different roles in the *in vivo* modulation of polyadenylation in *Escherichia coli*. *Mol Microbiol* 2000;**36**:982–94.
122. Masters M, Colloms MD, Oliver IR, He L, Macnaughton EJ, Charters Y. The pcnB gene of *Escherichia coli*, which is required for ColE1 copy number maintenance, is dispensable. *J Bacteriol* 1993;**175**:4405–13.
123. Raynal LC, Krisch HM, Carpousis AJ. The *Bacillus subtilis* nucleotidyltransferase is a tRNA CCA-adding enzyme. *J Bacteriol* 1998;**180**:6276–82.
124. Yue D, Maizels N, Weiner AM. CCA-adding enzymes and poly(A)polymerases are all members of the same nucleotidyltransferase superfamily: Characterization of the CCA-adding enzyme from the archael hyperthermophile *Sulfolobus shibatae*. *RNA* 1996;**2**:895–908.
125. Betat H, Rammelt C, Martin G, Mörl M. Exchange of regions between bacterial poly(A) polymerase and the CCA-adding enzyme generates altered specificiies. *Mol Cell* 2004;**15**:389–98.
126. Binns N, Masters M. Expression of the *Escherichia coli pcnB* gene is translationally limited using an inefficient start codon: A second chromosomal example of translation initiated at AUU. *Mol Microbiol* 2002;**44**:1287–98.
127. August JT, Ortiz PJ, Hurwitz J. Ribonucleic acid-dependent ribonucleotide incorporation. I. Purification and properties of the enzyme. *J Biol Chem* 1962;**237**:3786–93.
128. Sippel AE. Purification and characterization of adenosine triphosphate: Ribonucleic acid adenyltransferase from *Escherichia coli*. *Eur J Biochem* 1973;**37**:31–40.
129. Yehudai Reshelf S, Schuster C. Characterization of the *E. coli* poly(A) polymerase: Nucleotide specificity, RNA binding affinities and RNA structure dependence. *Nucleic Acids Res* 2000;**28**:1139–44.
130. Folichon M, Allemand F, Regnier P, Hajnsdorf E. Stimulation of poly(A) synthesis by *E. coli* poly(A)polymerase I is correlated with Hfq binding to poly(A) tails. *FEBS J* 2005;**272**:454–63.
131. Mohanty BK, Kushner SR. Polynucleotide phosphorylase functions both as a 3′–5′ exonuclease and a poly(A) polymerase in *Escherichia coli*. *Proc Natl Acad Sci USA* 2000;**97**:11966–71.
132. Feng Y, Cohen SN. Unpaired terminal nucleotides and 5′ monoposphorylation govern 3′ polyadenylation by *Escherichia coli* poly(A) polymerase I. *Proc Natl Acad Sci USA* 2000;**97**:6415–20.
133. Hajnsdorf E, Régnier P. Host factor HFq of *Escherichia coli* stimulates elongation of poly(A) tails by poly(A)polymerase I. *Proc Natl Acad Sci USA* 2000;**97**:1501–5.
134. Sano H, Feix G. Terminal riboadenylate transferase from *Escherichia coli*. *Eur J Biochem* 1976;**71**:577–83.
135. Zuo Y, Deutscher MP. Exoribonuclease superfamilies: Structural analysis and phylogenetic distribution. *Nucleic Acids Res* 2001;**29**:1017–26.
136. Ezraty B, Dahlgren B, Deutscher MP. The RNase Z homologue encoded by *Escherichia coli* elaC gene is RNase BN. *J Biol Chem* 2005;**280**:16542–5.

137. Cao G-J, Sarkar N. Poly(A) RNA in *Escherichia coli*: Nucleotide sequence at the junction of the lpp transcript and the polyadenylate moiety. *Proc Natl Acad Sci USA* 1992;**89**:7546–50.
138. Saira MI. Comparative sequence analysis of ribonucleases HII, III, II, PH and D. *Nucleic Acids Res* 1997;**25**:3187–95.
139. Spahr PF, Schlessinger D. Breakdown of messenger ribonucleic acid by a potassium-activated phosphodiesterase from *Escherichia coli*. *J Biol Chem* 1963;**238**:2251–3.
140. Shen V, Schlessinger D. In: P. Boyer, editor. *The Enzymes*. vol. XV Part B. New York: Academic Press; 1982. p. 501–15.
141. Deutscher MP, Reuven NB. Enzymatic basis for hydrolytic versus phosphorolytic mRNA degradation in *Escherichia coli* and *Bacillus subtilis*. *Proc Natl Acad Sci USA* 1991;**88**:3277–80.
142. Folichon M, Marujo PE, Arluison V, Le Derout J, Pellegrini O, Hajnsdorf E, et al. Fate of mRNA extremities generated by intrinsic termination: Detailed analysis of reactions catalyzed by ribonuclease II and poly(A) polymerase. *Biochimie* 2005;**87**:819–26.
143. Coburn GA, Mackie GA. Overexpression, purification and properties of *Escherichia coli* ribonuclease II. *J Biol Chem* 1996;**271**:1048–53.
144. Spickler C, Mackie GA. Action of RNase II and polynucleotide phosphorylase against RNAs containing stem-loops of defined structure. *J Bacteriol* 2000;**182**:2422–7.
145. Gupta RS, Kasai T, Schlessinger D. Purification and some novel properties of *Escherichia coli* RNase II. *J Biol Chem* 1977;**252**:8945–9.
146. Braun F, Hajnsdorf E, Régnier P. Polynucleotide phosphorylase is required for the rapid degradation of the RNase E-processed *rpsO* mRNA of *Escherichia coli* devoid of its 3' hairpin. *Mol Microbiol* 1996;**19**:997–1005.
147. Cheng Z-F, Deutscher MP. Purification and characterization of the *Escherichia coli* exoribonuclease RNase R. *J Biol Chem* 2002;**277**:21624–9.
148. Amblar M, Arraiano CM. A single mutation in *Escherichia coli* ribonuclease II inactivates the enzyme without affecting RNA binding. *FEBS J* 2005;**272**:2345.
149. Mott JE, Galloway JL, Platt T. Maturation of *Escherichia coli* tryptophan operon mRNA: Evidence for 3' exonucleolytic processing after rho-dependent termination. *EMBO J* 1985;**4**:1887–91.
150. Newbury SF, Smith NH, Robinson EC, Hiles ID, Higgins CF. Stabilization of translationally active mRNA by prokaryotic REP sequences. *Cell* 1987;**48**:297–310.
151. Pepe CM, Maslesa-Galic S, Simons RW. Decay of the IS10 antisense RNA by 3' exoribonucleases: Evidence that RNase II stabilizes RNA-OUT against PNPase attack. *Mol Microbiol* 1994;**13**:1133–42.
152. Zuo Y, Vincent HA, Zhang J, Wang Y, Deutscher MP, Malhotra A. Structural basis for processivity and single-strand specificity of RNase II. *Mol Cell* 2006;**24**:149–56.
153. Frazao C, McVey CE, Amblar M, Barbas A, Vonrhein C, Arraiano CM, et al. Unravelling the dynamics of RNA degradation by ribonuclease II and its RNA-bound complex. *Nature* 2006;**443**:110–4.
154. Barbas A, Matos RG, Amblar M, Lopez-Vinas E, Gomez-Puertas P, Arraiano CM. New insights into the mechanism of RNA degradation by ribonuclease II: Identification of the residue responsible for setting the RNase II end product. *J Biol Chem* 2008;**283**:13070–6.
155. Deutscher MP. Degradation of stable RNA in bacteria. *J Biol Chem* 2003;**278**:45041–4.
156. Amblar M, Barbas A, Fialho AM, Arraiano CM. Characterization of the functional domains of *Escherichia coli* RNase II. *J Mol Biol* 2006;**360**:921–33.
157. Grunberg-Manago M. In: W Cohn, editor. *Progress in nucleic acids research*. New York: Academic Press; 1963. p. 93–133.

158. Donovan WP, Kushner SR. Polynucleotide phosphorylase and ribonuclease II are required for cell viability and mRNA turnover in *Escherichia coli* K12. *Proc Natl Acad Sci USA* 1986;**83**:120–4.
159. Miczak A, Kaberdin VR, Wei C-L, Lin-Chao S. Proteins associated with RNase E in a multicomponent ribonucleolytic complex. *Proc Natl Acad Sci USA* 1996;**93**:3865–969.
160. Py B, Higgins CH, Krisch HM, Carpousis AJ. A DEAD-box RNA helicase in the *Escherichia coli* RNA degradosome. *Nature* 1996;**381**:169–72.
161. Carpousis AJ. The RNA degradosome of *Escherichia coli*: An mRNA-degrading machine assembled on RNase E. *Annu Rev Microbiol* 2007;**61**:71–87.
162. Symmons MF, Jones GH, Luisi BF. A duplicated fold is the structural basis for polynucleotide phosphorylase catalytic activity, processivity and regulation. *Structure* 2000;**8**:1215–26.
163. Symmons MF, Williams MG, Luisi BF, Jones GH, Carpousis AJ. Running rings around RNA: A superfamily of phosphate-dependent RNases. *TiBS* 2002;**27**:11–8.
164. Thang MN, Guschlbauer W, Zachau HG, Grunberg-Manago M. Degradation of transfer nucleic acid by polynucleotide phosphorylase. *J Mol Biol* 1967;**26**:403–21.
165. Kasai T, Gupta RS, Schlessinger D. Exoribonucleasesin wild-type *Escherichia coli* and RNase II-deficient mutants. *J Biol Chem* 1977;**252**:8950–6.
166. Vincent HA, Deutscher MP. Substrate recognition and catalysis by the exoribonuclease RNase R. *J Biol Chem* 2006;**281**:29769–75.
167. Cheng ZF, Deutscher MP. Quality control of ribosomal RNA mediated by polynucleotide phosphorylase and RNase R. *Proc Natl Acad Sci USA* 2003;**100**:6388–93.
168. Hajnsdorf E, Steier O, Coscoy L, Teysset L, Régnier P. Roles of RNase E, RNase II and PNPase in the degradation of the *rpsO* transcripts of *Escherichia coli*: Stabilizing function of RNase II and evidence for efficient degradation in an *ams-rnb-pnp* mutant. *EMBO J* 1994;**13**:3368–77.
169. Régnier P, Arraiano CM. Degradation of mRNA in bacteria: Emergence of ubiquitous features. *Bioessays* 2000;**22**:235–44.
170. Cheng Z-F, Zuo Y, Li Z, Rudd KE, Deutscher MP. The *vacB* gene required for virulence in *Shigella flexneri* encodes the exoribonuclease RNase R. *J Biol Chem* 1998;**273**:14077–80.
171. Coburn GA, Mackie GA. Degradation of mRNA in *Escherichia coli*: An old problem with some new twists. *Prog Nucl Acid Res Mol Biol* 1999;**62**:55–108.
172. Li Z, Deutscher MP. RNase E plays an essential role in the maturation of *Escherichia coli* tRNA precursors. *RNA* 2002;**8**:97–109.
173. Soderbom F, Svard SG, Kirsebom LA. RNase E cleavage in the 5′ leader of a tRNA precursor. *J Mol Biol* 2005;**352**:22–7.
174. Ow MC, Kushner SR. Initiation of tRNA maturation by RNase E is essential for cell viability in *E. coli*. *Genes Dev* 2002;**16**:1102–15.
175. Vanzo N, Li YS, Py B, Blum E, Higgins CF, Raynal LC, et al. Ribonuclease E organizes the protein interactions in the *Escherichia coli* RNA degradosome. *Genes Dev* 1998;**12**:2770–81.
176. Mackie GA. Ribonuclease E is a 5′-end-dependent endonuclease. *Nature* 1998;**395**:720–3.
177. Huang H, Liao J, Cohen SN. Poly(A) and poly(U)-specific RNA 3′ tail shortening by *E. coli* ribonuclease E. *Nature* 1998;**391**:99–102.
178. Walsh AP, Tock MH, Mallen MH, Kaberdin VR, von Gabain A, McDowall KJ. Cleavage of poly(A) tails on the 3′-end of RNA by ribonuclease E of *Escherichia coli*. *Nucleic Acids Res* 2001;**29**:1864–71.
179. Deutscher MP, Li Z. Exoribonucleases and their multiple roles in RNA metabolism. *Prog Nucleic Acid Res Mol Biol* 2001;**66**:67–105.
180. Li Z, Deutscher MP. The tRNA processing enzyme RNase T is essential for maturation of 5S RNA. *Proc Natl Acad Sci USA* 1995;**92**:6883–6.

181. Li Z, Pandit S, Deutscher MP. 3' exoribonucleolytic trimming is a common feature of the mauration of small, stable RNAs in *Escherichia coli*. *Proc Natl Acad Sci USA* 1998;**95**:2856–61.
182. Li Z, Pandit S, Deutscher MP. Maturation of 23S ribosomal RNA requires the exoribonuclease RNase T. *RNA* 1999;**5**:139–46.
183. Zuo Y, Deutscher MP. The physiological role of RNase T can be explained by its unusual substrate specificity. *J Biol Chem* 2002;**277**:29654–61.
184. Li Z, Reimers S, Pandit S, Deutscher MP. RNA quality control: Degradation of defective transfer RNA. *EMBO J* 2002;**21**:1132–8.
185. Mohanty BK, Kushner SR. Genomic analysis in *Escherichia coli* demonstrates differential roles for polynucleotide phosphorylase and RNase II in mRNA abundance and decay. *Mol Microbiol* 2003;**50**:645–58.
186. Worrall JA, Howe FS, McKay AR, Robinson CV, Luisi BF. Allosteric activation of the ATPase activity of the *Escherichia coli* RhlB RNA helicase. *J Biol Chem* 2008;**283**:5567–76.
187. Jasiecki J, Wegrzyn G. Growth-rate dependent RNA polyadenylation in *Escherichia coli*. *EMBO Rep* 2003;**4**:172–7.
188. Mohanty BK, Maples VF, Kushner SR. The Sm-like protein Hfq regulates polyadenylation dependent mRNA decay in *Escherichia coli*. *Mol Microbiol* 2004;**54**:905–20.
189. Raynal LC, Krisch HM, Carpousis AJ. Bacterial poly(A)polymerase: An enzyme that modulates RNA stability. *Biochimie* 1996;**78**:390–8.
190. Jones PG, VanBogelen RA, Neidhardt FC. Induction of proteins in response to low temperature in *Escherichia coli*. *J Bacteriol* 1987;**169**:2092–5.
191. Chen C, Deutscher MP. Elevation of RNase R in response to multiple stress conditions. *J Biol Chem* 2005;**280**:34393–6.
192. Goverde RL, Huis in't Veld JH, Kusters JG, Mooi FR. The psychrotrophic bacterium *Yersinia enterocolitica* requires expression of pnp, the gene for polynucleotide phosphorylase, for growth at low temperature (5 degrees C). *Mol Microbiol* 1998;**28**:555–69.
193. Luttinger A, Hahn J, Dubnau D. Polynucleotide phosphorylase is necessary for competence development in *Bacillus subtilis*. *Mol Microbiol* 1996;**19**:343–56.
194. Jarrige A-C, Mathy N, Portier C. PNPase autocontrols its expression by degrading a double-stranded structure in the *pnp* mRNA leader. *EMBO J* 2001;**20**:6845–55.
195. Portier C, Dondon L, Grunberg-Manago M, Régnier P. The first step in the functional inactivation of the *Escherichia coli* polynucleotide phosphorylase messenger is a ribonuclease III processing at the 5' end. *EMBO J* 1987;**6**:2165–70.
196. Robert-Le Meur M, Portier C. *E. coli* polynucleotide phosphorylase expression is autoregulated through an RNase III-dependent mechanism. *EMBO J* 1992;**11**:2633–41.
197. Cairrào F, Chora A, Zilhao R, Carpousis AJ, Arraiano CM. RNase II levels change according to the growth conditions: Characterization of *gmr*, a new *Escherichia coli* gene involved in the modulation of RNase II. *Mol Microbiol* 2001;**39**:1550–61.
198. Zilhao R, Cairrao F, Regnier P, Arraiano CM. PNPase modulates RNase II expression in *Escherichia coli*: Implications for mRNA decay and cell metabolism. *Mol Microbiol* 1996;**20**:1033–42.
199. Söderbom F, Wagner EGH. Degradation pathway of CopA, the antisense RNA that controls replication of plasmid R1. *Microbiology* 1998;**144**:1907–17.
200. Raynal LC, Carpousis AJ. Poly(A) polymerase I of *Escherichia coli*: Characterization of the catalytic domain, an RNA binding site and regions for the interaction with proteins involved in mRNA degradation. *Mol Microbiol* 1999;**32**:765–75.
201. Wahle E, Keller W. The biochemistry of polyadenylation. *Trends Biochem Sci* 1996;**21**:247–50.

202. Senear AW, Steitz JA. Site-specific interaction of Qb host factor and ribosomal protein S1 with Qb and R17 bacteriophage RNAs. *J Biol Chem* 1976;**251**:1902–12.
203. Zhang A, Altuvia S, Tiwari A, Argaman L, Hengge-Aronis R, Storz G. The oxyS regulatory RNA represses *rpoS* translation and binds the Hfq (HF-1) protein. *EMBO J* 1998;**17**:6061–8.
204. de Haseth PL, Uhlenbeck OC. Interaction of *Escherichia coli* host factor protein with oligoriboadenylates. *Biochemistry* 1980;**19**:6138–46.
205. Folichon M, Arluison V, Pellegrini O, Huntzinger E, Regnier P, Hajnsdorf E. The poly(A) binding protein Hfq protects RNA from RNase E and exoribonucleolytic degradation. *Nucleic Acids Res* 2003;**31**:7302–10.
206. Valentin-Hansen P, Eriksen M, Udesen C. The bacterial Sm-like protein Hfq: A key player in RNA transactions. *Mol Microbiol* 2004;**51**:1525–33.
207. Brennan RG, Link TM. Hfq structure, function and ligand binding. *Curr Opin Microbiol* 2007;**10**:125–33.
208. Gottesman S. Micros for microbes: Noncoding regulatory RNAs in bacteria. *Trends Genet* 2005;**21**:399–404.
209. Storz G, Opdyke JA, Zhang A. Controlling mRNA stability and translation with small, noncoding RNAs. *Curr Opin Microbiol* 2004;**7**:140–4.
210. Morita T, Maki K, Aiba H. RNase E based-ribonucleoprotein complexes: Mechanical basis of mRNA destabilization mediated by bacterial noncoding RNAs. *Genes Dev* 2005;**19**:2176–86.
211. Jasiecki J, Wegrzyn G. Localization of *Escherichia coli* poly(A) polymerase I in cellular membrane. *Biochem Biophys Res Commun* 2005;**329**:598–602.
212. Liou G-G, Jane W-N, Cohen SN, Lin N-S, Lin-Chao S. RNA degradosomes exist *in vivo* in *Escherichia coli* as multicomponent complexes associated with the cytoplasmic membrane *via* the N-terminal region of ribonuclease E. *Proc Natl Acad Sci USA* 2001;**98**:63–8.
213. Taghbalout A, Rothfield L. RNase E and the other constituents of the RNA degradosome are components of the bacterial cytoskeleton. *Proc Natl Acad Sci USA* 2007;**104**:1667–72.
214. Liou G-G, Chan H-Y, Lin C-S, Lin-Chao S. DEAD box RhlB helicase physically associates with exoribonuclease PNPase to degrade double stranded RNA independent of the degradosome asssembling region of RNase E. *J Biol Chem* 2002;**277**:41157.
215. Jasiecki J, Wegrzyn G. Phosphorylation of *Escherichia coli* poly(A) polymerase I and effects of this modification on the enzyme activity. *FEMS Microbiol Lett* 2006;**261**:118–22.
216. Feng Y, Huang H, Liao J, Cohen SN *Escherichia coli* poly(A) binding proteins that interact with components of degradosomes or impede RNA decay mediated by polynucleotide phosphorylase and RNase E. *J Biol Chem* 2001;**276**:31651–6.
217. Houseley J, LaCava J, Tollervey D. RNA-quality control by the exosome. *Nat Rev Mol Cell Biol* 2006;**7**:529–39.
218. Vanacova S, Wolf J, Martin G, Blank D, Dettwiler S, Friedlein A, *et al*. A new yeast poly(A) polymerase complex involved in RNA quality control. *PLoS Biol* 2005;**3**:e189.
219. Marujo PE, Braun F, Haugel-Nielsen J, Le Derout J, Arraiano CM, Regnier P. Inactivation of the decay pathway initiated at an internal site by RNase E promotes poly(A)-dependent degradation of the rpsO mRNA in *Escherichia coli*. *Mol Microbiol* 2003;**50**:1283–94.
220. Mohanty BK, Kushner SR. Polyadenylation of *Escherichia coli* transcripts plays an integral role in regulating intracellular levels of polynucleotide phosphorylase and RNase E. *Mol Microbiol* 2002;**45**:1315–24.
221. Joanny G, Le Derout J, Bréchemier-Baey D, Labas V, Vinh J, Régnier P, *et al*. Polyadenylation of a functional mRNA controls gene expression in *E. coli*. *Nucleic Acids Res* 2007;**35**:2494–502.
222. Urban JH, Vogel J. Two seemingly homologous noncoding RNAs act hierarchically to activate glmS mRNA translation. *PLoS Biol* 2008;**6**:e64.

223. Lisitsky I, Klaff P, Schuster G. Addition of destabilizing poly(A)-rich sequences to endonuclease cleavage sites during the degradation of chloroplast mRNA. *Proc Natl Acad Sci USA* 1996;**93:**13398–403.
224. Slomovic S, Portnoy V, Yehudai-Resheff S, Bronshtein E, Schuster G. Polynucleotide phosphorylase and the archaeal exosome as poly(A)-polymerases. *Biochim Biophys Acta* 2008;**1779:**247–55.
225. Rott R, Zipor G, Portnoy V, Liveanu V, Schuster G. RNA polyadenylation and degradation in cyanobacteria are similar to the chloroplast but different from *E. coli*. *J Biol Chem* 2003;**278:**15771–7.
226. Portnoy V, Schuster G. *Mycoplasma gallisepticum* as the first analyzed bacterium in which RNA is not polyadenylated. *FEMS Microbiol Lett* 2008;**283:**97–103.
227. Bralley P, Jones GH. cDNA cloning confirms the polyadenylation of RNA decay intermedites in *Streptomyces coelicolor*. *Microbiology* 2002;**148:**1421–5.
228. Yehudai-Resheff S, Portnoy V, Yogev S, Adir N, Schuster G. Domain analysis of the chloroplast polynucleotide phosphorylase reveals discrete functions in RNA degradation, polyadenylation, and sequence homology with exosome proteins. *Plant Cell* 2003;**15:**2003–19.
229. Wilusz CJ, Wilusz J. New ways to meet your (3′) end oligouridylation as a step on the path to destruction. *Genes Dev* 2008;**22:**1–7.
230. Lee K, Zhan X, Gao J, Qiu J, Feng Y, Meganathan R, et al. RraA. A protein inhibitor of RNase E activity that globally modulates RNA abundance in *E. coli*. *Cell* 2003;**114:**623–34.
231. Gao J, Lee K, Zhao M, Qiu J, Zhan X, Saxena A, et al. Differential modulation of *E. coli* mRNA abundance by inhibitory proteins that alter the composition of the degradosome. *Mol Microbiol* 2006;**61:**394–406.
232. Zhan X, Gao J, Jain C, Cieslewicz MJ, Swartz JR, Georgiou G. Genetic analysis of disulfide isomerization in *Escherichia coli*: Expression of DsbC is modulated by RNase E-dependent mRNA processing. *J Bacteriol* 2004;**186:**654–60.
233. Zhao M, Zhou L, Kawarasaki Y, Georgiou G. Regulation of RraA, a protein inhibitor of RNase E-mediated RNA decay. *J Bacteriol* 2006;**188:**3257–63.
234. Monzingo AF, Gao J, Qiu J, Georgiou G, Robertus JD. The X-ray structure of *Escherichia coli* RraA (MenG), a protein inhibitor of RNA processing. *J Mol Biol* 2003;**332:**1015–24.
235. Kanesaki T, Hamada T, Yonesaki T. Opposite roles of the dmd gene in the control of RNase E and RNase LS activities. *Genes Genet Syst* 2005;**80:**241–9.
236. Ueno H, Yonesaki T. Recognition and specific degradation of bacteriophage T4 mRNAs. *Genetics* 2001;**158:**7–17.
237. Zilhao R, Regnier P, Arraiano CM. The role of endonucleases in the expression of ribonuclease II in *Escherichia coli*. *FEMS Microbiol Lett* 1995;**130:**237–44.
238. Weber H, Pesavento C, Possling A, Tischendorf G, Hengge R. Cyclic-di-GMP-mediated signaling within the sigma network of *Escherichia coli*. *Mol Microbiol* 2006;**62:**1014–34.
239. Wassarman KM, Repoila F, Rosenow C, Storz G, Gottesman S. Identification of novel small RNAs using comparative genomics and microarrays. *Genes Dev* 2001;**15:**1637–51.
240. Rivas E, Klein RJ, Jones TA, Eddy SR. Computational identification of noncoding RNAs in *E. coli* by comparative genomics. *Current Biol* 2001;**11:**1369–73.
241. Chen S, Lesnik EA, Hall TA, Sampath R, Griffey RH, Ecker DJ, et al. A bioinformatics based approach to discover small RNA genes in the *Escherichia coli* genome. *BioSystems* 2002;**65:**157–77.
242. Hershberg R, Altuvia S, Margalit H. A survey of small RNA-encoding genes in *Escherichia coli*. *Nucleic Acids Res* 2003;**31:**1813–20.
243. Aiba H, Matsuyama S, Mizuno T, Mizushima S. Function of micF as an antisense RNA in osmoregulatory expression of the ompF gene in *Escherichia coli*. *J Bacteriol* 1987;**169:**3007–12.

244. Andersen J, Delihas N. micF RNA binds to the 5′ end of ompF mRNA and to a protein from *Escherichia coli*. *Biochemistry* 1990;**29**:9249–56.
245. Altuvia S, Zhang A, Argaman L, Tiwari A, Storz G. The *Escherichia coli* OxyS regulatory RNA represses fhlA translation by blocking ribosome binding. *EMBO J* 1998;**17**:6069–75.
246. Zhang A, Altuvia S, Storz G. The novel oxyS RNA regulates expression of the σ^s subunit of *Escherichia coli* RNA polymerase. *Nucl Acids Symp Ser* 1997;**36**:27–8.
247. Moller T, Franch T, Hojrup P, Keene D, Bächinger HP, Brennan RG, et al. HFq: A bacterial Sm-like protein that mediates RNA–RNA interaction. *Mol Cell* 2002;**9**:23–30.
248. Majdalani N, Cunning C, Sledjeski D, Elliott T, Gottesman S. DsrA RNA regulates translation of RpoS message by an anti-antisense mechanism, independent of its action as an antisilencer of transcription. *Proc Natl Acad Sci USA* 1998;**95**:12462–7.
249. Lease RA, Belfort M. A *trans*-acting RNA as a control switch in *Escherichia coli*: DsrA modulates function by forming alternative structures. *Proc Natl Acad Sci USA* 2000;**97**:9919–24.
250. Braun F, Le Derout J, Régnier P. Ribosomes inhibit an RNase E cleavage which induces the decay of the *rpsO* mRNA of *Escherichia coli*. *EMBO J* 1998;**17**:4790–7.
251. Muffler A, Traulsen DD, Fischer D, Lange R, Hengge-Aronis R. The RNA-binding protein HF-1 plays a global regulatory role which is largely, but not exclusively, due to its role in expression of the σ^s subunit of RNA polymerase in *Escherichia coli*. *J Bacteriol* 1997;**179**:297–300.
252. Takada A, Wachi M, Nagai K. Negative regulatory role of the *Escherichia coli hfq* gene in cell division. *Biochem Biophys Res Com* 1999;**266**:579–83.
253. Tsui H-CT, Leung H-CE, Winkler ME. Characterization of broadly pleiotropic phenotypes caused by an *hfq* insertion mutation in *Escherichia coli* K-12. *Mol Microbiol* 1994;**13**:35–49.
254. Ziolkowska K, Derreumaux P, Folichon M, Pellegrini O, Regnier P, Boni IV, et al. Hfq variant with altered RNA binding functions. *Nucleic Acids Res* 2006;**34**:709–20.
255. Vassilieva IM, Garber MB. The regulatory role of the HFq protein in bacterial cells. *Mol Biol* 2002;**36**:785–91.
256. Nogueira T, Springer M. Posttranscriptional control by global regulators of gene expression in bacteria. *Curr Opin Microbiol* 2000;**3**:154–8.
257. Zhang A, Wassarman KM, Rosenow C, Tjaden BC, Storz G, Gottesman S. Global analysis of small RNA and mRNA targets of Hfq. *Mol Microbiol* 2003;**50**:1111–24.
258. Sledjeski DD, Whitman C, Zhang A. HFq is necessary for regulation by the untranslated RNA DsrA. *J Bacteriol* 2001;**183**:1997–2005.
259. Zhang A, Wassarman KM, Ortega J, Steven AC, Storz G. The Sm-like HFq protein increases OxyS RNA interaction with target mRNAs. *Mol Cell* 2002;**9**:11–22.
260. Vanderpool CK, Gottesman S. Involvement of a novel transcriptional activator and small RNA in posttranscriptional regulation of the glucose phosphoenolpyruvate phosphotransferase system. *Mol Microbiol* 2004;**54**:1076–89.
261. Mikulecky PJ, Kaw MK, Brescia CC, Takach JC, Sledjeski DD, Feig AL. *Escherichia coli* Hfq has distinct interaction surfaces for DsrA, rpoS and poly(A) RNAs. *Nat Struct Mol Biol* 2004;**11**:1206–14.
262. Vecerek B, Rajkowitsch L, Sonnleitner E, Schroeder R, Blasi U. The C-terminal domain of *Escherichia coli* Hfq is required for regulation. *Nucleic Acids Res* 2008;**36**:133–43.
263. Arluison V, Derreumaux P, Allemand F, Folichon M, Hajnsdorf E, Regnier P. Structural modeling of the Sm-like protein Hfq from *Escherichia coli*. *J Mol Biol* 2002;**320**:705–12.
264. Kawamoto H, Koide Y, Morita T, Aiba H. Base-pairing requirement for RNA silencing by a bacterial small RNA and acceleration of duplex formation by Hfq. *Mol Microbiol* 2006;**61**:1013–22.

265. Brantl S. Regulatory mechanisms employed by *cis*-encoded antisense RNAs. *Curr Opin Microbiol* 2007;**10**:102–9.
266. Boisset S, Geissmann T, Huntzinger E, Fechter P, Bendridi N, Possedko M, *et al*. Staphylococcus aureus RNAIII coordinately represses the synthesis of virulence factors and the transcription regulator Rot by an antisense mechanism. *Genes Dev* 2007;**21**:1353–66.
267. Vytvytska O, Jakobsen JS, Balcunate G, Andersen JS, Baccarini M, von Gabain A. Host-factor I, Hfq, binds to *Escherichia coli ompA* mRNA in a growth rate-dependent fashion and egulates its stability. *Proc Natl Acad Sci USA* 1998;**95**:14118–23.
268. Massé E, Escorcia FE, Gottesman S. Coupled degradation of a small regulatory RNA and its mRNA targets in *Escherichia coli*. *Genes Dev* 2003;**17**:2374–83.
269. Tsui H-CT, Feng G, Winkler ME. Negative regulation of *mutS* and *mutH* repair gene expression by the Hfq and RpoS global regulators of *Escherichia coli* K-12. *J Bacteriol* 1997;**179**:7476–87.
270. Moll I, Afonyushkin T, Vytvytska O, Kaberdin VR, Blasi U. Coincident Hfq binding and RNase E cleavage sites on mRNA and small regulatory RNAs. *RNA* 2003;**9**:1308–14.
271. Geissmann TA, Touati D. Hfq, a new chaperoning role: Binding to messenger RNA determines access for small RNA regulator. *EMBO J* 2004;**23**:396–405.
272. Butland G, Peregrin-Alvarez JM, Li J, Yang W, Yang X, Canadien V, *et al*. Interaction network containing conserved and essential protein complexes in *Escherichia coli*. *Nature* 2005;**433**:531–7.
273. Kajitani M, Ishihama A. Identification and sequence determination of the host factor gene for bacteriophage Qb. *Nucleic Acids Res* 1991;**9**:1063–6.
274. Sukhodolets MV, Garges S. Interaction of *Escherichia coli* RNA polymerase with the ribosomal protein S1 and the Sm-like ATPase Hfq. *Biochemistry* 2003;**42**:8022–34.
275. Vecerek B, Moll I, Afonyushkin T, Kaberdin V, Bläsi U. Interaction of the RNA chaperone Hfq with mRNAs: Direct and indirect roles in iron metabolism of *Escherichia coli*. *Mol Microbiol* 2003;**50**:897–910.
276. Davis BM, Waldor MK. RNase E-dependent processing stabilizes MicX, a *Vibrio cholerae* sRNA. *Mol Microbiol* 2007;**65**:373–85.
277. Ueno H, Yonesaki T. Role of *Escherichia coli* HFq in late-gene silencing of bacteriophage T4 *dmd* mutant. *Genes Genet Syst* 2002;**77**:301–8.
278. Masse E, Arguin M. Ironing out the problem: new mechanisms of iron homeostasis. *Trends Biochem Sci* 2005;**30**:462–8.
279. Carpousis AJ. Degradation of targeted mRNAs in *Escherichia coli*: Regulation by a small antisense RNA. *Genes Dev* 2003;**17**:2351–5.
280. Afonyushkin T, Vecerek B, Moll I, Blasi U, Kaberdin VR. Both RNase E and RNase III control the stability of sodB mRNA upon translational inhibition by the small regulatory RNA RyhB. *Nucleic Acids Res* 2005;**33**:1678–89.
281. Prevost K, Salvail H, Desnoyers G, Jacques JF, Phaneuf E, Masse E. The small RNA RyhB activates the translation of shiA mRNA encoding a permease of shikimate, a compound involved in siderophore synthesis. *Mol Microbiol* 2007;**64**:1260–73.
282. Kimata K, Tanaka Y, Inada T, Aiba H. Expression of the glucose transporter gene, *ptsG*, is regulated at the mRNA degradation step in response to glycolytic flux in *Escherichia coli*. *EMBO J* 2001;**20**:3587–95.
283. Morita T, Mochizuki Y, Aiba H. Translational repression is sufficient for gene silencing by bacterial small noncoding RNAs in the absence of mRNA destruction. *Proc Natl Acad Sci USA* 2006;**103**:4858–3.
284. Morita T, Kawamoto H, Mizota T, Inada T, Aiba H. Enolase in the RNA degradosome plays a crucial role in the rapid decay of glucose transporter mRNA in the response to phosphosugar stress in *Escherichia coli*. *Mol Microbiol* 2004;**54**:1063–75.

285. Opdyke JA, Kang JG, Storz G. GadY, a small-RNA regulator of acid response genes in *Escherichia coli*. *J Bacteriol* 2004;**186**:6698–705.
286. Takada A, Umitsuki G, Nagai K, Wachi M. RNase E is required for induction of the glutamate-dependent acid resistance system in *Escherichia coli*. *Biosci Biotechnol Biochem* 2007;**71**:158–64.
287. Krinke L, Wulff DL. RNase III-dependent hydrolysis of lcII-O gene mRNA mediated by l OOP antisense RNA. *Genes Dev* 1990;**4**:2223–33.
288. Krinke L, Wulff DL. OOP RNA, produced from multicopy plasmids, inhibits lambda cII gene expression through an RNase III-dependent mechanism. *Genes Dev* 1987;**1**:1005–13.
289. Duhring U, Axmann IM, Hess WR, Wilde A. An internal antisense RNA regulates expression of the photosynthesis gene isiA. *Proc Natl Acad Sci USA* 2006;**103**:7054–8.
290. Waldbeser LS, Chen Q, Crosa JH. Antisense RNA regulation of the fatB iron transport protein gene in *Vibrio anguillarum*. *Mol Microbiol* 1995;**17**:747–56.
291. Blomberg P, Wagner EGH, Nordström K. Control of replication of plasmid R1: The duplex between the antisense RNA, Cop A, and its target, Cop T, is processed specifically *in vivo* and *in vitro* by RNase III. *EMBO J* 1990;**9**:2331–40.
292. Gerdes K, Nielsen A, Thorsted P, Wagner EG. Mechanism of killer gene activation. Antisense RNA-dependent RNase III cleavage ensures rapid turn-over of the stable hok, srnB and pndA effector messenger RNAs. *J Mol Biol* 1992;**226**:637–49.
293. Case CC, Simons EL, Simons RW. The IS10 transposase mRNA is destabilized during antisense RNA control. *EMBO J* 1990;**9**:1259–66.
294. Romeo T, Gong M, Liu MY, Brun-Zinkernagel AM. Identification and molecular characterization of csrA, a pleiotropic gene from *Escherichia coli* that affects glycogen biosynthesis, gluconeogenesis, cell size, and surface properties. *J Bacteriol* 1993;**175**:4744–55.
295. Sabnis NA, Yang H, Romeo T. Pleiotropic regulation of central carbohydrate metabolism in *Escherichia coli* via the gene csrA. *J Biol Chem* 1995;**270**:29096–104.
296. Liu MY, Yang H, Romeo T. The product of the pleiotropic *Escherichia coli* gene csrA modulates glycogen biosynthesis via effects on mRNA stability. *J Bacteriol* 1995;**177**:2663–72.
297. Liu MY, Romeo T. The global regulator CsrA of *Escherichia coli* is a specific mRNA-binding protein. *J Bacteriol* 1997;**179**:4639–42.
298. Liu MY, Gui G, Wei B, Preston JF, III, Oakford L, Yuksel U, et al. The RNA molecule CsrB binds to the global regulatory protein CsrA and antagonizes its activity in *Escherichia coli*. *J Biol Chem* 1997;**272**:17502–10.
299. Weilbacher T, Suzuki K, Dubey AK, Wang X, Gudapaty S, Morozov I, et al. A novel sRNA component of the carbon storage regulatory system of *Escherichia coli*. *Mol Microbiol* 2003;**48**:657–70.
300. Babitzke P, Romeo T. CsrB sRNA family: Sequestration of RNA-binding regulatory proteins. *Curr Opin Microbiol* 2007;**10**:156–63.
301. Wei BL, Brun-Zinkernagel AM, Simecka JW, Pruss BM, Babitzke P, Romeo T. Positive regulation of motility and flhDC expression by the RNA-binding protein CsrA of *Escherichia coli*. *Mol Microbiol* 2001;**40**:245–56.
302. Suzuki K, Babitzke P, Kushner S, Romeo T. Identification of a novel regulatory protein (CsrD) that targets the global regulatory RNAs CsrB and CsrC for degradation by RNase E. *Genes Dev* 2006;**20**:2605–17.
303. Heeb S, Kuehne SA, Bycroft M, Crivii S, Allen MD, Haas D, et al. Functional analysis of the posttranscriptional regulator RsmA reveals a novel RNA-binding site. *J Mol Biol* 2006;**355**:1026–36.
304. Pessi G, Williams F, Hindle Z, Heurlier K, Holden MT, Camara M, et al. The global posttranscriptional regulator RsmA modulates production of virulence determinants and N-acylhomoserine lactones in *Pseudomonas aeruginosa*. *J Bacteriol* 2001;**183**:6676–83.

305. Dulebohn D, Choy J, Sundermeier T, Okan N, Karzai AW. Trans-translation: The tmRNA-mediated surveillance mechanism for ribosome rescue, directed protein degradation, and nonstop mRNA decay. *Biochemistry* 2007;**46**:4681–93.
306. Yamamoto Y, Sunohara T, Jojima K, Inada T, Aiba H. SsrA-mediated *trans*-translation plays a role in mRNA quality control by facilitating degradation of truncated mRNAs. *RNA* 2003;**9**:408–18.
307. Mehta P, Richards J, Karzai AW. tmRNA determinants required for facilitating nonstop mRNA decay. *RNA* 2006;**12**:2187–98.
308. Sunohara T, Jojima K, Tagami H, Inada T, Aiba H. Ribosome stalling during translation elongation induces cleavage of mRNA being translated in *Escherichia coli*. *J Biol Chem* 2004;**279**:15368–75.
309. Sunohara T, Jojima K, Yamamoto Y, Inada T, Aiba H. Nascent-peptide-mediated ribosome stalling at a stop codon induces mRNA cleavage resulting in nonstop mRNA that is recognized by tmRNA. *RNA* 2004;**10**:378–86.
310. Hayes CS, Sauer RT. Cleavage of the A site mRNA codon during ribosome pausing provides a mechanism for translational quality control. *Mol Cell* 2003;**12**:903–11.
311. Pedersen K, Zavialov AV, Pavlov MY, Elf J, Gerdes K, Ehrenberg M. The bacterial toxin relE displays codon-specific cleavage of mRNAs in the ribosomal A site. *Cell* 2003;**112**:131–40.
312. Christensen SK, Gerdes K. RelE toxins from bacteria and Archaea cleave mRNAs on translating ribosomes, which are rescued by tmRNA. *Mol Microbiol* 2003;**48**:1389–400.
313. Christensen SK, Pedersen K, Hansen FG, Gerdes K. Toxin–antitoxin loci as stress-response-elements: ChpAK/MazF and ChpBK cleave translated RNAs and are counteracted by tmRNA. *J Mol Biol* 2003;**332**:809–19.
314. Ivanova N, Pavlov MY, Felden B, Ehrenberg M. Ribosome rescue by tmRNA requires truncated mRNAs. *J Mol Biol* 2004;**338**:33–41.
315. Karzai AW, Sauer RT. Protein factors associated with the ssrA-smpB tagging and ribosome rescue complex. *Proc Natl Acad Sci* 2001;**98**:3040–4.
316. Hong SJ, Tran QA, Keiler KC. Cell cycle-regulated degradation of tmRNA is controlled by RNase R and SmpB. *Mol Microbiol* 2005;**57**:565–75.
317. Richards J, Mehta P, Karzai AW. RNase R degrades non-stop mRNAs selectively in an SmpB-tmRNA-dependent manner. *Mol Microbiol* 2006;**62**:1700–12.
318. Winkler WC, Nahvi A, Roth A, Collins JA, Breaker RR. Control of gene expression by a natural metabolite-responsive ribozyme. *Nature* 2004;**428**:281–6.
319. Barrick JE, Breaker RR. The distributions, mechanisms, and structures of metabolite-binding riboswitches. *Genome Biol* 2007;**8**:R239.
320. Tucker BJ, Breaker RR. Riboswitches as versatile gene control elements. *Curr Opin Struct Biol* 2005;**15**:342–8.
321. Mandal M, Boese B, Barrick JE, Winkler WC, Breaker RR. Riboswitches control fundamental biochemical pathways in *Bacillus subtilis* and other bacteria. *Cell* 2003;**113**:577–86.
322. Barrick JE, Corbino KA, Winkler WC, Nahvi A, Mandal M, Collins J, et al. New RNA motifs suggest an expanded scope for riboswitches in bacterial genetic control. *Proc Natl Acad Sci USA* 2004;**101**:6421–6.
323. Collins JA, Irnov I, Baker S, Winkler WC. Mechanism of mRNA destabilization by the glmS ribozyme. *Genes Dev* 2007;**21**:3356–68.
324. Altman S, Wesolowski D, Guerrier-Takada C, Li Y. RNase P cleaves transient structures in some riboswitches. *Proc Natl Acad Sci USA* 2005;**102**:11284–9.
325. Deutscher MP. Degradation of RNA in bacteria: Comparison of mRNA and stable RNA. *Nucleic Acids Res* 2006;**34**:659–66.
326. Marcaida MJ, DePristo MA, Chandran V, Carpousis AJ, Luisi BF. The RNA degradosome: Life in the fast lane of adaptive molecular evolution. *Trends Biochem Sci* 2006;**31**:359–65.

327. Condon C. Maturation and degradation of RNA in bacteria. *Curr Opin Microbiol* 2007;**10**:271–8.
328. Sousa S, Marchand I, Dreyfus M. Autoregulation allows *Escherichia coli* RNase E to continuously adjust its synthesis to that of its substrates. *Mol Microbiol* 2001;**42**:867–78.
329. Bardwell JCA, Régnier P, Chen SM, Nakamura Y, Grunberg-Manago M, Court DL. Autoregulation of RNase III operon by mRNA processing. *EMBO J* 1989;**8**:3401–7.

The Role of 3'–5' Exoribonucleases in RNA Degradation

José M. Andrade,
Vânia Pobre, Inês J. Silva,
Susana Domingues, and
Cecília M. Arraiano

Instituto de Tecnologia Química e Biológica, Universidade Nova de Lisboa, Apartado 127, 2781-901 Oeiras, Portugal

I. Introduction	188
A. Dynamics of The RNA Degradation Machinery	188
B. 3'–5' Degradative Exoribonucleases	190
II. Polynucleotide Phosphorylase (PNPase)	193
A. Control of PNPase Expression	193
B. PNPase Activities	195
C. PNPase Complexes	195
D. PNPase Structure and Function	196
E. A Role for PNPase in Virulence	199
F. PNPase, An Exosome-Like Enzyme	199
III. RNase II	200
A. Control of RNase II Expression	200
B. RNase II Activity and RNA Degradation	201
C. RNase II Structure and Function	202
D. RNase II-like, Rrp44/Dis3: The Catalytic Subunit of The Eukaryotic Exosome	204
IV. RNase R	205
A. Control of RNase R Expression	206
B. RNA Degradation and Mode of Action	206
C. RNA Quality Control	207
D. Protein Quality Control	207
E. Stress and Stationary-Phase	208
F. A Role for RNase R in Virulence	209
G. Insights into The Structure and Function of RNase R	209
V. Oligoribonuclease	210
VI. Mechanisms of RNA Degradation	212
A. *pyr*F-*orf*F	212
B. *trx*A	212
C. *rps*O	214
D. *rps*T	214
E. *mal*EF	215
F. *omp*A	215
G. Small Noncoding RNAs	216
VII. Concluding Remarks	217
References	218

RNA degradation is a major process controlling RNA levels and plays a central role in cell metabolism. From the labile messenger RNA to the more stable noncoding RNAs (mostly rRNA and tRNA, but also the expanding class of small regulatory RNAs) all molecules are eventually degraded. Elimination of superfluous transcripts includes RNAs whose expression is no longer required, but also the removal of defective RNAs. Consequently, RNA degradation is an inherent step in RNA quality control mechanisms. Furthermore, it contributes to the recycling of the nucleotide pool in the cell. *Escherichia coli* has eight 3′–5′ exoribonucleases, which are involved in multiple RNA metabolic pathways. However, only four exoribonucleases appear to accomplish all RNA degradative activities: polynucleotide phosphorylase (PNPase), ribonuclease II (RNase II), RNase R, and oligoribonuclease. Here, we summarize the available information on the role of bacterial 3′–5′ exoribonucleases in the degradation of different substrates, highlighting the most recent data that have contributed to the understanding of the diverse modes of operation of these degradative enzymes.

I. Introduction

RNA degradation is far from being an unrestricted or random process. Instead, it is a controlled mechanism in which enzymes, together with RNA sequence and structure are important factors. Not surprisingly, RNases are key players in the regulation of gene expression. Different RNases may cooperate in the degradation of RNAs. They are divided into two main classes: endonucleases, proteins that cleave RNA internally and exoribonucleases that digest RNA one nucleotide at a time from one extremity. Bacterial RNases have been most extensively studied in *E. coli* where about 20 RNases have been identified so far. Such enzymes show either a high degree of specialization (with activities not performed by any other RNase) or a functional redundancy with other RNases. This is clearly evident when we analyze the eight exoribonucleases present in *E. coli*. While oligoribonuclease is an essential exoribonuclease that performs a very specialized function in the cell, all the others, despite having certain unique characteristics can be replaced by other exoribonucleases, at least to some extent. This review will focus on 3′–5′ exoribonucleases and their role in RNA degradation.

A. Dynamics of The RNA Degradation Machinery

The rate of RNA degradation plays a central role in establishing RNA levels (*1*, *2*). In *E. coli*, most mRNAs have half-lives ranging between 2 and 8 min, depending on the growth conditions (*3*, *4*). More stable RNAs include the noncoding rRNAs and tRNAs. Posttranscriptional regulation at the level of

RNA stability contributes to differential gene expression. Even different segments of a polycistronic RNA can be differentially regulated (5). Bacterial cells do not contain dedicated machineries to degrade the different classes of RNA; degradative enzymes function in both RNA processing and maturation pathways (6). Thus, the same set of enzymes has to cope with many different substrates and play different roles. For example, RNase II, a major exoribonuclease involved in mRNA degradation, is also able to participate in tRNA maturation (7); PNPase and RNase R, the other major contributors to mRNA decay, are also key enzymes in the degradation of stable RNA (7, 8). The RNA degradation machinery and consequently, RNA degradation pathways are dynamic and can change in response to different environmental stimuli. Exoribonucleases are a good example of these dynamics, as different stress conditions change the levels of some of these enzymes. RNases differ among themselves in substrate binding and catalytic activity, which may be advantageous when dealing with a multiplicity of substrates. Accordingly, the intrinsic characteristics of the different RNases afforded by their structural and functional features play a central role in establishing RNA degradation mechanisms.

RNA is far from being an innocent substrate waiting to be degraded. The sequence of the RNA can directly impact its turnover rate. RNA molecules can possess stabilizing elements, such as folded structural elements (particularly in G/C enriched sequences) that can present obstacles to ribonucleolytic activity by preventing the progression of exoribonucleases (9). Nucleotide modifications may protect the RNA and the 5′ phosphorylation state of the RNA molecule also influences its decay (10–12). Certain sequences make the transcript more susceptible to ribonucleolytic attack. Polyadenylation (the addition of poly(A) stretches at the 3′ RNA end) usually favors exonucleolytic activity (13, 14). In prokaryotes, transcription is coupled with translation and translating ribosomes can block nucleolytic activity and influence RNA stability (see the chapter by Dreyfus, in this volume). In fact, naked RNA is usually more susceptible to degradation (15). Moreover, premature arrest of translation usually results in the concomitant degradation of an mRNA, where exoribonuclease R plays the main role (16). Finally, RNA binding proteins may modulate transcript stability. For example, the RNA chaperone Hfq has been suggested to protect RNA from both endo- and exonucleolytic degradation (17).

In prokaryotes, it is commonly accepted that initiation of RNA decay is primarily endonucleolytic (see the chapter by Carpousis and colleagues, in this volume). Following this, exoribonucleases complete the decay process and rapidly eliminate intermediary fragments, explaining why we rarely see accumulation of RNA breakdown products. However, exoribonucleases can also promote degradation of full-length RNAs (18, 19). These enzymes can act in a 3′–5′ or 5′–3′ direction. 3′–5′ exoribonucleases seem to be more abundant with widespread representative members (20). Like their prokaryotic counterparts, 3'-5' exoribonucleases from archaea and eukarya are involved in the degradation of different RNA substrates. The archaeal and eukaryotic exosome, a multisubunit ring-shaped complex that contains multiple homologues of bacterial 3'-5' exoribonucleases, is a major factor in the degradation of deadenylated mRNAs (21, 22). The 5'-3' exoribonuclease Xrn1 is considered to be primarily responsible for eukaryotic mRNA decay, but no homologues of this enzyme have been found in archaea (23). *Bacillus subtilis* RNase J1 is at the moment the only identified 5'-3' exoribonuclease amongst prokaryotes (24) (see the chapter by Bechhofer, in this volume). However, RNase J1 is a ubiquitous enzyme, with homologues predicted in many other sequenced bacterial and archaeal genomes (25). Moreover, work by Hasenöhrl *et al.* further indicates that 5'-3' directional mRNA decay is a common pathway to all domains of life (26).

In *E. coli* (the reference genetic model organism by excellence), eight RNA 3'-5' exoribonucleases have been described: PNPase, RNase II, RNase R, RNase BN, RNase D, RNase PH, RNase T, and oligoribonuclease (Orn). Most are involved in stable RNA processing, acting in the trimming of inactive precursors leading to mature RNAs. However, only four exoribonucleases appear to accomplish all RNA degradative activity in the cell: PNPase, RNase II, RNase R, and oligoribonuclease. These four enzymes will be the target of this review (see Table I).

B. 3'-5' Degradative Exoribonucleases

E. coli 3'-5' degradative exoribonucleases have homologous counterparts in all domains of life, which brings into focus their importance in the ancient and critical pathways of RNA degradation. Most of them belong to the RNase II, DEDD, or PDX families of exoribonucleases (20, 27). They present a different linear domain organization and structure that may reflect the differences found in their catalytic activity (Fig. 1). The majority of the exoribonucleases (e.g., RNase II, RNase R, and oligoribonuclease) act hydrolytically releasing nucleoside monophosphates but others degrade RNA through a phosphorolytic mechanism releasing nucleoside diphosphates (e.g., PNPase). It has been suggested that the coexistence of these two mechanisms may be important for adaptation to environmental conditions (35). Exoribonucleases can show some specificity and even close homologues may act differently on a variety of substrates. For example, RNase R has a high affinity for rRNA, while RNase II greatly prefers poly(A) (36). Moreover, exoribonucleases can compete between themselves for access to the same RNA substrate. With the exception of oligoribonuclease, the other enzymes exhibit some degree of overlapping function and single mutants can survive.

TABLE I
3′–5′ Exonucleases Involved in RNA Degradation

Degradative 3′–5′ exonuclease	Family	Gene	kDa[a]	Structure	Catalytic mechanism	Substrate	Comments
PNPase	PDX	*pnp*	78	α_3 $(\alpha_3\beta_2)$[b]	Phosphorolytic	mRNA, rRNA, tRNA, small RNA	Homologues are found in archaeal and eukaryotic exosome. Degrades double-stranded RNA when in multiprotein complexes, for example, the degradosome. Cold shock protein. Involved in RNA quality control
RNase II	RNase II	*rnb*	73	α	Hydrolytic	mRNA, tRNA	Sensitive to RNA secondary structures. Can also protect RNA from degradation. Homologue protein is the only active catalytic subunit of eukaryotic exosome
RNase R	RNase II	*rnr*	92	α	Hydrolytic	mRNA, rRNA, tRNA, small RNA	Highly effective against RNA duplexes. Stress induced protein. Growth-phase regulated. Involved in RNA quality control
Oligoribonuclease	DEDD	*orn*	21	α_2	Hydrolytic	Oligoribonucleotides (2–5 mer)	Essential enzyme. Degrades the short oligoribonucleotides released from the degradative action of other exoribonucleases

[a] Molecular weight of the monomer.
[b] DEAD-box RNA helicase RhlB is the β subunit of PNPase.

Exonuclease structure

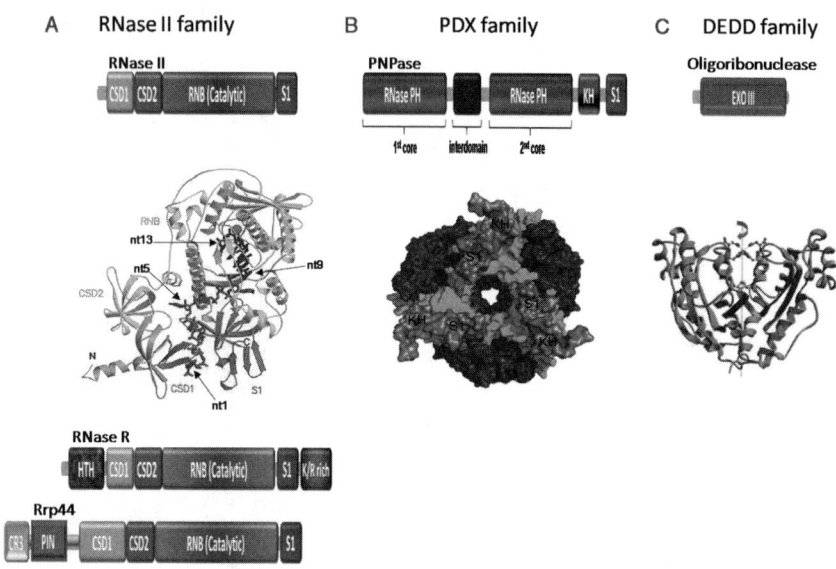

Fig. 1. Schematic representation of the domains found in the exoribonucleases from the RNase II/PDX/DEDD families and structures of representative members. (A) Top. Linear representation of RNase II domains: CSD1, CSD2, the central catalytic RNB, and S1. Middle. View of RNase II–RNA complex crystal structure, showing the distinct domains of the enzyme, the 13-mer bound RNA is represented and the Mg^{2+} ion is shown as a sphere (28). The other members of RNase II family are predicted to have additional domains (bottom): bacterial RNase R shows an additional helix-turn-helix at its N-terminal end and a charged C-terminal region rich in lysine and arginine residues (29); eukaryotic Rrp44 contains a predicted N-terminal with a PIN domain (shows a fold related to RNase H) (30) and a CR3 domain (cysteine rich region thought to be involved in protein–protein interaction) (31, 220). (B) Top. Schematic representation of PNPase (PDX family member) primary structure: two RNase PH folds (the first and second core) linked by an all α helical region (interdomain), followed by two RNA binding domains, KH and S1. Below. PNPase from S. antibioticus viewed along the threefold rotation axis (32). PNPase is a trimer, assemble in a doughnut-like shape, similar to that found in exosome. The S1 and KH domains are indicated. (C) Schematic representation of the primary structure of oligoribonuclease (member of the DEDD family of exoribonucleases). The ExoIII domain is common in a variety of other exoribonucleases, such as ribonuclease T and the epsilon subunit of DNA polymerase III. Below is a ribbon diagram of the crystal structure of the oligoribonuclease dimer from Xanthomonas campestris. The dotted line indicates the noncrystallographic 2-fold symmetry (33). Protein motifs were analysed using Pfam database (34).

Degradative exoribonucleases are processive and do not exhibit sequence specificity, but they differ in their ability to bind RNA and to degrade through extensive regions of secondary structure. Removal of structured RNAs emerges as a challenging degradation step. RNA duplexes usually act as the stabilizing elements within RNAs and are common in mRNAs, either the 5′ or 3′ untranslated regions (UTR) (4) or in intergenic regions of polycistronic transcripts, as REP (repetitive extragenic palindromic) sequences (37, 38). Furthermore, stable RNAs like rRNA, tRNA, and several small noncoding RNAs (sRNAs) are highly folded and pose an even greater challenge for degradation (39).

Exoribonucleases have to be controlled to avoid the elimination of RNA molecules still necessary to the cell; the regulation of exonucleolytic activity is thus extremely important. Access to the 3' end of RNA is the rate limiting event for exonucleolytic degradation. Exoribonucleases prefer a single-stranded region at the RNA 3' end to bind and initiate decay. Many RNAs have secondary structures with only a short, if any, 3' single-stranded region, hampering degradation. Polyadenylation of RNA restores the accessibility of RNA to exoribonucleases by providing them with an "on-ramp" or "toe-hold" (9, 40). However, even if a 3' linear stretch is provided and exoribonucleases are able to bind RNA, they may detach upon approaching stem-loops. The association of exoribonucleases in multiprotein complexes can be advantageous. E. coli PNPase associates with the RNA helicase B (RhlB), for example, which helps it to degrade folded RNAs (41, 42) and a related complex also exists in yeast (43–45). ATP-dependent RNA helicases are called "motor proteins" as they are able to unwind RNA duplexes, promoting exonucleolytic activity (21, 46). Remarkably, only RNase R is able to overcome highly folded structures on its own.

Here, we summarize the available information on bacterial 3'–5' exoribonucleases in RNA degradation, highlighting the most recent data that contributed to revealing the diverse mechanisms of these degradative enzymes. Most efforts, by far, to elucidate RNA degradation pathways have focussed on E. coli and we will mainly describe what is known in this model organism, although we will also include other organisms for comparison.

II. Polynucleotide Phosphorylase (PNPase)

The PDX family of exoribonucleases includes PNPase and RNase PH from bacteria, and the core of the exosome in archaea and eukaryotes (20, 27, 47). PNPase is involved in global mRNA decay, while the primary activity of RNase PH is the 3' maturation of tRNA precursors (48). PNPase is widely conserved from bacteria to plants and metazoans (20, 49). This enzyme was discovered in 1955 by Severo Ochoa and Marianne Grunberg-Manago (50). For this important discovery, Ochoa was rewarded with the Nobel Prize 4 years later.

A. Control of PNPase Expression

PNPase is encoded by the *pnp* gene that starts with an unusual UUG triplet and is located downstream of the *rpsO* gene (encoding ribosomal protein S15) (51). The *pnp* gene is transcribed from two promoters: one upstream of the *rpsO* gene and another upstream of *pnp* (51). *pnp* expression is negatively autoregulated at the posttranscriptional level by the concerted action of PNPase and RNase III (52). In an RNase III deficient strain there is a stabilization of the *pnp* mRNA, leading to 10-fold increased levels of PNPase (53).

The RNase III cleavage at the double-stranded region of the 5' end of the *pnp* mRNA leads to an RNA helix with a protruding 3' end that is preferentially removed by PNPase. Since PNPase is not able to interact with its message before it is processed at its 5' end, the absence of RNase III cleavage impedes the autogenous control of the *pnp* mRNA (52, 54). Thus, the stability of the *pnp* message is inversely correlated with the amount of active PNPase in the cell (55). Additionally, PNPase interaction with the *pnp* message prevents translation, further decreasing the level of PNPase (54). RNase II also regulates PNPase expression (56). In the absence of RNase II, PNPase levels are increased. Interestingly, PNPase overexpression also leads to a decrease in RNase II activity (see below), an intriguing example of cross-regulation (56). At what level RNase II regulates PNPase is still not known.

PNPase does not seem to be indispensable to *E. coli* at optimal temperature, unless either RNase II or RNase R is also missing (3, 57). Accumulation of rRNA fragments was observed in a strain lacking both PNPase and RNase R at nonpermissive temperature, while the absence of PNPase and RNase II led to mRNA accumulation (3, 8). Interestingly, an RNase II$^-$ RNase R$^-$ strain is viable, suggesting that PNPase can somehow replace their activity, at least under some growth conditions (19). PNPase is essential for growth at low temperatures (58–60). In fact, PNPase has been identified as a cold shock protein (61). Cold temperature induction of *pnp* expression occurs at several different posttranscriptional levels, including reversal of *pnp* autoregulation (62). During the cold-acclimation phase there is a 10-fold increase in *pnp* transcripts (60) and a 2-fold increase in PNPase protein level (63). Transcriptional activation does not appear to contribute significantly to the induction of *pnp* mRNA (60, 62, 63); the increase in *pnp* transcripts seems to be mainly due to their stabilization (60, 63). However, RNase III processing of *pnp* transcripts is not affected by the temperature downshift. Since the *pnp* mRNA is not efficiently translated during cold acclimation, the stabilization of the transcripts appears to compensate for this poor translation to maintain or increase the protein levels (60). After adaptation to the cold, *pnp* mRNA levels decrease.

PNPase levels are also affected by polyadenylation. When Poly(A) polymerase I (PAP I, encoded by the *pcnB* gene) is overexpressed, the availability of PNPase to carry out normal autoregulation drops significantly. It is likely that the increased levels of polyadenylated transcripts titrate out the amount of PNPase available to bind to the 5' end of its own transcripts (64).

B. PNPase Activities

PNPase catalyzes the processive 3′–5′ phosphorolytic degradation of RNA, releasing nucleoside diphosphates from the 3′ end. At low inorganic phosphate concentrations, the enzyme catalyzes the reverse reaction, that is, polymerization of single-stranded RNA from nucleoside diphosphates (65–67). PNPase also catalyzes the exchange reaction between the β-phosphate group of nucleoside diphosphates and free orthophosphate (68). In vivo, this enzyme is essentially devoted to the processive degradation of RNA, but is also responsible for adding the polyribonucleotide tails observed in E. coli mutants devoid of the main polyadenylating enzyme PAP I (69). Recent work demonstrated that ATP binds to PNPase and inhibits both its phosphorolytic and polymerization activities (70). Thus, PNPase-dependent RNA tailing and degradation occurs mainly at low ATP concentrations. Contrary to the homopolymeric poly(A) tails added by PAP I, the extensions synthesized by PNPase are heteropolymeric, containing all four nucleotides (69, 71).

In exponentially growing E. coli, more than 90% of the transcripts are polyadenylated. Rho-dependent transcription terminators were suggested to be modified by the polymerase activity of PNPase, whereas Rho-independent transcription terminators serve as polyadenylation signals for PAP I (72). In spinach chloroplasts, cyanobacteria and *Streptomyces coelicolor*, PNPase was suggested to be the main polyadenylating enzyme (73–75). On the other hand, data from *B. subtilis* suggested that an enzyme other than PNPase is in fact the main polyadenylating enzyme, as 3′ tails are still detected in a *pnp* mutant (76). The archaeal exosome, which is structurally very similar to PNPase (see below), has also been demonstrated to be responsible for the addition of heteropolymeric tails in *Sulfolobus* (77). Due to the ability of PNPase to carry out several distinct activities, the enzyme can be considered as a multifunctional protein. It is a pleiotropic regulator, involved in a number of different pathways of RNA degradation.

Although PNPase accounts for 10% of the exoribonucleolytic activity in E. coli extracts on poly(A) RNA (35), the enzyme seems to play a greater role in mRNA degradation than previously thought. Indeed, inactivation of PNPase leads to increased steady-state levels of more transcripts (17.3%) than inactivation of RNase II (7.3%) (78). The processive degradation of RNA by PNPase is stalled by double-stranded RNA structures (9). A minimal 3′ overhang of 7–10 unpaired ribonucleotides is required for an RNA molecule to be bound by PNPase (79, 80). The action of PNPase on folded RNAs can be stimulated by prior 3′ polyadenylation (9, 79, 81, 82).

C. PNPase Complexes

PNPase can be assembled into multienzyme complexes allowing the enzyme to degrade through extensive structured RNA. The degradosome is the main multiprotein complex that integrates PNPase. Besides PNPase, the major components of the degradosome are RNase E, an endoribonuclease that provides the scaffold for the entire structure, the DEAD-box RhlB helicase and enolase, a glycolytic enzyme (*46, 79, 83–85*) (see the chapter by Carpousis and colleagues, in this volume). The role of RNase E in degradosome assembly has been characterized in detail but little is known about the structural requirements of the other proteins (*86*). An alternative form of the degradosome assembles *in vivo* under conditions of cold shock. In this case, the helicase RhlB is replaced by CsdA, another DEAD-box helicase whose expression is induced by low temperatures (*87*).

PNPase can exist as a $\alpha_3\beta_2$ complex and the β-subunit has been shown to be RhlB (*41*). PNPase binds directly to the C-terminal region of RhlB independently of RNase E (*88*). The unwinding ability of RhlB probably enables PNPase to degrade certain double-stranded RNAs (*88*). PNPase also forms complexes with Hfq and PAP I. These complexes might be PNPase–Hfq–PAP I, PNPase–Hfq, or PNPase–PAP I (*89*). In an Hfq mutant, the biosynthetic activity of PNPase is enhanced, while the PAP I activity is significantly reduced, even though the levels of the two proteins remain unchanged (*89*).

D. PNPase Structure and Function

Pure PNPase is isolated from *E. coli* as a homotrimer of 78 kDa subunits (*90*). The X-ray crystal structure of the highly homologous *Streptomyces antibioticus* PNPase has been solved and recently the structure of *E. coli* PNPase was also solved (*91, 219*) (Fig. 1). Each subunit exhibits a five-domain structure composed of: (i) a structural core, formed by association of two homologous RNase PH domains (the first and the second core domains); (ii) an all α-helical domain that connects the two core domains, located at the bottom of the structure; and (iii) two C-terminal RNA-binding domains, KH and S1, on the top of the enzyme. These are two well-characterized examples of RNA-binding domains present in many enzymes involved in nucleic acid metabolism (*92, 93*). It has been suggested that the function of the PNPase S1 and KH domains in the degradosome is to bind and stabilize single-stranded RNA released by RhlB (*94*). The three subunits associate via trimerization interfaces of the core domains, forming a central channel where catalysis is thought to take place.

The second core RNase PH domain binds tungstate, a phosphate analogue, suggesting that the catalytic center is located in this domain (*91*). The first core domain was suggested to be involved in the additional guanosine pentaphosphate (pppGpp) synthetase activity of *S. antibioticus* PNPase, absent in the majority of PNPases. This region may also have been separately adapted as a second active site in PNPases, generally (*91*). Kinetic studies have established that the processivity of PNPase is in part due to the presence of distinct and separated RNA-binding sites. The two RNA-binding folds present in the C-terminal region of PNPase, although dispensable for detection of catalytic activity (94–96), could play an important role in this processivity. The trimer channel probably contributes both to processivity and to the regulation of PNPase activity by RNA structural elements (91).

The structure of PNPase represented a great advance in the identification of key conserved residues of the PNPase active site, structural elements that might be important for the catalytic activity and the arrangement of the accessory RNA-binding domains. However, our knowledge of the structure–function relationships in this complex enzyme is still fairly rudimentary. Cocrystallization with a substrate bound would be helpful to establish the exact catalytic mechanism of PNPase and how its high degree of processivity is achieved, as well as identifying the RNA binding sites and the RNA pathway inside the enzyme towards the catalytic center.

Several mutations have been introduced into fully phylogenetically conserved residues of E. coli PNPase, taking advantage of the structural information of S. antibioticus PNPase to predict their function (95). Most of the conserved residues are clustered in the second core domain (which contains the putative catalytic center) and in a small part of the first core domain. Only a few are located in the KH or S1 RNA-binding domains. Mutations in amino acids located around the tungstate binding site in the crystallographic structure abolish or severely decrease all catalytic activities of the enzyme, suggesting that these mutations affect the catalytic site directly (95).

Although the all α-helical domain is the least conserved of the five characteristic domains of PNPases (49, 95), it has also been implicated in the catalytic activity of E. coli PNPase (97). This domain connects the first and second core domains. In the homology model of E. coli PNPase (97), the adjacent residues, V304 and V305, are located in helix α8 of the all α-helical domain. Mutation of these residues to Ala and Asp, respectively, is thought to modify the reciprocal positioning of the two core domains, thus affecting the catalytic activity of the mutant enzyme (97). The authors of this study also isolated mutants in adjacent residues of helix α1 of the first core domain. These mutants exhibit a drastic reduction in catalytic activity. Analysis of the S. antibioticus structure (91) shows that helix α1 faces the putative phosphate binding site and could thus be a component of the catalytic center (97). These results, in agreement with previous mutagenesis data (95) and with recent studies on the spinach chloroplast and human PNPases (98, 99), suggest that the catalytic site of E. coli PNPase is probably composed of structural elements in both the first and second core domains. Interestingly, other mutations in the core region were analysed that do not affect phosphorolytic or polymerase activities, but rather RNA-binding is severely impaired. This is the case for a substitution of the conserved residue Gly454 by an aspartate in E. coli PNPase (86).

Mutations affecting the RNA-binding domains, either directly or indirectly, have limited effects on phosphorolysis. PNPase mutants lacking either the S1 or KH domain retain phosphorolytic activity, demonstrating that these RNA-binding domains are accessory, distinct from the domain(s) carrying the active center (94–96). However, determination of the kinetic parameters of the catalytic activities showed that these deletions result in dramatic increases in the K_m, the largest being observed in the double deletion mutant (96). RNA electrophoretic mobility shift assays demonstrated that the presence of both KH and S1 domains are required for a proper binding (96). However, the fact that the truncated derivatives still retain binding ability, although with much weaker affinities, indicates that the catalytic core has intrinsic RNA-binding activity (94).

PNPase mutants lacking the S1 or KH domain, or both, are probably unable to perform multiple cycles of RNA degradation (94). It was suggested that the truncated enzymes stall instead on their substrate, without releasing the degraded product, leading to a severe decrease in the number of molecules processed. A two-step model was proposed to explain these results, according to which the S1 and KH domains indirectly facilitate the activity of PNPase by promoting substrate binding and product release. First, the RNA molecule interacts weakly with the KH and S1 domains and strongly with the catalytic site, where it is subjected to phosphorolysis. This would proceed until a double-stranded helix is encountered by the enzyme. A key role for the S1 domain (acting in concert with the KH domain) would be the binding of another molecule followed by its placement into the active site, which would allow the displacement of the stalled molecule (94).

Superposition of the structures of the S1 domains of PNPase, RNase II, and the model of RNase R highlighted a similar overall structure of the three S1 domains (100). Indeed, the S1 domain of RNase II could be replaced by the S1 domains of PNPase or RNase R without abolishing the exonucleolytic activity of the enzyme. However, the S1 domains of the three enzymes are not equivalent. Strikingly, the S1 domain from PNPase was able to induce the trimerization of the RNase II–PNP hybrid protein, indicating that this domain might have a role in the biogenesis of PNPase multimers. A similar function was suggested for the S1 domain of RNase E (101). However, the role of S1 domain in the trimerization of PNPase is still elusive; while the S1 domain may be promoting the association of the monomers, the amino acids of this region are apparently not involved in the interaction between the monomers in the final PNPase structure (91).

As highlighted above, PNPase is essential for *E. coli* growth at low temperature, a phenomenon that has also been observed in *B. subtilis* and *Yersinia enterocolitica* (58, 59, 60, 102). Certain mutations of both the KH and S1 domains have been shown to confer a cold-sensitive phenotype (96, 97, 103).

Surprisingly, a plasmid expressing a truncated form of PNPase, missing the S1 domain, was able to complement the growth of a *pnp* deficient nonsense mutant at low temperature (*60*). Point mutations or short deletions in these domains seem to be more detrimental than their precise deletion (*96*).

Interestingly, although PNPase lacking the S1 or KH domain, or both, could still be assembled in the degradosome, the truncated forms were not able to participate in the helicase-dependent degradation of the highly structured REP element in the *mal*EF RNA substrate (*94*). Overall, these results suggest that the function of the S1 and KH domains of PNPase in the degradosome is vital at low temperature, when the stem-loop structures present in the target mRNAs are more stable and the efficiency of the degradosome is likely to be critical (*96*). The available data is still scarce and often difficult to interpret, however. PNPase function in the cold is still not clear, and the involvement of the other PNPase domains in the cold-sensitive phenotype has not yet been determined. Even though it is generally accepted that the catalytic activity of the enzyme is required for growth at low temperature, the available experimental data is still largely inconclusive. In this regard, it is surprising that a PNPase derivative, *pnp*-V304A-V305D (see above), whose phosphorolytic activity is strongly inhibited, complements the cold growth defect of the Δ*pnp-751* mutant (an in-frame deletion, encompassing about 95% of the *pnp* coding sequence) (*97*).

E. A Role for PNPase in Virulence

It has been suggested that PNPase modulates the expression of genes involved in bacterial adaptation to environmental changes. Disturbing this regulation leads to altered expression of selected virulence functions and altered pathogenesis (*104*). PNPase is involved in setting up the *Yersinia* type three secretion systems (TTSS) such that a maximal number of Yops (outer membrane protein virulence factors) are exported from the bacterium upon induction (*105*). PNPase modulates the *Yersinia* TTSS by affecting the steady-state levels of TTSS transcripts and controlling the secretion-rate through a mechanism that requires the RNA-binding surface of its S1 domain (*105, 106*). Conversely, in *Salmonella*, *pnp* mutants show an increased capacity to invade epithelial cells and to establish persistent infections in mice (*104*). The genes most affected by *pnp* mutation are contained in two pathogenicity islands, SPI 1 (containing genes for invasion) and SPI 2 (containing genes for intracellular growth) (*104*). The expression of these genes is normally reduced by PNPase.

F. PNPase, An Exosome-Like Enzyme

The structural and functional similarities between the archaeal exosome and the bacterial PNPase are remarkable (*107*) (see the chapter by Evguenieva-Hackenberg and Klug, in this volume). Crystallographic analysis of the archaeal exosome has revealed that the RNase PH, S1, and KH domains are organized into a quaternary structure that is quite similar to that of bacterial PNPase (*108*). Both form ring-like (doughnut-shaped) structures composed of six RNase PH domains surrounding a central channel that can accommodate a single-stranded RNA molecule. The active sites are located on one face of the molecule, where the central channel is open. The putative RNA-binding domains are positioned on the top of the hexamer, on the opposite side to the RNA-degrading sites. Even though the details of substrate recognition might differ, both the archaeal exosome and bacterial PNPase are thought to capture the substrate using their RNA-binding domains and lead the RNA molecule through the channel into the central cavity to reach the active sites (*109*).

The same core structure is present in the eukaryotic exosome: a ring-shaped structure produced by six different RNase PH-like polypeptides, stabilized by a trimer of proteins containing S1 or KH RNA-binding domains (*47, 110*). Surprisingly, although they preserve the doughnut-like shape of the core, the yeast and human exosomes, and possibly eukaryotic exosomes in general, do not display any phosphorolytic activity. The only exonucleolytic activity observed is hydrolytic (*110, 111*); see next section for details.

The emerging picture is that of an evolutionary conserved machine for RNA degradation, pointing to a common basis for RNA-degrading mechanisms in all three domains of life. PNPases and the archaeal and eukaryotic exosomes constitute a unique ring-shaped engine, heavily engaged in the $3'$–$5'$ degradation of RNA. The origin of this machine probably predates the divergence of bacteria, archaea, and eukaryotes.

III. RNase II

E. coli RNase II is the prototype of the widespread RNase II family of enzymes (*20, 27, 112, 113*). RNase II-like proteins are components of the exosome (*110, 111*) (see below), a complex of exoribonucleases involved in the maturation and turnover of RNA (*112*), in RNA interference (*114*), and in surveillance pathways that recognize and degrade aberrant RNAs (*43*).

A. Control of RNase II Expression

RNase II expression is differentially regulated at the transcriptional and posttranscriptional levels. The *E. coli rnb* gene encodes a monomer of approximately 73 kDa. A mutant with a complete deletion of the *rnb* gene is viable (*115, 116*). The *rnb* gene is transcribed from two promoters P1 and P2 and terminates in a Rho-independent terminator 10-nucleotides downstream of *rnb* stop codon (*115, 117, 118*). Both promoters are active in exponential cultures but have different expression levels; P2 is stronger than P1 and accounts for 75% of total expression (118). However, no transcription from the P2 promoter was observed *in vitro*. In *vivo* transcription initiation at P2 might be positively regulated by a cofactor that interacts directly with RNA polymerase and activates transcription (118).

PNPase was shown to modulate RNase II expression by degrading the *rnb* mRNA (56). RNase II expression is also affected at the posttranscriptional level by the endonucleases RNase III and RNase E. In the RNase III mutant, there is a decrease in RNase II levels and activity (119). However, RNase III action does not seem to affect *rnb* mRNA directly; instead the *pnp* mRNA is not properly cleaved in the RNase III mutant and PNPase levels are increased, which in turn affects RNase II expression (see above—PNPase expression). In the RNase E mutant, there is an increase in both levels of *rnb* mRNA and in RNase II activity and this effect is likely to be due to a direct involvement of RNase E in *rnb* mRNA degradation (119).

RNase II is, so far, the only exoribonuclease whose protein stability is known to be posttranslationally regulated. The Gmr (gene modulating RNase II) protein is encoded by a gene located downstream of *rnb* (120). In a Δ*gmr* mutant, there is a 3-fold increase in the RNase II levels and activity (120). RNase II protein turnover is slower in the Δ*gmr* strain (half-life > 2 h) compared to the wild-type (half-life < 60 min), that is, RNase II is more stable in the absence of *gmr* (120). The Gmr protein has a PAS domain, which can act as an environmental sensor detecting changes in growth conditions. It was also shown that RNase II levels are adjusted according to growth conditions, a phenomenon probably due to the action of Gmr (120).

B. RNase II Activity and RNA Degradation

RNase II is a single-stranded, sequence-independent, 3'-exoribonuclease that participates in the terminal stages of mRNA degradation. It is responsible for 90% of the hydrolytic activity on poly(A) RNA in *E. coli* crude extracts (35). It processively degrades RNA in the 3'–5' direction, yielding 5'-nucleoside monophosphates. When the substrate is shorter than ~12-nucleotides, the enzyme becomes distributive (121). The favorite substrate of RNase II is the homopolymer poly(A) and it can rapidly degrade some polyadenylated stretches necessary for degradation by other exoribonucleases *in vivo* (122–124). As a consequence, RNase II can paradoxically act as a protector of some RNAs from degradation by impairing the access of other exoribonucleases (122, 123, 125–128). Moreover, the protective role of RNase II is probably greater than previously thought. Mohanty and Kushner (78) have shown that in the absence of RNase II a large number (31%) of *E. coli* mRNAs are decreased, especially ribosomal protein genes, suggesting a major function for this enzyme in the protection of specific mRNAs through poly(A) tail removal.

Although the processive degradation activity of RNase II is easily blocked by stem-loop structures (9, 129), the enzyme is still able to bind to structured molecules. *In vitro*, both structured DNA and RNA oligos become susceptible to single-strand specific nucleases after RNase II binding (129). Interaction with a duplex presumably weakens or disrupts the double-stranded stem, making it more vulnerable. The fact that RNase II can disrupt short duplexes that can be further degraded by the cooperative action of another enzyme may be of metabolic significance (129).

Study of the mutant strain SK4803 (130) provided the first mutational data on the RNase II activity. This strain carries the *rnb296* allele that encodes an inactive RNase II enzyme. Amblar and Arraiano (131) demonstrated that a single amino acid substitution, D209N, led to complete inactivation of the enzyme without affecting its RNA-binding ability. This amino acid was also proposed to be involved in metal binding (Mg^{2+}) at the active site of the enzyme. These findings provided the first identification of a key residue for catalysis and supplied experimental evidence for the organization of this enzyme into independent functional domains.

C. RNase II Structure and Function

Sequence analysis had already identified the presence of distinctive conserved motifs in all the enzymes belonging to the RNase II family (27) (Fig. 1). These enzymes are large polypeptides predicted to contain one or more RNA-binding domains at the N-terminus. At this extremity, *E. coli* RNase II contains a sequence similar to the cold shock domain (CSD). This region is, however, the most variable both in length and in sequence among all proteins of the family (31, 132). At the C-terminus, the presence of an S1 RNA-binding domain has been proposed for all RNase II-like proteins. The catalytic ability has been assigned to a stretch of about 400 amino acids in the central region, termed the RNB domain (27). Multiple sequence alignments of this domain reveal the presence of four highly conserved sequence motifs (I–IV) containing some invariant carboxylate residues (27).

Experimental evidence concerning the role of these domains in *E. coli* RNase II activity was reported (132, 133). *In vitro* characterization of a set of deletion mutants established the independent nature of the different domains. Although the RNA-binding capability of the CSD and S1 domains was confirmed, the S1 domain seems to be more critical than the N-terminal CSD for the RNA-binding capacity of RNase II. Elimination of the N-terminal CSD resulted in an increase of the RNA-binding affinity of the enzyme for poly(A). This result suggested that the N-terminal CSD domain may carry out an important task in the control of mRNA degradation, by preventing the establishment of a tight complex between RNase II and the poly(A) tails on most prokaryotic transcripts. Such a strong interaction might restrain the movement on the poly(A) chain, thus diminishing the degradation efficiency (132). Interestingly, a mutant protein lacking both the N-terminal CSD and S1 can still binds and degrade RNA. A detailed analysis of its sequence pointed to the presence of an additional RNA-binding fold (OB-fold, named CSD2) in the N-terminal part of the RNB core (132).

Amblar et al. (100) performed functional studies on the S1 domain through the use of chimeric proteins, in which the RNase II S1 domain was replaced by the S1 domains from PNPase or RNase R. Both S1 domains were able to partially restore the RNA-binding ability and the exonucleolytic activity of an enzyme derivative lacking the S1 domain. Nevertheless, the DNA-binding capability of RNase II was decreased in the RNAII–R hybrid protein. As mentioned earlier, the RNase II–PNPase hybrid protein oligomerized as a trimer suggesting that the S1 domain from PNPase is involved in the biogenesis of multimers (see also PNPase structure and function).

Frazão et al. (28) reported the first X-ray crystallographic structure of the E. coli wild-type RNase II, as well as the structure of a point mutant (D209N) (see above) that fortuitously cocrystallized with a 13-nucleotide RNA molecule. The structure of E. coli RNase II was the first structure of an exoribonuclease from the RNase II family and completed the group portrait of the three 3′ exoribonuclease families, RNase II, DEDD, and PDX (113). The overall structure reveals four domains, as previously predicted by Amblar et al. (132). Three of these domains are RNA-binding (two CSDs and one S1 RNA-binding fold) and are grouped together on the bottom of the structure, forming an "anchor site". On the top of the structure is the active site, buried within the RNB catalytic domain (28).

The structure has unraveled crucial aspects of the mechanism of RNA degradation (Fig. 1). The RNA fragment interacts with the protein at two noncontiguous regions, the anchor and catalytic regions (121). Nucleotides 1–5, at the 5′-end of the RNA fragment, are located in the anchor region in a deep cleft between the two CSDs and the S1 domain. The final nucleotides 9–13 are located in a cavity deep within the RNB domain, stacked and clamped between conserved residues Phe358 and Tyr253. A 10-nucleotide fragment is the shortest RNA able to retain contacts with both anchor and catalytic regions. Since it has already been demonstrated that processivity is achieved by the presence of two independent sites for substrate binding (121), this fact explains why RNase II is processive on long RNA molecules but becomes distributive on substrates shorter than 10–15 nucleotides. The structure also explains the inability of RNase II to degrade double-stranded RNA. In fact, access to the catalytic pocket is restricted to single-stranded RNA by steric hindrance at its entrance.

RNase II only degrades RNA and not DNA. The RNA specificity is not for the nucleotide removed at the 3′ end, but for nucleotides that are just upstream of the terminal nucleotide (121). The structure revealed that the ribose rings of nucleotides, 8, 10, and 12 directly interact with conserved residues Glu390, Asp201, and Tyr313. These interactions might be responsible for the correct orientation of RNA at the catalytic site, thus allowing RNase II cleavage.

The model for RNA degradation involves one Mg^{2+} ion, four highly conserved aspartate residues (D201, D207, D209, and D210) in the catalytic pocket and the recruitment of a second Mg^{2+} for catalysis. However, these residues are not equivalent and their functions in RNA metabolism are distinct. Asp209 is the only essential residue, and Asp207 plays the least critical role for RNase II activity (134). As expected, mutations in Asp201 and Asp210 lead to a significant loss of RNase II activity. These residues have been proposed to coordinate the Mg^{2+} ion and their substitution would result in loss of one of the coordinations, affecting catalysis (134). After cleavage, as nucleotide 13 leaves the protein, the RNA translocates and the following nucleotide is prepared for the next cleavage event. When the RNA molecule is shorter than five nucleotides, the required packing of the bases can no longer occur, preventing the translocation of the RNA and generating a final end product of four nucleotides (28). Interestingly, changing Tyr253 to Ala, altered the smallest end product of degradation from 4 to 10-nucleotides (134). This led to the identification of the residue responsible for setting the size of the RNase II end-product. This mutation has been proposed to cause loosening of the RNA substrate at the catalytic site and, as a consequence, binding at the anchor region would be essential to keep the RNA attached to the protein and allow cleavage. Molecules shorter than 10-nucleotides would no longer be able to bind simultaneously to both sites and would thus be released.

The crystal structure of *E. coli* RNase II was also subsequently reported by Zuo and collegues (29). These authors noticed that a small opening in the active center cavity could likely serve as an exit port for the mononucleotide product. The fact that it lies in an acidic surface of the enzyme might preclude substrate entry via this path.

D. RNase II-like, Rrp44/Dis3: The Catalytic Subunit of The Eukaryotic Exosome

Rrp44/Dis3 protein, highly conserved from yeast to humans, is an essential RNase II-like 3′–5′ exoribonuclease that also processively hydrolyses single-stranded RNAs (Fig. 1). However, it also degrades RNAs containing secondary structures. This protein is part of the eukaryotic core exosome (30, 110, 112). Although six of the core exosome subunits have the bacterial RNase PH fold, recent reports have shown that the only exonucleolytic activity of the human and yeast exosome core resides in Rrp44 (110, 111). Lorentzen et al. (30) reported the crystal structure of yeast Rrp44ΔN in complex with single-stranded RNA. This protein lacked the predicted N-terminal PIN domain that is characteristic of Rrp44 orthologs, but absent from RNase II and RNase R. Recently it was demonstrated that the PIN domain has endonucleolytic activity both *in vivo* and *in vitro* (220). The activity of the PIN is sensitive to the phosphate status of the 5′end and this fact might help coordinate 5′ and 3′ degradation. The binding interactions at the catalytic site are essentially identical between Rrp44 and RNase II, supporting the view that these exoribonucleases share the same mechanism for 3′ binding and hydrolysis. The major difference between Rrp44 and RNase II lies in the relative orientation of the RNA-binding domains and the path taken by the RNA. Mutation of a key residue (Ala815) on the alternative RNA path had little or no effect on degradation of single-stranded RNA, but strongly affected the degradation of RNA duplexes. These results confirmed the importance of the different entry route of the RNA into the catalytic center for Rrp44's ability to unwind structured substrates (30). However, the crystal structure solved did not include the N-terminus of the protein, containing the PIN domain, which may also contribute to the unwinding of stable RNA molecules (30).

Interestingly, the RNase activity of the yeast Rrp44 is tightly regulated and decreases significantly upon association with the exosome core. A pseudoatomic model of the yeast exosome, based on EM images, was recently presented by Wang *et al.* (135). The model showed that the active site of Rrp44 is occluded inside the Rrp44 body and that the entrance to the recruitment channel is not directly accessible. The presence of the other core subunits was thus, proposed to downregulate the Rrp44 activity by sterically restricting the RNAs access to the nucleolytic active site. This modulation effect is especially notable for stem-loop containing RNA molecules and might be a regulatory mechanism that prevents unwanted degradation of stable RNA complexes (30).

Other RNase II homologues include the *Droshophila* Tazman, a developmentally regulated exoribonuclease (31). Mutations in RNase II-like enzymes have also been linked with abnormal chloroplast biogenesis (136), mitotic control, and cancer (137) and defects in kinetochore formation (138).

IV. RNase R

RNase R was initially identified as an enzyme merely responsible for residual hydrolytic activity in a mutant for RNase II (139). Only later was exoribonuclease R identified as the ~92 kDa protein product of the *rnr* gene (previously called *vac*B) (57). RNase R has since been shown to be quite important in RNA metabolism. It is highly effective against structured RNA and has recently been described to be relevant in RNA quality control, and in the processing and degradation of several RNA substrates (8, 16, 19, 80, 136, 140–142). RNase R belongs to the RNase II family of exoribonucleases (28, 113).

A. Control of RNase R Expression

The *rnr* gene is second in an operon together with *nsr*R (a transcriptional regulator), *rlm*B (rRNA methyltransferase), and *yjfI* (unknown function). Transcription is driven from a possible σ^{70} promoter upstream of *nsr*R (*57, 140*) but its transcriptional regulation remains unexplored. RNase R levels change according to different environmental stimuli. Like PNPase, RNase R is upregulated in cold shock and this was shown to be the consequence of a significant stabilization of the *rnr* transcripts at low temperatures (*140*). RNase R expression is controlled at the posttranscriptional level with the participation of other RNases. Processing of *rnr* transcripts is mainly achieved by RNase E, although RNase G may also participate (*143*). PNPase was also shown to participate in this regulation at the end of the acclimation phase (*140*). RNase R and RNase R RNase II double mutants are viable but a RNase R PNPase double mutant is lethal (*57, 140*). *E. coli* RNase R was shown to be important for growth at low temperatures and RNase R-deficient colonies are smaller especially in the cold (*140*). This enzyme is essential for survival at low temperatures in the psychrotrophic *Pseudomonas syringae* (*144*). Not surprisingly, being the only exoribonuclease present, RNase R is also essential in *Mycoplasma* (*20, 145*).

B. RNA Degradation and Mode of Action

RNase R acts as a monomer and requires a monovalent cation and Mg^{2+} to be active (*36, 142*). RNase R is a processive, sequence nonspecific 3′–5′ exoribonuclease that acts through a hydrolytic mechanism releasing nucleoside monophosphates. Degradation is processive as it does not accumulate intermediary fragments and, unlike RNase II, the final end product of digestion is a dinucleotide. The most distinctive feature of RNase R is its unusual effectiveness on RNA molecules with secondary structures, without the aid of a helicase activity. RNase R is able to degrade a duplex RNA provided there is a single-stranded 3′ overhang of more than 7-nucleotides in length, with optimal binding and maximal degradation rates obtained with a linear region of 10-nucleotides (*36, 142*). Unlike RNase II or PNPase, RNase R tightly binds the RNA and does not detach when approaching a double-stranded RNA region, proceeding very efficiently through even extensive secondary structures (*80, 142*). Recognition of the substrate requires ribose moieties because RNase R is unable to bind and degrade a DNA substrate (*142*). Together with the findings of *in vitro* and *in vivo* degradation of the highly structured rRNA and tRNA, it seems likely that the major role of RNase R *in vivo* is the degradation of structured RNAs (*7, 8, 36*).

RNase R has also emerged as an important novel contributor to mRNA degradation. The absence of both RNase R and PNPase results in the strong accumulation of REP-containing mRNA sequences. However, the presence of only one of these exoribonucleases is sufficient to remove such transcripts. Interestingly, RNase R is also thought to degrade the REP-free ompA transcript (19) (see below). This is, for now, the only reported example of a transcript stabilized specifically in the single absence of RNase R. The reason why RNase R drives degradation of the ompA mRNA is still unclear.

C. RNA Quality Control

RNase R plays a very important role in RNA quality control (7, 8), such as in the degradation of defective tRNAs (7, 142). Interestingly, this resembles the degradation of the presumably misfolded hypomodified $tRNA_i^{Met}$ by Rrp44 (146). RNase R, together with PNPase, eliminates aberrant fragments of 16S and 23S rRNA which accumulate to very high level in the absence of both enzymes, probably affecting ribosome maturation and assembly (8) (see the chapter by Deutscher, in this volume).

In *Mycoplasma*, RNase R has been shown to catalyze a broad range of RNA reactions including degradation and processing events (145). It has been suggested that *Mycoplasma* RNase R is also involved in RNA quality control mechanisms. Ribose 2'-O-methylation of stable RNA molecules, such as 23S rRNA, may act as a quality signal preventing degradation (145). Since RNase R has been shown to be sensitive to ribose modifications, RNA molecules that fail to be methylated may be considered defective and become substrates for RNase R.

D. Protein Quality Control

RNase R was copurified with a ribonucleoprotein complex that contained SmpB protein and the structured noncoding tmRNA/SsrA, a unique bifunctional RNA molecule that displays both tRNA and mRNA features (147). This complex coordinates a translational control process named *trans*-translation (for a comprehensive review see (148)). It recognizes and rescues stalled ribosomes on mRNAs lacking stop codons, translation of which may result in toxic truncated proteins (149). The SmpB·tmRNA complex directs the addition of a C-terminal tag to incomplete proteins that targets them for proteolysis. Moreover, the tmRNA system also targets the aberrant mRNAs for degradation by RNase R, thus avoiding the accumulation of defective proteins arising from new translational cycles (16).

The role of RNase R in the accuracy of gene expression is broadened by the finding that, in the absence of RNase R, the small stable SsrA/tmRNA is itself not properly processed, leading to defects in *trans*-translation (140). In fact, *rnr* mutants show significant defects in protein tagging for proteolysis (147). Under cold-shock conditions, the steady-state levels of the tmRNA precursor are increased 3-fold in the *rnr* mutant strain compared to wild-type cells (140). RNase R was also shown to affect tmRNA in other organisms, although in different ways. Interestingly, *Caulobacter crescentus* tmRNA is specifically degraded at a specific point in the cell cycle (*150*). On the other hand, tmRNA fragments, rather than its precursor or mature form, are stabilized in an *rnr* mutant in *P. syringae* (*144*). It has been suggested that the cold lethality phenotype associated with the *rnr* mutant in this strain may be caused by accumulation of these fragments.

E. Stress and Stationary-Phase

The large majority of studies on RNA degradation are performed under experimental laboratory conditions that clearly favor the analysis of cultures in the exponential phase of growth. However, in nature, bacteria have to cope with many different stresses and rarely live at such a high metabolic rate as those typically found during growth in rich media and at defined temperature. Instead, in the wild, bacteria can have drastically reduced active growth, and experience morphological and physiological changes that result in a state of dormancy known as stationary phase (*151, 152*). Adaptation to a challenging environment requires a rapid adjustment in RNA levels, requiring not only transcriptional regulation, but also fine-tuning control of RNA stability. In stationary phase, transcription is essentially controlled by the alternative sigma factor σ^s (encoded by *rpo*S), leading to the expression of stationary-phase-specific genes with the concomitant repression of the highly expressed genes of exponentially growing cells (*153*).

The degradative ribonucleolytic machinery in the cell also responds to environmental changes. For example, polyadenylation in stationary phase is characterized by the addition of heteropolymeric tails instead of the typically homopolymeric tails found in exponentially growing cells (*154*). RNase R is growth-phase regulated and its levels increase in the stationary phase of growth (*19*). This work demonstrated that RNase R can be a modulator of gene expression in stationary-phase cells (*19*). Remarkably, RNase R seems to play a central role in nature where bacteria spend most of the time in stationary phase.

Stationary-phase cells have to become resistant to a variety of environmental hardships including exposure to oxidative stress, abrupt temperature changes, and nutrient depletion. Factors that contribute to stress-resistance are crucial to cell adaptation. In fact, *E. coli* RNase R seems to be a general stress-induced protein whose levels are not only upregulated in stationary phase (2-fold), but also in response to heat shock (~2-fold) and cold shock (7–8-fold) (*19, 124, 140*). In mesophilic bacteria like *E. coli*, the cold shock response is triggered by a sudden temperature downshift (e.g., 37–10 °C), which causes the overexpression of a minor set of proteins concomitant with the transient repression of bulk protein synthesis (*155*). From the different stresses analyzed, cold shock treatment results by far in the highest upregulation of RNase R, to higher levels than the cold-induction of PNPase (63, 140). Such a large increase in the amount of RNase R suggests a highly important role for this exoribonuclease in these conditions.

In cold shock, some mRNAs become much more stabilized than others (140, 156, 157). This stabilization is believed to be highly dependent on the formation of secondary structures, which are less thermodynamically stable at higher temperatures. It seems reasonable to postulate that degradation of highly structured RNAs becomes even more important under such conditions. RNase R, with its remarkable intrinsic characteristics, seems to be the enzyme of choice to degrade such extensively structured RNAs. Indeed, RNase R overexpression complements the cold-sensitive phenotype of the cold shock helicase CsdA mutant (87, 158). Moreover, RNase R is the exoribonuclease that associates within the degradosome in the psychrotrophic bacterium *P. syringae* (159), instead of PNPase (42). RNase R seems thus to have a central role in the cellular adaptation to new stressful conditions. However, the mechanisms underlying RNase R upregulation in respsonse to different stresses are not known.

F. A Role for RNase R in Virulence

Stress resistance and virulence are intimately related. Many pathogenic bacteria have to cope with very harsh environments during the process of infection. Not surprisingly, the major regulators of stationary-phase physiology are also major regulators of virulence (160–162). RNase R has been implicated in the establishment of virulence in a growing number of pathogens including *Shigella flexneri*, enteroinvasive *E. coli*, and *Aeromonas hydrophila* (163, 164). RNase R-deficient bacteria have been shown to be less virulent than the wild-type parental strains. However, how this is achieved is still obscure. It has been suggested that RNase R may control the export of proteins involved in pathogenesis (163), as has been reported for PNPase (see PNPase section). However, the role of RNase R in disease is still controversial, as it does not seem to be a universal trait of pathogens. A recent report suggested that RNase R is dispensable for virulence of the mammalian intracellular pathogen *Brucella abortus* (165). Maybe differences in the mode of action of RNase R from *B. abortus* could explain this striking difference. Furthermore, as mentioned earlier, RNase R is the only exoribonuclease present in the pathogen *Mycoplasma* (145). Altogether, the available data suggests that bacterial RNase R may be attractive as a potential therapeutic agent, but clearly more studies are required.

G. Insights into The Structure and Function of RNase R

All exoribonucleases from the RNase II family share a similar modular organization of domains: an N-terminal region with two CSDs (CSD1 and CSD2), a central catalytic RNB, and a C-terminal S1 domain (Fig. 1). Based on the structure of RNase II, it has been proposed that the RNA-binding domains of RNase R can assume a more open disposition and direct the RNA towards the narrow catalytic pocket (28, 29, 134, 142). However, the widening of the anchoring site is not sufficient to explain the ability to unwind RNA duplexes. The catalytic domain probably only accommodates single-stranded RNA, explaining the requirement for a 3′ linear overhang to initiate degradation. What drives the unwinding of the RNA? It has been suggested that the energy released upon hydrolysis of the RNA can be used to force translocation of the substrate into the tight catalytic region causing it to unwind (142). However, we do not currently know how this may happen.

The RNA-binding region seems to play an important role and might somehow have the intrinsic ability to melt RNA duplexes. The RNase R S1 domain by itself does not seem to be the determinant for degradation of structured RNA (100). Could the uniqueness of RNase R rely on the other RNA binding motif, the CSD? Interestingly, an additional nucleic acid binding motif (Helix-turn-Helix) is predicted in the proximity of the CSD at the N-terminus. It is also possible that the catalytic domain holds the key for explaining RNase R action. Most of the residues that interact with the 3′ end of the RNA substrate are conserved throughout the RNase II family of exoribonucleases, suggesting a similar mechanism for hydrolysis amongst its members (30, 134). Structural differences might help explain the divergence in substrate recognition between these enzymes (30, 134).

The first model for RNase R was recently proposed (134) and structural differences with RNase II might help explaining the differences in substrate recognition between these enzymes (30, 134). The RNase R ortholog, Rrp44, can also digest structured RNA once a 3′ overhang is provided (30, 110, 146). It has been suggested that RNase R and Rrp44 might have a similar RNA alternative pathway, and this could be responsible for the degradation of double-stranded RNA. Nevertheless, the results did not lead to a clear explanation for the mechanism of RNase R (30). Clearly, resolution of RNase R structure is the key to fully understanding its remarkable mode of action.

V. Oligoribonuclease

The end products resulting from the degradation of PNPase, RNase II, and RNase R constitute a serious problem for cell viability, since these enzymes release RNA fragments 2–5 nucleotides in length whose accumulation may be deleterious (166) (see the chapter by Danchin, in this volume). The first time that the existence of an enzyme that degraded short oligoribonucleotides was suspected was in 1967 (167). Eight years later, Niyogi and Datta (168) called the enzyme "oligoribonuclease" because of its specificity for hydrolyzing short oligoribonucleotide chains (168). Genetic and biochemical studies established oligoribonuclease as a distinct enzyme from RNase D, BN, T, PH, and R. It has been reported to copurify with PNPase (169). The oligoribonuclease coding gene, *orn*, is essential in *E. coli* (166). From the known exoribonuclease genes in *E. coli* this is the only exoribonuclease required for cell viability (166).

Oligoribonuclease belongs to the DEDD family of exoribonucleases and contains a well-conserved ExoIII domain (20). It is a homodimeric ($\alpha 2$) enzyme and each monomer weighs $\sim$ 20 kDa (170). This enzyme produces mononucleotides and requires the presence of divalent cations (preferably Mn^{2+}) (168). Hydrolysis is processive in the 3'–5' direction, since the enzyme acts by degrading one oligoribonucleotide chain to completion before proceeding to the hydrolysis of another chain. Oligoribonuclease has a higher affinity for a 5-mer oligoribonucleotides than smaller substrates. However, the hydrolysis reaction rate also decreases with the increasing chain length (171). Longer RNA fragments are degraded to molecules of shorter size, which are then degraded at faster rates. The enzyme requires a free 3'-OH end, but is not sensitive to the 5' phosphorylation state of the RNA. Oligoribonuclease was shown to act very efficiently on oligonucleotides bearing 5'-ribonucleoside triphosphates. Similar results were obtained with oligonucleotides bearing 5'-ribonucleoside diphosphates and monophosphates (171).

Oligoribonuclease has close homologues in other organisms (170) such as the human protein Sfn, which exhibits a 3'–5' exoribonuclease activity on small single-stranded RNA and DNA oligomers (172), and the XC847 protein from the plant pathogen *Xanthomonas campestris* (33). Oligoribonucleases are inhibited by the nucleotide 3'-phosphoadenosine-5'-phosphate (pAp) that is generated in both prokaryotes and eukaryotes during the process of sulfur assimilation (173). *B. subtilis* does not have a homologue of this enzyme (173). However, a potential functional analogue of Orn, called YtqI, has been identified through its binding to pAp. *B. subtilis* YtqI can complement a conditional *orn* mutant in *E. coli* when expressed at similar levels. Orn and YtqI employ different mechanisms for the degradation of oligoribonucleotides, since YtqI shows *in vitro* preference for 3-mers, but is unable to act on a 5-mer substrate. In contrast with Orn and Sfn, YtqI is not inhibited by pAp and can degrade this nucleotide due to its pAp-phosphatase activity. Whereas Orn is essential in *E. coli*, YtqI is not essential in *B. subtilis*. This observation points to the existence of at least one more enzyme in *B. subtilis* able to degrade oligoribonucleotides.

The crystal structure of the XC847 oligoribonuclease from *X. campestris* has been solved (174) (Fig. 1). In this organism, one Mg^{2+} ion was found to be well coordinated with the active site carboxylate groups of Asp15, Glu17, and Asp166 (174). Only the preliminary X-ray characterization of the *E. coli* Orn structure has been reported (175). Although, it was initially proposed that oligoribonuclease does not attack deoxyribonucleotides (171), it was recently shown that Orn can degrade short DNA oligos like its human homologue Sfn. However, this degradation requires higher enzyme concentrations than the RNA-directed activity. Interestingly, this highlights a possible role for Orn in DNA repair, like its human counterpart (173). In RNA metabolism, oligoribonuclease acts as the "finishing enzyme" to degrade the oligoribonucleotides of 2–5 nucleotides in length to mononucleotides in a wide range of organisms.

VI. Mechanisms of RNA Degradation

The interplay between the different factors involved in RNA decay emphasizes the role of exoribonucleases in the degradation of many substrates. The construction of exoribonuclease mutants has been most helpful in deciphering the multiple RNA degradation pathways coexisting in a cell (3, 176, 177). The intrinsic characteristics of both the enzymes and RNA substrate seem to control the degradation of individual RNAs. However, some common features are evident from the analysis of different RNA degradation pathways. Polyadenylation emerges as an important factor controlling exonucleolytic activity (40). Perturbation of RNA structural features may also work as an efficient degradation signal. Relaxation of secondary structures may result in an easier accessibility of RNases, namely exposing the 3′ RNA end to exonucleolytic attack. In this section, we illustrate the most relevant examples of RNA degradation mechanisms in *E. coli* (Fig. 2).

A. *pyr*F-*orf*F

*pyr*F-*orf*F constitutes a dicistronic transcript of 1146 nucleotides in length that contains *pyr*F (orotidine-5′-monophosphate decarboxylase) and an open reading frame (*orf*F) encoding a polypeptide of unknown function (178–180). The full-length transcript is rapidly cleaved into a series of breakdown products (181). At least 18 endonucleolytic cleavage sites have been mapped throughout the full-length mRNA (181). The decay pattern observed in the wild-type and in a triple mutant ($rne\text{-}1^{ts}$ $pnp7^-$ $rnb500^{ts}$) is quite similar, but breakdown products are stabilized in the mutant, showing a role for RNase E, PNPase, and RNase II in this decay. The *pyr*F-*orf*F transcript can be degraded by more than one enzymatic pathway including both 5′–3′ and 3′–5′ decay. Which pathway is employed may be related to the particular context in which one or more of the decay-mediating factors has access to the mRNA.

B. *trx*A

The *trx*A gene, which encodes for thioredoxin, is transcribed as a monocistronic message of 493 nucleotides. Several *rne*, *pnp*, and/or *rnb* mutant strains were constructed in the study of the *trx*A decay (176). Northern and

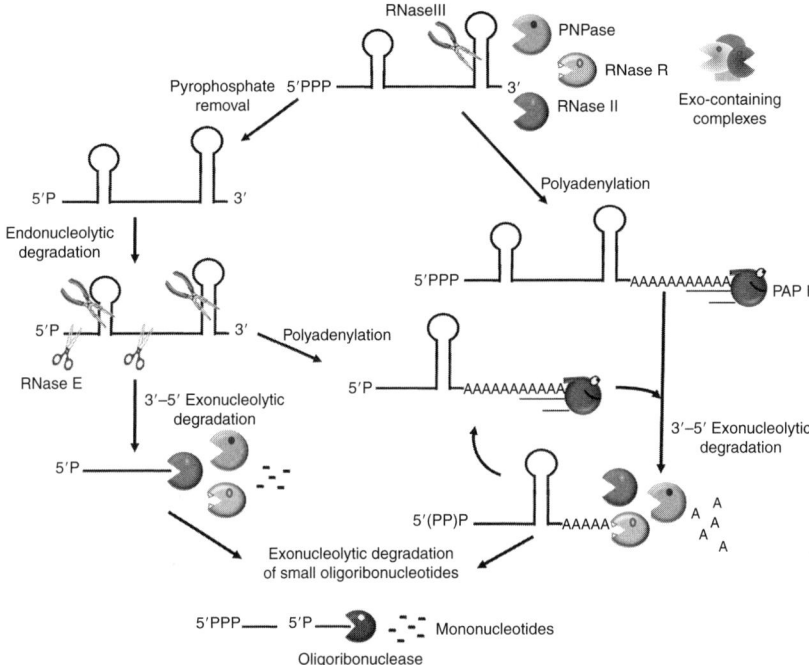

FIG. 2. Schematic representation of major RNA degradation pathways in *E. coli*. Primary transcripts often possess RNA secondary structures at the 5′ and/or 3′ extremities. RNase E endonucleolytic cleavage is responsible for the initiation of decay in the majority of transcripts. Pyrophosphate removal by RppH at the 5′ RNA end is the rate-limiting step for RNase E, which prefers a monophosphorylated 5′ end (*11, 12*). Whether this is a limiting step affecting the activity of other endonucleases (like RNase III) or exoribonucleases, is at the moment unknown. Following endonucleolytic cleavages, the linear transcripts are rapidly degraded by the 3′–5′ degradative exoribonucleases, RNase II, RNase R, and PNPase. A minor pathway in the cell is the exonucleolytic degradation of full-length transcripts. The short 3′ overhang usual found downstream of the transcription terminator can be extended by poly (A) polymerase (PAP I). Polyadenylation promotes RNA degradation by providing a toe-hold to which exoribonucleases can bind. Cycles of polyadenylation and exonucleolytic digestion can overcome RNA secondary structures. RNase R, unlike RNase II and PNPase, is efficient against highly structured RNAs. PNPase in association with other proteins, namely RNA helicases, can also unwind RNA duplexed. The small oligoribonucleotides (2–5 nucleotides) released by exoribonucleases are finally degraded to mononucleotides by oligoribonuclease.

S1 analysis showed that full-length transcripts are initially processed by endonucleolytic cleavages. Complete degradation of the initially cleaved transcripts occurs through possible additional endonucleolytic cleavages in the 3′–5′ direction, followed by exonucleolytic degradation by RNase II and PNPase. This was the first report of a progression of endonucleolytic cleavages in a 3′–5′

direction during the degradation of a full-length transcript. The triple RNase E^{ts} PNPase$^-$ RNase II^{ts} mutant shows discrete breakdown products of *trx*A mRNA that are dramatically stabilized compared to wild-type (*182*). Moreover, the *trx*A mRNA is still degraded despite the absence of RNase E, PNPase, and RNase II, indicating that other RNases are also involved. More recently, it was shown that increased polyadenylation led to a decrease in the half-life of *trx*A mRNA, showing a direct correlation between total *in vivo* poly (A) level and *trx*A mRNA stability (*14, 183*).

C. *rps*O

The *rps*O gene encodes the ribosomal protein S15 (*184, 185*). It is transcribed either as a monocistronic P1-t1 message terminated at the Rho-independent transcription terminator t1, or as a bicistronic *rps*O-*pnp* transcript (*51, 184, 185*). The intrinsic terminator stem-loop protects *rps*O mRNA from attack by exoribonucleases (*125*).

The stability of the *rps*O transcript is mainly controlled by RNase E, which catalyzes the initial nucleolytic cleavages at two sites, one located on each side of the transcriptional terminator (*125, 186, 187*). After RNase E cleavage, the mRNA lacking the 3′ terminal RNA secondary structure becomes an ideal substrate for PNPase (*187*). The RNA chaperone Hfq can protect the *rps*O mRNA from exonucleolytic degradation by PNPase and RNase II, and from cleavage by RNase E (*17*).

Exonucleolytic poly(A)-dependent degradation of *rps*O mRNA is stimulated when the primary pathway of decay mediated by RNase E is inactive (*13, 18, 128*). PAP I synthesizes destabilizing poly(A) tails downstream of the 3′ hairpin of the primary transcript, but also destabilizes breakdown fragments (*18, 188, 189*). Multiple rounds of polyadenylation and exonucleolytic digestion can overcome RNA secondary structures and complete the decay (*190–192*).

RNase II protects the full-length *rps*O mRNA from degradation by removing the poly(A) tails (*122*). It was recently shown that RNase R is the main enzyme involved in the poly(A)-dependent degradation of the *rps*O mRNA (*124*). Elongated *rps*O transcripts harboring poly(A) tails are specifically recognized by RNase R and strongly accumulate in the absence of this enzyme. The 3′ oligo(A)-extension may stimulate the binding of RNase R, allowing the complete degradation of the mRNA (*124*).

D. *rps*T

The *rps*T gene encodes the ribosomal protein S20. This gene is transcribed from two promoters (P1 and P2) and terminates at a Rho-independent terminator, leading to two monocistronic mRNA species, P1 with 447 nucleotides and P2 with 356 nucleotides (*193*). The first step in *rps*T mRNA decay is carried out by RNase E. Several lines of evidence suggest that this step is independent of polyadenylation. First, the cleavage of the *rps*T mRNA by RNase E *in vitro* is independent of ATP or PAP I (*191, 194*). Second, the absence of polyadenylation does not significantly affect either the steady-state levels or the stabilities of the *rps*T primary transcripts (*192*).

PAP I, PNPase, ATP, and phosphate are necessary to catalyze the degradation of the smaller intermediates generated by RNase E which retain the stable terminator stem-loop (*192*). A 147-nucleotide intermediate, called P0, strongly accumulates in strains lacking PAP I (*192, 195*). The poly(A)-dependent exoribonucleases PNPase and RNase R were shown to control the levels of this polyadenylated fragment (*124, 192, 195*). RNase II protects this fragment from exonucleolytic decay, as reported for the *rps*O mRNA (*122, 192*). In contrast, the steady-state levels of the full-length *rps*T transcripts are almost equal in exoribonuclease mutant strains compared to wild-type.

E. *mal*EF

The polycistronic *mal*EFG operon of *E. coli* encodes three proteins involved in the transport of maltodextrins. The *mal*EF intercistronic region contains two REP sequences (*5*) which protect the transcript from 3′–5′ exonucleolytic degradation (*37*). RNase E cleavage is required for *mal*E/*mal*F mRNA maturation (*195*). After RNase E activity, small RNA fragments with the *mal*E 3′REP-stabilizer arise. RNase R and PNPase are shown to have a major role in the degradation of these fragments (*80, 195*). PNPase degradation of *mal*EF transcript is only accomplished in the presence of RNase E and RhlB, indicating that the degradosome participates in this degradation (*94*). RhlB unwinds the folded RNA and passes it to PNPase (*195, 196*). PAP I facilitates degradation of the *mal*EF REP sequences which accumulate to high levels in ΔpcnB mutants (*195*).

F. *omp*A

OmpA is a major protein of *E. coli* outer membrane with important structural and stress-resistance roles. The *omp*A gene is transcribed as a monocistronic mRNA (*197*). It was previously demonstrated that *omp*A stability is growth rate-dependent and that shorter generation times in exponential phase corresponded to longer *omp*A mRNA half-lives (*198*). The *E. coli omp*A 5′ UTR contains stable stem-loops (*199*). An RNase E cleavage event in this region initiates the decay of this mRNA (*19, 200–202*). *In vitro* experiments showed that 30S ribosome subunits protect the *omp*A 5′ UTR from RNase E cleavage (*202*). Exonucleolytic degradation seems to account for the elimination of breakdown products. Indeed, the presence of only one of the exoribonucleases (RNase II, RNase R, or PNPase) may be sufficient to remove most of the decay intermediates (*80*). This is dependent on polyadenylation, as PAP I activity accelerates *omp*A mRNA decay (*14, 19, 72, 183*). Interestingly, exonucleolytic activity on the full-length *omp*A transcript was shown to be growth-phase regulated. PNPase and RNase R affect the stability of the *omp*A primary transcript only in the stationary phase. Fragments harboring intact 3' RNA ends accumulate in RNase R-deficient strains. Moreover, the increased levels of *omp*A mRNA found in the *rnr* mutant result in the upregulation of OmpA (19).

Recent studies demonstrated that the small noncoding RNA, MicA, is in fact the principal posttranscriptional regulator of the *omp*A expression (203). MicA RNA downregulates the steady-state levels of the *omp*A mRNA and the synthesis of OmpA protein (203, 204). Interestingly, the riboregulator MicA RNA is also under the control of exoribonucleases, namely in the stationary phase of growth (see below).

G. Small Noncoding RNAs

Exoribonucleases also have a very important role in the control of small noncoding RNAs (sRNAs). These RNAs have received considerable attention recently, since they play very important regulatory roles (205, 206). To better understand the action of sRNAs, it is important to study the processing and turnover of these molecules.

PNPase has emerged as a major exoribonuclease involved in the degradation of noncoding RNAs. In *E. coli*, the chromosomal encoded MicA and RybB sRNAs (which regulate outer membrane protein expression) are destabilized by PNPase, specifically in stationary phase (203, 207–209). This was shown to occur in a degradosome-independent manner. Interestingly, polyadenylation of MicA by PAP I appears not to be crucial for PNPase activity on this sRNA (209). Moreover, MicA decay appears to be independent of the target mRNA, since PNPase still degrades this sRNA in the absence of *omp*A. RNase II and RNase R appear not to be required (209). In *Salmonella typhimurium*, the sRNAs MicA, SraL, CsrB, and CsrC are also mainly degraded by PNPase. In the case of CsrB and CsrC, the absence of this exoribonuclease causes a change in degradation patterns, with the accumulation of several decay intermediates harboring the same 5' ends and different 3' ends (210).

Plasmid encoded antisense RNAs are major regulators of plasmid copy number. In almost all cases, these sRNAs are constitutively synthesized and rapidly turned over. The antisense RNA, CopA, inhibits replication of plasmid R1 by binding to a region located within the *rep*A mRNA, blocking the synthesis of the replication initiator protein RepA (211). The decay of CopA is initiated by an endonucleolytic cleavage (RNase E) followed by addition of poly(A) tail (212). As in other examples, the presence of RNase II decreases the rate of degradation of transcripts that are substrates of PNPase (122, 124).

RNAI is the copy number regulator of the plasmid ColE1. It hybridizes with the larger RNAII preventing its maturation to the primer used to initiate ColE1 replication (213). PNPase, PAP I, RNase E, and RNase III have been demonstrated to play roles in RNAI decay (*81, 214–216*). Two degradation pathways have been suggested for this RNA (*216*). The primary pathway starts with RNase E cleavage, followed by PAP I polyadenylation and PNPase-mediated degradation. The second mechanism begins with the adenylation of RNAI, followed by RNase III cleavage and subsequent exonucleolytic attack. In the absence of RNase E, RNase III, and PAP I, the antisense RNAI continues to disappear, showing that yet other enzymes are able to catalyze its decay.

Replication of ColE2 plasmid requires a plasmid encoded replication protein (Rep). RNAI from ColE2 controls *rep* expression by inhibition of translation (*217*). ColE2 RNAI degradation also starts with RNase E cleavage at a site close to the 5′ end. PAP I polyadenylates the 3′ ends of degradation intermediates and both RNase II and PNPase act in the further exonucleolytic degradation. A recent report (*218*) performed an interesting analysis of RNAI turnover in parallel with its *rep* mRNA target. The mechanism of *rep* mRNA degradation presents two main differences compared to RNAI degradation. Firstly, PAP I makes a minor contribution to the degradation of this mRNA, whereas it acts as a central player in the RNAI degradation. Secondly, RNase II seems to be the major exoribonuclease in *rep* mRNA decay, with PNPase playing only a minor role. In RNAI degradation, PNPase and RNase II act to similar extents.

VII. Concluding Remarks

The degradative exoribonucleases are major players in RNA metabolism. PNPase, RNase II, RNase R, and oligoribonuclease share some common features but they also exhibit unique characteristics. They not only have different catalytic activities but also have different requirements for substrate binding. Detailed genetic and biochemical studies have allowed us to gain insights into the properties of each enzyme and its interaction with its substrates. The most recent structural data have contributed to expanding this knowledge and to a better understanding of the mechanism of exoribonuclease action. The folding of the RNA substrate and 3′ end accessibility, together with the availability and intrinsic characteristics of the exoribonucleases, are the major features controlling exonucleolytic degradation. RNases are major regulators in the cell but they also have to be regulated. The stabilization of transcripts encoding RNases seems to be an important factor in their induction, but more studies are required to fully understand how the cell regulates the levels of the exoribonucleases.

The three-dimensional architecture of these enzymes represents an important advance for the design of further biochemical studies. A large amount of information has been gathered very recently concerning the RNase II family of enzymes. Surprisingly, the closely related exoribonucleases of these family exhibit important different properties. It has been proposed that structural differences between RNase II and Rrp44 may explain the different biochemical properties observed. However, these data have not yet clarified the underlying mechanism of the impressive RNase R activity against highly structured RNA. Clearly, the structure of RNase R will be a decisive contribution to the unveiling of the unique characteristics of this enzyme.

Recent advances in the study of degradative exoribonucleases have revealed many similarities between prokaryotes and eukaryotes. The structure of the core exosome is remarkably similar to the PNPase trimer. The catalytic subunit of the eukaryotic exosome is Rrp44, an RNase II-like exoribonuclease biochemically close to RNase R. Oligoribonuclease plays an essential role in the cell, removing the small oligoribonucleotide products resulting from the other exoribonuclease activities. The recent discovery of a 5'–3' exoribonuclease in *B. subtilis* raises the question of whether such an activity, typically found in eukaryotes, is present in other prokaryotes (24).

Degradative exoribonucleases are key players in the control of gene expression. The thriving RNA field still holds many unanswered questions. The continuous study of these diverse enzymes will surely guarantee a more detailed understanding of RNA metabolism.

REFERENCES

1. Régnier P, Arraiano CM. Degradation of mRNA in bacteria: Emergence of ubiquitous features. *Bioessays* 2000;**22**:235–44.
2. Arraiano CM, Maquat LE. Posttranscriptional control of gene expression: Effectors of mRNA decay. *Mol Microbiol* 2003;**49**:267–76.
3. Donovan WP, Kushner SR. Polynucleotide phosphorylase and ribonuclease II are required for cell viability and mRNA turnover in *Escherichia coli* K-12. *Proc Natl Acad Sci USA* 1986;**83**:120–4.
4. Bernstein JA, Khodursky AB, Lin PH, Lin-Chao S, Cohen SN. Global analysis of mRNA decay and abundance in *Escherichia coli* at single-gene resolution using two-color fluorescent DNA microarrays. *Proc Natl Acad Sci USA* 2002;**99**:9697–702.
5. Newbury SF, Smith NH, Higgins CF. Differential mRNA stability controls relative gene expression within a polycistronic operon. *Cell* 1987;**51**:1131–43.
6. Deutscher MP. Degradation of RNA in bacteria: Comparison of mRNA and stable RNA. *Nucleic Acids Res* 2006;**34**:659–66.
7. Li Z, Reimers S, Pandit S, Deutscher MP. RNA quality control: Degradation of defective transfer RNA. *EMBO J* 2002;**21**:1132–8.
8. Cheng ZF, Deutscher MP. Quality control of ribosomal RNA mediated by polynucleotide phosphorylase and RNase R. *Proc Natl Acad Sci USA* 2003;**100**:6388–93.
9. Spickler C, Mackie A. Action of RNases II and polynucleotide phosphorylase against RNAs containing stem-loops of defined structure. *J Bacteriol* 2000;**182**:2422–7.
10. Mackie GA. Ribonuclease E is a 5'-end-dependent endonuclease. *Nature* 1998;**395**:720–3.

11. Celesnik H, Deana A, Belasco JG. Initiation of RNA decay in *Escherichia coli* by 5' pyrophosphate removal. *Mol Cell* 2007;**27**:79–90.
12. Deana A, Celesnik H, Belasco JG. The bacterial enzyme RppH triggers messenger RNA degradation by 5' pyrophosphate removal. *Nature* 2008;**451**:355–8.
13. Hajnsdorf E, Braun F, Haugel-Nielsen J, Régnier P. Polyadenylylation destabilizes the *rpsO* mRNA of *Escherichia coli*. *Proc Natl Acad Sci USA* 1995;**92**:3973–7.
14. O'Hara EB, Chekanova JA, Ingle CA, Kushner ZR, Peters E, Kushner SR. Polyadenylylation helps regulate mRNA decay in *Escherichia coli*. *Proc Natl Acad Sci USA* 1995;**92**:1807–11.
15. Iost I, Dreyfus M. The stability of *Escherichia coli lacZ* mRNA depends upon the simultaneity of its synthesis and translation. *EMBO J* 1995;**14**:3252–61.
16. Richards J, Mehta P, Karzai AW. RNase R degrades nonstop mRNAs selectively in an SmpB-tmRNA-dependent manner. *Mol Microbiol* 2006;**62**:1700–12.
17. Folichon M, Arluison V, Pellegrini O, Huntzinger E, Régnier P, Hajnsdorf E. The poly(A) binding protein Hfq protects RNA from RNase E and exoribonucleolytic degradation. *Nucleic Acids Res* 2003;**31**:7302–10.
18. Marujo PE, Braun F, Haugel-Nielsen J, Le Derout J, Arraiano CM, Régnier P. Inactivation of the decay pathway initiated at an internal site by RNase E promotes poly(A)-dependent degradation of the *rpsO* mRNA in *Escherichia coli*. *Mol Microbiol* 2003;**50**:1283–94.
19. Andrade JM, Cairrão F, Arraiano CM. RNase R affects gene expression in stationary phase: Regulation of *ompA*. *Mol Microbiol* 2006;**60**:219–28.
20. Zuo Y, Deutscher MP. Exoribonuclease superfamilies: Structural analysis and phylogenetic distribution. *Nucleic Acids Res* 2001;**29**:1017–26.
21. Houseley J, LaCava J, Tollervey D. RNA-quality control by the exosome. *Nat Rev Mol Cell Biol* 2006;**7**:529–39.
22. Lorentzen E, Conti E. The exosome and the proteasome: Nano-compartments for degradation. *Cell* 2006;**125**:651–4.
23. Anantharaman V, Koonin EV, Aravind L. Comparative genomics and evolution of proteins involved in RNA metabolism. *Nucleic Acids Res* 2002;**30**:1427–64.
24. Mathy N, Benard L, Pellegrini O, Daou R, Wen T, Condon C. 5'–3' exoribonuclease activity in bacteria: Role of RNase J1 in rRNA maturation and 5' stability of mRNA. *Cell* 2007;**129**:681–92.
25. Even S, Pellegrini O, Zig L, Labas V, Vinh J, Brechemmier-Baey D, *et al*. Ribonucleases J1 and J2: Two novel endoribonucleases in *B. subtilis* with functional homology to *E. coli* RNase E. *Nucleic Acids Res* 2005;**33**:2141–52.
26. Hasenohrl D, Lombo T, Kaberdin V, Londei P, Blasi U. Translation initiation factor a/eIF2 (-gamma) counteracts 5'–3' mRNA decay in the archaeon *Sulfolobus solfataricus*. *Proc Natl Acad Sci USA* 2008;**105**:2146–50.
27. Mian IS. Comparative sequence analysis of ribonucleases HII, III, II PH and D. *Nucleic Acids Res* 1997;**25**:3187–95.
28. Frazão C, McVey CE, Amblar M, Barbas A, Vonrhein C, Arraiano CM, *et al*. Unravelling the dynamics of RNA degradation by ribonuclease II and its RNA-bound complex. *Nature* 2006;**443**:110–4.
29. Zuo Y, Vincent HA, Zhang J, Wang Y, Deutscher MP, Malhotra A. Structural basis for processivity and single-strand specificity of RNase II. *Mol Cell* 2006;**24**:149–56.
30. Lorentzen E, Basquin J, Tomecki R, Dziembowski A, Conti E. Structure of the active subunit of the yeast exosome core, Rrp44: Diverse modes of substrate recruitment in the RNase II nuclease family. *Mol Cell* 2008;**29**:717–28.
31. Cairrão F, Arraiano C, Newbury S. *Drosophila* gene tazman, an orthologue of the yeast exosome component Rrp44p/Dis3, is differentially expressed during development. *Dev Dyn* 2005;**232**:733–7.

32. Worrall JA, Luisi BF. Information available at cut rates: Structure and mechanism of ribonucleases. *Curr Opin Struct Biol* 2007;**17**:128–37.
33. Wu Y-Y, Chin K-H, Chou C-C, Lee C-C, Shr H-L, Gao FP, et al. Cloning, purification, crystallization and preliminary X-ray crystallographic analysis of XC847, a 30–50 oligoribonuclease from *Xanthomonas campestris*. *Acta Crystallogr. Sect. F* 2005;**61**:902–5.
34. Finn RD, Mistry J, Schuster-Bockler B, Griffiths-Jones S, Hollich V, Lassmann T, et al. Pfam: Clans, web tools and services. *Nucleic Acids Res* 2006;**34**:D247–51.
35. Deutscher MP, Reuven NB. Enzymatic basis for hydrolytic versus phosphorolytic mRNA degradation in *Escherichia coli* and *Bacillus subtilis*. *Proc Natl Acad Sci USA* 1991;**88**:3277–80.
36. Cheng ZF, Deutscher MP. Purification and characterization of the *Escherichia coli* exoribonuclease RNase R. Comparison with RNase II. *J Biol Chem* 2002;**277**:21624–9.
37. Higgins CF, Mclaren RS, Newbury SF. Repetitive extragenic palindromic sequences, mRNA stability and gene expression: Evolution by gene conversion? A review *Gene* 1988;**72**:3–14.
38. McLaren RS, Newbury SF, Dance GS, Causton HC, Higgins CF. mRNA degradation by processive 3′–5′ exoribonucleases *in vitro* and the implications for prokaryotic mRNA decay *in vivo*. *J Mol Biol* 1991;**221**:81–95.
39. Deutscher MP. Degradation of stable RNA in bacteria. *J Biol Chem* 2003;**278**:45041–4.
40. Dreyfus M, Régnier P. The poly(A) tail of mRNAs: Bodyguard in eukaryotes, scavenger in bacteria. *Cell* 2002;**111**:611–3.
41. Lin PH, Lin-Chao S. RhlB helicase rather than enolase is the β-subunit of the *Escherichia coli* polynucleotide phosphorylase (PNPase)-exoribonucleolytic complex. *Proc Natl Acad Sci USA* 2005;**102**:16590–5.
42. Carpousis AJ. The RNA degradosome of *Escherichia coli*: An mRNA-degrading machine assembled on RNase E. *Annu Rev Microbiol* 2007;**61**:71–87.
43. LaCava J, Houseley J, Saveanu C, Petfalski E, Thompson E, Jacquier A, et al. RNA degradation by the exosome is promoted by a nuclear polyadenylation complex. *Cell* 2005;**121**:713–24.
44. Vanácová S, Wolf J, Martin G, Blank D, Dettwiler S, Friedlein A, et al. A new yeast poly(A) polymerase complex involved in RNA quality control. *PLoS Biol* 2005;**3**:e189.
45. Wyers F, Rougemaille M, Badis G, Rousselle JC, Dufour ME, Boulay J, et al. Cryptic pol II transcripts are degraded by a nuclear quality control pathway involving a new poly(A) polymerase. *Cell* 2005;**121**:725–37.
46. Iost I, Dreyfus M. DEAD-box RNA helicases in *Escherichia coli*. *Nucleic Acids Res* 2006;**34**:4189–97.
47. Pruijn GJ. Doughnuts dealing with RNA. *Nat Struct Mol Biol* 2005;**12**:562–4.
48. Reuven NB, Deutscher MP. Substitution of the 3′ terminal adenosine residue of transfer RNA *in vivo*. *Proc Natl Acad Sci USA* 1993;**90**:4350–3.
49. Bermúdez-Cruz RM, Fernandez-Ramírez F, Kameyama-Kawabe L, Montañez C. Conserved domains in polynucleotide phosphorylase among eubacteria. *Biochimie* 2005;**87**:737–45.
50. Grunberg-Manago M, Oritz PJ, Ochoa S. Enzymatic synthesis of nucleic acidlike polynucleotides. *Science* 1955;**122**:907–10.
51. Portier C, Régnier P. Expression of the *rpsO* and *pnp* genes: Structural analysis of a DNA fragment carrying their control regions. *Nucleic Acids Res* 1984;**12**:6091–102.
52. Robert-Le Meur M, Portier C. *E. coli* polynucleotide phosphorylase expression is autoregulated through an RNase III-dependent mechanism. *EMBO J* 1992;**11**:2633–41.
53. Portier C, Dondon L, Grunberg-Manago M, Régnier P. The first step in the functional inactivation of the *Escherichia coli* polynucleotide phosphorylase messenger is a ribonuclease III processing at the 5′ end. *EMBO J* 1987;**6**:2165–70.

54. Robert-Le Meur M, Portier C. Polynucleotide phosphorylase of *Escherichia coli* induces the degradation of its RNase III processed messenger by preventing its translation. *Nucleic Acids Res* 1994;**22**:397–403.
55. Jarrige AC, Mathy N, Portier C. PNPase autocontrols its expression by degrading a double-stranded structure in the *pnp* mRNA leader. *EMBO J* 2001;**20**:6845–55.
56. Zilhão R, Cairrão F, Régnier P, Arraiano CM. PNPase modulates RNase II expression in *Escherichia coli*: Implications for mRNA decay and cell metabolism. *Mol Microbiol* 1996;**20**:1033–42.
57. Cheng ZF, Zuo Y, Li Z, Rudd KE, Deutscher MP. The *vacb* GENE required for virulence in *Shigella flexneri* and *Escherichia coli* encodes the exoribonuclease RNase R. *J Biol Chem* 1998;**273**:14077–80.
58. Luttinger A, Hahn J, Dubnau D. Polynucleotide phosphorylase is necessary for competence development in *Bacillus subtilis*. *Mol Microbiol* 1996;**19**:343–56.
59. Piazza F, Zappone M, Sana M, Briani F, Dehò G. Polynucleotide phosphorylase of *Escherichia coli* is required for the establishment of bacteriophage P4 immunity. *J Bacteriol* 1996;**178**:5513–21.
60. Zangrossi S, Briani F, Ghisotti D, Regonesi ME, Tortora P, Dehò G. Transcriptional and posttranscriptional control of polynucleotide phosphorylase during cold acclimation in *Escherichia coli*. *Mol Microbiol* 2000;**36**:1470–80.
61. Jones PG, VanBogelen RA, Neidhardt FC. Induction of proteins in response to low temperature in *Escherichia coli*. *J Bacteriol* 1987;**169**:2092–5.
62. Beran RK, Simons RW. Cold-temperature induction of *Escherichia coli* polynucleotide phosphorylase occurs by reversal of its autoregulation. *Mol Microbiol* 2001;**39**:112–25.
63. Mathy N, Jarrige AC, Robert-Le Meur M, Portier C. Increased expression of *Escherichia coli* polynucleotide phosphorylase at low temperatures is linked to a decrease in the efficiency of autocontrol. *J Bacteriol* 2001;**183**:3848–54.
64. Mohanty BK, Kushner SR. Polyadenylation of *Escherichia coli* transcripts plays an integral role in regulating intracellular levels of polynucleotide phosphorylase and RNase E. *Mol Microbiol* 2002;**45**:1315–24.
65. Godefroy T. Kinetics of polymerization and phosphorolysis reactions of *Escherichia coli* polynucleotide phosphorylase. Evidence for multiple binding of polynucleotide in phosphorolysis *Eur J Biochem* 1970;**14**:222–31.
66. Littauer UZ, Soreq H. Polynucleotide phosphorylase. In: P Boyer, Ed. *The Enzymes*. Vol. XV. New York: Academic Press. 1982; p. 517–53.
67. Sulewski M, Marchese-Ragona SP, Johnson KA, Benkovic SJ. Mechanism of polynucleotide phosphorylase. *Biochemistry* 1989;**28**:5855–64.
68. Grunberg-Manago M. Enzymatic synthesis of nucleic acids. *Prog Biophys Mol Biol* 1963;**13**:175–239.
69. Mohanty BK, Kushner SR. Polynucleotide phosphorylase functions both as a 3' right-arrow 5' exonuclease and a poly(A) polymerase in *Escherichia coli*. *Proc Natl Acad Sci USA* 2000;**97**:11966–71.
70. Del Favero M, Mazzantini E, Briani F, Zangrossi S, Tortora P, Dehò G. Regulation of *Escherichia coli* polynucleotide phosphorylase by ATP. *J Biol Chem* 2008;**283**:27355–9.
71. Slomovic S, Portnoy V, Yehudai-Resheff S, Bronshtein E, Schuster G. Polynucleotide phosphorylase and the archaeal exosome as poly(A)-polymerases. *Biochim Biophys Acta* 2008;**1779**:247–55.
72. Mohanty BK, Kushner SR. The majority of *Escherichia coli* mRNAs undergo posttranscriptional modification in exponentially growing cells. *Nucleic Acids Res* 2006;**34**:5695–704.
73. Yehudai-Resheff S, Hirsh M, Schuster G. Polynucleotide phosphorylase functions as both an exonuclease and a poly(A) polymerase in spinach chloroplasts. *Mol Cell Biol* 2001;**21**:5408–16.

74. Rott R, Zipor G, Portnoy V, Liveanu V, Schuster G. RNA polyadenylation and degradation in cyanobacteria are similar to the chloroplast but different from *Escherichia coli*. *J Biol Chem* 2003;**278**:15771–7.
75. Sohlberg B, Huang J, Cohen SN. The *Streptomyces coelicolor* polynucleotide phosphorylase homologue, and not the putative poly(A) polymerase, can polyadenylate RNA. *J Bacteriol* 2003;**185**:7273–8.
76. Campos-Guillen J, Bralley P, Jones GH, Bechhofer DH, Olmedo-Alvarez G. Addition of poly(A) and heteropolymeric 3' ends in *Bacillus subtilis* wild-type and polynucleotide phosphorylase-deficient strains. *J Bacteriol* 2005;**187**:4698–706.
77. Portnoy V, Evguenieva-Hackenberg E, Klein F, Walter P, Lorentzen E, Klug G, et al. RNA polyadenylation in archaea: Not observed in *Haloferax* while the exosome polynucleotidylates RNA in *Sulfolobus*. *EMBO Rep* 2005;**6**:1188–93.
78. Mohanty BK, Kushner SR. Genomic analysis in *Escherichia coli* demonstrates differential roles for polynucleotide phosphorylase and RNase II in mRNA abundance and decay. *Mol Microbiol* 2003;**50**:645–58.
79. Py B, Higgins CF, Krisch HM, Carpousis AJ. A DEAD-box RNA helicase in the *Escherichia coli* RNA degradosome. *Nature* 1996;**381**:169–72.
80. Cheng ZF, Deutscher MP. An important role for RNase R in mRNA decay. *Mol Cell* 2005;**17**:313–8.
81. Xu F, Cohen SN. RNA degradation in *Escherichia coli* regulated by 3' adenylation and 5' phosphorylation. *Nature* 1995;**374**:180–3.
82. Carpousis AJ, Vanzo NF, Raynal LC. mRNA degradation. A tale of poly(A) and multiprotein machines. *Trends Genet* 1999;**15**:24–8.
83. Carpousis AJ, Van Houwe G, Ehretsmann C, Krisch HM. Copurification of *E. coli* RNase E and PNPase: Evidence for a specific association between two enzymes important in RNA processing and degradation. *Cell* 1994;**76**:889–900.
84. Miczak A, Kaberdin VR, Wei CL, Lin-Chao S. Proteins associated with RNase E in a multicomponent ribonucleolytic complex. *Proc Natl Acad Sci USA* 1996;**93**:3865–9.
85. Vanzo NF, Li YS, Py B, Blum E, Higgins CF, Raynal LC, et al. Ribonuclease E organizes the protein interactions in the *Escherichia coli* RNA degradosome. *Genes Dev* 1998;**12**:2770–81.
86. Regonesi ME, Del Favero M, Basilico F, Briani F, Benazzi L, Tortora P, et al. Analysis of the *Escherichia coli* RNA degradosome composition by a proteomic approach. *Biochimie* 2006;**88**:151–61.
87. Prud'homme-Genereux A, Beran RK, Iost I, Ramey CS, Mackie GA, Simons RW. Physical and functional interactions among RNase E, polynucleotide phosphorylase and the cold-shock protein, CsdA: Evidence for a "cold shock degradosome." *Mol Microbiol* 2004;**54**:1409–21.
88. Liou GG, Chang HY, Lin CS, Lin-Chao S. DEAD box rhlb RNA helicase physically associates with exoribonuclease PNPase to degrade double-stranded RNA independent of the degradosome-assembling region of RNase E. *J Biol Chem* 2002;**277**:41157–62.
89. Mohanty BK, Maples VF, Kushner SR. The Sm-like protein Hfq regulates polyadenylation dependent mRNA decay in *Escherichia coli*. *Mol Microbiol* 2004;**54**:905–20.
90. Portier C. Quaternary structure of polynucleotide phosphorylase from *Escherichia coli*: Evidence of a complex between two types of polypeptide chains. *Eur J Biochem* 1975;**55**:573–82.
91. Symmons MF, Jones GH, Luisi BF. A duplicated fold is the structural basis for polynucleotide phosphorylase catalytic activity, processivity, and regulation. *Structure* 2000;**8**:1215–26.
92. Mattaj IW. RNA recognition: A family matter? *Cell* 1993;**73**:837–40.

93. Bycroft M, Hubbard TJ, Proctor M, Freund SM, Murzin AG. The solution structure of the S1 RNA binding domain: A member of an ancient nucleic acid-binding fold. *Cell* 1997;**88**:235–42.
94. Stickney LM, Hankins JS, Miao X, Mackie GA. Function of the conserved S1 and KH domains in polynucleotide phosphorylase. *J Bacteriol* 2005;**187**:7214–21.
95. Jarrige A, Brechemier-Baey D, Mathy N, Duche O, Portier C. Mutational analysis of polynucleotide phosphorylase from *Escherichia coli*. *J Mol Biol* 2002;**321**:397–409.
96. Matus-Ortega ME, Regonesi ME, Pina-Escobedo A, Tortora P, Dehò G, García-Mena J. The KH and S1 domains of *Escherichia coli* polynucleotide phosphorylase are necessary for autoregulation and growth at low temperature. *Biochim Biophys Acta* 2007;**1769**:194–203.
97. Briani F, Del Favero M, Capizzuto R, Consonni C, Zangrossi S, Greco C, et al. Genetic analysis of polynucleotide phosphorylase structure and functions. *Biochimie* 2007;**89**:145–57.
98. Yehudai-Resheff S, Portnoy V, Yogev S, Adir N, Schuster G. Domain analysis of the chloroplast polynucleotide phosphorylase reveals discrete functions in RNA degradation, polyadenylation, and sequence homology with exosome proteins. *Plant Cell* 2003;**15**:2003–19.
99. Sarkar D, Park ES, Emdad L, Randolph A, Valerie K, Fisher PB. Defining the domains of human polynucleotide phosphorylase (hPNPaseOLD-35) mediating cellular senescence. *Mol Cell Biol* 2005;**25**:7333–43.
100. Amblar M, Barbas A, Gomez-Puertas P, Arraiano CM. The role of the S1 domain in exoribonucleolytic activity: Substrate specificity and multimerization. *RNA* 2007;**13**:317–27.
101. Schubert M, Edge RE, Lario P, Cook MA, Strynadka NC, Mackie GA, et al. Structural characterization of the RNase E S1 domain and identification of its oligonucleotide-binding and dimerization interfaces. *J Mol Biol* 2004;**341**:37–54.
102. Goverde RL, Huis in't Veld JH, Kusters JG, Mooi FR. The psychrotrophic bacterium *Yersinia enterocolitica* requires expression of *pnp*, the gene for polynucleotide phosphorylase, for growth at low temperature (5 °C). *Mol Microbiol* 1998;**28**:555–69.
103. García-Mena J, Das A, Sánchez-Trujillo A, Portier C, Montanez C. A novel mutation in the KH domain of polynucleotide phosphorylase affects autoregulation and mRNA decay in *Escherichia coli*. *Mol Microbiol* 1999;**33**:235–48.
104. Clements MO, Eriksson S, Thompson A, Lucchini S, Hinton JC, Normark S, et al. Polynucleotide phosphorylase is a global regulator of virulence and persistency in *Salmonella enterica*. *Proc Natl Acad Sci USA* 2002;**99**:8784–9.
105. Rosenzweig JA, Weltman G, Plano GV, Schesser K. Modulation of Yersinia type three secretion system by the S1 domain of polynucleotide phosphorylase. *J Biol Chem* 2005;**280**:156–63.
106. Rosenzweig JA, Chromy B, Echeverry A, Yang J, Adkins B, Plano GV, et al. Polynucleotide phosphorylase independently controls virulence factor expression levels and export in *Yersinia* spp. *FEMS Microbiol Lett* 2007;**270**:255–64.
107. Symmons MF, Williams MG, Luisi BF, Jones GH, Carpousis AJ. Running rings around RNA: A superfamily of phosphate-dependent RNases. *Trends Biochem Sci* 2002;**27**:11–8.
108. Lorentzen E, Walter P, Fribourg S, Evguenieva-Hackenberg E, Klug G, Conti E. The archaeal exosome core is a hexameric ring structure with three catalytic subunits. *Nat Struct Mol Biol* 2005;**12**:575–81.
109. Lorentzen E, Dziembowski A, Lindner D, Seraphin B, Conti E. RNA channelling by the archaeal exosome. *EMBO Rep* 2007;**8**:470–6.
110. Liu Q, Greimann JC, Lima CD. Reconstitution, activities, and structure of the eukaryotic RNA exosome. *Cell* 2006;**127**:1223–37.
111. Dziembowski A, Lorentzen E, Conti E, Seraphin B. A single subunit, Dis3, is essentially responsible for yeast exosome core activity. *Nat Struct Mol Biol* 2007;**14**:15–22.

112. Mitchell P, Petfalski E, Shevchenko A, Mann M, Tollervey D. The exosome: A conserved eukaryotic RNA processing complex containing multiple 3'–5' exoribonucleases. *Cell* 1997;**91**:457–66.
113. Grossman D, van Hoof A. RNase II structure completes group portrait of 3' exoribonucleases. *Nat Struct Mol Biol* 2006;**13**:760–1.
114. Orban TI, Izaurralde E. Decay of mRNAs targeted by RISC requires XRN1, the Ski complex, and the exosome. *RNA* 2005;**11**:459–69.
115. Zilhão R, Caillet J, Régnier P, Arraiano CM. Precise physical mapping of the *Escherichia coli rnb* gene, encoding ribonuclease II. *Mol Gen Genet* 1995;**248**:242–6.
116. Piedade J, Zilhão R, Arraiano CM. Construction and characterization of an absolute deletion mutant of *Escherichia coli* ribonuclease II. *FEMS Microbiol Lett* 1995;**127**:187–93.
117. Zilhão R, Camelo L, Arraiano CM. DNA sequencing and expression of the gene *rnb* encoding *Escherichia coli* ribonuclease II. *Mol Microbiol* 1993;**8**:43–51.
118. Zilhão R, Plumbridge J, Hajnsdorf E, Régnier P, Arraiano CM. *Escherichia coli* RNase II: Characterization of the promoters involved in the transcription of *rnb*. *Microbiology* 1996;**142** (Pt. 2), 367–75.
119. Zilhão R, Régnier P, Arraiano CM. The role of endonucleases in the expression of ribonuclease II in *Escherichia coli*. *FEMS Microbiol Lett* 1995;**130**:237–44.
120. Cairrão F, Chora A, Zilhão R, Carpousis AJ, Arraiano CM. RNase II levels change according to the growth conditions: Characterization of *gmr*, a new *Escherichia coli* gene involved in the modulation of RNase II. *Mol Microbiol* 2001;**39**:1550–61.
121. Cannistraro VJ, Kennell D. The processive reaction mechanism of ribonuclease II. *J Mol Biol* 1994;**243**:930–43.
122. Marujo PE, Hajnsdorf E, Le Derout J, Andrade R, Arraiano CM, Régnier P. Rnase II removes the oligo(A) tails that destabilize the *rpsO* mRNA of *Escherichia coli*. *RNA* 2000;**6**:1185–93.
123. Mohanty BK, Kushner SR. Polynucleotide phosphorylase, RNase II and RNase E play different roles in the *in vivo* modulation of polyadenylation in *Escherichia coli*. *Mol Microbiol* 2000;**36**:982–94.
124. Andrade JM, Hajnsdorf E, Régnier P, Arraiano CM. The Poly(A)-dependent degradation of *rpsO* mRNA is primarily mediated by RNase R. *RNA* 2008, doi:10.1261/rna.1197309.
125. Hajnsdorf E, Steier O, Coscoy L, Teysset L, Régnier P. Roles of RNase E, RNase II and PNPase in the degradation of the *rpsO* transcripts of *Escherichia coli*: Stabilizing function of RNase II and evidence for efficient degradation in an *ams pnp rnb* mutant. *EMBO J* 1994;**13**:3368–77.
126. Pepe CM, Maslesa-Galic S, Simons RW. Decay of the IS10 antisense RNA by 3' exoribonucleases: Evidence that RNase II stabilizes RNA-OUT against PNPase attack. *Mol Microbiol* 1994;**13**:1133–42.
127. Coburn GA, Mackie GA. Overexpression, purification, and properties of *Escherichia coli* ribonuclease II. *J Biol Chem* 1996;**271**:1048–53.
128. Folichon M, Marujo PE, Arluison V, Le Derout J, Pellegrini O, Hajnsdorf E, et al. Fate of mRNA extremities generated by intrinsic termination: Detailed analysis of reactions catalyzed by ribonuclease II and poly(A) polymerase. *Biochimie* 2005;**87**:819–26.
129. Cannistraro VJ, Kennell D. The reaction mechanism of ribonuclease II and its interaction with nucleic acid secondary structures. *Biochem et Biophys Acta* 1999;**1433**:170–87.
130. Donovan WP, Kushner SR. Amplification of ribonuclease II (*rnb*) activity in *Escherichia coli* K-12. *Nucleic Acids Res* 1983;**11**:265–75.
131. Amblar M, Arraiano CM. A single mutation in *Escherichia coli* ribonuclease II inactivates the enzyme without affecting RNA binding. *FEBS J* 2005;**272**:363–74.
132. Amblar M, Barbas A, Fialho AM, Arraiano CM. Characterization of the functional domains of *Escherichia coli* RNase II. *J Mol Biol* 2006;**360**:921–33.

133. Arraiano CM, Barbas A, Amblar M. Characterizing ribonucleases *in vitro*: Examples of synergies between biochemical and structural analysis. *Methods Enzymol* 2008;**447**:131-160.
134. Barbas A, Matos RG, Amblar M, Lopez-Vinas E, Gomez-Puertas P, Arraiano CM. New insights into the mechanism of RNA degradation by ribonuclease II: Identification of the residue responsible for setting the RNase II end-product. *J Biol Chem* 2008;**283**:13070–6.
135. Wang HW, Wang J, Ding F, Callahan K, Bratkowski J, Butler JS, et al. Architecture of the yeast Rrp44-exosome complex suggests routes of RNA recruitment for 3'end processing. *Proc Nat Acad Sci USA* 2007;**104**:16844–9.
136. Bollenbach TJ, Lange H, Gutierrez R, Erhardt M, Stern DB, Gagliardi D. RNR1, a 3'–5' exoribonuclease belonging to the RNR superfamily, catalyzes 3' maturation of chloroplast ribosomal RNAs in *Arabidopsis thaliana*. *Nucleic Acids Res* 2005;**33**:2751–63.
137. Lim J, Kuroki T, Ozaki K, Kohsaki H, Yamori T, Tsuruo T, et al. Isolation of murine and human homologues of the fission-yeast dis3 + gene encoding a mitotic-control protein and its overexpression in cancer cells with progressive phenotype. *Cancer Res* 1997;**57**:921–5.
138. Murakami H, Goto DB, Toda T, Chen ES, Grewal SI, Martienssen RA, et al. Ribonuclease activity of Dis3 is required for mitotic progression and provides a possible link between heterochromatin and kinetochore function. *PLoS ONE* 2007;**2**:e317.
139. Kasai T, Gupta RS, Schlessinger D. Exoribonucleases in wild type *Escherichia coli* and RNase II-deficient mutants. *J Biol Chem* 1977;**252**:8950–6.
140. Cairrão F, Cruz A, Mori H, Arraiano CM. Cold shock induction of RNase R and its role in the maturation of the quality control mediator SsrA/tmRNA. *Mol Microbiol* 2003;**50**:1349–60.
141. Oussenko IA, Abe T, Ujiie H, Muto A, Bechhofer DH. Participation of 3'–5' exoribonucleases in the turnover of *Bacillus subtilis* mRNA. *J Bacteriol* 2005;**187**:2758–67.
142. Vincent HA, Deutscher MP. Substrate recognition and catalysis by the exoribonuclease RNase R. *J Biol Chem* 2006;**281**:29769–75.
143. Cairrão F, Arraiano CM. The role of endoribonucleases in the regulation of RNase R. *Biochem Biophys Res Commun* 2006;**343**:731–7.
144. Purusharth RI, Madhuri B, Ray MK. Exoribonuclease R in *Pseudomonas syringae* is essential for growth at low temperature and plays a novel role in the 3' end processing of 16 and 5 S ribosomal RNA. *J Biol Chem* 2007;**282**:16267–77.
145. Lalonde MS, Zuo Y, Zhang J, Gong X, Wu S, Malhotra A, et al. Exoribonuclease R in *Mycoplasma genitalium* can carry out both RNA processing and degradative functions and is sensitive to RNA ribose methylation. *RNA* 2007;**13**:1957–68.
146. Schneider C, Anderson JT, Tollervey D. The exosome subunit Rrp44 plays a direct role in RNA substrate recognition. *Mol Cell* 2007;**27**:324–31.
147. Karzai AW, Sauer RT. Protein factors associated with the SsrA.SmpB tagging and ribosome rescue complex. *Proc Natl Acad Sci USA* 2001;**98**:3040–4.
148. Richards J, Sundermeier T, Svetlanov A, Karzai AW. Quality control of bacterial mRNA decoding and decay. *Biochim Biophys Acta* 2008;**1779**(9), 574-82.
149. Karzai AW, Roche ED, Sauer RT. The SsrA–SmpB system for protein tagging, directed degradation and ribosome rescue. *Nat Struct Biol* 2000;**7**:449–55.
150. Hong SJ, Tran QA, Keiler KC. Cell cycle-regulated degradation of tmRNA is controlled by RNase R and SmpB. *Mol Microbiol* 2005;**57**:565–75.
151. Kolter R, Siegele DA, Tormo A. The stationary phase of the bacterial life cycle. *Annu Rev Microbiol* 1993;**47**:855–74.
152. Hengge-Aronis R. Interplay of global regulators and cell physiology in the general stress response of *Escherichia coli*. *Curr Opin Microbiol* 1999;**2**:148–52.
153. Ishihama A. Adaptation of gene expression in stationary phase bacteria. *Curr Opin Genet Dev* 1997;**7**:582–8.

154. Cao GJ, Sarkar N. Stationary phase-specific mRNAs in *Escherichia coli* are polyadenylated. *Biochem Biophys Res Commun* 1997;**239**:46–50.
155. Rodrigues DF, Tiedje JM. Coping with our cold planet. *Appl Environ Microbiol* 2008;**74**:1677–86.
156. Giangrossi M, Giuliodori AM, Gualerzi CO, Pon CL. Selective expression of the β-subunit of nucleoid-associated protein HU during cold shock in *Escherichia coli*. *Mol Microbiol* 2002;**44**:205–16.
157. Polissi A, De Laurentis W, Zangrossi S, Briani F, Longhi V, Pesole G, et al. Changes in *Escherichia coli* transcriptome during acclimatization at low temperature. *Res Microbiol* 2003;**154**:573–80.
158. Awano N, Xu C, Ke H, Inoue K, Inouye M, Phadtare S. Complementation analysis of the cold-sensitive phenotype of the *Escherichia coli csdA* deletion strain. *J Bacteriol* 2007;**189**:5808–15.
159. Purusharth RI, Klein F, Sulthana S, Jager S, Jagannadham MV, Evguenieva-Hackenberg E, et al. Exoribonuclease R interacts with endoribonuclease E and an RNA helicase in the psychrotrophic bacterium *Pseudomonas syringae* Lz4W. *J Biol Chem* 2005;**280**:14572–8.
160. Roop RM, II, Gee JM, Robertson GT, Richardson JM, Ng WL, Winkler ME. *Brucella* stationary-phase gene expression and virulence. *Annu Rev Microbiol* 2003;**57**:57–76.
161. Sonenshein AL. CodY, a global regulator of stationary phase and virulence in Gram-positive bacteria. *Curr Opin Microbiol* 2005;**8**:203–7.
162. Chaturongakul S, Raengpradub S, Wiedmann M, Boor KJ. Modulation of stress and virulence in *Listeria monocytogenes*. *Trends Microbiol* 2008;**16**:388–96.
163. Tobe T, Sasakawa C, Okada N, Honma Y, Yoshikawa M. Vacb, a novel chromosomal gene required for expression of virulence genes on the large plasmid of *Shigella flexneri*. *J Bacteriol* 1992;**174**:6359–67.
164. Erova TE, Kosykh VG, Fadl AA, Sha J, Horneman AJ, Chopra AK. Cold shock exoribonuclease R (VacB) is involved in *Aeromonas hydrophila* pathogenesis. *J Bacteriol* 2008;**190**:3467–74.
165. Miyoshi A, Rosinha GM, Camargo IL, Trant CM, Cardoso FC, Azevedo V, et al. The role of the *vacb* gene in the pathogenesis of *Brucella abortus*. *Microbes Infect* 2007;**9**:375–81.
166. Gosh S, Deutscher MP. Oligoribonuclease is an essential component of the mRNA decay pathway. *Proc Natl Acad Sci USA* 1999;**96**:4372–7.
167. Stevens A, Niyogi SK. Hydrolysis of oligoribonucleotides by an enzyme fraction from *Escherichia coli*. *Biochem Biophys Res Commun* 1967;**29**:550–5.
168. Niyogi SK, Datta AK. A novel oligoribonuclease of *Escherichia coli*. I. Isolation and properties. *J Biol Chem* 1975;**250**:7307–12.
169. Yu D, Deutscher MP. Oligoribonuclease is distinct from the other known exoribonucleases of *Escherichia coli*. *J Bacteriol* 1995;**177**:4137–9.
170. Zhang X, Zhu L, Deutscher MP. Oligoribonuclease is encoded by a highly conserved gene in the 3′–5′ exonuclease superfamily. *J Bacteriol* 1998;**180**:2779–81.
171. Datta AK, Niyogi K. A novel oligoribonuclease of *Escherichia coli*. II. Mechanism of action. *J Biol Chem* 1975;**250**:7313–9.
172. Nguyen LH, Erzberger JP, Root J, Wilson DM, III. The human homolog of *Escherichia coli* Orn degrades small single-stranded RNA and DNA oligomers. *J Biol Chem* 2000;**275**:25900–6.
173. Mechold U, Ogryzko V, Ngo S, Danchin A. Oligoribonuclease is a common downstream target of lithium-induced pAp accumulation in *Escherichia coli* and human cells. *Nucleic Acids Res* 2006;**34**:2364–73.

174. Chin KH, Yang CY, Chou CC, Wang AH, Chou SH. The crystal structure of XC847 from *Xanthomonas campestris*: A 3′–5′ oligoribonuclease of Dnaq fold family with a novel opposingly shifted helix. *Proteins* 2006;**65**:1036–40.
175. Fiedler TJ, Vincent HA, Zuo Y, Gavrialov O, Malhotra A. Purification and crystallization of *Escherichia coli* oligoribonuclease. *Acta Crystallogr D Biol Crystallogr* 2004;**60**:736–9.
176. Arraiano CM, Yancey SD, Kushner SR. Stabilization of discrete mRNA breakdown products in *ams pnp rnb* multiple mutants of *Escherichia coli* K-12. *J Bacteriol* 1988;**170**:4625–33.
177. Grunberg-Manago M. Messenger RNA stability and its role in control of gene expression in bacteria and phages. *Annu Rev Genet* 1999;**33**:193–227.
178. Donovan WP, Kushner SR. Cloning and physical analysis of the *pyrF* gene (coding for orotidine-5′-phosphate decarboxylase) from *Escherichia coli* K-12. *Gene* 1983;**25**:39–48.
179. Jensen KF, Larsen JN, Schack L, Sivertsen A. Studies on the structure and expression of *Escherichia coli pyrC*, *pyrD*, and *pyrF* using the cloned genes. *Eur J Biochem* 1984;**140**:343–52.
180. Turnbough CL Jr., Kerr KH, Funderburg WR, Donahue JP, Powell FE. Nucleotide sequence and characterization of the *pyrF* operon of *Escherichia coli* K12. *J Biol Chem* 1987;**262**:10239–45.
181. Arraiano CM, Cruz AA, Kushner SR. Analysis of the *in vivo* decay of the *Escherichia coli* dicistronic *pyrF-orfF* transcript: Evidence for multiple degradation pathways. *J Mol Biol* 1997;**268**:261–72.
182. Arraiano C, Yancey SD, Kushner SR. Identification of endonucleolytic cleavage sites involved in decay of *Escherichia coli trxA* mRNA. *J Bacteriol* 1993;**175**:1043–52.
183. Mohanty BK, Kushner SR. Analysis of the function of *Escherichia coli* poly(A) polymerase I in RNA metabolism. *Mol Microbiol* 1999;**34**:1094–108.
184. Régnier P, Portier C. Initiation, attenuation and RNase III processing of transcripts from the *Escherichia coli* operon encoding ribosomal protein S15 and polynucleotide phosphorylase. *J Mol Biol* 1986;**187**:23–32.
185. Régnier P, Grunberg-Manago M, Portier C. Nucleotide sequence of the *pnp* gene of *Escherichia coli* encoding polynucleotide phosphorylase. Homology of the primary structure of the protein with the RNA-binding domain of ribosomal protein S1. *J Biol Chem* 1987;**262**:63–8.
186. Régnier P, Hajnsdorf E. Decay of mRNA encoding ribosomal protein S15 of *Escherichia coli* is initiated by an RNase E-dependent endonucleolytic cleavage that removes the 3′ stabilizing stem and loop structure. *J Mol Biol* 1991;**217**:283–92.
187. Braun F, Hajnsdorf E, Régnier P. Polynucleotide phosphorylase is required for the rapid degradation of the RNase E-processed *rpsO* mRNA of *Escherichia coli* devoid of its 3′ hairpin. *Mol Microbiol* 1996;**19**:997–1005.
188. Haugel-Nielsen J, Hajnsdorf E, Régnier P. The *rpsO* mRNA of *Escherichia coli* is polyadenylated at multiple sites resulting from endonucleolytic processing and exonucleolytic degradation. *EMBO J* 1996;**15**:3144–52.
189. Hajnsdorf E, Régnier P. *E. coli* RpsO mRNA decay: RNase E processing at the beginning of the coding sequence stimulates poly(A)-dependent degradation of the mRNA. *J Mol Biol* 1999;**286**:1033–43.
190. Cohen SN. Surprises at the 3′ end of prokaryotic RNA. *Cell* 1995;**80**:829–32.
191. Coburn GA, Mackie GA. Differential sensitivities of portions of the mRNA for ribosomal protein S20 to 3′-exonucleases dependent on oligoadenylation and RNA secondary structure. *J Biol Chem* 1996;**271**:15776–81.
192. Coburn GA, Mackie GA. Reconstitution of the degradation of the mRNA for ribosomal protein S20 with purified enzymes. *J Mol Biol* 1998;**279**:1061–74.
193. Mackie GA, Parsons GD. Tandem promoters in the gene for ribosomal protein S20. *J Biol Chem* 1983;**258**:7840–6.

194. Mackie GA. Specific endonucleolytic cleavage of the mRNA for ribosomal protein S20 of *Escherichia coli* requires the product of the *ams* gene *in vivo* and *in vitro*. *J Bacteriol* 1991;**173**:2488–97.
195. Khemici V, Carpousis AJ. The RNA degradosome and poly(A) polymerase of *Escherichia coli* are required *in vivo* for the degradation of small mRNA decay intermediates containing REP-stabilizers. *Mol Microbiol* 2004;**51**:777–90.
196. Coburn GA, Miao X, Briant DJ, Mackie GA. Reconstitution of a minimal RNA degradosome demonstrates functional coordination between a 3′ exonuclease and a DEAD-box RNA helicase. *Genes Dev* 1999;**13**:2594–603.
197. von Gabain A, Belasco JG, Schottel JL, Chang AC, Cohen SN. Decay of mRNA in *Escherichia coli*: Investigation of the fate of specific segments of transcripts. *Proc Natl Acad Sci USA* 1983;**80**:653–7.
198. Nilsson G, Belasco JG, Cohen SN, von Gabain A. Growth-rate dependent regulation of mRNA stability in *Escherichia coli*. *Nature* 1984;**312**:75–7.
199. Emory SA, Bouvet P, Belasco JG. A 5′-terminal stem-loop structure can stabilize mRNA in *Escherichia coli*. *Genes Dev* 1992;**6**:135–48.
200. Melefors O, von Gabain A. Site-specific endonucleolytic cleavages and the regulation of stability of *E. coli ompA* mRNA. *Cell* 1988;**52**:893–901.
201. Arnold TE, Yu J, Belasco JG. mRNA stabilization by the *ompA* 5′ untranslated region: Two protective elements hinder distinct pathways for mRNA degradation. *RNA* 1998;**4**:319–30.
202. Vytvytska O, Moll I, Kaberdin VR, von Gabain A, Blasi U. Hfq (HF1) stimulates *ompA* mRNA decay by interfering with ribosome binding. *Genes Dev* 2000;**14**:1109–18.
203. Udekwu KI, Darfeuille F, Vogel J, Reimegard J, Holmqvist E, Wagner EG. Hfq-dependent regulation of ompA synthesis is mediated by an antisense RNA. *Genes Dev* 2005;**19**:2355–66.
204. Rasmussen AA, Eriksen M, Gilany K, Udesen C, Franch T, Petersen C, *et al*. Regulation of *ompA* mRNA stability: The role of a small regulatory RNA in growth phase-dependent control. *Mol Microbiol* 2005;**58**:1421–9.
205. Storz G, Opdyke JA, Zhang A. Controlling mRNA stability and translation with small, noncoding RNAs. *Curr Opin Microbiol* 2004;**7**:140–4.
206. Viegas SC, Arraiano CM. Regulating the regulators: How ribonucleases dictate the rules to control small non coding RNAs. *RNA Biology* 2008;**5**(4).
207. Guillier M, Gottesman S, Storz G. Modulating the outer membrane with small RNAs. *Genes Dev* 2006;**20**:2338–48.
208. Johansen J, Rasmussen AA, Overgaard M, Valentin-Hansen P. Conserved small noncoding RNAs that belong to the sigmae regulon: Role in down-regulation of outer membrane proteins. *J Mol Biol* 2006;**364**:1–8.
209. Andrade JM, Arraiano CM. PNPase is a key player in the regulation of small RNAs that control the expression of outer membrane proteins. *RNA* 2008;**14**:543–51.
210. Viegas SC, Pfeiffer V, Sittka A, Silva IJ, Vogel J, Arraiano CM. Characterization of the role of ribonucleases in *Salmonella* small RNA decay. *Nucleic Acids Res* 2007;**35**:7651–64.
211. Stougaard P, Molin S, Nordstrom K. RNAs involved in copy-number control and incompatibility of plasmid R1. *Proc Natl Acad Sci USA* 1981;**78**:6008–12.
212. Soderbom F, Binnie U, Masters M, Wagner EG. Regulation of plasmid R1 replication: PcnB and RNase E expedite the decay of the antisense RNA, copa. *Mol Microbiol* 1997;**26**:493–504.
213. Tomizawa J, Itoh T. Plasmid ColE1 incompatibility determined by interaction of RNA I with primer transcript. *Proc Natl Acad Sci USA* 1981;**78**:6096–100.
214. Lin-Chao S, Cohen SN. The rate of processing and degradation of antisense RNAI regulates the replication of ColE1-type plasmids *in vivo*. *Cell* 1991;**65**:1233–42.

215. Xu F, Lin-Chao S, Cohen SN. The *Escherichia coli pcnB* gene promotes adenylylation of antisense RNAI of ColE1-type plasmids *in vivo* and degradation of RNAI decay intermediates. *Proc Natl Acad Sci USA* 1993;**90**:6756–60.
216. Binnie U, Wong K, McAteer S, Masters M. Absence of RNase III alters the pathway by which RNAI, the antisense inhibitor of ColE1 replication, decays. *Microbiology* 1999;**145**(Pt. 11), 3089–100.
217. Takechi S, Yasueda H, Itoh T. Control of ColE2 plasmid replication: Regulation of Rep expression by a plasmid-coded antisense RNA. *Mol Gen Genet* 1994;**244**:49–56.
218. Nishio SY, Itoh T. Replication initiator protein mRNA of ColE2 plasmid and its antisense regulator RNA are under the control of different degradation pathways. *Plasmid* 2008;**59**:102–10.
219. Shi Z, Yang WZ, Lin-Chao S, Chak KF, Yuan HS. Crystal structure of *Escherichia coli* PNPase: Central channel residues are involved in processive RNA degradation. *RNA* 2008;**14**(11), 2361–71.
220. Schaeffer D, Tsanova B, Barbas A, Reis FP, Dastidar EG, Sanchez-Rotunno M, Arraiano CM, van Hoof A. The exosome contains domains with specific endoribonuclease, exoribonuclease and cytoplasmic mRNA decay activities. *Nat Struct Mol Biol* 2008; advance online publication, doi:10.1038/nsmb.1528.

Messenger RNA Decay and Maturation in *Bacillus subtilis*

> David H. Bechhofer
>
> Department of Pharmacology and Systems Therapeutics, Mount Sinai School of Medicine of New York University, New York, NY 10029

I. Introduction and Historical Perspective	232
II. The 5′ End	234
A. Focus on the 5′ End	234
B. RNase J	237
C. RNase J1 Activity: An Alternative Model for mRNA Decay	243
D. 5′ Stabilizers	244
E. Stability of mRNA Prediction (STOMP)	245
F. Pyrophosphatase Activity	250
G. Destabilizers	251
III. Body of The Message	251
A. Translation and Decay	251
B. Messenger RNA Length	253
C. Bs-RNase III	254
D. RNase M5, RNase P, and RNase Z	255
E. Differential Stability of Polycistronic mRNA	256
IV. The 3′ End	258
A. Transcription Terminator Structure	258
B. PNPase	259
C. Other 3′ Exoribonucleases	262
D. Polyadenylation	263
E. 3′-Terminal Fragment Turnover	264
V. Regulated mRNA Decay	264
VI. Conclusion	267
References	267

Our understanding of the ribonucleases that act to process and turn over RNA in *Bacillus subtilis*, a model Gram-positive organism, has increased greatly in recent years. This chapter discusses characteristics of *B. subtilis* ribonucleases that have been shown to participate in messenger RNA maturation and decay. Distinct features of a recently discovered ribonuclease, RNase J1, are reviewed, and are put in the context of a mechanism for the mRNA decay process in *B. subtilis* that differs greatly from the classical model developed for *E. coli*. This chapter is divided according to three parts of an mRNA—5′ end, body, and 3′ end—that could theoretically serve as sites for initiation of decay. How 5′-proximal elements affect mRNA half-life, and especially how these elements interface with RNase J1, forms the basis for a set of "rules" that may be useful in predicting mRNA stability.

I. Introduction and Historical Perspective

Bacillus subtilis is an accepted "model organism" for Gram-positive bacteria, just as *Escherichia coli* is a model for Gram-negative bacteria. It would seem that Jacques Monod's famous claim about the conservation of function between *E. coli* and the elephant should certainly apply to *E. coli* and *B. subtilis*. For basic aspects of gene expression, such as RNA transcription and protein translation, the high degree of similarity between these processes in *E. coli* and *B. subtilis* is what one would expect. However, researchers in the field of mRNA processing and decay encountered an unexpected (and not unwelcome) surprise a decade ago, when the *B. subtilis* genome first became available (1). Sequence homologues of several key enzymes involved in *E. coli* mRNA turnover were not apparent from the *B. subtilis* genome sequence. These included: RNase E, the endonuclease that is widely thought to be responsible for initiation of decay of most *E. coli* mRNAs, and which is essential in *E. coli* (see chapter by A.J. Carpousis in this volume); RNase II, a 3′-to-5′ exoribonuclease that is thought to be responsible for much of the turnover of RNA fragments generated by endonuclease cleavages (see chapters by A. Danchin and C. Arraiano in this volume); and oligoribonuclease, which is essential in *E. coli* and is responsible for conversion of oligonucleotides to mononucleotides (2), thus completing the turnover of mRNA (see chapter by C. Arraiano in this volume). The absence of sequence homologues for these *E. coli* enzymes in the *B. subtilis* genome suggested either a different mechanism for mRNA turnover altogether or, at the very least, a different set of proteins to accomplish this function. Those studying *B. subtilis* mRNA decay and processing were not going to reinvent the wheel.

Evidence for fundamental differences in RNA turnover between *E. coli* and *B. subtilis* was at hand long before publication of the *B. subtilis* genome. Twenty-five years earlier, Boyer and colleagues published back-to-back papers in which they followed incorporation of ^{18}O to determine the nature of RNA decay in *E. coli* and *B. subtilis* (3, 4). The conclusion was that RNA decay in *B. subtilis* was primarily phosphorolytic, that is, removal of a nucleotide unit from the end of a chain was by addition of phosphate, resulting in a nucleoside diphosphate product. RNA decay in *E. coli*, on the other hand, was primarily hydrolytic, that is, removal of a nucleotide unit was by addition of water, resulting in a nucleoside monophosphate product. Almost 20 years later,

B. subtilis mRNA decay

Deutscher and Reuven (5) showed that the hydrolytic RNase II is responsible for 90% of the degradative activity in *E. coli* cell extracts, while the phosphorolytic 3′-to-5′ exonuclease, polynucleotide phosphorylase (PNPase), played a minor role. In *B. subtilis* extracts, on the other hand, phosphorolytic activity predominated, and this was later attributed to the activity of the *B. subtilis* PNPase (6). Deutscher and Reuven speculated that the difference in how *E. coli* and *B. subtilis* degrade RNA reflects their respective environmental niches, with gut-dwelling *E. coli* occupying an energy-rich environment and soil-dwelling *B. subtilis* occupying an energy-poor environment. Phosphorolytic decay, and thereby the retention of phosphate bond energy in the products of RNA breakdown, would be particularly advantageous for *B. subtilis*.

A more recent indication of possible profound differences in how *E. coli* and *B. subtilis* handle turnover of mRNA has come to light with the discovery of the 5′-to-3′ exonuclease activity of *B. subtilis* RNase J1 (7). Such an activity had long been sought but had never been found in bacteria. The logic of employing a 5′ exonuclease in mRNA turnover is indisputable. As translation initiation regions (TIRs) are generally quite close to the 5′ end of an mRNA, attack on mRNA by modes other than 5′-to-3′ exonucleolytic activity are potentially wasteful or even deleterious. Initiation of decay by an endonuclease cleavage downstream of the TIR would generate an mRNA fragment that was capable of being bound by a ribosome but that would encode a prematurely terminated peptide product. Attack from the 3′ terminus, by a 3′-to-5′ exonuclease, would similarly leave mRNA fragments with intact TIRs and partial coding sequences. Two possible mechanisms could be used to avoid accumulation of unproductive mRNA fragments: One would be extremely rapid decay of the products of endonuclease cleavage. This is likely the case in *E. coli*, where one cannot easily observe mRNA decay intermediates, which suggests that turnover following initiation of decay by endonuclease cleavage occurs extremely rapidly, and there is likely built-in redundancy in the 3′-to-5′ exoribonucleases to ensure this is the case. The other mechanism would be processive exonuclease activity from the 5′ end. Since most mRNAs have a short leader region, this would functionally inactivate the mRNA almost as soon as decay was initiated. Such is the case with eukaryotic mRNA turnover, where 5′ decapping, which itself can be dependent on poly(A) tail shortening, allows the onset of degradation from the 5′ end by a 5′-to-3′ exonuclease (a number of reviews on eukaryotic mRNA decay have appeared in recent years, including (8–11)). That a 5′-to-3′ exonuclease activity has now been demonstrated in *B. subtilis*, an activity that does not appear to exist in *E. coli*, opens up new approaches to our understanding of the process of mRNA turnover in *B. subtilis*.

The familiar "all or none" pattern of mRNA decay is observed in *B. subtilis* as it is in other organisms. That is, probing for a particular mRNA by Northern blot analysis, especially if it is monocistronic, will reveal a single band representing the full-length transcript. Decay intermediates are rarely observed, unless RNA is isolated from a strain that is deficient in one or more ribonucleases. This pattern suggests that mRNA decay is first order, and depends on an initial step that is rate-determining; after initiation of decay commences, degradation of the mRNA occurs too rapidly to detect intermediates in the process. Thus, our discussion of mRNA characteristics and ribonuclease activities that determine stability will be divided according to three mRNA regions at which decay can theoretically initiate: the 5' end, the body of the message, and the 3' end.

In this review we focus exclusively on studies in *B. subtilis*, to cover this one model organism in a comprehensive manner. It is expected that findings for *B. subtilis* will be representative not only of other *Bacillus* species, but also of the many organisms that contain a ribonuclease complement similar to that of *B. subtilis*.

(*Technical note*: The discovery of rifampin, a specific inhibitor of bacterial RNA polymerase (*12*), can be viewed as the starting point for studies of mRNA decay in bacteria. Typically, rifampin is added to inhibit new transcription, and RNA is isolated at increasing times thereafter. This is followed by analysis (Northern blot, RT-PCR, microarray) of the percent of a particular mRNA that remains at times after rifampin addition. Interestingly, the use of rifampin for mRNA half-life studies in *B. subtilis* was criticized several years after its discovery (*13*), to the extent that these authors concluded that "rifampicin is too toxic to use as an antibiotic for assessing the lifetime of mRNA." However, others defended its use (*14*), and we have performed "rifampin recovery" experiments which demonstrate that incubation of *B. subtilis* cultures in 150 µg ml^{-1} rifampin for up to 1 h has no effect on the number of colony forming units when assayed after washing out the rifampin! Thus, it has become routine practice to measure mRNA half-life in *B. subtilis* after addition of rifampin.)

II. The 5' End

A. Focus on the 5' End

As mentioned above, the 5' end is the natural choice for initiation of decay. Attack at or near the 5' end will result, in most cases, in a rapid functional inactivation of a message whose TIR is located close to the 5' end. This avoids costly ribosome binding and initiation of translation on mRNA fragments that are unable to be translated into full-length protein. Early experiments with the stabilization of *ermC* and *ermA* mRNA by ribosome stalling near the 5' end (see Section V), suggested a key role for the 5' end in determining stability (*15–18*). The significance of the 5' region in determining the decay rate was also suggested by studies on growth-rate regulated stability of *sdh* mRNA (*19*). Such a role could be explained by the effect of the 5' end on the activity of a 5'-to-3' exoribonuclease. However, until recently no 5'-to-3' exoribonuclease had ever been identified in any prokaryote. This led to the proposal (16) that the activity responsible for initiation of decay in B. subtilis was a "5'-binding endonuclease," that is, an enzyme that cleaved mRNA endonucleolytically but that needed to access the mRNA via the 5' end, followed by looping or tracking to a recognition site. This proposal, which predated by many years the discovery of just such an activity for E. coli RNase E (20), was met with much skepticism, since "by definition, an endonuclease does not require a terminus to bind to an RNA molecule.... You simply cannot talk about a 5'-binding endonuclease" (quote from a review of a manuscript). In fact, such an activity was known for the RecBCD complex, which binds to an end of double-stranded DNA and makes endonucleolytic cleavages to produce DNA oligonucleotides (21). In any event, the 5'-binding endonuclease concept is now widely accepted also for RNA, and the activity of such an enzyme in E. coli—RNase E—is viewed as the primary mechanism for initiation of decay for many RNAs (22).

The model for mRNA turnover in E. coli is therefore as follows (Fig. 1A): RNase E binding at the 5' end is facilitated by removal of pyrophosphate to convert the 5'-triphosphate end to a 5'-monophosphate end (see Section II.F). Following 5'-end binding, RNase E loops to an internal cleavage site. Alternatively, RNase E can bind to an internal cleavage site directly, as is suggested in a number of cases (23–25). Endonuclease cleavage in the body of the message by RNase E generates an upstream fragment that has an unprotected 3' end and is degraded by RNase II, RNase R, or PNPase, and a downstream fragment that has a monophosphate 5' end. This monophosphate 5' end provides an excellent substrate for another round of binding and endonucleolytic cleavage (20, 26, 27). Decay of the 3'-terminal fragment, which contains a transcription terminator structure that is resistant to 3' exonuclease activity, depends on reiterative poly(A) addition and attack at the extended 3' end (28). See also chapter by E. Hajnsdorf and P. Regnier in this volume. The products of 3' exonuclease activity are monophosphate nucleosides and, at 5' termini, oligonucleotides that are 2–5 nucleotides (nts) long. These are degraded to mononucleotides by oligoribonuclease (2).

According to this model, endonuclease cleavage by RNase E at even a single site would result in rapid decay of most of the mRNA. Thus, the half-life of a message is determined by the rate of 5' binding and/or initial cleavage. As such, 5'-terminal or 5'-proximal elements that restrict access of RNase E confer stability to downstream RNA (29, 30).

In B. subtilis, the recent discovery of ribonucleases with similar 5'-end-dependent properties has had a profound influence on the study of B. subtilis mRNA processing and decay. We start with an in-depth discussion of these ribonucleases, which will set the stage for the subsequent account of elements at the 5' end that determine mRNA half-life.

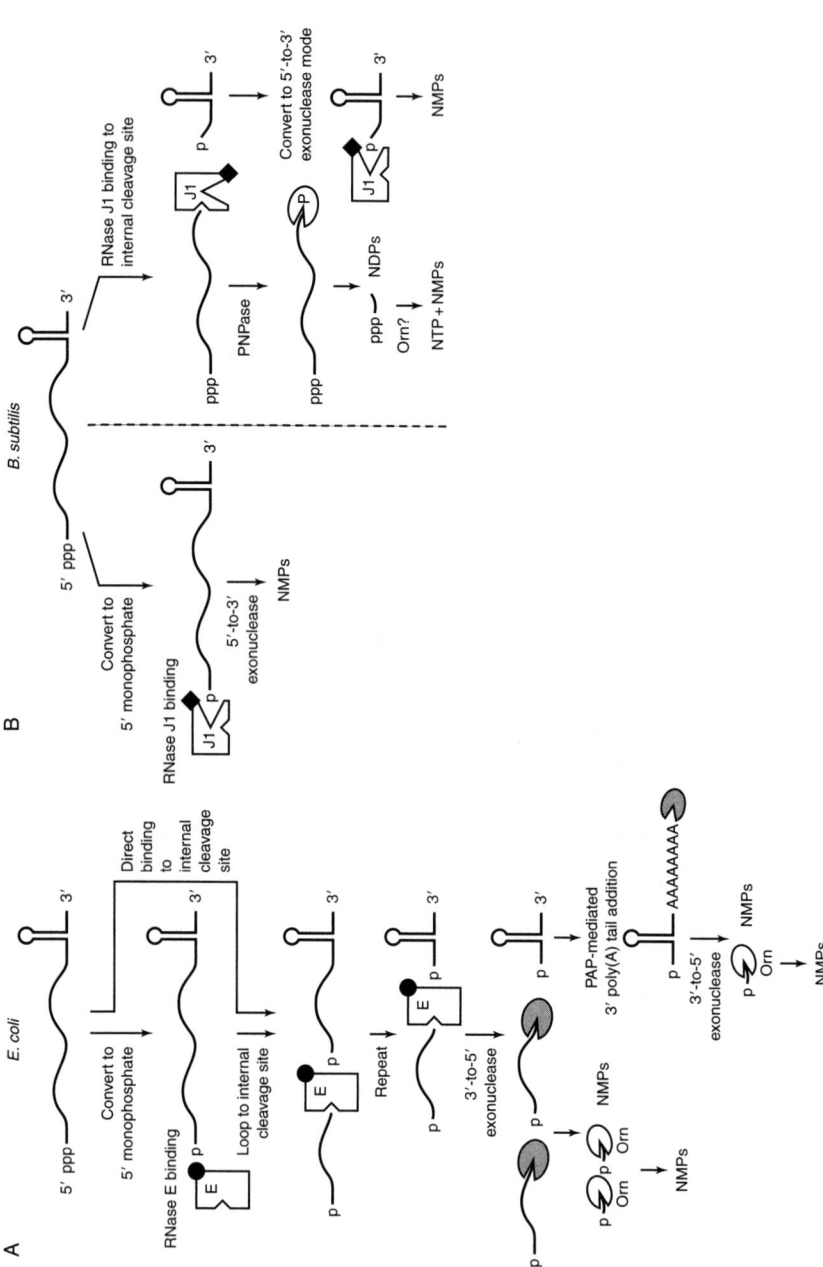

FIG. 1. General models for mRNA decay in (A) *E. coli* and (B) *B. subtilis*. Alternate pathways for *B. subtilis* mRNA decay are shown in *B* and are

B. RNase J

1. RNASE J ACTIVITIES

During the time when the substantial role of RNase E in *E. coli* mRNA decay was being clarified, attempts were made to identify a similar activity in *B. subtilis* by complementing an *E. coli* RNase E temperature-sensitive mutant with a library of shotgun-cloned *B. subtilis* DNA. These experiments were not successful. Nevertheless, the existence of a *B. subtilis* RNase E-like activity was suggested some time ago by the report that endonuclease cleavage at "site 2" in the leader region of *B. subtilis thrS* mRNA (encoding threonyl tRNA synthetase; Fig. 2) occurred at a similar site in *B. subtilis* and *E. coli*, and that processing at this site in *E. coli* was RNase E-dependent *(31)*. It took a number of years for the *B. subtilis* activity to be purified to the extent that the ribonuclease could be identified. In fact, two proteins with cleavage activity copurified from a ribosome-associated fraction *(32)*, and these were named RNase J1 (encoded by the *rnjA*, formerly *ykqC*, gene) and RNase J2 (encoded by the *rnjB*, formerly *ymfA*, gene). Thus, in addition to the 5'-to-3' exoribonuclease activity mentioned in Section I, RNase J1 and J2 also have endonuclease activity.

RNases J1 and J2 are both around 61 kDa and are 49% identical. RNase J1 is essential, while RNase J2 is not. A strain in which RNase J1 is expressed conditionally (IPTG controlled) and RNase J2 is deleted, shows an increase in global mRNA half-life (from 2.6 to 3.6 min) when RNase J1 expression levels are decreased by removing IPTG *(32)*. Deletion of the RNase J2 gene alone has no effect on mRNA half-life. While the difference in global mRNA half-life under full vs. limited RNase J1 expression is relatively small, this may be due to the continued presence of substantial amounts of RNase J1 protein, even in conditions where IPTG is not present, as shown by Western blot analysis in the Condon laboratory (unpublished results). In fact, the Western blot analysis

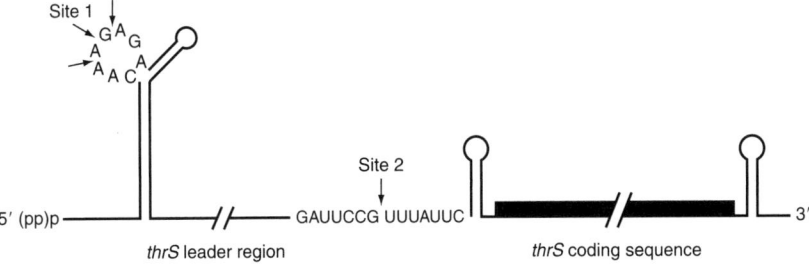

FIG. 2. Schematic of *thrS* mRNA, showing leader region and coding sequence. Sites 1 and 2 are the mapped 5' ends resulting from RNase J1 processing. Site 1 is now known to be the result of exonucleolytic decay from the initial 5' end, while site 2 is the result of endonucleolytic cleavage.

shows that the level of RNase J1 in the conditional strain, even in the presence of IPTG, is significantly less than in the wild-type strain. Thus, the difference in RNase J1 levels with vs. without IPTG is no more than three- to fourfold. In any event, the increase in global mRNA half-life in the RNase J1 conditional strain suggests a general role for RNase J1 in mRNA turnover.

Since most of the recent results in *B. subtilis* have been obtained with RNase J1, we focus on this activity, noting that *in vitro* results obtained with purified RNase J1 have, for the most part, been obtained also with purified RNase J2.

While the details of *thrS* gene expression are not pertinent to our discussion, it is of interest that RNase J1 endonuclease cleavage at site 2 in the *thrS* mRNA leader results in a downstream RNA product that is stabilized, relative to the full-length *thrS* mRNA (33). As the processing site is just upstream of a predicted stem-loop structure (Fig. 2), the downstream RNA product of RNase J1 cleavage is likely to be protected from further decay by virtue of this structure (see below for a detailed exposition of 5′ stabilizers). Thus, RNase J1 processing in this case contributes to positive regulation of *thrS* expression. Purified RNase J1 also cleaves the mRNA for *thrZ*, a minor threonyl tRNA synthetase, in a similar position (32). Endonuclease cleavages at such sites have been identified *in vivo* for a number of tRNA synthetase mRNAs (33), and these may also be due to RNase J1 activity. The suggestion is that this type of endonuclease cleavage is a common mechanism for achieving increased expression. RNase J1 would then function in gene regulation not only negatively as an initiator of mRNA decay, but also positively as an mRNA processing enzyme that creates a stable, translatable product.

2. Effect of 5′ Phosphorylation State

Although endonucleolytic cleavage at sites 1 and 2 in the *thrS* leader region (Fig. 2) was discussed in the paper that first described identification of RNase J1 and its characterization *in vitro* (32), in hindsight it is now evident that the site 1 "A-G box" cleavage is, in fact, not an endonuclease cleavage but the result of a block to 5′ exonuclease processivity (34). In fact, it was this "cleavage" that was shown to be sensitive to the phosphorylation state of the *thrS* mRNA 5′ end. Processing at the A-G box was much more efficient when the 5′ end of the substrate was monophosphorylated than when it was triphosphorylated. On the other hand, cleavage at site 2 was not significantly affected by the 5′ phosphorylation state (32). Similarly, we found recently that *trp* leader RNA with a 5′-triphosphate end is efficiently cleaved endonucleolytically by RNase J1 *in vitro* (35). Thus, while the effect of the 5′ phosphorylation state on RNase J1 activity is thought to mirror the effect on *E. coli* RNase E, which has a strong preference for a monophosphorylated 5′ end over a triphosphorylated 5′ end (20, 26), the consequences are different: the effect of the 5′ phosphorylation state is on the *endonuclease* activity of RNase E but on the *exonuclease* activity of RNase J1. This has ramifications for the mechanism of mRNA decay initiation in *E. coli* and *B. subtilis*, as will be discussed below.

The effect of 5' phosphorylation state on RNase J1 5'-to-3' exonuclease activity can be measured using uniformly labeled RNA and assaying accumulation of labeled UMP on high percentage polyacrylamide gels (7). It is found that the initial rate of RNase J1 exonuclease activity is three times faster on a substrate bearing a 5' monophosphate end, or a 5' hydroxyl end, than on the same substrate with a 5' triphosphate end. However, these experiments could greatly underestimate the effect of a 5'-triphosphate end on exonuclease activity, since release of labeled monophosphate nucleoside by exonuclease activity could occur following endonuclease cleavage of the RNA substrate, which would not reflect degradation starting at the 5' terminus. A different assay of RNase J1 exonuclease activity uses 5'-end-labeled RNA and measures activity by release of 5' terminal nucleotide, as assayed on thin-layer chromatography. In this case, the amount of release of the terminal nucleotide is at least eightfold greater when the terminal nucleotide is a monophosphate nucleoside than when it is a triphosphate nucleoside (35).

Evidence that RNase J1 requires a single-stranded 5' end to allow 5'-to-3' exonuclease activity comes from an experiment in which a complementary oligonucleotide was hybridized to the 5' end of a uniformly labeled substrate (7). This reduced the level of exonuclease activity substantially. As just mentioned, such an experiment also gives an underestimate of the effect of making the 5' end double stranded, since RNase J1 could cleave endonucleolytically downstream of the hybridized oligonucleotide and then revert to exonuclease activity. Hybridization of a 5'-targeted oligonucleotide could affect either one of the two RNase J1 activities. The better experiment, performed recently (C. Condon, personal communication), tests the effect of a 5'-targeted antisense oligonucleotide on release of labeled 5'-monophosphate nucleoside from the 5' end. Virtually no exonuclease activity is observed under these conditions.

Since the 5' phosphorylation state affects RNase J1 exonuclease activity, but not endonuclease activity, it would be of interest to ascertain whether RNase J1 endonuclease activity is at all dependent on binding at a 5' end. The best test of this would be to determine whether RNase J1 could cleave a circular RNA that contains a known cleavage site. A finding that RNAse J1 can indeed cleave circular RNA would suggest that access to internal cleavage sites on mRNA is not 5'-end-dependent, which would impact the model of initiation of mRNA decay in *B. subtilis* (discussed below).

It should be pointed out that analysis of RNase J1 endonuclease activity *in vitro* is complicated by the dual activity of this enzyme. For example, measuring the amount of an upstream fragment produced by RNase J1 endonuclease cleavage of a 5'-end labeled substrate will give an artificially low measurement if the labeled 5' end is also subject to exonuclease activity. Similarly, measuring the amount of the downstream fragment produced by RNase J1 endonuclease activity is difficult because the product will be sensitive to ongoing exonuclease activity. It would be most useful to obtain an RNase J1 mutant that has either 5' exonuclease or endonuclease activity, but not both.

3. STRUCTURE OF RNASE J

The sequence of RNase J1 reveals that it is a member of the β-CASP subfamily of zinc-dependent metallo-β-lactamases ("β-CASP" is named after other members of this subfamily: metallo-β-lactamase CPSF, Artemis, SnmI, Pso2). Structural information comes from work on the single RNase J enzyme of *Thermus thermophilus*, which is 61% similar to RNase J1 of *B. subtilis* and which cleaves *thrS* mRNA at the same site as RNase J1 (34). RNase J consists of three domains (Fig. 3): an N-terminal β-lactamase domain, a central β-CASP domain, and a C-terminal domain that is required for activity, possibly through dimerization. Deletion of the C-terminal domain severely affects activity, but it is not known whether this is due to lack of dimerization or due to an intrinsic role of the C-terminal domain in catalysis. Structural determination of RNase J in the presence of UMP revealed a binding site for the nucleotide that is immediately adjacent to the predicted catalytic site

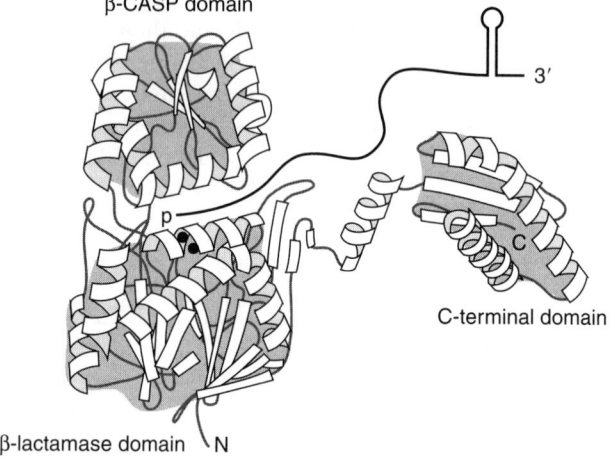

FIG. 3. Ribbon diagram of the three domains in *T. thermophilus* RNase J, based on the work of Putzer and colleagues (34). The enzyme is shown binding to the 5' monophosphate end of an RNA, whose path requires experimental verification. Black spheres in the β-lactamase domain represent zinc ions at the catalytic site.

(at which a pair of Zn^{2+} ions are coordinated), providing an explanation for the 5′-to-3′ exonuclease activity (Fig. 3). Furthermore, the architecture at this site does not seem to accommodate a 5′ triphosphate nucleoside, immediately suggesting an explanation for the ability of RNase J1 to discriminate between 5′ monophosphate and 5′ triphosphate ends. Interestingly, despite the lack of sequence homology, the overall structure of RNase E and RNase J share many common features, which may be related to their similar endonuclease activities.

Mutations in adjacent catalytic site residues results in severe impairment of both the 5′-to-3′ exonuclease and the endonuclease activities (34). This key finding indicates that there is only one catalytic site on the enzyme, and raises the difficult issue of how an enzyme that appears to be a 5′-binding ribonuclease, with its (apparently) single catalytic site adjacent to this 5′ binding site, can access and cleave endonucleolytically at distal internal sites. The 1-nt proximity of the 5′ binding site to the catalytic site is difficult to reconcile with a picture of RNase J binding to the 5′ end while the RNA loops around to correctly position an internal target site for endonuclease cleavage. Furthermore, if the exonucleolytic mode is clearly affected by the 5′ phosphorylation state, and if the endonucleolytic mode is initiated by 5′ binding, then one would expect endonuclease cleavage also to be affected by the 5′ phosphorylation state. This was not observed in the original study of RNase J1 and RNase J2 cleavage of *thrS* leader RNA (32), nor in a more recent study on *trp* leader RNA (35).

One could propose that RNase J1 has a second 5′-end binding pocket, located distal to the catalytic site, which can accommodate a triphosphorylated 5′ end. Binding of a 5′ end at the distal site could be followed by RNA looping such that the endonucleolytic cleavage site is positioned in the binding site that is adjacent to the catalytic site. This is similar to a model that has been proposed for RNase E (26, 36). According to this scenario, both the exonucleolytic and endonucleolytic activities of RNase J1 would be dependent on access to the 5′ end, and, if RNase J1 were indeed the key mRNA decay-initiating enzyme in *B. subtilis*, the phenomenon of 5′ stabilizers (described briefly in Section II.A and in more detail below) would be easily explained.

Alternatively, it may be that RNase J1 endonuclease cleavage is not dependent at all on 5′ binding, and the nature of the 5′ end restricts exonucleolytic decay but not internal cleavage. The limitation of RNase J1 endonuclease activity would be a function of the number of accessible recognition sites on the RNA (see below). An mRNA that had no available RNase J1 cleavage sites would need to rely on 5′-to-3′ exonuclease activity for turnover. An mRNA that had one or more RNase J1 cleavage sites would be susceptible to rapid initiation of decay, irrespective of the nature of the 5′ end. Although we and others have found that 5′ stabilizers are effective even for relatively long mRNA sequences (e.g., *lacZ* fusions, which are greater than 3 kb), it is possible that, in these cases, the 5' stabilizers are located upstream of sequences that fortuitously do not contain an accessible RNase J1 recognition site. More structural information on RNase J1 bound to RNA substrates, as well as a larger catalogue of RNase J1 recognition sites, will be needed to formulate hypotheses to address the issue of 5'-end-dependent activity.

4. RNase J1 Endonuclease Cleavage Sites

Sites of *in vivo* RNase J1 endonuclease cleavage have been mapped in five cases, and these are found in mRNA leader regions and in a stable RNA. They include: *infC* leader RNA (37), *thrS* and *thrZ* leader RNAs (32), *trp* leader RNA (35), and small cytoplasmic RNA or scRNA (38). The cleavages occur in single-stranded regions (in the case of scRNA, the single-stranded region is exposed after prior processing by Bs-RNase III) that are next to a downstream secondary structure, and, as shown in Table I, they occur mostly in AU-rich sequences. Superficially, these resemble *E. coli* RNase E cleavage sites in character, and the fact that *thrS* leader RNA is cleaved at the same site in *E. coli* and in *B. subtilis*, and that this cleavage is RNase E-dependent in *E. coli* (31), speaks for a conservation of site recognition. Indeed, as mentioned above, Putzer and colleagues document a similarity in the overall domain structures of *T. thermophilus* RNase J and *E. coli* RNase E (34). More definitive examples of RNase J1 endonuclease cleavage are needed to define the specificity more precisely. It should be noted that the activity of purified RNase J1 *in vitro* is not robust, and it has been suggested that cleavage by this enzyme *in vivo* may require an additional factor(s) (7). It should also be mentioned that, although *E. coli* RNase E appears to function in the context of a "degradosome" complex that includes PNPase and an RNA helicase, there is as yet no evidence for a degradosome complex in *B. subtilis*.

TABLE I
RNase J1 Endonuclease Cleavage Sites[a]

infC Leader	GUUGACCG-UACAUUU
thrS Leader	GAUUCCG-UUUAUUC
thrZ Leader	CCACGGG-U-U-AAUCA
trp Leader	CAUUAUG-U-U-UAUUC
scRNA	AUCAUCA-AAUUUUC

[a]Mapped cleavage sites are indicated by the dashes.

C. RNase J1 Activity: An Alternative Model for mRNA Decay

The discovery of the 5′-to-3′ exonuclease activity of *B. subtilis* RNase J1 (7) is truly a landmark event in the field of prokaryotic mRNA decay. RNase J1 is actually the first ribonuclease that is documented to have two enzymatic activities in a single polypeptide (34). The dual properties of RNase J1 allow one to propose an alternative model for mRNA decay in *B. subtilis*.

As described above, the now classical model of mRNA turnover in *E. coli* combines the action of a 5′-end-dependent endonuclease cleavage at a downstream site, generating (1) an upstream fragment with an unprotected 3′ end that is degraded rapidly by 3′-to-5′ exoribonucleases, and (2) a downstream fragment with a 5′ monophosphate end, which serves as the 5′ end for another round of binding and endonuclease cleavage (Fig. 1A). Assuming that RNase J1 is responsible for initiation of decay of a substantial fraction of *B. subtilis* mRNAs, an alternative model can be proposed (Fig. 1B) that involves RNase J1 in two possible pathways. For exonucleolytic decay (Fig. 1B, left), the 5′-triphosphate end of the initial transcription product needs to be converted first to a monophosphate (see Section II.F), after which RNase J1 can degrade the mRNA in the 5′-to-3′ direction. For initiation of decay by endonuclease cleavage (Fig. 1B, right), the 5′-triphosphorylated transcript acts as a substrate, and internal cleavage depends on the availability of RNase J1 recognition sites. These are accessed either by binding at the 5′ end and looping to the cleavage site or by direct binding to the internal site (see discussion above). The upstream RNA fragment generated by endonuclease cleavage is rapidly degraded by 3′-to-5′ exonuclease activity, primarily that of PNPase. The downstream RNA fragment, which has a 5′-monophosphate end, serves as a target either for additional RNase J1 endonuclease cleavage or for processive 5′-to-3′ exonucleolytic decay from the 5′ end. (In Section IV.E we discuss processivity of RNase J1 5′-to-3′ exonuclease activity on strong stem-loop structures.) This mechanism of mRNA turnover would presumably apply to any organism that has an RNase J type of activity.

The proposed alternative model for the role of RNase J1 in mRNA decay is attractive, since the "cut and chew" functions for the downstream fragments are combined in one enzyme. This raises the question whether RNase J1 is able to switch from endonucleolytic mode to exonucleolytic mode on the same RNA molecule, without release of the substrate. The structure of RNase J suggests that this may be likely. Since the 5′-monophosphate binding pocket is only one nt distance from the catalytic zinc ions (34), the 5′ monophosphate generated by endonuclease cleavage could immediately slip into the binding pocket to become subject to 5′ exonuclease activity. Such a mechanism would be highly efficient in turning over the downstream product of endonuclease cleavage, and, together with rapid PNPase degradation of the upstream product, would be consistent with the absence of detectable decay intermediates in a wild-type strain.

Such a mechanism for RNase J1 activity on downstream cleavage products would also explain an observation we made years ago in a mutant strain that is missing PNPase. In the PNPase-deficient strain, decay intermediates, which are virtually undetectable in the wild type, accumulate to a high degree (see Section IV.B). However, the overt difference in mRNA decay fragment accumulation between wild-type and PNPase mutant strains is only observed with a probe that is targeted to the 5' region of the message. Using a 3'-targeted probe, no difference is observed between wild-type and PNPase-deficient strains (39). This result can be explained by the dual action of RNase J1. Decay intermediates that contain the 5' end accumulate in the PNPase-deficient strain because they are the *upstream* products of RNase J1 cleavage, which may occur at several places in the body of the message. These need to be turned over by 3'-to-5' exonuclease activity, of which PNPase is the major one in *B. subtilis*. In the absence of PNPase, multiple upstream cleavage products accumulate. However, the *downstream* products of RNase J1 cleavage do not accumulate because they are either cleaved again by RNase J1 or degraded by the 5'-to-3' exonuclease activity of RNase J1.

D. 5' Stabilizers

As mentioned, the long-held assumption that the nature of the 5' end plays a major role in mRNA stability in *B. subtilis* came from early evidence of "5' stabilizers." These are defined as 5' elements that confer stability to downstream RNA, independent of the downstream RNA sequence or structure or translatability. Examples of such elements will be brought in the following discussion. The existence of 5' stabilizers suggests that many mRNAs are not degraded from the 3' end, likely because of the resistance of 3' transcription terminator structures to 3'-to-5' exoribonucleases. 5' stabilizers also suggest that a ribonuclease that initiates decay by cleaving endonucleolytically must access the message through the 5' end and not directly at its cleavage target site. For *E. coli* mRNA decay, the pervasive role of RNase E, which is a 5'-binding endonuclease, fits well with several observations of the protective effect of 5'-proximal secondary structure (22, 40, 41).

The discovery of RNase J1 5'-to-3' processive exonuclease activity in *B. subtilis* makes clear how, at least in this mode of mRNA processing, a 5' element could confer stability to long downstream sequences by interfering with decay from the 5' end. Any circumstance that makes the 5' end of an mRNA less recognizable as a single-stranded RNA binding site is predicted to increase the half-life of that message. These could include binding by a protein factor or ribosome close to the 5' end, formation of an intrinsic, 5'-terminal secondary structure, or the presence of a 5'-triphosphate terminus. Furthermore, any hindrance to movement of the 5'-binding enzyme at sites further downstream from the 5' end could block decay and increase mRNA half-life. For example, in the study that documented the 5'-to-3' exonuclease activity of RNase J1, Condon and colleagues demonstrated that the insertion of a particular 5'-stabilizer sequence, known as STAB-SD (42), and binding of a ribosome there blocked the processivity of RNase J1 (7).

If it is the case that the half-life of many *B. subtilis* mRNAs is determined by accessibility of RNase J1 to the 5' end, as well as the ability of RNase J1 to track along a message, one could hypothesize that certain 5'-proximal elements are good predictors of long or short mRNA half-life. In Section II.E, we present a framework for characterizing elements in the 5' end of *B. subtilis* mRNAs that allows predictions of mRNA stability.

E. Stability of mRNA Prediction (STOMP)

To facilitate a discussion of 5'-proximal elements that determine mRNA stability in *B. subtilis*, it is practical to talk about three categories of mRNA: (1) unstable: $t_{1/2} < 2$ min; (2) average stability: $t_{1/2} = 3$–8 min; and (3) extremely stable: $t_{1/2} > 15$ min. A microarray analysis of *B. subtilis* mRNA, which looked at decay rates of mRNAs encoded by about 1500 genes, showed that a large majority of mRNAs fall into the second category, with a half-life of 3–8 min (43). In an attempt to understand the instability of mRNAs that fall into the first category or the lower level of the second category versus the extreme stability of mRNAs that fall into the third category, Hambraeus and colleagues examined the 5' region of nine unstable mRNAs and nine stable mRNAs whose transcription start sites were known at the time. They were unable to establish "rules" that could be used to predict the stability of an mRNA based on the 5' region. It was noted that the relative strength of the interaction between the Shine-Dalgarno (SD) sequence and the 3' end of the 16S ribosomal RNA was not a predictor of mRNA stability.

We have published an extensive characterization of the effect of various 5'-proximal elements on the stability of Δ*ermC* mRNA, including the effect of unpaired 5'-terminal nucleotides, 5'-proximal secondary structure, and distance between 5' secondary structure and SD sequence (44). We used the information gained from the study of Δ*ermC* mRNA to explain the stability of 7 mRNAs from the Hambraeus *et al.* (43) microarray study, whose extreme stability was confirmed by Northern blotting. On the basis of our findings with Δ*ermC* mRNA, we can identify the following characteristics of the 5' region of an mRNA that can be used to predict, with some degree of confidence, the stability of an mRNA (Fig. 4).

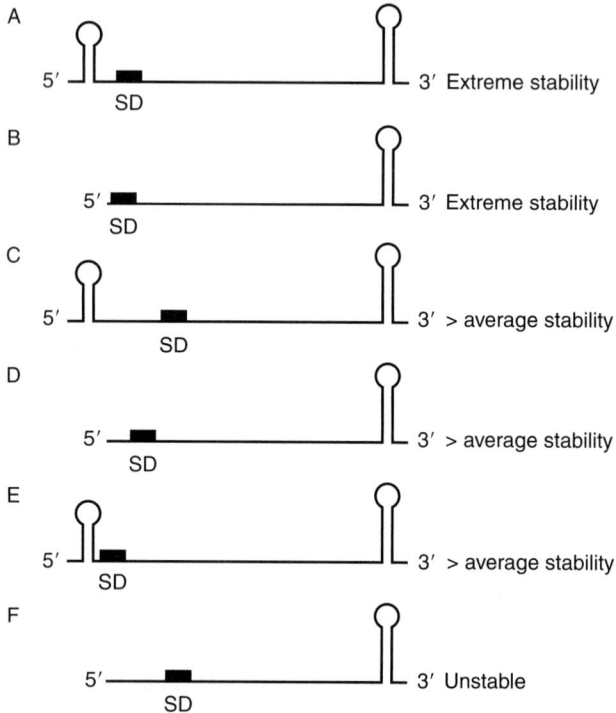

FIG. 4. Schematic diagrams of 5′ elements used in STOMP. SD, Shine-Dalgarno element.

Extremely stable mRNAs tend to have one of two arrangements in the 5′-proximal region (Fig. 4A and B): (1) a 5′-proximal secondary structure of reasonably low free energy (e.g., a ΔG_0 of -3.0 kcal mol^{-1} or lower), which begins 5 nts or less from the 5′ end, followed by an SD sequence that is a short distance away from the 3′ end of the secondary structure (but not too close to the secondary structure; that is, more than 5 nts distance); or (2) an SD sequence that is less than 15 nts from the 5′ end.

The rationale for how these characteristics can confer stability rests on the hypothesis that the 5′-end-dependent activity of RNase J1 is the major activity responsible for initiation of decay. As such, severe limitation of access to the 5′ end slows down the initiation of decay, producing an extremely stable mRNA. A 5′ terminus that is buried in, or very close to, a stable secondary structure (Fig. 4A) would inhibit binding of RNase J1 to the 5′ end, a prerequisite for initiating the decay process. Nevertheless, secondary structure formation would usually not be so strong as to completely preclude RNase J1 access, which would therefore bind *albeit* at a slower rate. For RNase J1 that is able to bind the 5′ end, the second element, an SD sequence located a short distance from the secondary structure, would come into play to block RNase J1 processivity—either its 5′-to-3′ processive exonuclease activity or perhaps tracking on the mRNA to an endonuclease cleavage site. Thus, the combination of 5′-proximal secondary structure to block access and a functional downstream SD sequence to block processivity would combine to put the mRNA into the extreme stability category. This model assumes that alternative pathways for decay (e.g., from the 3′ end or internal sites at which RNase J1 can bind without involvement of the 5′ end) are not prevalent in B. subtilis.

Even in the absence of strong 5′-terminal secondary structure, an SD sequence located close to the 5′ end (Fig. 4B) would confer extreme stability. We envision that a 16S rRNA-SD sequence interaction close to the 5′ end places the ribosome small subunit in such a way that the 5′ end of the mRNA is buried in ribosomal structure. Crystal structure data on a ribosome bound to mRNA shows that 15 nts upstream of the initiation codon are protected by the ribosome (45). The relatively large size of RNase J1 could contribute to difficulty in accessing the 5′ end of an mRNA that is in close contact with a ribosome.

Messenger RNAs in the high end of the average stability range (e.g., 6–8 min) may have 5′-terminal secondary structure but no proximal SD sequence (Fig. 4C), or may have an SD sequence located relatively close to the 5′ end (e.g., less than 20 nts), but not so close that ribosome binding shields the 5′ end (Fig. 4D). Neither one of these elements alone can confer stability in the extreme range but each can confer stability in the high-end of the average stability range. Another situation that would lead to some stability but not extreme stability would be the presence of 5′-terminal secondary structure with an SD sequence that was too close to the structure (Fig. 4E). The close proximity of these elements would cause inefficient formation of the secondary structure, due to nearby ribosome binding, or to inefficient ribosome binding, due to local secondary structure.

Messenger RNAs in the low end of the average stability range (e.g., 3–5 min) *or in the unstable category* would tend to have no predicted 5′-proximal secondary structure, and their SD sequence would be distant (e.g., more than 20 nts) from the 5′ end (Fig. 4F). These two characteristics would allow unhindered access of RNase J1 to the 5′ end, which could then initiate decay by either 5′-to-3′ exonuclease activity or endonuclease cleavage. Binding of a ribosome at a downstream SD sequence may not be able to block RNase J1 processivity enough to confer stability. On the other hand, the presence of a particularly strong SD sequence can be a stabilizer even if it is located well downstream of the 5′ end, as observed with the STAB-SD sequence (42), which has perfect complementarity to the final nine nucleotides of 16S rRNA.

We propose that features of the 5′ end of an mRNA are predictors of mRNA stability. The application of these "rules" is called "STOMP," for stability of mRNA prediction. Here, we use STOMP rules to extend our

TABLE II
Characteristics of Unstable and Extremely Stable mRNAs

	Gene name	Half-life (min)	nts From +1 to SD	5'-Terminal 2⁰ structure[a]	$\Delta G_0{}^b$	2⁰ Structure begins at	nts Between end of 2⁰ structure and RBS
1	argC	1	15	−			
2	dppA	3	22	−			
3	flgB	2	50	−			
4	sdhC	2	77	−			
5	odhA	2	98	+/−	−2.9	+1	67
6	pbpX	2	26	+/−	−1.5	+8	0
7	pdhA	4	51	+/−	−1.7	+2	27
8	qcrA	2	55	+/−	−1.0	+4	29
9	acoA	>15	13	−			
10	bofA	>15	9	−			
11	ftsH	>15	15	−			
12	gsiB	>15	9	−			
13	pbpE	>15	7	−			
14	sspA	>15	2	−			
15	yaaH	>15	7	−			
16	yxzE	>15	9	−			
17	yydF	>15	8	−			
18	aprE	>15	41	++	−9.0	+2	10
19	cotG	>15	27	+	−4.5	+1	12
20	opuE	>15	99	+++	−15.2	+1	59
21	sunA	>15	27	+	−3.8	+2	10
22	ybcO	>15	38	+++	−12.6	+1	8
23	abnA	>15	97	−			
24	ctsR	>15	25	−			
25	gltX	>15	27	−			
26	ybaN	>15	23	−			

[a] nts 1–40 were analyzed at the Zuker mfold site (http://frontend.bioinfo.rpi.edu/applications/mfold/cgi-bin/rna-form1.cgi).
[b] ΔG_0 value (kcal mol^{-1}) is for nts + 1 to 3 nts past the end of the predicted stem-loop structure.

previous analysis of seven stable mRNAs (44) to include 26 mRNAs (Table II) that have the following characteristics: (a) in the microarray study they are characterized either as unstable or as extremely stable; (b) the transcription start site is known (data obtained from the primary literature as documented in the DBTBS promoter database at http://dbtbs.hgc.jp); (c) they are either monocistronic or are the 5′-most cistron of a polycistronic mRNA. Eight of these mRNAs (1–8) are in the unstable or low end of average stability category, and 18 mRNAs (9–26) are in the extremely stable category. A number of relevant features of these mRNAs are shown in Table II, and can be summarized as follows.

Unstable mRNAs 1–4 have no 5′-terminal secondary structure and, except for the case of *argC*, their SD sequence is located more than 20 nts from the 5′ end. For unstable mRNAs 5–8, there is weak potential 5′-terminal secondary structure, and the location of the SD sequence relative to the 3′ end of these structures is either very close (*pbpX*) or at least 27 nts away.

Extremely stable mRNAs 9–17 have no predicted 5′-terminal secondary structure but their SD sequences are located 15 nts or less from the 5′ end. For one of these, *gsiB*, it has been demonstrated *in vivo* that the SD sequence is critical for stability (46). For extremely stable mRNAs 18–22, significant 5′-terminal secondary structure can be predicted, all of which begin either at the very 5′ terminus or 1 nt away. Four of these also have an SD sequence located 8–12 nts from the 3′ end of the predicted secondary structure, which we have found is an optimal distance (44). For *opuE* mRNA, the SD sequence is a considerable distance from the 5′-terminal structure. However, the extraordinary predicted strength of the 5′ structure is such that perhaps it hinders RNase J1 access to an extent that additional elements are not required to confer extreme stability.

In an interesting case of mRNA stabilization conferring a phenotype, a mutation in the *bmr3* gene (annotated as the *mdr* gene on the Subtilist web site) results in multidrug resistance (47). The mutation is a single nucleotide change that provides an additional base pairing in a 5′-terminal stem-loop structure, such that the free energy of the 5′ stem-loop formation goes from -5.2 to -9.5 kcal mol^{-1}. This mutation causes an increase in stability of *bmr3* mRNA from about 4 to greater than 15 min. Here, too, the SD sequence is located quite a distance from the end of the 5′-terminal secondary structure (22 nts), but there is a possibility of another stem-loop structure within this interval, and this may make the SD location irrelevant to stability.

Returning to Table II, mRNAs 23–26 are extremely stable, but they do not seem to conform to the STOMP rules. In the case of *abnA*, the very long leader region might be an indicator that a protein binds there, which is another way to confer stability (48). Thus, cases in which STOMP alone fails to account for the extreme stability of an mRNA may indicate that an additional factor, such as protein binding, is involved. It would be of interest to mutate the 5′-proximal nts of the remaining three extremely stable mRNAs, to determine what is required for their long half-life.

Detailed studies with $\Delta ermC$ mRNA have helped to understand how translational elements alone—in the absence of 5′-terminal secondary structure—confer relative stability to mRNAs (i.e., in the high end of average stability) (49). $\Delta ermC$ mRNA, with an SD sequence located 18 nts from the 5′ end, has a 6–8 min half-life. Relocating the SD sequence further away from the 5′ end (i.e., 28, 45, and 91 nts more) results in progressive shortening of half-life. The presence of a start codon is necessary to achieve the 8 min half-life; mutating the start codon to ACG or CCG results in a fourfold drop in half-life to around 2 min. This suggests that the interaction of the SD sequence with the 3′ end of 16S rRNA 3′ is not sufficient to provide stability, but that a tertiary complex (mRNA-16S rRNA-tRNA^{fmet}) is required. In fact, mutating the start codon has as big an effect on $\Delta ermC$ mRNA half-life as eliminating the SD sequence. The model that can be derived from these results is that ribosome binding and formation of a ternary complex near the 5′ end of a message competes with binding by RNase J1, resulting in an mRNA that has a high-end average stability.

It should be noted that, unlike in the case of $\Delta ermC$, where we find that ribosome binding *and* ternary complex formation are required to render $\Delta ermC$ mRNA somewhat stable, a downstream start codon is not required for stability conferred by STAB-SD, which consists of a "perfect" SD sequence, as mentioned above (42). Similarly, in the case of *aprE* (number 18 in Table I), where both a 5′ stem loop and a nearby SD sequence are needed for full stability, mutation of the stop codon has little effect on half-life (50).

The development of "rules" for predicting stability is valuable only to the extent that they can guide future research. The rules generate hypotheses that can facilitate decision-making on which elements to focus when determining the basis for stability or instability. However, it is clear that there is a lot to be learned about how *cis*- or *trans*-acting elements affect stability, and whether or not the access of RNase J1 to the 5′ end of an mRNA is the "bottom line" for most *B. subtilis* mRNAs.

F. Pyrophosphatase Activity

Belasco's group has recently published their findings on the role of *E. coli* pyrophosphatase activity at the 5′ end of a message in initiation of mRNA decay (51, 52). The model is that decay-initiating RNase E cleavage of many mRNAs depends on prior 5′-end binding, which is inhibited by a 5′ triphosphate end. Conversion of the 5′ triphosphate end to a 5′ monophosphate end by pyrophosphatase activity allows RNase E to bind and decay to begin with a downstream endonuclease cleavage. The *E. coli* enzyme responsible for pyrophosphatase activity on mRNA 5′ ends was recently identified as RppH (52).

As outlined in Section II.C, the endonuclease activity of RNase J1, which is not blocked by a 5′-triphosphate end, allows for initiation of decay without prior phosphatase activity. Thus, while attack of primary *B. subtilis* transcripts in the 5′-to-3′ exonucleolytic mode of RNase J1 would require prior pyrophosphate removal (Fig. 1B, left), initiation of mRNA decay by RNase J1 endonuclease activity would not be dependent on phosphohydrolase activity (Fig. 1B, right). It should be mentioned that *E. coli* RppH (formerly NudH) is a member of the NUDIX family of hydrolases (53), and at least six genes in *B. subtilis* are predicted to encode NUDIX proteins: *mutT, nudF, yjhB, ytbE, ytkD,* and *yvcI*. It remains to be determined whether pyrophosphatase activity in *B. subtilis* plays a role in mRNA decay, as it appears to in *E. coli*.

G. Destabilizers

There is an example in *B. subtilis* (as far as this author is aware, the only example) of a 5′ mRNA destabilizer. This is the case of the CIRCE element (CIRCE, controlling inverted repeat of chaperone expression), which is an inverted repeat sequence found in the 5′ region of class I heat shock genes (54). The CIRCE element in double-stranded DNA functions as a binding site for HrcA, a transcriptional repressor that is encoded by the first gene in the *dnaK* stress response operon, and this negatively regulates transcription of the heptacistronic *dnaK* operon and the bicistronic *groE* operon. Curiously, the same CIRCE sequence in RNA form, located near the 5′ end of either *hrcA* mRNA (55) or *groE* mRNA (56), functions as an mRNA destabilizer: The CIRCE stem-loop structure is predicted to have a low free energy of formation ($\Delta G_0 = -12.6$ kcal mol^{-1}) and begins just 5 nts from the 5′ end in both cases. For *hrcA* mRNA, the CIRCE secondary structure overlaps with the *hrcA* SD sequence, and, as mentioned above, this close distance could interfere with ribosome binding or secondary structure formation. When the CIRCE element is removed, the SD sequence is now located 11 nts from the 5′ end and the mRNA is more stable, so, according to STOMP rules, we can understand how the presence of CIRCE functions to destabilize *hrcA* mRNA. However, for *groE* mRNA, there is a 17-nt interval between the end of the CIRCE structure and the SD sequence. These characteristics, according to STOMP, would predict an extremely stable mRNA! It will be of interest to investigate the CIRCE destabilizer in more detail.

III. Body of The Message

A. Translation and Decay

Historically, there has been some controversy about whether a flow of ribosomes down the coding sequence of an mRNA is a necessary element in mRNA stability. Early experiments to assess the effect of ribosome flow on mRNA stability used various antibiotics that either froze ribosomes on the message or prevented initiation of translation, but there were likely indirect effects of the addition of antibiotics. In more recent experiments, a better assessment of the effect of translation on mRNA decay is achieved by the use of targeted mutations in either an SD sequence, an initiation codon, or introduction of a premature termination codon. The subject has been reviewed recently (57). Above, we discussed the effect on stability of 5′-proximal elements that are involved in translation initiation. Here, we address the issue of translation elongation and translational pausing.

It is safe to say that, in *B. subtilis*, translation of the body of the message is not a factor in determining half-life. This has been documented in detail for Δ*ermC* by the observation that introduction of premature termination codons, even immediately after the initiation codon, has no effect on mRNA half-life (49). Similar observations have been made for other *B. subtilis* mRNAs (42, 46, 50). These results suggest that the 5′-proximal TIR is the major determinant of stability, and that ribosome flow on the message is not required to protect against endonucleolytic cleavage. Furthermore, the suggestion is that ribosome flow on the message will not be protective, and that, once initiation of decay occurs, the presence of ribosomes will not prevent endonucleolytic cleavages from happening.

It has been found in one case that stalling of a ribosome in a coding sequence is concomitant with cleavage of the mRNA (58). This is the case for Δ*ermC* mRNA, upon which a ribosome that is bound by erythromycin stalls after translating nine codons (Fig. 5A) (59). Endonuclease cleavage occurs in the vicinity of the stalled ribosome and results in accumulation of a downstream

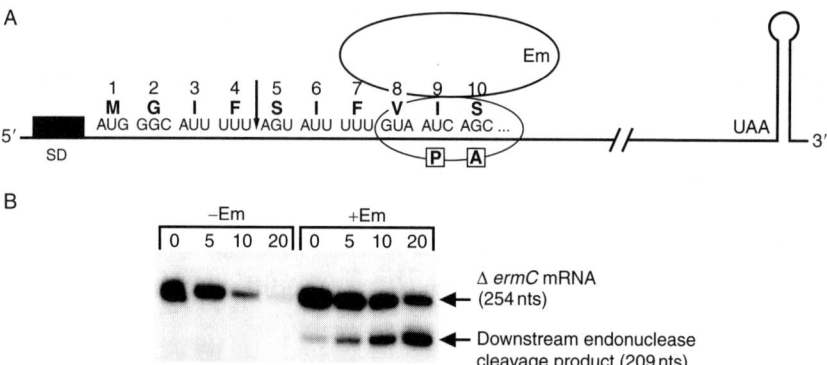

FIG. 5. Δ*ermC* mRNA processing. (A) Binding of erythromycin (Em) to a ribosome that is translating the *ermC* leader peptide coding sequence (first 10 codons shown), results in ribosome stalling with codon 9 in the ribosomal P site and codon 10 in the ribosomal A site. Downward arrow indicates site of cleavage by RNase J1. (B) Northern blot analysis of Δ*ermC* mRNA at times (min) after rifampin addition. The cleavage product accumulates with time (from (53)).

cleavage product (Fig. 5B). The mapped location of ΔermC cleavage site shows that this is not a case of "A-site cleavage" that has been documented in E. coli (60, 61). Recent work shows that the cleavage is catalyzed by RNase J1 (130). It remains to be determined whether this cleavage occurs only upon ribosome stalling or occurs constitutively but the downstream product is only observed when it is protected from rapid decay by the ribosome that is stalled near the 5' end.

B. Messenger RNA Length

Whether the length of an mRNA correlates with its half-life remains a matter of speculation. A recently published survey of correlation between length and stability in various organisms found no significant correlation for B. subtilis (62).

Despite the fact that longer mRNAs should provide more targets for a decay-initiating endonuclease, it has been found in several cases that 5' stabilizers in B. subtilis confer stability to long heterologous sequences, such as the E. coli lacZ coding sequence, which is about 3000 nts long. This suggests that protection at the 5' end is sufficient to confer stability independent of the length of, and therefore the number of potential cleavage targets in, the body of the message. In the absence of a 5' stabilizer, the mechanism for initiation of decay and subsequent extremely rapid turnover is such that one would likely not notice a difference in stability between an mRNA that had a few or many internal target cleavage sites.

On the other hand, it is possible to construct an mRNA that contains such a short sequence that, presumably, it provides no good target for a decay-initiating endonuclease. In an unpublished study, we have found that a small ΔermC mRNA construct around 125 nts in length, with an internal deletion that does not affect 5' or 3' elements, is twice as stable as the 254-nt ΔermC mRNA itself (J. Sharp and DHB, unpublished). The absolute size is not what matters but rather which sequences are deleted, since several other deletion constructs of the same small size are not significantly more stable than ΔermC mRNA. Thus, we hypothesize that the stability of the 125-nt RNA is due to the absence of good quality RNase J1 cleavage sites. By contrast, we find that trp leader RNA, which is only 140 nts long, has an extremely short half-life of about 1.5 min (63). This is despite the fact that about 20% of this RNA is in a 5-terminal stem-loop structure, the middle 40% of the RNA is bound by the TRAP protein complex, and a further 20% is in the 3' transcription terminator stem-loop structure, leaving only a small region of "accessible" single-stranded RNA between the TRAP binding site and the 3' terminator structure (Fig. 6A). We found that an RNase J1 target site is located in this single-stranded region,

A

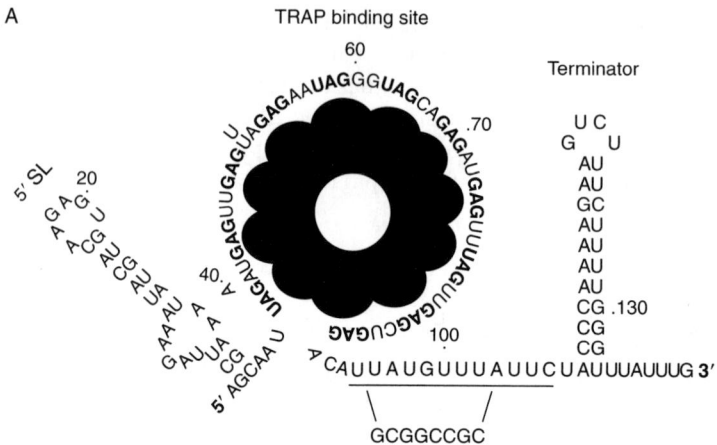

B

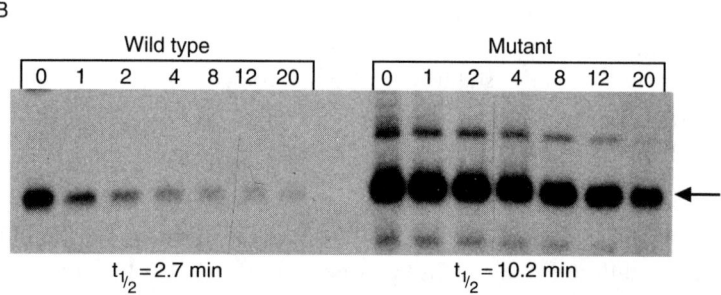

FIG. 6. Decay of *trp* leader RNA. (A) Schematic diagram of *trp* leader RNA bound by 11-mer TRAP complex in the presence of ample tryptophan. Binding of TRAP causes the terminator structure to form. 5′ SL, 5′ stem-loop structure. RNase J1 cleavage site at around nt 100 is underlined, and the mutation to change this AU-rich sequence to a GC-rich sequence is shown below. (B) Northern blot analysis of *trp* leader RNA decay at times (min) after rifampin addition, for RNA with wild-type and mutant RNase J1 cleavage site. Arrow at right indicates band that is full-length *trp* leader RNA. A fourfold increase in half-life is observed (from (61)).

and cleavage at this site results in rapid RNA turnover (35, 64). When this site is mutated from an AU-rich sequence to a GC-rich sequence (Fig. 6A), the stability of *trp* leader RNA increases fourfold (Fig. 6B).

C. Bs-RNase III

Cleavage at an internal target site by an endonuclease that does not require a 5′ end is a possible mechanism for initiation of decay. Indeed, several cases have been reported in which cleavage by the narrow-specificity endonuclease

RNase III of *E. coli* is the rate-limiting step for destabilization of an mRNA (65–67). The *B. subtilis* homologue of *E. coli* RNase III is Bs-RNase III, which is a 28 kDa protein that is 34% identical to the *E. coli* enzyme and that shows similar cleavage site specificity (68). Cleavage occurs at an internal loop sequence in a double-stranded structure that is formed by intramolecular base-pairing. The RNase III enzymes are part of a large family of endoribonucleases that include the siRNA and miRNA processing enzymes of eukaryotes.

The only known endogenous *B. subtilis* RNAs that are substrates for Bs-RNase III are 30S ribosomal precursor RNA (69) and scRNA (38, 70). An artificial construct containing a phage Bs-RNase III cleavage site in the body of an mRNA that bears the *ermC* 5′ stabilizer sequence is no longer stable upon addition of erythromycin (71). This demonstrates that the effect of a 5′ stabilizer is not a general one on the entire message but is specifically protective against binding at the 5′ end. Internal access to the message by way of a Bs-RNase III cleavage target overrides the 5′ stabilizing effect.

The fact that the *rncS* gene encoding Bs-RNase III is essential in *B. subtilis* (72) suggests that one or more critical mRNAs depend on Bs-RNase III for processing to give correct expression, and this may involve endonucleolytic cleavage to control mRNA stability. Rare cells that have a deletion of the *rncS* gene can be recovered, and presumably these have second-site mutations that compensate for the lack of Bs-RNase III. Such mutations have not been mapped to date.

Very recently, a second RNase III-like protein, called Mini-III, has been described in *B. subtilis*, which is involved in maturation of 23S rRNA (73). As is the case for Bs-RNase III, no endogenous mRNA substrates are known for this enzyme.

D. RNase M5, RNase P, and RNase Z

Do other endoribonucleases in *B. subtilis* function in mRNA turnover? RNase M5, whose major role is in the maturation of 5S rRNA (74), does not appear to have mRNA targets in *B. subtilis* (75).

RNase P is a ribonucleoprotein involved in maturation of tRNAs. In *E. coli*, RNase P, whose major role is in tRNA processing, is known to be involved in turnover of polycistronic mRNAs (76). A possible role for RNase P in *B. subtilis* mRNA processing *in vivo* has been reported recently (77). RNase P cleavage of the 5′ leader riboswitch region of *pbuE*, which encodes an adenine efflux protein and which is regulated by adenine binding, was observed *in vitro*. Reduction of RNase P expression *in vivo* resulted in an apparent destabilization of *pbuE* mRNA. The recent construction of conditional RNase P expressing strains, with mutants of both the protein component (78) and the RNA component (79), paves the way for observing possible activity of RNase P in mRNA decay.

RNase Z, another *B. subtilis* endoribonuclease, is required for 3' end formation of tRNAs that do not have an encoded 3' CCA sequence (*80*). The function of RNase Z in *E. coli*, whose tRNAs all have an encoded 3' CCA sequence, has recently been studied by Kushner and colleagues (*81*). Although an RNase Z deletion mutant had only a small effect on mRNA decay, the combination of an RNase Z deletion with an RNase E temperature-sensitive mutation had a large positive effect on the stability of several mRNAs that were tested. To date, a role, if any, of RNase Z in *B. subtilis* mRNA decay has not been reported.

E. Differential Stability of Polycistronic mRNA

Different cistrons in the same operon often need to be expressed at different levels. One way to ensure differential expression is to process a polycistronic mRNA such that a segment that encodes a protein needed in small amounts (e.g., a regulatory protein) becomes unstable, and segments that encode proteins needed in large amounts become more stable. There are three known examples of this form of posttranscriptional regulation in *B. subtilis*: the *dnaK* operon (*55*), the *gapA* operon (*82, 83*), and the *ilv-leu* operon (*84*). The organization and RNA products of these operons are shown schematically in Fig. 7 and are described in more detail below. The total number and steady-state level of transcripts is a function of both transcriptional and posttranscriptional regulation. Here we mention briefly the overall architecture of the operons and their RNA processing events, and the nature of the processing sites. An excellent discussion of these three cases can be found in Mader *et al.* (*84*).

The heptacistronic *dnaK* operon contains both an RNA processing site, in between the *hrcA* and *grpE* genes, and an internal transcription terminator located downstream of *dnaK* (Fig. 7A). The full-length transcript is 8.0 kb long. The first gene of the *dnaK* operon codes for the transcriptional repressor HrcA. An mRNA destabilizer is located at the 5' end of the *hrcA* mRNA, as has been described above in Section II.G, and this ensures a low steady-state level of *hrcA* mRNA. The downstream genes are expressed at higher levels than *hrcA*, and are not subject to destabilization, due to cleavage of operon mRNA at the processing site located between the *hrcA* and *grpE* genes. The processing site has been mapped to a putative secondary structure that contains a branched stem at which cleavage occurs (Fig. 7A). A second primary transcript arises from termination at the *dnaK* terminator, and this transcript is also processed at the *hrcA–grpE* processing site to give unstable upstream and stable downstream mRNAs. The stable downstream mRNA codes for the heat shock proteins GrpE and DnaK, which are present in higher amounts than other products of the operon.

A very similar processing mechanism occurs with the hexacistronic *gapA* operon (Fig. 7B). Here again, the first gene (*cggR*) codes for a transcriptional repressor protein. The neighboring gene, *gapA*, codes for glyceraldehyde-3-phosphate

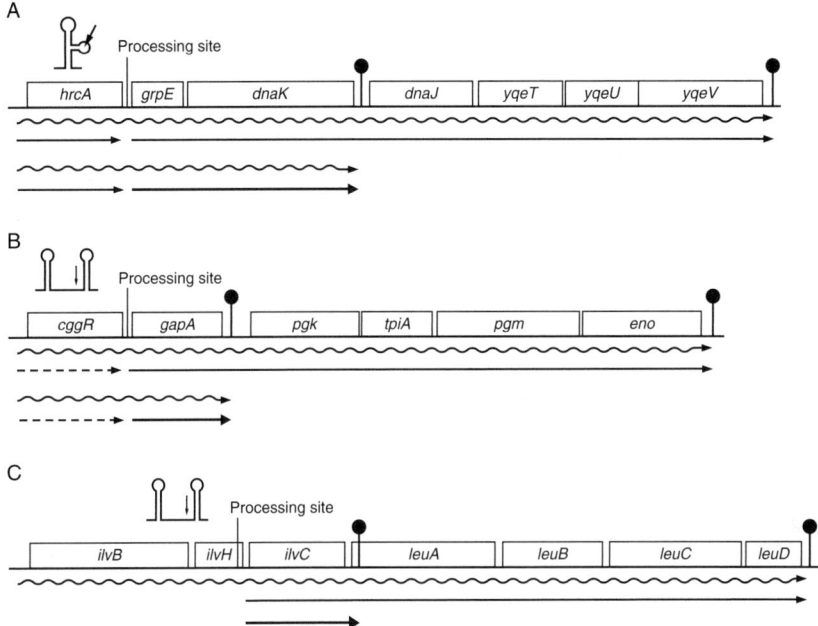

FIG. 7. Three examples of polycistronic mRNA processing in *B. subtilis*. Open rectangles are coding sequences. "Lollipops" are transcription terminator sequences. Location of the processing site for each operon is indicated by the vertical line, with the predicted secondary structure and site of endonuclease cleavage shown to the left. Wavy lines are initial transcripts, solid lines are processed transcripts, with the thickness indicating relative abundance. Arrows with dotted line in *B* indicate extremely unstable RNAs. Only the RNAs that are detected by Northern blotting are shown.

dehydrogenase, and more of the dehydrogenase enzyme than of the transcriptional regulatory protein is required in the cell. Differential expression of the *cggR* and downstream cistrons is achieved by cleavage of the sequence between the *cggR* and *gapA* coding sequences. The processing site is in a single-stranded region between two putative stem-loop structures (Fig. 7B). Cleavage here results in an extremely unstable upstream *cggR* mRNA (no detectable signal 2 min after rifampin addition) and a more stable downstream *gapA* mRNA with a half-life of 3.5 min. (There is also an internal promoter upstream of *pgk*. The transcript from this promoter is not shown in Fig. 7B.) Interestingly, the 5′ end of processed *gapA* mRNA has a 5′-terminal stem-loop structure, which starts at +3 and is predicted to form a stable structure ($\Delta G_0 = -9.9$ kcal mol^{-1}), as well as an SD sequence located 20 nts downstream of the stem loop. On the basis of STOMP rules, one might expect these two elements to confer extreme stability or at least (because the distance between the 5′-proximal structure and the SD sequence is a bit long) high-end average stability. However, it could be the case that STOMP rules that apply to primary transcription products, which bear a 5′ triphosphate end, do not necessarily apply to processed RNA fragments, which bear a 5′ monophosphate or hydroxyl end. Degradation of the downstream fragment may be RNase J1-dependent, and it has been shown in the case of *glmS* RNA (see Section V) that the product of ribozyme cleavage, which bears a 5′ hydroxyl, is efficiently degraded in an RNase J1-dependent manner, even though there is substantial 5′-proximal secondary structure predicted for this processed RNA (85).

The third example of a differentially processed operon mRNA is the heptacistronic *ilv-leu* operon (Fig. 7C). A processing site is located between *ilvH* and *ilvC*, which, by secondary structure prediction, resembles that of the *gapA* processing site. Cleavage at this site, combined with transcription termination downstream of *ilvC*, generates a relatively stable *ilvC* mRNA, resulting in a 10-fold greater amount of IlvC protein than the other proteins encoded by this operon.

The similarity of the posttranscriptional mechanism in these three cases suggests that it is likely to apply generally to polycistronic mRNA whose individual cistrons need to be expressed at different levels. To date, the identity of the endoribonuclease that cleaves at these processing sites is unknown, but RNase J1 is a prime candidate. With the availability of ribonuclease mutants that have been constructed in recent years, we can look forward to identification of the processing enzyme required to achieve differential stability.

IV. The 3′ End

A. Transcription Terminator Structure

The 3′ end of Rho-independent terminated transcripts consists of a stable stem-loop structure that, in most cases, will be a barrier to 3′-to-5′ exonuclease processivity (86, 87). It has been shown for *B. subtilis in vivo* and *in vitro* that strong secondary structure can block the action of PNPase, the major 3′ exonuclease in this organism (64, 88). Although *B. subtilis* does have a *rho* gene, there is little evidence for Rho-dependent transcription termination (89), and, unlike *E. coli*, *rho* is not an essential gene in *B. subtilis* (90). Therefore, it can be assumed that the overwhelming majority of *B. subtilis* transcripts are terminated in a Rho-independent manner and thus have a protected 3′ end.

Extensive study of 3′-end-initiated mRNA decay in *E. coli* has shown that the addition of a poly(A) tail by poly(A) polymerase (PAP) is necessary to provide a "toehold" for 3′ exonucleases to bind the 3′ end of messages (91). Reiterative 3′ degradation and polyadenylation is thought to overcome the structural barrier presented by the transcription terminator (28). Whether this is also the case in *B. subtilis* remains to be determined (see Sections IV.D and IV.E).

While the evidence is clear that the 5' end of a *B. subtilis* mRNA is dominant in determining stability, a "weak" 3' transcription terminator sequence that results in formation of a 3' end with less than typically strong secondary structure could allow decay of the message from the 3' end, most notably by a 3'-to-5' exonuclease such as PNPase. It can be predicted that replacement of a weak 3' end sequence with a very strong one, without a change in the 5'-proximal sequence, would increase stability. This was shown to be the case years ago, when the 3' end of the *B. thuringiensis cry* gene was found to act as a "retroregulator," that is, the presence of the *cry* transcription terminator at the 3' end of *penP* mRNA, which normally has a short, 2-min half-life in *B. subtilis*, resulted in a threefold increase in half-life (92). The *B. thuringiensis cry* transcription terminator is predicted to form a stem-loop structure with an extremely low free energy ($\Delta G_0 = -25.5$ kcal mol^{-1}). Thus, it is not surprising that replacement of the weaker *penP* terminator sequence with that of the *cry* gene could provide increased protection against processive decay from the 3' end, and thereby stabilize upstream sequences, as has been suggested (93). However, the structure that is predicted to form by the native transcription terminator of the *B. thuringiensis penP* gene is itself quite strong ($\Delta G_0 = -20.1$ kcal mol^{-1}), and one would not expect that strong versus stronger 3' end structures would make such a profound difference in upstream mRNA stability. Perhaps the difference in stability provided by these two 3' ends is related to the number of single-stranded nucleotides present following the 3'-terminal secondary structure. The *penP* transcript sequence following the predicted secondary structure contains a run of mostly Us (UUUUUGUUUU), which may be included in the transcript, providing a single-stranded "tail" for a 3'-to-5' exoribonuclease. The predicted secondary structure of the *cry* transcript, on the other hand, is followed by a heterogeneous sequence (ACUAUG...), and so only a short 3' single-stranded tail might be present in the terminated transcript. *In vitro* studies have shown that *B. subtilis* PNPase requires a significant 3' extension downstream of a transcription terminator to be able to bind to RNA (64).

B. PNPase

We mentioned above the earlier findings that PNPase, encoded by the *pnpA* gene, is the major 3'-to-5' exonuclease activity involved in *B. subtilis* mRNA turnover. PNPase is a member of the phosphate-dependent exoribonuclease (PDX) family, and close homologues are widely distributed in bacteria (94). *B. subtilis* PNPase is a large enzyme (molecular weight 77 kDa), and, by analogy to other PNPases, likely functions as a trimer (95). This 3'-to-5' exonuclease degrades RNA phosphorolytically, yielding nucleoside diphosphates as the product. Under conditions of excess nucleoside diphosphate, B. subtilis PNPase can undergo the reverse reaction and add nucleotides to the 3' end of an RNA substrate (96).

Despite playing a major role in mRNA decay, a *pnpA* deletion strain of B. subtilis is viable. The *pnpA* deletion does result in a number of diverse phenotypes, including cold sensitivity, filamentous growth, tetracycline sensitivity, and competence deficiency (6, 97). The basis for these phenotypes is not known, but likely reflects a requirement of PNPase activity to regulate expression of genes involved in these processes. Another phenotype of the PNPase-deficient strain is gross overexpression of the *trp* operon, even in the presence of ample tryptophan (63). We have shown that this is due to the inability of 3' exonucleases other than PNPase to degrade the *trp* leader RNA, which results in titration of the *trp* operon regulatory protein (TRAP) such that expression is no longer controlled. While the details of the *trp* system are not germane to this review, the concept that was demonstrated by these findings is likely of general significance, that is, that an RNA-binding regulatory protein needs to be released from its RNA target in a timely fashion, and degradation of the bound RNA by ribonucleases is one mechanism to achieve this release.

The *expected* phenotype of the PNPase-deficient strain is a defect in mRNA decay. Turnover of a number of monocistronic, small mRNAs are known to be severely affected by the absence of PNPase. In a wild-type strain, only full-length mRNA can be detected, while in the PNPase deletion strain, multiple decay intermediates are easily detected by a probe directed to the 5' end of the gene (39, 98). An example of the effect on mRNA decay of the PNPase deletion is *rpsO* mRNA, a monocistronic, 388-nt mRNA encoding ribosomal protein S15 (Fig. 8, lanes 1 and 2). In the absence of PNPase, a ~180-nt decay intermediate accumulates to a level that is higher than the native transcript. This observation is unlike the case in *E. coli*, where knocking out a single 3'-to-5' exoribonuclease does not lead to extensive accumulation of decay intermediates. Deletion of other 3' exoribonuclease genes results in even greater accumulation of decay intermediates (Fig. 8, lanes 5 and 6). These prominent decay intermediates have 3' ends located slightly downstream of predicted secondary or tertiary structures. It is hypothesized that these structures are easily degraded by PNPase but they block processivity of 3'-to-5' exoribonucleases other than PNPase that are present in the *pnpA* strain. The mechanism of PNPase processivity is somehow more efficient than that of other 3' exonucleases. The presence of PNPase alone, without any of the other B. subtilis 3'-to-5' exonucleases, gives an mRNA decay pattern similar to wild type (Fig. 8, lane 3).

It is interesting to note that the ability of 3'-to-5' exonucleases to proceed through secondary structure cannot necessarily be predicted based on the thermodynamic stability of the structure. For example, although PNPase

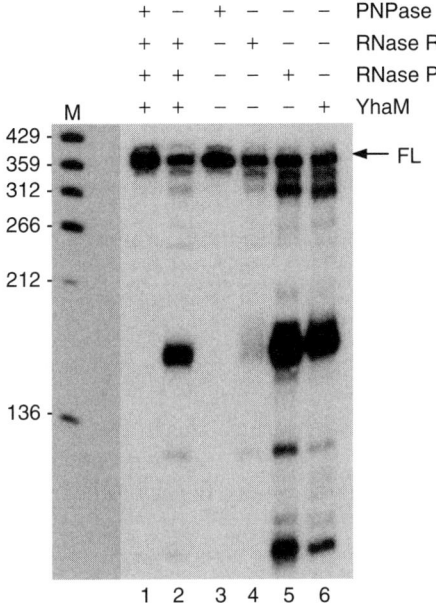

FIG. 8. Steady-state pattern of *rpsO* mRNA in wild-type and ribonuclease mutant strains. The probe was a 5′-end-labeled oligonucleotide complementary to sequences near the *rpsO* TIR. Migration of the full-length transcript is indicated by the arrow on the right. Marker lane (M) contains labeled DNA fragments with the sizes (in nts) indicated on the left (from (95), with permission from American Society for Microbiology).

clearly does a better job in degrading intermediate mRNA decay fragments than other exonucleases, a particular structure—termed "stem 2" RNA—is known to block PNPase while being susceptible to decay by other 3′-to-5′ exonucleases (88). It is not understood why this particular structure presents a barrier specifically to PNPase.

The viability of a *B. subtilis* PNPase-deficient strain, despite the massive accumulation of decay intermediates, is a puzzle. As these decay intermediates are detected by a probe that targets the 5′ end of the mRNA, they contain a functional TIR, which allows loading of ribosomes and initiation of translation. However, since these fragments do not contain a stop codon, the translating ribosome is not released for recycling and maintenance of the overall ribosome pool. We have shown that the *trans*-translation system of *B. subtilis* (99), which includes tmRNA (transfer-messenger RNA) encoded by the *ssrA* gene, is involved in releasing ribosomes from RNA fragments in the *pnpA* strain (98). Surprisingly, an *ssrA pnpA* double mutant strain is viable (DHB, unpublished). It remains to be determined how the maintenance of an adequate pool of free ribosomes is accomplished in such a strain.

Analysis of PNPase activity *in vitro* is an important step toward understanding RNA degradation mechanisms. In one such study, we found that PNPase binds poorly to the 3′ end of *trp* leader RNA (Fig. 6), which has a 3′-proximal secondary structure (formed by a transcription terminator sequence) followed by a single-stranded tail of 6 nts (*64*). When the 3′ tail of the substrate is extended to 17 single-stranded nts, PNPase is able to bind well. Perhaps, PAP activity *in vivo* provides such extended tails. However, even when bound, PNPase has difficulty degrading processively through the strong secondary structure that is provided by the transcription termination sequence. These observations support the hypothesis that PNPase does not degrade mRNAs from the native 3′ end, but relies on endonuclease cleavage in the body of the message as an entry site for rapid degradation. In the case of *trp* leader RNA, we have shown directly that initiation of decay occurs by RNase J1 cleavage (*35, 64*), and perhaps this is a general mechanism for many *B. subtilis* mRNAs. Thus, PNPase is not a decay-initiating enzyme but a secondary scavenger that operates once decay has been initiated by a primary processing enzyme.

C. Other 3′ Exoribonucleases

Other known 3′-to-5′ exoribonucleases in *B. subtilis* include RNase PH (*100*), RNase R (*101*), and YhaM (*102*). While all three of these exoribonucleases can participate in mRNA turnover in the absence of PNPase (*98*), this is likely not their primary function. RNase PH is, like PNPase, a phosphorolytic enzyme, and is a member of the PDX family of 3′ exoribonucleases (*94*). RNase PH is basically a smaller version (molecular weight of 26.5 kDa) of PNPase, without the large N-terminal domain and KH and S1 domains of PNPase. A medium resolution structure of *B. subtilis* RNase PH has been determined (*103*), and the functional form appears to be a hexamer. In the absence of the three other known *B. subtilis* 3′-to-5′ exoribonucleases, RNase PH can function in mRNA decay (*98*), but the major role of RNase PH appears to be in tRNA processing (*104*).

B. subtilis RNase R is a large (88 kDa) hydrolytic 3′-to-5′ exoribonuclease, a member of the RNR family (*94*), whose primary function in *B. subtilis* RNA metabolism is not known. *E. coli* RNase R has been shown to be important in mRNA decay (*105*) and is also the component of the *trans*-translation system that is responsible for degrading non-stop mRNAs (*106*). In the absence of PNPase, *B. subtilis* RNase R degrades mRNA that contains secondary structure more efficiently than the other remaining 3′-to-5′ exoribonucleases (Fig. 8, lane 4 vs. lanes 5 and 6) (*98*). In fact, a triple ribonuclease mutant that contains only RNase R (of the four known 3′-to-5′ exoribonucleases) handles mRNA decay better than when RNase R is present with RNase PH and YhaM.

We hypothesize that other RNases compete with RNase R for binding to 3′ single-stranded tails, and they may chew back these tails, which are necessary for efficient degradation by RNase R. *In vitro*, RNase R degrades through a strong secondary structure when the substrate includes a 3′-terminal, single-stranded extension (*101*). This may suggest a role for RNase R in mRNA turnover that relies on the activity of PAP to extend the 3′ end.

YhaM is the fourth known *B. subtilis* 3′-to-5′ exoribonuclease, and it was identified in extracts that were missing PNPase and RNase R (*102*). Interestingly, the overall domain structure of YhaM is unique to Gram-positive organisms. While YhaM can participate in mRNA turnover in the absence of other 3′ exonucleases, the true *in vivo* function of YhaM remains unexplored. YhaM has been implicated in DNA replication (*107*).

A mutant strain of *B. subtilis* has been constructed that is deleted for all four of the known 3′-to-5′ exoribonucleases: PNPase, RNase PH, RNase R, and YhaM (*98*). While the quadruple mutant strain is quite sick, the pattern of mRNA decay intermediates in this strain is similar to that in strains that are missing only PNPase. Experiments in *E. coli* show that inactivation of the two major 3′-to-5′ exoribonucleases, PNPase and RNase II, is lethal (*108*), and presumably 3′ exonuclease activity is essential in *B. subtilis* as well. Thus, the viability of the quadruple mutant strain suggests the existence of one or more unknown 3′ exonucleases. Whatever this additional activity is, it is not detectable to any significant degree in *B. subtilis* extracts that are deficient for the known 3′-to-5′ exonucleases. Perhaps *in vitro* conditions do not allow for activity of these other exonucleases, or they work only with other factors that are not present, or are not functional, in extracts.

D. Polyadenylation

That RNA 3′ polyadenylation occurs in *B. subtilis* RNA has been known for many years, as first demonstrated by Sarkar and colleagues (*109, 110*). Recently, it was shown that poly(A) and heteropolymeric tails are added in *B. subtilis* to both translated and untranslated RNAs (*111*). Poly(A) addition occurred even in a PNPase deletion strain, suggesting that PNPase, which can function as a PAP in *E. coli* (*112*), is not required for polyadenylation in *B. subtilis*. The only *B. subtilis* gene with considerable homology to *E. coli* PAP, the *cca* gene, was shown to be a nucleotidyl transferase and not a PAP (*113*). Thus, while the existence of a *B. subtilis* PAP is clear from the documented presence of non-templated 3′ extensions, the identity of the protein(s) responsible for this activity is unknown, and it is not yet possible to construct the mutant strains that will allow one to determine whether a deficiency in polyadenylation affects mRNA turnover.

E. 3′-Terminal Fragment Turnover

A longstanding question in the field of mRNA decay can be called the "end degradation problem," which is how 3′-terminal fragments get turned over despite their intrinsic resistance to 3′-to-5′ exonuclease activity. Such structures are likely to exist at the 3′ end of many *E. coli* mRNAs and virtually all *B. subtilis* mRNAs, since Rho-independent transcription termination is not known to occur in this organism. These structures are not substrates for endonuclease cleavage by RNase E or RNase J1, which seem to recognize single-stranded RNA. As mentioned before in Section II.A and shown in Fig. 1A, the solution to the end degradation problem in *E. coli* is thought to involve repeated addition of a 3′ poly(A) extension and subsequent degradation.

Recent data from our laboratory suggest that the solution to the end degradation problem in *B. subtilis* is much simpler, and involves RNase J1 itself. Experiments using 3′-end-labeled RNA oligonucleotides that contain a strong, 3′-proximal secondary structure show that RNase J1 5′ exonuclease activity is capable of degrading through secondary structure to some extent *in vitro* (35). This may be even more efficient *in vivo*, where other factors may be involved in RNase J1 activity, as is suggested by the weak activity of purified RNase J1 (7). Thus, we envision that, after cleaving endonucleolytically at its 3′-most target site, RNase J1 binds to the 5′ end of the downstream fragment and degrades the remaining mRNA in the 5′-to-3′ direction. Direct evidence for this mode of 3′-end fragment turnover *in vivo* has recently been obtained. In conditions where cellular RNase J1 levels were lowered, an accumulation of the 3′-terminal fragments of *trp* leader RNA, *rpsO* mRNA, and Δ*ermC* mRNA was observed, suggesting that RNase J1 is required for the rapid turnover of these fragments (35).

V. Regulated mRNA Decay

Early studies on *B. subtilis* mRNA decay were concerned with possible differences in mRNA stability in cells in logarithmic growth phase versus cells in post-logarithmic growth phase or sporulation. It was shown that sporulation was dependent on ongoing mRNA synthesis, since addition of rifampin inhibited stages of sporulation; thus, mRNA was generally as unstable during post-logarithmic growth as during logarithmic growth (114, 115). Nevertheless, growth-rate dependent stability for specific mRNAs can be demonstrated. The half-life for mRNA encoded by the succinate dehydrogenase (*sdh*) operon, an mRNA that is preferentially expressed during vegetative growth, decreases about sixfold (2.6–0.4 min) when cells enter stationary phase (116). The half-life of mRNA encoded by the subtilisin gene (*aprE*), an mRNA that is preferentially expressed during stationary phase, is longer in stationary phase than during vegetative growth (*117*). A later report claims that the rate of degradation of the *xynA* transcript is increased in response to stress (*118*).

These reports do not address a mechanism for how the half-life of a particular mRNA is regulated. One could imagine several possibilities:

(1) Regulation in *cis*, at the level of the mRNA itself. For example, use of an alternative transcription start site, or processing of a primary transcript, which results in a different 5′-terminal sequence or structure that makes the mRNA more or less susceptible to initiation of decay. The first example of control of *B. subtilis* mRNA stability by a regulated change in *cis* has recently been described (*85*). This is the metabolite-sensing, riboswitch-controlled *glmS* mRNA, already mentioned above (Section III.E). *glmS* mRNA half-life decreases when ribozyme cleavage occurs at a 5′-proximal site in the *glmS* leader region. The *glmS* ribozyme activity is activated by binding of glucosamine-6-phosphate (GlcN6P), the metabolite produced by GlcN6P synthase, which is encoded by the *glmS* gene itself—an elegant feedback loop. GlcN6P-induced cleavage leaves a 5′ hydroxyl end (as opposed to the triphosphate native 5′ end), which is a target for RNase J1-mediated decay. Thus, altering the character of the 5′ end of the mRNA by metabolite-regulated endonuclease cleavage appears to be the mechanism for controlling *glmS* mRNA half-life.

(2) Regulation in *trans*, that is, binding of a factor to a 5′ site that protects mRNA from decay. There is much evidence for this mechanism in *B. subtilis*. The earliest report on regulation of mRNA stability in *trans* involved stalling of a ribosome near the 5′ end of the mRNA encoded by the *ermC* gene. Expression of *ermC*, which encodes a ribosomal RNA methylase that confers erythromycin (Em) resistance, is induced by Em itself in a translational attenuation mechanism. Em-induced ribosome stalling on a peptide coding sequence in the leader region causes rearrangement of the *ermC* leader RNA to allow high-level translation of the downstream methylase coding sequence (Fig. 9) (*119, 120*). Stabilization of *ermC* mRNA is concomitant with Em-induced translation (*121*). This stabilization depends on translation of the leader peptide coding sequence, but not on translation of the body of the *ermC* message (*15*). The data suggest that ribosome stalling near the 5′ end of *ermC* mRNA confers stability by blocking access of a 5′-end-dependent ribonuclease that normally initiates mRNA decay, and this is likely to be RNase J1. Extending the 5′ end of *ermC* mRNA away from the ribosome binding site results in a lack of Em-induced stability (*16*). Similar results were obtained by Weisblum and colleagues, using the mRNA encoded by the *ermA* gene (*17, 18*). These studies were

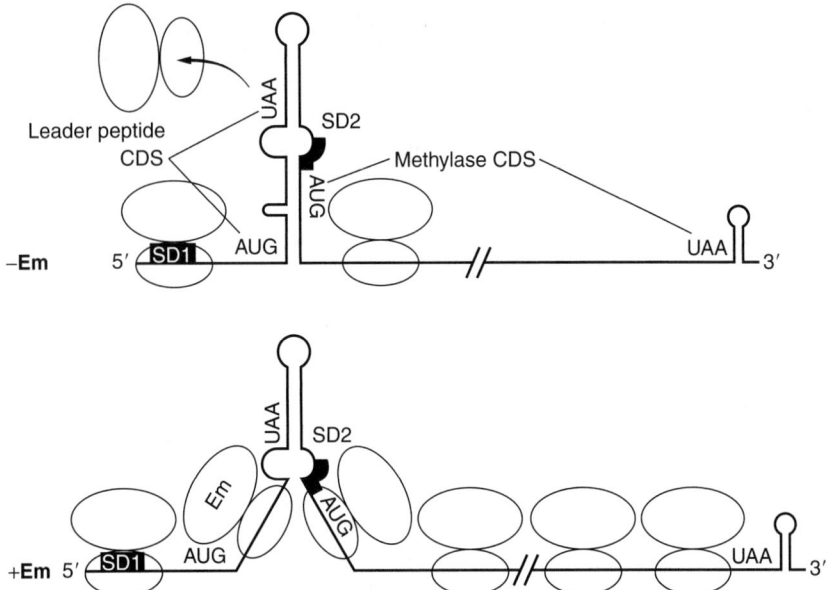

FIG. 9. Induction of *ermC* expression by Em. In the absence of Em, ribosomes translate the leader peptide without stalling, and the *ermC* leader RNA secondary structure restricts ribosome binding at SD2, the ribosome binding site for translation of the rRNA methylase coding sequence (CDS). Addition of Em causes a ribosome that is translating the leader peptide to stall, resulting in opening up of the leader RNA stem structure and allowing free access to SD2. Backup of stalled ribosomes near the 5′ end results in a 10-fold increase in mRNA half-life.

the first in *B. subtilis* to point to the 5′ end of an mRNA as being critical for determining mRNA half-life. It should be noted that addition of tetracycline to a *B. subtilis* culture also results in stabilization of several (and perhaps many) mRNAs, which is likely due to a similar effect of tetracycline-induced ribosome pausing near the 5′ end (*122*).

A similar *trans* mechanism, dependent on binding of a protein rather than a ribosome, explains the regulated stability of the monocistronic *glpD* mRNA. *glp* genes are involved in glycerol metabolism, and are regulated transcriptionally and posttranscriptionally. At the transcriptional level, GlpP protein (activated by the presence of glyceraldehyde-3-phosphate) plays the role of an antiterminator protein, allowing high-level transcription of several *glp* genes. This increase in gene expression is supplemented by posttranscriptional regulation in which binding of activated GlpP to a 5′-proximal inverted repeat sequence results in a 10-to-15-fold stabilization of *glpD* mRNA (*50, 123*). It is likely that GlpP binding near the 5′ end of *glpD* mRNA decreases the accessibility of the 5′ end to RNase J1.

Further examples of regulated stability are the cold shock genes *cspB* and *cspC*. Upon temperature downshift from 37 to 14 °C, there is about a 30-fold increase in the stability of *cspB* and *cspC* mRNAs (*124*). A mechanism for how this is achieved in *B. subtilis* has not been described; however, a similar cold-shock induced phenomenon occurs in *E. coli*, and a model for stabilization based on increased ribosome density at the lower temperature has recently been proposed (*125*).

(3) *Regulation of ribonuclease activity.* Stability could be controlled by modification of the level or specificity of one or more ribonucleases involved in mRNA turnover. While there is ample evidence in *E. coli* for regulation of RNase E activity (*126–128*), there has been no report of such regulation in *B. subtilis*.

VI. Conclusion

The last ten years have seen an explosion in our understanding of *B. subtilis* ribonuclease content and function. The earlier assumption that understanding the process of mRNA decay in *E. coli* would give a good picture of this process in other bacterial species no longer holds true. Bacteria have distinct machineries to solve the problem of rapid mRNA turnover, which is likely needed for rapid changes in gene expression profile that enable timely responses to changes in growth conditions. Many questions about *B. subtilis* mRNA decay need attention, and we have mentioned some of these along the way. On the basis of the overt differences depicted in Fig. 1 for the mechanisms of mRNA turnover in the model Gram-negative and Gram-positive bacteria, we can expect that interesting developments are on the way.

References

1. Kunst F, Ogasawara N, Moszer I, Albertini AM, Alloni G, Azevedo V, et al. The complete genome sequence of the gram-positive bacterium *Bacillus subtilis*. *Nature* 1997;**390**:249–56.
2. Ghosh S, Deutscher MP. Oligoribonuclease is an essential component of the mRNA decay pathway. *Proc Natl Acad Sci USA* 1999;**96**:4372–7.
3. Chaney SG, Boyer PD. Incorporation of water oxygens into intracellular nucleotides and RNA. II. Predominantly hydrolytic RNA turnover in *Escherichia coli*. *J Mol Biol* 1972;**64**:581–91.
4. Duffy JJ, Chaney SG, Boyer PD. Incorporation of water oxygens into intracellular nucleotides and RNA. I. Predominantly non-hydrolytic RNA turnover in *Bacillus subtilis*. *J Mol Biol* 1972;**64**:565–79.
5. Deutscher MP, Reuven NB. Enzymatic basis for hydrolytic versus phosphorolytic mRNA degradation in *Escherichia coli* and *Bacillus subtilis*. *Proc Natl Acad Sci USA* 1991;**88**:3277–80.

6. Wang W, Bechhofer DH. Properties of a *Bacillus subtilis* polynucleotide phosphorylase deletion strain. *J Bacteriol* 1996;**178**:2375–82.
7. Mathy N, Benard L, Pellegrini O, Daou R, Wen T, Condon C. 5′-to-3′ Exoribonuclease activity in bacteria: Role of RNase J1 in rRNA maturation and 5′ stability of mRNA. *Cell* 2007;**129**:681–92.
8. Fritz DT, Bergman N, Kilpatrick WJ, Wilusz CJ, Wilusz J. Messenger RNA decay in Mammalian cells: the exonuclease perspective. *Cell Biochem Biophys* 2004;**41**:265–78.
9. Meyer S, Temme C, Wahle E. Messenger RNA turnover in eukaryotes: Pathways and enzymes. *Crit Rev Biochem Mol Biol* 2004;**39**:197–216.
10. Newbury SF. Control of mRNA stability in eukaryotes. *Biochem Soc Trans* 2006;**34**:30–4.
11. Wilusz CJ, Wilusz J. Bringing the role of mRNA decay in the control of gene expression into focus. *Trends Genet* 2004;**20**:491–7.
12. Hartmann G, Honikel KO, Knusel F, Nuesch J. The specific inhibition of the DNA-directed RNA synthesis by rifamycin. *Biochim Biophys Acta* 1967;**145**:843–4.
13. Coote JG, Wood DA, Mandelstam J. Lethal effect of rifampicin in *Bacillus subtilis* as a complicating factor in the assessment of the lifetime of messenger ribonucleic acid. *Biochem J* 1973;**134**:263–70.
14. Brown S, Coleman G. Stability of rapidly labelled messenger ribonucleic acid in *Bacillus amyloliquefaciens* during the phases of minimum and maximum extracellular enzyme formation. *J Mol Biol* 1975;**96**:335–44.
15. Bechhofer DH, Dubnau D. Induced mRNA stability in *Bacillus subtilis*. *Proc Natl Acad Sci USA* 1987;**84**:498–502.
16. Bechhofer DH, Zen KH. Mechanism of erythromycin-induced ermC mRNA stability in *Bacillus subtilis*. *J Bacteriol* 1989;**171**:5803–11.
17. Sandler P, Weisblum B. Erythromycin-induced stabilization of ermA messenger RNA in *Staphylococcus aureus* and *Bacillus subtilis*. *J Mol Biol* 1988;**203**:905–15.
18. Sandler P, Weisblum B. Erythromycin-induced ribosome stall in the ermA leader: A barricade to 5′-to-3′ nucleolytic cleavage of the ermA transcript. *J Bacteriol* 1989;**171**:6680–8.
19. Melin L, Friden H, Dehlin E, Rutberg L, von Gabain A. The importance of the 5′-region in regulating the stability of sdh mRNA in *Bacillus subtilis*. *Mol Microbiol* 1990;**4**:1881–9.
20. Mackie GA. Ribonuclease E is a 5′-end-dependent endonuclease. *Nature* 1998;**395**:720–3.
21. Myers RS, Stahl FW. Chi and the RecBC D enzyme of *Escherichia coli*. *Annu Rev Genet* 1994;**28**:49–70.
22. Coburn GA, Mackie GA. Degradation of mRNA in *Escherichia coli*: An old problem with some new twists. *Prog Nucleic Acid Res Mol Biol* 1999;**62**:55–5108.
23. Arnold TE, Yu J, Belasco JG. mRNA stabilization by the ompA 5′ untranslated region: Two protective elements hinder distinct pathways for mRNA degradation. *RNA* 1998;**4**:319–30.
24. Baker KE, Mackie GA. Ectopic RNase E sites promote bypass of 5′-end-dependent mRNA decay in *Escherichia coli*. *Mol Microbiol* 2003;**47**:75–88.
25. Marujo PE, Braun F, Haugel-Nielsen J, Le Derout J, Arraiano CM, Regnier P. Inactivation of the decay pathway initiated at an internal site by RNase E promotes poly(A)-dependent degradation of the rpsO mRNA in *Escherichia coli*. *Mol Microbiol* 2003;**50**:1283–94.
26. Jiang X, Belasco JG. Catalytic activation of multimeric RNase E and RNase G by 5′-monophosphorylated RNA. *Proc Natl Acad Sci USA* 2004;**101**:9211–6.
27. Spickler C, Stronge V, Mackie GA. Preferential cleavage of degradative intermediates of rpst mRNA by the *Escherichia coli* RNA degradosome. *J Bacteriol* 2001;**183**:1106–9.
28. Coburn GA, Mackie GA. Reconstitution of the degradation of the mRNA for ribosomal protein S20 with purified enzymes. *J Mol Biol* 1998;**279**:1061–74.
29. Bouvet P, Belasco JG. Control of RNase E-mediated RNA degradation by 5′-terminal base pairing in *E. Coli*. *Nature* 1992;**360**:488–91.

30. Emory SA, Bouvet P, Belasco JG. A 5′-terminal stem-loop structure can stabilize mRNA in *Escherichia coli*. *Genes Dev* 1992;**6**:135–48.
31. Condon C, Putzer H, Luo D, Grunberg-Manago M. Processing of the *Bacillus subtilis* thrS leader mRNA is RNase E-dependent in *Escherichia coli*. *J Mol Biol* 1997;**268**:235–42.
32. Even S, Pellegrini O, Zig L, Labas V, Vinh J, Brechemmier-Baey D, *et al*. Ribonucleases J1 and J2: Two novel endoribonucleases in *B. subtilis* with functional homology to *E. coli* RNase E. *Nucleic Acids Res* 2005;**33**:2141–52.
33. Condon C, Putzer H, Grunberg-Manago M. Processing of the leader mRNA plays a major role in the induction of thrS expression following threonine starvation in *Bacillus subtilis*. *Proc Natl Acad Sci USA* 1996;**93**:6992–7.
34. Li de la Sierra-Gallay I, Zig L, Jamalli A, Putzer H. Structural insights into the dual activity of RNase J. *Nat Struct Mol Biol* 2008;**15**:206–12.
35. Deikus G, Condon C, Bechhofer DH. Role of *Bacillus subtilis* RNase J1 endonuclease and 5′-exonuclease activities in trp leader RNA turnover. *J Biol Chem* 2008;**283**:17158–67.
36. Callaghan AJ, Marcaida MJ, Stead JA, McDowall KJ, Scott WG, Luisi BF. Structure of *Escherichia coli* RNase E catalytic domain and implications for RNA turnover. *Nature* 2005;**437**:1187–91.
37. Choonee N, Even S, Zig L, Putzer H. Ribosomal protein L20 controls expression of the *Bacillus subtilis* infC operon via a transcription attenuation mechanism. *Nucleic Acids Res* 2007;**35**:1578–88.
38. Yao S, Blaustein JB, Bechhofer DH. Processing of *Bacillus subtilis* small cytoplasmic RNA: Evidence for an additional endonuclease cleavage site. *Nucleic Acids Res* 2007;**35**:4464–73.
39. Bechhofer DH, Wang W. Decay of ermC mRNA in a polynucleotide phosphorylase mutant of *Bacillus subtilis*. *J Bacteriol* 1998;**180**:5968–77.
40. Carpousis AJ. The RNA degradosome of *Escherichia coli*: An mRNA-degrading machine assembled on RNase E. *Annu Rev Microbiol* 2007;**61**:71–87.
41. Kushner SR. mRNA decay in *Escherichia coli* comes of age. *J Bacteriol* 2002;**184**:4658–65.
42. Agaisse H, Lereclus D. STAB-SD: A Shine-Dalgarno sequence in the 5′ untranslated region is a determinant of mRNA stability. *Mol Microbiol* 1996;**20**:633–43.
43. Hambraeus G, von Wachenfeldt C, Hederstedt L. Genome-wide survey of mRNA half-lives in *Bacillus subtilis* identifies extremely stable mRNAs. *Mol Genet Genomics* 2003;**269**:706–14.
44. Sharp JS, Bechhofer DH. Effect of 5′ proximal elements on decay of a model mRNA in *Bacillus subtilis*. *Mol Microbiol* 2005;**57**:484–95.
45. Yusupova GZ, Yusupov MM, Cate JH, Noller HF. The path of messenger RNA through the ribosome. *Cell* 2001;**106**:233–41.
46. Jurgen B, Schweder T, Hecker M. The stability of mRNA from the gsiB gene of *Bacillus subtilis* is dependent on the presence of a strong ribosome binding site. *Mol Gen Genet* 1998;**258**:538–45.
47. Ohki R, Tateno K. Increased stability of bmr3 mRNA results in a multidrug-resistant phenotype in *Bacillus subtilis*. *J Bacteriol* 2004;**186**:7450–5.
48. Glatz E, Nilsson RP, Rutberg L, Rutberg B. A dual role for the *Bacillus subtilis* glpD leader and the GlpP protein in the regulated expression of glpD: antitermination and control of mRNA stability. *Mol Microbiol* 1996;**19**:319–28.
49. Sharp JS, Bechhofer DH. Effect of translational signals on mRNA decay in *Bacillus subtilis*. *J Bacteriol* 2003;**185**:5372–9.
50. Hambraeus G, Karhumaa K, Rutberg B. A 5′ stem-loop and ribosome binding but not translation are important for the stability of *Bacillus subtilis* aprE leader mRNA. *Microbiology* 2002;**148**:1795–803.
51. Celesnik H, Deana A, Belasco JG. Initiation of RNA decay in *Escherichia coli* by 5′ pyrophosphate removal. *Mol Cell* 2007;**27**:79–90.

52. Deana A, Celesnik H, Belasco JG. The bacterial enzyme rppH triggers messenger RNA degradation by 5′ pyrophosphate removal. *Nature* 2008;**451**:355–8.
53. McLennan AG. The Nudix hydrolase superfamily. *Cell Mol Life Sci* 2006;**63**:123–43.
54. Schumann W. The *Bacillus subtilis* heat shock stimulon. *Cell Stress Chaperones* 2003;**8**:207–17.
55. Homuth G, Mogk A, Schumann W. Post-transcriptional regulation of the *Bacillus subtilis* dnaK operon. *Mol Microbiol* 1999;**32**:1183–97.
56. Yuan G, Wong SL. Regulation of groE expression in *Bacillus subtilis*: The involvement of the sigma A-like promoter and the roles of the inverted repeat sequence (CIRCE). *J Bacteriol* 1995;**177**:5427–33.
57. Deana A, Belasco JG. Lost in translation: The influence of ribosomes on bacterial mRNA decay. *Genes Dev* 2005;**19**:2526–33.
58. Drider D, DiChiara JM, Wei J, Sharp JS, Bechhofer DH. Endonuclease cleavage of messenger RNA in *Bacillus subtilis*. *Mol Microbiol* 2002;**43**:1319–29.
59. Vazquez-Laslop N, Thum C, Mankin AS. Molecular mechanism of drug-dependent ribosome stalling. *Mol Cell* 2008;**30**:190–202.
60. Hayes CS, Sauer RT. Cleavage of the A site mRNA codon during ribosome pausing provides a mechanism for translational quality control. *Mol Cell* 2003;**12**:903–11.
61. Sunohara T, Jojima K, Tagami H, Inada T, Aiba H. Ribosome stalling during translation elongation induces cleavage of mRNA being translated in *Escherichia coli*. *J Biol Chem* 2004;**279**:15368–75.
62. Feng L, Niu DK. Relationship between mRNA stability and length: An old question with a new twist. *Biochem Genet* 2007;**45**:131–7.
63. Deikus G, Babitzke P, Bechhofer DH. Recycling of a regulatory protein by degradation of the RNA to which it binds. *Proc Natl Acad Sci USA* 2004;**101**:2747–51.
64. Deikus G, Bechhofer DH. Initiation of decay of *Bacillus subtilis* trp leader RNA. *J Biol Chem* 2007;**282**:20238–44.
65. Bardwell JC, Regnier P, Chen SM, Nakamura Y, Grunberg-Manago M, Court DL. Autoregulation of RNase III operon by mRNA processing. *EMBO J* 1989;**8**:3401–7.
66. Portier C, Dondon L, Grunberg-Manago M, Regnier P. The first step in the functional inactivation of the *Escherichia coli* polynucleotide phosphorylase messenger is a ribonuclease III processing at the 5′ end. *EMBO J* 1987;**6**:2165–70.
67. Regnier P, Grunberg-Manago M. Cleavage by RNase III in the transcripts of the met Y-nus-A-infb operon of *Escherichia coli* releases the tRNA and initiates the decay of the downstream mRNA. *J Mol Biol* 1989;**210**:293–302.
68. Mitra S, Bechhofer DH. Substrate specificity of an RNase III-like activity from *Bacillus subtilis*. *J Biol Chem* 1994;**269**:31450–6.
69. Wang W, Bechhofer DH. *Bacillus subtilis* RNase III gene: Cloning, function of the gene in *Escherichia coli*, and construction of *Bacillus subtilis* strains with altered rnc loci. *J Bacteriol* 1997;**179**:7379–85.
70. Oguro A, Kakeshita H, Nakamura K, Yamane K, Wang W, Bechhofer DH. *Bacillus subtilis* RNase III cleaves both 5′- and 3′-sites of the small cytoplasmic RNA precursor. *J Biol Chem* 1998;**273**:19542–7.
71. DiMari JF, Bechhofer DH. Initiation of mRNA decay in *Bacillus subtilis*. *Mol Microbiol* 1993;**7**:705–17.
72. Herskovitz MA, Bechhofer DH. Endoribonuclease RNase III is essential in *Bacillus subtilis*. *Mol Microbiol* 2000;**38**:1027–33.
73. Redko Y, Bechhofer DH, Condon C. Mini-III, an unusual member of the RNase III family of enzymes, catalyses 23S ribosomal RNA maturation in B. Subtilis. *Mol Microbiol* 2008;**68**:1096–106.

74. Sogin ML, Pace NR. In vitro maturation of precursors of 5S ribosomal RNA from *Bacillus subtilis*. *Nature* 1974;**252**:598–600.
75. Condon C, Rourera J, Brechemier-Baey D, Putzer H. Ribonuclease M5 has few, if any, mRNA substrates in *Bacillus subtilis*. *J Bacteriol* 2002;**184**:2845–9.
76. Li Y, Altman S. A specific endoribonuclease, RNase P, affects gene expression of polycistronic operon mRNAs. *Proc Natl Acad Sci USA* 2003;**100**:13213–8.
77. Seif E, Altman S. RNase P cleaves the adenine riboswitch and stabilizes pbuE mRNA in *Bacillus subtilis*. *RNA* 2008;**14**:1237–43.
78. Gossringer M, Kretschmer-Kazemi Far R, Hartmann RK. Analysis of RNase P protein (rnpA) expression in *Bacillus subtilis* utilizing strains with suppressible rnpA expression. *J Bacteriol* 2006;**188**:6816–23.
79. Wegscheid B, Condon C, Hartmann RK. Type A and B RNase P RNAs are interchangeable in vivo despite substantial biophysical differences. *EMBO Rep* 2006;**7**:411–7.
80. Pellegrini O, Nezzar J, Marchfelder A, Putzer H, Condon C. Endonucleolytic processing of CCA-less tRNA precursors by RNase Z in *Bacillus subtilis*. *EMBO J* 2003;**22**:4534–43.
81. Perwez T, Kushner SR. RNase Z in *Escherichia coli* plays a significant role in mRNA decay. *Mol Microbiol* 2006;**60**:723–37.
82. Ludwig H, Homuth G, Schmalisch M, Dyka FM, Hecker M, Stulke J. Transcription of glycolytic genes and operons in *Bacillus subtilis*: Evidence for the presence of multiple levels of control of the gapA operon. *Mol Microbiol* 2001;**41**:409–22.
83. Meinken C, Blencke HM, Ludwig H, Stulke J. Expression of the glycolytic gapA operon in *Bacillus subtilis*: Differential syntheses of proteins encoded by the operon. *Microbiology* 2003;**149**:751–61.
84. Mader U, Hennig S, Hecker M, Homuth G. Transcriptional organization and posttranscriptional regulation of the *Bacillus subtilis* branched-chain amino acid biosynthesis genes. *J Bacteriol* 2004;**186**:2240–52.
85. Collins JA, Irnov I, Baker S, Winkler WC. Mechanism of mRNA destabilization by the glmS ribozyme. *Genes Dev* 2007;**21**:3356–68.
86. McLaren RS, Newbury SF, Dance GS, Causton HC, Higgins CF. mRNA degradation by processive 3′–5′ exoribonucleases in vitro and the implications for prokaryotic mRNA decay in vivo. *J Mol Biol* 1991;**221**:81–95.
87. Spickler C, Mackie GA. Action of RNase II and polynucleotide phosphorylase against RNAs containing stem-loops of defined structure. *J Bacteriol* 2000;**182**:2422–7.
88. Farr GA, Oussenko IA, Bechhofer DH. Protection against 3′-to-5′ RNA decay in *Bacillus subtilis*. *J Bacteriol* 1999;**181**:7323–30.
89. Shimotsu H, Henner DJ. Construction of a single-copy integration vector and its use in analysis of regulation of the trp operon of *Bacillus subtilis*. *Gene* 1986;**43**:85–94.
90. Quirk PG, Dunkley EA, Jr, Lee P, Krulwich TA. Identification of a putative *Bacillus subtilis* rho gene. *J Bacteriol* 1993;**175**:8053.
91. Dreyfus M, Regnier P. The poly(A) tail of mRNAs: Bodyguard in eukaryotes, scavenger in bacteria. *Cell* 2002;**111**:611–3.
92. Wong HC, Chang S. Identification of a positive retroregulator that stabilizes mRNAs in bacteria. *Proc Natl Acad Sci USA* 1986;**83**:3233–7.
93. Hess GF, Graham RS. Efficiency of transcriptional terminators in *Bacillus subtilis*. *Gene* 1990;**95**:137–41.
94. Zuo Y, Deutscher MP. Exoribonuclease superfamilies: Structural analysis and phylogenetic distribution. *Nucleic Acids Res* 2001;**29**:1017–26.
95. Symmons MF, Jones GH, Luisi BF. A duplicated fold is the structural basis for polynucleotide phosphorylase catalytic activity, processivity, and regulation. *Structure* 2000;**8**:1215–26.

96. Mitra S, Hue K, Bechhofer DH. In vitro processing activity of *Bacillus subtilis* polynucleotide phosphorylase. *Mol Microbiol* 1996;**19**:329–42.
97. Luttinger A, Hahn J, Dubnau D. Polynucleotide phosphorylase is necessary for competence development in *Bacillus subtilis*. *Mol Microbiol* 1996;**19**:343–56.
98. Oussenko IA, Abe T, Ujiie H, Muto A, Bechhofer DH. Participation of 3′-to-5′ exoribonucleases in the turnover of *Bacillus subtilis* mRNA. *J Bacteriol* 2005;**187**:2758–67.
99. Wiegert T, Schumann W. Ssra-mediated tagging in *Bacillus subtilis*. *J Bacteriol* 2001;**183**:3885–9.
100. Craven MG, Henner DJ, Alessi D, Schauer AT, Ost KA, Deutscher MP, et al. Identification of the rph (RNase PH) gene of *Bacillus subtilis*: Evidence for suppression of cold-sensitive mutations in *Escherichia coli*. *J Bacteriol* 1992;**174**:4727–35.
101. Oussenko IA, Bechhofer DH. The yvaJ gene of *Bacillus subtilis* encodes a 3′-to-5′ exoribonuclease and is not essential in a strain lacking polynucleotide phosphorylase. *J Bacteriol* 2000;**182**:2639–42.
102. Oussenko IA, Sanchez R, Bechhofer DH. *Bacillus subtilis* YhaM, a member of a new family of 3′-to-5′ exonucleases in gram-positive bacteria. *J Bacteriol* 2002;**184**:6250–9.
103. Harlow LS, Kadziola A, Jensen KF, Larsen S. Crystal structure of the phosphorolytic exoribonuclease RNase PH from *Bacillus subtilis* and implications for its quaternary structure and tRNA binding. *Protein Sci* 2004;**13**:668–77.
104. Wen T, Oussenko IA, Pellegrini O, Bechhofer DH, Condon C. Ribonuclease PH plays a major role in the exonucleolytic maturation of CCA-containing tRNA precursors in *Bacillus subtilis*. *Nucleic Acids Res* 2005;**33**:3636–43.
105. Cheng ZF, Deutscher MP. An important role for RNase R in mRNA decay. *Mol Cell* 2005;**17**:313–8.
106. Richards J, Mehta P, Karzai AW. RNase R degrades non-stop mRNAs selectively in an SmpB-tmRNA-dependent manner. *Mol Microbiol* 2006;**62**:1700–12.
107. Noirot-Gros MF, Dervyn E, Wu LJ, Mervelet P, Errington J, Ehrlich SD, et al. An expanded view of bacterial DNA replication. *Proc Natl Acad Sci USA* 2002;**99**:8342–7.
108. Donovan WP, Kushner SR. Polynucleotide phosphorylase and ribonuclease II are required for cell viability and mRNA turnover in *Escherichia coli* K-12. *Proc Natl Acad Sci USA* 1986;**83**:120–4.
109. Gopalakrishna Y, Sarkar N. Characterization of polyadenylate-containing ribonucleic acid from *Bacillus subtilis*. *Biochemistry* 1982;**21**:2724–9.
110. Gopalakrishna Y, Sarkar N. The synthesis of DNA complementary to polyadenylate-containing RNA from *Bacillus subtilis*. *J Biol Chem* 1982;**257**:2747–50.
111. Campos-Guillen J, Bralley P, Jones GH, Bechhofer DH, Olmedo-Alvarez G. Addition of poly (A) and heteropolymeric 3′ ends in *Bacillus subtilis* wild-type and polynucleotide phosphorylase-deficient strains. *J Bacteriol* 2005;**187**:4698–706.
112. Mohanty BK, Kushner SR. Polynucleotide phosphorylase functions both as a 3′ right-arrow 5′ exonuclease and a poly(A) polymerase in *Escherichia coli*. *Proc Natl Acad Sci USA* 2000;**97**:11966–71.
113. Raynal LC, Krisch HM, Carpousis AJ. The *Bacillus subtilis* nucleotidyltransferase is a tRNA CCA-adding enzyme. *J Bacteriol* 1998;**180**:6276–82.
114. Leighton T. Further studies on the stability of sporulation messenger ribonucleic acid in *Bacillus subtilis*. *J Biol Chem* 1974;**249**:7808–12.
115. Leighton TJ, Doi RH. The stability of messenger ribonucleic acid during sporulation in *Bacillus subtilis*. *J Biol Chem* 1971;**246**:3189–95.
116. Melin L, Rutberg L, von Gabain A. Transcriptional and posttranscriptional control of the *Bacillus subtilis* succinate dehydrogenase operon. *J Bacteriol* 1989;**171**:2110–5.

117. Resnekov O, Rutberg L, von Gabain A. Changes in the stability of specific mRNA species in response to growth stage in *Bacillus subtilis*. *Proc Natl Acad Sci USA* 1990;**87**:8355–9.
118. Allmansberger R. Degradation of the *Bacillus subtilis* xynA transcript is accelerated in response to stress. *Mol Gen Genet* 1996;**251**:108–12.
119. Dubnau D. Translational attenuation: The regulation of bacterial resistance to the macrolide-lincosamide-streptogramin B antibiotics. *CRC Crit Rev Biochem* 1984;**16**:103–32.
120. Weisblum B. Inducible resistance to macrolides, lincosamides and streptogramin type B antibiotics: the resistance phenotype, its biological diversity, and structural elements that regulate expression—a review. *J Antimicrob Chemother* 1985;**16**(Suppl. A), 63–90.
121. Shivakumar AG, Hahn J, Grandi G, Kozlov Y, Dubnau D. Posttranscriptional regulation of an erythromycin resistance protein specified by plasmic pe194. *Proc Natl Acad Sci USA* 1980;**77**:3903–7.
122. Wei Y, Bechhofer DH. Tetracycline induces stabilization of mRNA in *Bacillus subtilis*. *J Bacteriol* 2002;**184**:889–94.
123. Glatz E, Persson M, Rutberg B. Antiterminator protein GlpP of *Bacillus subtilis* binds to glpD leader mRNA. *Microbiology* 1998;**144**(Pt. 2), 449–56.
124. Kaan T, Jurgen B, Schweder T. Regulation of the expression of the cold shock proteins CspB and CspC in *Bacillus subtilis*. *Mol Gen Genet* 1999;**262**:351–4.
125. Hankins JS, Zappavigna C, Prud'homme-Genereux A, Mackie GA. Role of RNA structure and susceptibility to RNase E in regulation of a cold shock mRNA, cspA mRNA. *J Bacteriol* 2007;**189**:4353–8.
126. Cairrao F, Chora A, Zilhao R, Carpousis AJ, Arraiano CM. RNase II levels change according to the growth conditions: Characterization of gmr, a new *Escherichia coli* gene involved in the modulation of RNase II. *Mol Microbiol* 2001;**39**:1550–61.
127. Lee K, Zhan X, Gao J, Qiu J, Feng Y, Meganathan R, *et al.* RraA. A protein inhibitor of RNase E activity that globally modulates RNA abundance in *E. Coli*. *Cell* 2003;**114**:623–34.
128. Gao J, Lee K, Zhao M, Qiu J, Zhan X, Saxena A, *et al.* Differential modulation of *E. Coli* mRNA abundance by inhibitory proteins that alter the composition of the degradosome. *Mol Microbiol* 2006;**61**:394–406.
129. Mechold U, Fang G, Ngo S, Ogryzko V, Danchin A. YtqI from *Bacillus subtilis* has both oligoribonuclease and pap-phosphatase activity. *Nucleic Acids Res* 2007;**35**:4552–61.
130. Yao S, Blaustein J, Bechhofer DH. Erythromycin-induced ribosome stalling and RNase J1-mediated mRNA processing in *Bacillus subtilis*. *Mol Microbiol* 2008;**69**:1439–49.

RNA Degradation in Archaea and Gram-Negative Bacteria Different from *Escherichia coli*

ELENA EVGUENIEVA-
HACKENBERG AND
GABRIELE KLUG

*Institut für Mikrobiologie und
Molekularbiologie, University of Giessen,
Heinrich-Buff-Ring 26-32, D-35392
Giessen, Germany*

I. Introduction	276
II. RNA Degradation Mechanisms in Bacteria and Eukarya	277
A. Stabilizing and Destabilizing RNA Elements	278
B. Mechanisms for RNA Degradation in Bacteria	278
C. Mechanisms for RNA Degradation in Eukarya	285
III. Features of mRNA in Archaea	287
A. Structural Features of Archaeal mRNA	287
B. Half-Lives of mRNAs in Archaea	288
C. Changes in The Half-Lives of mRNAs in Response to Environmental Stimuli	291
IV. Archaeal Proteins with Unexpected Endoribonucleolytic Activities	293
A. Small DNA-Binding Proteins from *Sulfolobus* with RNase Activity	294
B. Two Different Dehydrogenases from *Sulfolobus* with RNase Activity	295
C. Archaeal IF5A from *H. salinarum* NRC-1 Shows Endoribonucleolytic Activity	297
V. Archaeal Proteins with Similarities to Bacterial or Eukaryotic Proteins Involved in RNA Degradation—A Short Overview	299
VI. The Archaeal Exosome	300
A. Composition of The Archaeal Exosome	300
B. Structure and Mechanism of The Archaeal Exosome	302
VII. RNA Degradation in Archaea	308
A. RNA Degradation in The Exosome-Harboring *S. solfataricus*	308
B. RNA Degradation in Archaea Without Exosome	309
VIII. Concluding Remarks	310
References	311

Exoribonucleolytic and endoribonucleolytic activities are important for controlled degradation of RNA and contribute to the regulation of gene expression at the posttranscriptional level by influencing the half-lives of specific messenger RNAs. The RNA half-lives are determined by the characteristics of the RNA substrates and by the availability and the properties of the involved proteins—ribonucleases and assisting polypeptides. Much is known about RNA degradation in Eukarya and Bacteria, but there is limited information about RNA-degrading enzymes and RNA destabilizing or stabilizing elements in the domain of the Archaea. The recent progress in the understanding of the structure and function of the archaeal exosome, a protein complex with RNA-degrading and RNA-tailing capabilities, has given some first insights into the mechanisms of RNA degradation in the third domain of life and into the evolution of RNA-degrading enzymes. Moreover, other archaeal RNases with degrading potential have been described and a new mechanism for protection of the 5′-end of RNA in Archaea was discovered. Here, we summarize the current knowledge on RNA degradation in the Archaea. Additionally, RNA degradation mechanisms in *Rhodobacter capsulatus* and *Pseudomonas syringae* are compared to those in the major model organism for Gram-negatives, *Escherichia coli*, which dominates our view on RNA degradation in Bacteria.

I. Introduction

From the elongation of the RNA chain during synthesis, to its degradation by ribonucleases (RNases), RNA undergoes a variety of highly specific interactions with numerous proteins. These interactions are necessary for the maturation of precursor molecules or determine the availability of the different RNA species in the cell and, in this way, influence gene expression. While the different aspects of RNA degradation in Eubacteria and Eukarya have been extensively studied, there is still not much information about these processes and their regulation in Archaea, the third domain of life.

Archaea are prokaryotic organisms, which are phylogenetically distinct from Eubacteria and Eukarya (1). Morphologically they are similar to Eubacteria, but biochemically they show similarities to both Eubacteria (e.g., some metabolic pathways) and Eukarya (e.g., replication and transcription mechanisms), and also exhibit unique features (e.g., the composition of the cytoplasmic membrane and the cell wall) (recently summarized in (2)). In addition to their wide distribution in soil and water of moderate climatic regions, many Archaea are extremophiles and live at very high temperature, under high pressure, in acidic environments, or in almost saturated salt lakes. Moreover, the methanogenic Archaea are the only methane producing organisms on earth. The extremophilic Archaea are fascinating research subjects and it is still mysterious how they maintain cellular integrity and function using biomolecules that are in general similar to those of other organisms. Many of the archaeal protein complexes are simplified, presumably ancestral versions of their eukaryotic counterparts, can be handled relatively easily, and are therefore suited for biochemical studies. Unfortunately, Archaea still cannot be genetically manipulated easily. For approximately two decades, shuttle vectors were available only for halophilic Archaea (3), but in the last few years, genetic systems for the major model organisms of Archaea, including *Sulfolobus* species, were developed (4–6).

When investigating Archaea, our assumptions are usually based on the experience and knowledge accumulated from the other two domains of life: Bacteria and Eukarya. From the research on RNA decay in Bacteria and Eukarya, we know that the features of the RNA molecules determine their longevity (reviewed in (7–9)) and that the presence of certain enzymes and enzyme complexes in the cell is indicative of the corresponding degradation mechanisms (10). The availability of many completed genome sequences and of large protein data bases, together with the progress in bioinformatic analyses, enables a straightforward and efficient way for experimental comparison of the RNA degradation mechanisms in the three domains of life. However, we should keep in mind that Archaea may exhibit unique and unexpected features.

Similarly, our view on the mechanisms for RNA decay in Bacteria is strongly biased by the data obtained on the Gram-negative *Escherichia coli* and the Gram-positive *Bacillus subtilis* as model organisms (8). The limited investigation of other organisms belonging to the genera *Rhodobacter*, *Pseudomonas*, *Salmonella*, and *Streptomyces* (reviewed in (11), see also (12)) have shown both similarities and differences to the two major model organisms of the bacterial domain. Generally, the organisms investigated from all three domains of life so far seem to exhibit similar strategies for RNA stabilization and destabilization to perform quality control, maturation, and turnover of RNA. However, even in phylogenetically related groups of organisms like Gram-negative Bacteria or methanogenic Archaea, considerable differences in the mechanisms of RNA decay exist (see below).

We will first give a brief overview of RNA stability determinants and enzymes involved in RNA degradation in Bacteria and Eukarya, to allow us to compare the mechanisms for RNA degradation in Archaea with those operating in the other two domains of life.

II. RNA Degradation Mechanisms in Bacteria and Eukarya

An RNA can be cleaved internally by endoribonucleases or can be degraded nucleotide by nucleotide from the end of the molecule by exoribonucleases, in either the 5′–3′ or 3′–5′ direction. Certain RNA features interfere with RNase activities and are therefore considered as stabilizing elements; attributes of RNA that are recognized by the degrading enzymes are called destabilizing elements. Bacterial and eukaryotic messenger RNAs (mRNAs) have different stability determinants, which can be part of the primary transcript or added posttranscriptionally.

A. Stabilizing and Destabilizing RNA Elements

In Bacteria, mRNAs are often polycistronic and are not subjected to extensive modifications like the monocistronic mRNAs in Eukarya, which are usually spliced, capped, and polyadenylated. The methylguanosine cap at the 5′-end and the long ($\sim$ 200 nt) poly(A) tails at the 3′-end of eukaryotic mRNAs have stabilizing effects. The triphosphate at the 5′-end of bacterial primary transcripts and stem-loop structures originating from Rho-independent terminators at the 3′-end are also stabilizing structures. Additionally, stem-loop structures in the 5′-untranslated region of bacterial mRNA have a stabilizing effect. The only posttranscriptional modification of bacterial mRNA is its nontemplated polynucleotidylation at the 3′-end. The resulting poly(A)- or adenine-rich tails are usually short (up to 50 nt (13, 14)) and are destabilizing elements (see chapters by Hajnsdorf and Regnier, and Stern and Schuster in this volume). Similarly, certain eukaryotic RNA molecules can be primed for degradation in the nucleus by the addition of short poly(A) tails to their 3′-end (Fig. 1) (reviewed in (8, 9)).

The internal portion of RNA molecules can also harbor stability determinants. Endonucleolytic cleavage sites in bacterial mRNAs and AU-rich elements (AREs) in the 3′-noncoding region of eukaryotic mRNAs are destabilizing elements (9). Finally, stem-loops in bacterial mRNAs have a stabilizing effect. In polycistronic mRNAs, they contribute to posttranscriptional regulation of gene expression due to differential stability of mRNA segments (Fig. 1) (reviewed in (7)).

B. Mechanisms for RNA Degradation in Bacteria

1. RNA Degradation in *E. Coli* and *B. Subtilis*: Selected Features

In the best investigated model organism, the Gram-negative bacterium *E. coli*, a set of RNases and RNase-containing protein complexes, some of them with overlapping functions, are responsible for RNA degradation (see chapter by Carpousis in this volume). The removal of the stabilizing 5′-triphosphate is performed by the enzyme RppH, a pyrophosphatase, which marks mRNA for rapid degradation (15, 16). The essential endoribonuclease RNase E senses the 5′-end and cleaves monophosphorylated RNAs with much higher efficiency than triphosphorylated RNAs (17). Although *in vitro* experiments have shown that RNase E interacts with the 5′-monophosphate and scans the mRNA for internal cleavage sites (specific AU-rich single-stranded regions) in the 3′–5′

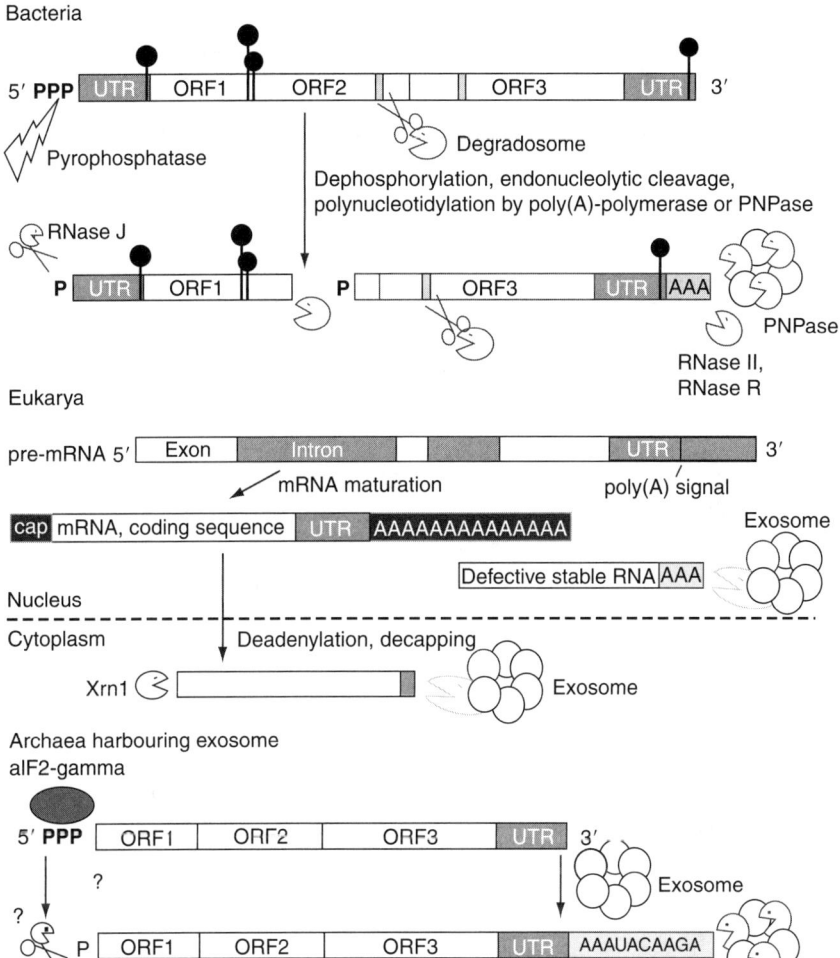

FIG. 1. Schematic overview of RNA degradation in the three domains of life: Bacteria, Eukarya, and Archaea. Endoribonucleases are symbolized by scissors, exoribonucleases by a "pacman." The black hairpin structures represent stabilizing mRNA secondary structures. For further details, see the text.

direction in vitro (18), a net 5′–3′ progression is generally observed in vivo (19, 20). The resulting RNA fragments are strongly destabilized and degraded by the help of endoribonucleases, tailing enzymes, RNA chaperones, RNA helicases, and exoribonucleases. These proteins can be organized in different complexes (8).

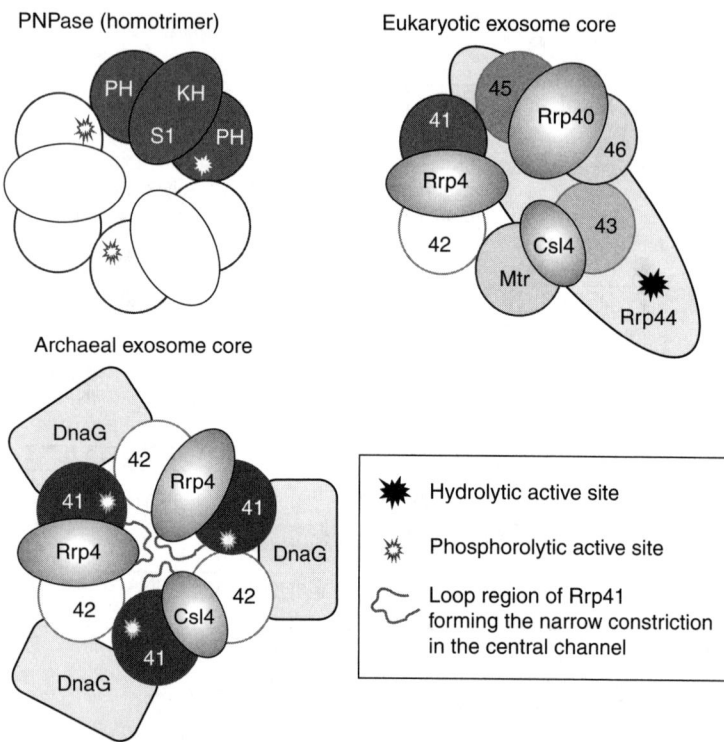

FIG. 2. Schematic overview of the structure of PNPase and the core structure of the exosome in Eukarya and Archaea (a view from the top). One of the monomers of the homotrimeric PNPase is shown in dark gray, the other two monomers are in white. Each monomer contains two RPD domains, an S1, and a KH domain. PH, RPD domain; S1, KH, the corresponding RNA-binding domains; 41–46, Rrp41–Rrp46; Mtr, Mtr3. The exact location of the archaeal DnaG protein in the exosome is not known, but it is present in similar amounts to Rrp41 and Rrp42 in coimmunoprecipitated protein complexes.

Exoribonucleolytic degradation in E. coli proceeds in the 3′–5′ direction. E. coli harbors three exoribonucleases with degradative functions: the hydrolytic RNases, RNase II and RNase R; and the phosphorolytic RNase, PNPase (reviewed in (21)). Each monomer of the homotrimeric bacterial PNPase contains two RNase PH domains (RPDs), only one of which is catalytically active, and the RNA-binding domains S1 and KH. The six RPDs of the trimer are arranged in a ring, on the top of which are positioned the S1 and KH domains (Fig. 2) (reviewed in (22)). Both RNase II and RNase R harbor two cold-shock domains (CSD) containing the RNA-binding OB fold in their N-terminal portions, a catalytically active RNB domain in the central region, and an S1 domain, which also contains an OB fold, at the C-terminus of the polypeptide (23, 24).

The three exoribonucleases cannot degrade RNA all the way to a single remaining mononucleotide; their final degradation products are oligonucleotides of 2–5 bases, which are degraded to completion by oligoribonuclease (25). For successful interaction with RNA, all three exoribonucleases need a single-stranded 3′-end region with a minimal length of 7–8 nt. Often, tailing of the substrate is necessary prior to exoribonucleolytic degradation. Polyadenylation of 3′-ends corresponding to Rho-independent terminators or to sites of endonucleolytic cleavage is performed by poly(A) polymerase, while Rho-dependent 3′-ends of RNA appear to be preferentially polynucleotidylated by PNPase in a reverse reaction to phosphorolysis (26). Interestingly, PNPase and the RNA chaperone Hfq were copurified with His-tagged poly(A) polymerase from *E. coli*. In this protein complex, Hfq possibly melts strong secondary structures at the 3′-end enabling their successful polyadenylation and subsequent exoribonucleolytic degradation (27).

RNase R is the only exoribonuclease capable of degrading structured RNA by itself (28). RNase II and PNPase stop degradation ∼ 8 nt downstream of a stable stem-loop. In this respect, it should be noticed that fast degradation of poly(A) tails by RNase II may actually lead to mRNA stabilization (29). To perform degradation of structured RNA, PNPase forms a complex with the DEAD-box helicase RhlB, which unwinds RNA helices in an ATP-dependent manner (30). Additionally, PNPase and RhlB are subunits of the *E. coli* degradosome, an RNA-degrading protein complex, which is organized by RNase E (31, 32).

The catalytic domain of RNase E is localized in the N-terminal half of the polypeptide. The unstructured C-terminal half is the scaffold for the assembly of the degradosome. The major subunits of this complex are RNase E, PNPase, RhlB, and enolase (Fig. 3) (32). Within the degradosome, the glycolytic enzyme enolase plays a role in the regulation of the stability of a specific mRNA in response to phosphosugar stress (33). The C-terminal region of RNase E is not only important for the assembly of the degradosome, but also for the interaction with Hfq, which binds small noncoding RNAs (ncRNAs) with regulatory functions. In this alternative protein complex, Hfq mediates the interaction of the ncRNAs with their mRNA targets, which are then degraded in an RNase E-dependent manner (34). As mentioned above, Hfq also influences the polyadenylation of RNA and was detected in a complex with poly(A) polymerase and PNPase (27).

The accumulated data show that dynamic protein complexes of different composition are responsible for RNA decay in *E. coli* depending on the environmental conditions and the physiological requirements. In this respect, it is important to note that RNase E activity is controlled by two inhibitors, RraA and RraB, which are expressed under different conditions and influence the composition of the degradosome (35, 36). Additionally, it is noteworthy that

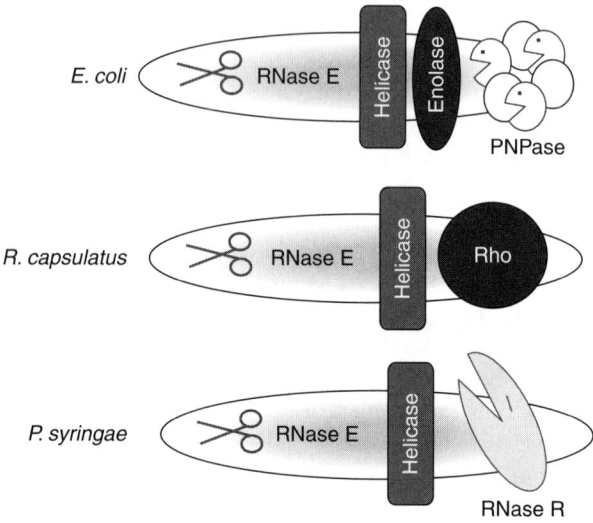

Fig. 3. Schematic representation of the composition of the degradosome in *E. coli*, *R. capsulatus*, and *P. syringae*.

the degradosome forms cytoskeleton-like structures via its RNase E and RhlB subunits *in vivo*, pointing to spatial organization of RNA degradation in the cell (37, 38).

In addition to RNase E, the endoribonucleases RNase G, which shows homology to the N-terminal RNase E region and has similar substrate requirements, and the double-strand-specific endoribonuclease RNase III participate in degradation of some mRNAs in *E. coli* (21). RNase E, which is considered to be the major degrading endonuclease in *E. coli*, is present in the Gram-positive *Streptomyces*, but not in *B. subtilis*, implying different mechanisms of RNA degradation in *B. subtilis*. Two related proteins (RNase J1 and J2), which are not orthologous to RNase E, but exhibit similar endonucleolytic activity, were identified in *B. subtilis*, and surprisingly, they turned out to also have 5′–3′ exoribonucleolytic activity. This was the first example of exoribonucleolytic degradation in 5′–3′ direction in Bacteria. Interestingly, RNase J homologs, although not present in gamma-Proteobacteria like *E. coli*, are widely distributed in other Bacteria and in Archaea (39, 40).

The above model for RNA degradation in *E. coli* is widely accepted. However, it was pointed out that this model does not explain the results of experiments for determination of 5′-ends arising due to endoribonucleolytic cleavages in the cell. The majority of the 5′-ends of RNA fragments were found to carry 5'-OH and to begin with an adenine, consistent with endonucleolytic degradation by an RNase A-type ribonuclease. Therefore, it was postulated that an unidentified enzyme other than RNase E, which produces monophosphorylated 5'-ends, is the major degrading endonuclease in *E. coli* (41).

Further aspects of RNA decay in Bacteria like the action of toxin–antitoxin systems (42) (see chapter by Inouye in this volume) and the nonstop mRNA decay, in which the tmRNA plays a key role (43, 44), are not described here. Since some of the RNases from the toxin–antitoxin systems produce RNA fragments with 5'-OH groups and an adenine at the 5'-end, they may be responsible for the 5'-ends detected in *E. coli* (45). Degradation of stable RNAs is probably performed in similar way to mRNA degradation (see chapter by Deutscher in this volume).

2. Degradation of mRNA in the Alpha-Proteobacterium *Rhodobacter capsulatus*

The role of mRNA degradation in the regulated expression of bacterial photosynthesis genes was intensely studied using the *puf* mRNA in *Rhodobacter capsulatus*. The influence of oxygen tension on the decay of *puf* mRNA in *R. capsulatus* was one of the first examples showing the importance of environmental factors on the stability of prokaryotic transcripts. The *puf* operon encodes pigment-binding proteins of the light-harvesting complex I (LHI) and the reaction center (RC) of the photosynthetic apparatus of *R. capsulatus*. These proteins are expressed under low oxygen tension, and individual segments of the polycistronic *puf mRNA* display extremely different half-lives contributing to the molar ratio of LHI and RC proteins required to form functional photosynthetic complexes (7, 46–49).

It was shown that several stabilizing (stem-loop structures) and destabilizing (RNase E cleavage sites) elements in specific positions of the primary 3.2 kb *puf* transcript (half-life of less than 0.5 min) are necessary for the nucleolytic formation of the 2.7 kb *pufBALMX* (half-life of 8 min) and 0.5 kb *pufBA* mRNA (half-life of 33 min) segments. The decay rate of the 2.7 kb transcript is increased under conditions of high oxygen tension (half-life of 3.5 min), and it was shown that endonucleolytic cleavages in a specific region of the transcript are the oxygen-dependent steps inactivating this message (7, 49). The endonucleolytic cleavages are performed by RNase E, which is part of the *R. capsulatus* degradosome. The composition of this protein complex differs from that of the *E. coli* degradosome (Fig. 3) (50).

The degradosome of *R. capsulatus*, partially purified in a way similar to that described for the *E. coli* degradosome (ammonium sulfate fractionation and sulfopropyl-sepharose chromatography in the presence of protein inhibitors), contains RNase E, the transcription termination factor Rho, two DEAD-box RNA helicases (65 and 74 kDa), a few unidentified proteins, and PNPase in very low amounts. Fractionation of the purified proteins through glycerol density gradient revealed the presence of heterogeneous complexes: a complex containing RNase E, the 75 kDa helicase and an unidentified 47 kDa protein, and a larger, 240 kDa complex containing RNase E, the 65 kDa helicase, Rho and an unidentified protein of 36 kDa. In contrast to the *E. coli* degradosome, PNPase is not a major component of the degradosome of *R. capsulatus*, but it was detected by Western blot analysis and by a PNPase-specific functional assay in the degradosome-containing glycerol density gradient fractions. The 75 kDa DEAD-box helicase and the 36 kDa protein are loosely associated with the protein complex(es) and are not copurified by immunoprecipitation with RNase E-specific antibodies. The strongly interacting *R. capsulatus* proteins that coimmunoprecipitated with RNase E were Rho, the 65 kDa DEAD-box helicase, and the 47 kDa polypeptide (50).

The role of Rho (a hexameric RNA–DNA helicase involved in transcription termination in *E. coli*) in the context of the degradosome is not known, but it may link RNA transcription to RNA degradation or processing. Interestingly, the amount of Rho in the degradosome was strongly increased under high oxygen tension, while the amount of the two DEAD-box RNA helicases was differently influenced. Moreover, the degradosome isolated from cells grown under aerobic conditions displayed higher degradation activity *in vitro* than the degradosome from semiaerobically grown cells, leading to the conclusion that the composition and activity of the *R. capsulatus* degradosome vary under different environmental conditions (51).

RNase E of *R. capsulatus* is a large polypeptide (118 kDa) with a conserved N-terminal portion (50). Most probably, the C-terminal region serves as a scaffold for complex assembly, as was shown for *E. coli* degradosome. The differences in the composition of the RNase E-based protein complexes in *E. coli* and *R. capsulatus* suggest that formation of RNA-degrading protein complexes is important for bacterial cells, but in evolution, different organisms adopted different mechanisms for processing and regulated degradation of RNA. This can be seen not only on the level of the proteins involved in RNA metabolism, but also on the RNA level. For example, the cold-shock-dependent stabilization of the *R. capsulatus cspA* mRNA is not regulated by rate-limiting cleavage in the 5'-UTR like in *E. coli*, although the transcript is cleaved endonucleolytically by RNase E of the *R. capsulatus* degradosome (52). Another example is the finding that the CIRCE element, an inverted DNA repeat known to be involved in the temperature-dependent regulation of heat shock genes, has a different function in *R. capsulatus* (53). It is not involved in the temperature-dependent transcription of the *groESL* genes as in most other bacteria, or in the specific destabilization of transcripts as in *B. subtilis* (54). Instead, it exerts a slight stabilizing effect on the *groESL* mRNA. It was found that deletion of CIRCE results in rapid recognition of an RNase E cleavage site in the U-rich region of the Rho-independent terminator of transcription, providing the first hints at the degradation mechanism of the *groESL* transcript by the action of RNase E and the degradosome in *R. capsulatus* (53).

3. THE DEGRADOSOME OF THE GAMMA-PROTEOBACTERIUM *PSEUDOMONAS SYRINGAE* Lz4W

The psychrophilic bacterium, *Pseudomonas syringae* Lz4W, contains a degradosome complex of 240 kDa, in which RNase E is associated with the hydrolytic exoribonuclease RNase R and the DEAD-box helicase RhlE (Fig. 3). The protein complex was purified by the procedure described above, which was also used for the isolation of the degradosomes from *E. coli* and *R. capsulatus*. Several other proteins were detected in the degradosome-containing fractions of the glycerol density gradient, including the DEAD-box helicase SrmB. However, specific interaction in a protein complex was confirmed only for RNase E, RNase R, and RhlE: RNase E and RNase R were coimmunoprecipitated with RNase E-specific antibodies, and all three proteins were copurified in a pull-down assay (55).

E. coli RNase R and the homologous eukaryotic protein Rrp44 are capable of degrading stem-loop-containing RNA by themselves and this raised the question about the role of RhlE in the degradosome of the Antarctic *P. syringae*. Besides the possibility that RhlE cooperates with RNase E in remodeling RNA substrates to reveal RNase E cleavage sites (56), a helpful role for RhlE for an efficient, ATP-dependent exoribonucleolytic degradation of structured transcripts by the degradosome was shown *in vitro*. *P. syringae* Lz4W possesses PNPase in addition to RNase R, but phosphorolytic activity was not detected in the degradosome preparations from this organism, showing that PNPase is not part of the degradosome (55).

Little is known about RNA degradation in the cold-adapted *P. syringae* Lz4W. It was shown that RNase R is essential for growth at 4 °C, although it is not cold inducible like in *E. coli*, and that the growth defect of an *rnr* mutant cannot be repaired by overexpression of PNPase. RNase R was found to be involved in 3'-end maturation of 16S and 5S rRNA, and in tmRNA turnover (57). The data show that the precise functions of RNases and RNase-containing complexes may differ in different organisms. However, the comparison of the degradosomes in phylogenetically distant bacteria shows that the interaction between endonucleases, exoribonucleases, and RNA helicases is important for bacterial cells and was retained during evolution.

C. Mechanisms for RNA Degradation in Eukarya

The eukaryotic RNA decay mechanisms are mentioned here very briefly, and only exosome-mediated degradation is described in more detail. The exosome, a protein complex with 3'–5' exoribonucleolytic activity, is present in the nucleus and in the cytoplasm, where it interacts with different proteins that enable its participation in a variety of different RNA maturation and degradation pathways (reviewed recently in (9)). In the nucleus, defective rRNA and tRNA molecules are targeted for degradation by protein complexes called TRAMP, which add short destabilizing poly(A) tails and mediate the interaction of the exosome with the tailed substrates (58). Incorrectly processed mRNAs, which carry a binding site for the nuclear RNA-binding protein NRD1, are also targets for degradation by the exosome (Fig. 1) (59).

In the cytoplasm, after deadenylation (degradation of the stabilizing long poly(A) tails at the 3′-end, which are protected by poly(A)-binding proteins) and decapping (removal of the methylguanosine cap from the 5′-end), eukaryotic mRNAs are degraded exoribonucleolytically in the 5′–3′ direction, by the RNase Xrn1, or in the 3′–5′ direction, by the exosome (Fig. 1) (reviewed in (9)). Proteins involved in the 5′–3′ decay pathway are concentrated in cytoplasmic processing bodies (P-bodies) (60). mRNAs containing destabilizing AREs (adenine-rich elements) in the 3′-untranslated region are rapidly deadenylated. Although the exosome is recruited by ARE-binding proteins (61), both the 3′–5′ and 5′–3′ exoribonucleolytic pathways are important for the decay of ARE-containing RNAs (62). ARE-containing mRNAs accumulate in the cytoplasm in discrete granular structures distinct from the P-bodies, together with poly(A)-specific RNase, the exosome, and ARE-binding proteins (63). Histone mRNAs, which do not carry stabilizing poly(A) tails, are oligouridylated, decapped, and degraded simultaneously from the 5′-end and from the 3′-end (64).

The mechanisms described above are responsible for general mRNA turnover. In addition, specific mRNAs are rapidly degraded in the process of RNA interference (RNAi) which silences gene expression posttranscriptionally. Double-stranded RNA molecules are first cleaved by Dicer, which belongs to the RNase III family of proteins. The resulting 21–22 nt small interfering RNAs (siRNA) are incorporated into the RNA-induced silencing complex (RISC), which cleaves mRNAs complementary to siRNAs (65). After the endonucleolytic cleavage, the distal RNA fragment is degraded by Xrn1 and the proximal fragment is degraded by the exosome with the help of a protein complex named Ski (Ski2, Ski3, and Ski8), without undergoing deadenylation or decapping (66). The Ski complex is important for the cytoplasmic functions of the exosome, and the interaction between the two protein complexes is mediated by the protein Ski7 (67).

The bypassing of deadenylation and/or decapping is also typical for the degradation of aberrant mRNAs. The specialized degradation pathways involved are termed mRNA surveillance pathways and include nonsense-mediated mRNA decay (translation termination occurs too soon), nonstop mRNA decay (translation termination does not occur), and no-go mRNA decay (translation elongation stalls). Nonsense mRNA is degraded exonucleolytically in the 5′–3′ by Xrn1, as well as in the 3′–5′ direction by the exosome together with the Ski proteins. These enzyme complexes may also participate in the no-go mRNA decay pathway, after an endonucleolytic cleavage by the Dom34 protein. Nonstop mRNA is degraded in the 3′–5′ direction by the exosome with the help of the Ski proteins (68–70).

The core of the eukaryotic exosome, which is present in the nucleus and in the cytoplasm, consists of 10 essential subunits (Rrp4, Rrp40–Rrp46, Mtr3, and Csl4). The six RPD-containing polypeptides of the eukaryotic exosome (Rrp41, Rrp42, Rrp43, Rrp45, Rrp46, and Mtr3) form a hexameric ring, on the top of which bind the S1 and KH domain-containing subunits Rrp4, Rrp40, and Csl4. This nine-subunit exosome shows structural similarity to PNPase, but is catalytically inactive. The hydrolytic RNase activity of the core eukaryotic exosome is due to its 10th subunit, Rrp44, which belongs to the *rnr* family of proteins, together with bacterial RNases II and R (Fig. 2) (9, 71–73). Consistent with differences in the spatial arrangement of the RNA-binding domains CSD1, CSD2, and S1 in Rrp44 and RNase II, Rrp44 can degrade structured RNA, resembling bacterial RNase R more than RNase II (23, 24). Rrp44 binds at the bottom of the hexameric ring of the exosome, and this binding probably regulates the activity of the exosome and its substrate accessibility (74, 75).

III. Features of mRNA in Archaea

Despite the prokaryotic structure of the cells, some molecular mechanisms that operate in Archaea show more similarities to those in Eukarya. Little is known about RNA decay in Archaea, but has become clear that different archaeal groups have quite different strategies for RNA degradation.

A. Structural Features of Archaeal mRNA

Archaeal mRNA is more similar to bacterial than to eukaryotic mRNA: it is essentially free of introns, is often polycistronic, is not modified by the addition of a methylguanosine cap at the 5′-end, and does not have long stabilizing poly(A) tails at the 3′-end (2, 74, 76). Instead, it carries a triphosphate as a stabilizing structure at the 5′-end (77), and, in some but not all Archaea, heteropolymeric adenine (A)-rich tails with a likely destabilizing function at the 3′-end (78, 79). Archaea lacking an exosome have no RNA tails at all—these halophilic and some methanogenic Archaea are the only organisms found so far in which RNA does not undergo posttranscriptional modification at the 3′-end (Fig. 1). This finding was described by different groups and was analyzed by different methods—3′-end labeling of RNA and digestion with RNases A and T1, oligo-d(T)-primed reverse transcription PCR (RT-PCR), as well as circularization of RNA and subsequent RT-PCR (78–80). Nothing is known about internal stabilizing or destabilizing mRNA elements in Archaea. Stability determinants, together with the general turnover mechanisms in a specific organism, influence the half-lives of different mRNAs, and are thus important factors in the regulation of gene expression.

B. Half-Lives of mRNAs in Archaea

To understand the contribution of RNA-degrading mechanisms to the posttranscriptional regulation of gene expression in Archaea, it was necessary to determine the half-lives of different archaeal mRNAs and to analyze whether their stability changed under different conditions. It is well known that bacteria control gene expression at the posttranscriptional level by changing the half-lives of specific mRNA species in response to environmental stimuli (7). Bacterial mRNAs are generally unstable—their half-lives are in the range of seconds to minutes, enabling a quick response to changing environmental conditions (7, 81, 82). The half-life of eukaryotic mRNAs ranges from tens of minutes to hours (83, 84) and it has been assumed that the half-lives of mRNAs correlate with the doubling time of the corresponding cells.

A prerequisite for RNA half-life determination is the efficient arrest of transcription. In Bacteria, this is achieved by the addition of rifampicin, which blocks the function of RNA polymerase. Archaeal RNA polymerase is of a eukaryotic type and is not sensitive to rifampicin (85, 86). The first determinations of the stability of archaeal mRNAs were performed with the mesophilic methanogene *Methanococcus vannielii*. RNA transcription was inhibited by addition of bromoethanesulfonate (BES) or removal of H_2, and the half-lives of the analyzed transcripts were 7, 15, 30, and 57 min, respectively, suggesting that mRNAs in Archaea are long lived. Puromycin, pseudomonic acid, or virginiamycin did not inhibit transcription (87).

Further attempts to inhibit transcription in Archaea led to the use of the DNA-intercalating reagent actinomycin D (88). It was shown that actinomycin D successfully stops incorporation of uridine in *Sulfolobus*, *Haloferax*, and *Halobacterium* (89–92), strongly suggesting that this antibiotic is useful for half-life measurements. The longevity of several transcripts was determined by Northern blot analysis for *Sulfolobus solfataricus* and *Haloferax mediterranei* (89, 90). In these two studies, different mRNAs encoding transcription factors and metabolic enzymes were analyzed. The following approximate half-lives were reported for *S. solfataricus* mRNAs: 6, 13, 25, 37, 54 min and at least 2 h, which were in the range reported for *M. vannielii* (87, 89) In the study with *H. mediterranei*, the half-lives of mRNAs encoding gas-vesicle proteins (*gvp*) from two operons, including genes for structural proteins and transcriptional regulators were determined. The half-lives of four mRNA segments were between 5 and 12 min, and only the mRNA for the major structural protein GvpA was much longer, 40 or 80 min, depending on the growth conditions (90).

Finally, RNA decay in *S. solfataricus*, *S. acidocaldarius*, and *Halobacterium salinarum* was studied on a global level, by microarray analysis. Surprisingly, the median half-lives of archaeal mRNAs turned out to be rather short: 5 min in the thermoacidophilic crenarchaea *S. solfataricus* and *S. acidocaldarius*, and 9.4 min in the halophilic euryarchaeon *H. salinarum*. For selected transcripts, the microarray data were confirmed by Northern blotting and quantitative reverse-transcriptase PCR (qRT-PCR) (*91, 92*).

In the transcriptome studies, 2064 genes of *S. solfataricus*, 1582 genes of *S. acidocaldarius*, and 1717 genes of *H. salinarum* were analyzed (*91, 92*). The half-lives of mRNAs varied between 2 and more than 20 min in the two *Sulfolobus* species, and between 5 and more than 18 min in *H. salinarum*. About 8% of the *Sulfolobus* mRNAs displayed half-lives longer than 20 min, and the half-lives of 50% of the mRNAs were in the range of 4–8 min. The mRNAs of *H. salinarum* showed less variation—the half-lives of 79% of the mRNAs were between 8 and 12 min, and a total of 30 mRNAs showed half-lives longer than 18 min. Messages with very short half-lives—less than 5 min in *H. salinarum* or less than 2 min in *Sulfolobus*—were not detected.

The GC content of the chromosome of *H. salinarum* (67.9%) is higher than that of its megaplasmids (57.9% and 59.2%). The mean half-life of the plasmid-derived mRNAs was shorter (8.7 min) than the mean half-life of the chromosome-derived mRNAs (10.6 min). The plasmid-derived mRNAs showed broader distribution of different half-lives—only 31% of them showed half-lives between 8 and 9 min, while 83% of the chromosome-derived mRNAs showed half-lives in this range. It is not clear whether these differences are related to the function of the genes located on the different replicons or are due to the different GC content.

In *Sulfolobus*, there was a correlation between the length of a transcript and its stability—the longer the mRNA, the shorter its half-life. No such correlation was found in *Halobacterium*, in Bacteria, or in Eukarya. Not surprisingly, there was a correlation between gene function and mRNA stability. In *Halobacterium* and in *Sulfolobus*, the most unstable transcripts were those encoding proteins involved in energy metabolism, amino acid metabolism, transcription, and signal transduction, while those related to central intermediary metabolism and lipid metabolism had longer half-lives. Interestingly, mRNAs for translational regulators in *Halobacterium* and for some transcriptional regulators in *Sulfolobus* were found to have significantly longer half-lives than the mRNAs for the corresponding basal transcriptional and translational machinery, suggesting differential regulation of gene expression at the posttranscriptional level. Generally, mRNAs of highly expressed genes (defined as high steady-state mRNA level) were found to be the most unstable and were enriched for essential functions. This should be helpful for quick adaptation to changing conditions. Additionally, it was pointed out that short half-life and a low number of translation events per mRNA contribute to a minimization of undesired protein fluctuations in the cell, and this is important for essential proteins and for subunits of protein complexes (for more detailed discussion, see (92)).

It is still not clear whether the median mRNA half-life correlates with the prokaryotic or eukaryotic nature of the organism, with its generation time, or with the environmental parameter of its habitat. In *Halobacterium*, the median half-life is 9.4 min, about two times longer than the half-live of a typical bacterial mRNA (*81, 82*), but is still two times shorter than the half-life of a typical mRNA in the eukaryote *Saccharomyces cerevisiae* (*84*). However, the median mRNA half-life in *Sulfolobus* of 5 min is similar to that of bacterial mRNAs. Thus, although *Sulfolobus* has a doubling time of about 4–6 h, the longevity of its mRNAs is similar to that of *E. coli*, which has a doubling time of 20 min. *Halobacterium* has a similar doubling time to *Sulfolobus*, and the question arises whether prokaryotic mRNAs are generally more unstable than eukaryotic mRNAs. However, the relative stability of the mRNAs in *Halobacterium* might be due to the high salt concentrations in its habitat and in the cytoplasm. Alternatively, or in addition, the mRNA stability in *Sulfolobus* might be generally lower than in other Archaea because of its hyperthermophilic nature. Under high temperatures of $\sim 80\,°C$, a high degree of intrinsic RNA instability is expected.

To date, a global analysis of the longevity of archaeal mRNAs has only been performed for these extremophiles, and it is not clear whether the results are representative of mesophilic Archaea, which are widely distributed in the environment. In addition, the obvious differences that exist between the mRNA degradation mechanisms in *Sulfolobus* and *Halobacterium* should be taken into account: *Sulfolobus* harbors an exosome and tailed RNAs, while *Halobacterium* does not (see below (78)). These differences may have a general effect on mRNA half-life. It should also be mentioned that the arrest in uridine incorporation after addition of actinomycin D mainly reflects an arrest of ribosomal RNA synthesis. Low levels of continuing mRNA transcription, which cannot be detected, would result in an overestimation of half-lives (discussed in (93)). However, even if there is a general overestimation of the median mRNA half-life in Archaea or an overestimation of the half-lives of specific mRNAs, the available data demonstrate that there are clear differences in the half-lives of different mRNAs. Thus, mRNA-specific stability elements must exist that determine the rate of decay of particular mRNAs and influence their steady-state levels.

Although the median half-life in the two *Sulfolobus* species was very similar, there was no correlation between the half-lives of mRNAs encoding homologous proteins (92). In line with this, the posttranscriptional regulation of the expression of the gas-vesicle protein (*gvp*) genes in *H. salinarum* and in

H. mediterranei seems to be quite different (*91*). The *gvp* genes are transcribed from two operons: *gvp*ACNO and *gvp*DEFGHIJKLM. GvpA is the major structural protein, which constitutes the wall of the vesicle, together with the less abundant GvpC. GvpD and GvpE are the repressor and activator of the transcription of the *gvp* operons, respectively. Gas vesicles are formed in stationary growth phase under high salt conditions, and the expression of the individual genes is growth phase and salt dependent. In *H. mediterranei*, the full-length transcripts of both operons can be detected at different growth stages when 25% salt is present in the medium—the *gvp*ACNO message is detected at stationary growth phase, while the *gvp*DEFGHIJKLM message is detected in exponential cultures. Segments of the polycistronic mRNAs containing different open reading frames (ORFs) are not equally abundant, as determined by Northern blot, with the first ORF of each operon being present in the highest amount (*94–97*).

The regulation of the steady-state amount of different *gvp* mRNAs is different in *H. salinarum* and in *H. mediterranei* (*90, 91*). The half-life of all evaluated *gvp* mRNAs in *H. salinarum* were between 8 and 9 min. Thus, the greater abundance of the *gvp*A message compared to the *gvp*N and *gvp*O messages in this organism is most likely due to a higher transcription rate at the *gvp*A promoter and premature termination downstream of the *gvp*A ORF, preventing the transcription of the whole operon. Whether premature termination of transcription occurs in *H. mediterranei* is not clear, but it was shown that differential stability of mRNA segments contributes to the posttranscriptional regulation of *gvp* genes in this organism. This is demonstrated by the finding that in the stationary phase, in the presence of 25% salt in the growth medium, the half-life of *gvp*A (80 min) is considerably longer than the half-lives of the *gvp*ACNO, the *gvp*D, and the *gvp*DE transcripts (between 5 and 11 min). This longer half-life obviously contributes to the higher steady-state level of *gvp*A mRNA and to the greater production of the GvpA polypeptide in comparison to the other Gvp proteins under conditions of gas-vesicle formation.

Thus, different, even closely related archaeal species exhibit different strategies for regulation of homologous genes, with more emphasis on transcriptional or posttranscriptional mechanisms, and probably also on translational regulation, depending on the case.

C. Changes in The Half-Lives of mRNAs in Response to Environmental Stimuli

It has been well established that, in Bacteria, the half-lives of specific mRNAs change in response to the changes in the environment, contributing to the regulation of gene expression. The prokaryotic nature of the archaeal cells, their lifestyle—most of them are free-living unicellular organisms—and the structural similarities between archaeal and bacterial mRNAs lead to the expectation that Archaea also exploit the regulation of mRNA stability for adaptation to changing conditions. The pressure to use this strategy for regulation is even stronger in Archaea than in Bacteria, because mRNAs are often leaderless in the Archaea (*80, 98, 99*). In *Sulfolobus* as well as in *Halobacterium*, many mRNAs do not possess untranslated 5′-regions (5′-UTRs), and thus cannot be translationally regulated by antisense RNAs overlapping the Shine–Dalgarno (SD) sequences or by alternative secondary structures in the region of translational initiation, regulatory mechanisms commonly found in Bacteria (*100*). In *Sulfolobus*, many small noncoding RNAs have been described, which are complementary to the translation initiation region (*101*), but this type of regulation seems to apply to distal ORFs in polycistronic mRNAs rather than the leaderless monocistronic mRNAs or the proximal ORFs of polycistronic messages.

The question whether RNA stability in Archaea differs under different growth conditions was addressed in a study with *H. mediterranei* (*100*). In stationary growth phase, in presence of 25% salt in the medium, the half-life of the *gvp*A mRNA was 80 min. Interestingly, when only 18% salt was present in the medium, the half-life of *gvp*A mRNA in the stationary phase was only 39 min. Thus, under conditions leading to less production of gas vesicles (low salt), the *gvp*A message is destabilized. This clearly shows that in *H. mediterranei*, the gene for the major structural protein of the gas vesicles, GvpA, is regulated on posttranscriptional level in response to environmental stimuli.

The stability of the *H. mediterranei gvp*A mRNA was also analyzed in the gas-vesicle-deficient species *Haloferax volcanii*. In late exponential growth phase, in presence of 25% salt, the half-life of *gvp*A mRNA was 160 min, while at 18% salt it was only 90 min, showing a similar tendency toward destabilization under low salt conditions, like in *H. mediterranei*. Since only *gvp*A mRNA showed changed stability under different salt concentrations and other *gvp* mRNAs did not, this effect cannot be attributed to generally higher RNase activities in media with low salt concentration. Instead, it must be due to destabilization of protective structures on *gvp*A mRNA in low salt environment. As mentioned above, *H. salinarum* does not increase the amount of its *gvp*A mRNA by increasing its stability when gas vesicles are formed. It is not clear whether *gvp*A mRNA of *H. mediterranei* harbors particular stabilizing elements that are missing in the message of *H. salinarum* (*91*).

A change in half-life was also detected at different growth stages under high salt conditions for a fragment of the *gvp*D mRNA. During early exponential growth, the signal for the 2.0 kb *gvp*DE mRNA is more intense than the signal of the 0.45 kb *gvp*D segment in *H. mediterranei*, whereas during stationary growth phase the reverse is true. The half-life of these RNAs at different growth stages was measured in *H. volcanii*, where they were transcribed from a *gvp*DE construct under the control of the salt-independent *fdx* promoter from *H. salinarum*. The half-life of the 2.0 kb *gvp*DE transcript was between 9 and 11 min in early exponential, late exponential, and stationary growth phases. In contrast, the 0.45 kb *gvp*D segment was not detectable in early exponential phase, had a half-life of 9 min in late exponential phase, and the half-life increased to 21 min in stationary phase.

As observed for the *gvp*A mRNA, the half-lives of the *gvp*DE transcripts were about twofold longer in *H. volcanii* than in *H. mediterranei*. In stationary growth phase under high salt conditions, the 2.0 kb *gvp*DE transcript had a half-life of 5.5 min (compared to 10 min in *H. volcanii*) and the 0.45 kb *gvp*D segment of mRNA had a half-life of 10 min (compared to 21 min in *H. volcanii*). Thus, either RNA is degraded generally more slowly in *H. volcanii*, or mRNA transcription is not as efficiently inhibited by actinomycin D in this archaeon. Despite these differences, a stabilization mechanism specific for the 0.45 kb *gvp*D segment seems to exist, which operates in two different *Haloferax* species. In summary, the data show that under different physiological conditions, the longevity of specific mRNAs changes in Archaea. Thus, differential gene expression is also regulated posttranscriptionally in the third domain of life, but nothing is known about the underlying mechanisms yet.

IV. Archaeal Proteins with Unexpected Endoribonucleolytic Activities

The availability of many sequenced genomes allows the identification of archaeal proteins with similarity to already characterized RNases from the other two domains of life. However, Archaea-specific RNases cannot be found using this approach. If such RNases can be detected in cell-free extracts, they can be purified by protein fractionation, however. The specificity of the test system used is important in such approaches. If the RNase activity is highly specific and the substrate is known, as in the case of the tRNA-processing endonucleases RNase P and RNase Z, and the splicing endonuclease, the corresponding protein can be enriched and identified (*102–104*). However, the cleavage sites of many important RNases cannot be easily predicted. In addition to the use of incorrect substrates, inappropriate or suboptimal reaction conditions may lead to failure to detect RNase activity. Another problem is the high risk of contaminations with highly active ribonucleases like RNase A, leading to false-positive results.

Several attempts were undertaken to identify archaeal endoribonucleases involved in RNA degradation by purification from cell-free extracts using classical chromatography methods. Various proteins were detected and characterized that cleave RNA endonucleolytically *in vitro*, but which have other obviously important roles in the cell: several small DNA-binding proteins from *S. solfataricus* and *S. acidocaldarius*, two dehydrogenases from *S. solfataricus*, and the halobacterial aIF-5A protein. Their properties are briefly summarized here, although the physiological relevance of these proteins with respect to RNA metabolism is not clear.

A. Small DNA-Binding Proteins from *Sulfolobus* with RNase Activity

The first attempt to isolate RNases from *S. solfataricus* led to the purification of three small basic proteins. Two of them, named p2 (Sso7d) and p3 (8.5-kDa RNase), showed significant similarity to 7 kDa DNA-binding proteins like Sso7d from *S. solfataricus* and its homolog from *S. acidocaldarius*, Sac7d. These DNA-binding proteins increase the melting temperature of dsDNA by 40 °C and are regarded as archaeal histones (105).

The RNase properties of Sso7d and the 8.5-kDa RNase were the same. The two proteins did not require divalent cations for activity and were capable of cleaving yeast tRNA but not homopolymers. The recombinant proteins showed the same activity as the proteins directly purified from *S. solfataricus* (106). However, another group reported that the RNase activity of Sso7d is due to copurification of RNase A present in the growth medium (107). To demonstrate the intrinsic RNase activity of Sso7d, individual amino acid residues were exchanged (E35L and K12L). The RNase activity was strongly abolished, although the overall structure of the protein was not changed, as judged from NMR thermal denaturation profiles, leading to the conclusion that Sso7d is indeed an RNase (108).

Although Sso7d is the smallest known enzyme, it interacts with DNA and RNA via different surfaces—its RNase activity is not abolished in the presence of DNA. The preferred cleavage sites were mapped in double-stranded and in single-stranded RNA regions without sequence specificity, but they correspond to intrinsically unstable bonds: the same digestion pattern was detected after nonspecific alkaline degradation of the RNA substrate and after incubation with Sso7d. Upon incubation of RNA with the enzyme, the most intrinsically unstable bonds are attacked first, but after prolonged incubation the substrate is completely destroyed, presumably to mononucleotides or oligonucleotides (108).

Although the amino acid sequence of Sso7d does not contain histidine residues, its catalytic mechanism seems to resemble that of RNase A. It was not investigated whether RNase A produces similar digestion pattern on tRNA or whether the activity of Sso7d is abolished by an RNase A inhibitor. It was pointed out that Sso7d might function as an RNase *in vivo* despite its low activity (K_m of 5 µM), because it is highly abundant (estimated minimal intracellular concentration of 1 mM (*108*)). The conditions for optimal activity are compatible with a physiological function as RNase—the enzyme showed pH–activity profiles with an optimum in the range 6.7–7.6 and thermostability up to 85 °C, consistent with the hyperthermophilic nature of *S. solfataricus* (*108*).

A similar basic protein of 9 kDa with RNase activity and DNA-binding properties was purified from *S. acidocaldarius* and named SaRD. The RNase activity was not affected in the presence of different divalent cations at concentrations up to 1 mM. After N-terminal sequencing of the protein, a degenerate oligodeoxyribonucleotide was designed to identify the corresponding gene by Southern hybridization. Two separate genes were detected, cloned, and the corresponding proteins were overproduced in *E. coli*. They were identified as the DNA-binding proteins Sac7d and Sac7e. SaRD was shown to be identical to Sac7e and while this protein showed RNase activity, Sac7d did not. Moreover, an amino acid substitution was shown to be crucial for its activity, indicating that the activity is not due to spurious RNase contamination (*109, 110*).

B. Two Different Dehydrogenases from *Sulfolobus* with RNase Activity

Curiously, the isolation of two ribonucleolytically active dehydrogenases, which cleave preferentially between C and A in single-stranded RNA regions, was the result of an attempt to purify an archaeal dsRNA-specific enzyme with RNase III-like properties (*111*). Protein fractions from *S. solfataricus* cell-free extracts were monitored for RNase III-like activity using a short, well-characterized substrate of RNase III from *E. coli* (called N26) and appropriate reaction conditions. Since RNase III is magnesium dependent, 10 mM $MgCl_2$ was used in the activity assays. After six fractionation steps, during which an endoribonuclease activity associated with major peaks in the elution profiles was followed, two proteins were detected in the fractions with the highest RNase activity. They were identified as aspartate-semialdehyde dehydrogenase (ASD) and acyl-CoA dehydrogenase (A-CoA).

The final elution fraction containing the native ASD and A-CoA proteins produced two different N26 RNA cleavage patterns, depending on the presence or absence of $MgCl_2$ in the reaction buffer. The recombinant His-tagged ASD and A-CoA proteins of *S. solfataricus*, which were isolated from *E. coli*, did not cleave RNA at 10 mM $MgCl_2$, but produced the magnesium-independent N26 RNA cleavage pattern. It is not clear whether the RNase properties of the endogenous ASD and A-CoA proteins differ from the properties of the recombinant proteins or whether the final elution fractions from *S. solfataricus* contained two different types of RNases.

The interaction of the recombinant proteins with RNA was characterized using several different rRNA-derived *in vitro* transcripts. The interest in the RNase activity of dehydrogenases was based on reports that eukaryotic dehydrogenases can bind RNA and that certain RNases share homology with dehydrogenases (*112*, *113*). The RNase activity of the two *S. solfataricus* proteins was indistinguishable with respect to their RNA cleavage patterns and their salt (50 mM KCl), pH (neutral pH), and temperature optima ($\sim 60\ ^\circ$C). Primer extension analysis revealed that preferential cleavage occurs in single-stranded regions between pyrimidine and A-residues, preferentially between C- and A-residues. The RNase activities were not dependent on divalent cations, were inhibited at ion concentrations higher than 5 mM $MgCl_2$ or 150 mM KCl, and were abolished by the RNase A inhibitor RNasin (Promega). The last finding pointed to possible contaminations with RNases. The doubts were supported by the finding that spurious amounts of RNase A (in femtomolar concentrations) produced very similar RNA cleavage patterns. The dehydrogenases were used in nanomolar concentrations in the activity assays (*111*).

To show that the RNase activity is an intrinsic feature of the dehydrogenase polypeptides, truncated protein variants were tested. The catalytic center for RNase activity of the ASD polypeptide was localized in the first 205 N-terminal amino acids, which contains a predicted Rossmann fold (NAD-binding site), whereas a polypeptide consisting of the 1440 C-terminal amino acids was not active. Further subcloning revealed that the RNase active site was located in the first 73 N-terminal amino acids, which comprise the first mononucleotide-binding site of the putative Rossmann fold.

Further evidence for the RNase activity of the dehydrogenases was obtained from assays in which their activity and that of RNase A were differently affected by the presence of tRNA and ssDNA in the reaction mixtures: while the presence of tRNA or ssDNA inhibited RNA degradation by the dehydrogenases, the RNase A activity was enhanced. To exclude the possibility that the excess of dehydrogenase polypeptides changes the behavior of contaminant RNase A in the reaction mixture, control reactions were performed with pure RNase A in femtomolar concentration together with dehydrogenases in nanomolar concentration. In presence of ssDNA, RNase A still cleaved the radioactively labeled RNA substrate efficiently, strongly suggesting that the RNase activity of the dehydrogenases is similar to but distinct from that of RNase A.

Although it was reported that the binding of RNA by the eukaryotic glyceraldehyde 3-phosphate dehydrogenase (GAPDH) is abolished in the presence of NAD^+, this compound did not affect the RNase activity of the archaeal ASD and A-CoA dehydrogenases, or of the tested eukaryotic and bacterial GAPDHs. This strongly suggests that the amino acid residues involved in NAD$^+$ binding are different from those responsible for RNase activity, and that the RNase active site is located at different surface than the NAD-binding site (111).

C. Archaeal IF5A from *H. salinarum* NRC-1 Shows Endoribonucleolytic Activity

To identify enzymes involved in RNA processing and degradation in *H. salinarum*, a biochemical screen for such activities was performed (114). Different *in vitro* transcripts were used as substrates: an *in vitro*-synthesized transcript derived from the *puf* mRNA of *R. capsulatus*, which contains single-stranded regions with RNase E cleavage sites (47), and structured transcripts derived from bacterial pre-rRNA (115). Surprisingly, it was found that RNA remains stable for hours when incubated at 37 or 42 °C with cell-free extracts from *H. salinarum* under physiologically relevant salt conditions (3–4.5 M KCl) in buffer containing 10 mM MgCl$_2$. RNase activity was detected only at KCl or NaCl concentrations below 200 mM (in "low salt buffer") without addition of MgCl$_2$ (114). This is consistent with recent data (103) that showed that, *in vitro*, the optimal reaction conditions for RNase Z from the halophilic archaeon *H. volcanii* are at 10 mM KCl.

One protein which was purified to homogeneity (as judged from silver stained gels) and displayed RNase activity was identified by mass spectroscopy as a homolog of the eukaryotic translation initiation factor 5A (eIF-5A) and was named archaeal (a) IF-5A (114). Its RNase properties were characterized using His-tagged variants purified from *H. salinarum* and from *E. coli*. Additionally, truncated polypeptides and mutant variants were tested. Most probably, the protein isolated from *H. salinarum* carries a unique posttranslational modification, hypusination, which is not present in the variant purified from *E. coli*. The hypusination is typical for this protein, and is restricted to Eukarya and Archaea—the enzyme which carries out this modification, deoxyhypusine synthase, has so far only been found in these two domains of life. It was shown that eIF-5A efficiently binds structured RNA containing certain motifs and that this interaction with RNA is hypusine dependent (116). Both the genes for eIF-5A and deoxyhypusine synthase are essential in yeast (117, 118). The interest in aIF-5A was further supported by studies on temperature-sensitive eIF-5A mutants indicating an involvement of this protein in mRNA turnover (119).

All preparations of the full-length aIF-5A led to identical RNA cleavage patterns showing that hypusination is not necessary for RNA cleavage *in vitro* (114). Divalent cations were not needed for cleavage, MgCl$_2$ concentrations above 5 mM and salt concentrations (KCl or NaCl) above 300 mM inhibited cleavage; maximal activity was observed at 120 mM KCl. The effect of polyamines on RNA cleavage was also tested because polyamines are known to bind RNA and hypusine is a modified lysine with structural contribution from the polyamine spermidine (summarized in (120)). At low concentrations (1 mM for putrescine, 0.5 mM for spermidine, 10 mM for ornithine), the polyamines increased RNase activity, whereas at higher concentrations (20, 4, and 120 mM, respectively) they inhibited RNase activity. The major cleavage sites were located in single-stranded regions between C and A. These characteristics strongly resemble those of RNase A and of the two archaeal dehydrogenases with RNase properties (see SectionIV.B.).

The two separated domains of the protein did not show any RNase activity and the exchange of one amino acid residue (glutamic acid at position 117 was replaced by alanine), or four residues (arginine–lysine at positions 72, 73 were exchanged to glycine–alanine and arginine–lysine at positions 122, 123 were exchanged to alanine–glycine), strongly reduced RNase activity. These results showed that the RNase activity is an intrinsic property of the aIF-5A protein. Similarly to the RNase activity of the dehydrogenases, the addition of tRNA at concentrations above 25 ng/μl inhibited RNase activity of aIF-5A (114), whereas RNase A activity was increased in presence of tRNA (111). aIF-5A forms oligomers with RNase activity and hypusination stabilizes the binding of aIF-5A to RNA. Moreover, it turned out that the eukaryotic eIF-5A also shows hypusine-dependent RNase activity, in addition to its RNA-binding capability (114).

aIF-5A contains two domains connected by a short hinge. The C-terminal domain is similar to CspA (the major cold-shock protein of *E. coli*, which is an RNA chaperone) and harbors an oligomer-binding (OB) fold, often found in sugar- and nucleotide-binding proteins. This fold is probably involved in RNA binding by aIF-5A. The N-terminal domain contains the hypusine residue, a KOW fold, implicated in nucleic acid binding, and a SH3-like motif, probably involved in protein–protein interactions. According to the proposed structural model, it is likely that an RNA molecule running through the central cleft between the two domains is in contact with all the amino acid residues that were shown to be crucial for the RNase activity of aIF-5A (114).

In summary, the attempts to purify novel RNase activities from Archaea resulted in the isolation of very different proteins. We observed that prolonged incubation of RNA with various proteins (including two dehydrogenases from *Sulfolobus* and the aIF-5A from *Halobacterium*), in the absence of divalent cations and RNase inhibitors, leads to cleavages of intrinsically unstable bonds between a pyrimidine and an adenine in single-stranded regions. The specificity of the small DNA-binding proteins from *Sulfolobus* seems to be different from RNase A-type cleavages, because they do not cleave preferentially between a pyrimidine and an adenine (108). In all cases, mutagenesis experiments suggested that the observed RNase activities are not due to contaminating RNases, but their physiological relevance is unclear.

V. Archaeal Proteins with Similarities to Bacterial or Eukaryotic Proteins Involved in RNA Degradation—A Short Overview

The proteins involved in RNA metabolism are among those that are evolutionarily the most conserved. Consequently, archaeal orthologs of bacterial or eukaryotic proteins involved in RNA degradation can be predicted using the completed genomic sequences of representatives of all three domains of life and the appropriate bioinformatic tools. In this way, several genes potentially encoding RNA-degrading proteins in Archaea were identified (10).

The analysis of completed genomes revealed that most Archaea harbor genes for potential subunits of an archaeal exosome-like protein complex. In 2001, Koonin et al. (121) published that three orthologs of eukaryotic exosomal subunits (Rrp41, Rrp42, and Rrp4) are encoded side by side in a highly conserved archaeal superoperon, together with ribosomal proteins, proteasomal subunits, RNase P proteins, and others. A fourth ortholog of a eukaryotic exosome subunit, Csl4, was found to be encoded in another operon in Archaea. In recent years, the existence of an archaeal exosome was experimentally demonstrated (122, 123). The core of this protein complex was found to resemble bacterial PNPase both structurally and functionally (124, 125). Surprisingly, the genes for the four different exosomal subunits were not found in the genomes of *Halobacterium* and *Methanothermobacter*, although the superoperon was preserved (121). This finding suggests that alternative mechanisms for RNA degradation exist in Archaea.

Further, genes for proteins with limited homology to RNase R and RNase E have been detected in Archaea (10). The RNase R-like protein from *Haloferax* was confirmed to be an exoribonuclease (79). Genes for RNase R orthologs were found in the halophilic Archaea but not in the methanogenic Archaea lacking the exosome (79). Although biochemical and serological evidence for an RNase E-like activity in a halophilic archaeon was published a while ago, the corresponding polypeptide was not identified (126). The archaeal protein with similarity to a part of the N-terminal RNase E region was shown to bind strongly and specifically AU-rich RNA, but catalytic activity has not been detected so far (127). Additionally, Archaea harbor genes for proteins showing homology to the recently discovered RNase J, which acts both as an RNase E-like endonuclease and a 5′–3′ exoribonuclease in *Bacillus* (39, 40). Consistent with this, evidence for RNA decay in the 5′–3′ direction, in addition to decay in the 3′–5′ direction, was recently reported (77). The archaeal RNase J ortholog (which is identical to the *mbl* RNAse proposed by Koonin to be an exosomal subunit in Archaea (121)), is still not characterized. The RNase J-like proteins are the only candidates for degrading exoribonucleases in methanogenic Archaea which lack the exosome. Archaeal orthologs of eukaryotic 5′–3′ exoribonucleases, such as Xrn1, have not been predicted so far.

RNase III-like proteins are also not encoded in the Archaea (10), suggesting that Archaea do not exploit the mechanisms of RNAi for gene silencing. However, the identification of archaeal Argonaute proteins led to the proposal that RNAi-related mechanisms may exist in the third domain of life (128). BLAST analysis shows, however, that specific groups of Argonaute proteins can be found in unrelated Archaea and Bacteria, while related species often do not harbor Argonaute homologs. For example, Argonaute homologs can be found by BLAST analysis in bacterial genera like *Aquifex*, *Geobacter*, and *Mesorhizobium* (but not in the related *Sinorhizobium*) as well as in the archaeon *Haloarcula* (but not in the related *Halobacterium*). An Argonaute homolog can be detected in *Sulfolobus tokodaii* but not in *S. solfataricus*. This is indicative for spread of such proteins by horizontal transfer, most probably by viruses and/or plasmids, and argues against a fundamental function of the archaeal Argonaute proteins in gene silencing. The identification of small noncoding RNAs, which are complementary to the translational start of specific mRNAs, suggested that posttranscriptional gene silencing is based on antisense RNAs in Archaea (101). Subsequent ribonucleolytic events are not excluded, however.

Finally, archaeal genomes do not encode poly(A) polymerase, and RNA tailing is limited to species containing the exosome, which obviously exhibits a dual function in Archaea (78, 79).

VI. The Archaeal Exosome

The existence of an archaeal exosome *in vivo* has so far been demonstrated for *Sulfolobus* and *Methanothermobacter* (122, 123). The recombinant exosomes of *Archaeoglobus* and *Pyrococcus* have also been reconstituted and studied *in vitro* (125, 129).

A. Composition of The Archaeal Exosome

The first experimental evidence for the existence of an exosome-like complex in Archaea was presented for *S. solfataricus* (122). Coimmunoprecipitation of proteins using a cell-free extract and polyclonal antibodies against the predicted recombinant subunit of the exosome *Sso*Rrp41 resulted in the isolation of a protein complex containing the four orthologs of the eukaryotic subunits of the exosome (Rrp4, Rrp41, Rrp42, and Csl4), in addition to the archaeal protein annotated as DnaG and low amounts of Cpn (the thermosome).

After separation of the cell-free extract by glycerol density gradient ultracentrifugation, *Sso*Rrp41 was detected by Western blotting in high molecular weight fractions corresponding to complexes with a sedimentation coefficients of 11S ($\sim$ 240 kDa) and 30S–50S (together with the ribosomal subunits). Coimmunoprecipitation of the exosome from the different fractions of the glycerol density gradient revealed that intact exosome (containing the subunits SsoRrp41, SsoRrp42, SsoRrp4, SsoDnaG, and SsoCsl4 in stoichiometric amounts) was present only in the 30S–50S fractions. Additionally, a Cdc48 homolog and low amounts of Cpn copurified with the exosome from those fractions. In contrast, the exosome from the 240-kDa fractions contained a degraded form of SsoRrp4 and less SsoDnaG than either SsoRrp41 or SsoRrp42. It also contained the conserved hypothetical protein Sso16p of unknown function and a low amount of Cpn. SsoCdc48 and SsoCsl4 were not detectable. The reason for the detection of the exosome in 30S–50S fractions is not clear. It was proposed that the archaeal exosome is involved in rRNA maturation (a function confirmed for the eukaryotic exosome (130)), and therefore is found in association with the ribosomal subunits. The significance of the copurification of SsoCdc48 and Sso16p with the exosome in separate glycerol density gradient fractions and the role of these two proteins is not clear. Cpn and Cdc48 have chaperone properties and may stick nonspecifically to the protein complex, or they may be involved in the maintenance of its integrity (122, 131).

SsoDnaG was shown to be a core subunit of the exosome of S. solfataricus. The possibility that SsoDnaG binds nonspecifically to the protein complex was excluded by the findings that coimmunoprecipitation of the exosome with SsoRrp41-directed and with SsoDnaG-directed antibodies results in the purification of protein complexes with very similar stoichiometry, and that depletion of SsoRrp41 from the cell extract is paralleled by SsoDnaG depletion (131). Obviously, the archaeal DnaG proteins are not involved in DNA replication like the bacterial primase DnaG. Instead, Archaea harbor a eukaryotic-type primase, and the proteins annotated as DnaG are subunits of the exosome with so far unknown role in RNA metabolism. Their annotation was based on the topoisomerase-primase (TOPRIM) domain, which is found not only in bacterial primases and in topoisomerases, but also in endoribonucleases. Moreover, archaeal DnaG proteins contain a domain found in RNA helicases and a domain specific for this group of archaeal polypeptides. Thus, their domain composition suggests RNA helicase or endoribonuclease function (discussed in (122, 131)). It is possible that DnaG (and the exosome) in Archaea participate in 5S rRNA maturation, similarly to the TOPRIM domain-containing RNase M5 in B. subtilis (132). Difficulties in the recovery of soluble recombinant SsoDnaG impaired so far the investigation of its physiological role (131).

Although recombinant exosomes from several Archaea have been characterized, *Methanothermobacter thermautotrophicus* is the only species besides *S. solfataricus* in which the existence of the protein complex was experimentally verified. The exosome of *M. thermautotrophicus* was separated by blue native gel electrophoresis followed by SDS-PAGE. The core subunits Rrp4, Rrp41, Rrp42, and DnaG were detected together in a complex of 900 kDa, along with a homomultimer of the splicing endonuclease (123). The splicing endonuclease cleaves bulge–helix–bulge motifs in double-stranded RNA and is involved in rRNA processing (*104, 133*). This strongly suggests that this enzyme and the exosome participate in RNA processing in a coordinated manner.

B. Structure and Mechanism of The Archaeal Exosome

Based on similarities between bacterial PNPase and the subunits of the exosome in Eukarya and Archaea, it was proposed that the archaeal proteins Rrp41 and Rrp42 are arranged in a hexameric ring, to which Rrp4 and Csl4 bind, forming the core archaeal exosome (*22*). The presence of two different RPD-containing subunits in the proposed hexameric ring of the archaeal exosome strongly suggested that this protein complex degrades RNA in a phosphorolytic manner like PNPase. Based on previous reports that particular recombinant subunits of the eukaryotic exosome degrade RNA (*71, 134*), it was expected that at least one of the subunits of the archaeal exosome should exhibit RNase activity separately. However, none of the recombinant orthologs of the exosomal subunits from Archaea (Rrp41, Rrp42, Rrp4, and Csl4) showed any RNase activity alone. Reconstitution experiments revealed that complex formation is necessary for the activity of the exosome; the hexameric ring is sufficient for activity, but this activity is modulated in the presence of the S1 proteins in the nine-subunit exosome (*124, 125, 131*). These findings demonstrate the interdependence of the exosomal subunits in respect to the function of the protein complex. They also exclude the possibility that RNase activity in the preparations of recombinant archaeal exosomes is due to contaminating *E. coli* proteins. The elimination of contaminating proteins originating from the overexpression host is easy in the case of exosomes from hyperthermophilic Archaea, since they are active at high temperatures between 60 and 80 °C, while *E. coli* RNases with similar phosphorolytic activity (PNPase and RNase PH) are readily inactivated at these temperatures.

In the past few years, the exosomes from several hyperthermophilic Archaea were reconstituted using recombinant subunits, and their crystal structures were determined. Moreover, the three-dimensional structures of exosomes together with RNA or NDP substrates were resolved. In these studies, the effect of point mutations in exosomal subunits on the interaction with substrates was analyzed, giving important information about the interaction of RNA with specific amino acid residues in the central channel of the hexameric core of the exosome, and about the mechanism for RNA degradation (*124, 125, 135–137*).

1. THE HEXAMERIC RING OF THE EXOSOME AND ITS INTERACTION WITH RNA

The hexameric ring, built by three Rrp41/Rrp42 dimers, is the minimal catalytic subunit of the archaeal exosome that possesses phosphorolytic activity (Fig. 2). Its structure strongly resembles the structure of the hexameric ring built from the two RPD domains present in each polypeptide of the homotrimeric bacterial PNPase. The phosphate-binding sites in the archaeal hexamer were visualized using a phosphate-mimicking ion and structure-guided mutations localized the active sites in the Rrp41 subunit. However, formation of the hexamer is necessary for RNA degradation and polymerization, because Rrp42 participates in substrate binding. The three active sites are located in the central channel near the bottom of the pore. RNA is bound in a cleft between Rrp41 and Rrp42 and is degraded releasing ribonucleotide 5′-diphosphates (NDPs). The phosphorolytic mode of degradation was shown by the reversal of the reaction: in the presence of nucleoside diphosphates (NDPs), the Rrp41–Rrp42 hexamer polymerizes RNA. Additionally, the NDPs produced during phosphorolytic degradation were experimentally shown by thin layer chromatography (*78, 124, 125, 131, 135–137*).

In the nine-subunit exosome, three S1 domain-containing proteins, which can be represented by Rrp41 and/or Csl4, bind on the top of the hexameric ring (Fig. 4; (*125*)). The narrow opening of the central channel on the top side of the hexamer is the entry site for an RNA substrate, although the active sites are located in the vicinity of the wider opening on the opposite, bottom side (*135*). The entry pore includes a narrow constriction (neck) of 8–10 Å formed by loops of the Rrp41 subunits (Figs. 2 and 4). The constriction ensures that only single-stranded RNA can be threaded through to reach one of the active sites on the bottom of the hexamer. This was demonstrated using RNA oligoribonucleotides containing a stable stem-loop structure at the 5′-end followed by poly(A) tails of different length in degradation assays with the hexameric ring of the *S. solfataricus* exosome. Only substrates carrying tails of 10 nt or more were degraded and a tail of ∼ 9 nt remained intact, in accordance with crystallographic data showing that the distance from the opening on the top of the hexamer to the active site is spanned by ∼ 9 nt (*135*).

The experiments described above confirm that the single-stranded RNA substrate follows a path from the top to the bottom of the hexamer, even in absence of the S1-containing subunits of the exosome. This can be explained by the electrostatic surface of the hexamer, which is negatively charged at the bottom and on the side (excluding interactions with RNA in these regions) and is positively charged at the entry pore and at the central channel down to the active sites (*125*). Thus, the only hexamer surfaces suitable for RNA binding are those of the entry site and of the phosphorolytic chamber. The hydrophobic surfaces on the top, which are the sites of interaction of the hexamer with the S1 subunits, do not impair RNA degradation by the hexamer, but the degradation by the nine-subunit exosome is more efficient (see below).

In the central channel of the hexamer, the two most distal nucleotides (N1 and N2, numbering from the 3′-end) reach the active site of an Rrp41 subunit (Fig. 4). The nucleotides N1–N4 are bound in a cleft of an

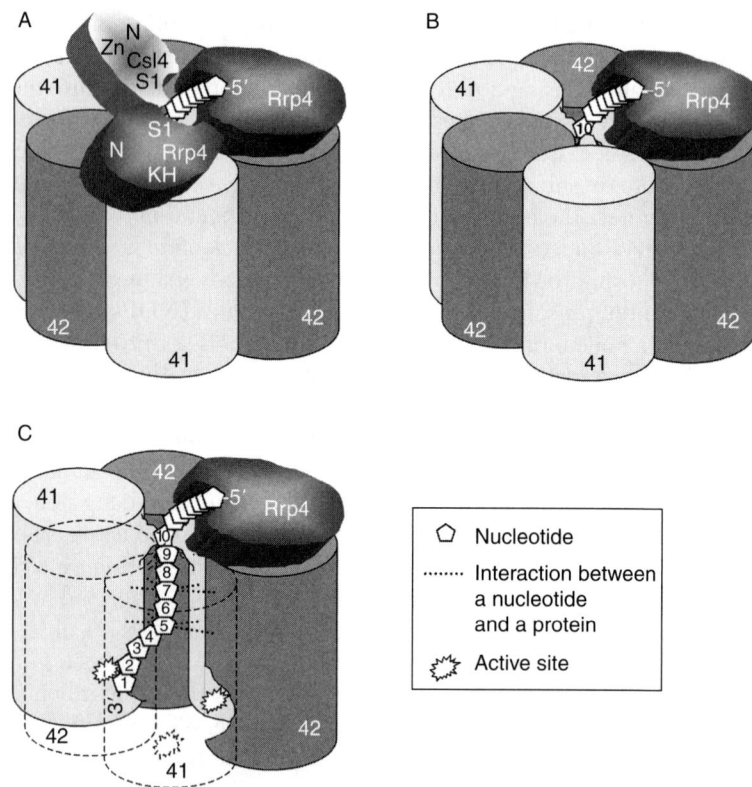

FIG. 4. Schematic representation of the structure of the archaeal exosome with bound RNA. N, N-terminal domain; S1, KH, the corresponding RNA-binding domains; Zn, Zn-ribbon domain, 41, Rrp41; 42, Rrp42. (A) The nine-subunit exosome with the protruding 5′-end of an RNA substrate. RNA is bound to the S1 domain of one of the S1-containing subunits of the exosome, in this case Rrp4. (B) A view on the neck in the central channel of the hexamer. This narrow constriction formed by loops of the three Rrp41 subunits interacts with the 10th nucleotide as numbered from the 3′-end of the substrate. In comparison to (A), two S1 subunits (Rrp4 and Csl4) are missing. (C) A view into the central channel of the hexamer.

Rrp41–Rrp42 dimer by ionic interactions with a ladder of arginine residues (*135, 136*). Arginine side chains contributed by both Rrp41 and Rrp42 are involved, explaining the importance of complex formation for enzymatic activity. Mutation of these arginines to glutamates abolishes RNA degradation. The substrate binding is performed mainly by electrostatic interactions with the phosphate groups of the ribose–phosphate backbone of the substrate. Importantly, interactions with 2′-OH of the sugar provide RNA specificity to the hexamer (*135*).

The next nucleotide, N5, interacts with two Rrp41–Rrp42 dimers, and this interaction is essential for degradation of a 15-mer poly(A) RNA (Fig. 4) (*137*). Following the path of RNA/protein interaction to the top of the hexamer, N7 also interacts with two dimers and N10 interacts with all three dimers in the neck structure at the top of the ring (*137*). Mutations in this Rrp4 region also abolish RNA degradation by the hexamer (*125, 136*). The (sequence-nonspecific) interaction of N7 and N10 with the exosome is mediated by the base and not by the phosphate–ribose backbone. The N7- and N10-binding sites are flexible structures of the exosome, which are stabilized upon RNA binding (*137*). Thus, from the active site to the neck structure, the channel of the hexameric ring is spanned by 10 nucleotides of single-stranded RNA in extended conformation (Fig. 4). The nature of the interactions between RNA and amino acid residues in the central channel of the hexamer suggests a sequence-nonspecific binding and degradation. However, a preference for AU- and U-rich RNA was reported for the hexameric RPD-core of the eukaryotic exosome and for PNPase (*138*).

After phosphorolysis, the rNDP product is not released simply by diffusion. Instead, a conformational change at the N1-binding site leads to controlled (active) release of the product through a conserved side channel in concert with the entry of an inorganic phosphate ion. The RNA substrate is then translocated in a way that the nucleotide located at 3′-end is positioned in the N1-binding site (*137*). The polymerization of RNA seems to follow these steps in the opposite direction.

2. THE NINE-SUBUNIT EXOSOME OF ARCHAEA

The spatial structure of the nine-subunit core of the archaeal exosome is similar to the structure of bacterial PNPase and of the nine-subunit exosome in Eukarya. When reconstituted *in vitro*, the nine-subunit exosome comprises three Rrp4 polypeptides or three Csl4 polypeptides, or both proteins in a stoichiometry of 2:1 or 1:2, in addition to the hexameric ring composed of Rrp41 and Rrp42 (Figs. 2 and 4A). The two isoforms of the *Archaeoglobus fulgidus* exosome, containing either Rrp4 or Csl4, were crystallized and their three-dimensional structure was determined, as well as the structure of the Rrp4-containing nine-subunit exosome of *S. solfataricus* (*125, 136*).

The Rrp4 (contains an N-terminal domain and the RNA-binding S1 and KH domains) and Csl4 (contains an N-terminal domain and the RNA-binding S1 and Zn-ribbon domains) subunits of the archaeal exosome bind on the top of the hexameric ring and form a flat, trimeric cap (Fig. 4A). Both Rrp4 and Csl4 strongly bind to Rrp41 with their N-terminal regions. The structure of Rrp41, which harbors the active site, is slightly different in the two isoforms of the

A. *fulgidus* exosome (*125*). Such structural plasticity of Rrp41 is compatible with the observation that different isoforms of the nine-subunit exosome differ in their activity (*125, 131*).

The S1 domains of the trimeric cap interact with Rrp41 and Rrp42, and are located at the center of the complex, in contrast to the position of the N-terminal (NT), KH, and Zn-ribbon domains at the periphery. The concave sites of the three S1 domains form the perimeter of the entry pore of the nine-subunit exosome. The KH domain of an Rrp4 polypeptide packs to the convex side of its S1 domain and also interacts with Rrp41 and Rrp42 similarly to the S1 domain. The Zn-ribbon domain of a Csl4 polypeptide also shares a hydrophobic interface with its S1 domain, but its position on the top of the nine-subunit exosome is not identical to the position of the KH domain. Instead, the Zn-ribbon domain is located between two adjacent S1 domains. According to the structural data, heterotrimeric caps containing Rrp4 and Csl4 can be formed (*125*).

The hydrophobic surfaces of the nine domains of the trimeric cap are involved in protein–protein interactions between the individual domains and between the cap and the hexamer. The remaining cap surface represents the top of the nine-subunit exosome and is suitable for interaction with RNA substrates and accessory protein factors, which may help to unwind RNA secondary structures, to guide the exosome to different substrates, or to regulate its activity, as has been shown for the eukaryotic exosome (*9*). Since the archaeal exosome exhibits a dual function in the cell as an RNA degradation and an RNA polynucleotidylation machine (*78*), interacting proteins may be involved in the regulation of the two different activities. The results of coimmunoprecipitation experiments suggest that nine-subunit exosomes of different composition exist *in vivo* (*122, 131*), although the predominant composition of the trimeric caps *in vivo*—homotrimeric caps or heterotrimeric caps of variable Rrp4 to Csl4 stoichiometry—is not known.

The existence of two different S1 subunits in exosomes with different RNA-binding domains and different electrostatic surfaces leads to the expectation that the particular isoforms of the exosome may interact with different macromolecules. It was pointed out that, in accordance with these assumptions, a mutation affecting the interaction between the Zn-ribbon domain and the S1 domain of the eukaryotic Csl4 abolishes degradation of mRNA but not processing of rRNA (*67, 125*). The top surface of the *A. fulgidus* Rrp4-exosome is positively charged throughout, with highest positive charge in the region of the S1-pore, in contrast to the top surface of the *A. fulgidus* Csl4-exosome, at which solely the electrostatic surface of its S1-pore is highly positively charged. In both isoforms of the exosome, the greatest positive charge is located on the concave sites of the S1 domains (S1-pore), which obviously represents a major RNA-binding site (*125*). According to the proposed model, a recruited RNA substrate strongly interacts with the S1-pore and is threaded through the neck on the top of the hexamer into the central phosphorolytic chamber, until the 3'-most nucleotide (N1) reaches an active site. This model is supported by the three-dimensional structure of the *S. solfataricus* Rrp4-exosome with a stalled RNA substrate (*136*). In contrast to the well-characterized binding sites for RNA in the central channel of the hexameric ring (*135, 137*), the RNA-binding sites outside the hexamer are still unknown.

In agreement with the above model, mobility shift experiments with a 30-mer poly(A) RNA demonstrated that the substrate is bound stronger by the nine-subunit exosomes of *A. fulgidus* than by the hexameric ring (*125*). Additionally, it was shown that the affinity of the *S. solfataricus* Rrp4-exosome for a 7-mer poly(A) RNA is 40-fold higher than that of the hexameric ring and at least 300-fold higher than the binding affinity of monomeric Rrp4. Thus, the Rrp4 polypeptide has strikingly low intrinsic RNA-binding affinity, despite the presence of the S1 and KH domains; however, its affinity strongly increases in the context of the protein complex and contributes to proper interaction with the substrate. This is shown also by the finding that the Rrp4-exosome binds the 7-mer poly(A) RNA with a stoichiometry of 1:1, but the hexamer binds this substrate in stoichiometry of one protein complex to three RNA molecules, in accordance to the presence of three active sites in the phosphorolytic chamber (*139*). RNA degradation is strongly increased when nine-subunit exosomes are compared to the hexamer, as demonstrated with 30-mer poly(A) RNA and with a heteropolymeric 90 nt RNA (*129, 131*). These data demonstrate the interdependence of the exosomal subunits for interactions with substrates—only the properly formed, physiologically relevant complex of at least nine-subunit is capable of efficient substrate binding and degradation.

Although it was previously published that eukaryotic Rrp4 shows hydrolytic activity (*71, 134*), the archaeal Rrp4 proteins are not catalytically active, either alone or in the context of the exosome; the same is true for the archaeal Csl4 proteins. These experimental findings are in agreement with the crystallographic data, which do not point to catalytic centers with hydrolytic activities in these polypeptides (*125, 131*).

It was suggested that the Rrp4-exosome of *A. fulgidus* has a fixed arrangement, because the Rrp4 subunits interact with the hexamer via very hydrophobic surfaces. Only small local movements of the S1 subunits in this exosome were proposed to mediate the movement of the substrate into the central channel during degradation nucleotide by nucleotide (*125*). However, in the Rrp4-exosome of *S. solfataricus*, only the N-terminal domain seems to be fixed, and the S1 and KH domains, which interact closely with each other and form a single structural unit, seem to be rather flexible structures (*136*). The S1 and KH domains of the bacterial PNPase were not well resolved in the crystal structure, probably because they are also flexible (*140*). Moreover, the structure of a heteromeric exosome containing the Rrp41 and Rrp42 subunits from *Pyrococcus horikoshii* and the Rrp4 subunits from *Pyrococcus abyssi* was studied in solution by small angle X-ray scattering. According to the model proposed in this study, the Rrp4 subunits are attached to the hexameric core as extended and flexible arms, which may bind RNA and direct it to the central channel (*129*).

The structural arrangement of the 10th core subunit of the archaeal exosome, DnaG (Fig. 2), as well as the identity of the exosomal subunits that interact with the splicing endonuclease and with the proteins Cdc48 and 16p in the exosomes of different Archaea are not known.

VII. RNA Degradation in Archaea

So far, only RNA-degrading protein complexes or proteins from *S. solfataricus* (belonging to Crenarchaeota) and from *H. volcanii* (belonging to Euryarchaeota) have been characterized functionally, and substantial differences were found in their mechanisms of RNA degradation in 3′–5′ direction (*78*, *131*). The discovery of a protein factor protecting triphosphorylated 5′-ends of mRNA in Archaea from decay in 5′–3′ direction revealed unexpected similarities to RNA degradation in Eukarya and Bacteria. Moreover, the analysis of posttranscriptional modifications at the 3′-end of RNA in different Archaea in addition to genomic data suggests that at least three different types of RNA-degrading mechanisms operate in the third domain of life (*78*, *79*).

A. RNA Degradation in The Exosome-Harboring *S. solfataricus*

The accumulated data strongly suggest that the exosome is very important for RNA degradation in Archaea, such as *S. solfataricus*, harboring this protein complex, since it is a 3′–5′ exoribonuclease and is the polynucleotidylation enzyme in these organisms (Fig. 1). The recombinant archaeal exosome can degrade RNA and can perform the reverse reaction, polyadenylation of poly(A) RNA (*78*, *124*, *125*, *129*, *137*). Consistent with this, the native coimmunoprecipitated exosome of *S. solfataricus* performs these reactions *in vitro*, and it was shown that the exoribonucleolytic and the polyadenylation activities of cell-free extracts of *S. solfataricus* can be assigned to the exosome (*78*, *131*). Moreover, the genome of *S. solfataricus* does not contain genes for other predicted 3′–5′ exoribonucleases with degradative function and thus, the exosome seems to be the major 3′–5′ exoribonuclease in this archaeon (*121*). This assumption is supported by the similarities of this protein complex to the eukaryotic exosome and to PNPase (*9*, *141*, *142*). In addition, the exosome is the most likely polynucleotidylating machine in *S. solfataricus*, since no other proteins with such a predicted function are encoded in the genome, and since, in different Archaea, the presence of the exosome has been correlated with the presence of posttranscriptionally added, heteropolymeric adenine-rich RNA tails (78, 79).

Sequences of posttranscriptionally added RNA tails from different exosome-containing Archaea were determined and tailed RNA molecules were identified as truncated mRNAs and rRNA molecules. The truncated molecules, which were tailed by the exosome, are probably the products of premature transcription arrest or endonucleolytic cleavage, or may have arisen due to a premature stop in the exonucleolytic decay from the 3′-end. Based on bacterial models, RNA tailing is most probably an event preceding and enhancing rRNA decay and therefore these data suggest a function of the exosome in targeting mRNA and rRNA fragments for degradation. In one case, a tailed precursor of 16S rRNA was also detected, suggesting a function of the exosome in rRNA maturation (78, 79).

In *S. solfataricus* and probably in many other Archaea, mRNA decay proceeds not only in 3′–5′ direction, but, in absence of triphosphate at the 5′-end, also in 5′–3′ direction (77). Messages with a triphosphate at the 5′-end are protected by aIF2-gamma, which is a trimeric translation initiation factor with an additional function—it binds to the 5′-triphosphate and resembles the eukaryotic cap-complex. This was shown recently in experiments demonstrating that aIF2-gamma binds the 5′-triphosphate of any RNA *in vitro* and that it protects the 5′-end of mRNA from degradation *in vitro* during incubation with S100 extracts. When a synthetic mRNA prebound with aIF2-gamma was incubated with the extract and was then analyzed in Northern hybridization with radiolabeled probes complementary to the 5′-end or to the 3′-end, the signal intensity corresponding to the 3′-end decreased to 5% of the input RNA after 60 min of incubation, while those corresponding to the 5′-end decreased to only 70% of input. Without prebinding of aIF2-gamma, the 3′-end decayed similarly, and the signal corresponding to the 5′-end decreased to 20% of input, consistent with a protective function of aIF2-gamma. Moreover, 10-fold overproduction of aIF2-gamma in *S. solfataricus* stabilizes bulk mRNA and the 5′-end segment of a specific mRNA *in vivo* (77). Most probably, the degradation in 3′–5′ direction is performed by the exosome. The degradation in 5′–3′ direction may be performed by the archaeal RNase J homolog (39, 40) (named also mbl RNase (121)), which is still not characterized in Archaea.

B. RNA Degradation in Archaea Without Exosome

The genomes of halophilic and many methanogenic Archaea do not contain the genes for the orthologs of eukaryotic exosomal subunits (Rrp4, Rrp41, Rrp42, and Csl4), suggesting that different RNA degradation mechanisms operate in these Archaea (121). This assumption is strengthened by the finding that Archaea without exosomes do not exhibit any posttranscriptional RNA modification—they do not have any RNA tails (78, 79). However, in almost all eukaryotic and bacterial cells, RNA tailing has an important role in the mechanisms of RNA degradation.

It is not clear how RNA decay is performed in Archaea without an exosome, but they seem to represent two groups. Halophilic Archaea contain an RNase R homolog, which was reported to be a hydrolytic exoribonuclease with an essential function (79). In these Archaea and in *Mycoplasma*, bacteria with minimal genome, the RNase R homolog is the only obvious candidate for an exoribonuclease with degradative function (143). Methanogenic Archaea without an exosome do not harbor genes for archaeal RNase R (79). If they do not have another, so far undiscovered Archaea-specific 3′–5′ exoribonuclease, their RNA degradation may depend solely on the putative RNase J (39), and thus RNA degradation in these organisms may proceed only in 5′–3′ direction (40). It should be noted that all Archaea contain the gene for the protein annotated as DnaG, which is part of the exosome in Archaea harboring this protein complex (144). Thus, the archaeal DnaG proteins seem to have an extraordinarily important role in RNA metabolism.

VIII. Concluding Remarks

Although the mechanisms for RNA degradation in Archaea are still not as well understood as in Eukarya and in Bacteria, the available data suggest that similar strategies are exploited in all three domains of life. Obviously, all organisms protect the 5′- and the 3′-end of RNA from interaction with degrading enzymes and enzymatic complexes and they appear to use (with few exceptions) the posttranscriptional addition of short poly(A) or (A)-rich tails for destabilization of the molecules. On the protein level, most prokaryotes use phosphorolytic enzymes (PNPase and the archaeal exosome) with a dual function—3′–5′ RNA degradation releasing NDPs in an energy saving way and RNA synthesis at the 3′-end using NDPs. Bacterial PNPase and the archaeal exosome are structurally and functionally very similar, and share a striking similarity to the eukaryotic exosome, although the nine-subunit core of the latter is catalytically inactive. However, its 10th hydrolytically active subunit shows similarity to RNase R, which is also present in Bacteria and in Archaea. The importance of this enzyme is underlined by the finding that RNase R seems to be the only enzyme responsible for RNA degradation in 3′–5′ direction in some organisms, and the only exoribonuclease capable of degrading structured RNA without help of RNA helicases. However, many ribonucleolytically active proteins and protein complexes interact with helicases unwinding

RNA secondary structures. The data suggest that in the last universal common ancestor, a PNPase/exosome-like protein complex existed, in parallel with ancestral RNase R and RNA helicase polypeptides.

REFERENCES

1. Woese CR, Fox GE. Phylogenetic structure of the prokaryotic domain: The primary kingdoms. *Proc Natl Acad Sci USA* 1977;**74**:5088–90.
2. Kletzin A. General characteristics and important model organisms. In: R Cavicchioli, editor. Archaea—molecular and cellular biology. Washington, DC: ASM Press; 2007. p. 14–92.
3. Holmes ML, Dyall Smith ML. A plasmid vector with a selectable marker for halophilic archaebacteria. *J Bacteriol* 1990;**172**:756–61.
4. Jonuscheit M, Martusewitsch E, Stedman KM, Schleper C. A reporter gene system for the hyperthermophilic archaeon *Sulfolobus solfataricus* based on a selectable and integrative shuttle vector. *Mol Microbiol* 2003;**48**:1241–52.
5. Lucas S, Toffin L, Zivanovic Y, Charlier D, Moussard H, Forterre P, et al. Construction of a shuttle vector for, and spheroplast transformation of, the hyperthermophilic archaeon *Pyrococcus abyssi*. *Appl Environ Microbiol* 2002;**68**:5528–36.
6. Zhang JK, White AK, Kuettner HC, Boccazzi P, Metcalf WW. Directed mutagenesis and plasmid-based complementation in the methanogenic archaeon *Methanosarcina acetivorans* C2A demonstrated by genetic analysis of proline biosynthesis. *J Bacteriol* 2002;**184**:1449–54.
7. Rauhut R, Klug G. mRNA degradation in bacteria. *FEMS Microbiol Rev* 1999;**23**:353–70.
8. Condon C. Maturation and degradation of RNA in bacteria. *Curr Opin Microbiol* 2007;**10**:271–8.
9. Houseley J, LaCava J, Tollervey D. RNA-quality control by the exosome. *Nat Rev Mol Cell Biol* 2006;**7**:529–39.
10. Anantharaman V, Koonin EV, Aravind L. Comparative genomics and evolution of proteins involved in RNA metabolism. *Nucleic Acids Res* 2002;**30**:1427–64.
11. Marcaida MJ, DePristo MA, Chandran V, Carpousis AJ, Luisi BF. The RNA degradosome: Life in the fast lane of adaptive molecular evolution. *Trends Biochem Sci* 2006;**31**:359–65.
12. Viegas SC, Pfeiffer V, Sittka A, Silva IJ, Vogel J, Arraiano CM. Characterization of the role of ribonucleases in *Salmonella* small RNA decay. *Nucleic Acids Res* 2007;**35**:7651–64.
13. Hajnsdorf E, Braun F, Haugel Nielsen J, Regnier P. Polyadenylylation destabilizes the rpsO mRNA of *Escherichia coli*. *Proc Natl Acad Sci USA* 1995;**92**:3973–7.
14. O'Hara EB, Chekanova JA, Ingle CA, Kushner ZR, Peters E, Kushner SR. Polyadenylylation helps regulate mRNA decay in *Escherichia coli*. *Proc Natl Acad Sci USA* 1995;**92**:1807–11.
15. Deana A, Celesnik H, Belasco JG. The bacterial enzyme RppH triggers messenger RNA degradation by 5′ pyrophosphate removal. *Nature* 2008;**451**:355–8.
16. Celesnik H, Deana A, Belasco JG. Initiation of RNA decay in *Escherichia coli* by 5′ pyrophosphate removal. *Mol Cell* 2007;**27**:79–90.
17. Mackie GA. Ribonuclease E is a 5′-end-dependent endonuclease. *Nature* 1998;**395**:720–3.
18. Feng Y, Vickers TA, Cohen SN. The catalytic domain of RNase E shows inherent 3′ to 5′ directionality in cleavage site selection. *Proc Natl Acad Sci USA* 2002;**99**:14746–51.
19. Baker KE, Mackie GA. Ectopic RNase E sites promote bypass of 5′-end-dependent mRNA decay in *Escherichia coli*. *Mol Microbiol* 2003;**47**:75–88.
20. Goodrich AF, Steege DA. Roles of polyadenylation and nucleolytic cleavage in the filamentous phage mRNA processing and decay pathways in *Escherichia coli*. *RNA* 1999;**5**:972–85.

21. Deutscher MP. Degradation of RNA in bacteria: Comparison of mRNA and stable RNA. *Nucleic Acids Res* 2006;**34**:659–66.
22. Symmons MF, Williams MG, Luisi BF, Jones GH, Carpousis AJ. Running rings around RNA: A superfamily of phosphate-dependent RNases. *Trends Biochem Sci* 2002;**27**:11–8.
23. Frazao C, McVey CE, Amblar M, Barbas A, Vonrhein C, Arraiano CM, et al. Unravelling the dynamics of RNA degradation by ribonuclease II and its RNA-bound complex. *Nature* 2006;**443**:110–4.
24. Lorentzen E, Basquin J, Tomecki R, Dziembowski A, Conti E. Structure of the active subunit of the yeast exosome core, Rrp44: Diverse modes of substrate recruitment in the RNase II nuclease family. *Mol Cell* 2008;**29**:717–28.
25. Ghosh S, Deutscher MP. Oligoribonuclease is an essential component of the mRNA decay pathway. *Proc Natl Acad Sci USA* 1999;**96**:4372–7.
26. Mohanty BK, Kushner SR. The majority of *Escherichia coli* mRNAs undergo post-transcriptional modification in exponentially growing cells. *Nucleic Acids Res* 2006;**34**:5695–704.
27. Mohanty BK, Maples VF, Kushner SR. The Sm-like protein Hfq regulates polyadenylation dependent mRNA decay in *Escherichia coli*. *Mol Microbiol* 2004;**54**:905–20.
28. Cheng ZF, Deutscher MP. Purification and characterization of the *Escherichia coli* exoribonuclease RNase R. Comparison with RNase II. *J Biol Chem* 2002;**277**:21624–9.
29. Marujo PE, Hajnsdorf E, Le Derout J, Andrade R, Arraiano CM, Regnier P. RNase II removes the oligo(A) tails that destabilize the rpsO mRNA of *Escherichia coli*. *RNA* 2000;**6**:1185–93.
30. Liou GG, Chang HY, Lin CS, Lin Chao S. DEAD box RhlB RNA helicase physically associates with exoribonuclease PNPase to degrade double-stranded RNA independent of the degradosome-assembling region of RNase E. *J Biol Chem* 2002;**277**:41157–62.
31. Carpousis AJ, Van Houwe G, Ehretsmann C, Krisch HM. Copurification of *E. coli* RNAase E and PNPase: Evidence for a specific association between two enzymes important in RNA processing and degradation. *Cell* 1994;**76**:889–900.
32. Carpousis AJ. The RNA degradosome of *Escherichia coli*: An mRNA-degrading machine assembled on RNase E. *Annu Rev Microbiol* 2007;**61**:71–87.
33. Morita T, Kawamoto H, Mizota T, Inada T, Aiba H. Enolase in the RNA degradosome plays a crucial role in the rapid decay of glucose transporter mRNA in the response to phosphosugar stress in *Escherichia coli*. *Mol Microbiol* 2004;**54**:1063–75.
34. Morita T, Maki K, Aiba H. RNase E-based ribonucleoprotein complexes: Mechanical basis of mRNA destabilization mediated by bacterial noncoding RNAs. *Genes Dev* 2005;**19**:2176–86.
35. Lee K, Zhan X, Gao J, Qiu J, Feng Y, Meganathan R, et al. RraA: A protein inhibitor of RNase E activity that globally modulates RNA abundance in *E. coli*. *Cell* 2003;**114**:623–34.
36. Gao J, Lee K, Zhao M, Qiu J, Zhan X, Saxena A, et al. Differential modulation of *E. coli* mRNA abundance by inhibitory proteins that alter the composition of the degradosome. *Mol Microbiol* 2006;**61**:394–406.
37. Taghbalout A, Rothfield L. RNaseE and RNA helicase B play central roles in the cytoskeletal organization of the RNA degradosome. *J Biol Chem* 2008;**283**:13850–5.
38. Taghbalout A, Rothfield L. RNaseE and the other constituents of the RNA degradosome are components of the bacterial cytoskeleton. *Proc Natl Acad Sci USA* 2007;**104**:1667–72.
39. Even S, Pellegrini O, Zig L, Labas V, Vinh J, Brechemmier Baey D, et al. Ribonucleases J1 and J2: Two novel endoribonucleases in *B. subtilis* with functional homology to *E. coli* RNase E. *Nucleic Acids Res* 2005;**33**:2141–52.
40. Mathy N, Benard L, Pellegrini O, Daou R, Wen T, Condon C. 5′-to-3′ exoribonuclease activity in bacteria: Role of RNase J1 in rRNA maturation and 5′ stability of mRNA. *Cell* 2007;**129**:681–92.

41. Kennell D. Processing endoribonucleases and mRNA degradation in bacteria. *J Bacteriol* 2002;**184**:4645–57.
42. Condon C. Shutdown decay of mRNA. *Mol Microbiol* 2006;**61**:573–83.
43. Mehta P, Richards J, Karzai AW. tmRNA determinants required for facilitating nonstop mRNA decay. *RNA* 2006;**12**:2187–98.
44. Richards J, Mehta P, Karzai AW. RNase R degrades non-stop mRNAs selectively in an SmpB–tmRNA-dependent manner. *Mol Microbiol* 2006;**62**:1700–12.
45. Zhang J, Zhang Y, Zhu L, Suzuki M, Inouye M. Interference of mRNA function by sequence-specific endoribonuclease PemK. *J Biol Chem* 2004;**279**:20678–84.
46. Klug G, Adams CW, Belasco J, Doerge B, Cohen SN. Biological consequences of segmental alterations in mRNA stability: Effects of deletion of the intercistronic hairpin loop region of the *Rhodobacter capsulatus* puf operon. *EMBO J* 1987;**6**:3515–20.
47. Fritsch J, Rothfuchs R, Rauhut R, Klug G. Identification of an mRNA element promoting rate-limiting cleavage of the polycistronic puf mRNA in *Rhodobacter capsulatus* by an enzyme similar to RNase E. *Mol Microbiol* 1995;**15**:1017–29.
48. Heck C, Rothfuchs R, Jager A, Rauhut R, Klug G. Effect of the pufQ–pufB intercistronic region on puf mRNA stability in *Rhodobacter capsulatus*. *Mol Microbiol* 1996;**20**:1165–78.
49. Klug G. Endonucleolytic degradation of puf mRNA in *Rhodobacter capsulatus* is influenced by oxygen. *Proc Natl Acad Sci USA* 1991;**88**:1765–9.
50. Jager S, Fuhrmann O, Heck C, Hebermehl M, Schiltz E, Rauhut R, et al. An mRNA degrading complex in *Rhodobacter capsulatus*. *Nucleic Acids Res* 2001;**29**:4581–8.
51. Jager S, Hebermehl M, Schiltz E, Klug G. Composition and activity of the *Rhodobacter capsulatus* degradosome vary under different oxygen concentrations. *J Mol Microbiol Biotechnol* 2004;**7**:148–54.
52. Jager S, Evguenieva Hackenberg E, Klug G. Temperature-dependent processing of the cspA mRNA in *Rhodobacter capsulatus*. *Microbiology (Reading, England)* 2004;**150**:687–95.
53. Jager S, Jager A, Klug G. CIRCE is not involved in heat-dependent transcription of groESL but in stabilization of the mRNA 5′-end in *Rhodobacter capsulatus*. *Nucleic Acids Res* 2004;**32**:386–96.
54. Homuth G, Mogk A, Schumann W. Post-transcriptional regulation of the *Bacillus subtilis* dnaK operon. *Mol Microbiol* 1999;**32**:1183–97.
55. Purusharth RI, Klein F, Sulthana S, Jager S, Jagannadham MV, Evguenieva Hackenberg E, et al. Exoribonuclease R interacts with endoribonuclease E and an RNA helicase in the psychrotrophic bacterium *Pseudomonas syringae* Lz4W. *J Biol Chem* 2005;**280**:14572–8.
56. Khemici V, Poljak L, Toesca I, Carpousis AJ. Evidence *in vivo* that the DEAD-box RNA helicase RhlB facilitates the degradation of ribosome-free mRNA by RNase E. *Proc Natl Acad Sci USA* 2005;**102**:6913–8.
57. Purusharth RI, Madhuri B, Ray MK. Exoribonuclease R in *Pseudomonas syringae* is essential for growth at low temperature and plays a novel role in the 3′ end processing of 16 and 5 S ribosomal RNA. *J Biol Chem* 2007;**282**:16267–77.
58. LaCava J, Houseley J, Saveanu C, Petfalski E, Thompson E, Jacquier A, et al. RNA degradation by the exosome is promoted by a nuclear polyadenylation complex. *Cell* 2005;**121**:713–24.
59. Vasiljeva L, Buratowski S. Nrd1 interacts with the nuclear exosome for 3′ processing of RNA polymerase II transcripts. *Mol Cell* 2006;**21**:239–48.
60. Sheth U, Parker R. Decapping and decay of messenger RNA occur in cytoplasmic processing bodies. *Science* 2003;**300**:805–8.
61. Chen CY, Gherzi R, Ong SE, Chan EL, Raijmakers R, Pruijn GJ, et al. AU binding proteins recruit the exosome to degrade ARE-containing mRNAs. *Cell* 2001;**107**:451–64.

62. Stoecklin G, Mayo T, Anderson P. ARE-mRNA degradation requires the 5′–3′ decay pathway. *EMBO Rep* 2006;**7**:72–7.
63. Lin WJ, Duffy A, Chen CY. Localization of AU-rich element-containing mRNA in cytoplasmic granules containing exosome subunits. *J Biol Chem* 2007;**282**:19958–68.
64. Mullen TE, Marzluff WF. Degradation of histone mRNA requires oligouridylation followed by decapping and simultaneous degradation of the mRNA both 5′ to 3′ and 3′ to 5′. *Genes Dev* 2008;**22**:50–65.
65. Meister G, Tuschl T. Mechanisms of gene silencing by double-stranded RNA. *Nature* 2004;**431**:343–9.
66. Orban TI, Izaurralde E. Decay of mRNAs targeted by RISC requires XRN1, the Ski complex, and the exosome. *RNA* 2005;**11**:459–69.
67. van Hoof A, Staples RR, Baker RE, Parker R. Function of the ski4p (Csl4p) and Ski7p proteins in 3′-to-5′ degradation of mRNA. *Mol Cell Biol* 2000;**20**:8230–43.
68. Doma MK, Parker R. Endonucleolytic cleavage of eukaryotic mRNAs with stalls in translation elongation. *Nature* 2006;**440**:561–4.
69. Chang YF, Imam JS, Wilkinson MF. The nonsense-mediated decay RNA surveillance pathway. *Annu Rev Biochem* 2007;**76**:51–74.
70. Mitchell P, Tollervey D. An NMD pathway in yeast involving accelerated deadenylation and exosome-mediated 3′→5′ degradation. *Mol Cell* 2003;**11**:1405–13.
71. Mitchell P, Petfalski E, Shevchenko A, Mann M, Tollervey D. The exosome: a conserved eukaryotic RNA processing complex containing multiple 3′→5′ exoribonucleases. *Cell* 1997;**91**:457–66.
72. Liu Q, Greimann JC, Lima CD. Reconstitution, activities, and structure of the eukaryotic RNA exosome. *Cell* 2006;**127**:1223–37.
73. Dziembowski A, Lorentzen E, Conti E, Seraphin B. A single subunit, Dis3, is essentially responsible for yeast exosome core activity. *Nat Struct Mol Biol* 2007;**14**:15–22.
74. Brown JW, Reeve JN. Polyadenylated, noncapped RNA from the archaebacterium *Methanococcus vannielii*. *J Bacteriol* 1985;**162**:909–17.
75. Wang HW, Wang J, Ding F, Callahan K, Bratkowski MA, Butler JS, et al. Architecture of the yeast Rrp44 exosome complex suggests routes of RNA recruitment for 3′ end processing. *Proc Natl Acad Sci USA* 2007;**104**:16844–9.
76. Brown JW, Reeve JN. Polyadenylated RNA isolated from the archaebacterium *Halobacterium halobium*. *J Bacteriol* 1986;**166**:686–8.
77. Hasenohrl D, Lombo T, Kaberdin V, Londei P, Blasi U. Translation initiation factor a/eIF2(-gamma) counteracts 5′ to 3′ mRNA decay in the archaeon *Sulfolobus solfataricus*. *Proc Natl Acad Sci USA* 2008;**105**:2146–50.
78. Portnoy V, Evguenieva Hackenberg E, Klein F, Walter P, Lorentzen E, Klug G, et al. RNA polyadenylation in Archaea: Not observed in *Haloferax* while the exosome polynucleotidylates RNA in *Sulfolobus*. *EMBO Rep* 2005;**6**:1188–93.
79. Portnoy V, Schuster G. RNA polyadenylation and degradation in different Archaea; roles of the exosome and RNase R. *Nucleic Acids Res* 2006;**34**:5923–31.
80. Brenneis M, Hering O, Lange C, Soppa J. Experimental characterization of *Cis*-acting elements important for translation and transcription in halophilic archaea. *PLoS Genet* 2007;**3**:e229.
81. Bernstein JA, Khodursky AB, Lin PH, Lin Chao S, Cohen SN. Global analysis of mRNA decay and abundance in *Escherichia coli* at single-gene resolution using two-color fluorescent DNA microarrays. *Proc Natl Acad Sci USA* 2002;**99**:9697–702.
82. Hambraeus G, von Wachenfeldt C, Hederstedt L. Genome-wide survey of mRNA half-lives in *Bacillus subtilis* identifies extremely stable mRNAs. *Mol Genet Genomics* 2003;**269**:706–14.

83. Raghavan A, Bohjanen PR. Microarray-based analyses of mRNA decay in the regulation of mammalian gene expression. *Brief Funct Genomic Proteomic* 2004;**3**:112–24.
84. Wang Y, Liu CL, Storey JD, Tibshirani RJ, Herschlag D, Brown PO. Precision and functional specificity in mRNA decay. *Proc Natl Acad Sci USA* 2002;**99**:5860–5.
85. Prangishvilli D, Zillig W, Gierl A, Biesert L, Holz I. DNA-dependent RNA polymerase of thermoacidophilic archaebacteria. *Eur J Biochem* 1982;**122**:471–7.
86. Langer D, Hain J, Thuriaux P, Zillig W. Transcription in archaea: Similarity to that in eucarya. *Proc Natl Acad Sci USA* 1995;**92**:5768–72.
87. Hennigan AN, Reeve JN. mRNAs in the methanogenic archaeon *Methanococcus vannielii*: Numbers, half-lives and processing. *Mol Microbiol* 1994;**11**:655–70.
88. Reich E, Franklin RM, Shatkin AJ, Tatum EL. Effect of actinomycin D on cellular nucleic acid synthesis and virus production. *Science* 1961;**134**:556–7.
89. Bini E, Dikshit V, Dirksen K, Drozda M, Blum P. Stability of mRNA in the hyperthermophilic archaeon *Sulfolobus solfataricus*. *RNA (New York, NY)* 2002;**8**:1129–36.
90. Jager A, Samorski R, Pfeifer F, Klug G. Individual gvp transcript segments in *Haloferax mediterranei* exhibit varying half-lives, which are differentially affected by salt concentration and growth phase. *Nucleic Acids Res* 2002;**30**:5436–43.
91. Hundt S, Zaigler A, Lange C, Soppa J, Klug G. Global analysis of mRNA decay in *Halobacterium salinarum* NRC-1 at single-gene resolution using DNA microarrays. *J Bacteriol* 2007;**189**:6936–44.
92. Andersson AF, Lundgren M, Eriksson S, Rosenlund M, Bernander R, Nilsson P. Global analysis of mRNA stability in the archaeon *Sulfolobus*. *Genome Biol* 2006;**7**:R99.
93. Klug G, Evguenieva-Hackenberg E, Omer AD, Dennis PP, Marchfelder A. RNA processing. In: R Cavicchioli, editor. Archaea—molecular and cellular biology. Washington, DC: ASM Press; 2007. p. 158–74.
94. Englert C, Horne M, Pfeifer F. Expression of the major gas vesicle protein gene in the halophilic archaebacterium *Haloferax mediterranei* is modulated by salt. *Mol Gen Genet* 1990;**222**:225–32.
95. Englert C, Kruger K, Offner S, Pfeifer F. Three different but related gene clusters encoding gas vesicles in halophilic archaea. *J Mol Biol* 1992;**227**:586–92.
96. Roder R, Pfeifer F. Influence of salt on the transcription of the gas-vesicle genes of *Haloferax mediterranei* and identification of the endogenous transcriptional activator gene. *Microbiology* 1996;**142**:1715–23.
97. Hofacker A, Schmitz KM, Cichoncyzk A, Sartorius Neef S, Pfeifer F. GvpE- and GvpD-mediated transcription regulation of the p-gvp genes encoding gas vesicles in *Halobacterium salinarum*. *Microbiology* 2004;**150**:1829–38.
98. Slupska MM, King AG, Fitz Gibbon S, Besemer J, Borodovsky M, Miller JH. Leaderless transcripts of the crenarchaeal hyperthermophile Pyrobaculum aerophilum. *J Mol Biol* 2001;**309**:347–60.
99. Benelli D, Maone E, Londei P. Two different mechanisms for ribosome/mRNA interaction in archaeal translation initiation. *Mol Microbiol* 2003;**50**:635–43.
100. Aiba H. Mechanism of RNA silencing by Hfq-binding small RNAs. *Curr Opin Microbiol* 2007;**10**:134–9.
101. Tang TH, Polacek N, Zywicki M, Huber H, Brugger K, Garrett R, *et al*. Identification of novel non-coding RNAs as potential antisense regulators in the archaeon *Sulfolobus solfataricus*. *Mol Microbiol* 2005;**55**:469–81.
102. Darr SC, Pace B, Pace NR. Characterization of ribonuclease P from the archaebacterium *Sulfolobus solfataricus*. *J Biol Chem* 1990;**265**:12927–32.
103. Schierling K, Rosch S, Rupprecht R, Schiffer S, Marchfelder A. tRNA 3′ end maturation in archaea has eukaryotic features: The RNase Z from *Haloferax volcanii*. *J Mol Biol* 2002;**316**:895–902.

104. Kleman Leyer K, Armbruster DW, Daniels CJ. Properties of *H. volcanii* tRNA intron endonuclease reveal a relationship between the archaeal and eucaryal tRNA intron processing systems. *Cell* 1997;**89**:839–47.
105. Fusi P, Tedeschi G, Aliverti A, Ronchi S, Tortora P, Guerritore A. Ribonucleases from the extreme thermophilic archaebacterium *S. solfataricus*. *Eur J Biochem* 1993;**211**:305–10.
106. Fusi P, Grisa M, Tedeschi G, Negri A, Guerritore A, Tortora P. An 8.5-kDa ribonuclease from the extreme thermophilic archaebacterium *Sulfolobus solfataricus*. *FEBS Lett* 1995;**360**:187–90.
107. Oppermann UC, Knapp S, Bonetto V, Ladenstein R, Jornvall H. Isolation and structure of repressor-like proteins from the archaeon *Sulfolobus solfataricus*. Co-purification of RNase A with Sso7c. *FEBS Lett* 1998;**432**:141–4.
108. Shehi E, Serina S, Fumagalli G, Vanoni M, Consonni R, Zetta L, et al. The Sso7d DNA-binding protein from *Sulfolobus solfataricus* has ribonuclease activity. *FEBS Lett* 2001;**497**:131–6.
109. Kulms D, Schafer G, Hahn U. SaRD, a new protein isolated from the extremophile archaeon *Sulfolobus acidocaldarius*, is a thermostable ribonuclease with DNA-binding properties. *Biochem Biophys Res Commun* 1995;**214**:646–52.
110. Kulms D, Schafer G, Hahn U. Overproduction of Sac7d and Sac7e reveals only Sac7e to be a DNA-binding protein with ribonuclease activity from the extremophilic archaeon *Sulfolobus acidocaldarius*. *Biol Chem* 1997;**378**:545–51.
111. Evguenieva Hackenberg E, Schiltz E, Klug G. Dehydrogenases from all three domains of life cleave RNA. *J Biol Chem* 2002;**277**:46145–50.
112. Nagy E, Rigby WF. Glyceraldehyde-3-phosphate dehydrogenase selectively binds AU-rich RNA in the NAD(+)-binding region (Rossmann fold). *J Biol Chem* 1995;**270**:2755–63.
113. Baker ME, Grundy WN, Elkan CP. Spinach CSP41, an mRNA-binding protein and ribonuclease, is homologous to nucleotide-sugar epimerases and hydroxysteroid dehydrogenases. *Biochem Biophys Res Commun* 1998;**248**:250–4.
114. Wagner S, Klug G. An archaeal protein with homology to the eukaryotic translation initiation factor 5A shows ribonucleolytic activity. *J Biol Chem* 2007;**282**:13966–76.
115. Evguenieva Hackenberg E, Klug G. RNase III processing of intervening sequences found in helix 9 of 23S rRNA in the alpha subclass of Proteobacteria. *J Bacteriol* 2000;**182**:4719–29.
116. Xu A, Jao DL, Chen KY. Identification of mRNA that binds to eukaryotic initiation factor 5A by affinity co-purification and differential display. *Biochem J* 2004;**384**:585–90.
117. Sasaki K, Abid MR, Miyazaki M. Deoxyhypusine synthase gene is essential for cell viability in the yeast *Saccharomyces cerevisiae*. *FEBS Lett* 1996;**384**:151–4.
118. Schnier J, Schwelberger HG, Smit McBride Z, Kang HA, Hershey JW. Translation initiation factor 5A and its hypusine modification are essential for cell viability in the yeast *Saccharomyces cerevisiae*. *Mol Cell Biol* 1991;**11**:3105–14.
119. Valentini SR, Casolari JM, Oliveira CC, Silver PA, McBride AE. Genetic interactions of yeast eukaryotic translation initiation factor 5A (eIF5A) reveal connections to poly(A)-binding protein and protein kinase C signaling. *Genetics* 2002;**160**:393–405.
120. Wolff EC, Kang KR, Kim YS, Park MH. Posttranslational synthesis of hypusine: Evolutionary progression and specificity of the hypusine modification. *Amino Acids* 2007;**33**:341–50.
121. Koonin EV, Wolf YI, Aravind L. Prediction of the archaeal exosome and its connections with the proteasome and the translation and transcription machineries by a comparative-genomic approach. *Genome Res* 2001;**11**:240–52.
122. Evguenieva Hackenberg E, Walter P, Hochleitner E, Lottspeich F, Klug G. An exosome-like complex in *Sulfolobus solfataricus*. *EMBO Rep* 2003;**4**:889–93.
123. Farhoud MH, Wessels HJ, Steenbakkers PJ, Mattijssen S, Wevers RA, van Engelen BG, et al. Protein complexes in the archaeon *Methanothermobacter thermautotrophicus* analyzed by blue native/SDS-PAGE and mass spectrometry. *Mol Cell Proteomics* 2005;**4**:1653–63.

124. Lorentzen E, Walter P, Fribourg S, Evguenieva Hackenberg E, Klug G, Conti E. The archaeal exosome core is a hexameric ring structure with three catalytic subunits. *Nat Struct Mol Biol* 2005;**12**:575–81.
125. Buttner K, Wenig K, Hopfner KP. Structural framework for the mechanism of archaeal exosomes in RNA processing. *Mol Cell* 2005;**20**:461–71.
126. Franzetti B, Sohlberg B, Zaccai G, von Gabain A. Biochemical and serological evidence for an RNase E-like activity in halophilic Archaea. *J Bacteriol* 1997;**179**:1180–5.
127. Kanai A, Oida H, Matsuura N, Doi H. Expression cloning and characterization of a novel gene that encodes the RNA-binding protein FAU-1 from *Pyrococcus furiosus*. *Biochem J* 2003;**372**:253–61.
128. Parker JS, Roe SM, Barford D. Crystal structure of a PIWI protein suggests mechanisms for siRNA recognition and slicer activity. *EMBO J* 2004;**23**:4727–37.
129. Ramos CR, Oliveira CL, Torriani IL, Oliveira CC. The Pyrococcus exosome complex: Structural and functional characterization. *J Biol Chem* 2006;**281**:6751–9.
130. Allmang C, Kufel J, Chanfreau G, Mitchell P, Petfalski E, Tollervey D. Functions of the exosome in rRNA, snoRNA and snRNA synthesis. *EMBO J* 1999;**18**:5399–410.
131. Walter P, Klein F, Lorentzen E, Ilchmann A, Klug G, Evguenieva Hackenberg E. Characterization of native and reconstituted exosome complexes from the hyperthermophilic archaeon *Sulfolobus solfataricus*. *Mol Microbiol* 2006;**62**:1076–89.
132. Allemand F, Mathy N, Brechemier-Baey D, Condon C. The 5S rRNA maturase, ribonuclease M5, is a Toprim domain family member. *Nucleic Acids Res* 2005;**33**:4368–76.
133. Tang TH, Rozhdestvensky TS, d'Orval BC, Bortolin ML, Huber H, Charpentier B, et al. RNomics in Archaea reveals a further link between splicing of archaeal introns and rRNA processing. *Nucleic Acids Res* 2002;**30**:921–30.
134. Chekanova JA, Dutko JA, Mian IS, Belostotsky DA. *Arabidopsis thaliana* exosome subunit AtRrp4p is a hydrolytic 3′→5′ exonuclease containing S1 and KH RNA-binding domains. *Nucleic Acids Res* 2002;**30**:695–700.
135. Lorentzen E, Conti E. Structural basis of 3′ end RNA recognition and exoribonucleolytic cleavage by an exosome RNase PH core. *Mol Cell* 2005;**20**:473–81.
136. Lorentzen E, Dziembowski A, Lindner D, Seraphin B, Conti E. RNA channelling by the archaeal exosome. *EMBO Rep* 2007;**8**:470–6.
137. Navarro MV, Oliveira CC, Zanchin NI, Guimaraes BG. Insights into the mechanism of progressive RNA degradation by the archaeal exosome. *J Biol Chem* 2008;**283**:14120–31.
138. Anderson JR, Mukherjee D, Muthukumaraswamy K, Moraes KC, Wilusz CJ, Wilusz J. Sequence-specific RNA binding mediated by the RNase PH domain of components of the exosome. *RNA* 2006;**12**:1810–6.
139. Oddone A, Lorentzen E, Basquin J, Gasch A, Rybin V, Conti E, et al. Structural and biochemical characterization of the yeast exosome component Rrp40. *EMBO Rep* 2007;**8**:63–9.
140. Symmons MF, Jones GH, Luisi BF. A duplicated fold is the structural basis for polynucleotide phosphorylase catalytic activity, processivity, and regulation. *Structure* 2000;**8**:1215–26.
141. Buttner K, Wenig K, Hopfner KP. The exosome: A macromolecular cage for controlled RNA degradation. *Mol Microbiol* 2006;**61**:1372–9.
142. Lorentzen E, Conti E. The exosome and the proteasome: Nano-compartments for degradation. *Cell* 2006;**125**:651–4.
143. Zuo Y, Deutscher MP. Exoribonuclease superfamilies: Structural analysis and phylogenetic distribution. *Nucleic Acids Res* 2001;**29**:1017–26.
144. Makarova KS, Koonin EV. Evolutionary and functional genomics of the Archaea. *Curr Opin Microbiol* 2005;**8**:586–94.

The Making of tRNAs and More – RNase P and tRNase Z

ROLAND K. HARTMANN*,
MARKUS GÖSSRINGER*,
BETTINA SPÄTH[†],
SUSAN FISCHER[†], AND
ANITA MARCHFELDER[†]

*Philipps-Universität Marburg, Institut für Pharmazeutische Chemie, Marbacher Weg 6, D-35037 Marburg, Germany
[†]Universität Ulm, Biologie II, 89069 Ulm, Germany

I. Introduction... 320
   A. Number and Organization of tRNA Genes................................. 320
   B. tRNA Genes Encoding CCA 322
   C. Role of RNase P and tRNase Z for Cell Viability 322
II. Processing by RNase P 324
   A. Architecture of RNase P... 324
   B. Substrate Recognition by Bacterial RNase P – Results from
      In Vitro and In Vivo Studies.. 328
   C. RNase P – Exceptions to the Rule in Bacteria and Archaea.............. 330
   D. RNase P in Organelles 332
   E. Expression of RNase P 332
   F. The Versatility of RNase P: Non-tRNA Substrates 337
   G. RNase P – Association with Cellular Components and Oligomeric State . 345
   H. Life with RNase P.. 346
III. Removal of the tRNA 3′-Trailer 347
   A. The Endonucleolytic Removal of tRNA 3′-Trailers is Catalyzed
      by tRNase Z 348
   B. Concerted Processing by Endo- and Exonucleases....................... 355
   C. Do Archaea Use Exclusively Pathway 1 (tRNase Z)?...................... 357
   References... 358

Transfer-RNA (tRNA) molecules are essential players in protein biosynthesis. They are transcribed as precursors, which have to be extensively processed at both ends to become functional adaptors in protein synthesis. Two endonucleases that directly interact with the tRNA moiety, RNase P and tRNase Z, remove extraneous nucleotides on the molecule's 5′- and 3′-side, respectively. The ribonucleoprotein enzyme RNase P was identified almost 40 years ago and is considered a vestige from the "RNA world". Here, we present the state of affairs on prokaryotic RNase P, with a focus on recent findings on its role in RNA metabolism. tRNase Z was only identified 6 years ago, and we do not yet have a comprehensive understanding of its function. The current knowledge on prokaryotic tRNase Z in tRNA 3′-processing is reviewed here. A second, tRNase Z-independent pathway of tRNA 3′-end maturation involving 3′-exonucleases will also be discussed.

I. Introduction

tRNAs are the central nucleic acid adaptors in the decoding of genetic information at the ribosome. The tRNA fraction can contribute up to 20% of the total RNA of a bacterial cell (1), and the tRNA concentration within an *Escherichia coli* cell has been estimated to be 0.5 mM (200,000 molecules per cell; (2)). tRNAs are generally transcribed with additional 5′- and 3′-sequences which have to be removed to yield a functional molecule (Fig. 1). Two endonucleolytic activities play a central role in tRNA maturation: RNase P, a ribonucleoprotein enzyme, generates the mature 5′-ends; tRNase Z cleaves CCA-less pre-tRNAs after the discriminator nucleotide (the discriminator is the unpaired nucleotide 5′ to the CCA sequence), thus providing the substrate for CCA addition by tRNA nucleotidyltransferase (Fig. 1). If the pre-tRNA contains a CCA triplet, 3′-processing is usually performed by exonucleases instead of tRNase Z (Fig. 1). Moreover, a substantial number of tRNA introns require excision in the Archaea. After a brief introduction to prokaryotic tRNA genes, we will focus on the two tRNA-processing endonucleases that interact directly with the tRNA body, RNase P, and tRNase Z, discussing aspects of their structure, function, biogenesis, and evolution.

A. Number and Organization of tRNA Genes

E. coli and *Bacillus subtilis* are the main bacterial model systems in which tRNA-processing has been studied in detail. The *B. subtilis* genome encodes 86 tRNA genes, organized in 21 transcription units, including single genes, tRNA genes as part of rRNA operons and tRNA gene clusters comprising up to 21 tRNA genes (1). An identical number of tRNA genes is encoded in the *E. coli* K12 genome, including single genes, tRNA genes within rRNA operons and tRNA operons encompassing 2–4 and up to 7 (*metT*) consecutive tRNA genes (3, 4). It is noteworthy that 60 tRNA genes in *B. subtilis* are associated with rRNA operons, 54 of which are located immediately downstream of 5S rRNA genes.

tRNA PROCESSING

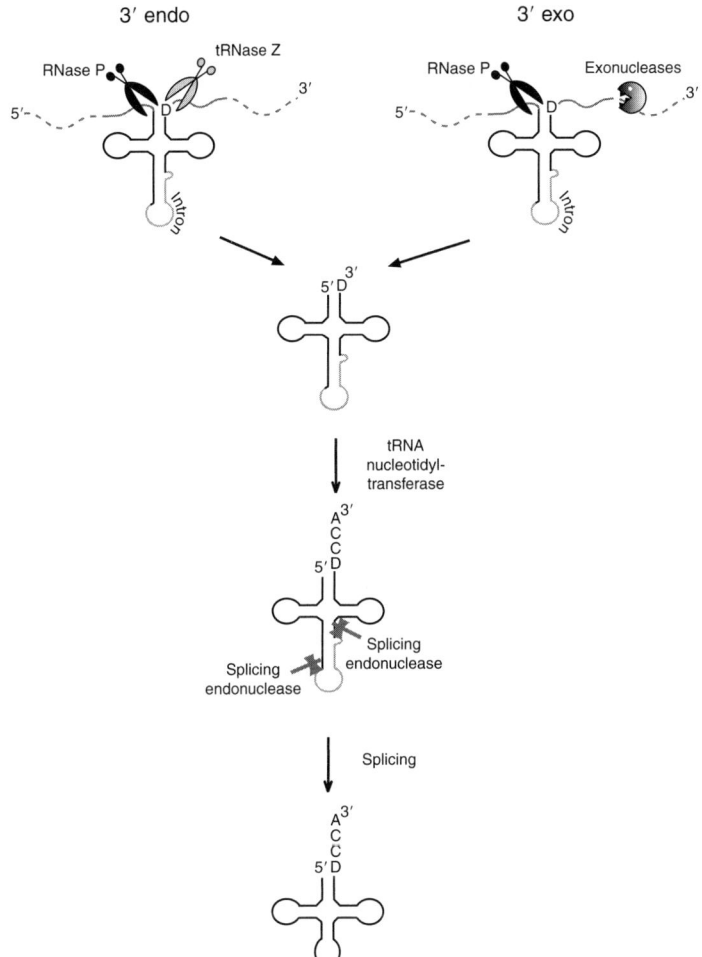

FIG. 1. The two tRNA-processing pathways. In both pathways, tRNA 5'-end maturation is catalyzed by RNase P. tRNA 3'-end maturation in pathway 1 (left) involves the endonuclease tRNase Z, whereas the final 3'-processing steps in pathway 2 are catalyzed by exonucleases (right). After removal of the 3'-trailer, the tRNA nucleotidyltransferase adds the CCA triplet. Introns in archaeal pre-tRNAs are removed by the splicing endonuclease. D, the discriminator nucleotide preceding CCA.

Only 14 tRNA genes (four downstream of 5S rRNA genes) are part of rRNA transcription units in *E. coli*, suggesting that these two model bacteria may use different strategies to regulate tRNA transcription (*1*). In Archaea, clustering of tRNA genes occurs as well. For example, 39 tRNA genes have been identified in the Euryarchaeon *Methanothermobacter thermautotrophicus* ΔH, encompassing 10 or 11 apparently single-gene transcriptional units, eight operons with two tRNA genes each, and two operons with five tRNA genes (5). A tRNA^{Ala} is encoded in the two rRNA operons, and three tRNA transcripts contain introns in their anticodon loop, with one (tRNA^{Pro}) harboring a second intron (5). Introns in tRNAs are widespread among Archaea, occurring at various positions in the tRNA; they vary in size from 11 to 44 nt and contain bulge–helix–bulge structural motifs (6, 7). Introns within tRNAs are rarely found in bacteria, where their position is restricted to the anticodon loop and which belong to another class of RNA structures, the much larger group I or II introns (7, 8). A peculiarity is found in the small hyperthermophilic archaeon *Nanoarchaeum equitans* (see also Section II.C): here, five tRNA species are encoded as separate 5′- and 3′-halves. Annealing of the separately transcribed tRNA 5′- and 3′-halves is supported by sequence extensions that base pair to form an intron-like structure in the anticodon arm. These *trans*-introns contain bulge–helix–bulge motifs that are removed by the tRNA splicing endonuclease of *N. equitans* (9).

B. tRNA Genes Encoding CCA

The 3′-CCA terminus of tRNAs is a key determinant for bacterial RNase P and tRNase Z. Whereas a 3′-CCA motif is not encoded in eukaryotic tRNA genes, this varies among Bacteria and Archaea (Table I). For example, all tRNA genes encode CCA in *E. coli* and *Aeropyrum pernix* (archaeon), none codes for CCA in *Borrelia burgdorferi* (bacterium) or *Archaeoglobus fulgidus* (archaeon), and about two-third of the tRNA genes encode CCA in *B. subtilis* or the archaeon *Methanocaldococcus jannaschii* (7).

C. Role of RNase P and tRNase Z for Cell Viability

The protein (encoded by the *rnpA* gene) and RNA (encoded by *rnpB*) subunits of bacterial RNase P are both essential for cell growth. This was initially inferred from *E. coli* strains carrying temperature-sensitive *rnpA* and *rnpB* mutant alleles (10–13). The essential character of both subunits was later substantiated using an *E. coli rnpB* deletion strain (14), and *E. coli* and *B. subtilis* mutant strains with suppressible *rnpB* (15, 16) or *rnpA* (17) expression.

tRNase Z (also termed RNase Z) is essential in organisms where 3′-end maturation of some CCA-less tRNA precursors depends on endonucleolytic cleavage immediately 3′ to the discriminator (see Section III). In a *B. subtilis* mutant strain, the chromosomal tRNase Z gene ($rnz = yqjK$) was put under control of an IPTG-dependent promoter. The removal of IPTG from the culture medium resulted in complete growth inhibition, demonstrating that the tRNase Z gene is essential in *B. subtilis* (18).

tRNA PROCESSING   323

TABLE I
DISTRIBUTION OF RIBONUCLEASES INVOLVED IN tRNA 3'-PROCESSING THROUGHOUT BACTERIA
AND ARCHAEA AND DISTRIBUTION OF CCA-ENCODING GENES

Organism	Z	E, G, EG	PH	T	+CCA/ $\sum$ tRNA genes
I. Bacteria					
Actinobacteria					
Mycobacterium tuberculosis[a]	–	EG	PH	–	16/45[b]
Chlamydiae					
Chlamydia trachomatis[a]	Z	EG	–	–	37/41[b]
Cyanobacteria					
Synechocystis sp. PCC 6803[a]	Z	EG	–	–	0/42[b]
Firmicutes					
Bacillales					
Bacillus subtilis[a]	Z	–	PH	–	59/86[c]
Clostridiales					
Clostridium perfringens[d]	Z	G	PH	–	93/93[b]
Lactobacillales					
Lactococcus lactis[a]	Z	–	–	–	0/61[b]
Mollicutes					
Mycoplasma genitalium[a]	–	–	–	–	36/36[b]
Ureaplasma urealyticum[a]	–	–	–	–	30/30[b]
Others					
Aquifex aeolicus[a]	–	EG	PH	–	32/43[b]
Deinococcus radiodurans[a]	Z	–	PH	–	47/48[b]
Thermotoga maritima[a]	Z	EG	–	–	45/46[b]
Proteobacteria					
α-proteobacteria					
Rickettsia prowazekii[a]	–	EG	PH	–	32/32[b]
β-proteobacteria					
Neisseria meningitidis[a]	–	E, G	PH	–	57/59[b]
γ-proteobacteria					
Escherichia coli[a]	Z	E, G	PH	T	86/86[e]
Haemophilus influenza[a]	–	E, G	PH	T	56/56[b]
Pseudomonas aeruginosa[a]	–	E, G	PH	T	60/62[b]
Vibrio cholerae[a]	–	E, G	PH	T	97/97[b]
δ - and ε - proteobacteria					
Campylobacter jejuni[a]	–	–	–	–	43/43[b]
Helicobacter pylori[a]	–	–	–	–	36/36[b]

(continued)

TABLE I (continued)

Organism	Z	E, G, EG	PH	T	+CCA/ $\sum$ tRNA genes
Spirochaetes					
Borrelia burgdorferi[a]	Z	–	–	–	0/33[b]
Treponema pallidum[a]	Z	–	–	–	1/45[b]
II. Archaea					
Euryarchaeota					
Methanopyrus kandleri[f]	Z	–	PH	–	29/34[b]
Methanosarcina mazei Go1[f]	Z	–	PH	–	8/57[g]
Methanocaldococcus jannaschii[a]	Z	–	–	–	25/36[b]
Haloferax volcanii[f]	Z	–	–	–	0/51[g]
Halobacterium sp.[a]	Z	–	–	–	0/47[b]
Methanothermobacter thermautotrophicus[a]	Z	–	PH	–	1/39[b]
Archaeoglobus fulgidus[a]	Z	–	PH	–	0/46[b]
Pyrococcus abyssi[a]	Z	–	PH	–	44/46[b]
Pyrococcus furiosus[f]	Z	–	PH	–	46/46[g]
Thermoplasma acidophilum[a]	Z	–	PH	–	1/46[b]
Crenarchaeota					
Sulfolobus solfataricus[a]	Z	–	PH	–	1/46[b]
Aeropyrum pernix[a]	Z	–	PH	–	46/46[b]
Pyrobaculum aerophilum[f]	Z	–	PH	–	22/46[b]
Nanoarchaeota					
Nanoarchaeum equitans[f]	Z	–	PH	–	0/43[h]

Z, tRNase Z; PH, RNase PH; E, RNase E; G, RNase G; EG, RNase E/G (E/G: RNases similar in size to RNase G (approx. 500 amino acids) but equally homologous to RNase G and to the N-terminal part of RNase E); –, no homologues were found.

[a] Data are from (201).
[b] Data are from (7).
[c] Data are from (18).
[d] Data are from (197).
[e] Data are from (194).
[f] Homologues of ribonucleases were identified using BLAST searches with *E. coli* sequences against the prokaryotic genome sequences.
[g] CCA data are from http://archaea.ucsc.edu.
[h] Data are from (9).

II. Processing by RNase P

A. Architecture of RNase P

The ribonucleoprotein enzyme RNase P (EC 3.1.26.5) universally removes 5′-leader sequences from precursor tRNAs (5′-pre-tRNAs) to generate the mature tRNA 5′-ends. The enzyme harbors a single RNA subunit of common ancestry among all three domains of life, as inferred from ubiquitously conserved features in the structural core of the RNA (19). The RNA subunit (P RNA, 350–400 nucleotides) of bacterial RNase P is an efficient catalyst *in vitro* in the absence of its single protein cofactor (approx. 14 kDa; (20)). The very low RNA-alone activity of archaeal, eukaryal, and organellar P RNAs (21–23) correlates with the presence of multiple protein subunits, usually 4 in Archaea and 9–10 in Eukarya (Fig. 2).

Two architectural subtypes of bacterial RNase P RNAs (Fig. 3) have been found in nature: the phylogenetically prevailing type A (for ancestral), represented by *E. coli* RNase P RNA, and type B (for *Bacillus*), the latter being confined to the low G + C Gram-positive Bacteria, such as *B. subtilis* (24). Outside this subphylum, the RNase P RNA of the green nonsulfur bacterium *Thermomicrobium roseum* is the only reported case of an RNase P RNA that has acquired most of the type B-specific structural elements by independent convergent evolution (25) or by lateral *rnpB* gene transfer.

Bacterial RNase P RNAs consist of two independent folding domains (26), termed the specificity (S) and catalytic (C) domains (Fig. 3). The S domain interacts with the T-arm module of pre-tRNAs and thus increases substrate affinity and confers specificity for tRNAs and tRNA-like molecules. The C domain includes all structural elements forming the active site and provides the binding interface for the single protein subunit, which is dispensable *in vitro* but essential *in vivo* (see above). The bacterial RNase P protein binds roughly within the P2–P3–P4 region of the RNA subunit (Fig. 3) (27–30) and increases substrate affinity through interaction with the 5'-leader (31–33), which entails tighter binding of key metal ions involved in substrate positioning and catalysis (34). Binding of the protein to the RNA induces conformational changes in both subunits (35). According to covariation and crosslinking analyses (25, 36–40) as well as crystal structures (41, 42), much of the catalytically proficient overall RNA fold of bacterial RNase P RNAs results from a set of long-range tertiary interactions, some of which orient the C and S domains toward each other (Fig. 3). Three-dimensional models of bacterial RNase P RNA–tRNA complexes are shown in Fig. 4. For more information on the structure, biochemistry, and phylogeny of bacterial RNase P, the reader is referred to recent reviews (19, 43).

Based on current knowledge, archaeal RNase P enzymes are composed of a single RNA and usually four different protein subunits (Rpp21, Rpp29, Rpp30, and Pop5; (44–46)). Archaeal RNase P RNAs have been grouped into two structural classes, type A and type M. Type A RNAs are structurally related to bacterial-type A RNase P RNAs, but display a variety of deviations from the bacterial consensus, among them a lack of P18 and P13/14 (47) (see Fig. 3). Archaeal-type M RNAs are structurally more simplified, and show resemblance to eukaryal RNase P RNAs. In contrast to type M RNAs, residual RNA-alone activity has been observed for archaeal-type A RNAs, albeit at low levels and

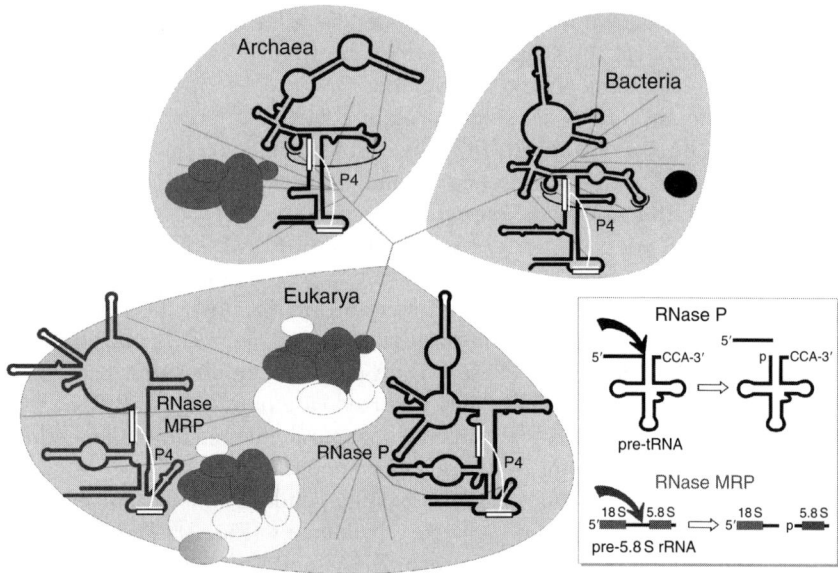

FIG. 2. RNase P family tree. RNase P, present in all three domains of life, consists of an RNA subunit (black line drawing of the 2D structure) that forms a ribonucleoprotein complex with one protein in Bacteria (black oval), typically four in Archaea (dark gray ovals), or nine to ten in Eukarya (white and gray ovals). For each of the four archaeal proteins, a homolog is present in eukaryal RNase P (dark gray ovals). All RNase P RNA subunits have in common the same basic architecture of their core (lower part of the line drawings) and share conserved nucleotides clustering in and around helix P4 (white boxes connected by the white line) (203) in the catalytic core, indicating that all RNase P RNAs stem from a common progenitor; RNA helix P4 harbors a cluster of metal ion-binding sites and is close to the substrate acceptor stem in enzyme–substrate complexes (204). Eukaryal RNase P has a sibling, RNase MRP (mitochondrial RNA processing, gray line drawing). Both ribonucleoprotein enzymes share the majority of their protein subunits as well as the conservation within their RNA core structure. Thin black lines in the RNase P structural sketches depict base-pairing interactions of nucleotides distant in the 2D presentations. RNase MRP protein subunits unique to RNase MRP are shown as ovals shaded in light gray (45, 205). Cleavage by RNase P and RNase MRP results in 3′-hydroxyl and 5′-phosphate (indicated by "p") ends of the products. The major substrate for the Eukarya-specific RNase MRP is precursor 5.8S rRNA; the cleavage site is located several nucleotides upstream of the mature 5′-end. RNase MRP was originally reported to cleave mammalian mitochondrial RNA transcripts to generate the primer for heavy-strand DNA replication (206); RNase MRP is further involved in the degradation of a B-type cyclin mRNA in yeast, an activity important for cell cycle progression (207, 208).

limited to conditions of very high ionic strength (21). A major cause of the low activity of archaeal RNase P RNAs in the absence of protein cofactors seems to be the low functionality of archaeal S domains (21, 48). For more details on the architecture of archaeal (and eukaryal) RNase P, the review by Walker and Engelke (45) is recommended.

tRNA PROCESSING

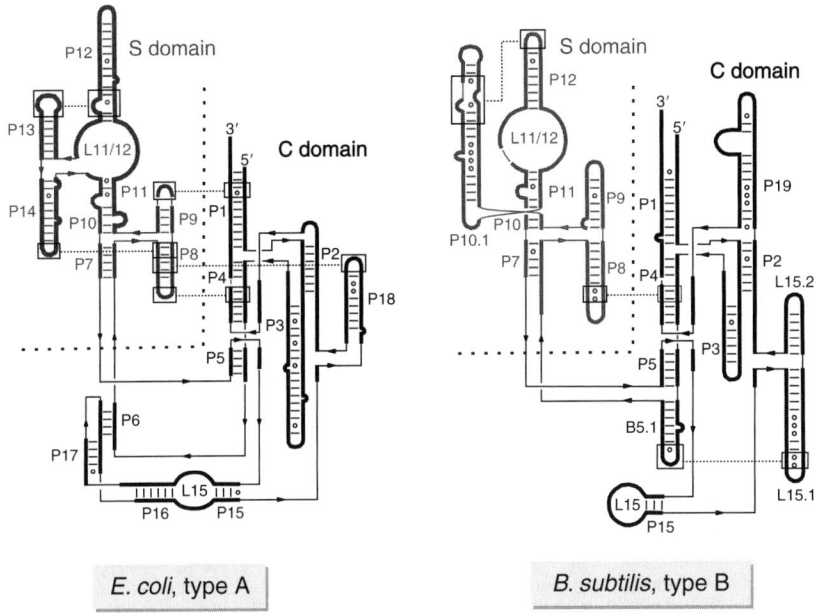

FIG. 3. Secondary structure illustrations (29) of *E. coli* (type A, left) and *B. subtilis* (type B, right) RNase P RNAs. C domain, catalytic domain; S domain, specificity domain; P, helical (paired) elements; L, loop region. Tertiary interactions are indicated by gray dotted lines. Type A RNAs usually form three loop–helix interdomain contacts (L8–P4, L9–P1, L18–P8) (39, 40, 42, 209, 210); in type B RNAs, P5.1 interacts with P15.1 (41) and the L9–P1 interaction may also form (39).

Crystal structures are available for each type of archaeal P protein (reviewed in (19)). Wilson et al. (49) saw pronounced similarity between archaeal Pop5 and the bacterial P protein in terms of type and arrangement of secondary structure elements. Yet, the functional relevance of this observation remains unclear, since archaeal Pop5 proteins, either alone or coexpressed with Rpp30, are unable to functionally replace the bacterial P protein in *B. subtilis* (50). Yeast two-hybrid (51, 52) and NMR analyses (49) have provided evidence for formation of heterodimers Pop5/Rpp30 and Rpp21/29 even in the absence of the RNA subunit. Protein pairs Pop5/Rpp30 and, to a lesser extent, Rpp21/29 can activate archaeal RNase P RNA from the hyperthermophilic archaeon *Pyrococcus furiosus* independently of each other, and both protein pairs reduced the Mg^{2+} requirements for efficient processing. Pop5/Rpp30 also stimulated activity of the C domain alone. High catalytic proficiency required the presence of all four archaeal P protein subunits. The current picture is that Rpp21/29 bind the S and Pop5/Rpp30 the C domain of archaeal RNase P RNA as a prerequisite for optimal cooperation between the two domains (53).

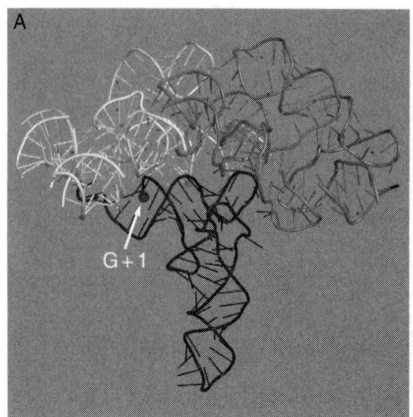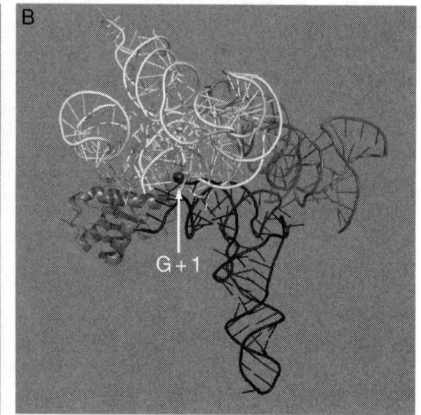

FIG. 4. Models of bacterial RNase P RNA-tRNA complexes. (A) *Thermotoga maritima* (type A) RNase P RNA in complex with mature tRNA based on the RNase P RNA crystal structure ((42); 2A2E.pdb). (B) Complex of *Bacillus stearothermophilus* (type B) RNase P RNA, mature yeast tRNA^{Phe} and the *B. subtilis* RNase P protein according to (30); this model is deposited at "The RNase P Database" (211). C domains are shown in white, S domains in grey and tRNAs in black; the phosphates of the tRNA 5′-end (G + 1) at the RNase P cleavage site are highlighted as dark gray spheres and indicated by white arrows. The protein subunit is shown as a ribbon cartoon with α-helices and β-strands in grey. See also colored online version of the figure.

B. Substrate Recognition by Bacterial RNase P – Results from *In Vitro* and *In Vivo* Studies

Apart from interactions between the 5′-leader and the protein subunit, and between the T-arm and the S domain, bacterial RNase P enzymes recognize a set of elements near the cleavage site in the pre-tRNA substrate. These include the 5′-leader nucleotide immediately upstream of the cleavage site (−1, preferentially a U-residue), the 5′-terminal nucleotide of mature tRNA (+1, usually a G-residue), the 3′-CCA terminus and the discriminator nucleotide (+73) preceding CCA (Fig. 5). For further details, the reader is referred here to the following reviews (54–56).

RNase P has to act with similar efficiency on pre-tRNAs carrying unrelated 5′-flanking sequences and which differ in the structural details of their mature domains. Pre-tRNA substrates do not always include the full set of recognition elements mentioned above. For example, *E. coli* initiator tRNA^{Met} carries a C–A mismatch at the end of the acceptor stem, one leucine isoacceptor has a G–U wobble pair at this position, and several tRNAs do not have a G-residue at +1. Such deviations from the consensus (Fig. 5) weaken the interaction with RNase P RNA (57). Harris and coworkers (57) provided evidence that the *E. coli* RNase P holoenzyme binds all 5′-pre-tRNAs with uniform affinity and

tRNA PROCESSING

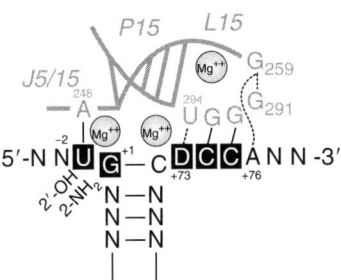

Fig. 5. Contacts between the C domain of *E. coli* RNase P RNA (gray) with pre-tRNA at and near the cleavage site (between nt −1 and +1). Key substrate nucleotides involved in this interaction are highlighted. D = A, G, or U; solid lines, Watson–Crick base pairs; dotted lines, other interactions. Multiple Mg^{2+} ions are involved in this interaction, but their exact position is unknown. For further information, see (55, 56). The 2′-OH of U − 1 and the 2-NH_2 at G + 1 are important functional groups in the interaction.

has apparently achieved this capacity during evolution by combining suboptimal tRNA structures with 5′-leader sequences that confer high pre-tRNA affinity, similar to uniform substrate binding by elongation factor Tu (58). In contrast to the holoenzyme, substrate recognition by *E. coli* RNase P RNA is highly sensitive to structural deviations in the vicinity of the cleavage site, entailing large affinity losses (57). Thus, an important biological function of the bacterial RNase P protein is to offset differences in pre-tRNA structure such that binding and catalysis are uniform.

Another aspect that merits discussion is the fact that certain bacterial 5′-pre-tRNAs need to be processed at a site shifted by 1 nt to generate a mature tRNA product with an 8-bp instead of a 7-bp long acceptor stem (tRNA^{His}, tRNA^{SeCys}) (59, 60). For example, *E. coli* 5′-pre-tRNA^{His} has a U-residue at −2 and a G-residue at −1 that forms an additional bp with C + 73. This prevents formation of the contact between nt +73 and U294 in RNase P RNA (Fig. 5) and channels the enzyme to another reaction pathway that leads to cleavage 1 nt upstream. This relay-type switching of cleavage site upon disruption of a subset of contacts illustrates the flexibility of RNA and might explain why RNase P has retained an RNA subunit in evolution.

1. ROLE OF THE CCA INTERACTION *IN VIVO*

The two C-residues of the tRNA 3′-CCA motif form Watson–Crick base pairs with two G-residues in L15 (15, 61–64) (Fig. 5). L15 is an internal loop in type A RNAs and an apical loop in type B RNAs (Fig. 3). Cyanobacterial RNase P RNAs are an exception to this rule and apparently do not form this interaction, as they lack the two conserved G-residues in L15 (65, 66).

Recently, the role of the CCA interaction was studied *in vivo* in *E. coli* and *B. subtilis rnpB* mutant strains, using *rnpB* alleles in which each of the pairing G-residues in L15 was changed to a C-residue to disrupt base pairing (15, 64). In both organisms, the mutant RNase P RNA alleles (*rnpBC292* or *rnpBC293* in *E. coli*, *rnpBC258* or *rnpBC259* in *B. subtilis*) caused severe growth defects. Compensatory mutations in 3'-CCA (3'-G_{74}CA) of model substrates fully restored the catalytic performance of the C293 (*E. coli*) or C259 (*B. subtilis*) mutant RNase P RNAs, demonstrating that Watson–Crick base pairing is the sole constraint at this position. In contrast, a corresponding substrate mutation (3'-CG_{75}A) only incompletely rescued the defects of the C292 and C258 mutant RNase P RNAs, showing that the identity of G292 (*E. coli*) and G258 (*B. subtilis*) is crucial for full enzyme function. The *in vivo* defect of the mutant RNase P RNA was inferred to lie primarily in the recruitment of catalytically relevant Mg^{2+}, with a possible contribution from altered RNA folding. The best coherence between *in vitro* and *in vivo* phenotypes was seen at Mg^{2+} concentrations as low as 2 mM (15, 64).

One-third of the tRNA genes in *B. subtilis* do not encode the 3'-CCA triplet. Based on the importance of the CCA interaction for *B. subtilis* RNase P, one might predict that the enzyme would act inefficiently on CCA-less 5'-pre-tRNAs and thus act only after 3'-end processing by tRNase Z and CCA addition have occurred. On the other hand, *B. subtilis* tRNase Z activity inversely correlates with the length of tRNA 5'-extensions, which led to the proposal that processing by RNase P precedes that of tRNase Z in cases where CCA-less tRNA transcripts have longer 5'-extensions (>30 nt) (18). This issue was addressed in a tRNase Z mutant strain, where CCA-less 3'-pre-tRNAs accumulate under nonpermissive conditions (18). Surprisingly, a 5'-RACE analysis for two such 3'-pre-tRNAs (*trnSL–Ala1* and *trnSL–Val2*) revealed transcripts with mature 5'-ends exclusively (64). Thus, *B. subtilis* RNase P is able to cleave CCA-less ptRNAs *in vitro* and *in vivo*, albeit with reduced efficiency. This then raises the question why mutations in the CCA-binding site of RNase P RNA cause severe effects on cell growth. Possible explanations are a global deceleration in pre-tRNA maturation or a severe inhibition of 5'-maturation for a particular pre-tRNA subset. The finding that RNase P cleaves CCA-less 3'-pre-tRNAs in *B. subtilis* (64) suggests that RNase P acts before tRNase Z on CCA-less tRNA transcripts with long 5'-leaders.

C. RNase P – Exceptions to the Rule in Bacteria and Archaea

RNase P, first discovered in *E. coli* almost 40 years ago (67), is found in Bacteria, Archaea, Eukarya, and eukaryotic organelles (44). In the vast majority of bacterial genomes, the genes encoding the RNA (*rnpB*) and the protein subunit (*rnpA*) have been identified. The *rnpB* gene is positionally volatile among Bacteria, and the *rnpA* gene is usually part of a transcription unit with the gene for ribosomal protein L34 (*rpmH*). In rare cases, such as in the Firmicutes *Lactococcus lactis* or *Streptococcus pyogenes*, *rpmH* is separated from *rnpA* and transferred to a position approximately 2-kb downstream of *rnpA* (44).

With respect to RNase P, there is one remarkable exception among Bacteria: in *Aquifex aeolicus*, candidate genes for *rnpB* and *rnpA* could not be identified (68) despite intense efforts in several research groups (e.g., (69)). This situation seems to be unique to *A. aeolicus* and possibly very close relatives (70), since canonical *rnpB* and *rnpA* genes have been identified in other *Aquificales* (71). Two groups recently reported the partial purification of RNase P-like tRNA 5'-end maturation activity from *A. aeolicus* extracts (70, 72), with protein component(s) apparently being sensitive to thiol oxidation. At present, it is not clear and a matter of debate whether the *A. aeolicus* enzyme contains an RNA subunit or not (70, 72).

A definite exception to the rule of RNase P ubiquity is found in the archaeal domain. The "nanosized" hyperthermophilic archaeon *N. equitans*, which grows attached to the surface of a specific archaeal host, *Ignicoccus hospitalis* (73, 74), is devoid of RNase P and synthesizes leaderless tRNAs (75). The *N. equitans* genome encodes the machinery for information processing and repair, but lacks genes for lipid, cofactor, amino acid, or nucleotide biosyntheses. It is the smallest microbial genome (491 kbp) sequenced to date, with one of the highest coding densities (95% (76)). "Life without RNase P" has been made possible in *N. equitans* by the conserved positioning of all tRNA promoters such that the primary transcripts are 5'-mature tRNAs bearing a 5'-triphosphate terminus (75). Since tRNA gene transcription is initiated with a purine in *N. equitans*, tRNA^{Tyr}, which carries a cytosine at position 1 in other archaea, has a 5'-terminus extended by a single G-residue. This unorthodox tRNA was shown to be a substrate for the cognate *N. equitans* aminoacyl-tRNA synthetase (TyrRS), whereas a tRNA variant lacking this extra 5'-G-residue was a poor substrate for the same enzyme. Likewise, in the initiator tRNA^{Met} of *N. equitans*, the A1–U72 bp required for the interaction between the initiator tRNA and initiation factor aIFγ in other archaea (7, 77), is shifted to position $-1/+73$ to form an 8-bp acceptor stem, and this tRNA is charged efficiently by its cognate MetRS (75). Evidently, the TyrRS and MetRS enzymes of *N. equitans* have coevolved with their unusual tRNA substrates. The primary driving force for elimination of tRNA 5'-end maturation may have been the pressure for genome compaction. The hyperthermophilic crenarchaeon *Pyrobaculum aerophilum* (78), a fast-evolving organism like *N. equitans*, might represent an intermediate evolutionary state toward abandoning transcription of 5'-pre-tRNAs and eventually loss of RNase P; although most tRNA genes in

P. aerophilum contain a promoter at a conserved distance to direct transcription of leaderless tRNAs, a few promoter positions suggest the synthesis of pre-tRNAs with short 5'-leaders (75). Consistently, a conventional RNase P is not detectable in *P. aerophilum* and only one RNase P protein subunit (Rpp29) could be identified in its genome (44).

D. RNase P in Organelles

Since organelles have a bacterial origin, a short detour into tRNA 5'-end maturation in organelles seems appropriate. While a bacterial-type RNase P RNA is still encoded in some chloroplast genomes, as well as in the mitochondrial genomes of some lower eukaryotes (44, 79), the bacterial-like RNase P protein subunit (*rnpA*) is not identifiable in any of the known organellar genomes. Furthermore, RNA-alone activity, a hallmark of bacterial RNase P RNAs, has not been demonstrated for any of the *bona fide* organellar RNase P RNAs, with one exception: the RNase P RNA of the cyanelle of *Cyanophora paradoxa* was shown to have weak but specific RNA-alone activity (23), despite a higher protein content than in bacterial RNase P enzymes (80, 81). The cyanelle, a photosynthetic organelle of *C. paradoxa*, occupies an intermediate position in plastid evolution, combining features of free-living cyanobacteria and modern chloroplasts. The increase in protein content observed for the cyanelle RNase P (81) reflects a general trend in organellar RNase P evolution. For example, *Saccharomyces cerevisiae* mitochondrial RNase P is composed of an RNA and a 105 kDa protein subunit (82), unrelated to the 14 kDa protein subunit found in bacteria. Very recently, the human mitochondrial RNase P was identified to be composed exclusively of three protein subunits and no RNA subunit (83). There is also evidence for the lack of an RNA component in the tRNA 5'-end maturation activity of spinach chloroplasts (84, 85).

E. Expression of RNase P

The bacterial cell has to adapt to different environmental conditions. Varying needs for mature tRNAs are expected to lead to changes in the level of expression of tRNA-processing enzymes like RNase P. Furthermore, the 1:1 stoichiometry of the RNA and protein component in bacterial RNase P necessitates an adjustment of the expression rate of the genes encoding both subunits, *rnpA* and *rnpB*. Coordinated synthesis of RNase P components is even more complex for the multisubunit enzymes in Archaea and Eukarya. For historical reasons, the RNase P RNA of *E. coli* will be termed "M1 RNA" and that of *B. subtilis*, "P RNA," in the following.

1. Expression of RNase P RNA in E. coli and B. subtilis

In *E. coli*, three potential promoters (P_1, P_2, P_3) were found immediately upstream of the *rnpB* gene (86, 87). Though transcripts derived from all three promoters are detectable *in vivo*, the proximal P_1 promoter is the most active one (87, 88); P_2 and P_3 may yet play a role under certain metabolic conditions. In *E. coli*, four open reading frames (*garPLRK*) are located upstream of *rnpB*, in the same transcriptional orientation. Earlier *in vivo* studies suggested that the upstream promoter preceding *garPLRK* can also direct transcripts that include the M1 RNA and terminate at the *rnpB* terminator, and from which functional M1 RNA is released (89). However, according to a recent study, these long transcripts are primarily degraded in an RNase E-dependent manner, rather than being processed to functional M1 RNA (90).

Under conditions of stress like nutrient starvation, to conserve energy, *E. coli* initiates a global change in the cellular metabolism, termed the stringent response. The first regulatory response to amino acid starvation is the synthesis of guanosine tetraphosphate (ppGpp) by the *relA* gene product, which results in a sudden decrease in stable RNA synthesis. The P_1 promoter of *rnpB* encodes a G/C-rich discriminator motif between the -10 region and the transcription start site that has been implicated in the regulation of stringently controlled promoters (88, 91). The synthesis of M1 RNA was shown to be growth-rate dependent, but the question of whether M1 RNA expression is stringently controlled is still a matter of debate (91–93).

In *B. subtilis*, a potential *rnpB* promoter is positioned correctly for transcription initiation at the 5′-terminal G(+1) residue of mature P RNA (94). A guanosine at the transcription start site is a hallmark of σ^A-dependent promoters that are downregulated in response to amino acid starvation in *B. subtilis* (95). This promoter is overlapped by a putative second promoter lying 15 nucleotides upstream, although a P RNA transcript containing 15 extra nucleotides at its 5′-end has not been detected (94). The fact that both promoters show some deviations from the -10 and -35 consensus sequences in *B. subtilis* may explain the relatively low levels of expression of P RNA in *B. subtilis* (20–200 copies per cell; (94, 96)).

2. Expression of the Bacterial RNase P Protein

In Bacteria, the genes for the small ribosomal protein L34 (*rpmH*) and the RNase P protein (*rnpA*) are adjacent to each other in the majority of bacterial genomes (44), suggesting cotranscription, as has been shown for *E. coli*. Here, *rpmH* and *rnpA* constitute an operon with one minor and two major promoters preceding the *rpmH* gene (97). In Northern blot analyses, three mRNA species derived from the *rpmH–rnpA* operon were detected, the two shorter ones lacking the *rnpA* cistron and the longer and much less abundant one including it (98). *E. coli* cells contain a 60- to 80-fold molar excess of ribosomes over M1 RNA (92). Likewise, in *B. subtilis* the number of ribosomes per cell was estimated to exceed that of RNase P by a factor of 30–300 (94, 96). This implies that the ribosomal protein L34, encoded by the same transcription unit as *rnpA*, is produced in excess over RNase P protein in bacteria such as *E. coli* and *B. subtilis*. Different strategies of how bacteria achieve an unequal production of L34 and the RNase P protein have been discussed (17, 99). In several bacteria, hairpin structures are encoded in the *rpmH–rnpA* operon, either in the intergenic region (e.g., *B. subtilis*) or in the 5′-portion of the *rnpA* gene (e.g., *E. coli*). On the transcriptional level, such hairpin structures could contribute to transcriptional polarity by causing premature termination of transcription downstream of *rpmH*. This mechanism could explain why the level of *rpmH* mRNA exceeds that of *rnpA* by a factor of ~90 in exponentially growing *B. subtilis* cells (17). Another interesting regulatory mechanism was proposed by Feltens *et al.* (99) for thermophilic bacteria of the genus *Thermus*, where the *rnpA* gene completely overlaps the *rpmH* gene in a different reading frame. This results in the synthesis of extended RNase P proteins. The distance between the purine-rich ribosomal binding site and the start codons of *rpmH* and *rnpA* is different and seems to be suboptimal for *rnpA*. This may cause less abundant translational initiation of the *rnpA* reading frame than that of *rpmH*. Furthermore, several codons are present in the *rnpA* region that are rarely used in highly expressed *Thermus* genes. This could lead to a slower rate of translation and possibly a higher frequency of abortive translation for ribosomes translating in the *rnpA* reading frame relative to those translating in the *rpmH* frame. Similarly, the codon usage of *rnpA* in *E. coli* does not correspond to that of the most abundant proteins (100).

3. Processing of Bacterial RNase P RNA Transcripts

The processing of RNase P RNA transcripts differs considerably in the two main bacterial model systems, *E. coli* and *B. subtilis*, which parallels differences in their repertoire of ribonucleases. For example, some of the key enzymes of RNA metabolism in *E. coli*, such as RNase E, RNase II and oligoribonuclease, are missing in *B. subtilis* (101).

In *E. coli*, mature M1 RNA isolated from wild-type cells is present as a single band of 377 nucleotides, indicating efficient processing. The major *rnpB* primary transcript is initiated at the proximal promoter P_1 to yield the mature 5′-end (position +1) of M1 RNA, with the majority of transcripts (>94%; (102)) terminating at the first terminator, T_1 (103). This primary transcript (pM1 RNA) of 413 nucleotides (104) carries a 3′-extension of ~36 nucleotides, including the termination stem-loop (86, 104). A putative RNase E recognition sequence (5′-GAUUU) is located immediately downstream of the mature 3′-end. It has been suggested that the major processing pathway is initiated by

RNase E cleavage at this site, leaving 1 or 2 extra nucleotides at the 3′-terminus of M1 RNA (*89, 105, 106*) that are subsequently trimmed to the mature 3′-end at position +377 by exoribonucleases. Li *et al*. (*107*) showed that the exoribonuclease RNase T is the major activity in this final maturation step. At first glance, the biological significance of the M1 RNA maturation process is not apparent since M1 RNA with extra nucleotides at either the 3′-terminus or its 5′-terminus is catalytically active (*108, 109*). However, inhibition of RNase E cleavage at the aforementioned 5′-GAUUU site led to severe growth defects, demonstrating that 3′-end processing of M1 RNA is required for cell viability (*110*). It was further shown that 3′-maturation protects M1 RNA primary transcripts from poly(A)-dependent degradation by converting an unstable transcript into a stable form (*110*). The authors proposed that the primary transcript of M1 RNA may enter two distinct pathways: (i) processing to form stable M1 RNA by RNase E cleavage at positions +378/+379 and exoribonuclease trimming to generate the mature 3′-end at position +377 and (ii) polyadenylation by poly(A) polymerase I and degradation. This model is consistent with the finding that M1 RNA precursors contain nonencoded adenylate residues at the 3′-terminus in some exoribonuclease-deficient strains (*111*).

Model transcripts of M1 RNA, containing 270 nucleotides of natural upstream sequence (upRNA) were found to be primarily degraded *in vivo*, rather than processed to mature M1 RNA (*90*). In contrast to 5′-mature M1 RNA transcripts initiated at the P_1 promoter (pM1 RNA), upRNA was cleaved within the M1 RNA moiety by the catalytic domain of RNase E *in vitro*, suggesting a role for RNase E in the degradation of upRNA. The difference in susceptibility to RNase E may be explained by the preference of RNase E for unstructured regions present in the 5′-portion of upRNA. The cleavage products generated, which have 5′-monophosphates, may then be even more efficient substrates for subsequent cleavage by RNase E within the mature M1 RNA moiety. In contrast, the 5′-end of pM1 RNA is part of a stem structure and carries a 5′-triphosphate, unfavorable for RNase E cleavage. Thus, it appears that RNase E exerts different effects on *rnpB* expression depending on the nature of the M1 RNA primary transcript. Pan and coworkers (*112*) showed that RNA polymerase pauses after about one-third of M1 RNA has been transcribed. This seems to allow the 5′-portion of the M1 RNA to fold into a transient nonnative structure that later facilitates interaction with sequence elements in the 3′-portion of M1 RNA to achieve the native fold of the full-length RNA. The extra 5′-sequences in upRNA may also interfere with the formation of this nonnative structure, retard M1 RNA folding and result in an expanded time window for RNase E to access the M1 RNA portion in upRNA transcripts.

As outlined in Section II.E.1, transcription of *B. subtilis rnpB* is either initiated at the 5′-terminal G(+1) residue of the mature P RNA or ∼15 nucleotides upstream (*94*). A potential transcription terminator located between 20 and 50 nucleotides downstream of the coding sequence for the mature P RNA (94) suggests that P RNA is synthesized as a precursor that requires processing, at least at the 3'-end. The processing mechanism for P RNA proposed by Loria and Pan (113) is exceptional up to now. The authors observed that addition of the RNase P protein to precursor P RNA with extra 5'- and 3'-sequences results in efficient autolytic maturation of the 5'- and 3'-ends *in vitro*. The 5'- and 3'-processing events are independent of each other and several intermediates with heterogeneous 3'-ends are produced over the course of the reaction. Autolytic processing *in vitro* produces the precise 5'-end of the mature *B. subtilis* P RNA. However, the 3'-end of the final *in vitro* product (at nt 397) was 4 nucleotides shorter than the reported 3'-end of the native mature P RNA at nt 401 (94). Thus, the physiological role of autolytic processing is not yet clear. For *E. coli* RNase P, autolytic *in vitro* cleavage within the 5'-extension of M1 RNA has also been observed in the presence of the protein cofactor (89), in line with the general capacity of bacterial RNase P holoenzymes to cleave single-stranded RNA substrates (114).

4. Coordinated Expression of *rnpA* and *rnpB*

Studies with *E. coli* mutant strains encoding temperature-sensitive RNase P proteins (A49, ts241) demonstrated that cell growth at the nonpermissive temperature results in a substantial reduction of M1 RNA levels compared to wild-type cells (115). This deficiency was not the result of inefficient M1 RNA processing since no larger precursors of this RNA were detected. These findings suggested a link between cellular M1 RNA levels and the assembly of functional RNase P holoenzymes. Similar observations were made with another mutant strain of *E. coli* in which the chromosomal *rnpB* expression is switched off in the absence of arabinose (15). Complementation of this strain with low copy number plasmid-encoded *rnpB* genes demonstrated that the *rnpB*C293 (G to C mutation in the tRNA 3'-CCA-binding sequence; see Section II.B.1) mutant allele is unable to substitute for the native *rnpB* gene under nonpermissive growth conditions. However, simultaneous overexpression of the *rnpA* gene encoding the RNase P protein partially restored the growth defect, consistent with the idea that the processing defects were alleviated by increases in the cellular amount of RNase P holoenzyme. Indeed, for *B. subtilis* it was recently shown that P protein overexpression results in an increase in plasmid-encoded steady-state levels of P RNA (64), and, as observed for *E. coli* (15), P protein overexpression rescued the otherwise lethal phenotype of *rnpB* mutant alleles that disrupt base pairing with tRNA 3'-CCA (64).

As mentioned above, there is evidence that *rnpB* primary transcripts can undergo either 3′-end maturation or oligoadenylation resulting in degradation (*110*). A knockout of poly(A) polymerase (*pcnB*) blocked the degradation pathway and consequently increased the level of 3′-precursor M1 RNA, but did not lead to increases in the levels of mature M1 RNA. The authors proposed that mature M1 RNA may only stably accumulate when complexed with its protein subunit. Such a mechanism may explain why RNase P protein overexpression results in increases in steady-state levels of RNase P RNA (*64*), and, by inference, in elevated RNase P holoenzyme concentrations.

F. The Versatility of RNase P: Non-tRNA Substrates

The common function of RNase P enzymes is the removal of 5′-leader sequences from precursor tRNA molecules. In *E. coli*, two physiologically important stable RNAs, tmRNA (=10Sa RNA) (*116*) and 4.5S RNA (*117, 118*), add to the number of substrates processed by RNase P. RNase P recognizes both substrates via RNA modules that mimic the acceptor stem and T-arm of tRNAs; tmRNA plays an important role in the rescue of ribosomes stalled on truncated mRNAs and the degradation of the incomplete polypeptides associated with such ribosomes (*119*), while 4.5S RNA is part of the signal recognition particle (SRP) that targets ribosomes to the plasma membrane (*120, 121*). Other non-tRNA substrates of *E. coli* RNase P include several phage-induced regulatory RNAs, such as the phi80-induced M_3 RNA (*122*), CI RNA from satellite phage P4 (*123*), and the C4 repressor RNA of bacteriophages P1/P7 (*124*). Studies demonstrating specific processing of small RNA duplex substrates (*125, 126*) and even single-stranded RNA oligonucleotides by bacterial RNase P (*114*) suggested the enzyme might have a wide spectrum of substrates in addition to tRNAs. Indeed, RNase P cleavage sites were further identified in riboswitches and intergenic regions of polycistronic operons in *E. coli*, *B. subtilis*, and *Salmonella typhimurium* (see Table II). Microarray analysis of an *E. coli* strain carrying a temperature-sensitive mutation in the protein component of RNase P revealed that RNase P affects the expression of several di- or polycistronic operons (*tna*, *secG*, *rbs*, and *his*). RNase P cleavage in the 5′-untranslated region or intergenic regions of these transcripts seems to negatively affect the abundance of the RNA fragments downstream of these cleavage sites (*127*). A similar finding was reported for the lactose operon of *E. coli*, where an RNase P cleavage site was identified between *lacY* and *lacA*; the cleavage product containing *lacA* was rapidly degraded (*128*), explaining the natural polarity between *lacZ* and *lacA* expression. RNase P cleavage sites in leader, intergenic or regulatory regions are not always associated with the classical "single-strand/helix junction" motif mimicking a tRNA acceptor stem with a single-stranded 5′-leader (*127, 129, 130*). *In vitro* enzymatic probing of several of these novel substrates confirmed to some extent the

TABLE II
Non-tRNA Substrates of Bacterial RNase P

RNA substrate	Gene function	Cleavage in vivo?	Cleavage in vitro?	% Cleavage efficiency[a]	Substrate structure	Experimental strategies to characterize RNase P cleavage	Effect of RNase P cleavage on 3'-product	Ref.
Bacteriophage Φ80: M$_3$ RNA	Immediate early gene	Unknown	Yes[a]	10[b]	Classical RNase P motif	In vitro cleavage assay Cleavage site mapping (in vitro) End-group determination In vitro deletion analysis of the substrate In vivo analysis (RNase P mutant strain)		(122)
E. coli: 4.5 S RNA	Protein secretion/targeting of cytoplasmic membrane proteins	Yes	Yes	~90	Classical RNase P motif	In vitro cleavage assay Cleavage site mapping (in vitro) End-group determination In vivo analysis (RNase P mutant strain) Kinetic analysis of substrate mutants		(117, 118)
E. coli: tmRNA	Rescue of stalled ribosomes ("trans-translation")	Yes	ND	ND	Classical RNase P motif	In vivo analysis and cleavage site mapping (RNase P mutant strain)		(116)
S. typhimurium: his operon (hisC/hisB)	Histidine biosynthesis	Yes	Yes/no[c]	ND	Classical RNase P motif	In vivo analysis[d] In vitro cleavage assay[e] In vitro translation analysis with mutated transcripts	Stabilization	(134)

Name	Function				Motif	Methods	Effect	Ref.
Bacteriophages P1 and P7: C4 antisense RNA	Establishment and maintenance of lysogeny	Yes	Yes	ND	Classical RNase P motif	*In vitro* cleavage assay; Cleavage site mapping (*in vitro* and *in vivo*); Dependence of RNase P cleavage *in vivo* on "CCA" interaction		(124)
Satellite phage P4: CI RNA	P4 immunity factor	Yes	ND	ND	Classical RNase P motif	Cleavage site mapping (*in vivo*); *In vivo* analysis (infection of RNase P mutant strain with P4); Northern blot analysis		(123)
E. coli: *tna* operon (*tnaL–tnaA*)	Tryptophanase operon: tryptophan metabolism	Yes	Yes	4	No classical RNase P motif	Microarray analysis of RNase P mutant strain; *In vitro* cleavage assay; Cleavage site mapping (*in vitro* and *in vivo*); End-group determination; Structure probing (enzymatic); Kinetic analysis; Cleavage assays with M1 RNA mutants; Northern blot analysis; *tnaA* activity assay	Downregulation	(127)
E. coli: *tna* operon (*tnaA–tnaB*)		ND	Yes	13	Classical RNase P motif	Microarray analysis of RNase P mutant strain; *In vitro* cleavage assay	Downregulation	
E. coli: *secG* operon	*secG*: protein export	ND	Yes	0.02	Classical RNase P motif	Cleavage site mapping (*in vitro*); End-group determination	Downregulation	

(*continued*)

TABLE II (continued)

RNA substrate	Gene function	Cleavage in vitro?	Cleavage in vitro?	% Cleavage efficiency[a]	Substrate structure	Experimental strategies to characterize RNase P cleavage	Effect of RNase P cleavage on 3'-product	Ref.
E. coli: rbs operon (rbsB–rbsK)	rbs operon: D-ribose transport	ND	Yes	0.52	Classical RNase P motif/no classical RNase P motif[g]	Structure probing (enzymatic) Kinetic analysis Tertiary structure analysis (cleavage assays with M1 RNA mutants)	Downregulation	
E. coli: his operon (hisL/hisG)	his operon: histidine biosynthesis	ND	Yes	0.02	No classical RNase P motif		Downregulation	
E. coli: his operon (hisC/hisB)		ND	Yes	2	No classical RNase P motif		Downregulation	
E. coli: lac operon (lacY/lacA)	lac operon: transport and metabolism of lactose	Yes	Yes	10	Classical RNase P motif	*In vitro* cleavage assay Cleavage site mapping (*in vitro* and *in vivo*) End-group determination Structure probing (enzymatic and mutational analysis) Kinetic analysis *In vivo* analysis (using RNase P mutant strain/reporter gene assay) Western blot analysis Quantitative RT-PCR	Downregulation	(128)

E. coli: coenzyme B$_{12}$ riboswitch-btuB	btuB: cobalamin outer membrane transporter	Yes	Yes	10	Classical RNase P motif[h]	In vitro cleavage assay[i] Cleavage site mapping (in vitro) End-group determination Reporter gene assay in RNase P mutant strain Kinetic analysis Cleavage site analysis (base mutations in the substrate) Northern blot analysis	Downregulation	(129)
B. subtilis: coenzyme B$_{12}$ riboswitch-ywrC(BAK) operon	ywrC: encodes putative metal-binding protein	ND	Yes	ND	No classical RNase P motif[f]	In vitro cleavage assay[i] End-group determination[j]	ND	
B. halodurans: OLE RNA	Unknown (forms complex with product of the downstream gene BH2780 in vitro)	ND	Yes	ND	Classical RNase P motif/no classical RNase P motif[k]	In vitro cleavage assay (incl. chimeric RNase P)[l] Cleavage site mapping End-group determination Reporter gene assay in RNase P- and RNase E-deficient strains Northern blot analysis	Downregulation/ no effect[m]	(130)

(continued)

TABLE II (continued)

RNA substrate	Gene function	Cleavage in vivo?	Cleavage in vitro?	% Cleavage efficiency[a]	Substrate structure	Experimental strategies to characterize RNase P cleavage	Effect of RNase P cleavage on 3′-product	Ref.
B. subtilis: adenine riboswitch-pbuE	pbuE: encodes adenine efflux pump	Yes[n]	Yes	ND	Classical RNase P motif[h]	In vitro cleavage assay[o] Cleavage site mapping (in vitro) End-group determination Kinetic analysis Reporter gene assay in RNase P mutant strain	Stabilization	(212)

The two dashed lines indicate that the experimental strategies listed in between were applied to the *E. coli tna*, *secG*, *rbs*, and *his* operon.
ND, not determined.
[a]% cleavage efficiency relative to pre-tRNA substrate (RNase P holoenzyme, multiple turnover).
[b]Using a cellular RNase P preparation.
[c]Cleavage with S30 cellular extract from *E. coli*; no cleavage with purified RNase P holoenzyme.
[d]Analysis in *S. typhimurium*, RNase P mutant *E. coli* strains (A49/ts709), and other RNase mutant *E. coli* strains.
[e]Cleavage assays performed with S30 and S100 cellular extract from *E. coli* (A49/ts709) and reconstitution of extracts with purified *E. coli* RNase P holoenzyme.
[f]Predicted structure.
[g]Two cleavage sites in the intergenic region (*rbsB–rbsK*); only one corresponds to the classical RNase P motif.
[h]Predicted structure, validated by mutational analysis.
[i]*In vitro* cleavage using *E. coli* RNase P holoenzyme in the presence and absence of coenzyme B₁₂.
[j]*In vitro* cleavage using *E. coli* RNase P holoenzyme.
[k]Predicted structure; cleavage site recognized by *E. coli* and *B. subtilis* RNase P corresponds to classical RNase P motif; cleavage site recognized by partially purified RNase P from *B. halodurans* does not correspond to classical RNase P motif.
[l]*In vitro* cleavage using *E. coli* RNase P (M1 RNA and holoenzyme) and *B. subtilis* RNase P (P RNA and holoenzyme) and partially purified RNase P from *B. halodurans*.
[m]Conflicting results from reporter gene assay (β-galactosidase) in *E. coli* strains (A49/wild-type) versus Northern blot analysis in *B. subtilis*.
[n]Two cleavages sites.
[o]*In vitro* cleavage assay with different substrate variants, in the presence and absence of adenine.

computer-predicted secondary structures, which lack any obvious similarity to canonical RNase P substrates (127). Yet, their authentic structures in the context of long cellular primary transcripts and the structural elements recognized by RNase P are hard to assess. Riboswitches undergo structural changes in response to metabolite binding and the transient structures captured by RNase P are even more difficult to predict. Selective cleavage of the ligand-free or ligand-bound riboswitch conformer by RNase P might introduce an intriguing regulatory layer to control the expression of RNA portions downstream of the RNase P cleavage site. However, in the documented cases (the coenzyme B_{12} riboswitches of *E. coli* and *B. subtilis*, and the *pbuE* adenine riboswitch of *B. subtilis*; see Table II), a dependence on the presence of the ligand was not observed.

Theoretically, RNase P cleavage within polycistronic mRNAs could either up- or downregulate the level of the downstream cleavage products. In the case of canonical substrates, RNase P cleavage occurs immediately 5' to a helical structure, releasing the downstream cleavage product with a 5'-terminal stem-loop structure. The degradation of RNAs is usually initiated by endonucleolytic RNase E cleavage (131). Since RNase E cleaves RNAs with unpaired nucleotides at the 5'-end more effectively than those with 5'-terminal base pairing (132, 133), removal of a 5'-single-stranded sequence by RNase P may protect the downstream cleavage fragment of an mRNA from RNase E-induced degradation. The precedence for this has been found in the polycistronic *his* transcript of *S. typhimurium*. Here, a fragment that is released by RNase E cleavage downstream of the *hisC* cistron is stabilized by subsequent RNase P-catalyzed removal of single-stranded sequences from its 5'-end (134). However, stabilization owing to RNase P cleavage is certainly not a general rule, as RNase P cleavage between the *lacY* and *lacA* cistrons in *E. coli* also leaves a downstream fragment with a 5'-terminal hairpin, which is yet rapidly degraded (128). RNase E has a preference for 5'-monophosphorylated RNA fragments, and RNase P cleavage generates such 5'-ends and may thus trigger RNase E-mediated degradation of transcript portions 3' to the cleavage site. Thus, depending on the context, RNase P cleavage can have stabilizing or destabilizing effects.

What is the overall physiological relevance of RNase P cleavages in mRNA transcripts compared with 5'-maturation of tRNA, 4.5S RNA, and tmRNA, taking into account that RNase P tends to cleave its mRNA substrates at least one order of magnitude less efficiently than its tRNA substrates (127, 128) (Table II)? Some of these RNase P cleavages in mRNA transcripts might simply represent "off-target" effects, which are tolerated because they do not have severe consequences for the cell. However, in other cases where RNase P cleavages proved beneficial to the cell, these processing events may have been integrated into mechanisms that fine-tune gene expression. One such example may be the polar expression of the *lac* operon discussed above (*128*). There is also evidence for a focus of *E. coli* RNase P function in phosphoribosyl pyrophosphate (PRPP) metabolism and protein secretion, where the enzyme processes several RNA targets (*127*).

To shed more light on the question how important these "side jobs" of RNase P are compared to 5'-pre-tRNA processing, it is instructive to look at *in vivo* complementation studies. RNase P RNAs and proteins of *E. coli* and *B. subtilis* can replace each other *in vivo*, and even a single copy of the *E. coli rnpB* gene inserted into the *B. subtilis* chromosome supports growth of *B. subtilis* under conditions where expression of its native *rnpB* gene is suppressed (*15, 16*). Likewise, a phylogenetically broad spectrum of bacterial *rnpA* genes can replace the *B. subtilis rnpA* gene *in vivo* (*50*). These findings indicate that bacterial RNase P components are widely interchangeable with respect to their central function in tRNA metabolism. However, this may not necessarily pertain to the processing of non-tRNA substrates that are usually bound with lower affinity by the enzyme. A documented example is the noncoding OLE RNA from *Bacillus halodurans*, for which cleavage site selection *in vitro* varied with the type of RNase P protein (*E. coli* or *B. subtilis*) (*130*). This observation could suggest that heterologous or hybrid enzymes, although able to replace the homologous RNase P enzyme and support bacterial growth, may not be able to cleave a subset of non-tRNA substrates or at least not at their authentic sites in the foreign host bacterium. Such cleavage events might then be identifiable as of lower importance for cell survival.

A recurrent theme in RNase P research has been the observation that the bacterial RNase P holoenzyme acts much more efficiently on non-tRNA substrates than the RNA subunit alone. Thus, the protein subunit, in addition to increasing the affinity for 5'-pre-tRNA (*31*), enables the enzyme to act on a broader spectrum of substrate structures (*118, 127, 128, 135*). One may speculate that the increased protein content in RNase P from Archaea and Eukarya could have further expanded the number of substrates recognized or the number of other factors interacting with RNase P. Evidence for such an increased functional complexity comes from the recent finding that human nuclear RNase P associates with the chromatin of tRNA and 5S rRNA genes in proliferating cells and is required for the transcription of those genes by RNA polymerase III (*136*). The recruitment of human RNase P to chromatin and its integral role as an RNA polymerase III transcription factor may be mediated by the protein subunits since all RNase P proteins tested so far are associated with the chromatin of these RNA genes. It will be interesting to see if also archaeal RNase P has functions beyond tRNA processing.

G. RNase P – Association with Cellular Components and Oligomeric State

There is evidence for specialized reaction areas within bacterial cells, such as certain membrane areas, the nucleoid, multienzyme complexes, storage granules, and cytoskeletal elements (137, 138). Estimates on the RNase P content in bacterial cells such as *B. subtilis* range from 20 to 200 copies per cell (94, 96). An association of RNase P with certain cellular components or structures could be a mechanism to locally concentrate the ribonucleoprotein for the sake of increased enzymatic efficacy.

Indeed, it has been reported that *B. subtilis* RNase P specifically binds to the 30S ribosomal subunit, with evidence that a dimeric form (see below) of RNase P prevails in such complexes (96). The physiological role of RNase P association with 30S subunits is not yet clear. Interestingly, the presence of 30S subunits impaired RNase P activity with a non-tRNA substrate, but not with a pre-tRNA substrate. Association with 30S subunits may thus put a leash on RNase P to prevent "off-target" cleavage of non-tRNA substrates (96). Another possibility is that 30S association may guide RNase P to ribosome-bound mRNA substrates that require cleavage before translation can occur. Purification of *E. coli* RNase P by salt washing of ribosome fractions also suggested an association with ribosomes in this bacterium (67). In addition, the RNase P protein subunit, when overexpressed in *E. coli*, binds to ribosomes and is quantitatively recovered in the pellet of a 100.000g centrifugation (139).

Apirion and coworkers (140) further provided evidence that RNase P and RNase E activities sediment as a complex and accumulate immediately above the ribosome pellet upon ultracentrifugation of cell extracts prepared from *E. coli*. Moreover, both RNases were reported to cosediment in a sucrose density gradient of cell extracts. In a later study, endoribonucleases RNase III, RNase E, and RNase P were found to fractionate primarily with the cell membrane (141). The cosedimentation as particles larger than their known individual sizes and their association with membrane fractions implied a closer interaction between RNase P and RNase E. This finding may also point to a connection between RNase P and the RNA degradosome in *E. coli*, a multi-protein complex composed of RNase E, PNPase, RhlB, and enolase (131); see also chapter by A. J. Carpousis and colleagues in this volume.

A link between RNase P and the ribosome may also exist in Archaea. It was recently shown for the hyperthermophilic archaeon *Pyrococcus horikoshii* that the putative ribosomal protein L7Ae forms a specific complex with RNase P RNA and elevates the temperature for optimum activity of the RNase P holoenzyme (142). This raises the possibility that L7Ae represents a fifth protein subunit of archaeal RNase P in some hyperthermophilic archaea.

B. subtilis RNase P was shown to occur *in vitro* in both a monomeric (one RNA and one protein subunit) and a dimeric form (two RNA and two protein subunits) (*143*). In the absence of substrate, the dimer–monomer equilibrium is sensitive to monovalent ions and the total holoenzyme concentration. Formation of the dimer is strongly favored at lower monovalent salt (0.1 M NH_4Cl), whereas the monomer prevails at 0.8 M NH_4Cl (*144*). Enzyme–substrate complexes with monomeric pre-tRNA always contain the RNase P monomer. However, in the presence of a tandem tRNA precursor and low monovalent salt (0.1 M NH_4Cl), the dimeric enzyme form is favored (*144*). Since *B. subtilis* lacks RNase E, which is essential for tRNA processing in *E. coli* (*145*), and because the majority of its tRNAs are organized in clustered tRNA operons, the dimeric RNase P holoenzyme may facilitate the processing of multicistronic precursor tRNA transcripts. In contrast to *B. subtilis* RNase P, only weak dimer formation and a propensity to form aggregates has been observed for the *E. coli* holoenzyme (*33, 143*).

H. Life with RNase P

The example of *N. equitans* (see Section II.C) illustrates that tRNA 5′-end maturation is not an absolutely required functional step in tRNA biogenesis. To achieve this state, the protein synthesis machinery of *N. equitans* apparently had to undergo only subtle evolutionary adaptations, such as changes in the mode of recognition by aminoacyl-tRNA synthetases in the cases of $tRNA^{Tyr}$ and initiator $tRNA^{Met}$. The presence of substantial levels of 5′-triphosphorylated tRNAs in *N. equitans* further suggests that such tRNAs participate in protein synthesis. Likewise, *E. coli* ribosomes were shown to accept tRNAs that carry a 5′-triphosphate instead of the canonical 5′-monophosphate terminus in a coupled *in vitro* transcription–translation system (*146*). Although *N. equitans* provides the "proof of principle" that an organism is capable of thriving without RNase P, it is the result of a very specialized evolution, not representative of the vast majority of living organisms. The genome of *N. equitans* is extremely condensed and major metabolic functions are provided by the archaeal host of *N. equitans* (see Section II.C). Several reasons come to mind to explain why RNase P has been retained in the bulk of organisms throughout evolution. The enzyme permits a more flexible tRNA promoter placement, thus generally favoring genome rearrangements and contributing to their plasticity and evolution. RNase P has certainly favored tRNA gene duplications, since the presence of the enzyme made it possible to release functional tRNA moieties from bi- or multicistronic tRNA transcripts derived from loci that underwent tRNA gene duplication events. Also, archaeal viruses and conjugative plasmids encoding certain integrase proteins were found to use primarily tRNA genes as integration sites (*147*). Considering that the resulting genomic rearrangement could change the distance between the tRNA gene and its promoter, such events would jeopardize the synthesis of functional tRNA if there were no RNase P activity in the cell. Thus, the structure and variability of present-day genomes, as well as the pace of their evolution, would have been inconceivable without RNase P.

III. Removal of the tRNA 3′-Trailer

For some stable RNAs, such as ribosomal RNA, minor 3′-end heterogeneities are compatible with RNA function (148). In contrast, maturation of the tRNA 3′-end has to occur with uncompromised accuracy: for CCA-containing precursors, all nucleotides downstream of CCA have to be removed, while 3′-processing of CCA-less tRNAs has to take place immediately 3′ to the discriminator nucleotide, to permit addition of the CCA triplet by tRNA nucleotidyltransferase as a prerequisite for tRNA aminoacylation.

While the mechanism of maturation of the tRNA 5′-end is conserved throughout all domains of life, tRNA 3′-processing differs from domain to domain and even from organism to organism. Two main processing modes have been described: (1) one-step maturation by direct endonucleolytic cleavage at the tRNA 3′-end (pathway 1; Fig. 6A) and (2) multistep maturation involving endo- and exonucleases (pathway 2; Fig. 6B). The one-step endonucleolytic pathway 1 is always catalyzed by the enzyme tRNase Z (EC 3.1.26.11),

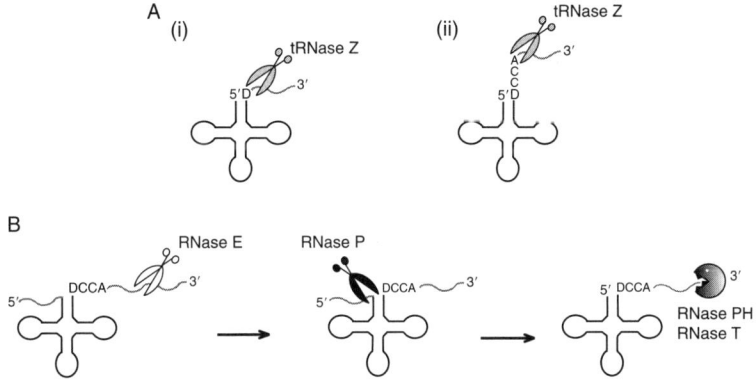

Fig. 6. tRNA 3′-processing pathways in prokaryotes. Two main pathways have been observed for tRNA 3′-processing in prokaryotes. (A) (i) Pathway 1 is a single-step reaction involving an endonucleolytic cut by tRNase Z 3′ to the discriminator nucleotide (D). (ii) The tRNase Z from *Thermotoga maritima* is an exception, since the enzyme can also cleave 3′ to the CCA sequence. (B) Pathway 2 is a multistep reaction involving an endonucleolytic cut downstream of the tRNA 3′-end and a subsequent nucleolytic digestion by several exonucleases after the removal of the 5′-leader sequence by RNase P. In *E. coli*, the main endonuclease cleaving the 3′-trailer is RNase E and the most important exonucleases are RNase PH and RNase T.

whereas players in the multistep pathway 2 include several exonucleases which can be redundant. Some organisms use solely the endonucleolytic pathway 1, others exclusively the multistep pathway 2. Yet other organisms normally employ pathway 1 but are able to utilize pathway 2 as a backup if pathway 1 is not functional, and vice versa. If either pathway is the only available mode of tRNA 3′-processing, then this pathway becomes essential (149, 150).

One criterion for pathway preference seems to be the presence of the terminal CCA in the tRNA primary transcript. Precursor tRNAs which contain the terminal CCA are more often matured by pathway 2. Another criterion is the length and structure of the tRNA 3′-trailer, which tends to be extended and to encompass structured regions if the tRNA moiety is embedded in multicistronic primary transcripts (see Section I). In case of single tRNA gene transcription units in Archaea, tRNA 3′-trailers are usually rather short and may be easily removed by exonucleases. In bacteria, single tRNA genes are transcribed as pre-tRNAs containing 3′-stem-loop structures, which likely necessitate endonucleolytic cleavage 5′ to the stem-loop. In case of multicistronic transcripts, endonucleolytic processing downstream of the tRNA is necessary. This cleavage may occur either directly 3′ to the discriminator (pathway 1) or further downstream (pathway 2).

A. The Endonucleolytic Removal of tRNA 3′-Trailers is Catalyzed by tRNase Z

Detection of endonucleolytic processing of the 3′-trailer of tRNAs was first described in 1979 (151, 152). However, the protein and gene sequences of the corresponding endonuclease were not identified until 2002, marking a breakthrough in the investigation of tRNA 3′-end maturation (153). Based on the sequence of the founding enzyme, wheat tRNase Z, numerous homologs from all three domains of life (Bacteria, Archaea, and Eukarya) were identified (153, 154). Synonyms for tRNase Z include RNase Z, 3′-tRNase, 3′-pre-tRNase, ElaC, ZiPD, and RNase BN. tRNase Z enzymes exist in two versions: a short form (280–360 amino acids in length) present in all three domains of life, and a long version (750–930 amino acids), so far found only in eukaryotes. All short tRNase Z enzymes analyzed to date form homodimers (155–159).

Here, we will concentrate on the short tRNase Z enzymes of Bacteria and Archaea. The current state of knowledge on eukaryotic tRNase Z has been reviewed recently (154, 160, 161).

1. tRNase Z Is a Member of the Metallo-β-Lactamase Family

The metallo-β-lactamase (MBL) family is characterized by the β-lactamase fold, consisting of an α–β/β–α structure (162). Members of the MBL family process a wide variety of substrates, most of which possess an ester linkage and a negative charge. They include, among others, class B β-lactamases (substrates: β-lactams), glyoxalase II (substrates: S-D-lactoylglutathiones), cAMP phosphodiesterases (substrate: cAMP), and the β-CASP proteins (derived from metallo-β-lactamase, CPSF, Artemis, SNM1, Pso2) acting on nucleic acids (162–164).

CPSF and tRNase Z Have Similar Features. Three members of the β-CASP subgroup of the MBLs are ribonucleases: RNase J1, Int11, and CPSF-73 (163). RNase J1 is involved in *thrS* mRNA leader and 16S rRNA processing (165–167), Int11 is a subunit of the integrator complex, which is involved in snoRNA 3'-end processing (168), and CPSF-73 is a subunit of the human cleavage and polyadenylation specificity factor (CPSF) (169, 170). CPSF-73 was only recently shown to be the endonuclease responsible for the cleavage in the 3'-UTR of mRNAs that precedes polyadenylation in eukaryotes (171). The structure of human CPSF-73 has been solved, confirming the existence of the β-lactamase domain (amino acids 1–208 and amino acids 395–460) and the β-CASP domain (171). Both CPSF-73 and tRNase Z have a high affinity for Zn^{2+} ions and show the same metal ion-binding mode which differs from canonical metallo-β-lactamases (171). Although tRNase Z does not contain the complete β-CASP domain it contains the characteristic β-CASP motifs A to C. Because of the functional and partial structural similarities of CPSF-73-like proteins and tRNase Z enzymes, one could consider placing tRNase Z into the β-CASP subgroup of MBL proteins.

The Exosite Is a Unique Structural Element. In addition to the metallo-β-lactamase domain, tRNase Z enzymes possess a structural element called the *exosite* (Fig. 7) (172). This feature is unique among the MBL family members and thus a tRNase Z-specific element. The exosite of tRNase Z appears as a flexible arm which protrudes from the main protein body. It takes part in the binding of pre-tRNA, but its absence has little effect on catalytic activity with the small substrate bis(*p*-nitrophenyl) phosphate (bpNPP) (Fig. 8C) (172). There are two different types of exosites: the glycine/proline-rich ZiPD exosite (*E. coli, B. subtilis, M. jannaschii* tRNase Z) and the TM-type exosite (*Thermotoga maritima* tRNase Z) which contains a cluster of 4–5 basic amino acid residues instead of the glycine/proline motif (172) (Fig. 7).

2. CRYSTAL STRUCTURES OF SHORT tRNASE Z ENZYMES

The X-ray structures of three bacterial tRNase Z enzymes have been solved to date, those from *B. subtilis* (BsuTrz (155)), *E. coli* (EcoTrz (173)), and *T. maritima* (TmaTrz (156)). All three enzymes share a similar overall geometry with equal numbers of α-helices and β-sheets. They form homodimers and the monomeric subunits are arranged in head-to-head fashion (173). The

```
HvoTrz    (139) HRTA--SVGYALVEDDRPGRFDREKAEELGVPVGPAFGRLHAGEDVELEDGTVVRSEQVV
PfuTrz    (138) HGIP--ALGYVFKEKDRRGNFDLEKIKSLGLTPGPWMKELEKRKIIKIGE-RVIRLSEVT
MjaTrz    (143) HGIP--SYAYIFKEIKKP-RLDIEKAKKLGVKIGPDLKKLKNGEAVKNIYGEIIKPEYVL
BsuTrz    (140) HGVE--AFGYRVQEKDVPGSLKADVLKEMNIPPGPVYQKIKKGETVTLEDGRIINGNDFL
EcoTrz    (141) HPLE--CYGYRIEEHDKPGALNAQALKAAGVPPGPLFQELKAGKTITLEDGRQINGADYL
TmaTrz    (134) HVSSEVSFGYHIFEVRR--KL---KKEFQGLD-SKEISRL------VKEKGR-----DFV
AthTrzS1  (133) HVIQ--SQGYVVYSTKY--KL---KKEYIGLS-GNEIKNLKVSG-VEI--------TDSI
CPSF-73   (158) HVLG--AAMFMI--|------------------------------------------EIA

HvoTrz    (197) GDPRPGRTVVYTGD
PfuTrz    (195) GPKKRGAKVVYTGD
MjaTrz    (200) LPPKKGFCLAYSGD
BsuTrz    (198) EPPKKGRSVVFSGD
EcoTrz    (199) AAPVPGKALAIFGD
TmaTrz    (177) TEEYHKKVLTISGD
AthTrzS1  (176) ITP-|---EVAFTGD
CPSF-73   (171) ----|-GVKLLYTGD
```

FIG. 7. Alignment of tRNase Z exosites and CPSF-73. tRNase Z sequences beginning with motif 3 (H) and ending with motif 4 (D) were aligned with the corresponding sequence of human CPSF-73; the box marks the tRNase Z exosite. The bacterial tRNase Z enzymes of *B. subtilis* (BsuTrz) and *E. coli* (EcoTrz) and the archaeal tRNase Z proteins of *H. volcanii* (HvoTrz), *P. furiosus* (PfuTrz), and *M. jannaschii* (MjaTrz) share a ZiPD exosite (G/P-rich motif, underlined and highlighted in bold). The Tma exosite (basic residues shown in bold) which is present in *T. maritima* Trz and the eukaryotic *A. thaliana* Trz[S1] is shown for comparison (see also (154)). The exosite is missing in CPSF-73.

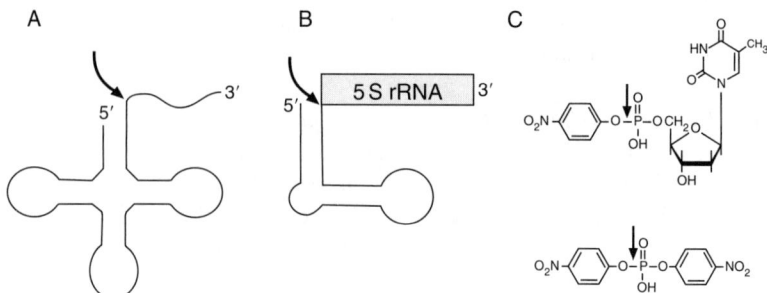

FIG. 8. tRNase Z substrates. (A) The key substrates of tRNase Z are pre-tRNAs with their conserved overall fold. (B) *H. volcanii* pre-5S rRNA contains a tRNA-like structure in the 5′-leader sequence and is processed by tRNase Z (149). (C) The small chromogenic phosphodiester model substrates thymidine-5′-*p*-nitrophenyl phosphate (TpNPP) and bis(*p*-nitrophenyl) phosphate (bpNPP) are processed efficiently by *E. coli* tRNase Z and less efficiently by other Trz proteins. The tRNase Z cleavage sites are indicated by arrows.

monomers concertedly build an active site cleft which is capable of accommodating single-stranded RNA (Fig. 9). The exosite of EcoTrz and BsuTrz is similar in length and structure, but a superposition of both tRNase Z structures shows a 20° difference in the angle between the exosites and the

tRNA PROCESSING

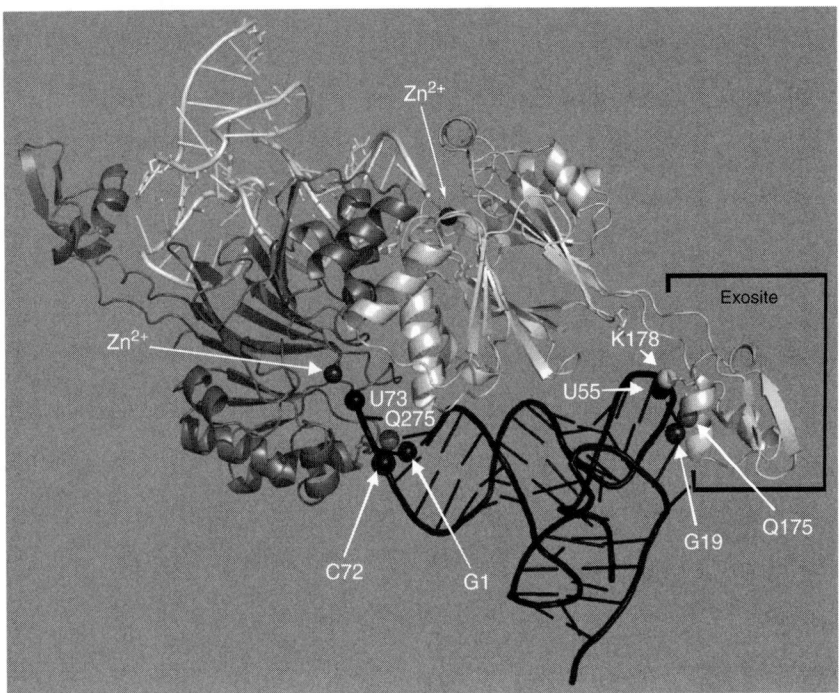

FIG. 9. Crystal structure of the *B. subtilis* tRNase Z-tRNA complex. The BsuTrz dimer complexed with two molecules tRNA^{Thr} (molecule 1: black, lower right; molecule 2: white, upper left), generated by crystallographic symmetry (174); protein monomers are shown as ribbon presentations in light and dark gray, respectively. Key functionalities of the tRNA-tRNase Z interaction are illustrated as spheres for the tRNA molecule in black; contacts to the BsuTrz subunit that provides the active site: the 3′-oxygen of U73, the phosphorous atom of C72, the N7 of G1 in hydrogen bonding distance to the amide nitrogen of Q275; contacts to the exosite of the second protein subunit: the 3′-oxygen of U55 interacting with the ε-amino group of K178 in the exosite, and the O6 of G19 accepting an H-bond from the backbone imino nitrogen of Q175. The two Zn^{2+} ions are shown as well. See also colored online version of the figure.

corresponding protein cores. This difference clearly illustrates the high degree of flexibility of the exosite, which seems to be a necessary property for substrate binding (*173*). Comparison of the catalytic sites of the three proteins reveals major differences, particularly with respect to the number and positions of metal ions, but these likely reflect the crystallographic conditions rather than functional differences between the proteins. In the crystal structure of the BsuTrz homodimer, the catalytic core of monomer A contains two Zn^{2+} ions, whereas a distortion of one metal-binding site inhibits Zn^{2+} binding to both metal-binding sites in monomer B (*155, 173*). TmaTrz contains one Zn^{2+} ion per monomer (*156*). Among the three enzymes, only the structure of EcoTrz showed two fully loaded catalytic sites, which may have resulted from Zn^{2+} addition during crystallization (173). To date, crystal structures for archaeal and eukaryotic tRNase Z enzymes are not available.

Structure of tRNase Z Bound to tRNA. The crystal structure of BsuTrz in complex with a $tRNA^{Thr}$ revealed a monomer of tRNase Z complexed to 52 nt of the tRNA in the asymmetric unit (174). A functional dimer was created by crystallographic symmetry that suggests that both subunits of the dimer can bind tRNA molecules simultaneously. No electron density was detectable for the anticodon arm and the 3'-extension, but all nucleotides of the T-arm and acceptor stem were visible in the protein–tRNA complex. Former data indicating that T-arm and acceptor stem are involved in tRNA recognition are confirmed by the observation that the only contacts between the tRNase Z protein and the tRNA are with nucleotides from the T-arm and acceptor stem (Fig. 9) (175, 176). The T-arm and acceptor stem are clamped between the flexible exosite of one subunit and the helix $\alpha 7$ of the other and "bear a distinct resemblance to a ski boot in its bindings" (177). The recognition of tRNA substrates occurs mainly via their sugar–phosphate backbone, which is expected in view of the wide variety of pre-tRNA substrates with different primary sequences, but conserved tertiary structures. Two positively charged motifs in tRNase Z play a central role in tRNA binding, one located near the active site and the second in the exosite. The G1–C72 pair and the discriminator nucleotide U73 represent the major contact sites with the adjoining monomeric subunit that contains the corresponding catalytic site (174). Two base-specific contacts with the highly conserved guanosine residues at positions 1 and 19 were also identified (Fig. 9) (174).

Inhibitory Effects on the tRNase Z Activity by 5'-Extensions and the CCA Motif. Some tRNase Z enzymes are inhibited by the presence of the CCA motif (18, 178), while others are not (179, 180). The presence of long 5'-extensions also inhibits (158) or slows down (158, 181, 182) tRNase Z activity, with the exception of the enzyme from the hyperthermophilic archaeon *P. furiosus*, PfuTrz (158). Crystallographic data of BsuTrz help to explain the inhibitory effects of these elements. Condon and colleagues (155) compared the structure of BsuTrz, which cannot cleave substrates carrying a 3'-CCA, with TmaTrz, which cleaves 3' to the CCA motif. They proposed that the loop between $\beta 1$ and $\beta 2$ in *B. subtilis* tRNase Z could be responsible for the inhibitory effect of CCA-containing tRNA precursors since *T. maritima* tRNase Z lacks this loop. In BsuTrz and in EcoTrz, this loop covers the active site which results in a potential "collision" of the loop with the amino group of C_{74}.

There is a narrow channel just wide enough for single-stranded RNA which leads from the catalytic center to the exterior of the protein. This channel, which is proposed to be the exit path for the 3'-extension of pre-tRNAs, may explain why 5'-extensions can inhibit tRNase Z enzymes; the longer the 5'-extension of the tRNA, the higher is the probability that 5'-leader and 3'-trailer sequences base pair and thus prevent threading of the single-stranded 3'-trailer into the channel (155).

3. tRNase Z Substrates

Precursor tRNAs are the key substrates of the tRNase Z enzyme (Fig. 8A). All recombinant tRNase Z enzymes investigated to date show tRNA 3'-processing activity in vitro (18, 150, 153, 158, 172, 178, 180, 183–186). Generally, tRNase Z enzymes cleave the pre-tRNA molecule 3' to the discriminator leaving a 3'-hydroxyl group on the 3'-end of the tRNA and a 5'-phosphate on the released trailer. In addition, E. coli tRNase Z efficiently cleaves the small chromogenic substrates bpNPP and thymidine-5'-p-nitrophenyl phosphate (TpNPP) (Fig. 8C) (157). Although these substrates are nonphysiological, they are good model substrates for investigation of the phosphodiesterase activity. Other short tRNase Z enzymes are also able to cleave bpNPP albeit with reduced efficiency (150, 158, 160, 187). Recently, the involvement of tRNase Z in pre-5S rRNA cleavage in *Haloferax volcanii* was reported (149). Analysis of the sequence upstream of the 5S rRNA revealed the presence of a mini-tRNA structure (Fig. 8B); the 5S rRNA moiety seems to mimic a pre-tRNA trailer, explaining why tRNase Z can process this substrate (149). This particular 3'-trailer (the 5S rRNA) folds into a complex structure which includes base pairing between the 5'- and 3'-termini of the 5S rRNA portion only a few nucleotides downstream of the mini-tRNA structure. It would be interesting to see whether HvoTrz forms a wider channel than BsuTrz at the catalytic site to accommodate the 5S rRNA double strand. The mini-tRNA structure was also processed by tRNase Z enzymes from other organisms (149). An involvement of the E. coli tRNase Z in mRNA processing was recently shown by Kushner and colleagues (178). In this case, canonical tRNase Z recognition elements could not be identified in the RNA substrate.

4. Bacterial tRNase Z Enzymes

The first bacterial tRNase Z enzyme for which tRNA 3'-processing activity was demonstrated was BsuTrz (18). BsuTrz is required for maturation of pre-tRNAs lacking a CCA motif and is thus essential for cell viability. Downregulation of BsuTrz expression leads to an accumulation of CCA-less pre-tRNAs in vivo (18). A CCA triplet 3' to the discriminator base inhibits the BsuTrz activity in vitro, the most inhibiting element being the first C-residue of CCA (18). As mentioned previously, the CCA motif is often encoded in prokaryotic tRNA genes, but the proportion of tRNAs with and without CCA triplet varies between organisms (Table I). Since about two thirds of the *B. subtilis* tRNA genes encode CCA, their 3′-maturation was predicted to be achieved via an alternative pathway. The nature of this alternative pathway in *B. subtilis* was just recently revealed (see Section III.B).

In *Synechocystis*, the CCA triplet is generally not encoded in the tRNA genes, and the recombinant SspTrz cleaves pre-tRNAs 3′ to the discriminator. However, pre-tRNAs with 3′-trailers harboring partial CCA motifs (C and CC) are cleaved *in vitro* by SspTrz 3′ to C and CC, respectively, whereas CCA-containing pre-tRNAs are not cleaved (*150*). Interestingly, although the CCA-containing pre-tRNAs are not cleaved by SspTrz they are bound by the enzyme. The properties of the *Synechocystis* tRNase Z may be an adaptation to the mode of CCA addition in this organism. *Synechocystis* has evolved two separate nucleotidyltransferases for addition of CCA, one adding both C-residues and the second adding the terminal A (*150*).

T. maritima is an organism with only a single CCA-less pre-tRNA, whereas all other tRNA genes encode the CCA motif. Nashimoto and colleagues (*185*) showed that TmaTrz cleaves CCA-containing pre-tRNAs immediately 3′ to the CCA motif and the single pre-tRNA without CCA 3′ to the discriminator base. Thus, TmaTrz uses this motif as a guide sequence to redirect the cleavage site, rather than being inhibited by the CCA triplet.

E. coli tRNase Z, also known as Elac, ZiPD (*157, 188*), or RNase BN (*189*), was initially identified as a zinc-dependent phosphodiesterase hydrolyzing the chromogenic substrate bpNPP and therefore called ZiPD (*157*). In contrast to its *Bacillus* counterpart, EcoTrz is not essential for cell viability (*188*). Later, the tRNA 3′-processing activity of EcoTrz was demonstrated *in vitro* (*172, 178, 185*) and *in vivo* (*178*). Schilling *et al.* (*172*) showed that EcoTrz needs its exosite for binding of pre-tRNAs but not for hydrolysis of bpNPP. Since all *E. coli* tRNA genes encode CCA, EcoTrz does not play a major role in tRNA maturation in this organism (see Section II.B). Recent results indicate a role for EcoTrz in mRNA decay (*178*). EcoTrz cleaves the *rpsT* mRNA encoding ribosomal protein S20 at locations distinct from those cleaved by RNase E, suggesting a role for tRNase Z in RNA metabolism, which may be even more pronounced in organisms lacking RNase E (*178*). Another possibility could be that EcoTrz works as a backup 3′-maturation enzyme for pre-tRNAs with incorrect nucleotides incorporated into the CCA triplet (*173*). Such a function could be useful under conditions of high mutation rates or high tRNA turnover (*173*). *E. coli* RNase BN, which was first described as having such a kind of 3′-maturation backup function in 1975 (*190*), has recently been shown to be encoded by the tRNase Z gene (*189*).

5. ARCHAEAL tRNASE Z PROTEINS

Early work on tRNA 3′-processing in Archaea, using crude protein extracts (*181*), showed that tRNA 3′-processing in this domain of life occurs via an endonucleolytic pathway. The tRNase Z gene from *M. jannaschii* was one of the first to be identified (153), and homologs are present in all archaeal genomes sequenced to date, including the recently sequenced korarchaeotal genome of *Candidatus Korarchaeum cryptofilum* (191). In vitro tRNA 3′-processing activity has been shown for tRNase Z proteins from *P. aerophilum* (185), *Thermoplasma acidophilum* (185), *H. volcanii* (HvoTrz) (149, 158), and *P. furiosus* (PfuTrz) (158). HvoTrz is an essential enzyme, since a knockout of this gene is lethal (149). As mentioned above, HvoTrz was identified as the 5′-end maturation enzyme of *H. volcanii* 5S rRNA. HvoTrz is only active at low salt concentrations *in vitro*, which is unexpected in view of the high intracellular KCl concentrations (2–4 M) in this organism. Electron microscopy excluded the presence of intracellular compartments that might provide areas of lower salt concentrations (149). The thermophilic tRNase Z PfuTrz has a broader substrate spectrum than HvoTrz. PfuTrz cleaves intron-containing tRNA precursors, 5′-extended pre-tRNAs and bpNPP and the glyoxalase II substrate, S-D-lactoylglutathione (SLG) (158). PfuTrz is the only tRNase Z known so far which has such a broad substrate spectrum and which is able to process a substrate of an enzyme that belongs to a different subclass of the MBL family. All MBL enzymes share a common ancient structural motif, which has provided the basis for grouping these different enzymatic activities as part of the MBL family. The finding that PfuTrz is able to cleave substrates of different MBL family members suggests that this enzyme might be closely related to an ancestral MBL enzyme which had a broader substrate spectrum than most of the MBL enzymes of today.

B. Concerted Processing by Endo- and Exonucleases

1. tRNA 3′ END MATURATION IN *E. coli*

E. coli and *B. subtilis* are the best studied model systems for tRNA 3′-end maturation by pathway 2 (Fig. 1). Analysis of tRNA 3′-end maturation in *E. coli* began 30 years ago (10, 192, 193) and is the most extensively studied. In *E. coli*, where all tRNA genes encode the CCA sequence (194), 3′-end maturation begins for most pre-tRNAs (at least for 59 out of 86 (195)) with an endonucleolytic cleavage by RNase E several nucleotides downstream of the tRNA, followed by 5′-end processing catalyzed by RNase P (195, 196). The remaining 3′-trailer nucleotides are subsequently removed by exonucleases (Fig. 10A). Four exoribonucleases are able to catalyze this trimming reaction: RNases II, D, T, and PH, but two of them, RNase T and RNase PH, are most effective. For long 3′-trailers it has been observed that, after the endonucleolytic cleavage by RNase E, exonucleases PNPase and RNase II first remove several nucleotides before RNase P removes the 5′-leader (Fig. 10B). Subsequently, RNases PH and T remove the remaining nucleotides (196). A survey of RNase E

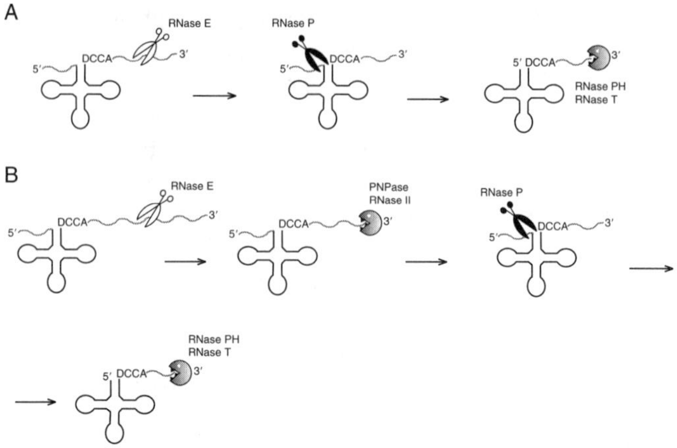

FIG. 10. Processing of long and short 3′-trailers in *E. coli*. Most 3′-trailers are first cleaved endonucleolytically by RNase E. (A) Pre-tRNAs with short 3′-trailers are subsequently processed at the 5′-end by RNase P before the remaining nucleotides are removed by RNase PH and RNase T. (B) Long trailers are first digested by exonucleases PNPase or RNase II before RNase P processes the 5′-end. Only then are the final nucleotides removed by RNase PH and RNase T.

recognition sites in pre-tRNA 3′-trailer sequences revealed an A/U-rich element in all *E. coli* tRNA precursors, except for tRNAs encoded by the *valT, valZ, valV*, and *valY* loci (197). It seems that A/U-rich elements are conserved in bacteria that possess RNase E and tRNA 3′-processing exonucleases, suggesting that RNA sequences and processing nucleases have coevolved (197). It is very likely that bacteria encoding RNase E and an *E. coli*-like set of 3′-exonucleases will utilize pathway 2 (Fig. 6B) for tRNA 3′-end maturation. Kushner and colleagues (198) showed that two tRNA clusters, *valV–valW* and *leuQ–leuP–leuV*, can be matured independently of RNase E in *E. coli*. The intergenic spacer between *valV* and *valW* consists of the tetranucleotide sequence UCCU, which is not recognized by RNase E. The dicistronic precursor is processed by RNase P and the remaining 3′-trailer is removed directly by RNases PH and T. The intergenic spacers in the *leuQ–leuP–leuV* cluster contain potential RNase E motifs further downstream but not in the immediate vicinity of the CCA triplets. In this case also, RNase P cleavage releases the tRNAs and the remaining 3′-trailers are removed directly by exonucleases (198).

E. coli strains with simultaneous deletion of genes for RNases II, Z, D, T, and PH are not viable, but cells survive when only one of these genes is active. The substrate requirements of the tRNA 3′-processing exonucleases can be figuratively described as a healthy appetite for single-stranded unstructured RNA and a heartily dislike for CCA.

2. tRNA 3'-END MATURATION IN B. SUBTILIS

In *B. subtilis*, the 27 CCA-less tRNA precursors are matured at the 3'-end by tRNase Z (18), and pathway 2 is predicted to be utilized for the CCA-containing pre-tRNAs (Fig. 6). However, an RNase E-like endonuclease to initiate 3'-end maturation as in *E. coli* has not been found in *B. subtilis*. For the exonucleolytic removal of the 3'-trailer of the CCA-containing pre-tRNAs, several candidates exist in *B. subtilis*: RNases PH, R, YhaM, and PNPase. Condon and colleagues (199) showed RNase PH to be the main exonuclease in this process, effectively removing the 3'-trailer up to one nucleotide downstream of the CCA. Deletion of the RNase PH gene results in accumulation of tRNAs carrying a few extra nucleotides at their 3'-ends. This is in contrast to *E. coli* where deletion of the RNase PH gene affected only one out of 11 tRNAs analyzed (200). Removal of the last nucleotide by *B. subtilis* RNase PH is slow, suggesting that this reaction is catalyzed by another enzyme. This is reminiscent of the situation in *E. coli* where the last nucleotide has to be removed by RNase T, which preferentially removes the nucleotide 3' to the CCA. However, a functional homolog of RNase T has not yet been identified in *B. subtilis*. In the absence of RNase PH, maturation of CCA-containing pre-tRNAs still occurs to some extent in *B. subtilis*, raising the possibility that one of the other exoribonucleases identified in this organism might be involved (199). However, neither single deletion of the gene coding for RNase R or YhaM, nor simultaneous deletion of all three exonucleases, RNase PH, R, and YhaM, had a substantial effect on tRNA 3'-processing. Even a strain with RNases PH, R, YhaM, and PNPase simultaneously deleted still showed a significant level of tRNA maturation. Thus, at least one tRNA 3'-end maturation activity in *B. subtilis* has remained unidentified.

C. Do Archaea Use Exclusively Pathway 1 (tRNase Z)?

In Archaea, the distribution of tRNA genes with and without an encoded CCA triplet is similar to that of bacteria (Table I). In both domains, there are organisms encoding tRNAs exclusively without CCA (e.g., *B. burgdorferi* and *Halobacterium salinarum*), or entirely with CCA (e.g., *E. coli* and *P. furiosus*), while others have a mixture of both (e.g., *B. subtilis* and *M. jannaschii*). Thus, one would expect archaea to also possess two 3'-processing pathways, an endonucleolytic and a mixed endonucleolytic/exonucleolytic pathway (Fig. 6). However, pathway 2 has not been identified in Archaea so far. Interestingly, all archaeal organisms sequenced up to date encode a tRNase Z gene enabling them to process the tRNA 3'-ends endonucleolytically. Since the tRNase Z from the bacterium *T. maritima* has been shown to process CCA-containing tRNAs immediately 3' to the CCA (185), it is possible that archaeal tRNase Z enzymes generally process both CCA-less and CCA-containing pre-tRNAs.

An interesting example is the tRNase Z from *P. aerophilum*, which contains CCA-less and CCA-containing tRNA genes, but processes exclusively CCA-containing pre-tRNAs *in vitro* (*185*). If this reflects the *in vivo* situation, then the organism requires additional (endonucleolytic or exonucleolytic) enzymatic activities or cofactors for tRNA 3'-end formation of CCA-less pre-tRNAs.

Archaea encode only one homolog of the known bacterial exonucleases, mostly RNase PH (often two copies of it), or RNase R in haloarchaea, whereas no homologs of the bacterial RNases II, D, T, and PNPase could be identified (*201*). The presence of a homolog of the exonuclease RNase PH would allow archaea to employ this exonuclease for tRNA 3'-maturation. It is yet unlikely that archaea encode only a single exonuclease; other exonucleases may remain to be discovered or proteins unrelated to known exonucleases may have acquired additional functions as RNases, as has been shown for the elongation factor aIF-5A, for example (*202*).

Acknowledgments

Financial support from the VolkswagenStiftung to A. M. and the Deutsche Forschungsgemeinschaft to A. M. and R. K. H. is greatly acknowledged. We thank Liang Tong for helpful discussion and Dagmar K. Willkomm for critical reading of the manuscript. We apologize to all colleagues who contributed to the field of tRNA processing in prokaryotes and whose papers have not been cited owing to space limitations.

References

1. Dittmar KA, Mobley EM, Radek AJ, Pan T. Exploring the regulation of tRNA distribution on the genomic scale. *J Mol Biol* 2004;**337**:31–47.
2. Goodsell DS. Inside a living cell. *Trends Biochem Sci* 1991;**16**:203–6.
3. Fournier MJ, Ozeki H. Structure and organization of the transfer ribonucleic acid genes of *Escherichia coli* K-12. *Microbiol Rev* 1985;**49**:379–97.
4. Inokuchi H, Yamao F. Structure and expression of prokaryotic tRNA genes. *In* tRNA: Structure, biosynthesis and function, (D Söll and U. RajBhandary, Eds.). ASM Press, Washington, DC, 1995; p. 17–30.
5. Smith DR, Doucette-Stamm LA, Deloughery C, Lee H, Dubois J, Aldredge T, *et al*. Complete genome sequence of *Methanobacterium thermoautotrophicum* deltaH: Functional analysis and comparative genomics. *J Bacteriol* 1997;**179**:7135–55.
6. Marck C, Grosjean H. Identification of BHB splicing motifs in intron-containing tRNAs from 18 archaea: Evolutionary implications. *RNA* 2003;**9**:1516–31.
7. Marck C, Grosjean H. tRNomics: Analysis of tRNA genes from 50 genomes of Eukarya, Archaea, and Bacteria reveals anticodon-sparing strategies and domain-specific features. *RNA* 2002;**8**:1189–232.
8. Vogel J, Hess WR. Complete 5' and 3' end maturation of group II intron-containing tRNA precursors. *RNA* 2001;**7**:285–92.

9. Randau L, Münch R, Hohn MJ, Jahn D, Söll D. *Nanoarchaeum equitans* creates functional tRNAs from separate genes for their 5'- and 3'-halves. *Nature* 2005;**433**:537–41.
10. Schedl P, Primakoff P, Roberts J. Processing of *E. coli* tRNA precursors. *Brookhaven Symp Biol* 1975;**26**:53–76.
11. Kirsebom LA, Baer MF, Altman S. Differential effects of mutations in the protein and RNA moieties of RNase P on the efficiency of suppression by various tRNA suppressors. *J Mol Biol* 1988;**204**:879–88.
12. Shiraishi H, Shimura Y. Mutations affecting two distinct functions of the RNA component of RNase P. *EMBO J* 1986;**5**:3673–9.
13. Baer MF, Wesolowski D, Altman S. Characterization *in vitro* of the defect in a temperature-sensitive mutant of the protein subunit of RNase P from *Escherichia coli*. *J Bacteriol* 1989;**171**:6862–6.
14. Waugh DS, Pace NR. Complementation of an RNase P RNA (*rnpB*) gene deletion in *Escherichia coli* by homologous genes from distantly related eubacteria. *J Bacteriol* 1990;**172**:6316–22.
15. Wegscheid B, Hartmann RK. The precursor tRNA 3'-CCA interaction with *Escherichia coli* RNase P RNA is essential for catalysis by RNase P *in vivo*. *RNA* 2006;**12**:2135–48.
16. Wegscheid B, Condon C, Hartmann RK. Type A and B RNase P RNAs are interchangeable *in vivo* despite substantial biophysical differences. *EMBO Rep* 2006;**7**:411–7.
17. Gößringer M, Kretschmer-Kazemi Far R, Hartmann RK. Analysis of RNase P protein (*rnpA*) expression in *Bacillus subtilis* utilizing strains with suppressible *rnpA* expression. *J Bacteriol* 2006;**188**:6816–23.
18. Pellegrini O, Nezzar J, Marchfelder A, Putzer H, Condon C. Endonucleolytic processing of CCA-less tRNA precursors by RNase Z in *Bacillus subtilis*. *EMBO J* 2003;**22**:4534–43.
19. Evans D, Marquez SM, Pace NR. RNase P: Interface of the RNA and protein worlds. *Trends Biochem Sci* 2006;**31**:333–41.
20. Guerrier-Takada C, Gardiner K, Marsh T, Pace N, Altman S. The RNA moiety of ribonuclease P is the catalytic subunit of the enzyme. *Cell* 1983;**35**:849–57.
21. Pannucci JA, Haas ES, Hall TA, Harris JK, Brown JW. RNase P RNAs from some Archaea are catalytically active. *Proc Natl Acad Sci USA* 1999;**96**:7803–8.
22. Kikovska E, Svärd SG, Kirsebom LA. Eukaryotic RNase P RNA mediates cleavage in the absence of protein. *Proc Natl Acad Sci USA* 2007;**104**:2062–7
23. Li D, Willkomm DK, Schön A, Hartmann RK. RNase P of the *Cyanophora paradoxa* cyanelle: A plastid ribozyme. *Biochimie* 2007;**89**:1528–38.
24. Haas ES, Banta AB, Harris JK, Pace NR, Brown JW. Structure and evolution of ribonuclease P RNA in Gram-positive bacteria. *Nucleic Acids Res* 1996;**24**:4775–82.
25. Haas ES, Brown JW. Evolutionary variation in bacterial RNase P RNAs. *Nucleic Acids Res* 1998;**26**:4093–9.
26. Loria A, Pan T. Domain structure of the ribozyme from eubacterial ribonuclease P. *RNA* 1996;**2**:551–63.
27. Biswas R, Ledman DW, Fox RO, Altman S, Gopalan V. Mapping RNA–protein interactions in ribonuclease P from *Escherichia coli* using disulfide-linked EDTA-Fe. *J Mol Biol* 2000;**296**:19–31.
28. Rox C, Feltens R, Pfeiffer T, Hartmann RK. Potential contact sites between the protein and RNA subunit in the *Bacillus subtilis* RNase P holoenzyme. *J Mol Biol* 2002;**315**:551–60.
29. Tsai HY, Masquida B, Biswas R, Westhof E, Gopalan V. Molecular modeling of the three-dimensional structure of the bacterial RNase P holoenzyme. *J Mol Biol* 2003;**325**:661–75.
30. Buck AH, Kazantsev AV, Dalby AB, Pace NR. Structural perspective on the activation of RNAse P RNA by protein. *Nat Struct Mol Biol* 2005;**12**:958–64.

31. Kurz JC, Niranjanakumari S, Fierke CA. Protein component of *Bacillus subtilis* RNase P specifically enhances the affinity for precursor-tRNA^{Asp}. *Biochemistry* 1998;**37**:2393–400.
32. Rueda D, Hsieh J, Day-Storms JJ, Fierke CA, Walter NG. The 5' leader of precursor tRNA^{Asp} bound to the *Bacillus subtilis* RNase P holoenzyme has an extended conformation. *Biochemistry* 2005;**44**:16130–9.
33. Buck AH, Dalby AB, Poole AW, Kazantsev AV, Pace NR. Protein activation of a ribozyme: The role of bacterial RNase P protein. *EMBO J* 2005;**24**:3360–8.
34. Kurz JC, Fierke CA. The affinity of magnesium binding sites in the *Bacillus subtilis* RNase P×pre-tRNA complex is enhanced by the protein subunit. *Biochemistry* 2002;**41**:9545–58.
35. Guo X, Campbell FE, Sun L, Christian EL, Anderson VE, Harris ME. RNA-dependent folding and stabilization of C5 protein during assembly of the *E. coli* RNase P holoenzyme. *J Mol Biol* 2006;**360**:190–203.
36. Chen JL, Nolan JM, Harris ME, Pace NR. Comparative photocross-linking analysis of the tertiary structures of *Escherichia coli* and *Bacillus subtilis* RNase P RNAs. *EMBO J* 1998;**17**:1515–25.
37. Haas ES, Morse DP, Brown JW, Schmidt FJ, Pace NR. Long-range structure in ribonuclease P RNA. *Science* 1991;**254**:853–6.
38. Haas ES, Brown JW, Pitulle C, Pace NR. Further perspective on the catalytic core and secondary structure of ribonuclease P RNA. *Proc Natl Acad Sci USA* 1994;**91**:2527–31.
39. Massire C, Jaeger L, Westhof E. Phylogenetic evidence for a new tertiary interaction in bacterial RNase P RNAs. *RNA* 1997;**3**:553–6.
40. Massire C, Jaeger L, Westhof E. Derivation of the three-dimensional architecture of bacterial ribonuclease P RNAs from comparative sequence analysis. *J Mol Biol* 1998;**279**:773–93.
41. Kazantsev AV, Krivenko AA, Harrington DJ, Holbrook SR, Adams PD, Pace NR. Crystal structure of a bacterial ribonuclease P RNA. *Proc Natl Acad Sci USA* 2005;**102**:13392–7.
42. Torres-Larios A, Swinger KK, Krasilnikov AS, Pan T, Mondragon A. Crystal structure of the RNA component of bacterial ribonuclease P. *Nature* 2005;**437**:584–7.
43. Kazantsev AV, Pace NR. Bacterial RNase P: A new view of an ancient enzyme. *Nat Rev Microbiol* 2006;**4**:729–40.
44. Hartmann E, Hartmann RK. The enigma of ribonuclease P evolution. *Trends Genet* 2003;**19**:561–9.
45. Walker SC, Engelke DR. Ribonuclease P: The evolution of an ancient RNA enzyme. *Crit Rev Biochem Mol Biol* 2006;**41**:77–102.
46. Ellis JC, Barnes J, Brown JW. Is Alba an RNase P subunit? *RNA Biol* 2007;**4**:169–72.
47. Harris JK, Haas ES, Williams D, Frank DN, Brown JW. New insight into RNase P RNA structure from comparative analysis of the archaeal RNA. *RNA* 2001;**7**:220–32.
48. Pulukkunat DK, Gopalan V. Studies on *Methanocaldococcus jannaschii* RNase P reveal insights into the roles of RNA and protein cofactors in RNase P catalysis. *Nucleic Acids Res* 2008;**36**:4172–80.
49. Wilson RC, Bohlen CJ, Foster MP, Bell CE. Structure of *Pfu* Pop5, an archaeal RNase P protein. *Proc Natl Acad Sci USA* 2006;**103**:873–8.
50. Gößringer M, Hartmann RK. Function of heterologous and truncated RNase P proteins in *Bacillus subtilis*. *Mol Microbiol* 2007;**66**:801–13.
51. Hall TA, Brown JW. Interactions between RNase P protein subunits in archaea. *Archaea* 2004;**1**:247–54.
52. Kifusa M, Fukuhara H, Hayashi T, Kimura M. Protein–protein interactions in the subunits of ribonuclease P in the hyperthermophilic archaeon *Pyrococcus horikoshii* OT3. *Biosci Biotechnol Biochem* 2005;**69**:1209–12.
53. Tsai HY, Pulukkunat DK, Woznick WK, Gopalan V. Functional reconstitution and characterization of *Pyrococcus furiosus* RNase P. *Proc Natl Acad Sci USA* 2006;**103**:16147–52.

54. Kurz JC, Fierke CA. Ribonuclease P: A ribonucleoprotein enzyme. *Curr Opin Chem Biol* 2000;**4**:553–8.
55. Harris ME, Christian EL. Recent insights into the structure and function of the ribonucleoprotein enzyme ribonuclease P. *Curr Opin Struct Biol* 2003;**13**:325–33.
56. Kirsebom LA. RNase P RNA mediated cleavage: Substrate recognition and catalysis. *Biochimie* 2007;**89**:1183–94.
57. Sun L, Campbell FE, Zahler NH, Harris ME. Evidence that substrate-specific effects of C5 protein lead to uniformity in binding and catalysis by RNase P. *EMBO J* 2006;**25**:3998–4007.
58. LaRiviere FJ, Wolfson AD, Uhlenbeck OC. Uniform binding of aminoacyl-tRNAs to elongation factor Tu by thermodynamic compensation. *Science* 2001;**294**:165–8.
59. Orellana O, Cooley L, Söll D. The additional guanylate at the 5' terminus of *Escherichia coli* tRNA^{His} is the result of unusual processing by RNase P. *Mol Cell Biol* 1986;**6**:525–9.
60. Burkard U, Söll D. The unusually long amino acid acceptor stem of *Escherichia coli* selenocysteine tRNA results from abnormal cleavage by RNase P. *Nucleic Acids Res* 1988;**16**:11617–24.
61. Kirsebom LA, Svärd SG. Base pairing between *Escherichia coli* RNase P RNA and its substrate. *EMBO J* 1994;**13**:4870–6.
62. Svärd SG, Kagardt U, Kirsebom LA. Phylogenetic comparative mutational analysis of the base-pairing between RNase P RNA and its substrate. *RNA* 1996;**2**:463–72.
63. Busch S, Kirsebom LA, Notbohm H, Hartmann RK. Differential role of the intermolecular base-pairs G292-C(75) and G293-C(74) in the reaction catalyzed by *Escherichia coli* RNase P RNA. *J Mol Biol* 2000;**299**:941–51.
64. Wegscheid B, Hartmann RK. *In vivo* and *in vitro* investigation of bacterial type B RNase P interaction with tRNA 3'-CCA. *Nucleic Acids Res* 2007;**35**:2060–73.
65. Hess WR, Fingerhut C, Schön A. RNase P RNA from *Prochlorococcus marinus*: Contribution of substrate domains to recognition by a cyanobacterial ribozyme. *FEBS Lett* 1998;**431**:138–42.
66. Pascual A, Vioque A. Functional reconstitution of RNase P activity from a plastid RNA subunit and a cyanobacterial protein subunit. *FEBS Lett* 1999;**442**:7–10.
67. Robertson HD, Altman S, Smith JD. Purification and properties of a specific *Escherichia coli* ribonuclease which cleaves a tyrosine transfer ribonucleic acid presursor. *J Biol Chem* 1972;**247**:5243–51.
68. Swanson RV. Genome of *Aquifex aeolicus*. *Methods Enzymol* 2001;**330**:158–69.
69. Li Y, Altman S. In search of RNase P RNA from microbial genomes. *RNA* 2004;**10**:1533–40.
70. Marszalkowski M, Willkomm DK, Hartmann RK. 5'-End maturation of tRNA in *Aquifex aeolicus*. *Biol Chem* 2008;**389**:395–403.
71. Marszalkowski M, Teune JH, Steger G, Hartmann RK, Willkomm DK. Thermostable RNase P RNAs lacking P18 identified in the *Aquificales*. *RNA* 2006;**12**:1915–21.
72. Lombo TB, Kaberdin VR. RNA processing in *Aquifex aeolicus* involves RNase E/G and an RNase P-like activity. *Biochem Biophys Res Commun* 2008;**366**:457–63.
73. Huber H, Hohn MJ, Rachel R, Fuchs T, Wimmer VC, Stetter KO. A new phylum of Archaea represented by a nanosized hyperthermophilic symbiont. *Nature* 2002;**417**:63–7.
74. Jahn U, Gallenberger M, Paper W, Junglas B, Eisenreich W, Stetter KO, Rachel R, Huber H. *Nanoarchaeum equitans* and *Ignicoccus hospitalis*: new insights into a unique, intimate association of two archaea. *J Bacteriol* 2008;**190**:1743–50.
75. Randau L, Schröder I, Söll D. Life without RNase P. *Nature* 2008;**453**:120–3.
76. Waters E, Hohn MJ, Ahel I, Graham DE, Adams MD, Barnstead M, et al. The genome of *Nanoarchaeum equitans*: Insights into early archaeal evolution and derived parasitism. *Proc Natl Acad Sci USA* 2003;**100**:12984–8.

77. Farruggio D, Chaudhuri J, Maitra U, RajBhandary UL. The A1 × U72 base pair conserved in eukaryotic initiator tRNAs is important specifically for binding to the eukaryotic translation initiation factor eIF2. *Mol Cell Biol* 1996;**16**:4248–56.
78. Fitz-Gibbon ST, Ladner H, Kim UJ, Stetter KO, Simon MI, Miller JH. Genome sequence of the hyperthermophilic crenarchaeon *Pyrobaculum aerophilum*. *Proc Natl Acad Sci USA* 2002;**99**:984–9.
79. Seif E, Cadieux A, Lang BF. Hybrid *E. coli*—mitochondrial ribonuclease P RNAs are catalytically active. *RNA* 2006;**12**:1661–70.
80. Baum M, Cordier A, Schön A. RNase P from a photosynthetic organelle contains an RNA homologous to the cyanobacterial counterpart. *J Mol Biol* 1996;**257**:43–52.
81. Cordier A, Schön A. Cyanelle RNase P: RNA structure analysis and holoenzyme properties of an organellar ribonucleoprotein enzyme. *J Mol Biol* 1999;**289**:9–20.
82. Morales MJ, Dang YL, Lou YC, Sulo P, Martin NC. A 105-kDa protein is required for yeast mitochondrial RNase P activity. *Proc Natl Acad Sci USA* 1992;**89**:9875–9.
83. Holzmann J, Frank P, Löffler E, Bennett KL, Gerner C, Rossmanith W. RNase P without RNA: Identification and functional reconstitution of the human mitochondrial tRNA processing enzyme. *Cell* 2008;**135**:462–74.
84. Wang MJ, Davis NW, Gegenheimer P. Novel mechanisms for maturation of chloroplast transfer RNA precursors. *EMBO J* 1988;**7**:1567–74.
85. Thomas BC, Li X, Gegenheimer P. Chloroplast ribonuclease P does not utilize the ribozyme-type pre-tRNA cleavage mechanism. *RNA* 2000;**6**:545–53.
86. Reed RE, Baer MF, Guerrier-Takada C, Donis-Keller H, Altman S. Nucleotide sequence of the gene encoding the RNA subunit (M1 RNA) of ribonuclease P from *Escherichia coli*. *Cell* 1982;**30**:627–36.
87. Motamedi H, Lee Y, Schmidt FJ. Tandem promoters preceding the gene for the M1 RNA component of *Escherichia coli* ribonuclease P. *Proc Natl Acad Sci USA* 1984;**81**:3959–63.
88. Lee Y, Ramamoorthy R, Park CU, Schmidt FJ. Sites of initiation and pausing in the *Escherichia coli rnpB* (M1 RNA) transcript. *J Biol Chem* 1989;**264**:5098–103.
89. Lundberg U, Altman S. Processing of the precursor to the catalytic RNA subunit of RNase P from *Escherichia coli*. *RNA* 1995;**1**:327–34.
90. Ko JH, Han K, Kim Y, Sim S, Kim KS, Cho B, Lee K, Lee Y. Dual function of RNase E for control of M1 RNA biosynthesis in *Escherichia coli*. *Biochemistry* 2008;**47**:762–70.
91. Jung YH, Lee Y. *Escherichia coli rnpB* promoter mutants altered in stringent response. *Biochem Biophys Res Commun* 1997;**230**:582–6.
92. Dong H, Kirsebom LA, Nilsson L. Growth rate regulation of 4.5 S RNA and M1 RNA the catalytic subunit of *Escherichia coli* RNase P. *J Mol Biol* 1996;**261**:303–8.
93. Park JW, Jung Y, Lee SJ, Jin DJ, Lee Y. Alteration of stringent response of the *Escherichia coli rnpB* promoter by mutations in the −35 region. *Biochem Biophys Res Commun* 2002;**290**:1183–7.
94. Reich C, Gardiner KJ, Olsen GJ, Pace B, Marsh TL, Pace NR. The RNA component of the *Bacillus subtilis* RNase P. Sequence, activity, and partial secondary structure. *J Biol Chem* 1986;**261**:7888–93.
95. Krasny L, Tiserova H, Jonak J, Rejman D, Sanderova H. The identity of the transcription +1 position is crucial for changes in gene expression in response to amino acid starvation in *Bacillus subtilis*. *Mol Microbiol* 2008;**69**:42–54.
96. Barrera A, Pan T. Interaction of the *Bacillus subtilis* RNase P with the 30S ribosomal subunit. *RNA* 2004;**10**:482–92.
97. Hansen FG, Hansen EB, Atlung T. The nucleotide sequence of the *dnaA* gene promoter and of the adjacent *rpmH* gene, coding for the ribosomal protein L34, of *Escherichia coli*. *EMBO J* 1982;**1**:1043–8.

98. Panagiotidis CA, Drainas D, Huang SC. Modulation of ribonuclease P expression in *Escherichia coli* by polyamines. *Int J Biochem* 1992;**24**:1625–31.
99. Feltens R, Gößringer M, Willkomm DK, Urlaub H, Hartmann RK. An unusual mechanism of bacterial gene expression revealed for the RNase P protein of *Thermus* strains. *Proc Natl Acad Sci USA* 2003;**100**:5724–9.
100. Hansen FG, Hansen EB, Atlung T. Physical mapping and nucleotide sequence of the *rnpA* gene that encodes the protein component of ribonuclease P in *Escherichia coli*. *Gene* 1985;**38**:85–93.
101. Condon C. Maturation and degradation of RNA in bacteria. *Curr Opin Microbiol* 2007;**10**:271–8.
102. Lee YM, Lee Y, Park CU. Transcription termination in the M1 RNA gene of *Escherichia coli*. *Korean Biochem J* 1989;**22**:276–81.
103. Reed RE, Altman S. Repeated sequences and open reading frames in the 3′ flanking region of the gene for the RNA subunit of *Escherichia coli* ribonuclease P. *Proc Natl Acad Sci USA* 1983;**80**:5359–63.
104. Sakamoto H, Kimura N, Shimura Y. Processing of transcription products of the gene encoding the RNA component of RNase P. *Proc Natl Acad Sci USA* 1983;**80**:6187–91.
105. Gurevitz M, Jain SK, Apirion D. Identification of a precursor molecular for the RNA moiety of the processing enzyme RNase P. *Proc Natl Acad Sci USA* 1983;**80**:4450–4.
106. Kim S, Kim H, Park I, Lee Y. Mutational analysis of RNA structures and sequences postulated to affect 3′ processing of M1 RNA, the RNA component of *Escherichia coli* RNase P. *J Biol Chem* 1996;**271**:19330–7.
107. Li Z, Pandit S, Deutscher MP. 3′ exoribonucleolytic trimming is a common feature of the maturation of small, stable RNAs in *Escherichia coli*. *Proc Natl Acad Sci USA* 1998;**95**:2856–61.
108. Guerrier-Takada C, Altman S. Catalytic activity of an RNA molecule prepared by transcription *in vitro*. *Science* 1984;**223**:285–6.
109. Guerrier-Takada C, Altman S. M1 RNA with large terminal deletions retains its catalytic activity. *Cell* 1986;**45**:177–83.
110. Kim KS, Sim S, Ko JH, Lee Y. Processing of m1 RNA at the 3′ end protects its primary transcript from degradation. *J Biol Chem* 2005;**280**:34667–74.
111. Li Z, Pandit S, Deutscher MP. Polyadenylation of stable RNA precursors *in vivo*. *Proc Natl Acad Sci USA* 1998;**95**:12158–62.
112. Wong TN, Sosnick TR, Pan T. Folding of noncoding RNAs during transcription facilitated by pausing-induced nonnative structures. *Proc Natl Acad Sci USA* 2007;**104**:17995–8000.
113. Loria A, Pan T. The 3′ substrate determinants for the catalytic efficiency of the *Bacillus subtilis* RNase P holoenzyme suggest autolytic processing of the RNase P RNA *in vivo*. *RNA* 2000;**6**:1413–22.
114. Hansen A, Pfeiffer T, Zuleeg T, Limmer S, Ciesiolka J, Feltens R, Hartmann RK. Exploring the minimal substrate requirements for *trans*-cleavage by RNase P holoenzymes from *Escherichia coli* and *Bacillus subtilis*. *Mol Microbiol* 2001;**41**:131–43.
115. Motamedi H, Lee K, Nichols L, Schmidt FJ. An RNA species involved in *Escherichia coli* ribonuclease P activity. Gene cloning and effect on transfer RNA synthesis *in vivo*. *J Mol Biol* 1982;**162**:535–50.
116. Komine Y, Kitabatake M, Yokogawa T, Nishikawa K, Inokuchi H. A tRNA-like structure is present in 10Sa RNA, a small stable RNA from *Escherichia coli*. *Proc Natl Acad Sci USA* 1994;**91**:9223–7.
117. Bothwell AL, Garber RL, Altman S. Nucleotide sequence and *in vitro* processing of a precursor molecule to *Escherichia coli* 4.5 S RNA. *J Biol Chem* 1976;**251**:7709–16.

118. Peck-Miller KA, Altman S. Kinetics of the processing of the precursor to 4.5 S RNA, a naturally occurring substrate for RNase P from *Escherichia coli*. *J Mol Biol* 1991;**221**:1–5.
119. Keiler KC. Biology of *trans*-translation. *Annu Rev Microbiol* 2008;**62**:133–51.
120. Herskovits AA, Bochkareva ES, Bibi E. New prospects in studying the bacterial signal recognition particle pathway. *Mol Microbiol* 2000;**38**:927–39.
121. Ulbrandt ND, Newitt JA, Bernstein HD. The *E. coli* signal recognition particle is required for the insertion of a subset of inner membrane proteins. *Cell* 1997;**88**:187–96.
122. Bothwell AL, Stark BC, Altman S. Ribonuclease P substrate specificity: Cleavage of a bacteriophage phi80-induced RNA. *Proc Natl Acad Sci USA* 1976;**73**:1912–6.
123. Forti F, Sabbattini P, Sironi G, Zangrossi S, Deho G, Ghisotti D. Immunity determinant of phage–plasmid P4 is a short processed RNA. *J Mol Biol* 1995;**249**:869–78.
124. Hartmann RK, Heinrich J, Schlegl J, Schuster H. Precursor of C4 antisense RNA of bacteriophages P1 and P7 is a substrate for RNase P of *Escherichia coli*. *Proc Natl Acad Sci USA* 1995;**92**:5822–6.
125. McClain WH, Guerrier-Takada C, Altman S. Model substrates for an RNA enzyme. *Science* 1987;**238**:527–30.
126. Forster AC, Altman S. External guide sequences for an RNA enzyme. *Science* 1990;**249**:783–6.
127. Li Y, Altman S. A specific endoribonuclease, RNase P, affects gene expression of polycistronic operon mRNAs. *Proc Natl Acad Sci USA* 2003;**100**:13213–8.
128. Li Y, Altman S. Polarity effects in the lactose operon of *Escherichia coli*. *J Mol Biol* 2004;**339**:31–9.
129. Altman S, Wesolowski D, Guerrier-Takada C, Li Y. RNase P cleaves transient structures in some riboswitches. *Proc Natl Acad Sci USA* 2005;**102**:11284–9.
130. Ko JH, Altman S. OLE RNA, an RNA motif that is highly conserved in several extremophilic bacteria, is a substrate for and can be regulated by RNase P RNA. *Proc Natl Acad Sci USA* 2007;**104**:7815–20.
131. Carpousis AJ. The RNA degradosome of *Escherichia coli*: An mRNA-degrading machine assembled on RNase E. *Annu Rev Microbiol* 2007;**61**:71–87.
132. Mackie GA, Genereaux JL. The role of RNA structure in determining RNase E-dependent cleavage sites in the mRNA for ribosomal protein S20 *in vitro*. *J Mol Biol* 1993;**234**:998–1012.
133. Bouvet P, Belasco JG. Control of RNase E-mediated RNA degradation by 5′-terminal base pairing in *E. coli*. *Nature* 1992;**360**:488–91.
134. Alifano P, Rivellini F, Piscitelli C, Arraiano CM, Bruni CB, Carlomagno MS. Ribonuclease E provides substrates for ribonuclease P-dependent processing of a polycistronic mRNA. *Genes Dev* 1994;**8**:3021–31.
135. Liu F, Altman S. Differential evolution of substrates for an RNA enzyme in the presence and absence of its protein cofactor. *Cell* 1994;**77**:1093–100.
136. Reiner R, Ben-Asouli Y, Krilovetzky I, Jarrous N. A role for the catalytic ribonucleoprotein RNase P in RNA polymerase III transcription. *Genes Dev* 2006;**20**:1621–35.
137. Gitai Z. The new bacterial cell biology: Moving parts and subcellular architecture. *Cell* 2005;**120**:577–86.
138. Gitai Z. Diversification and specialization of the bacterial cytoskeleton. *Curr Opin Cell Biol* 2007;**19**:5–12.
139. Vioque A, Arnez J, Altman S. Protein–RNA interactions in the RNase P holoenzyme from *Escherichia coli*. *J Mol Biol* 1988;**202**:835–48.
140. Jain SK, Pragai B, Apirion D. A possible complex containing RNA processing enzymes. *Biochem Biophys Res Commun* 1982;**106**:768–78.
141. Miczak A, Srivastava RA, Apirion D. Location of the RNA-processing enzymes RNase III, RNase E and RNase P in the *Escherichia coli* cell. *Mol Microbiol* 1991;**5**:1801–10.

142. Fukuhara H, Kifusa M, Watanabe M, Terada A, Honda T, Numata T, Kakuta Y, Kimura M. A fifth protein subunit Ph1496p elevates the optimum temperature for the ribonuclease P activity from *Pyrococcus horikoshii* OT3. *Biochem Biophys Res Commun* 2006;**343**:956–64.
143. Fang XW, Yang XJ, Littrell K, Niranjanakumari S, Thiyagarajan P, Fierke CA, Sosnick TR, Pan T. The *Bacillus subtilis* RNase P holoenzyme contains two RNase P RNA and two RNase P protein subunits. *RNA* 2001;**7**:233–41.
144. Barrera A, Fang X, Jacob J, Casey E, Thiyagarajan P, Pan T. Dimeric and monomeric *Bacillus subtilis* RNase P holoenzyme in the absence and presence of pre-tRNA substrates. *Biochemistry* 2002;**41**:12986–94.
145. Ow MC, Kushner SR. Initiation of tRNA maturation by RNase E is essential for cell viability in *E. coli*. *Genes Dev* 2002;**16**:1102–15.
146. Noren CJ, Anthony-Cahill SJ, Suich DJ, Noren KA, Griffith MC, Schultz PG. *In vitro* suppression of an amber mutation by a chemically aminoacylated transfer RNA prepared by runoff transcription. *Nucleic Acids Res* 1990;**18**:83–8.
147. She Q, Shen B, Chen L. Archaeal integrases and mechanisms of gene capture. *Biochem Soc Trans* 2004;**32**:222–6.
148. Li Z, Deutscher MP. The tRNA processing enzyme RNase T is essential for maturation of 5S RNA. *Proc Natl Acad Sci USA* 1995;**92**:6883–6.
149. Hölzle A, Fischer S, Heyer R, Schütz S, Zacharias M, Walther P, Allers T, Marchfelder A. Maturation of the 5S rRNA 5′ end is catalyzed *in vitro* by the endonuclease tRNase Z in the archaeon *H. volcanii*. *RNA* 2008;**14**:928–37.
150. Ceballos-Chavez M, Vioque A. Sequence-dependent cleavage site selection by RNase Z from the cyanobacterium *Synechocystis* sp. PCC 6803. *J Biol Chem* 2005;**280**:33461–9.
151. Garber RL, Gage LP. Transcription of a cloned *Bombyx mori* tRNA2Ala gene: Nucleotide sequence of the tRNA precursor and its processing *in vitro*. *Cell* 1979;**18**:817–28.
152. Hagenbüchle O, Larson D, Hall GI, Sprague KU. The primary transcription product of a silkworm alanine tRNA gene: Identification of *in vitro* sites of initiation, termination and processing. *Cell* 1979;**18**:1217–29.
153. Schiffer S, Rösch S, Marchfelder A. Assigning a function to a conserved group of proteins: the tRNA 3′-processing enzymes. *EMBO J* 2002;**21**:2769–77.
154. Vogel A, Schilling O, Späth B, Marchfelder A. The tRNase Z family of proteins. Physiological functions, substrate specificity and structural properties. *Biol Chem* 2005;**386**:1253–64.
155. de la Sierra-Gallay IL, Pellegrini O, Condon C. Structural basis for substrate binding, cleavage and allostery in the tRNA maturase RNase Z. *Nature* 2005;**433**:657–61.
156. Ishii R, Minagawa A, Takaku H, Takagi M, Nashimoto M, Yokoyama S. Crystal structure of the tRNA 3′ processing endoribonuclease tRNase Z from *Thermotoga maritima*. *J Biol Chem* 2005;**280**:14138–44.
157. Vogel A, Schilling O, Niecke M, Bettmer J, Meyer-Klaucke W. ElaC encodes a novel binuclear zinc phosphodiesterase. *J Biol Chem* 2002;**277**:29078–85.
158. Späth B, Schubert S, Lieberoth A, Settele F, Schütz S, Fischer S, Marchfelder A. Two archaeal tRNase Z enzymes: similar but different. *Arch Microbiol* 2008;**25**:25.
159. Späth B, Kirchner S, Vogel A, Schubert S, Meinlschmidt P, Aymanns S, Nezzar J, Marchfelder A. Analysis of the functional modules of the tRNA 3′ endonuclease (tRNase Z). *J Biol Chem* 2005;**280**:35440–7.
160. Späth B, Canino G, Marchfelder A. tRNase Z: The end is not in sight. *Cell Mol Life Sci* 2007;**64**:2404–12.
161. Ceballos M, Vioque A. tRNase Z. *Protein Pept Lett* 2007;**14**:137–45.
162. Aravind L. An evolutionary classification of the metallo-beta-lactamase fold proteins. *In Silico Biol* 1999;**1**:69–91.

163. Callebaut I, Moshous D, Mornon JP, de Villartay JP. Metallo-beta-lactamase fold within nucleic acids processing enzymes: The beta-CASP family. *Nucleic Acids Res* 2002;**30**:3592–601.
164. Daiyasu H, Osaka K, Ishino Y, Toh H. Expansion of the zinc metallo-hydrolase family of the beta-lactamase fold. *FEBS Lett* 2001;**503**:1–6.
165. Britton RA, Wen T, Schaefer L, Pellegrini O, Uicker WC, Mathy N, Tobin C, Daou R, Szyk J, Condon C. Maturation of the 5' end of *Bacillus subtilis* 16S rRNA by the essential ribonuclease YkqC/RNase J1. *Mol Microbiol* 2007;**63**:127–38.
166. Mathy N, Benard L, Pellegrini O, Daou R, Wen T, Condon C. 5'-to-3' exoribonuclease activity in bacteria: Role of RNase J1 in rRNA maturation and 5' stability of mRNA. *Cell* 2007;**129**:681–92.
167. Even S, Pellegrini O, Zig L, Labas V, Vinh J, Brechemmier-Baey D, Putzer H. Ribonucleases J1 and J2: Two novel endoribonucleases in *B. subtilis* with functional homology to *E. coli* RNase E. *Nucleic Acids Res* 2005;**33**:2141–52.
168. Baillat D, Hakimi MA, Naar AM, Shilatifard A, Cooch N, Shiekhattar R. Integrator, a multiprotein mediator of small nuclear RNA processing, associates with the C-terminal repeat of RNA polymerase II. *Cell* 2005;**123**:265–76.
169. Dominski Z. Nucleases of the metallo-beta-lactamase family and their role in DNA and RNA metabolism. *Crit Rev Biochem Mol Biol* 2007;**42**:67–93.
170. Keller W, Minvielle-Sebastia L. A comparison of mammalian and yeast pre-mRNA 3'-end processing. *Curr Opin Cell Biol* 1997;**9**:329–36.
171. Mandel CR, Kaneko S, Zhang H, Gebauer D, Vethantham V, Manley JL, Tong L. Polyadenylation factor CPSF-73 is the pre-mRNA 3'-end-processing endonuclease. *Nature* 2006;**444**:953–6.
172. Schilling O, Späth B, Kostelecky B, Marchfelder A, Meyer-Klaucke W, Vogel A. Exosite modules guide substrate recognition in the ZiPD/ElaC protein family. *J Biol Chem* 2005;**280**:17857–62.
173. Kostelecky B, Pohl E, Vogel A, Schilling O, Meyer-Klaucke W. The crystal structure of the zinc phosphodiesterase from *Escherichia coli* provides insight into function and cooperativity of tRNase Z-family proteins. *J Bacteriol* 2006;**188**:1607–14.
174. Li de la Sierra-Gallay I, Mathy N, Pellegrini O, Condon C. Structure of the ubiquitous 3' processing enzyme RNase Z bound to transfer RNA. *Nat Struct Mol Biol* 2006;**13**:376–7.
175. Nashimoto M, Tamura M, Kaspar RL. Minimum requirements for substrates of mammalian tRNA 3' processing endoribonuclease. *Biochemistry* 1999;**38**:12089–96.
176. Schiffer S, Helm M, Théobald-Dietrich A, Giegé R, Marchfelder A. The plant tRNA 3' processing enzyme has a broad substrate spectrum. *Biochemistry* 2001;**40**:8264–72.
177. Redko Y, Li de Lasierra-Gallay I, Condon C. When all's zed and done: The structure and function of RNase Z in prokaryotes. *Nat Rev Microbiol* 2007;**5**:278–86.
178. Perwez T, Kushner SR. RNase Z in *Escherichia coli* plays a significant role in mRNA decay. *Mol Microbiol* 2006;**60**:723–37.
179. Minagawa A, Takaku H, Takagi M, Nashimoto M. A novel endonucleolytic mechanism to generate the CCA 3' termini of tRNA molecules in *Thermotoga maritima*. *J Biol Chem* 2004;**279**:15688–97.
180. Schiffer S, Rösch S, Marchfelder A. Recombinant RNase Z does not recognize CCA as part of the tRNA and its cleavage efficieny is influenced by acceptor stem length. *Biol Chem* 2003;**384**:333–42.
181. Schierling K, Rösch S, Rupprecht R, Schiffer S, Marchfelder A. tRNA 3' end maturation in archaea has eukaryotic features: The RNase Z from *Haloferax volcanii*. *J Mol Biol* 2002;**316**:895–902.

182. Pellegrini O, Nezzar J, Marchfelder A, Putzer H, Condon C. Endonucleolytic processing of CCA-less tRNA precursors by RNase Z in *Bacillus subtilis*. *EMBO J* 2003;**22**:4534–43.
183. Takaku H, Minagawa A, Takagi M, Nashimoto M. A candidate prostate cancer susceptibility gene encodes tRNA 3' processing endoribonuclease. *Nucleic Acids Res* 2003;**31**:2272–8.
184. Dubrovsky EB, Dubrovskaya VA, Levinger L, Schiffer S, Marchfelder A. *Drosophila* RNase Z processes mitochondrial and nuclear pre-tRNA 3' ends *in vivo*. *Nucleic Acids Res* 2004;**32**:255–62.
185. Minagawa A, Takaku H, Takagi M, Nashimoto M. A novel endonucleolytic mechanism to generate the CCA 3' termini of tRNA molecules in *Thermotoga maritima*. *J Biol Chem* 2004;**279**:15688–97.
186. Zareen N, Yan H, Hopkinson A, Levinger L. Residues in the conserved His domain of fruit fly tRNase Z that function in catalysis are not involved in substrate recognition or binding. *J Mol Biol* 2005;**350**:189–99.
187. Späth B, Settele F, Schilling O, D'Angelo I, Vogel A, Feldmann I, Meyer-Klaucke W, Marchfelder A. Metal requirements and phosphodiesterase activity of tRNase Z enzymes. *Biochemistry* 2007;**46**:14742–50.
188. Schilling O, Ruggeberg S, Vogel A, Rittner N, Weichert S, Schmidt S, Doig S, Franz T, Benes SC, Andrews SC, Baum M, Meyer-Klaucke W. Characterization of an *Escherichia coli* elaC deletion mutant. *Biochem Biophys Res Commun* 2004;**320**:1365–73.
189. Ezraty B, Dahlgren B, Deutscher MP. The RNase Z homologue encoded by *Escherichia coli* elaC gene is RNase BN. *J Biol Chem* 2005;**280**:16542–5.
190. Seidman JG, Schmidt FJ, Foss K, McClain WH. A mutant of *Escherichia coli* defective in removing 3' terminal nucleotides from some transfer RNA precursor molecules. *Cell* 1975;**5**:389–400.
191. Elkins JG, Podar M, Graham DE, Makarova KS, Wolf Y, Randau L, Hedlund BP, Brochier-Armanet V, Kunin V, Anderson I, Lapidus A, Goltsman E, Barry K, Koonin EV, Hugenholtz P, Kyrpides G, Wanner G, Richardson P, Keller M, Stetter KO. A korarchaeal genome reveals insights into the evolution of the Archaea. *Proc Natl Acad Sci USA* 2008;**105**:8102–7.
192. Bikoff EK, Gefter ML. *In vitro* synthesis of transfer RNA. I. Purification of required components. *J Biol Chem* 1975;**250**:6240–7.
193. Bikoff EK, LaRue BF, Gefter ML. *In vitro* synthesis of transfer RNA. II. Identification of required enzymatic activities. *J Biol Chem* 1975;**250**:6248–55.
194. Blattner FR, Plunkett G, III, Bloch CA, Perna NT, Burland V, Riley M, Collado-Vides J, Glasner CK, Rode CK, Mayhew GF, Gregor J, Davis NW, Kirkpatrick HA, Goeden MA, Rose B, Mau B, Shao Y. The complete genome sequence of *Escherichia coli* K-12. *Science* 1997;**277**:1453–74.
195. Ow MC, Kushner SR. Initiation of tRNA maturation by RNase E is essential for cell viability in *E. coli*. *Genes Dev* 2002;**16**:1102–15.
196. Li Z, Deutscher MP. RNase E plays an essential role in the maturation of *Escherichia coli* tRNA precursors. *RNA* 2002;**8**:97–109.
197. Li Z, Gong X, Joshi VH, Li M. Co-evolution of tRNA 3' trailer sequences with 3' processing enzymes in bacteria. *RNA* 2005;**11**:567–77.
198. Mohanty BK, Kushner SR. Ribonuclease P processes polycistronic tRNA transcripts in *Escherichia coli* independent of ribonuclease E. *Nucleic Acids Res* 2007;**35**:7614–25.
199. Wen T, Oussenko IA, Pellegrini O, Bechhofer DH, Condon C. Ribonuclease PH plays a major role in the exonucleolytic maturation of CCA-containing tRNA precursors in *Bacillus subtilis*. *Nucleic Acids Res* 2005;**33**:3636–43.
200. Li Z, Deutscher MP. Maturation pathways for *E. coli* tRNA precursors: A random multienzyme process *in vivo*. *Cell* 1996;**86**:503–12.

201. Condon C, Putzer H. The phylogenetic distribution of bacterial ribonucleases. *Nucleic Acids Res* 2002;**30**:5339–46.
202. Wagner S, Klug G. An archaeal protein with homology to the eukaryotic translation initiation factor 5A shows ribonucleolytic activity. *J Biol Chem* 2007;**282**:13966–76.
203. Marquez SM, Harris JK, Kelley ST, Brown JW, Dawson SC, Roberts EC, Pace NR. Structural implications of novel diversity in eucaryal RNase P RNA. *RNA* 2005;**11**:739–51.
204. Christian EL, Smith KM, Perera N, Harris ME. The P4 metal binding site in RNase P RNA affects active site metal affinity through substrate positioning. *RNA* 2006;**12**:1463–7.
205. Xiao S, Scott F, Fierke CA, Engelke DR. Eukaryotic ribonuclease P: A plurality of ribonucleoprotein enzymes. *Annu Rev Biochem* 2002;**71**:165–89.
206. Chang DD, Clayton DA. A novel endoribonuclease cleaves at a priming site of mouse mitochondrial DNA replication. *EMBO J* 1987;**6**:409–17.
207. Gill T, Cai T, Aulds J, Wierzbicki S, Schmitt ME. RNase MRP cleaves the CLB2 mRNA to promote cell cycle progression: Novel method of mRNA degradation. *Mol Cell Biol* 2004;**24**:945–53.
208. Gill T, Aulds J, Schmitt ME. A specialized processing body that is temporally and asymmetrically regulated during the cell cycle in *Saccharomyces cerevisiae*. *J Cell Biol* 2006;**173**:35–45.
209. Brown JW, Nolan JM, Haas ES, Rubio MA, Major F, Pace NR. Comparative analysis of ribonuclease P RNA using gene sequences from natural microbial populations reveals tertiary structural elements. *Proc Natl Acad Sci USA* 1996;**93**:3001–6.
210. Marszalkowski M, Willkomm DK, Hartmann RK. Structural basis of a ribozyme's thermostability: P1–L9 interdomain interaction in RNase P RNA. *RNA* 2008;**14**:127–33.
211. Brown JW. The Ribonuclease P Database. *Nucleic Acids Res* 1999;**27**:314.
212. Seif E, Altman S. RNase P cleaves the adenine riboswitch and stabilizes *pbuE* mRNA in *Bacillus subtilis*. *RNA* 2008;**14**:1237–43.

Maturation and Degradation of Ribosomal RNA in Bacteria

Murray P. Deutscher

Department of Biochemistry and Molecular Biology, University of Miami Miller School of Medicine, Miami, Florida

I. Introduction 369
II. Early Studies of rRNA Maturation and Degradation 370
III. Current Understanding of rRNA Processing 373
  A. 16S rRNA 374
  B. 23S rRNA 378
  C. 5S rRNA 380
  D. Other Factors Involved in rRNA Processing 382
IV. Degradation of rRNA 383
  A. Misassembly and rRNA Quality Control 384
  B. RNase I-Mediated rRNA Degradation 386
  C. rRNA Degradation as a Response to Poor Environmental Conditions 386
  References 388

Ribosomal RNAs are the major components of ribosomes and are responsible for their catalytic activity. The three bacterial rRNAs (16S, 23S, and 5S) are cotranscribed as a single molecule that must be converted to the mature, functioning species through a series of nucleolytic processing events and base and sugar modifications that occur in the context of the assembling ribosome. One focus of this review is to examine the reactions that lead from the rRNA precursor to the mature species and to describe the ribonucleases (RNases) that carry out these processing reactions. rRNA, although usually stable in growing cells, also can be degraded if its assembly into ribosomes is aberrant or in response to certain stress conditions, such as starvation. The second focus of this review is to describe these degradative reactions, the RNases that carry them out, and the conditions that initiate the turnover process.

I. Introduction

Ribosomes are complex ribonucleoprotein (RNP) particles whose primary function is to serve as the site and catalyst for protein biosynthesis (1). Ribosomes are comprised of two subunits; in prokaryotes, these are 30S and 50S particles, which during translation join together to form the 70S functioning ribosome. The mass of the ribosome is approximately two-thirds RNA and one-third protein. The smaller 30S subunit contains a single 16S rRNA molecule and 21 proteins, while the 50S subunit contains one molecule each of 23S rRNA and 5S rRNA and 33 proteins. In recent years, with the availability of X-ray structures of prokaryotic ribosomes and subunits, considerable attention has been devoted to detailed analysis of ribosome structure and function. These topics will not be dealt with here, but for a recent review see reference (2).

In order to function properly, the many proteins and the RNAs of each ribosomal subunit must come together accurately in an ordered assembly process. Concomitant with ribosome assembly, the rRNAs, which are initially synthesized as precursor molecules, must undergo nucleolytic processing to generate the mature molecules found in the functioning particles. In addition, during maturation, the RNA molecules are extensively modified, primarily by base and sugar methylations and by conversion of specific uridines to pseudo-uridines. One major focus of this review is an examination of the nucleolytic processing events by which rRNA precursors are converted to their mature forms, and to describe, as far as is known, the ribonucleases (RNases) that catalyze these maturation reactions. Base and sugar modification of rRNA will not be discussed.

As might be expected from the complexity of ribosome biogenesis, errors in assembly can occur due to misfolded rRNAs, misordered addition of ribosomal proteins or improper conformational rearrangements. Such altered ribosomes appear to be subject to a quality control process that leads to their elimination *in vivo*. Likewise, under certain stress conditions or upon exposure of cells to certain agents, ribosomes can be extensively degraded. For each of these situations, we are still in the early stages of understanding how the process occurs, what determines when an rRNA will be degraded, and which RNases catalyze these degradative reactions. These areas are the second major focus of this review.

II. Early Studies of rRNA Maturation and Degradation

In this section, I will present a summary of the more important observations and conclusions that emerged from early studies of rRNA maturation and degradation.

The rRNA genes in most eubacteria are organized into operons (3–5). In *Escherichia coli*, seven rRNA operons are present, all with similar structure. The gene for 16S rRNA is near the 5′ end of the operon, followed by the gene for 23S rRNA and the 5S rRNA is nearest the 3′ end. In *E. coli* rRNA operons, a gene for at least one tRNA is located in the spacer region between 16S and 23S rRNA, and depending on the operon, tRNA genes may also be present downstream of 5S rRNA. Sequence analysis revealed that complementary regions flanked both the mature 16S and 23S rRNA sequences, providing the opportunity for extensive base pairing in the RNA transcript (6, 7). The importance of these secondary structures, as well as strong evidence for cotranscription of all the RNAs encoded by the operon became apparent from studies of a mutant strain deficient in the endoribonuclease, RNase III, an enzyme that specifically cleaves double-stranded RNA (8). A very similar gene organization is found in the 10 rRNA operons of *Bacillus subtilis* (4, 5).

In the absence of RNase III, cells accumulated a 30S rRNA molecule containing all the sequences in the operon (9). This molecule can be cleaved by RNase III *in vitro* within the flanking double-stranded regions to generate products (17S and p23S) containing the sequences of mature 16S and 23S rRNAs as well as additional residues at their 3′ and 5′ ends (6, 7). These studies not only provided evidence that rRNAs were cotranscribed, but they also identified the first rRNA processing enzyme, RNase III. Subsequent work showed that 30S rRNA molecules never exist in wild-type cells; rather, the first RNase III cleavage in the double-stranded region flanking 16S rRNA occurs as soon as this structure can form, and before transcription of the operon is completed (10). Since the RNase III cleavages do not lead to mature 16S or 23S rRNA, these early studies also indicated that additional RNase action would be required to complete the maturation process.

Other studies provided additional, important information about rRNA processing. Further examination of the RNase III-deficient strain revealed that functional ribosomes could still be made even in the absence of this endoribonuclease. However, under these conditions, normal maturation of 23S rRNA does not occur, although mature 16S rRNA can be made (11). These findings indicated that other endoribonucleases could separate the 23S and 16S rRNA regions and that additional RNase action could generate partially or fully matured RNA molecules. Inasmuch as mature 16S rRNA is made, these data imply that the RNases that normally generate the 5′ and 3′ termini of this molecule can work independently of RNase III action.

Some understanding of the nature of these enzymes also was obtained. An endoribonuclease activity was identified (termed RNase M16) that participated in maturation of the 5′ end of 16S rRNA (12). A mutant strain deficient in this enzyme accumulated an RNA molecule of 16.3S, smaller than the original 17S precursor generated by RNase III action, but still containing ~60 extra nucleotides at the 5′ end. In light of more recent work indicating that 5′ maturation is a two-step process (see the following text), these early observations are now completely understandable. The 16.3S processing intermediate contained a mature 3′ terminus indicating that a separate enzyme is involved in 3′ processing, and such an activity was partially purified (13). However, the identity of the 3′ processing RNase remains a mystery.

Additional early work provided insights into 5S rRNA maturation. In *E. coli*, following RNase III action, 5S rRNA is released as a 9S precursor molecule derived from the 3′ part of the rRNA transcript (14, 15). In addition to the 5S rRNA sequence, extra 5' residues from the spacer between the 23S and 5S rRNAs, and extra 3' residues, which may contain downstream tRNAs, are also present. The 9S precursor is not normally observed because it is rapidly converted to a 5S precursor containing three extra residues at each end by the essential endoribonuclease, RNase E (16). However, the 9S molecule accumulates in an RNase E temperature-sensitive mutant when placed at a nonpermissive temperature (14–16). Although the enzymes involved were not identified, it was suggested that the three extra residues present at each end of the final 5S precursor are removed by exoribonucleases since intermediates with one or two extra residues are observed in cells treated with chloramphenicol (17, 18). Subsequent work has borne this out, at least for 3' maturation of 5S rRNA (see the following text).

In B. subtilis, 5S rRNA maturation was found to proceed by a different route. In this organism, a single enzyme, RNase M5, cleaves extra residues from both the 3' and 5' ends of the 5S rRNA precursor (19). Interestingly, action of the catalytic subunit of RNase M5 is dependent upon binding of a second protein to the 5S rRNA precursor, ribosomal protein L18 (known at the time as BL16), which is normally associated with 5S rRNA in the ribosome (20). Thus, the true substrate of RNase M5 is an RNP particle (21). Cleavage by RNase M5 *in vitro* occurs in a double-stranded region of the 5S precursor with the cuts staggered by one nucleotide. Apparently, cleavages on the two strands occur simultaneously as intermediates processed at only one end were not found (19). This would differ from 5S RNA maturation in E. coli in which the two ends appear to be acted on independently. However, in more recent work (22), evidence for independent cleavages of the two ends of 5S RNA by RNase M5 was obtained.

Among the most important conclusions to emerge from these early studies is that rRNA maturation events generally occur in the context of an RNP particle, namely the assembling ribosome. In fact, some early studies suggested that protein synthesis by the newly-made ribosome is necessary to complete the final steps of the rRNA maturation process (reviewed in 20). The close connection between rRNA processing and ribosome assembly was supported by the isolation of numerous assembly mutants in which the rRNAs were not matured completely. While the coupling of these two processes makes sense biologically, it has made *in vitro* analysis of rRNA processing extremely challenging. Not only are the potential RNP substrates difficult to prepare or isolate, but in most cases, one does not even know which preribosomal intermediate might be the appropriate substrate for a particular rRNA processing reaction. Consequently, essentially all advances in identifying rRNA maturation enzymes have come, not from *in vitro* biochemistry with rRNA or rRNP substrates, but from analysis of mutant strains lacking RNases already identified through other functions.

rRNA maturation and degradation

Inasmuch as ribosomes are generally stable in growing bacterial cells, relatively few early studies were devoted to rRNA degradation. Nevertheless, it became clear, early on, that under certain stress conditions rRNA could be extensively degraded (reviewed in 21). These instances are related primarily to conditions of slow or no growth or of damage to the cell membrane. Thus, during starvation, nutritional downshift, stationary phase, or medium-imposed slow growth, rRNA may be degraded. Such degradation affects not only pre-existing ribosomes, but also newly synthesized rRNA if there are insufficient ribosomal proteins to assemble the rRNA into ribosomes. For example, during very slow growth, as much as 70% of newly synthesized rRNA does not make it into ribosomes and apparently is degraded (23) since free rRNA does not accumulate (24). These observations emphasize the importance of balanced synthesis of rRNA and ribosomal proteins.

A large number of agents, such as antibiotics, detergents, and heavy metals also promote rRNA degradation (25). What these agents have in common is that they lead to membrane damage and allow the periplasmic enzyme, RNase I, to enter the cell. Since, RNase I is able to act on intact ribosomes, its entry can result in massive rRNA degradation. This scenario is supported by the observation that 3' mononucleotides, which are derived solely from the action of RNase I, are products of rRNA degradation under these conditions. Moreover, in mutant strains lacking RNase I, this rRNA breakdown does not occur. Based on these, and other observations, it is clear that sequestering RNase I in the periplasm, away from cellular RNA, is important for cell viability.

Early attempts to identify RNases responsible for rRNA degradation were hampered by the unavailability of mutant strains entirely deficient for the enzymes that were known at that time, and of course, by the lack of knowledge of all the RNases present in a cell (26–28). Nevertheless, some important observations were made. RNase I was shown to play a role, but primarily at very elevated temperatures (45–50 °C), conditions that would lead to membrane damage and RNase I entry. Evidence was also obtained for the involvement of polynucleotide phosphorylase (PNPase) and for a hydrolytic exoribonuclease, probably distinct from RNase II (28). In light of current knowledge (see the following text), these observations were remarkably accurate.

III. Current Understanding of rRNA Processing

In recent years, our understanding of rRNA maturation has progressed both as a consequence of the discoveries of new RNases and the generation of mutant strains lacking these enzymes, and from an expansion of information about RNases initially identified for their roles in other processes. However, considering the amount of time elapsed since the early studies described above,

TABLE I
RNASES PARTICIPATING IN rRNA MATURATION IN *E. COLI* AND *B. SUBTILIS*

	RNase	
Substrate	*E. coli*	*B. subtilis*
Initial 30S transcript	RNase III	RNase III
16S precursor		
5'	RNase E, RNase G[a]	RNaseJ1
3'	Unknown	Unknown
23S precursor		
5'	Unknown	Mini-III
3'	RNase T	Mini-III
5S precursor		
5'	RNase E, unknown[a]	RNase M5
3'	RNase E, RNase T[a]	RNase M5

[a]When two RNases listed, first one acts followed by the second which generates the mature terminus.

progress has been relatively slow, and much remains unknown. In this section, our current knowledge of the maturation of 16S, 23S, and 5S rRNAs will be described (Table I).

A. 16S rRNA

In *E. coli*, the action of RNase III on rRNA gene transcripts leads to production of a 17S precursor of 16S rRNA (Fig. 1). This initial cleavage product contains an additional 115 nucleotides (nt) at its 5' end and 33 nt at the 3' end that must be removed to generate the mature RNA molecule (6, 7). As noted, early studies (12) had identified a mutant strain, termed BUMMER, that accumulated a 16.3S intermediate containing only ∼60 extra 5' residues as well as a mature 3' terminus suggesting that conversion of the 17S precursor to mature 16S rRNA was at least a two-step process. In addition, an activity was described, termed RNase M16, which converted the 16.3S RNA in 30S subunits or 70S ribosomes to mature RNA *in vitro*. However, the relation of that activity to the catalog of known *E. coli* RNases remained unclear.

A major advance in our understanding of 16S rRNA maturation came with the discovery that two endoribonucleases were required to process the 5' end of 16S rRNA (29, 30). One was RNase E, already known to be an important participant in both mRNA degradation and 5S rRNA maturation (16). The second enzyme was a homolog of the N-terminal, active region of RNase E, originally termed CafA (31), and now renamed RNase G (29, 30). In the absence of either protein, the rate of 5' processing is slowed dramatically, and

rRNA maturation and degradation

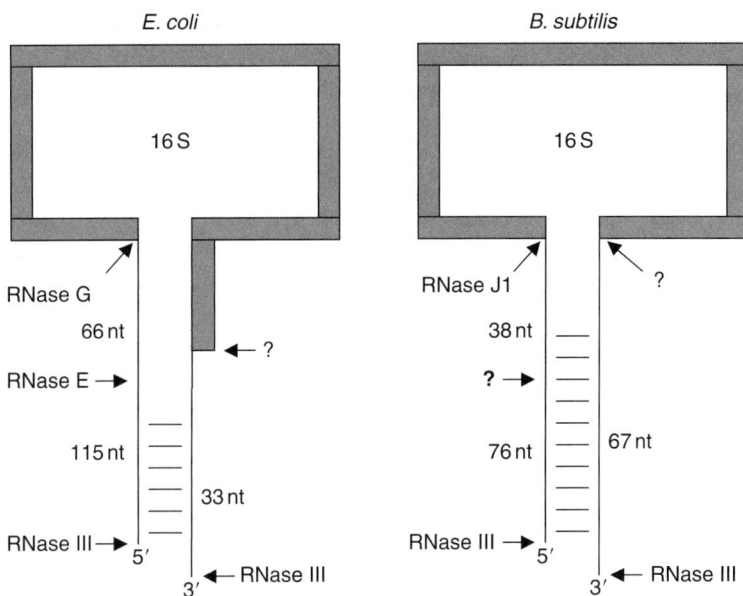

FIG. 1. Diagram of 16S rRNA maturation in *E. coli* and *B. subtilis*. Mature 16S sequences are shown as filled in area, and precursor sequences as a line. Enzymes acting at different positions are named, where known, and denoted with a **question mark (?)**, if unknown. The number of precursor residues present, prior to the action of the RNase immediately above, are indicated.

when both enzymes are missing, 5′ processing is blocked completely (29). Detailed examination using site-directed RNase H cleavage, which allowed analysis to single nucleotide resolution, revealed that in the absence of RNase E, a mature 5′ terminus could be generated by RNase G, although much more slowly. In contrast, in a strain lacking RNase G, the major product that accumulates contains 66 extra 5′ residues (29), exactly as was found in the BUMMER strain (12). Over time, a small amount of RNA with a mature 5′ terminus and a product with four or five extra 5′ residues can be generated by the RNase E still present in this strain. Complementation analysis confirmed that the BUMMER strain lacked active RNase G, and sequence analysis showed that the *rng* gene in the BUMMER strain contained an 11-bp deletion at nucleotide residues 729–739 leading to a truncated RNase G protein of ∼30 kDa compared to the normal 55 kDa (30).

The *in vivo* observations of the roles of RNases E and G in 5′ maturation were confirmed by *in vitro* analysis (29). Ribosome substrates containing rRNA with 115 and/or 66 extra 5′ residues were isolated from *rne rng* or *rng* mutant strains, respectively, and were treated with extracts from wild type, *rng* mutant or RNase G overexpressing strains, or with purified RNase E. These experiments showed that RNase G converted the +66-nt intermediate to the mature 5' terminus, whereas it acted poorly on the precursor with 115 extra residues. Based on these observations, it is clear that RNase M16 is RNase G.

These studies of 5' processing also provided some information about 3' processing of 16S rRNA. Again, using the RNase H cleavage procedure, but with a different complementary oligonucleotide, it was possible to examine the status of the 3' terminus in strains lacking RNase E and/or RNase G (29). In wild type cells, removal of the 33 extra 3' residues occurs rapidly. This is also true for the *rng* mutant. However, the single or double mutant strains deficient in RNase E process the 3' terminus more slowly. Thus, while 3' maturation can proceed in the complete absence of 5' maturation, it is more efficient when RNase E has already acted at the 5' terminus. In addition, since no intermediates with less than 33 extra 3' nucleotides were observed even when 3' processing was slowed, it is likely that maturation at the 3' terminus of 16S rRNA occurs by a single endonucleolytic cleavage.

Based on all of these observations, a model can be proposed for 16S rRNA maturation in *E. coli*. RNase III cleavage of the growing rRNA transcript releases the portion containing the 16S rRNA as a 17S precursor with extra residues at each end. This RNA is present in a preribosomal particle, and all subsequent processing events occur in this context. Cleavage by RNase E then shortens the 5' precursor sequence from 115 to 66 nt. The resulting 16.3S intermediate is also missing the 33-nt 3' precursor sequence. This has led to the suggestion that the 3' sequence might be removed first (30). However, since the absence of RNase E slows 3' processing, it is likely that the preferred route is to first cleave on the 5' side. In the 17S precursor, 26 of the 33 extra 3' residues are able to base pair with 5' precursor-specific nucleotides to form a terminal stem. Once RNase E cleavage occurs, this stem can no longer form, and 3' maturation could be facilitated. Confirmation of this idea must await identification of the 3' processing enzyme.

At the present time, almost nothing is known about 3' processing of 16S rRNA in *E. coli*. One activity identified early on was not characterized sufficiently to relate it to currently known *E. coli* RNases (13). As noted earlier, the maturation apparently is endonucleolytic, but neither RNase III, E, nor G appears to be involved (29). Likewise, the absence of the known *E. coli* exoribonucleases, alone or in various combinations, does not affect 3' processing (unpublished observations). Thus, it appears either that the RNase responsible is an as yet undiscovered enzyme or that some overlapping combination of known RNases is involved, but they all have not yet been simultaneously removed by mutation.

In another eubacterium, *B. subtilis*, processing of the 16S rRNA proceeds by a dramatically different mechanism (Fig. 1). *B. subtilis* lacks orthologs of RNases E and G (reviewed in 32), indicating that some other RNase would have to be involved in the 5' maturation process. The likely candidate is an enzyme, termed RNase J1, a product of the *ykqC* gene, now renamed *rnjA* (33). This enzyme was originally found to be an endoribonuclease that affected mRNA stability (34). However, more recent work revealed that depletion of RNase J1 also led to the accumulation of 16S rRNA precursors, implicating this enzyme in the processing of the rRNA as well (33). The closely-related enzyme, RNase J2, does not appear to be involved.

Primer extension analysis showed that the 5' ends of the precursors in the RNase J1 mutant corresponded well to both an RNase III cleavage site identified previously at +76 and to a site at +38 (33). Why there are two ends is not understood. As in *E. coli*, the accumulated 16S precursors present in the mutant strain were incorporated into 30S and 70S ribosomes, and into small polysomes, suggesting that they could function *in vivo*. The data also suggested that maturation occurs subsequent to 30S subunit assembly. To eliminate the possibility that the action of RNase J1 was indirect, such as in ribosome assembly rather than due to its RNase activity, an active site histidine residue was altered. Analysis of cells carrying this mutated enzyme showed that the catalytic activity of RNase J1 was required for 16S rRNA 5' processing *in vivo* (33).

In vitro analysis confirmed that purified RNase J1 could act on 70S ribosomes containing the precursor 16S rRNA, establishing that its action is direct (33). Most interestingly, subsequent *in vitro* analysis indicated that RNase J1 acted on 16S rRNA precursors as a 5'–3' exoribonuclease that released 5' mononucleotides (35). However, this activity was relatively weak and at extended times of incubation actually degraded the 16S rRNA precursor. Since these experiments utilized naked RNA that had been transcribed *in vitro*, and therefore carried no modifications, it is possible that the RNase acted aberrantly. Nevertheless, based on the *in vivo* analysis, it is clear that RNase J1 participates in 16S rRNA maturation in *Bacillus*. Moreover, the *in vitro* studies suggest that the mode of action of RNase J1, also being a 5'–3' exoribonuclease, may be unique among prokaryotic RNases. What is not yet clear is whether it acts on the +76 as well as the +38 molecule (36).

Although the enzymes required for 3' maturation of 16S rRNA in *E. coli* and *B. subtilis* remain to be elucidated, recent work in the organism *Pseudomonas syringae* has provided some provocative information regarding this process (37). In this organism, the 3'–5' exoribonuclease, RNase R, which heretofore had been considered a degradative enzyme (38), appears to be responsible for maturation of the 3' ends of 16S and 5S rRNAs. The *rnr* gene encoding RNase R was found to be required for normal growth and viability of this organism at 4 °C (37), although in contrast to *E. coli* (39, 40), the enzyme is not cold inducible. Most interestingly, the *rnr* mutant strain accumulates precursors of 16S and 5S rRNAs at 4 °C, and these molecules are incorporated into ribosomes. The 16S rRNA precursors that accumulate in the *rnr* strain contain 36 extra 3' nt up to the RNase III cleavage site in this organism. Some molecules also are unprocessed at the 5' end, but the significance of this observation is not yet understood (37). It might suggest that removal of the extra 3' residues facilitates 5' processing or that processing at the two termini is somehow coordinated. Further work is also required to prove that RNase R acts directly to remove 3' residues, and that it removes precursor residues right up to the mature 3' terminus. If the RNase R of *P. syringae* is as processive as that of *E. coli*, and is also able to digest structured RNAs (41, 42), some mechanism would be required to stop its action at the mature 3' end, perhaps proteins of the 30S subunit.

B. 23S rRNA

In *E. coli*, RNase III-mediated cleavage of the rRNA transcript generates a 23S rRNA precursor containing three or seven extra 5' residues and seven to nine extra 3' residues (43) (Fig. 2). Earlier studies had suggested that the 5' residues would be removed endonucleolytically, whereas processing at the 3' end would be carried out by an exoribonuclease. In the intervening years,

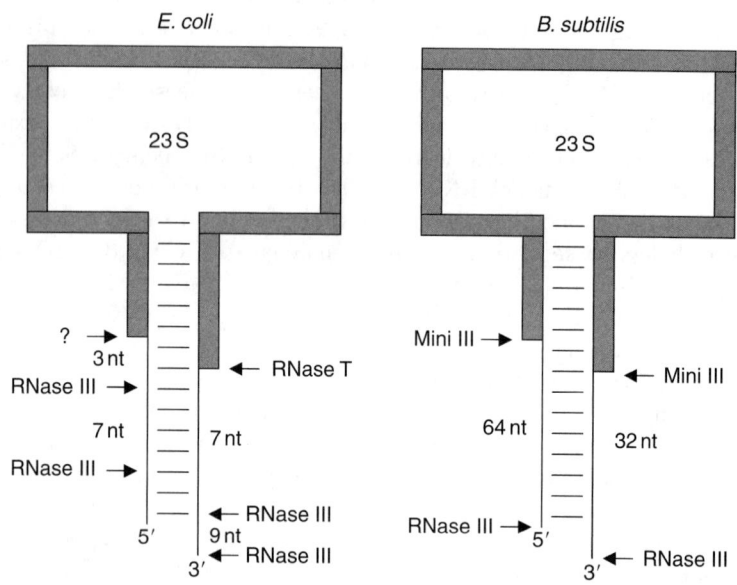

FIG. 2. Diagram of 23S rRNA maturation in *E. coli* and *B. subtilis*. Mature 23S sequences are shown as filled in area, and precursor sequences as a line. Enzymes acting at different positions are named, where known, and denoted with a **question mark** (?), if unknown. The number of precursor residues present, prior to the action of the RNase immediately above, are indicated.

almost no progress has been made in our understanding of the 5′ maturation process. Preliminary data from our laboratory (Lövgren and Deutscher, unpublished observations) with a partially purified activity supports the conclusion that removal of the extra 5′ nucleotides is due to an endoribonuclease as no partially shortened products are observed. However, since a processive 5′–3′ exoribonuclease would behave similarly, additional purification and product analysis will be necessary to conclusively establish an endonucleolytic mechanism.

On the other hand, 3′ maturation of 23S rRNA is now known to require the exoribonuclease, RNase T (44). In the absence of this enzyme, mutant cells accumulate precursors with extra 3′ residues, and little, if any, mature product is made. Other exoribonucleases may help to shorten the full 3′ extension, but RNase T is required for removal of the last few residues. Nevertheless, even in the absence of RNase T, 23S rRNA products with extra 3′ residues are incorporated into ribosomes, and cell growth is affected only slightly, indicating that final 3′ maturation is not required for functioning ribosomes.

Confirmation of a direct role for RNase T in 23S rRNA maturation came from in vitro studies (44). Purified RNase T rapidly and completely converted the 23S precursors in mutant ribosomes to the mature size. In contrast, naked 23S precursor was converted much more slowly and more product 1 nt shorter than the mature form was generated. These observations demonstrated a direct role for RNase T, and showed that processing is more efficient and more accurate when the rRNA precursor is in the context of the ribosome. The fact that maturation occurs so efficiently with fully assembled ribosomes also suggests that final 3′ maturation is a very late event in ribosome biogenesis. It is also interesting to note that 5′ processing of 23S rRNA is unaffected by events at the 3′ end. Thus, even though the 3′ and 5′ ends of 23S rRNA are paired, maturation of the 5′ end is an independent process.

As with 16S rRNA, final maturation of 23S rRNA in *B. subtilis* differs from that in *E. coli* (Fig. 2). A recently discovered member of the RNase III family of enzymes, named Mini-III, was shown to be responsible for maturation of both the 5′ and 3′ termini of *B. subtilis* 23S rRNA (45). In the absence of this enzyme, cells accumulate 23S precursors and aberrantly matured forms of 23S rRNA, the latter arising from secondary pathways that apparently utilize exoribonucleases. Although Mini-III can act on naked RNA, its action is much more rapid on 50S ribosome subunits containing the 23S precursor. Thus, as with most other rRNA maturation enzymes, processing is greatly facilitated by the context of the ribosome.

Mini-III is a dimeric enzyme, as is RNase III, and this structure presumably enables it to simultaneously cleave both sides of the double-stranded stalk flanking the mature 23S rRNA in the precursor molecule (45). Interestingly, although the action of Mini-III has much in common with RNase III, it lacks the double-stranded RNA binding domain present in other members of the RNase III family. Mini III has a distinct RNA binding motif, and a model of its structure suggests its mode of binding to RNA differs from that of RNase III. However, its catalytic mechanism is proposed to be similar to that of RNase III. *In vivo*, the two enzymes appear to have nonoverlapping functions.

The Mini-III enzyme has been identified in 110 sequenced prokaryotic genomes, amounting to almost 20% of the total. However, its distribution is quite narrow, existing primarily in Firmicutes and Cyanobacteria. It is absent from Proteobacteria and from Archaea (*45*).

In some bacterial species, maturation of 23S rRNA proceeds in an unusual manner. Thus, in *Salmonella*, intervening sequences are present which are removed by the action of RNase III (*46*). Inasmuch as the resulting fragments are not re-ligated, the mature 23S rRNA exists as fragments. In certain α-proteobacteria, an internal transcribed spacer is removed from the 5′ end of 23S rRNA. Cleavages by RNase III and RNase E are involved, but additional unknown RNases are necessary to generate the mature 5′ terminus (*47*).

C. 5S rRNA

5S rRNA is located the most 3′ of the three rRNAs in the primary transcript. In *E. coli*, RNase III cleavage releases the 5S rRNA and adjacent sequences as a 9S precursor (*14*) that is subsequently cleaved by RNase E (*16*) (Fig. 3). The resulting product still retains three extra nucleotides at each end that must be removed to generate the mature 5S rRNA. At present, we do not know how the three residues at the 5′ terminus are removed. However, the characteristics of the reaction with 5S rRNA are essentially identical to those observed with the 23S rRNA substrate (*44*). An RNase E/G type enzyme has also been shown recently to participate in 5S RNA maturation in the hyperthermophilic bacterium, *Aquifex acolicus* (*48*).

E. coli mutant strains lacking RNase T produce no mature 5S rRNA (*49*). In the absence of this enzyme, a processing intermediate containing primarily two extra 3′ residues and a mature 5′ end accumulates, and this molecule is assembled into functioning ribosomes. Thus, 5′ maturation of 5S rRNA is independent of events at the 3′ terminus, and the presence of extra 3′ residues has little effect on ribosome function. *In vitro*, purified RNase T accurately and efficiently removes the extra 3′ residues from ribosomes derived from RNase T-deficient cells to generate the mature 3′ end, and no further nucleotide removal occurs up to 1 h of incubation. The same experiment carried out with a naked RNA substrate proceeds more slowly, and the final product lacks the normal 3′ terminal residue of mature 5S rRNA (*49*). These data re-emphasize the importance of the ribosome context for efficient and accurate rRNA maturation.

rRNA maturation and degradation

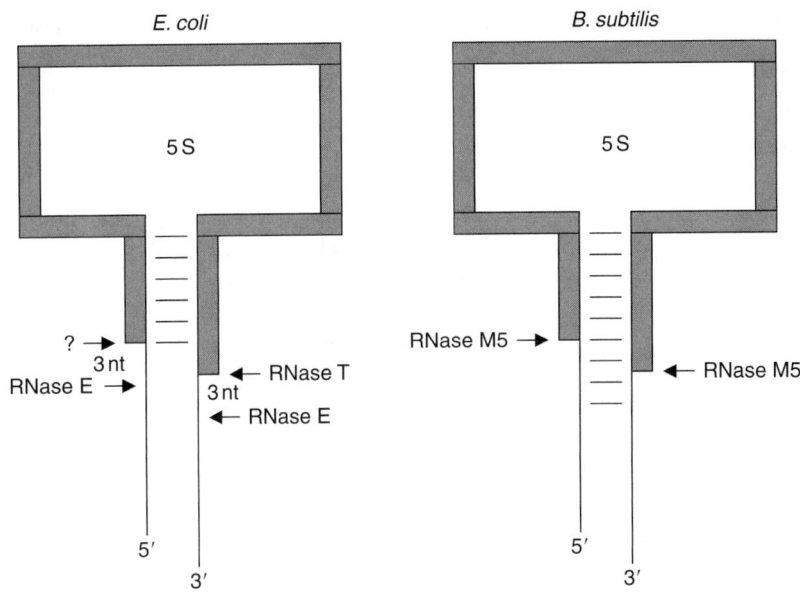

FIG. 3. Diagram of 5S rRNA maturation in *E. coli* and *B. subtilis*. Mature 5S sequences are shown as filled in area, and precursor sequences as a line. Enzymes acting at different positions are named, where known, and denoted with a **question mark (?)**, if unknown. The number of precursor residues present, prior to the action of the RNase immediately above, are indicated.

From this discussion, it is obvious that processing of the 3′ termini of 23S and 5S rRNA are strikingly similar. In both cases, only RNase T from among the eight *E. coli* exoribonucleases functions to remove the last 3′ residues to generate the mature 3′ terminus (*44, 49*). This is understandable because only RNase T is able to efficiently trim residues close to a double-stranded stem (*50*). In mature 5S rRNA, there is only a single unpaired 3′ residue following the stem (*49*), and in 23S rRNA, only two unpaired residues (*44*). So, in each case, trimming off the last precursor specific residues would require action essentially adjacent to a stem. For both 23S and 5S rRNA, other exoribonucleases may shorten the 3′ precursor sequences in the absence of RNase T, but for removal of the last few residues, RNase T is essential.

The similarity in 3′ maturation of 23S and 5S rRNAs extends even further (*44, 49*). Processing of each molecule appears to be a very late event, such that fully assembled 70S ribosomes, which must be functioning *in vivo* because RNase T-deficient cells grow almost normally (*51*), are substrates for RNase T *in vitro*. In addition, in each assembled RNA molecule, the extra 3′ residues are accessible to RNase T suggesting that they are not buried within the structure of the ribosome, but are exposed, consistent with the positions of the mature 3′

ends in ribosome crystal structures (52). Yet, in each case, only the correct number of residues are removed by RNase T (44, 49). Consequently, we can conclude that only these residues are exposed in the assembled ribosome or that, once the mature RNA is made, the structure of the ribosome changes to protect against further action by RNase T. The fact that RNase T acts distributively, rather than processively, would allow such a mechanism (53).

The similar nature of the 3′ processing raises the interesting question of how closely related the 5′ maturation of 23S and 5S rRNA might be. We already know that both precursor molecules carry short 5′ extensions adjacent to a double-stranded stem (44, 50). Likewise, in both cases, 5′ maturation proceeds independently of completion of 3′ maturation. Based on these observations, it would not be surprising if the same enzyme were responsible for 5′ maturation of both 23S and 5S rRNA. Interestingly, in *P. syringae*, RNase R appears to participate in the processing of the 3′ end of 5S rRNA (37), although it is not yet clear whether it removes the last few precursor residues adjacent to the mature terminus.

As discussed earlier, 5S rRNA maturation in *B. subtilis* is carried out by a single enzyme, RNase M5, which cleaves in a double-stranded region to generate both the mature 5′ and 3′ termini (19) (Fig. 3). The similarity to the recently described RNase Mini-III that acts on 23S rRNA is evident (45), so that in this organism as well, there are striking similarities between 23S and 5S rRNA maturation. Additionally, neither Mini-III nor RNase M5 are essential enzymes, and functioning ribosomes can be made containing precursors to both 23S and 5S rRNA (45, 54), a situation identical to that in *E. coli*. Microarray analysis of *B. subtilis* lacking RNase M5 indicated that the enzyme has little, if any, action on mRNAs in this organism (55). Thus, while RNase M5 is distributed widely, particularly among Gram-positive bacteria (54), its role, so far, is limited to the nonessential cleavage of 5S rRNA precursors.

D. Other Factors Involved in rRNA Processing

The foregoing discussion has emphasized the pathways of rRNA processing and the RNases that carry out these processes. However, as has been mentioned repeatedly, essentially all rRNA maturation reactions occur in the context of the assembling ribosome or in association with at least some ribosomal proteins (43). It is most likely that ribosome biogenesis involves an ordered series of events that include addition of ribosomal proteins, cleavage and trimming of rRNA precursors, and rRNA modifications. Each step probably facilitates the accuracy and efficiency of the following step until the ribosome subunit has been accurately assembled. That is not to say that some events could not be bypassed or that alternative pathways might not be possible. We have already seen that ribosomes may function even though certain normal rRNA processing events have not taken place, and it is also known that ribosomes may function in the absence of certain rRNA modifications. Nevertheless, a preferred assembly pathway for ribosomes likely exists, and deviations from it lead to ribosomes that do not function optimally or are affected by temperature or other stress conditions. If the misassembly is very dramatic, the ribosome may, in fact, be degraded, and that process will be discussed in the following section as RNA quality control.

The strongest evidence for this framework comes from the very large number of mutations that affect ribosome biogenesis, and the fact that in many of these mutant strains the rRNAs are not fully matured (1, 2). As a consequence, it is clear that rRNA maturation can be affected by many factors distinct from the RNases that actually carry out the processing steps. A detailed discussion is beyond the scope of this review, but these factors include individual ribosomal proteins, RNA helicases, rRNA modification enzymes, chaperones, and certain GTPases (2). In almost all cases, it is not clear how these mutations ultimately affect rRNA maturation. Some may be direct effects, and others indirect. However, if ribosome assembly is considered as the ordered series of events described earlier, it would not be at all surprising that a misplaced ribosomal protein, the lack of an RNA modification or the absence of an RNA rearrangement could influence an RNA processing event. For example, RNA cleavage sites might be blocked or altered and RNA trimming reactions might not stop at the appropriate residue leading to precursors that cannot be processed or which are processed inaccurately. So, even when all the rRNA processing RNases have been identified, and the rRNA maturation reactions described, much work will still be required to completely understand how rRNA processing coordinates with ribosome assembly.

IV. Degradation of rRNA

Ribosomes and rRNAs generally are extremely stable in growing cells, but a number of distinct circumstances can lead to rRNA degradation (reviewed in 21). One such situation, which probably occurs at a low level throughout growth, is a quality control process that removes grossly misassembled ribosomes that arise due to errors in the complex process of ribosome biogenesis. Misassembled ribosomes may also be present due to exogenous agents that interfere with the assembly process. Secondly, rRNA may be degraded as a consequence of damage to the cell membrane by any of a variety of chemicals. Such damage may allow entry into the cell of the nonspecific endoribonuclease, RNase I, which can then attack even intact ribosomes and initiate rRNA breakdown. Finally, rRNA may be degraded as a physiological response to poor environmental conditions that lead to slowed growth and diminished protein synthesis. While the initiating mechanism may differ in each of these

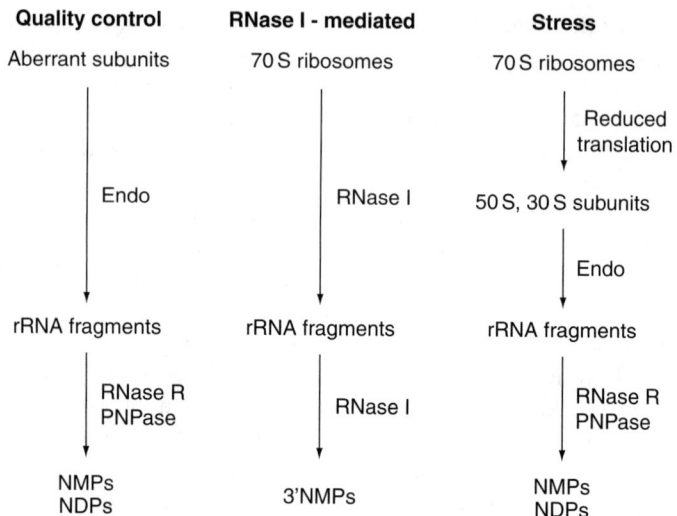

FIG. 4. Proposed modes of ribosome degradation. Three different mechanisms of ribosome turnover are suggested, each with distinct initial substrates. In quality control, the substrates are aberrant or misassembled ribosome subunits; RNase I that has leaked into the cell due to membrane damage can act even on intact 70S ribosomes; and during stress, the decrease in translation leads to an accumulation of 30S and 50S subunits that are the substrates. In each case, cleavages are initiated by an endoribonuclease, either RNase I or an as yet unknown RNase (s). The rRNA fragments generated are further broken down by the exoribonucleases, PNPase and RNase R to NDP's and NMP's, respectively. When RNase I is present, it can also act on the rRNA fragments, but its final products would be 3' NMP's.

situations, the actual degradative machinery may overlap or even be identical. At present, we are in the early stages of understanding the pathways and RNases responsible for rRNA degradation; our current knowledge in this area is discussed in this section (Fig. 4).

A. Misassembly and rRNA Quality Control

Defective ribosomes have the potential to interfere with the function of their normal counterparts. Hence, the presence of mechanisms to remove them is not unexpected; however, direct evidence for such a process was obtained only recently. Several hints that an rRNA quality control process might exist were obtained over the years from studies of mutations in rRNA. For example, many deletion mutants of *E. coli* 23S rRNA derived from a plasmid-encoded gene are stably incorporated into ribosomal particles, whereas certain others are not (56). In particular, mutations which remove the 5' or 3' terminal regions of 23S rRNA are not found in RNP particles. A similar situation was observed as a consequence of mutations in the 5' leader region of 16S rRNA even though this region is removed during maturation (57). In each of these instances, the mutations in rRNA presumably led to a ribosome subunit that is assembled improperly, rendering the constituent rRNA susceptible to degradation. One can imagine similar effects due to mutations in ribosomal proteins. Indeed, a mutation which alters ribosomal protein S10 also leads to substantial degradation of 16S rRNA (58).

More direct evidence for a rRNA quality control process comes from studies of on *E. coli* mutant strain deficient in the exoribonucleases, PNPase and RNase R (59). Such a strain is normally nonviable; however, it can be studied if the null *rnr* mutation is combined with a temperature-sensitive mutation in *pnp*. As expected, upon elevation of the temperature to 42 °C, this strain ceases growth after a few generations and rapidly loses viability. Examination of RNAs at various times after the shift to 42 °C revealed a decrease in 16S and 23S rRNA and a massive accumulation of large fragments derived from each of these molecules, amounting to about half of the total rRNA. No fragments were observed at 31 °C, a temperature at which PNPase is active. The full length 16S and 23S rRNA species both contained mature 5' and 3' termini indicating that the absence of PNPase and RNase R activities did not affect rRNA maturation, as would be expected from what is known about this process in *E. coli*. Ribosome analysis and pulse-chase experiments showed both the disappearance of preexisting ribosomes and the inability to fully assemble new ones. Fragments of rRNA appeared within 5 min of the pulse.

The finding that fragments of 16S and 23S rRNA accumulate in cells deficient in PNPase and RNase R, but not in cells containing either enzyme (59), indicates that the fragments normally are eliminated by these RNases, and explains why they had not been observed previously. Nevertheless, they do show that rRNA, a presumed stable molecule, does turn over. The identity of the enzyme(s) that generate the fragments is not known, but the disappearance of both preexisting and newly-assembled particles demonstrates that both can be substrates. These observations can be explained by a model in which the preexisting ribosomes become substrates for degradation as a result of cessation of growth (to be discussed later), while the newly-made rRNA is degraded because these ribosomes do not fully assemble. It is likely that the latter situation occurs because the accumulating rRNA fragments sequester ribosomal proteins leaving an insufficient amount for the assembly of new ribosomes. Eventually, after a few generations under these conditions, the cells lose viability.

Misassembly of ribosome subunits can be enhanced by certain antibiotics (reviewed in 21). For example, erythromycin interferes with 50S subunit assembly and neomycin does the same to 30S subunits. In the presence of these agents, there is increased degradation of rRNA. As a result, it was predicted that cells depleted of an RNase involved in the degradative process would show enhanced sensitivity to the antibiotics (60). A test of this hypothesis using cells treated with azithromycin indicated that PNPase and RNase II participate in degradation of 23S rRNA. Unfortunately, mutant strains deficient in RNase R, previously implicated in rRNA degradation (59), were not examined.

Why does misassembly of ribosome subunits lead to rRNA degradation? The most obvious explanation is that rRNA normally is protected in fully assembled subunits against the action of endoribonucleases, whereas some part of the RNA is accessible when the subunit is not intact. However, an alternative explanation, which will be described in detail below, is that misassembled subunits cannot function in protein synthesis, and therefore, do not form 70S ribosomes, and perhaps, it is actually the 70S ribosome which is resistant to RNase action.

B. RNase I-Mediated rRNA Degradation

Intact 70S ribosomes are generally resistant to the action of RNases. The major exception to this conclusion is the nonspecific endoribonuclease, RNase I (61). Consequently, it is imperative that active RNase I not be present inside cells, and accordingly, most RNase I is present in the periplasm of Gram-negative cells. However, a small pool of RNase I is present intracellularly, although presumably in an inactive state. How it is kept inactive is unclear; presumably, the reducing environment of the cell plays a role, and the possibility of an RNase I inhibitor, analogous to the RNase inhibitor in eukaryotic cells, also needs to be considered.

A large number of agents or conditions promote the breakdown of cellular RNA, which consists primarily of rRNA. In many cases, these agents or conditions affect the cell membrane, which could allow the entry of periplasmic RNase I into the cytoplasm resulting in the extensive RNA turnover observed. A variety of studies (reviewed in 21) strongly support the conclusion that RNase I is involved. Most direct, are experiments which show that degradation does not occur in RNase I$^-$ strains under the particular conditions used. Thus, it is clear that RNase I can be a mediator of rRNA degradation in certain special situations. What is not clear is whether RNase I serves any role in rRNA breakdown under usual physiological conditions, and if so, how its entry into the cell or activation in the cytoplasm might be modulated.

C. rRNA Degradation as a Response to Poor Environmental Conditions

Cells adjust their ribosome concentration to growth rate such that fast growing cells may have 10-fold more ribosomes than slow growing ones (62). As cells encounter conditions of nutritional deprivation and their growth rate

rRNA maturation and degradation

slows or ceases, excess ribosomes may be degraded (reviewed in 21). Studies of this phenomenon in the laboratory generally involve rapid and dramatic changes of medium which do not reflect the situation in nature in which the depletion of nutrients would be expected to be gradual. In the latter case, cells may initially be able to adjust their ribosome complement by stopping ribosome synthesis and diluting out the excess ribosomes through cell division. At longer times of nutrient deprivation, ribosome degradation would eventually occur as well. In the laboratory situation, ribosome degradation seems to be the norm, and although not completely reflective of the process as it occurs in the natural environment, it does provide information on the capabilities and enzymatic machinery present in cells that carry out the degradative process.

Important early studies provided support for a model in which rRNA in ribosomes would first be attacked by an endoribonuclease, suggested to be RNase I, followed by exonuclease action on the resulting oligonucleotides to release mononucleotides (26). The exonucleases involved were initially suggested to be PNPase and RNase II (26), but subsequent studies raised questions about the involvement of RNase II, and pointed to a different hydrolytic enzyme (28). Inasmuch as many RNases had not yet been identified at the time of these early studies, the enzymology of the process was necessarily incomplete. For example, it is now likely that RNase I involvement is restricted to high temperature when that enzyme can enter the cell from the periplasm. However, the identity of the endoribonuclease(s) that initiates ribosome breakdown at 37 °C has still not been determined conclusively, although *in vitro* studies have shown that RNases E and G are able to carry out such cleavages (Zundel and Deutscher, unpublished observations). One report has suggested that the endoribonuclease, RNase LS, may participate in the degradation of 23S rRNA as a fragment of this molecule accumulated in its absence (63)

With regard to the exoribonucleases that participate in rRNA degradation, it is now known that only PNPase, in association with an RNA helicase, or RNase R are able to digest highly structured RNAs (42). In fact, as noted earlier, fragments of rRNA accumulate in the absence of these enzymes during rRNA quality control. Therefore, it is extremely likely that the same two exoribonucleases would participate in the rRNA degradation caused by nutrient deprivation. In addition, studies *in vitro* have confirmed that PNPase and RNase R can act on ribosome sububits (Zundel and Deutscher, unpublished observations).

A major question surrounding the degradation of ribosomes under stress conditions is how the process is initiated. How do ribosomes that previously were stable become destabilized such that they are susceptible to the action of RNases? Formally, there appear to be only two possibilities: (*1*) a new RNase is activated or a preexisting RNase is elevated, or (*2*) the structure of the ribosome is altered to render it sensitive to RNases already present. The simplest explanation that would allow ribosome degradation to be built into the system when cells enter slow or no growth would be to have the actual substrates be the 30S and 50S ribosome subunits. As growth slows, fewer ribosomes would be engaged in translation and would, therefore, remain as subunits. Structural information supports this idea since it is now known that the subunit interfaces are primarily RNA (64). These regions would be protected in the translating 70S ribosome, but would become exposed in the free subunits, providing a readily accessible site for RNase action. The simplicity of this model makes it very appealing, and would allow cells to remove excess ribosomes because those that are not in use are by their very structure substrates for degradation.

The fact that ribosomes may turn over under certain conditions, although recognized for many years, has not generally been appreciated, partly due to the terminology of rRNA being considered a "stable RNA." However, this area of investigation is now receiving increased attention. Future work will be directed towards conclusively identifying the RNases that participate in the process and defining the sites at which the rRNAs are cleaved. These studies, together with those that support regulation of the process (e.g., RNase R, a known participant in rRNA degradation, increases dramatically under conditions in which rRNA is destabilized (39, 40)), will greatly clarify our understanding of this important metabolic pathway of rRNA.

Acknowledgments

Research from the author's laboratory was made possible by support from the National Institutes of Health. I thank many of my colleagues who, over the years, have contributed to this work.

References

1. Noller HF, Nomura M. Ribosomes. In: Neidhardt FC, editor. *Escherichia coli and Salmonella*, 2nd ed. Washington, DC: ASM Press; 1996. p. 167–86.
2. Kaczanowska M, Ryden-Aulin M. Ribosome biogenesis and the translation process in *Escherichia coli*. *Microbiol Mol Biol Rev* 2007;**71**:477–94.
3. Brasius J, Dull TJ, Sleeter D, Noller HF. Gene organization and primary structure of a ribosomal RNA operon from *Escherichia coli*. *J Mol Biol* 1981;**148**:107–27.
4. LaFauci G, Widom RL, Eisner RL, Jarvis ED, Rudner R. Mapping of rRNA genes with integrable plasmids in *Bacillus subtilis*. *J Bacteriol* 1986;**165**:204–14.
5. Jarvis ED, Widom RL, LaFauci G, Setoguchi Y, Richter IR, Rudner R. Chromosomal organization of rRNA operons in *Bacillus subtilis*. *Genetics/Society* 1988;**120**:625–35.
6. Young RA, Steitz JA. Complementary sequences 1700 nucleotides apart form a ribonuclease III cleavage site in *Escherichia coli* ribosomal precursor RNA. *Proc Natl Acad Sci USA* 1978;**75**:3593–7.

7. Bram RJ, Young RA, Steitz JA. The ribonuclease III site flanking 23S sequences in the 30S ribosomal precursor RNA of *E. coli*. *Cell* 1980;**19**:393–401.
8. Robertson HD, Webster RE, Zinder ND. Purification and properties of ribonuclease III from *Escherichia coli*. *J Biol Chem* 1968;**243**:82–91.
9. Dunn JJ, Studier FW. T7 early RNAs and *Escherichia coli* ribosomal RNAs are cut from large precursors *in vivo* by ribonuclease III. *Proc Natl Acad Sci USA* 1973;**70**:3296–300.
10. Gegenheimer P, Apirion D. *Escherichia coli* ribosomal ribonucleic acids are not cut from an intact precursor molecule. *J Biol Chem* 1975;**250**:2407–9.
11. King CK, Schlessinger D. S1 nuclease mapping analysis of ribosomal RNA processing in wild type and processing deficient *Escherichia coli*. *J Biol Chem* 1983;**258**:12034–42.
12. Dahlberg AE, Dahlberg JE, Lund E, Tokimatsu H, Rabson AB, Calvert PC, *et al*. Processing of the 5′ end of *Escherichia coli* 16S ribosomal RNA. *Proc Natl Acad Sci USA* 1978;**75**:3598–602.
13. Hayes F, Vasseur M. Processing of the 17-S *Escherichia coli* precursor RNA in the 27S preribosomal particle. *Eur J Biochem* 1976;**61**:433–42.
14. Ghora BK, Apirion D. Structural analysis and *in vitro* processing to p5 rRNA of 9S RNA molecule isolated from an *rne* mutant of *E. coli*. *Cell* 1978;**15**:1055–66.
15. Ghora BK, Apirion D. Identification of a novel RNA molecule in a new RNA processing mutant of *Escherichia coli* which contains 5S rRNA sequences. *J Biol Chem* 1979;**254**:1951–6.
16. Misra TK, Apirion D. RNase E, an RNA processing enzyme from *Escherichia coli*. *J Biol Chem* 1979;**254**:11154–9.
17. Feunteun J, Jordan BK, Monier R. Study of the maturation of 5S rRNA precursors in *Escherichia coli*. *J Mol Biol* 1972;**70**:465–74.
18. Szeberenyi J, Tomcsanyi T, Apirion D. Maturation of the 3′ end of 5-S ribosomal RNA from *Escherichia coli*. *Eur J Biochem* 1985;**149**:113–8.
19. Sogin ML, Pace B, Pace NR. Partial purification and properties of a ribosomal RNA maturation endonuclease from *Bacillus subtilis*. *J Biol Chem* 1977;**252**:1350–7.
20. Stahl DA, Pace B, Marsh T, Pace NR. The ribonucleoprotein substrate for a ribosomal RNA-processing nuclease. *J Biol Chem* 1984;**259**:11448–53.
21. Pace B, Stahl DA, Pace NR. The catalytic element of a ribosomal RNA-processing complex. *J Biol Chem* 1984;**259**:11454–8.
22. Allemand F, Mathy N, Brechemier-Bacy D, Condon C. The 5S rRNA maturase, ribonuclease M5, is a toprim domain family member. *Nucleic Acids Res* 2005;**33**:4368–76.
23. Norris TE, Koch AL. Effects of growth rate on the relative rates of synthesis of messenger, ribosomal and transfer RNA in *Escherichia coli*. *J Mol Biol* 1972;**64**:633–49.
24. Lindahl L. Intermediates and time kinetics of the *in vivo* assembly of *Escherichia coli* ribosomes. *J Mol Biol* 1975;**92**:15–37.
25. Deutscher MP. Degradation of stable RNA in bacteria. *J Biol Chem* 2003;**278**:45041–4.
26. Kaplan R, Apirion D. The involvement of ribonuclease I, ribonuclease II, and polynucleotide phosphorylase in the degradation of stable ribonucleic acid during carbon starvation in *Escherichia coli*. *J Biol Chem* 1974;**249**:149–51.
27. Kaplan R, Apirion D. The fate of ribosomes in *Escherichia coli* starved for a carbon source. *J Biol Chem* 1975;**250**:1854–63.
28. Cohen L, Kaplan R. Accumulation of nucleotides by starved *Escherichia coli* cells as a probe for the involvement of ribonucleases in ribonucleic acid degradation. *J Bacteriol* 1977;**129**:651–7.
29. Li Z, Pandit S, Deutscher MP. RNase G (Caf A protein) and RNase E are both required for the 5(maturation of 16S ribosomal RNA). *EMBO J* 1999;**18**:2878–85.

30. Wachi M, Umitsuki G, Shimizu M, Takada A, Nagai K. Escherichia coli cafA gene encodes a novel RNase, designated as RNase G, involved in processing of the 5' end of 16S rRNA. Biochem Biophys Res Commun 1999;**259**:483–8.
31. Wachi M, Umitsuki G, Nagai K. Functional relationship between Escherichia coli RNase E and cafA protein. Mol Gen Genet 1997;**253**:515–9.
32. Condon C. Maturation and degradation of RNA in bacteria. Curr Opin Microbiol 2007;**10**:1–8.
33. Britton RA, Wen T, Shaefer L, Pellegrini O, Vicker WC, Mathy N, et al. Maturation of the 5' end of Bacillus subtilis 16S rRNA by the essential ribonuclease YkqC/RNase J1. Mol Microbiol 2007;**63**:127–38.
34. Even S, Pellegrini O, Zig L, Labas V, Vinh J, Brechemmier-Baey D, et al. Ribonucleases J1 and J2: Two novel endoribonucleases in B. subtilis with functional homology to E. coli RNase E. Nucleic Acids Res 2005;**33**:2141–52.
35. Mathy N, Benard L, Pellegrini O, Daov R, Wen T, Condon C. 5'–3' exoribonuclease activity in bacteria: role of RNase J1 in rRNA maturation and 5' stability of mRNA. Cell 2007;**129**:681–92.
36. de la Sierra-Gallay IL, Zig L, Jamalli A, Putzer H. Structural insights into the dual activity of RNase J. Nal Struct Mol Biol 2008;**2**:206–12.
37. Purusharth RI, Madhuri B, Ray MK. Exoribonuclease R in Pseudomonas syringae is essential for growth at low temperature and plays a novel role in the 3' end processing of 16S and 5S ribosomal RNA. J Biol Chem 2007;**282**:16267–77.
38. Deutscher MP. Degradation of RNA in bacteria: Comparison of mRNA and stable RNA. Nucleic Acids Res 2006;**34**:659–66.
39. Cairrao F, Cruz A, Mori H, Arraiano CM. Cold shock induction of RNase R and its role in the maturation of the quality control mediator SsrA/ tmRNA. Mol Microbiol 2003;**50**:1349–60.
40. Chen C, Deutscher MP. Elevation of RNase R in response to multiple stress conditions. J Biol Chem 2003;**280**:34393–6.
41. Cheng Z, Deutscher MP. Purification and characterization of the Escherichia coli exoribonuclease RNase R: comparison with RNase II. J Biol Chem 2002;**277**:21624–9.
42. Cheng ZF, Deutscher MP. An important role for RNase R in mRNA decay. Mol Cell 2005;**17**:313–8.
43. King TC, Sirdeskmukh R, Schlessinger D. Nucleolytic processing of ribonucleic acid transcripts in prokaryotes. Microbiol Rev 1986;**50**:428–51.
44. Li Z, Pandit S, Deutscher MP. Maturation of 23S ribosomal RNA requires the exoribonuclease RNase T. RNA 1999;**5**:139–46.
45. Redko Y, Bechhofer DH, Condon C. Mini-III, an unusual member of the RNase III family of enzymes, catalyzes 23S ribosomal RNA maturation in B. subtilis. Mol Microbiol 2008;**68**:1096–106.
46. Burgin AB, Parados K, Lane DJ, Pace NR. The excision of intervening sequences from Salmonella 23S ribosomal RNA. Cell 1990;**60**:405–14.
47. Klein F, Evguenieva-Hackenberg E. RNase E is involved in 5'-end 23S rRNA processing in α-proteobacteria. Biochem Biophys Res Commun 2002;**299**:780–6.
48. Lombo TB, Kaberdin VR. RNA processing in Aquifex aeolicus involves RNase E/G and an RNase P-like activity. Biochem Biophys Res Commun 2008;**366**:457–63.
49. Li Z, Deutscher MP. The tRNA processing enzyme RNase T is essential for maturation of 5S rRNA. Proc Natl Acad Sci USA 1995;**92**:6883–6.
50. Li Z, Pandit S, Deutscher MP. 3' exoribonucleolytic trimming is a common feature of the maturation of small stable RNAs in Escherichia coli. Proc Natl Acad Sci USA 1998;**95**:2856–61.

51. Padmanabha KP, Deutscher MP. RNase T affects *Escherichia coli* growth and recovery from metabolic stress. *J Bacteriol* 1991;**173**:1376–81.
52. Schuwirth BS, Borovinskaya MA, Hau CW, Zhang W, Vila-Sanjurjo A, Holton JM, et al. Structures of the bacterial ribosome at 3.5 Å resolution. *Science* 2005;**310**:827–34.
53. Deutscher MP, Marlor CW. Purification and characterization of *Escherichia coli* RNase T. *J Biol Chem* 1985;**260**:7067–71.
54. Condon C, Brechemier-Baev D, Grunberg-Manago M, Putzer H. Identification of the gene encoding the 5S ribosomal RNA maturase in *Bacillus subtilis*: Mature 5S rRNA is dispensable for ribosome function. *RNA* 2001;**7**:242–53.
55. Condon C, Rourera J, Brechemier-Baev D, Putzer H. Ribonuclease M5 has few, if any, mRNA substrates in *Bacillus subtilis*. *J Bacteriol* 2002;**184**:2845–9.
56. Liiv A, Tenson T, Remme J. Analysis of the ribosome large subunit assembly and 23S rRNA stability *in vivo*. *J Mol Biol* 1996;**263**:396–410.
57. Schaferkordt J, Wagner R. Effects of base change mutations within an *Escherichia coli* ribosomal RNA leader region on rRNA maturation and ribosome formation. *Nucleic Acids Res* 2001;**29**:3394–403.
58. Kuwano M, Taniguchi H, Ono M, Endo H, Ohnishi Y. An *Escherichia coli* K12 mutant carrying altered ribosomal protein (S10). *Biochem Biophys Res Commun* 1997;**75**:156–62.
59. Cheng ZF, Deutscher MP. Quality control of ribosomal RNA mediated by polynucleotide phosphorylase and RNase R. *Proc Natl Acad Sci USA* 2003;**100**:6388–93.
60. Silvers JA, Champney WS. Accumulation and turnover of 23S ribosomal RNA in azithromycin-inhibited ribonuclease mutant strains of *Escherichia coli*. *Arch Microbiol* 2005;**184**:66–7.
61. Shen V, Schlessinger D. RNases I, II, and IV of *Escherichia coli*. In: Boyer PD, editor. *The Enzymes*, Vol. XV. part B. New York: Academic Press: 1982. P. 501–15.
62. Bremer H, Dennis PP. Modulation of chemical composition and other parameters of the cell by growth rate. In: Neidhardt FC, editor. *Escherichia coli and Salmonella*, 2nd ed. Washington, DC: ASM Press; 1996. p. 1553–69.
63. Otsuka Y, Yonesaki T. A novel doribonuclease, Rnase 1S, in *Escherichia coli*. *Genetics/Society* 2005;**169**:13–20.
64. Rackham O, Wang K, Chin JW. Functional epitopes at the ribosome subunit interface. *Nat Chem Biol* 2006;**2**:254–8.

RNA Polyadenylation and Decay in Mitochondria and Chloroplasts

GADI SCHUSTER[*] AND
DAVID STERN[†]

[*]*Department of Biology, Technion—Israel Institute of Technology, Haifa 32000, Israel*

[†]*Boyce Thompson Institute for Plant Research, Tower Rd., Ithaca, New York 14853*

I. Introduction	394
II. Polyadenylation of RNA	395
A. The Stable Poly(A) Tail of Nucleus-Encoded mRNA	395
B. The Polyadenylation-Stimulated Degradation Pathway	395
C. Is Polyadenylation A Required Step in The Degradation Pathway?	397
III. The Enzymes	399
A. Endoribonucleases	399
B. Exoribonucleases	404
C. The Family of Poly(A)-Adding Enzymes	408
IV. RNA Degradation and Polyadenylation in Chloroplasts	408
V. RNA Degradation and Polyadenylation in Mitochondria	409
A. Plant Mitochondria	409
B. Yeast Mitochondria: RNA Metabolism Without Polyadenylation	410
C. Trypanosome Mitochondria: Both Stable and Unstable Poly(A) Tails	411
D. Animal Mitochondria	412
VI. Conclusions and Perspectives	413
References	414

Mitochondria and chloroplasts were originally acquired by eukaryotic cells through endosymbiotic events and retain their own gene expression machinery. One hallmark of gene regulation in these two organelles is the predominance of posttranscriptional control, which is exerted both at the gene-specific and global levels. This review focuses on their mechanisms of RNA degradation, and therefore mainly on the polyadenylation-stimulated degradation pathway. Overall, mitochondria and chloroplasts have retained the prokaryotic RNA decay system, despite evolution in the number and character of the enzymes involved. However, several significant differences exist, of which the presence of stable poly(A) tails, and the location of PNPase in the intermembrane space in animal mitochondria, are perhaps the most remarkable. The known and predicted proteins taking part in polyadenylation-stimulated degradation pathways are described, both in chloroplasts and four mitochondrial types: plant, yeast, trypanosome, and animal.

Abbreviations

PAP	poly(A) polymerase
PNPase	polynucleotide phosphorylase
rNTr	ribonucleotidyl transferase
RNase E	ribonuclease E
RNase J	ribonuclease J

I. Introduction

Chloroplasts and mitochondria originated from what are considered to be the most successful symbiotic events to have occurred over the last 1.5 billion years (1–3). It is assumed that the chloroplast originated from a cyanobacterial ancestor, while mitochondria arose from a α-proteobacterium. In both symbiotic events, a prokaryotic organism entered the eukaryotic precursor and this was followed by extensive gene transfer from the organelle to the nuclear genome. Today, of the thousands of proteins present in organelles, only a very limited number remain encoded by the organellar genome. For example, only 13 proteins are encoded in the human mitochondrial genome, and about 90 in the chloroplasts of *Arabidopsis*. Still, organelles harbor a complete gene expression system, which includes DNA replication and maintenance, transcription, posttranscriptional, translational, and posttranslational activities, *albeit* much of it nucleus-encoded. The posttranscriptional components include splicing, editing, 3′, 5′, and intercistronic processing, and the addition of stable poly(A) tails in the case of animal mitochondria. Several of these events can modulate RNA half-life, which is a very significant factor in the regulation of organellar gene expression (4, 5).

This chapter primarily concerns the polyadenylation and degradation of organellar RNAs, as it is a central organellar process derived from prokaryotes. However, a special characteristic of animal and trypanosome mitochondria is also discussed, namely the presence of stable poly(A) tails, which have not been found in prokaryotes. The enzymes involved in RNA degradation, including exo and endoribonucleases will be described, as well as the various incarnations of the polyadenylation-stimulated degradation pathway found in organelles and prokaryotic organisms.

II. Polyadenylation of RNA

A. The Stable Poly(A) Tail of Nucleus-Encoded mRNA

The addition of a stable poly(A) tail to the 3′ end of nucleus-encoded mRNA (excluding histone mRNAs) is a well-defined and long-known phenomenon in eukaryotes (Fig. 1) (6). Historically, the observations that mRNA was mostly retained on an oligo(dT) column, and that cDNA could be obtained by reverse transcription of total RNA using an oligo(dT) primer, paved the way for the discovery of a stable poly(A) tail at the 3′ end. Further biochemical analysis showed that following transcription by RNA polymerase II, mRNA is cleaved and polyadenylated by a high molecular weight complex consisting of several proteins. The stable poly(A) tail functions in the transport of mRNA from the nucleus to the cytoplasm and in translation initiation. In addition, it is significantly shortened during the initial steps of RNA degradation (7–10). However, whether or not it is required for stability and/or determines the half-life of the transcript is still a matter of debate. It is assumed that in the nucleus and cytoplasm, the stable poly(A) tail is fully bound by the poly(A) binding protein (11) (Fig. 1).

B. The Polyadenylation-Stimulated Degradation Pathway

In a somewhat opposing manner to the function of the stable poly(A) tail of nucleus-encoded mRNA, the prokaryotic/organellar poly(A) tail usually functions to tag the RNA molecule for rapid exonucleolytic degradation. This phenomenon was first identified in *E. coli* (see Chapter in this volume by Hajnsdorf and Regnier), but is now well-known in all kingdoms of life including

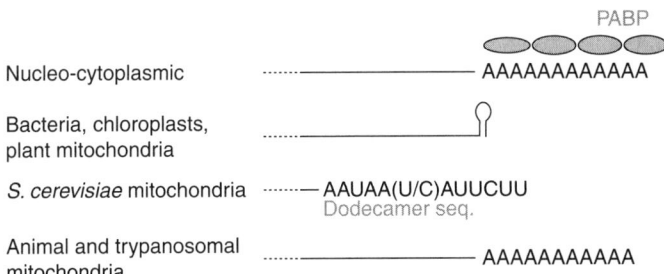

FIG. 1. The 3′ ends of mature mRNAs in various systems. The details of each are discussed in the text. "Dodecamer seq." is an encoded tag found at the end of yeast mitochondrial transcripts.

prokaryotes, archaea and organelles, and the nucleus of eukaryotic cells. Indeed, it has been found in all organisms analyzed to date, excluding the few that are described below.

In bacteria and chloroplasts, the initial event in the mRNA degradation pathway is generally thought to be an endonucleolytic cleavage (Fig. 2), preceded in some cases in *E. coli*, and probably other bacteria, by the RppH-catalyzed removal of 5' pyrophosphate (12, 13). Following the initial cleavage, a wave of endonucleolytic cleavages may degrade the RNA into many fragments. These fragments are then digested by exonucleases, with or without preceding polyadenylation (Fig. 2). The different endoribonucleases that may be involved are described below. Because in some situations none of these candidate endoribonucleases have been found, for example, in plant and human mitochondria, it may be that the endonucleolytic cleavage does not take place. In these cases, RNA degradation may begin directly at the polyadenylation/exonucleolytic step (14).

As mentioned earlier, following endonucleolytic cleavage the RNA fragment can be polyadenylated and exonucleolytically degraded (Fig. 2). Therefore, unlike stable nuclear poly(A) tails, the polyadenylation in this context is transient. Accordingly, progressive RT-PCR amplification methods are required to detect these tails. The enzymes performing the polyadenylation step are polynucleotide phosphorylase (PNPase) and several poly(A) polymerases (PAPs) of the nucleotidyltransferase (Ntr) family. The tails can be either homopolymeric, composed exclusively of adenosines (A), or heteropolymeric, composed of the four nucleotides, with adenosine being the most abundant (poly(A)-rich tails) (Fig. 3). "Chimeric" tails where part is heteropolymeric and part homopolymeric, were also recently observed in the chloroplasts of different plants (Larum and Schuster, manuscript in preparation). Generally, homopolymeric tails are produced by PAP, while heteropolymeric tails result from PNPase activity functioning in synthetic rather than degradation mode. The pervasiveness and transience of such poly(A) tails is remarkable, and raises the question of why RNA fragments are elongated as a prelude to their degradation.

The answer is that unstable poly(A) tails are believed to serve as a platform or runway for exonucleases to bind the 3' end of the RNA and degrade it in the 3'–5' direction. It is very likely that the addition of the tail enables the exonuclease to digest RNA even with stem-loops and other structures that normally function as efficient barriers to exonucleases. It is also possible that this step is not built on a single polyadenylation and processive exonucleolytic degradation event, but rather on repeated cycles of polyadenylation and degradation. That is, whenever a normally processive exoribonucleases is stalled by an RNA structure, it dissociates and a new polyadenylation event occurs, adding a platform for a new molecule of the exonuclease, and perhaps modifying the secondary structure in order to weaken it.

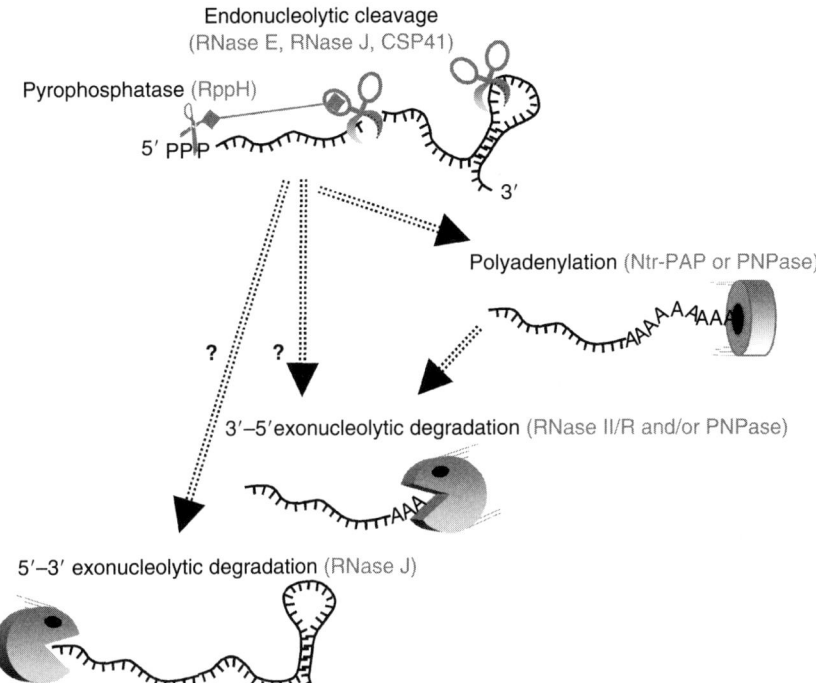

Fig. 2. The polyadenylation-stimulated RNA degradation pathway. The stages of polyadenylation-stimulated RNA turnover are: (A) endonucleolytic cleavage, (B) polyadenylation, and (C) exonucleolytic digestion. The first endonucleolytic cleavage is believed to be performed by RNase E in *E. coli* and related bacteria. In *E. coli*, it was recently shown that the removal of 5' end pyrophosphate by RppH in some cases precedes and stimulates the RNase E cleavage. RNase J has been implicated in this function in *Bacillus subtilis*. CSP41 is an endonuclease present in the chloroplast and may be also involved in the initial cleavage. The polyadenylation is performed by Ntr-PAP, producing homopolymeric poly(A) tails or by PNPase, producing heteropolymeric poly(A)-rich tails. In hyperthermophilic and several methanogenic archaea, the heteropolymeric tails are synthesized by the archaeal exosome. The 3'–5' exonucleolytic degradation step is carried out by PNPase and RNase II/R in bacteria and organelles. Dashed lines with a question mark indicate possible pathways and shortcuts that yet have to be shown to take place. The 5'–3' exonucleolytic degradation is predicted to be carried out by RNase J in organisms in which it is present.

C. Is Polyadenylation A Required Step in The Degradation Pathway?

The response to this question seems to be a bit complicated. On the one hand, in organisms where the polyadenylation-stimulated degradation pathway takes place, it is very difficult to knock out the polyadenylating enzymes while still retaining viability. For example, attempts to knock out both PNPase and

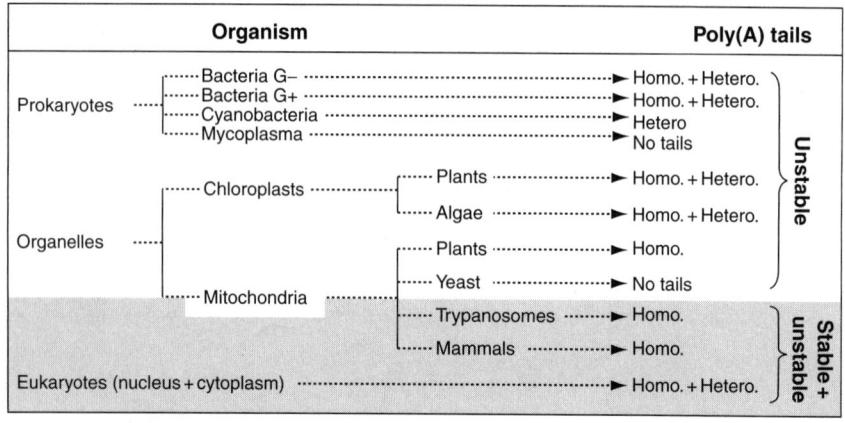

Fig. 3. Poly(A) tails in different organisms and organelles. The distribution of homopolymeric (Homo.), heteropolymeric (Hetero.), stable, and unstable tails between different organisms and organelles. "No tails" indicates that no polyadenylation takes place in the organism/organelle.

PAP in *E. coli* resulted in a significantly slower growth rate and a high rate of reversion (15). Therefore, temperature sensitive mutants are often used (16). Deletion of the only polyadenylating enzyme, PNPase, in cyanobacteria, resulted in a lethal phenotype (17, 18). While in *B. subtilis* there is only one gene encoding an Ntr type protein, which has been shown not be active as PAP, both hetero and homopolymeric poly(A)-tails are present (19, 20). A PNPase deletion mutant is viable, and the tails in this strain are homopolymeric (20). Therefore, if indeed the *B. subtilis* Ntr protein is not PAP, a new type of PAP that is not a member of the Ntr family, could be hypothesized. In addition, the inhibition of polyadenylation in a lysed-chloroplast system resulted in the accumulation of endonucleolytic cleavage products, a result that is similar to that obtained by the inhibition of exoribonucleases (21). Taken together, this limited set of experimental data suggests that in organisms and organelles in which poly(A)-stimulated degradation pathway takes place, its absence or inhibition leads to lethality or a growth defect.

On the other hand, several organisms and organelles have been described in which RNA is degraded without polyadenylation. These include yeast mitochondria, halophilic, and several methanogenic archaea, and the small-genome parasitic bacterium *Mycoplasma* (Fig. 3) (14, 22–25). In these systems there may be a more pronounced function of RNA helicases, which would fulfill the role of destabilizing secondary structures. It is an interesting question as to whether during evolution these organisms/organelles lost the polyadenylation process or simply never possessed it. In either case, the evolutionary pressure leading to the present-day situation remains elusive. For example, halophilic archaea, which live at a very high salt concentration and normal temperature, that is, a condition which favors the formation of RNA secondary structure, degrade RNA without polyadenylation (23). However, the hyperthermophilic group, which lives at a very high temperature where the RNA is not expected to be highly folded, utilizes the polyadenylation-stimulated degradation pathway (22, 23).

Taken together, the polyadenylation-stimulated RNA degradation pathway is composed of several steps including a possible initial endonuclease cleavage, followed by additional cleavages, polyadenylation of the cleavage products, and exonucleolytic degradation. Polyadenylation seems to provide a platform for the processive exoribonucleases and helps them overcome highly structured RNA barriers. This pathway (with some variations) evolved and is present in all organisms and organelles but a limited few. Where the pathway is present, it appears to be very important for normal growth.

III. The Enzymes

A. Endoribonucleases

1. RNase E

RNase E is an endoribonuclease found in many bacteria, some algae and archaea, as well as in higher plants, where it is predicted to be localized in the chloroplast (Fig. 4). It plays an important role in the processing and degradation of RNA in *E. coli*. RNase E was discovered in *E. coli* as an rRNA maturation enzyme (28) and was later shown to be involved in the processing of numerous other RNAs, including the antisense regulator of *E. coli* plasmid replication, RNAI; the precursor of M1 RNA, which is the catalytic subunit of the RNase P; tRNAs; and small noncoding regulatory RNAs and their targets (29–34). In addition, RNase E alters the stability of total RNA as well as numerous specific transcripts (35–38). Moreover, the enzyme concentration in the cell is regulated by a feedback loop, in which RNase E controls the stability of its own mRNA (39–41).

The *E. coli* version of this protein is essential for cell viability and contains 1061 amino acids in two distinct domains, an amino-terminal catalytic region and a carboxy-terminal region. The latter serves as a scaffold for assembling the degradosome, a high molecular weight complex that also contains PNPase, RNA helicase B (Rhl B), and the glycolytic enzyme enolase (42–45). This degradosome complex, however, is not present in cyanobacteria or spinach chloroplasts (18, 46, 47). RNase E cleaves single-stranded RNA with a preference for A/U-rich sequences (37, 48). RNase G is another *E. coli* endonuclease possessing about 50% sequence similarity to the RNase E catalytic region.

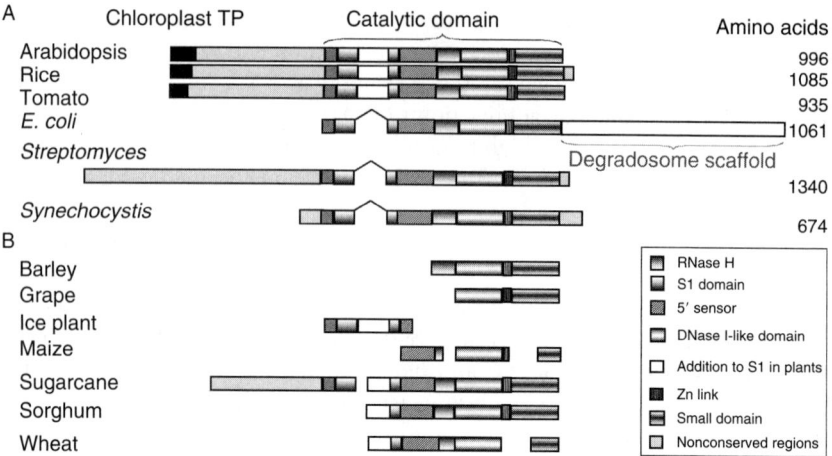

FIG. 4. Plant RNase E proteins. (A) The amino acid sequences of *Arabidopsis*, rice, and tomato RNase E homologues were aligned to those of the *E. coli*, *Streptomyces*, and *Synechocystis*. Regions of significant homology are shown as patterned boxes, with catalytic subdomains designated accordingly (26). The N-terminus of the plant proteins includes sequences shown in *Arabidopsis* and predicted in other plants to constitute a chloroplast TP (27). In addition, the plant proteins contain N-terminal extensions of several hundred amino acids that are not homologous between species, as well as a stretch of about 120 amino acids inside the S1 domain, which is not present in any bacterial sequence. (B) Plant ESTs (Expressed Sequence Tags). Related to RNase E, with the position of each domain in the full-length sequences indicated. When more than one EST was found, the ESTs are shown on one line aligned to the full-length sequence of *Arabidopsis*. Sequence accession numbers are as follows: *Escherichia coli*, P21513; *Arabidopsis thaliana*, NP_850987; rice (*Oryza sativa*), NP_001061542; tomato (*Lycopersicon esculentum*) (27); *Streptomyces*, NP_626836; *Synechocystis*, NP_439978; barley (*Hordeum vulgare*), TC141965; grape (*Vitis vinifera*), TC42553; ice plant (*Carpobrotus edulis*), TC6253; maize (*Zea mays*), TC308943, TC049636, TC284127; sugarcane (*Saccharum officinarum*), TC51842, TC68873; sorghum (*Sorghum bicolor*), TC106341; wheat (*Triticum aestivum*), TC270983, TC271418.

RNase G has overlapping, but nonidentical, cleavage specificity with RNase E (*49*, *50*). Both enzymes were consequently combined into a newly named family of RNase E/G proteins. The cleavage activity of the family members depends on the number of phosphates at the 5′ end: RNA containing one phosphate is a much better substrate than RNA with three phosphates or no phosphate (*12*, *51–53*). (See also Chapter in this volume by Carpousis *et al.*). The structure of the catalytic portion of *E. coli* RNase E has been solved, revealing its mode of action and the mechanism by which the number of phosphates at the 5′ end affects activity (*26*, *54*).

Genes and/or ESTs encoding RNase E/G-like proteins have been found in many bacteria, cyanobacteria, red and green algae, and the nuclear genomes of higher plants, but not in eukaryotes lacking chloroplasts (Fig. 4) (*27*, *55*, *56*). The classification of RNase E-like polypeptides into several groups based on

the domain architecture has been proposed (55). In many bacteria other than
E. coli, as well as in the nuclear genomes of several green algae and higher
plants, only one member of the RNase E/G group is encoded, and it is generally
termed RNase E.

The *Arabidopsis* RNase E, encoded by the At2g04270 locus, has recently
been characterized (27, 57). Since the T-DNA null insertion mutant for this
protein cannot grow without adding sucrose to the medium (57), *Arabidopsis*
RNase E may be required for chloroplast development, similar to its requirement for viability in *E. coli*. Analysis of the N-terminal 63 amino acids revealed
a canonical chloroplast transit peptide (TP) that likely directs the cytoplasmically translated protein into this organelle. Indeed, experimental analysis verified the localization of this protein to the chloroplast (27), and its absence can
be correlated with reduced accumulation of some chloroplast transcripts (57).
Similar proteins are also encoded by the nuclear genomes of perhaps all other
plants (Fig. 4), but apparently not in *Chlamydomonas*.

The carboxyl half of the plant RNase E homologues contains the multi-domain catalytic region and is similar to the amino-terminal region of *E. coli*
RNase E, both in sequence and domain architecture (26). Interestingly, the S1
domain of the plant proteins, which is important for RNA cleavage activity,
contains a uniformly located insertion of 121 nonconserved amino acids (Fig. 4).
A similar insertion in the S1 domain was described before for the RNase E of α-
proteobacteria (55). The endonucleolytic activities of the catalytic portions of
the *E. coli* and *Arabidopsis* RNase E proteins were found to be very similar.
Both were sensitive to the number of phosphates at the 5′ end and to substrate
secondary structure. In both enzymes, replacing the two highly conserved lysine
residues at positions 546 and 552 (*E. coli* residues 106 and 112), located in the S1
domain, significantly reduced catalytic activity. Therefore, the catalytic domains
of the prokaryotic and chloroplast RNase E have apparently retained very
similar properties despite their long evolutionary separation.

Although the sensitivity of cleavage activity to the number of phosphates
located at the 5′ end of the transcripts is conserved in chloroplast RNase E,
this is based on the activity of the catalytic domain without the plant-specific
amino-terminal extension. While the 5′ ends of bacterial mRNAs correspond
mainly to the transcription initiation site and therefore contain three phosphates,
in chloroplasts mRNAs are often processed at their 5′ ends, meaning that they
would have monophosphate, which would *a priori* be sensitive to RNase E
cleavage. However, as best studied in *Chlamydomonas*, but perhaps also true in
higher plants, chloroplast mRNAs are often protected from degradation by
nucleus-encoded proteins that specifically bind the 5′ end (56, 58–60). It would
appear, then, that the apparently poor cleavage activity of the chloroplast enzyme
on triphosphorylated substrates likely protects primary transcripts, but not their
processed derivatives, from undesired degradation.

2. RNase J

Many organisms lack an RNase E homologue, suggesting that another endoribonuclease is responsible for endonucleolytic processing and turnover. Recently, the purification and identification of two novel *B. subtilis* endoribonucleases, RNases J1 and J2, was described (*61*). These RNases, like the tRNA 3′ processing endonuclease RNase Z, belong to the zinc-dependent metallo β-lactamase group and *in vitro* assays suggest that they are functionally homologous to RNase E, since they have similar substrate specificity in terms of cleavage site selection in AU-rich single-stranded regions (*61, 62*). Indeed, the *B. subtilis thrS* leader mRNA, which is a substrate of RNase J, is cleaved at the same site by RNase E when it is expressed in *E. coli* (*63*).

Genes encoding RNase J homologues are widespread in eubacteria, archaea, algae, and higher plants. Although they appear to replace RNase E in many organisms such as *Chlamydomonas*, others such as *Synechocystis* and higher plants encode both types of enzyme. The *Chlamydomonas* and *Arabidopsis* nuclear genomes each contain a single *RNJ* gene (EU518648-EU518649 and At5g63420, respectively), and the N-terminus of the *Arabidopsis* gene product targets GFP to chloroplasts in transient assays (Bollenbach and Stern, unpublished data).

Surprisingly, analysis of *B. subtilis* RNase J revealed both endonuclease and 5′–3′ exonuclease activity, making it the first 5′–3′ exonuclease discovered in prokaryotes (Fig. 2) (*64*). If the analysis of green algal and higher plant RNase J proteins demonstrates chloroplast localization and 5′–3′ exoribonuclease activity, it may be possible that this is the enzyme responsible for the net 5′–3′ exonucleolytic activity that has been characterized in *Chlamydomonas* chloroplasts (*65–67*). The interplay of the endo and exonuclease activities of this protein in RNA processing and/or degradation in bacteria and possibly the chloroplast awaits further study. Moreover, the division of labor between RNase E and RNase J in cyanobacteria and higher plant chloroplasts, where both enzymes appear to be present, will be interesting to decipher.

The observation that RNase J is essential for embryo development in *Arabidopsis*—plants heterozygous for a T-DNA insertion in the *RNJ* coding sequence produce siliques containing aborted embryos (http://www.seedgenes.org)—suggests a nonredundancy with RNase E. This phenomenon may be related to a function in 16S rRNA maturation as was recently reported for *B. subtilis* RNase J (*68*).

3. CSP41

CSP41a (chloroplast stem-loop binding protein, 41 kDa) and CSP41b are widespread, highly conserved endoribonucleases, which are unique to photosynthetic organisms. The photosynthetic bacteria *Synechocystis* sp. PCC6803 and *Nostoc* sp. PCC7120 encode only a CSP41b homologue, whereas plant and algal nuclear genomes encode both CSP41a and CSP41b. Phylogenetic and motif analyses have shown that CSP41a and CSP41b are paralogs of a cyanobacterial ancestor that diverged from a bacterial epimerase/dehydratase (69, 70).

CSP41a was first purified from spinach chloroplasts as a *petD*-specific RNA-binding protein and a nonspecific endoribonuclease (71, 72). Spinach CSP41a was shown to cleave synthetic stem-loop-containing *petD*, *psbA*, and *rbcL* RNAs, and could cleave arbitrary single-stranded RNAs (72). This suggested that it could initiate turnover of chloroplast transcripts by endonucleolytic cleavage, the first step in the poly(A)-stimulated turnover pathway (Fig. 2). In vitro measurements of tobacco chloroplast mRNA degradation rates in CSP41a-deficient plants showed a 7-fold, 2-fold, and 5-fold decrease in the rates of *rbcL*, *psbA*, and *petD* transcript turnover, respectively (73), suggesting that CSP41a may participate broadly in chloroplast mRNA turnover. Recent analysis of a CSP41b mutant led to the suggestion that it functions to process 23S rRNA (74).

Most chloroplast open reading frames encode inverted repeat (IR) sequences in their 3′ untranslated regions that can fold into stable stem-loop structures. Prior research has shown that these IRs act as processing determinants and protect upstream sequences against 3′–5′ exonucleolytic degradation (75). As mentioned earlier, CSP41 has no sequence specificity, but displays a substrate preference for stem-loop containing RNAs from *petD*, *psbA*, and *rbcL in vitro* (72). This property would make CSP41 a candidate for RNA maturation leading to turnover (73). The analysis suggests that CSP41 has broad substrate specificity, and that stem-loop structure is a major determinant of CSP41 cleavage rates and transcript half-life in the chloroplast.

4. RNASE P AND RNASE Z

Ribonuclease P (RNase P) is an endoribonuclease that processes the 5′ leader sequence of precursor tRNA (Fig. 5). In bacteria, RNase P is a small ribonucleoprotein complex consisting of a catalytic RNA and a protein cofactor (76). In human cells, a highly purified nuclear RNase P has at least ten distinct protein subunits associated with a single RNA species, the H1 RNA (77–79). In addition, a subset of these protein subunits is shared with RNase MRP (80), a mitochondrial and ribosomal RNA-processing ribonucleoprotein (81, 82). However, it is not known if these protein subunits are also shared with the mitochondrial form of human RNase P, a ribonucleoprotein particle shown to have an RNA moiety that is identical to H1 RNA (83). RNase P is an essential enzyme present in all organisms, except in some archaea that produce leaderless tRNAs (81, 84). There is still a debate concerning the type of RNase P in mitochondria and chloroplasts and the extent to which the organellar form contains the catalytic RNA subunit (83, 85, 86).

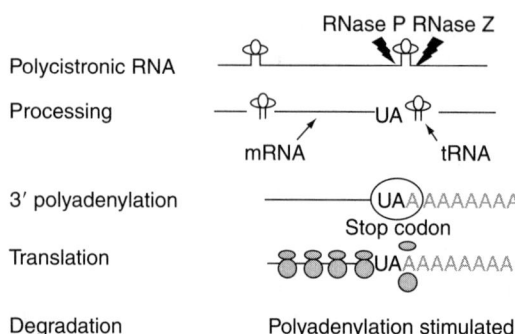

Fig. 5. RNA processing in animal mitochondria. The genome is transcribed into polycistronic RNAs in which the mRNAs are punctuated by tRNAs, which are processed as detailed in the main text. These tRNAs are cleaved at the 5' and 3' ends by the endoribonucleases RNase P and RNase Z, respectively. Several mRNAs contain incomplete translational stop codons composed of only U or UA instead of UAA. The addition of a stable poly(A) tail at the 3' end creates the complete stop codon. The mRNA is then translated, and eventually degraded by the polyadenylation-stimulated degradation pathway.

RNase Z is a member of a highly conserved family of metallo-β-lactamase proteins that is found in prokaryotes, eukaryotes, and archaea (87, 88). This endoribonuclease is involved in the processing of tRNA precursors lacking an encoded CCA terminus at their 3' end (88, 89). However, it is generally not active on tRNA precursors that contain a chromosomally encoded CCA (90). RNase Z cleavage of tRNA precursors generates substrates for tRNA nucleotidyl transferase-catalyzed addition of CCA to produce functional tRNA molecules (88).

B. Exoribonucleases

1. PNPase (Polynucleotide Phosphorylase)

PNPase (EC 2.7.7.8) was discovered during studies of biological phosphorylation in *Azotobacter vinelandii* (91), and was later characterized in the context of its role in *E. coli* RNA synthesis (92). In fact, PNPase was the first enzyme shown to catalyze the synthesis of polynucleotides from ribonucleotides; unlike RNA polymerases, PNPase catalyzes this reaction in a template-independent manner.

As a phosphorylase, PNPase catalyzes both processive 3'–5' degradation and RNA polymerization, and participates in the degradation, processing, and polyadenylation of RNA in bacteria and organelles (4, 25, 93–96). PNPase has also been reported to be a global regulator of virulence and persistency in *Salmonella enterica* (97), and its activity in some way regulates both chloroplast isoprenoid metabolism (98) and the ability of *Chlamydomonas* to survive phosphate starvation (99). Recent work provided evidence of a possible function of PNPase in the dedifferentiation process of human cancer cells, a possible involvement in apoptosis and a role in the protection of cellular RNA from oxidizing damage (*100, 101*). Unlike the situation in bacteria and plant organelles, where PNPase is directly involved in RNA metabolism, the human PNPase was recently shown to be localized to the mitochondrial intermembrane space. Therefore, it is not directly engaged in degrading and polymerizing mitochondrial transcripts, since these are located in the matrix (*102–104*).

Analysis of the processing and accumulation of chloroplast or mitochondrial transcripts in plants in which the corresponding PNPase was depleted, revealed dramatic changes in the 3′ end processing and accumulation of polyadenylated transcripts related to ribosomal, messenger, and transfer RNAs (*105–108*). Interestingly, these effects were not observed when the expression of the single PNPase of *Chlamydomonas reinhardtii* was down-regulated, although the accumulation of chloroplast transcripts seen in wild-type cells upon phosphate starvation, was not observed in PNPase-deficient cells (*99*).

Even though the human PNPase is located in the intermembrane space where no RNA is known to be present, the enzyme is phosphorolytically active when expressed as a recombinant protein (*109*). Moreover, siRNA-mediated knockdown of PNPase expression in human cells significantly affected the polyadenylation of mitochondrial transcripts, as well as ATP generation and other mitochondrial activities (*104*). Since mitochondrial transcripts were affected in the knockdown cells, it was suggested that PNPase phosphorolytic activity in the intermembrane space is important for proper mitochondrial functioning, perhaps by fine-tuning the phosphate and nucleotide concentrations (*102, 104, 110*). Therefore, the effects on polyadenylation and processing of mitochondrial transcripts would be indirect. This hypothesis suggests that there is a substrate for PNPase in the intermembrane space. However, since no RNA has yet been located in this compartment, one might speculate that there is a hitherto unknown substrate for this enzyme. In addition, since no gene encoding a protein related to the RNase II/R family with a mitochondrial TP has been identified in the human genome (see below), this would suggest that we do not yet have a candidate for a human mitochondrial matrix exoribonuclease.

Genes encoding PNPase homologues have been identified in almost all bacteria and eukaryotes, with the exception of *Mycoplasma*, trypanosomes, and yeast (*25*). Furthermore, there is no PNPase in archaea, although the hyperthermophiles and some methanogenic archaea contain an exosome that is structurally and enzymatically very similar to PNPase (*23, 25, 111*). The primary structures of PNPases encoded in bacteria and in the nuclear genomes of plants and mammals comprise five domains: two N-terminal core domains homologous to the *E. coli* phosphorylase RNase PH, separated by an α-helical domain, and two C-terminal RNA-binding domains (KH and S1) (*112–114*).

Structural analysis of PNPase from the bacterium *Streptomyces antibioticus* revealed a homotrimeric complex surrounding a central channel that can accommodate a single-stranded RNA molecule (112, 113).

Protein sequence alignments and structural observations revealed that bacterial and chloroplast PNPases are evolutionarily related to archaeal and eukaryotic exosomes. The exosome functions in 3′–5′ RNA degradation, RNA processing, and quality control of gene expression in the cytoplasm and nucleus of eukaryotic cells (8). It is comprised of 9–11 proteins, including six that are related to RNase PH, three related to the S1 and KH RNA-binding domains, and two others related to the hydrolytic ribonucleases, RNase II and RNase D. Overall, the core 9-subunit exosome is structurally similar to trimeric PNPase (96, 114–118). Therefore, PNPase, and the archaeal and eukaryotic exosomes represent functionally and evolutionarily conserved machines for 3′–5′ exonucleolytic degradation. Nevertheless, while the archaeal exosome is very similar to PNPase and is phosphorolytically active, the yeast and human exosome complexes are not (115, 117). Instead, the yeast and human exosomes degrade RNA only hydrolytically, and perhaps retain their circular shape as a result of evolutionary pressure for RNA binding and/or structural features (119).

2. RNase II/R

The RNR exoribonuclease family members, which are typified by *E. coli* RNase II and RNase R, are hydrolytically processive 3′–5′ exoribonucleases that release 5′ monophosphate nucleotides. These enzymes are widely distributed among eukaryotes, eubacteria, *Mycoplasma*, and the archaea. While most eukaryotic nuclear genomes encode at least three RNR homologues, some prokaryotic genomes encode only a single RNR-like enzyme and exceptional cases, such as *Mycoplasma*, encode a single RNR homologue as the only exoribonuclease (24, 120). The halophilic archaea also encode an RNR homologue, while hyperthermophiles and several methanogens contain the archaeal exosome (22, 23). Interestingly, no homologue of RNase II/R could be detected in those methanogens that do not contain the archeal exosome, or in the human mitochondrial matrix (22, 96). The *Arabidopsis* nuclear genome encodes three homologues, including RNR1, which is both plastid and mitochondria-localized, and RNR2 and RNR3, which are localized to the nucleus and cytosol, and are therefore putative exosome subunits (107, 121–123). Homozygous *Arabidopsis* T-DNA mutants for RNR1 can germinate only in the present of sucrose and the maturation of the 3′ ends of the 23S, 16S, and 5S rRNAs is impaired, while mRNAs appear to be unaffected (121). *Chlamydomonas* appears to encode two RNR members, RNB1, and RNB2. The N-terminus of the former does not target YFP significantly to an organelle, suggesting it is cytosolic and a good candidate for the exosome. RNB2, by the same criterion, is localized to the chloroplast (99).

In *E. coli*, the RNR family members differ in their abilities to remain processive through secondary structures (see Chapter in this volume by Arraiano and colleagues). For example, RNase II becomes distributive near stem-loops and is eventually inhibited by them, while RNase R can melt secondary structures (*124*). Therefore, although both enzymes are nonspecific exonucleases in *E. coli*, RNase II is more active on single-stranded homopolymeric transcripts such as poly(A), while RNase R has a preference for rRNAs (*124*).

An RNase II crystal structure has recently shed light on the catalytic activity and substrate specificity of RNR enzymes (*125, 126*). RNase II folds into four domains comprising two N-terminal RNA-binding moieties, a central catalytic domain, and a C-terminal S1-like RNA binding region. The N- and C-terminal domains form a clamp atop the catalytic domain, which funnels the ssRNA substrate into a narrow channel that houses the active site. Although the domain structure and sequence motifs are highly conserved among RNR family members, it is thought that differences in the clamp arrangement and thus RNA binding properties play an important role in regulating the activity on transcripts containing secondary structures.

Chloroplast/mitochondrial RNR1 is inhibited by secondary structures when assayed *in vitro* (*107, 121*). This is consistent with the fact that it participates in the processing of precursor RNAs, in particular the 3′ ends of rRNAs. Since both rRNAs and mature mRNAs often contain terminal stem-loops in both organelles (Fig. 1), any degradative action of RNR1 would require prior endonucleolytic cleavage and polyadenylation, or recruitment of an RNA helicase. The latter mechanism is employed by yeast mitochondrial Dss1, an RNase R homologue that digests secondary structures by complexing with a helicase. It should be noted that there is no PNPase in yeast mitochondria, thus Dss1 is the only exonuclease so far identified in that organelle (*127*).

RNase II, RNase R, and PNPase, which represent the major exoribonuclease activities in *E. coli*, have significantly different substrate specificities and catalytic properties *in vitro*, but share overlapping functions *in vivo*. In *Synechocystis*, there is a single RNase II/R homologue. In addition, PNPase functions as the only polyadenylation enzyme (in addition to its function in degradation). Accordingly, deletion of *Synechocystis* PNPase- or RNase II/R-encoding genes, unlike the situation in *E. coli* (*128*), leads to loss of viability (*18*). Similarly, since there is no PNPase in yeast mitochondria, deletion of the RNase II/R homologue *DSS1* leads to mitochondrial dysfunction and eventually to the loss of its genome (*127*).

Plant chloroplast PNPase and RNR1 catalyze distinguishable reactions *in vivo*, but may functionally overlap. Repression of the gene encoding chloroplast PNP, for example, leads to defects in mRNA and 23S rRNA 3′ processing, but plants are viable and grow on soil (*105*). Similar observations were made for the mitochondrial enzyme, although the growth phenotype is much stronger (*108*). In contrast, *rnr1* null mutants are defective in rRNA but not in mRNA processing (121, 129). RNR1 mutants are marginally viable on soil, owing to a dependence on RNR1 for chloroplast development in cotyledons, and perhaps an effect on mitochondrial mRNA metabolism (107).

C. The Family of Poly(A)-Adding Enzymes

In eukaryotes, a stable poly(A) tail is added to almost all mRNAs during the transcription termination process, initiated by the cleavage of the nascent RNA chain (130). The addition of a stable poly(A) tail is accomplished by the so-called canonical PAP. In addition, a diverse family of related enzymes polyadenylate transcripts in various systems, mostly as part of the polyadenylation-stimulated degradation pathway. These enzymes belong to the ribonucleotidyl transferase (rNTr) family that catalyzes the nontemplated addition of homopolymeric adenosine tails to the 3′ hydroxyl group of RNAs and tags them for degradation (131, 132).

The rNTr superfamily includes the PAPs mentioned earlier, as well as terminal uridylyl transferases, poly(U)-polymerases and the ubiquitous CCA-adding enzymes (CCAtrs) responsible for the synthesis or repair of the 3′ terminal sequence of tRNA molecules (131). rNTr-PAPs are very similar in protein sequence to CCAtrs. However, one motif, which forms a predicted β-loop near the catalytic center, has been identified that seems to be PAP-specific (133). rNTr-PAPs are found in the β, γ, and δ subdivisions of the proteobacteria and several *Chlamydiales* and *Spirochaetales*, but not in Gram-positive bacteria or bacteria that diverged before the Gram-positives (133). In addition, no homologues can be detected in archaea. In plants, chloroplasts and mitochondria are thought to host NTr-PAP homologues encoded in the nucleus and several rNTr-PAPs were recently identified bioinformatically (133). For example, bioinformatic analysis of the *Chlamydomonas* genome revealed eight NTR/noncanonical PAPs, of which three are predicted to be located in the chloroplast (134). Similar numbers of NTr-PAPs were identified in human cells, where one was found to be mitochondrial (132, 135–137).

Although PNPase has been found to be primarily responsible for poly(A) addition in spinach chloroplasts (138), the presence of homopolymeric poly(A) tails in *Arabidopsis* and other plant chloroplasts suggests the activity of an NTr-PAP. Genes encoding putative NTr-PAPs were identified in the *Arabidopsis* nuclear genome and await experimental validation (133).

IV. RNA Degradation and Polyadenylation in Chloroplasts

In principle, the RNA metabolic pathways in the chloroplast were retained from their prokaryotic ancestors and therefore, the elucidation of the polyadenylation-stimulated degradation pathway in *E. coli* has paved the way for defining this process in the chloroplast. However, chloroplast-specific variations have been observed, as discussed at the beginning of this chapter. Thus, when standard methods like oligo(dT)-primed reverse transcription PCR and others were used for the detection of nonabundant and truncated transcripts decorated with posttranscriptionally added tails, these studies surprisingly revealed heteropolymeric, poly(A)-rich tails, the first observation of such tails in any organism (139). Similar heteropolymeric tails, produced mainly by PNPase or the archaeal exosome, were later discovered in bacteria, archaea, and human cells (Fig. 3) (96).

Further studies revealed that, as in *E. coli*, both PNPase and Ntr-PAP are polyadenylating enzymes in the chloroplast, but the contribution to the tail population may differ between different plants. For example, while most of the tails in *Arabidopsis* chloroplasts are homopolymeric, suggesting a major contribution by Ntr-PAP, the majority of tails in spinach are heteropolymeric, suggesting a major contribution by PNPase (138) (Larum and Schuster, manuscript in preparation). It is interesting to note that PNPase is exclusively responsible for polyadenylation in cyanobacteria, which is considered the closest bacterial relative of the evolutionary ancestor of the chloroplast (18). Therefore, it may be suggested that Ntr-PAP evolved in the chloroplast following endosymbiosis. Since both PNPase and Ntr-PAP are active in polyadenylation in both *E. coli* and the chloroplast, it may also be suggested that the conversion of the Ntr-CCA enzyme to Ntr-PAP occurred more than once in the evolution of bacteria and organelles (96). Alternatively, it could be suggested that PAP was lost during cyanobacterial evolution.

The enzymes involved in the polyadenylation-stimulated degradation pathway in the chloroplast (Fig. 2) are as follows: for the initial endonucleolytic cleavage step, any of the three known endoribonucleases, RNase E, RNase J, and CSP41a/b, or perhaps the three of these together. The second step of polyadenylation is performed by PNPase (heteropolymeric tails) and perhaps Ntr-PAP (homopolymeric tails). The third step of the exonucleolytic digestion is carried out by the phosphorylase activity of PNPase and/or the hydrolytic enzyme RNase II/R. It is possible that the enzyme oligoribonuclease degrades the residual oligomers, as it does in *E. coli* (see Chapter in this volume by Danchin). This possibility is supported by the finding that homologues of the *E. coli* enzyme are encoded in both the *Arabidopsis* and *Chlamydomonas* genomes (134).

V. RNA Degradation and Polyadenylation in Mitochondria

A. Plant Mitochondria

The RNA degradation system in plant mitochondria resembles that of bacteria and chloroplasts in the sense that transcripts are not decorated with stable poly(A) tails. Therefore, the poly(A) tails are destabilizing. Both PNPase and RNase II/R are present. Nonabundant polyadenylated transcripts were observed in several plants and their quantity significantly increased when the expression of the mitochondrial PNPase was impaired (*107, 140–143*). Indeed, the accumulation of truncated polyadenylated transcripts in PNPase-deficient plant mitochondria prominently reveals the normal action of the polyadenylation-stimulated degradation pathway. However, there are two major differences between plant mitochondria and chloroplasts. The first is that unlike bacteria and chloroplasts, heteropolymeric tails have not yet been detected in mitochondria, suggesting that plant mitochondrial PNPase does not work as a polymerase *in vivo* and that the tails are produced by a yet-to-be identified PAP. The second difference is that no endoribonuclease of the RNase E or RNase J type has been identified in mitochondria, although they do contain tRNA processing enzymes. Therefore, the question is whether there is any mRNA endonuclease. The observation that RNase E and RNase J are restricted to bacteria and photosynthetic organisms, where they bear predicted chloroplast TPs, argues against this, but a mitochondrial localization, or dual localization, cannot be ruled out. Alternatively, a role for the tRNA processing enzymes RNase P and/or RNase Z could be hypothesized. In summary, we speculate that the plant mitochondrial polyadenylation-stimulated degradation pathway consists simply of polyadenylation and exoribonucleolytic digestion.

B. Yeast Mitochondria: RNA Metabolism Without Polyadenylation

Yeast mitochondria are the only organelles known to metabolize RNA without polyadenylation (*14, 127, 144*). This was found to be the case both in *Saccharomyces cerevisiae* and *Schizosaccharomyces pombe*, suggesting that it might be a general phenomenon of fungal mitochondria. Because no *PNP* gene is present in yeast, it is conceivable that during evolution yeast simultaneously lost the *PNP* gene and mitochondrial polyadenylation (Fig. 3) (*114, 145*).

A protein complex consisting of an exoribonuclease belonging to the RNR family and an RNA helicase, defined as the mtExo or the mitochondrial degradosome, has been identified in *S. cerevisiae* mitochondria (*127*). In the absence of PNPase, this complex might be exclusively responsible for exonucleolytic activity in this organelle. This would resemble the case of halophilic archaea and the small genome parasitic bacteria *Mycoplasma*, which also lack RNA polyadenylation, but retain an RNR homologue presumably responsible for exoribonucleolytic activity (*23, 24*). The 3′ ends of yeast mitochondrial mRNAs are characterized by a conserved dodecamer sequence that is encoded in the mitochondrial genome, and is believed to be bound by a specific protein that may protect the 3′ end from exonucleolytic degradation (*14*) (Fig. 1).

C. Trypanosome Mitochondria: Both Stable and Unstable Poly(A) Tails

Trypanosomes are among the earliest branching eukaryotes and their mitochondria are of great interest both because of the unique arrangement of the mitochondrial DNA and the novel posttranscriptional events that govern gene expression. The best-known of these is the massive editing of most protein-coding transcripts, *via* the insertion or deletion of uridines (*146*). The addition of poly(A) tails was detected in these organelles and intriguingly, it seems that both constitutively stable and unstable poly(A) tails are present, depending on the editing stage of the particular transcript (*147*).

Similar to the stable 3′ end tails that characterize animal mitochondrial transcripts, two fractions of short tails composed of several adenosines (oligo (A)), and long tails (poly(A)), were identified. However, unlike human mitochondria where the tails are exclusively composed of adenosines (*104, 148*), in *Trypanosoma brucei* mitochondria the long tails contain many uridines and are therefore considered poly(A/U) extensions (*149*). The short poly(A) tails were found to be required and sufficient for maintaining the abundance of partially edited, fully edited and unedited mRNAs in mitochondria (*149*).

A PAP (KPAP1) was recently identified and characterized in *T. brucei* mitochondria (*149*). This PAP was found to be essential for parasite viability and mitochondrial function, and is engaged in a polyadenylation complex that also includes pentatricopeptide repeat (PPR) proteins (*149*). PPR proteins are characterized by multiple repeats of about 35 amino acids and are involved in the posttranscriptional regulation of gene expression, mainly in organelles (*150, 151*).

The coexistence of stable and unstable poly(A) tails in the same organelle was also observed in mammalian mitochondria and yeast and mammalian nuclei (Fig. 3) (*148, 152–154*). This implies the presence of a mechanism that can distinguish between stabilizing and destabilizing tails because in many cases, this will result in opposite functions, for example, exonucleolytic degradation as opposed to stability determination and translation. Unraveling the details of this molecular mechanism is one of the present challenges.

In terms of exoribonucleases, an RNR-type enzyme, but not PNPase, has been identified in *T. brucei* mitochondria (*155*). Experiments analyzing RNA degradation in protein extracts and organellar systems revealed higher degradation activity for polyadenylated RNA than nonpolyadenylated molecules, as well as an important role for UTP polymerization in this process (*156–158*). Therefore, by analogy to RNA editing, the polyadenylation, UTP-polymerization, and degradation of RNA in this organelle appear to have adopted unique characteristics.

D. Animal Mitochondria

The 16.6 kb circular mammalian mitochondrial genome encodes two rRNAs, 22 tRNAs, and 13 protein components of the oxidative phosphorylation complexes (159–163). Unlike plant mitochondrial genomes, their mammalian counterparts have been extensively condensed. They lack introns and aside from one regulatory region, the so-called D-loop, intergenic sequences are absent or limited to a few bases. Both rRNA and tRNA molecules are unusually small. In several cases, genes encode only partial translation termination codons, which become functional only after posttranscriptional polyadenylation, as described below (14, 163).

Both DNA strands (termed H and L) are fully transcribed, resulting in polycistronic RNA molecules, which are then endonucleolytically processed to produce mRNAs, rRNAs, and tRNAs (Fig. 5). The transcripts are then polyadenylated, producing functional stop codons. Transfer RNAs are subjected to the addition of the CCA motif and the modification of internal nucleotides (162–165). The intergenic regions of the L-strand primary transcript are believed to be rapidly degraded. RNA degradation is a key component of mitochondrial gene regulation, as it is also required to eliminate aberrant transcripts (14, 25, 159, 166). Our mechanistic knowledge of human mitochondrial RNA degradation is very limited, the most significant difference with most other organellar systems being the presence of stable poly(A) tails at the mature 3' ends of mRNA (Figs. 1 and 5).

This stable poly(A) tail was described more than 30 years ago (160, 167, 168) and it has been proposed to determine transcript stability, perhaps in conjunction with a putative poly(A)-binding protein (14, 169). Yet, the only established function is the completion of translational stop codons where the encoded one is incomplete (170). Although PNPase is present in human mitochondria, the homopolymeric poly(A) nature of the tails suggest synthesis by a PAP (171). As mentioned earlier, PNPase was recently localized to the intermembrane space, whereas RNA metabolism occurs in the matrix (110, 136, 137). In cells where the expression of this PAP was drastically reduced by RNAi, polyadenylation still occurred, but tail length was reduced from an average of 43 to 8 adenosines (104, 137). This result suggests that the residual PAP can still produce the oligoadenylated tails, or that more than one enzyme is responsible for the polyadenylation activity in human mitochondria.

If one assumes that the degradation of mammalian mitochondrial RNA is mechanistically similar to other organellar and prokaryotic pathways, one would expect to be able to find truncated, low-abundance polyadenylated fragments in these mitochondria, in addition to the full-length RNAs with stable poly(A) tails at their 3' ends. Indeed, analysis using oligo(dT)-primed reverse transcriptase PCR of human mitochondrial RNA, from both cancer cell lines and primary fibroblasts, revealed many such molecules derived from each gene that was analyzed, including mRNA, rRNA, and tRNA (*148*). Furthermore, a bioinformatic tool developed to search the human EST database for cDNAs corresponding to polyadenylated truncated human mitochondrial RNAs, was successful in finding hundreds of such ESTs (*148*). The resulting ESTs represented the entire human mitochondrial transcriptome, including the L-strand intergenic regions.

In all systems for which this has been investigated, there is a strict correlation between the presence of truncated polyadenylated RNA molecules and the prokaryotic/organellar polyadenylation-stimulated RNA degradation mechanism. Therefore, this internal polyadenylation is most likely part of the RNA degradation process, meaning that in this respect, mammalian mitochondria stay true to their prokaryotic origin (*96, 148*). In this light, the discovery that PNPase is located in the intermembrane space came as a surprise, since it suggests that unlike other systems, it is not directly involved in RNA metabolism (*102, 103*). However, as described earlier, when human PNPase is expressed in bacteria, it is active as a phosphorylase, and reducing its amount in the cell by siRNA drastically affects polyadenylation and ATP production as well as other mitochondrial processes, most likely by indirect means (*102, 104, 109*). As discussed in the section on PNPase, these observations suggest that PNPase fine-tunes the nucleotide concentration in mitochondria and maintains mitochondrial homeostasis, with mRNA metabolism being one of the processes influenced by this activity. Constitutive knockdown of PNPase in human cell lines demonstrated transcript-dependent effects on mitochondrial mRNA processing and polyadenylation (*104*). These effects, which included abnormal 5′ and 3′end processing and fluctuations in the lengths of poly(A) tails, did not seem to influence mitochondria mRNA abundance, the polypeptide synthesis rate, or protein accumulation. Since polyadenylation of the *cox1* transcript was abolished in this experiment, the results demonstrated that at least in this case, a stable poly(A) tail is not required for stabilization or translation initiation (*104*).

How do animal mitochondria differentiate between stable polyadenylation and degradation-inducing poly(A) tails? Is there a second polyadenylating enzyme? Is there an initial endonucleolytic cleavage, and if so, what is the enzyme involved? In the absence of PNPase in the matrix and the lack of a member of the RNase II/R family with a mitochondrial targeting peptide, what is the exonuclease that degrades mitochondrial transcripts, if there is one at all? These questions are currently being investigated and promise to reveal an evolutionarily unique outcome.

VI. Conclusions and Perspectives

Based on our current knowledge of RNA degradation/polyadenylation pathways in various bacteria, archaea, yeast, plants, and animals, a broader view of their evolution has been achieved. In addition, the power of comparative genomics to understand the origin of complex RNA degradation pathways is evident. Continuing and broader investigations will reveal different combinations of enzymes, as well as the interplay between stable and unstable poly(A) tails, which in turn will help establish the role of each in a given organism.

Acknowledgments

This work was supported by Binational Scientific Foundation (BSF) (2005184) and Binational Agriculture Research and Development Foundation (BARD) (IS-3605-04CR) awards to D.B.S. and G.S., and by an Israel Science Foundation (ISF) (266/05) award to G.S.

References

1. Hoffmeister M, Martin W. Interspecific evolution: Microbial symbiosis, endosymbiosis and gene transfer. *Environ Microbiol* 2003;**5**:641–9.
2. Dyall SD, Brown MT, Johnson PJ. Ancient invasions: From endosymbionts to organelles. *Science* 2004;**304**:253–7.
3. Gould SB, Waller RF, McFadden GI. Plastid evolution. *Annu Rev Plant Biol* 2008;**59**:491–517.
4. Bollenbach T, Schuster G, Portnoy V, Stern D. Polyadenylation, processing and degradation of chloroplast RNA. *Top Curr Genet* 2007;**19**:175–211.
5. Leister D, Schneider A. From genes to photosynthesis in *Arabidopsis thaliana*. *Int Rev Cytol* 2003;**228**:31–83.
6. Edmonds M. A history of poly A sequences: From formation to factors to function. *Prog Nucleic Acid Res Mol Biol* 2002;**71**:285–389.
7. Doma MK, Parker R. RNA quality control in eukaryotes. *Cell* 2007;**131**:660–8.
8. Houseley J, LaCava J, Tollervey D. RNA-quality control by the exosome. *Nat Rev Mol Cell Biol* 2006;**7**:529–39.
9. Garneau NL, Wilusz J, Wilusz CJ. The highways and byways of mRNA decay. *Nat Rev Mol Cell Biol* 2007;**8**:113–26.
10. Isken O, Maquat LE. Quality control of eukaryotic mRNA: Safeguarding cells from abnormal mRNA function. *Genes Dev* 2007;**21**:1833–56.
11. Kuhn U, Wahle E. Structure and function of poly(A) binding proteins. *Biochim Biophys Acta* 2004;**1678**:67–84.
12. Celesnik H, Deana A, Belasco JG. Initiation of RNA decay in *Escherichia coli* by 5′ pyrophosphate removal. *Mol Cell* 2007;**27**:79–90.
13. Deana A, Celesnik H, Belasco JG. The bacterial enzyme RppH triggers messenger RNA degradation by 5′ pyrophosphate removal. *Nature* 2008;**451**:355–8.
14. Gagliardi D, Stepien PP, Temperley RJ, Lightowlers RN, Chrzanowska-Lightowlers ZM. Messenger RNA stability in mitochondria: Different means to an end. *Trends Genet* 2004;**20**:260–7.
15. Reuven NB, Zhou Z, Deutscher MP. Functional overlap of tRNA nucleotidyltransferase, poly (A) polymerase I, and polynucleotide phosphorylase. *J Biol Chem* 1997;**272**:33255–9.
16. Hajnsdorf E, Braun F, Haugel-Nielsen J, Regnier P. Polyadenylation destabilizes the *rpsO* mRNA of *Escherichia coli*. *Proc Natl Acad Sci USA* 1995;**92**:3973–7.

17. Kushner SR. mRNA decay in prokaryotes and eukaryotes: Different approaches to a similar problem. *IUBMB Life* 2004;**56**:585–94.
18. Rott R, Zipor G, Portnoy V, Liveanu V, Schuster G. RNA polyadenylation and degradation in cyanobacteria are similar to the chloroplast but different from *Escherichia coli*. *J Biol Chem* 2003;**278**:15771–7.
19. Raynal LC, Krisch HM, Carpousis AJ. The *Bacillus subtilis* nucleotidyltransferase is a tRNA CCA-adding enzyme. *J Bacteriol* 1998;**180**:6276–82.
20. Campos-Guillen J, Bralley P, Jones GH, Bechhofer DH, Olmedo-Alvarez G. Addition of poly (A) and heteropolymeric 3′ ends in *Bacillus subtilis* wild-type and polynucleotide phosphorylase-deficient strains. *J Bacteriol* 2005;**187**:4698–706.
21. Lisitsky I, Klaff P, Schuster G. Blocking polyadenylation of mRNA in the chloroplast inhibits its degradation. *Plant J* 1997;**12**:1173–8.
22. Portnoy V, Schuster G. RNA polyadenylation and degradation in different archaea; roles of the exosome and RNase R. *Nucleic Acids Res* 2006;**34**:5923–31.
23. Portnoy V, Evguenieva-Hackenberg E, Klein F, Walter P, Lorentzen E, Klug G, et al. RNA polyadenylation in archaea: Not observed in *Haloferax* while the exosome polyadenylates RNA in *Sulfolobus*. *EMBO Rep* 2005;**6**:1188–93.
24. Portnoy V, Schuster G. *Mycoplasma gallisepticum* as the first analyzed bacterium in which RNA is not polyadenylated. *FEMS Microbiol Lett* 2008;**283**:97–103.
25. Slomovic S, Portnoy V, Liveanu V, Schuster G. RNA polyadenylation in prokaryotes and organelles; different tails tell different tales. *Crit Rev Plant Sci* 2006;**25**:65–77.
26. Callaghan AJ, Marcaida MJ, Stead JA, McDowall KJ, Scott WG, Luisi BF. Structure of *Escherichia coli* RNase E catalytic domain and implications for RNA turnover. *Nature* 2005;**437**:1187–91.
27. Schein A, Sheffy-Levin S, Glaser F, Schuster G. The RNase E/G-type endoribonuclease of higher plants is located in the chloroplast and cleaves RNA similarly to the *E. coli* enzyme. *RNA* 2008;**14**:1057–68.
28. Ghora BK, Apirion D. Structural analysis and *in vitro* processing to p5 rRNA of a 9S RNA molecule isolated from an rne mutant of *E. coli*. *Cell* 1978;**15**:1055–66.
29. Afonyushkin T, Vecerek B, Moll I, Blasi U, Kaberdin VR. Both RNase E and RNase III control the stability of sodB mRNA upon translational inhibition by the small regulatory RNA RyhB. *Nucleic Acids Res* 2005;**33**:1678–89.
30. Morita T, Maki K, Aiba H. RNase E-based ribonucleoprotein complexes: Mechanical basis of mRNA destabilization mediated by bacterial noncoding RNAs. *Genes Dev* 2005;**19**:2176–86.
31. Udekwu KI, Darfeuille F, Vogel J, Reimegard J, Holmqvist E, Wagner EG. Hfq-dependent regulation of OmpA synthesis is mediated by an antisense RNA. *Genes Dev* 2005;**19**:2355–66.
32. Kaberdin VR, Chao YH, Lin-Chao S. RNase E cleaves at multiple sites in bubble regions of RNA I stem loops yielding products that dissociate differentially from the enzyme. *J Biol Chem* 1996;**271**:13103–9.
33. Li Z, Deutscher MP. RNase E plays an essential role in the maturation of *Escherichia coli* tRNA precursors. *RNA* 2002;**8**:97–109.
34. Ow MC, Kushner SR. Initiation of tRNA maturation by RNase E is essential for cell viability in *E. coli*. *Genes Dev* 2002;**16**:1102–15.
35. Ono M, Kuwano M. A conditional lethal mutation in an *Escherichia coli* strain with a longer chemical lifetime of messenger RNA. *J Mol Biol* 1979;**129**:343–57.
36. Arraiano CM, Yancey SD, Kushner SR. Stabilization of discrete mRNA breakdown products in ams pnp rnb multiple mutants of *Escherichia coli* K-12. *J Bacteriol* 1988;**170**:4625–33.
37. Mackie GA. Secondary structure of the mRNA for ribosomal protein S20: Implications for cleavage by ribonuclease E. *J Biol Chem* 1992;**267**:1054–61.

38. Hajnsdorf E, Braun F, Haugel-Nielsen J, Le Derout J, Regnier P. Multiple degradation pathways of the rpsO mRNA of Escherichia coli. RNase E interacts with the 5′ and 3′ extremities of the primary transcript. Biochimie 1996;78:416–24.
39. Diwa A, Bricker AL, Jain C, Belasco JG. An evolutionarily conserved RNA stem-loop functions as a sensor that directs feedback regulation of RNase E gene expression. Genes Dev 2000;14:1249–60.
40. Sousa S, Marchand I, Dreyfus M. Autoregulation allows Escherichia coli RNase E to adjust continuously its synthesis to that of its substrates. Mol Microbiol 2001;42:867–78.
41. Ow MC, Liu Q, Mohanty BK, Andrew ME, Maples VF, Kushner SR. RNase E levels in Escherichia coli are controlled by a complex regulatory system that involves transcription of the rne gene from three promoters. Mol Microbiol 2002;43:159–71.
42. Py P, Higgins CF, Krisch HM, Carpousis AJ. A DEAD-box RNA helicase in the Escherichia coli degradosome. Nature 1996;381:169–72.
43. Miczak A, Kaberdin VR, Wei CL, Lin-Chao S. Proteins associated with RNase E in a multicomponent ribonucleolytic complex. Proc Natl Acad Sci USA 1996;93:3865–9.
44. Vanzo NF, Li YS, Py B, Blum E, Higgins CF, Raynal LC, et al. Ribonuclease E organizes the protein interactions in the Escherichia coli RNA degradosome. Genes Dev 1998;12:2770–81.
45. Carpousis AJ. The RNA degradosome of Escherichia coli: An mRNA-degrading machine assembled on RNase E. Annu Rev Microbiol 2007;61:71–87.
46. Kaberdin VR, Miczak A, Jakobsen JS, Lin-Chao S, McDowall KJ, von Gabain A. The endoribonucleolytic N-terminal half of Escherichia coli RNase E is evolutionarily conserved in Synechocystis sp. and other bacteria but not the C-terminal half, which is sufficient for degradosome assembly. Proc Natl Acad Sci USA 1998;95:11637–42.
47. Baginsky S, Shteiman-Kotler A, Liveanu V, Yehudai-Resheff S, Bellaoui M, Settlage RE, et al. Chloroplast PNPase exists as a homo-multimer enzyme complex that is distinct from the Escherichia coli degradosome. RNA 2001;7:1464–75.
48. Cohen SN, McDowall KJ. RNase E: Still a wonderfully mysterious enzyme. Mol Microbiol 1997;23:1099–106.
49. Ow MC, Perwez T, Kushner SR. RNase G of Escherichia coli exhibits only limited functional overlap with its essential homologue, RNase E. Mol Microbiol 2003;49:607–22.
50. Lee K, Bernstein J, Cohen S. RNase G complementation of rne null mutation identifies functional interrelationships with RNase E in Escherichia coli. Mol Microbiol 2002;43:1445–56.
51. Mackie GA. Ribonuclease E is a 5′-end-dependent endonuclease. Nature 1998;395:720–3.
52. Jiang X, Diwa A, Belasco JG. Regions of RNase E important for 5′-end-dependent RNA cleavage and autoregulated synthesis. J Bacteriol 2000;182:2468–75.
53. Tock MR, Walsh AP, Carroll G, McDowall KJ. The CafA protein required for the 5′-maturation of 16 S rRNA is a 5′-end-dependent ribonuclease that has context-dependent broad sequence specificity. J Biol Chem 2000;275:8726–32.
54. Worrall JA, Luisi BF. Information available at cut rates: Structure and mechanism of ribonucleases. Curr Opin Struct Biol 2007;17:128–37.
55. Lee K, Cohen SN. A Streptomyces coelicolor functional orthologue of Escherichia coli RNase E shows shuffling of catalytic and PNPase-binding domains. Mol Microbiol 2003;48:349–60.
56. Bollenbach TJ, Schuster G, Stern DB. Cooperation of endo- and exoribonucleases in chloroplast mRNA turnover. Prog Nucleic Acid Res Mol Biol 2004;78:305–37.
57. Mudd EA, Sullivan S, Gisby MF, Mironov A, Kwon CS, Chung WI, et al. A 125 kDa RNase E/G-like protein is present in plastids and is essential for chloroplast development and autotrophic growth in Arabidopsis. J Exp Bot 2008;59:2597–610.
58. Mayfield SP. Chloroplast gene regulation: Interaction of the nuclear and chloroplast genomes in the expression of photosynthetic proteins. Curr Opin Cell Biol 1990;2:509–13.

59. Barkan A, Goldschmidt-Clermont M. Participation of nuclear genes in chloroplast gene expression. *Biochimie* 2000;**82**:559–72.
60. Rochaix J-D. Posttranscriptional steps in the expression of chloroplast genes. *Annu Rev Cell Biol* 1992;**8**:1–28.
61. Even S, Pellegrini O, Zig L, Labas V, Vinh J, Brechemmier-Baey D, et al. Ribonucleases J1 and J2: Two novel endoribonucleases in *B. subtilis* with functional homology to *E. coli* RNase E. *Nucleic Acids Res* 2005;**33**:2141–52.
62. de la Sierra-Gallay IL, Zig L, Jamalli A, Putzer H. Structural insights into the dual activity of RNase J. *Nat Struct Mol Biol* 2008;**15**:206–12.
63. Condon C, Putzer H, Luo D, Grunberg-Manago M. Processing of the *Bacillus subtilis* thrS leader mRNA is RNase E-dependent in *Escherichia coli*. *J Mol Biol* 1997;**268**:235–42.
64. Mathy N, Benard L, Pellegrini O, Daou R, Wen T, Condon C. 5′–3′ exoribonuclease activity in bacteria: Role of RNase J1 in rRNA maturation and 5′ stability of mRNA. *Cell* 2007;**129**:681–92.
65. Drager RG, Girard-Bascou J, Choquet Y, Kindle KL, Stern DB. In vivo evidence for 5′–3′ exoribonuclease degradation of an unstable chloroplast mRNA. *Plant J* 1998;**13**:85–96.
66. Drager RG, Higgs DC, Kindle KL, Stern DB. 5′–3′ exoribonucleolytic activity is a normal component of chloroplast mRNA decay pathways. *Plant J* 1999;**19**:521–31.
67. Hicks A, Drager RG, Higgs DC, Stern DB. An mRNA 3′ processing site targets downstream sequences for rapid degradation in *Chlamydomonas* chloroplasts. *J Biol Chem* 2002;**277**:3325–33.
68. Britton RA, Wen T, Schaefer L, Pellegrini O, Uicker WC, Mathy N, et al. Maturation of the 5′ end of *Bacillus subtilis* 16S rRNA by the essential ribonuclease YkqC/RNase J1. *Mol Microbiol* 2007;**63**:127–38.
69. Baker ME, Grundy WN, Elkan CP. Spinach CSP41, an mRNA-binding protein and ribonuclease, is homologous to nucleotide-sugar epimerases and hydroxysteroid dehydrogenases. *Biochem Biophys Res Commun* 1998;**248**:250–4.
70. Yamaguchi K, Beligni MV, Prieto S, Haynes PA, McDonald WH, Yates JR, III, et al. Proteomic characterization of the *Chlamydomonas reinhardtii* chloroplast ribosome. Identification of proteins unique to the 70S ribosome. *J Biol Chem* 2003;**278**:33774–85.
71. Yang J, Schuster G, Stern DB. CSP41, a sequence-specific chloroplast mRNA binding protein, is an endoribonuclease. *Plant Cell* 1996;**8**:1409–20.
72. Yang J, Stern DB. The spinach chloroplast endoribonuclease CSP41 cleaves the 3′-untranslated region of *petD* mRNA primarily within its terminal stem-loop structure. *J Biol Chem* 1997;**272**:12874–80.
73. Bollenbach TJ, Tatman DA, Stern DB. CSP41a, a multifunctional RNA-binding protein, initiates mRNA turnover in tobacco chloroplasts. *Plant J* 2003;**36**:842–52.
74. Beligni MV, Mayfield SP. *Arabidopsis thaliana* mutants reveal a role for CSP41a and CSP41b, two ribosome-associated endonucleases, in chloroplast ribosomal RNA metabolism. *Plant Mol Biol* 2008;**67**:389–401.
75. Stern DB, Gruissem W. Control of plastid gene expression: 3′ inverted repeats act as mRNA processing and stabilizing elements, but do not terminate transcription. *Cell* 1987;**51**:1145–57.
76. Kazantsev AV, Pace NR. Bacterial RNase P: A new view of an ancient enzyme. *Nat Rev Microbiol* 2006;**4**:729–40.
77. Gopalan V, Vioque A, Altman S. RNase P: Variations and uses. *J Biol Chem* 2002;**277**:6759–62.
78. Jarrous N, Reiner R. Human RNase P: A tRNA-processing enzyme and transcription factor. *Nucleic Acids Res* 2007;**35**:3519–24.
79. Torres-Larios A, Swinger KK, Pan T, Mondragon A. Structure of ribonuclease P—a universal ribozyme. *Curr Opin Struct Biol* 2006;**16**:327–35.

80. Evans D, Marquez SM, Pace NR. RNase P: Interface of the RNA and protein worlds. *Trends Biochem Sci* 2006;**31**:333–41.
81. Walker SC, Engelke DR. Ribonuclease P: The evolution of an ancient RNA enzyme. *Crit Rev Biochem Mol Biol* 2006;**41**:77–102.
82. Clayton DA. Molecular biology: A big development for a small RNA. *Nature* 2001;**410**:29–31.
83. Puranam RS, Attardi G. The RNase P associated with HeLa cell mitochondria contains an essential RNA component identical in sequence to that of the nuclear RNase P. *Mol Cell Biol* 2001;**21**:548–61.
84. Randau L, Schroder I, Soll D. Life without RNase P. *Nature* 2008;**453**:120–3.
85. Rossmanith W, Karwan RM. Characterization of human mitochondrial RNase P: Novel aspects in tRNA processing. *Biochem Biophys Res Commun* 1998;**247**:234–41.
86. Thomas BC, Li X, Gegenheimer P. Chloroplast ribonuclease P does not utilize the ribozyme-type pre-tRNA cleavage mechanism. *RNA* 2000;**6**:545–53.
87. Schurer H, Schiffer S, Marchfelder A, Morl M. This is the end: Processing, editing and repair at the tRNA 3'-terminus. *Biol Chem* 2001;**382**:1147–56.
88. Schiffer S, Rosch S, Marchfelder A. Assigning a function to a conserved group of proteins: The tRNA 3'-processing enzymes. *EMBO J* 2002;**21**:2769–77.
89. Dubrovsky EB, Dubrovskaya VA, Levinger L, Schiffer S, Marchfelder A. Drosophila RNase Z processes mitochondrial and nuclear pre-tRNA 3' ends *in vivo*. *Nucleic Acids Res* 2004;**32**:255–62.
90. Pellegrini O, Nezzar J, Marchfelder A, Putzer H, Condon C. Endonucleolytic processing of CCA-less tRNA precursors by RNase Z in *Bacillus subtilis*. *EMBO J* 2003;**22**:4534–43.
91. Grunberg-Manago M, Oritz PJ, Ochoa S. Enzymatic synthesis of nucleic acid-like polynucleotides. *Science* 1955;**122**:907–10.
92. Littauer UZ, Soreq H. Polynucleotide phosphorylase. *In* The enzymes, (PD Boyer, Ed.), p. 15. Academic Press, New York.
93. Grunberg-Manago M. Messenger RNA stability and its role in control of gene expression in bacteria and phages. *Annu Rev Genet* 1999;**33**:193–227.
94. Littauer UZ, Grunberg-Manago M. Polynucleotide phosphorylase. *In* The encyclopedia of molecular biology, (TE Creighton, Ed.), pp. 1911–8. Wiley, New York.
95. Jarrige A, Brechemier-Baey D, Mathy N, Duche O, Portier C. Mutational analysis of polynucleotide phosphorylase from *Escherichia coli*. *J Mol Biol* 2002;**321**:397–409.
96. Slomovic S, Portnoy V, Yehudai-Resheff S, Bronshtein E, Schuster G. Polynucleotide phosphorylase and the archaeal exosome as poly(A)-polymerases. *Biochim Biophys Acta* 2008;**1779**:247–55.
97. Clements MO, Eriksson S, Thompson A, Lucchini S, Hinton JC, Normark S, et al. Polynucleotide phosphorylase is a global regulator of virulence and persistency in *Salmonella enterica*. *Proc Natl Acad Sci USA* 2002;**99**:8784–9.
98. Sauret-Gueto S, Botella-Pavia P, Flores-Perez U, Martinez-Garcia JF, San Roman C, Leon P, et al. Plastid cues posttranscriptionally regulate the accumulation of key enzymes of the methylerythritol phosphate pathway in *Arabidopsis*. *Plant Physiol* 2006;**141**:75–84.
99. Yehudai-Resheff S, Zimmer SL, Komine Y, Stern DB. Integration of chloroplast nucleic acid metabolism into the phosphate deprivation response in *Chlamydomonas reinhardtii*. *Plant Cell* 2007;**19**:1023–38.
100. Sarkar D, Park ES, Barber GN, Fisher PB. Activation of double-stranded RNA dependent protein kinase, a new pathway by which human polynucleotide phosphorylase (hPNPase(old-35)) induces apoptosis. *Cancer Res* 2007;**67**:7948–53.
101. Wu J, Li Z. Human polynucleotide phosphorylase reduces oxidative RNA damage and protects HeLa cell against oxidative stress. *Biochem Biophys Res Commun* 2008;**372**:288–92.

102. Chen HW, Rainey RN, Balatoni CE, Dawson DW, Troke JJ, Wasiak S, et al. Mammalian polynucleotide phosphorylase is an intermembrane space RNase that maintains mitochondrial homeostasis. *Mol Cell Biol* 2006;**26**:8475–87.
103. Rainey RN, Glavin JD, Chen HW, French SW, Teitell MA, Koehler CM. A new function in translocation for the mitochondrial i-AAA protease Yme1: Import of polynucleotide phosphorylase into the intermembrane space. *Mol Cell Biol* 2006;**26**:8488–97.
104. Slomovic S, Schuster G. Stable PNPase RNAi silencing: Its effect on the processing and adenylation of human mitochondrial RNA. *RNA* 2008;**14**:310–23.
105. Walter M, Kilian J, Kudla J. PNPase activity determines the efficiency of mRNA 3′-end processing, the degradation of tRNA and the extent of polyadenylation in chloroplasts. *EMBO J* 2002;**21**:6905–14.
106. Perrin R, Lange H, Grienenberger JM, Gagliardi D. AtmtPNPase is required for multiple aspects of the 18S rRNA metabolism in *Arabidopsis thaliana* mitochondria. *Nucleic Acids Res* 2004;**32**:5174–82.
107. Perrin R, Meyer EH, Zaepfel M, Kim YJ, Mache R, Grienenberger JM, et al. Two exoribonucleases act sequentially to process mature 3′-ends of atp9 mRNAs in *Arabidopsis* mitochondria. *J Biol Chem* 2004;**279**:25440–6.
108. Holec S, Lange H, Kuhn K, Alioua M, Borner T, Gagliardi D. Relaxed transcription in *Arabidopsis* mitochondria is counterbalanced by RNA stability control mediated by polyadenylation and polynucleotide phosphorylase. *Mol Cell Biol* 2006;**26**:2869–76.
109. Portnoy V, Palnizky G, Yehudai-Resheff S, Glaser F, Schuster G. Analysis of the human polynucleotide phosphorylase (PNPase) reveals differences in RNA binding and response to phosphate compared to its bacterial and chloroplast counterparts. *RNA* 2008;**14**:297–309.
110. Chen HW, Koehler CM, Teitell MA. Human polynucleotide phosphorylase: Location matters. *Trends Cell Biol* 2007;**17**:600–8.
111. Lorentzen E, Walter P, Fribourg S, Evgunieva-Hackenberg E, Klug G, Conti E. The archaeal exosome core is a hexameric ring structure with three catalytic subunits. *Nat Struct Mol Biol* 2005;**12**:575–81.
112. Symmons MF, Jones GH, Luisi BF. A duplicated fold is the structural basis for polynucleotide phosphorylase catalytic activity, processivity, and regulation. *Structure* 2000;**8**:1215–26.
113. Symmons MF, Williams MG, Luisi BF, Jones GH, Carpousis AJ. Running rings around RNA: A superfamily of phosphate-dependent RNases. *Trends Biochem Sci* 2002;**27**:11–8.
114. Yehudai-Resheff S, Portnoy V, Yogev S, Adir N, Schuster G. Domain analysis of the chloroplast polynucleotide phosphorylase reveals discrete functions in RNA degradation, polyadenylation, and sequence homology with exosome proteins. *Plant Cell* 2003;**15**:2003–19.
115. Dziembowski A, Lorentzen E, Conti E, Seraphin B. A single subunit, Dis3, is essentially responsible for yeast exosome core activity. *Nat Struct Mol Biol* 2007;**14**:15–22.
116. Hernandez H, Dziembowski A, Taverner T, Seraphin B, Robinson CV. Subunit architecture of multimeric complexes isolated directly from cells. *EMBO Rep* 2006;**7**:605–10.
117. Liu Q, Greimann JC, Lima CD. Reconstitution, activities, and structure of the eukaryotic RNA exosome. *Cell* 2006;**127**:1223–37.
118. Lin-Chao S, Chiou NT, Schuster G. The PNPase, exosome and RNA helicases as the building components of evolutionarily-conserved RNA degradation machines. *J Biomed Sci* 2007;**14**:523–32.
119. Lorentzen E, Basquin J, Tomecki R, Dziembowski A, Conti E. Structure of the active subunit of the yeast exosome core, Rrp44: Diverse modes of substrate recruitment in the RNase II nuclease family. *Mol Cell* 2008;**29**:717–28.
120. Zuo Y, Deutscher MP. Exoribonuclease superfamilies: Structural analysis and phylogenetic distribution. *Nucleic Acids Res* 2001;**29**:1017–26.

121. Bollenbach TJ, Lange H, Gutierrez R, Erhardt M, Stern DB, Gagliardi D. RNR1, a 3′–5′ exoribonuclease belonging to the RNR superfamily, catalyzes 3′ maturation of chloroplast ribosomal RNAs in *Arabidopsis thaliana*. *Nucleic Acids Res* 2005;**33**:2751–63.
122. Chekanova JA, Gregory BD, Reverdatto SV, Chen H, Kumar R, Hooker T, et al. Genome-wide high-resolution mapping of exosome substrates reveals hidden features in the *Arabidopsis* transcriptome. *Cell* 2007;**131**:1340–53.
123. Lange H, Holec S, Cognat V, Pieuchot L, Le Ret M, Canaday J, et al. Degradation of a polyadenylated rRNA maturation by-product involves one of the three RRP6-like proteins in *Arabidopsis thaliana*. *Mol Cell Biol* 2008;**28**:3038–44.
124. Cheng ZF, Deutscher MP. Purification and characterization of the *Escherichia coli* exoribonuclease RNase R. Comparison with RNase II. *J Biol Chem* 2002;**277**:21624–9.
125. Frazao C, McVey CE, Amblar M, Barbas A, Vonrhein C, Arraiano CM, et al. Unravelling the dynamics of RNA degradation by ribonuclease II and its RNA-bound complex. *Nature* 2006;**443**:110–4.
126. Zuo Y, Vincent HA, Zhang J, Wang Y, Deutscher MP, Malhotra A. Structural basis for processivity and single-strand specificity of RNase II. *Mol Cell* 2006;**24**:149–56.
127. Dziembowski A, Piwowarski J, Hoser R, Minczuk M, Dmochowska A, Siep M, et al. The yeast mitochondrial degradosome. Its composition, interplay between RNA helicase and RNase activities and the role in mitochondrial RNA metabolism. *J Biol Chem* 2003;**278**:1603–11.
128. Donovan WP, Kushner SR. Polynucleotide phosphorylase and ribonuclease II are required for cell viability and mRNA turnover in *Escherichia coli* K-12. *Proc Natl Acad Sci USA* 1986;**83**:120–4.
129. Kishine M, Takabayashi A, Munekage Y, Shikanai T, Endo T, Sato F. Ribosomal RNA processing and an RNase R family member in chloroplasts of *Arabidopsis*. *Plant Mol Biol* 2004;**55**:595–606.
130. Danckwardt S, Hentze MW, Kulozik AE. 3′ end mRNA processing: Molecular mechanisms and implications for health and disease. *EMBO J* 2008;**27**:482–98.
131. Martin G, Keller W. RNA-specific ribonucleotidyl transferases. *RNA* 2007;**13**:1834–49.
132. Martin G, Doublie S, Keller W. Determinants of substrate specificity in RNA-dependent nucleotidyl transferases. *Biochim Biophys Acta* 2008;**1779**:206–16.
133. Martin G, Keller W. Sequence motifs that distinguish ATP(CTP):tRNA nucleotidyl transferases from eubacterial poly(A) polymerases. *RNA* 2004;**10**:899–906.
134. Zimmer SL, Fei Z, Stern DB. Genome-based analysis of *Chlamydomonas reinhardtii* exoribonucleases and poly(A) polymerases predicts unexpected organellar and exosomal features. *Genetics/Society* 2008;**179**:125–36.
135. Wilusz CJ, Wilusz J. New ways to meet your (3′) end oligouridylation as a step on the path to destruction. *Genes Dev* 2008;**22**:1–7.
136. Nagaike T, Suzuki T, Katoh T, Ueda T. Human mitochondrial mRNAs are stabilized with polyadenylation regulated by mitochondria-specific poly(A) polymerase and polynucleotide phosphorylase. *J Biol Chem* 2005;**280**:19721–7.
137. Tomecki R, Dmochowska A, Gewartowski K, Dziembowski A, Stepien PP. Identification of a novel human nuclear-encoded mitochondrial poly(A) polymerase. *Nucleic Acids Res* 2004;**32**:6001–14.
138. Yehudai-Resheff S, Hirsh M, Schuster G. Polynucleotide phosphorylase functions as both an exonuclease and a poly(A) polymerase in spinach chloroplasts. *Mol Cell Biol* 2001;**21**:5408–16.
139. Lisitsky I, Klaff P, Schuster G. Addition of poly(A)-rich sequences to endonucleolytic cleavage sites in the degradation of spinach chloroplast mRNA. *Proc Natl Acad Sci USA* 1996;**93**:13398–403.
140. Gagliardi D, Leaver CJ. Polyadenylation accelerates the degradation of the mitochondrial mRNA associated with cytoplasmic male sterility in sunflower. *EMBO J* 1999;**18**:3757–66.

141. Lupold DS, Caoile AG FS, Stern DB. Polyadenylation occurs at multiple sites in maize mitochondrial *cox2* mRNA and is independent of editing status. *Plant Cell* 1999;**11**:1565–78.
142. Kuhn J, Tengler U, Binder S. Transcript lifetime is balanced between stabilizing stem-loop structures and degradation-promoting polyadenylation in plant mitochondria. *Mol Cell Biol* 2001;**21**:731–42.
143. Gagliardi D, Perrin R, Marechal-Drouard L, Grienenberger JM, Leaver CJ. Plant mitochondrial polyadenylated mRNAs are degraded by a 3'–5'-exoribonuclease activity, which proceeds unimpeded by stable secondary structures. *J Biol Chem* 2001;**276**:43541–7.
144. Schafer B, Hansen M, Lang BF. Transcription and RNA-processing in fission yeast mitochondria. *RNA* 2005;**11**:785–95.
145. Leszczyniecka M, DeSalle R, Kang DC, Fisher PB. The origin of polynucleotide phosphorylase domains. *Mol Phylogenet Evol* 2004;**31**:123–30.
146. Simpson L, Sbicego S, Aphasizhev R. Uridine insertion/deletion RNA editing in trypanosome mitochondria: A complex business. *RNA* 2003;**9**:265–76.
147. Kao CY, Read LK. Opposing effects of polyadenylation on the stability of edited and unedited mitochondrial RNAs in *Trypanosoma brucei*. *Mol Cell Biol* 2005;**25**:1634–44.
148. Slomovic S, Laufer D, Geiger D, Schuster G. Polyadenylation and degradation of human mitochondrial RNA: The prokaryotic past leaves its mark. *Mol Cell Biol* 2005;**25**:6427–35.
149. Etheridge RD, Aphasizheva I, Gershon PD, Aphasizhev R. 3' adenylation determines mRNA abundance and monitors completion of RNA editing in *T. brucei* mitochondria. *EMBO J* 2008;**27**:1596–608.
150. Lurin C, Andres C, Aubourg S, Bellaoui M, Bitton F, Bruyere C, et al. Genome-wide analysis of *Arabidopsis* pentatricopeptide repeat proteins reveals their essential role in organelle biogenesis. *Plant Cell* 2004;**16**:2089–103.
151. Delannoy E, Stanley WA, Bond CS, Small ID. Pentatricopeptide repeat (PPR) proteins as sequence-specificity factors in posttranscriptional processes in organelles. *Biochem Soc Trans* 2007;**35**:1643–7.
152. Lacava J, Houseley J, Saveanu C, Petfalski E, Thompson E, Jacquier A, et al. RNA degradation by the exosome is promoted by a nuclear polyadenylation complex. *Cell* 2005;**121**:713–24.
153. Vanacova S, Wolf J, Martin G, Blank D, Dettwiler S, Friedlein A, et al. A new yeast poly(A) polymerase complex involved in RNA quality control. *PLoS Biol* 2005;**3**:e189.
154. Wyers F, Rougemaille M, Badis G, Rousselle JC, Dufour ME, Boulay J, et al. Cryptic Pol II transcripts are degraded by a nuclear quality control pathway involving a new poly(A) polymerase. *Cell* 2005;**121**:725–37.
155. Penschow JL, Sleve DA, Ryan CM, Read LK. TbDSS-1, an essential *Trypanosoma brucei* exoribonuclease homolog that has pleiotropic effects on mitochondrial RNA metabolism. *Eukaryot Cell* 2004;**3**:1206–16.
156. Ryan CM, Militello KT, Read LK. Polyadenylation regulates the stability of *Trypanosoma brucei* mitochondrial RNAs. *J Biol Chem* 2003;**278**:32753–62.
157. Ryan CM, Read LK. UTP-dependent turnover of *Trypanosoma brucei* mitochondrial mRNA requires UTP polymerization and involves the RET1 TUTase. *RNA* 2005;**11**:763–73.
158. Etheridge RD, Aphasizheva I, Gershon PD, Aphasizhev R. 3' adenylation determines mRNA abundance and monitors completion of RNA editing in *T. brucei* mitochondria. *EMBO J* 2008;**27**:1596–608.
159. Taanman JW. The mitochondrial genome: Structure, transcription, translation and replication. *Biochim Biophys Acta* 1999;**1410**:103–23.
160. Attardi G, Schatz G. Biogenesis of mitochondria. *Ann Rev Cell Biol* 1988;**4**:289–333.
161. Anderson S, Bankier AT, Barrell BG, de Bruijn MH, Coulson AR, Drouin J, et al. Sequence and organization of the human mitochondrial genome. *Nature* 1981;**290**:457–65.

162. Ojala D, Montoya J, Attardi G. tRNA punctuation model of RNA processing in human mitochondria. *Nature* 1981;**290**:470–4.
163. Montoya J, Ojala D, Attardi G. Distinctive features of the 5′-terminal sequences of the human mitochondrial mRNAs. *Nature* 1981;**290**:465–70.
164. Montoya J, Gaines GL, Attardi G. The pattern of transcription of the human mitochondrial rRNA genes reveals two overlapping transcription units. *Cell* 1983;**34**:151–9.
165. Levinger L, Morl M, Florentz C. Mitochondrial tRNA 3′ end metabolism and human disease. *Nucleic Acids Res* 2004;**32**:5430–41.
166. King MP, Attardi G. Posttranscriptional regulation of the steady-state levels of mitochondrial tRNAs in HeLa cells. *J Biol Chem* 1993;**268**:10228–37.
167. Hirsch M, Penman S. Mitochondrial polyadenylic acid-containing RNA: Localization and characterization. *J Mol Biol* 1973;**80**:379–91.
168. Amalric F, Merkel C, Gelfand R, Attardi G. Fractionation of mitochondrial RNA from HeLa cells by high-resolution electrophoresis under strongly denaturing conditions. *J Mol Biol* 1978;**118**:1–25.
169. Temperley RJ, Seneca SH, Tonska K, Bartnik E, Bindoff LA, Lightowlers RN, et al. Investigation of a pathogenic mtDNA microdeletion reveals a translation-dependent deadenylation decay pathway in human mitochondria. *Hum Mol Genet* 2003;**12**:2341–8.
170. Bobrowicz AJ, Lightowlers RN, Chrzanowska-Lightowlers Z. Polyadenylation and degradation of mRNA in mammalian mitochondria: A missing link? *Biochem Soc Trans* 2008;**36**:517–9.
171. Piwowarski J, Grzechnik P, Dziembowski A, Dmochowska A, Minczuk M, Stepien PP. Human polynucleotide phosphorylase, hPNPase, is localized in mitochondria. *J Mol Biol* 2003;**329**:853–7.

Killer and Protective Ribosomes

MARC DREYFUS

Expression Génétique Microbienne (CNRS UPR9073), IBPC, 11 rue P. et M. Curie 75005 Paris, France

I. Introduction	424
II. Nucleolytic *Versus* Non-nucleolytic Inactivation of mRNAs	426
A. Definitions	426
B. Translation and Non-nucleolytic Inactivation	427
C. A New Case for Non-nucleolytic Inactivation: The Action of sRNAs	427
D. Possible Physical Separation of Translation and mRNA Decay	428
E. What is the Global Importance of Non-nucleolytic Inactivation?	430
III. Translation–Degradation Interplay: Conceptual Ambiguities and Technical Caveats	430
A. Direct or Translation-Mediated Changes of mRNA Stability	430
B. Practical Aspects: How to Measure mRNA Stability Without Affecting It?	433
C. The *lacZ* Gene, or the Importance of Comparing Comparable Items	435
IV. Killer Ribosomes	436
A. Historical Background	436
B. The *daa* Operon	437
C. ssrA Tagging as a Tracer for Killer Ribosomes	438
D. Looking for the Nuclease	441
E. Ambiguous Lessons from Gram-Positive Organisms	442
F. Some Pending Questions	445
V. Protective Ribosomes	446
A. mRNAs are Generally Protected by Ribosomes	446
B. Mechanism of RNase E Cleavage: The 5′ Tethering and Internal Entry Pathways	447
C. Local *Versus* Distal Protection	450
D. A Specific Role of RNase E C-Terminal Half in the Degradation of Untranslated mRNA?	454
E. Transcription, Translation, or RNase E Preferences: The Source of Directionality	454
F. Protective Ribosomes in *B. subtilis*	455
VI. Concluding Remarks	457
References	458

In prokaryotes, translation influences mRNA decay. The breakdown of most *Escherichia coli* mRNAs is initiated by RNase E, a 5′-dependent endonuclease. Some mRNAs are protected by ribosomes even if these are located far upstream of cleavage sites ("protection at a distance"), whereas others require direct shielding of these sites. I argue that these situations reflect different modes of interaction of RNase E with mRNAs. Protection at a distance is most impressive in Bacilli, where ribosomes can protect kilobases of unstable downstream sequences. I propose that this protection reflects the role in mRNA decay of RNase J1, a $5' \rightarrow 3'$ exonuclease with no *E. coli* equivalent. Finally, recent years have shown that besides their protective role, ribosomes can also cleave their mRNA under circumstances that cause ribosome stalling. The endonuclease associated with this "killing" activity, which has a eukaryotic counterpart ("no-go decay"), is not characterized; it may be borne by the distressed ribosome itself.

I. Introduction

Since early studies on *Escherichia coli* revealed the metabolic instability of mRNA (1), the possibility that this instability might be further modulated by translation has frequently been considered. However, the magnitude of this modulation and even its nature—protective or killer ribosomes?—has long remained obscure. Nevertheless, the field has become less descriptive and more interpretative over the years, as our understanding of mRNA decay itself has improved. Early claims that translating ribosomes can degrade their own message in *E. coli* were rapidly challenged, and for more than 20 years it was considered that translation, on the contrary, had only a protective effect against degradation. This belief was based on the observation that circumstances that repress the translation of individual mRNAs also generally destabilize them, and *vice versa*. After the discovery, in the early 1990s, of RNase E as the endonuclease that initiates the degradation of most mRNAs (see chapter by A.J. Carpousis and colleagues in this volume), a widely accepted view was that the translation-mediated stabilization of mRNAs reflects a direct, presumably steric, competition between ribosomes and RNase E. Yet, long-standing observations have suggested that at least some mRNAs are inactivated by non-nucleolytic events that precede degradation. No direct competition between translation and degradation can take place in these cases. The recent observation that RNase E is not spread throughout the cytosol like the ribosomes, but is bound to the inner membrane is intriguing in this regard. It raises the possibility that translation and mRNA degradation are to some extent separated in space, and that mRNAs are inactivated non-nucleolytically by passing from one "compartment" to the other. Oddly, we still do not know how widespread non-nucleolytic inactivation is among *E. coli* mRNAs.

Several recent discoveries have further obscured the simplest case of a direct competition between ribosomes and RNase E. Among these discoveries is the realization that RNase E, although an endonuclease, prefers substrates carrying a 5' monophosphate end. Since genuine mRNAs bear a 5' triphosphate extremity, pyrophosphate removal, which is catalyzed by the newly discovered

RNA pyrophosphohydrolase RppH, presumably constitutes an inconspicuous but essential step in the degradation of many mRNAs. The presence of ribosomes near the 5′ end might inhibit pyrophosphate removal or the recognition of the 5′ monophosphate by RNase E, which may explain why certain mRNAs can be protected without requiring the direct presence of ribosomes over the RNase E cleavage sites ("protection at a distance"). Finally and perhaps most remarkably, "killer ribosomes" have made an unexpected come-back during the last decade, and it is now believed that almost any circumstance that results in ribosome stalling can cause mRNA cleavage, either by the ribosome itself or by unknown associated endonucleases. This remarkable phenomenon, which is distinct from the action of the known toxins (see chapter by Y. Yamaguchi and M. Inouye in this volume), may be important for bulk mRNA decay particularly under stress conditions.

While the *E. coli* field was moving ahead, significant information became available on the translation–degradation interplay in other bacteria, most notably *Bacillus subtilis*. Early observations, in the late 1980s and in the 1990s, were again purely descriptive. A salient feature was that, in *B. subtilis*, a single stalled ribosome could protect kilobases of downstream mRNA that would otherwise have been unstable. This protection at a distance was much more prominent than in *E. coli*. Recently, it has become clear that this quantitative difference reflects a fundamental divergence in the way mRNAs are degraded in *E. coli* and *B. subtilis*. In contrast to transcription or translation, the mechanism of mRNA decay has not been conserved over the 3-billion years of prokaryotic evolution. In particular, an enzyme that appears to be the major player of mRNA decay in *B. subtilis*, the newly discovered RNase J1/2, has no ortholog in *E. coli*. This enzyme possess both endonuclease and 5′→3′ exonuclease activities (see chapter by D. Bechhofer in this volume), and the impressive protection at a distance observed in *B. subtilis* may reflect a prevalence of this latter activity, which for long has been thought to be confined to eukaryotes. At present, we cannot exclude that yet other decay mechanisms, and therefore other styles of translation–degradation interplay, predominate in distant bacterial phyla.

From these brief considerations, it should be clear that, in its widest sense, the relationship between translation and degradation is quite intricate with few general rules. Yet, insofar as posttranscriptional events have a global bearing on the regulation of gene expression (2), it is also an essential and inescapable facet of prokaryotic physiology. Various aspects of the translation–degradation interplay have been reviewed recently (3–5). In particular, with Susan Joyce, I presented a critical view of the literature up to the spring of 2002 (3). In the present update, I have chosen to de-emphasize our former discussion on kinetic models for the competition between ribosomes and RNase E; readers interested in this question should refer to the former version. Conversely, I now discuss topics—evidence for killer ribosomes, possible bearing of translation on new decay mechanisms identified in *E. coli*, interpretation of the protection at a distance for Gram-positive organisms, new information on nucleolytic *versus* non-nucleolytic inactivation, etc.—for which recent progress has been most spectacular.

II. Nucleolytic *Versus* Non-nucleolytic Inactivation of mRNAs

A. Definitions

The "functional" lifetime of an mRNA is the time during which it can support protein synthesis, whereas its "physical" or "chemical" lifetime (often simply called "stability" here) is the time during which it remains physically intact, that is, immune to nuclease attack. Many mRNAs are presumably inactivated by an endo- or exonucleolytic degradation that prevents further translation. In this situation, which is hereafter referred to as "nucleolytic" inactivation, the two definitions of mRNA lifetime coincide. Nucleolytic inactivation obviously occurs when the mRNA is cleaved by the translating ribosome itself ((6, 7) see Section IV), or by an endonuclease that cleaves the mRNA between ribosomes or ahead of the leading ribosome (8, 9), or within elements that are outside the coding sequence but are essential for translation (e.g., the Shine–Dalgarno (SD) sequence ((10, 11); see chapter by Marc Uzan in this volume). In contrast, other mRNAs are known to lose their ability to support protein synthesis because of non-nucleolytic events, with degradation only occurring subsequently. This "non-nucleolytic" inactivation can be any process that effectively prevents further ribosome loading. An obvious and classical example is the binding of a translational repressor provided this binding is virtually irreversible ((12); see also the case of sRNAs below). Other processes that can result in non-nucleolytic inactivation are the progressive formation of secondary structures that irreversibly sequester the ribosome binding site (RBS) (13) or even the binding of the RNase that will subsequently degrade the mRNA (e.g., RNase E), provided this binding is distinguishable from the cleavage step and can prevent ribosome loading on its own (for discussion see Section V(B) in (14)). Other, recently identified mechanisms for non-nucleolytic inactivation are evoked below. In these cases, the functional lifetime will be *shorter* than the physical lifetime, although in practice the two lifetimes may be very close if the translationally inactivated mRNA becomes prone to nucleolytic degradation. This latter situation can be distinguished from *bona fide* nucleolytic inactivation by inhibiting the RNases involved in decay: in the case of non-nucleolytic inactivation, RNase inhibition will extend the physical lifetime without affecting the functional lifetime, so that the two become distinguishable. Non-nucleolytic inactivation was recognized early in *E. coli* (*15–17*) and somewhat more recently in *B. subtilis* (*18, 19*).

B. Translation and Non-nucleolytic Inactivation

From the viewpoint of gene expression, the distinction between nucleolytic and non-nucleolytic inactivation is essential: in the former case, mRNA degradation is a direct player in gene expression—the higher the stability, the higher the expression—whereas in the latter case, it is just a scavenging process. The distinction is also important, albeit on a more subtle register, for the interpretation of the translation–degradation interplay. In the case of nucleolytic inactivation, any correlation between translation and stability reflects a direct effect of ribosomes on mRNA cleavage; as such, it can provide interesting information on the mechanism of cleavage itself (for practical examples, see Section V). In contrast, in the case of non-nucleolytic inactivation, such correlation will only indicate that translation can facilitate or (more frequently) inhibit the inactivation process. Consequently, it may give information on the mechanism of inactivation, but not on that of the subsequent degradation. As a practical example, the *thrS* mRNA of *E. coli*, encoding threonyl tRNA synthetase (ThrS), can be inactivated non-nucleolytically by the binding of a translational repressor, the ThrS protein itself. ThrS competes directly with the initiating ribosome (*20*). At the onset of its synthesis, the *thrS* mRNA may bind either the ribosome or the repressor; in the former case, it gets translated normally, whereas in the latter it is destroyed so quickly that it is not even detectable (*12*). In the presence of mRNA mutations that favor translation over repression, the proportion of translated mRNAs increases. There is more mRNA in the cell, that is, mRNA appears globally more stable. However, those mRNA molecules that bind the repressor are still immediately degraded, as before. Thus, favoring translation tells us nothing about the mechanism of degradation of the repressed mRNA molecules, because they remain untranslated anyhow (*12*).

C. A New Case for Non-nucleolytic Inactivation: The Action of sRNAs

Among the translational repressors, *trans*-encoded small regulatory RNAs (sRNAs) deserve a special mention from the viewpoint of the mechanism of mRNA inactivation. In *E. coli*, many of these newly discovered regulators negatively modulate gene expression at a posttranscriptional level in response to environmental stress. They usually bind their target mRNAs in the RBS region with the help of the RNA chaperone protein Hfq, repressing translation and generally also destabilizing the mRNA (for reviews, see (*21, 22*) and the chapter by E. Hajnsdorf and P. Regnier in this volume). In several cases, it has been observed that, under repression conditions, mutations that affect RNase E activity stabilize the target mRNA chemically without relieving repression (23, 24), as expected for non-nucleolytic inactivation (see above). Yet, the Aiba group has obtained evidence that at least in some cases mRNA degradation is not secondary to translational repression—as in "classical" non-nucleolytic inactivation—but rather can occur *simultaneously* with it. When iron is scarce, the sRNA RyhB accumulates, repressing the translation of the *sodB* mRNA and favoring its degradation; the same holds true for the sRNAs SgrS and its target, the *ptsG* mRNA, in response to elevated phosphosugar levels (25). Now, biochemical work shows that both sRNAs can form a ribonucleoproteic complex with the RNA chaperone Hfq and RNase E, with Hfq bridging the sRNAs and the non-catalytic C-terminal half (CTH) of RNase E (26). It is then easy to understand that, by binding to the RBS of the *sodB* and *ptsG* mRNAs, the sRNAs not only repress translation, but also promote immediate mRNA degradation by directly recruiting RNase E (25). Consistent with this model, the removal of the nonessential CTH of RNase E is enough to stabilize both the *sodB* and *ptsG* mRNAs under repression conditions (24). Interestingly, the sRNA–Hfq–RNase E complex would be functionally equivalent to the eukaryotic RISC complex that mediates RNA interference (24). Now, since the binding of the sRNA to the RBS is enough to inactivate the mRNA, why is it necessary to simultaneously destroying it? It is worth noting that, in the case of the RyhB sRNA, the degradation of the target mRNA is coupled to that of the sRNA itself (27). Thus, in addition to scavenging the inactivated target mRNA, the immediate degradation process ensures that the sRNA acts stoichiometrically rather than catalytically; thereby, it remains a transient species whose action ceases once its synthesis is shut-off.

To my knowledge, the existence of RISC-like complexes associating sRNAs and RNase E has not yet been independently confirmed, and there is no doubt that alternative mechanisms exist whereby sRNA-mediated translational repression can be coupled to mRNA degradation (see discussion in (5)). Yet, it is intriguing that several other sRNAs (and in particular the porine regulators MicA, MicC, and MicF) resemble RyhB and SgrS in requiring the integrity of the CTH for the destabilization of their target mRNAs (23). It is thus possible that the mechanism of action of RyhB and SgrS will prove general, and that many mRNAs are simultaneously repressed *and* degraded following sRNA binding.

D. Possible Physical Separation of Translation and mRNA Decay

Implicit in our thinking on the translation–degradation interplay is the notion that transcription, translation, and mRNA decay all occur in the same cellular compartment so that they can influence each other. This assumption is based on the fact that the prokaryotic cell lacks a nuclear membrane; yet, this feature does not rule out some degree of compartmentalization. Particularly intriguing in this respect is the recent discovery that RNase E is not localized randomly in the bacterial cytosol, but attached to the cytoplasmic side of the inner membrane (28–30). As far as transcription and translation are concerned, there is much evidence that both processes are colocalized and tightly coupled in E. coli. Indeed, ribosomes start attaching to RBSs as soon as they have been synthesized (31–33), and transcription often stops prematurely when the leading ribosome lags behind the RNA polymerase or when the nascent transcript remains untranslated altogether (see (34, 35) and references therein). Because RNA polymerase localizes in the nucleoid and ribosomes in the cytosol (cf. (36) for the visualization of these molecules in B. subtilis), it is likely, then, that proficient transcription is confined to the nucleoid–cytosol interface and that RNA polymerase molecules within the nucleoid body are inactive (for the existence of a large pool of DNA-bound, inactive RNA polymerase molecules, see (37)).

Now, what about colocalization of translation and mRNA decay? Images of the E. coli cell provide no evidence that the nucleoid sticks to the inner membrane, where RNase E is located. This, in turn, suggests that the degradation of mRNAs might be separated geographically and temporally from their synthesis and utilization. Such a situation occurs in eukaryotes where cytoplasmic mRNA decay takes place in dedicated loci, the "P-bodies" (38). This idea implies that many E. coli mRNAs could be inactivated non-nucleolytically, the inactivating event being the migration from the transcription/translation compartment to the degradation compartment. I do not believe that this is the case, however, because it is clear that for at least some mRNAs, RNase E-mediated degradation starts well before mRNA synthesis is complete (8, 39–42). Thus, at least in these cases, transcription and mRNA decay must occur simultaneously. To reconcile these apparently conflicting observations, I note that the size of an mRNA molecule is similar to the dimensions of the bacterial cell (if extended, a 1000 nt RNA molecule may reach 0.35 μm in length; (43)), so that even relatively short nascent mRNAs could span the gap between the nucleoid and the inner membrane. Alternatively, movements of the nucleoid within the cell may be fast, so that any region of the nucleoid–cytosol interface could contact the membrane within seconds. Finally, it is possible that two populations of RNase E exist in the cell, a major membrane-bound fraction with a structural role, and a minor cytoplasmic fraction dedicated to mRNA decay. By tentatively comparing the amount of RNase E present in the cell with the rate of synthesis of its substrates, we have speculated that the enzymatic turnover of RNase E is globally quite low (44), which might suggest that only a fraction of it is actually involved in RNA processing and decay. The nature of the possible structural role of RNase E, which has sometimes been evoked (45), remains elusive.

E. What is the Global Importance of Non-nucleolytic Inactivation?

Given the above considerations, it is important to estimate the fraction of mRNAs that are non-nucleolytically inactivated. In *E. coli*, mutations that affect bulk mRNA stability can provide a clue. The conditional *rne1* or *rne3071* mutations, which at the non-permissive temperature of 43 °C result in the inactivation of RNase E, stabilize bulk mRNA physically (by about three- to sixfold) without an equivalent increase in functional lifetime (*46, 47*). These observations led to the perception that many mRNAs that decay *via* an RNase E-controlled process are inactivated non-nucleolytically rather than nucleolytically. Still these experiments showed a significant increase in the functional lifetime of bulk mRNA, implying that a subpopulation of mRNAs are inactivated nucleolytically; moreover, the very significance of these observations is questionable because they do not correspond to steady-state growth. In contrast, the *rne131* mutation (formally called *smbB131*; (*48*)), which removes the nonessential CTH of RNase E, does not impair steady-state growth at 37 °C. In this mutant, the functional and physical lifetimes of bulk mRNA increase by the same factor (about twofold (*49, 50*)). On this basis, we concluded that nucleolytic inactivation is the rule rather than the exception, at least for those mRNAs that are stabilized by CTH removal (*49*). Supporting evidence was also obtained by artificially depleting the cellular RNase E pool. We found that a 10-fold reduction in RNase E concentration does not compromise steady-state growth, although the growth rate is halved. Under these conditions, the functional lifetime of bulk mRNA increases significantly (twofold) (*44*). Altogether, these data suggest that a large fraction of all mRNAs, presumably the majority, are inactivated nucleolytically by RNase E cleavage, at least under conditions of balanced growth in rich medium. In the following sections, mRNAs are assumed to be inactivated nucleolytically unless otherwise stated. Yet, the possibility that non-nucleolytic inactivation prevails under certain conditions must be kept in mind (see also Section V(C)).

III. Translation–Degradation Interplay: Conceptual Ambiguities and Technical Caveats

A. Direct or Translation-Mediated Changes of mRNA Stability

In *E. coli*, the interpretation of cis-acting mutations that alter mRNA stability (or of any other experimental changes that affect this stability) has been confused for years by the translation-degradation interplay. Do these mutations affect stability directly (e.g., by introducing or removing endonuclease cleavage sites), or indirectly, *via* an effect on translation? At first sight, it seems straightforward to decide whether translation is affected or not, by measuring the "frequency of translation initiation" (FTI) (*41*). This parameter is usually defined as the ratio between the rate of synthesis of a protein and the steady-state concentration of its mRNA (or, more correctly, of the fraction of this mRNA that is functionally active (*51*)). At the microscopic level, the FTI is inversely proportional to the spacing of ribosomes on the mRNA. Now, let us consider the effect of a mutation that decreases translation initiation in the case of a very tight coupling between translation and degradation. Any delay in loading the next ribosome will allow RNase access and cause the immediate onset of mRNA decay of that mRNA molecule (Fig. 1A and B). The remaining visible mRNA, if one were to look at electron micrographs for example, will therefore appear just as closely packed with ribosomes as in the wild-type situation, even though translation initiation has been impaired—otherwise it would not have survived. The measurable FTI will appear to be the same, but the measured mRNA lifetime will be shorter. This situation is illustrated in Fig. 1B, where the inefficiently translated mRNA starts decaying after only a few ribosomes have been loaded. Formally, it is impossible to distinguish this situation from one in which the mutation primarily favors mRNA cleavage rather than impairing translation: in both cases, ribosome packing (that is, FTI) remains invariant and only stability decreases (compare Fig. 1B and C). The *lacZ* gene is a good illustration of this situation: in this case, mutations that affect translation initiation have been found to impact mRNA stability much more severely than ribosome packing (*41, 42, 52*).

In practice, many cases are not really that ambiguous. For instance, the introduction of a cleavage site for a well-defined endonuclease (e.g., RNase III (*53*) or RNase E (*54*)) is likely to destabilize the mRNA directly, rather than indirectly by adventitiously decreasing translation. Likewise, the 5' ends of many mRNAs carry hairpins, which are known to hinder the action of the pyrophosphatase RppH *in vitro* and subsequently to impede the entry of RNase E or RNase G ((*55*); see "5' tethering pathway" below). Therefore, the destabilization often observed upon removing these hairpins *in vivo* (*56*) likely reflects a facilitated attack by these endonucleases, rather than a decrease in efficiency of translation. These situations correspond to Fig. 1C. Conversely mutations which affect Shine–Dalgarno sequences or initiation codons or remove or introduce stop codons, and are associated with an effect on mRNA stability (*41, 56–61*) are likely to do so *via* their effect on translation, that is, they clearly correspond to the situation in Fig. 1B. However, in several other cases, the ambiguity described above was recognized but not solved. Cho and Yanofsky characterized mutations in the 5'UTR of the *trpE* gene that affected mRNA stability, but they could not decide whether the affected parameter was

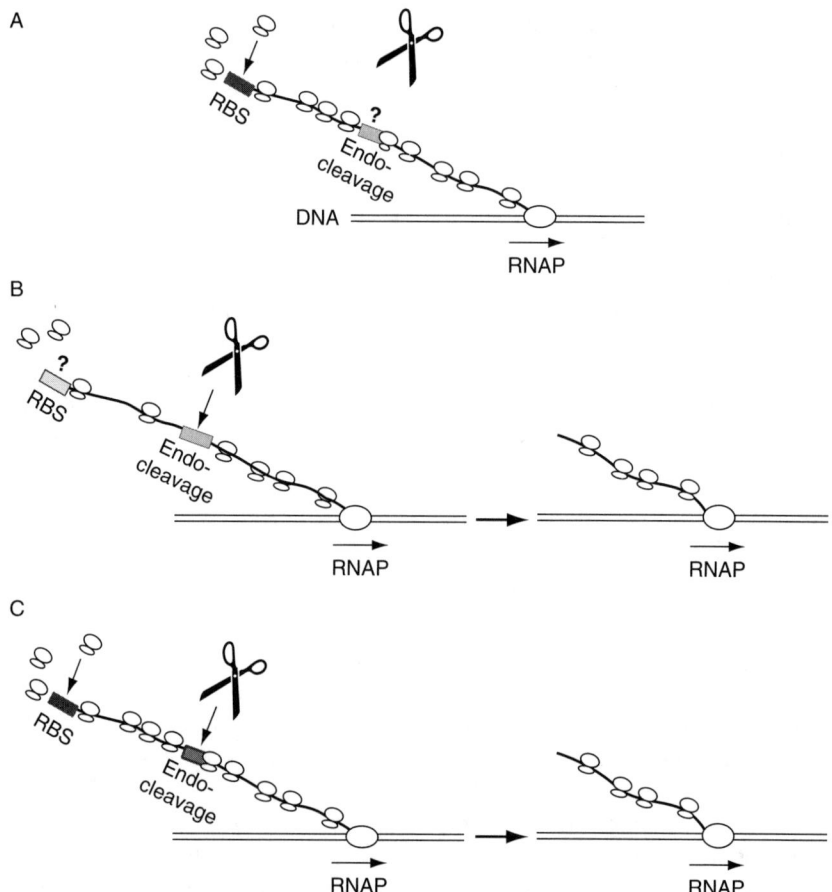

FIG. 1. Cartoon showing the effect of a mutation that either decreases the efficiency of translation initiation (B) or increases endonuclease susceptibility (C), when translation and degradation are tightly coupled. In (A) the translation and degradation rates represent the wild-type situation. Rectangular boxes symbolize ribosome binding site (RBS) or an endonuclease cleavage sites, as indicated; a dark-colored box means that the RBS or the endonuclease cleavage site is efficient, a light-colored box that it is less efficient. The endonuclease (scissors) is assumed to cleave the mRNA when the delay between two successive initiations exceeds a certain threshold. Translation initiation being a stochastic process, this event statistically occurs earlier if translation is inefficient (B) rather than efficient (A). The lifetime of the mRNA will therefore be reduced in (B). However, before cleavage occurs, the successive initiations must have been close enough to prevent cleavage, hence the equal ribosome packing ahead of the cleavage site in (A) and (B). In (C), cleavage occurs earlier than in (A) because the cleavage site is recognized more efficiently by the nuclease. It is notable that (C) is indistinguishable from (B) from the viewpoint of both ribosome packing and mRNA stability. RNAP stands for RNA polymerase.

primarily stability or translation (62). When studying mutations downstream of the *lacZ* RBS that affected mRNA stability, Petersen first favored a direct effect on stability (63, 64), but subsequently considered the possibility that at least in some cases the primary target was, in fact, translation (51). Lundberg et al. observed that the stability of the *ompA* mRNA increased with growth rate whereas the FTI does not, and concluded that these stability changes were unrelated to translation (65). However, in revisiting this question recently using in particular *in vitro* approaches, two groups reached the opposite conclusion, that is, that the parameter affected by growth rate is in fact translation (66, 67). More generally, *in vitro* studies performed under conditions where translation–degradation coupling does not take place have been found useful in solving the above ambiguity. For example, several r-proteins repress the expression of their own operons at a posttranscriptional level, and the corresponding mRNAs are destabilized *in vivo* under repression conditions (68, 69). However, *in vitro* studies show unambiguously that repression operates at the translational level, and therefore that destabilization is secondary to the absence of translation (70). The same hierarchy appears to hold true for the autoregulation of the chaperone Hfq, which also takes place at posttranscriptional level (71). Similarly, using an *in vitro* approach, McCormick et al. showed that a series of 5′UTR deletions that affected the stability of a hybrid S10′-′*lacZ* mRNA *in vivo*, had in fact a direct effect on translation, with stability being affected only secondarily (72). The reverse situation is illustrated by the *rnc* mRNA, which is destabilized after RNase III processing. Again, using an *in vitro* approach, Matsunaga et al. were able to show that the primary effect of this processing is to increase the susceptibility of the mRNA to an endonuclease (presumably RNase E), rather than to repress translation (73).

B. Practical Aspects: How to Measure mRNA Stability Without Affecting It?

The stability of an mRNA can be measured in a number of ways. Some of them do not require that cell growth be interrupted. For instance, stability can be assessed by comparing the steady-state concentration of the mRNA with its rate of synthesis; the latter can be measured directly by hybridization after pulse-labeling the RNA (62) or deduced from the expression of a suitable reporter gene (74). Alternatively, if the synthesis of the mRNA is controlled by an inducible promoter, its stability can be inferred from the kinetics of mRNA appearance or disappearance after transcription is switched on or off, respectively (64, 68, 75). However, by far the most commonly used method consists of blocking cellular transcription with suitable drugs (e.g., rifampicin) and subsequently following the decay of the mRNA of interest. This method is the only applicable one in the case of genome-wide studies (e.g., see (50, 76)).

However, since the steady-state growth is effectively interrupted at the time of measurement, it can yield grossly distorted results due in part to translation–degradation interplay. After rifampicin treatment, many mRNAs remain apparently stable during extensive periods of time before fast decay begins (54, 57, 72, 75, 77). It has been proposed that under these conditions free ribosomes accumulate rapidly due to their release from fast-decaying mRNA; this accumulation, in turn, transiently enhances the translation of the surviving mRNAs and hence their stability (72, 77). Conversely, the subsequent fast-decay phase may be accelerated by the increase in the availability of free RNase E, after the *de novo* synthesis of its substrates has ceased (54). The promoter–proximal region of the *spc* mRNA has been shown to be *immediately* destabilized following rifampicin addition (75). In normally growing cells, the stability of this mRNA is down-regulated by the r-protein S8, the repressor of the *spc* mRNA. To explain their results, the authors proposed that, in the presence of rifampicin, the synthesis of 16S rRNA, the primary target of S8, stops before that of S8 protein itself; the pool of free S8 protein then expands, ultimately resulting in repression and destabilization of the *spc* mRNA (75). More generally, it has been repeatedly reported that the lifetime of mRNAs encoding r-proteins (that is, the L11–L1 mRNA (69), the S13–S4 mRNA (68), or the L20 mRNA (54)) cannot be reliably measured in the presence of rifampicin because of the progressive accumulation of the cognate repressors.

The above examples illustrate how treatments that interrupt growth can have unforeseen effects on mRNA stability. A further example is illustrated by the action of translation inhibitors. Drugs such as chloramphenicol, tetracycline, or fusidic acid, which block the movement of translating ribosomes, generally stabilize *E. coli* mRNAs, whereas puromycin or kasugamycin, which strip ribosomes off mRNAs, destabilize them (see Section IV(A)). Traditionally, these effects are interpreted as evidence that translating ribosomes can protect mRNA against nucleases. However, the following experiments show that, regardless of their mode of action, translational inhibitors also affect the activity of the nucleases themselves. We have studied the effect of translational inhibitors on the stability of *untranslated* mRNAs, that is, the *lacZ* mRNA lacking an RBS (precautions were taken to eliminate polarity; see Section V(C)) and RNAI, a small untranslated RNA that first undergoes RNase E cleavage near the 5' end, and then trimming by PNPase assisted by poly(A) polymerase (78, 79). Whatever their mode of action, all inhibitors tested markedly stabilized these untranslated RNAs. Moreover, in the case of RNAI, both the RNase E cleavage and PNPase-mediated pathways were inhibited. The proposed interpretation is based on two known facts. First, both RNase E and PNPase autoregulate their synthesis by degrading their own mRNAs (40, 80); therefore they are presumably never present in excess in the cell. Second, rRNA synthesis is boosted after a translational block, and rRNA becomes unstable,

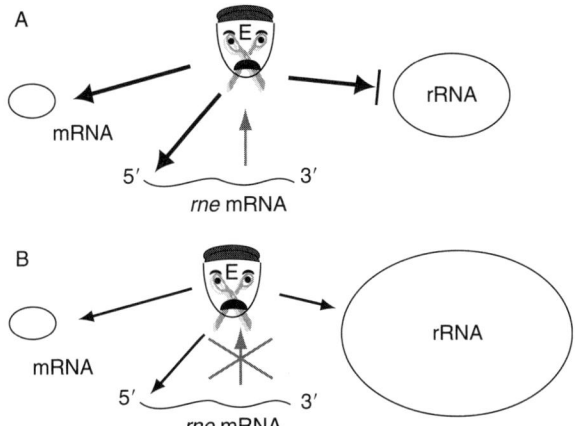

FIG. 2. Bulk mRNA stabilization after a translational block: a plausible model. (A) In growing cells, RNase E (E) is synthesized from the *rne* mRNA (vertical arrow) and used for the degradation of mRNAs (thick black arrows), whereas the degradation of rRNA is blocked after limited processing (barred arrow). The free pool of RNase E is limited because it degrades its own mRNA (*rne*), autoregulating its production. After a translational block (B), the synthesis of rRNA is boosted and the newly rRNA is prone to RNase E attack (thin arrow) because it cannot assemble with r-proteins. In the meantime RNase E synthesis is discontinued (cross on the vertical arrow). We assume that RNase E is then titrated, hence the stabilization of the mRNAs (thinner arrows compared to (A)).

presumably because it can no longer assemble with r-proteins (*81*). RNase E and PNPase presumably participate in this degradation (*82*). We therefore hypothesized that RNase E and PNPase become permanently titrated with excess substrate when translation is blocked, explaining mRNA stabilization (*83*) (Fig. 2). Where the above model correct, the prediction was that it should be possible to titrate RNase E or PNPase—and thus stabilize mRNAs—even in the absence of a translational block, simply by inducing the synthesis of a highly expressed substrate in growing cells. However, titration and hence mRNA stabilization should then be transient, because the pools of RNase E or PNPase are now free to expand in response to the stabilization of their own mRNAs. These predictions were confirmed experimentally (*44*).

C. The *lacZ* Gene, or the Importance of Comparing Comparable Items

From the viewpoint of the translation–degradation interplay, no prokaryotic gene has been more extensively studied than the *E. coli lacZ* gene (see (*8, 34, 41, 51, 72, 84, 85*) and references therein), often with divergent results. Starting from the genuine chromosomal *lac* operon, we introduced a variety of mutations in the *lacZ* RBS that decrease its efficiency (that is, by altering the Shine–Dalgarno sequence or initiation codon, or by creating local secondary structures), and we recorded the effect of these mutations on the stability of the mRNA. To this end, we compared the steady-state level of the mRNA in each case with its rate of synthesis (*41, 52, 86*). Whatever the mutation used, the stability decreased steadily as translation declined, until it became invariant at very low translation levels. The steady-state mRNA level continues to decrease in this translation range, however, due to transcriptional effects ("polarity"), and we have proposed a plausible interpretation for the shift from stability to polarity effects (*41*). Several reports in the literature are consistent with these results; those which do not can possibly be explained by alterations in the 5' region of the mRNA that may affect stability on their own, or by other experimental differences (see (*34, 41, 72*) for discussion). However, one of these reports warrants a special mention given our forthcoming discussion (see Section V(C)). It was reported that the presence of a ribosome lingering over the first few codons of *lacZ* can stabilize the whole mRNA (*84*). However, only the stability of the 5' mRNA region was probed in this study. Small fragments encompassing this region are known to accumulate in the cell, presumably because they are relatively stable; when translation is inefficient, they are much more abundant than the full-length message itself (*85, 87–89*) (see Fig. 7 below). Moreover, in the complete absence of translation, no synthesis of the full-length message can be observed unless special precautions are taken to eliminate polarity (*41, 72, 87, 88*). Under the latter conditions, we have observed that the presence of translating ribosomes over the beginning of the coding sequence cannot protect the entire mRNA (*89*), although fragments corresponding to the translated region are indeed stabilized (see Section V(C) and Fig. 7 below).

IV. Killer Ribosomes

A. Historical Background

In the late 1960s, biochemical work on *E. coli* suggested that translating ribosomes, or a subpopulation of them, may play an active role in mRNA degradation, *via* an associated 5'→3' exonuclease activity that was named RNase V (*90, 91*). For a while, this idea seemed consistent with the finding that drugs such as chloramphenicol, tetracycline, or fusidic acid, which block the movement of translating ribosomes, also block bulk mRNA degradation (*51, 90, 92*; see Section III(B)). However, subsequent work failed to confirm the existence of a ribosome-bound 5'→3' exoncleolytic activity (*93, 94*); rather, "RNase V" was suggested to be nothing but contaminating RNase II, an abundant 3'→5' exonuclease that largely exists in a soluble form (see chapter by C. Arraiano and colleagues in this volume). Moreover, other inhibitors of protein synthesis, such as kasugamycin or puromycin, which act by stripping ribosomes off mRNAs, were found to *destabilize*, rather than stabilize bulk mRNA (*51*, *59*, *92*, *95*). This latter observation led to the opposing view that ribosomes generally exert a protecting effect on mRNAs. According to this view, stabilization of mRNAs by ribosome-stalling drugs would reflect shielding of nuclease-sensitive sites by stalled ribosomes. We now know that effects mediated by translation inhibitors are difficult to interpret because of the indirect action of these inhibitors on the degradation machinery itself (see Section III(B)). Nevertheless, for three decades, the prevailing view was that the presence of ribosomes invariably improves mRNA stability, or is at least neutral to it. This view has been amply supported by experiments on many individual mRNAs (see Section V).

Yet, in 1998, the Moseley group produced the first clear hint that translation can *facilitate* the endonucleolytic cleavage of an mRNA, that of the *E. coli daa* operon. Since then, several groups, including those of Drs Sauer, Bouloc, and— particularly—Aiba, have considerably extended these observations, showing that cleavage occurs in a variety of situations that result in ribosome stalling and that it does not generally involve any previously identified nucleases. The simplest explanation is that the cleavage is mediated by the ribosome itself ("killer ribosome"). The recent discovery of this remarkable phenomenon highlights the persisting difficulty of establishing dogmas in the field of mRNA catabolism, even with an organism as thoroughly studied as *E. coli*.

B. The *daa* Operon

The polycistronic *daaA-E* mRNA, which encodes the F1845 fimbriae, is processed—presumably endonucleolytically—within a small open reading frame (*daaP*) which lies between the penultimate and last genes of the operon (*96*). This processing likely contributes to the differential expression of individual *daa* genes: the downstream fragment, which encodes the fimbrial subunit, is much more stable than the upstream fragment encoding regulatory or accessory proteins. This processing is independent of RNase E or RNase III (*96*). Moreover, in contrast to all current observations at that time, it was found to require the presence of translating ribosomes over the *daaP* sequence. The amino acid sequence of DaaP *per se* was important for processing, suggesting that the nascent polypeptide is somehow involved (*97*). Later studies indicated that processing requires the synthesis of a particular tripeptide (GPP) near the end of the *daaP* gene and presumably took place immediately afterward, when the second proline codon was in the P-site (Fig. 3A). As judged from the 5' end of the downstream fragment, cleavage occurred 12 nt upstream of this P-site codon regardless of the local nucleotide sequence (*6*). Since this position is well within the mRNA region that is protected from nucleases by the ribosome, the

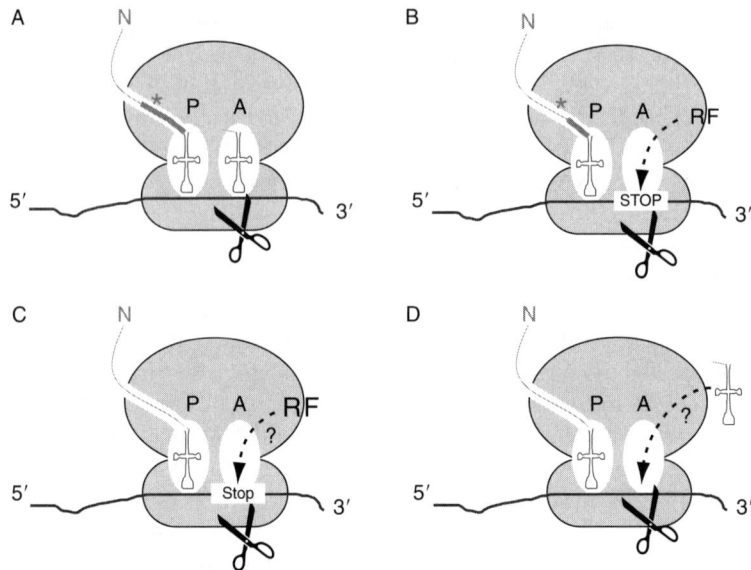

FIG. 3. In *E. coli*, ribosome stalling causes mRNA cleavage. (A) The synthesis of certain stretches of amino acids (thick grey line, red on the web version) cause stalling and subsequently mRNA cleavage by an unknown endonuclease (scissors; note that cleavage does not necessarily occur at the A-site codon only (see Fig. 5)) (6, 7, 119). Presumably, stalling results from an interaction (star) of the nascent peptide with the ribosome exit tunnel (118). 'N' stands for the N-terminal end of the nascent peptide. (B) same as (A), except that cleavage occurs during translation termination. In the examples described (108, 112), the amino acid stretches that cause stalling and cleavage at a stop codon are shorter than in (A). RF: release factor. (C) Same as (B), except that there is no stall-promoting sequence: rather, stalling is due to depletion of release factors or to mutations that reduce their activity (113). (D) Cleavage of coding sequences also occurs when the mRNA encounters a stretch of rare codons (115), or a codon whose cognate amino acid is missing (120), suggesting that cleavage occurs whenever the ribosome stalls on the mRNA, independently of any particular interaction between the nascent polypeptide and the ribosome.

authors hypothesized that the cleavage agent may be nothing but the ribosome itself, or a ribosome-bound nuclease. Possibly, the newly synthesized GPP tripeptide somehow modifies the translating ribosome, converting it into a "killer" form. It should be noted, however, that the endonucleolytic nature of the cleavage was not rigorously established since only the downstream processed mRNA fragment was observed along with the unprocessed mRNA: the upstream fragment remained undetectable.

C. ssrA Tagging as a Tracer for Killer Ribosomes

Further progress came from *trans*-translation studies. *trans*-Translation, discovered in 1996 by R. Sauer and colleagues (98), allows the rescuing of ribosomes that reach the 3' end of an mRNA lacking a stop codon. Examples of such "nonstop" mRNAs are products of premature transcription termination (*99*), or normal mRNAs that have been severed from their stop codon by exo- or endonucleases (*100*). The product of the *ssrA* gene, tmRNA, and its associated protein partner SmpB, then interact with the empty A-site of the distressed ribosome, adding a noncoded alanine to the nascent polypeptide and then switching the ribosome to an *ssrA*-encoded ORF that ends with a *bona fide* stop codon. This ORF encodes a tag that targets the released protein for degradation, and the ribosome itself is recycled (for recent reviews see (*101, 102*)).

In 1999, Roche and Sauer reported that the stalling of ribosomes over a cluster of rare codons yielded an incomplete, *ssrA*-tagged polypeptide (*103*). Situations that resulted in slow translation termination—e.g., the presence of an inefficient stop codon or certain arrangements of codons just before the stop signal—were also found to result in *ssrA*-tagging (*104–107*). In these examples, *trans*-translation took place on an *a priori* intact mRNA, and so the following question arose: was tmRNA able to recognize a stalled ribosome with a codon present in its A-site, or was the mRNA first cleaved to generate a nonstop mRNA, the "normal" substrate for *trans*-translation? Roche and Sauer were unable to detect a 3′ truncated mRNA and concluded that the first hypothesis must be correct. However, in 2003, the same group observed that the *ybeL* mRNA, whose product is *ssrA*-tagged after the last sense codon, was cleaved at the stop codon or very close to it, and concluded that the mRNA must first be converted into a nonstop mRNA before tagging occurs (*108*) (Fig. 3B). Intriguingly, cleavage occurred when a proline codon was in the P-site, as for the *daaP* mRNA. In addition, it was independent of the RelE toxin, which is known to mediate A-site cleavage at specific codons (*109*), or of any of the other known toxin and antitoxin (TA) pairs (*108*) (see chapter by Y. Yamaguchi and M. Inouye in this volume for more information on TA pairs). The truncated *ybeL* mRNA was barely detectable in an *ssrA*$^+$ background and we now understand why: by removing the stalled ribosomes, tmRNA accelerates the exonucleolytic attack of the nonstop mRNA (*100*). In most cases, the exonuclease implicated in this scavenging process seems to be RNase R (*110*), which had been known for some time to associate with tmRNA and SmpB (*111*).

Subsequently, the case for an endonuclease activity associated with distressed ribosomes, yielding a nonstop mRNA and ultimately tmRNA intervention, received considerable support. The use of variant tmRNAs that stabilize tagged proteins, so that the exact tagging point can be determined, has been useful in this regard. Sunohara *et al.* confirmed the work of Hayes and Sauer in demonstrating that the mRNA of an engineered *crp* gene whose last two codons encoded certain XP dipeptides, was cleaved at the stop codon, that is, with the proline codon in the P site. Concomitantly, the completed polypeptide was tagged (*112*) (Fig. 3B). This study brought the first strong evidence that cleavage was actually endonucleolytic, since both the upstream and downstream fragments could be detected. Interestingly, cleavage was apparently more efficient in the presence of a functional tmRNA than in its absence, suggesting that tmRNA already interacts with the ribosome during the cleavage step, that is, before tagging. Somewhat later, the same group demonstrated that mRNA cleavage (and polypeptide tagging) can occur at a stop codon regardless of the nature of the preceding codons, provided the activity of the cognate release factor (RF) was low (Fig. 3C). Presumably, it is ribosome stalling *per se*, whether caused by RF scarcity or by the particular structure of the finished peptide (other C-terminal sequences can be problematic in addition to certain XP dipeptides; see (*104, 107*)), that caused cleavage (*113*). Interestingly, in this last study, cleavage occurred not only at the stop codon (in the A site), but also upstream and downstream of it. The upstream cleavage (that is, 15 nt upstream of the P-site) was located approximately at the same position as in the *daaP* case (see above). In hindsight, these results are also quite reminiscent of an early report of Björnsson and Isaksson (*114*). These authors observed mRNA processing 13 nt upstream of an inefficient stop codon, but the possibility that ribosome might actively participate in this cleavage was not considered at that time (*114*).

While mRNA cleavage occurred during inefficient translation termination in the above examples, several groups also observed that ribosome pausing within coding sequences can result in mRNA cleavage and polypeptide tagging. Thus, clusters of rare arginine codons were found to result in mRNA cleavage. At least three such codons were required, and cleavage was abolished by the overexpression of the cognate tRNAs (*115*) (Fig. 3D). While the accumulation of rare codons is an artificial situation not normally found in *E. coli* mRNAs, genuine pausing sites also exist within coding sequences. The SecM ("secretion monitor") polypeptide, which is cotranslationally translocated to the periplasm, possesses such a translational "arrest sequence" near its C-terminal end. This arrest sequence plays a regulatory function: when SecM is efficiently exported to the periplasm, ribosome arrest is rapidly overcome by the translocation force, whereas when it is not exported, the ribosome stalls there durably. This stalling, in turn, favors the translation of the downstream *secA* gene encoding the translocation ATPase, presumably by stabilizing a structure where the *secA* RBS is accessible ("translation attenuation") (*116–118*). Ribosome stalling occurs because of the interaction of a particular 17-amino acid sequence of SecM with the exit tunnel of the ribosome; most important for stalling are the five last amino acids (IRAGP) of this sequence (*7, 118*). Interrupting translation just after the proline codon does not affect the arrest, which suggests that stalling occurs when the proline codon is in the P-site, as in the cases mentioned above. Two groups have studied the effect of the SecM arrest on mRNA integrity, either within the genuine *secM–secA* operon (*119*), or within an artificial construct (7). In both cases, mRNA cleavage was observed, together with tmRNA tagging of the incomplete polypeptide (Fig. 3A). Thus, like *daaP*, *secM* is a genuine *E. coli* gene whose polypeptide sequence causes internal mRNA cleavage. However, cleavage in this case occurred at multiple sites, yielding upstream fragments that end within the A-site of the stalled ribosome (that is, just after the proline codon), but also 12–15 nt further upstream (like in *daaP*) and 12–15 nt further downstream (7). Interestingly, only one 5′ extremity, corresponding to the most upstream cleavage, was detected for the downstream fragment. The possible reason for this difference is discussed below (see Section IV(F))

In the above examples, the ribosome stalling that causes mRNA cleavage is due to special sequence features in the nascent polypeptides (*secM*, *daaP*, *ybeL*), to artificially introduced clusters of rare codons, or to particular genetic backgrounds (e.g., mutations that reduce RF availability). However, very recently, the Aiba group produced evidence that several mRNAs—and perhaps all of them—can be internally cleaved following starvation for an amino acid, such as serine (*120*). Shortage of serine was achieved by exposing the cells to serine hydroxamate, a serine analog. Although the cleavage sites were not mapped precisely, they seem to be located at or near the hungry codons, consistent with the view that it is the stalling of ribosomes that causes cleavage (Fig. 3D). Most importantly, mRNA cleavage was still observed in *E. coli* strains lacking all five characterized toxin–antitoxin (TA) pairs; moreover, cleavage occurred at serine hydroxamate concentrations that were far too low to induce the synthesis of the RelE toxin, and it was also independent of ppGpp. The incomplete polypeptides are subsequently tagged and degraded, which could be a valuable source of amino acids for reorienting the metabolic activity of the cell. Generalizing their results, the authors showed that amino acid downshift in general also resulted in mRNA cleavage, although in this case they did not show that cleavage was independent of the known toxins. In summary, this work suggests that, by reducing the pool of translatable mRNAs and by recycling ribosomes and amino acids, TA-independent mRNA cleavage constitutes a genuine response to amino acid shortage. Intriguingly, these are also the major functions attributed to TA pairs (*109*, *121*). Presumably, future work will clarify the respective importance of TA-dependent and TA-independent mRNA cleavage during amino acid starvation; the severity of the starvation is likely to be important here.

D. Looking for the Nuclease

How do stalled ribosomes kill mRNAs? The endonuclease activity might be an intrinsic property of the translating ribosome itself: RNA is thermodynamically unstable, and structural work has revealed intimate contacts between the mRNA and the ribosome, some of which may catalyze cleavage (20, 122). Yet, in a purified in vitro system, stalled ribosomes do not cleave their mRNAs (123) even in the presence of tmRNA that stimulates cleavage in vivo (see above). Alternatively, then, the endonuclease activity may be brought (or activated) by a dedicated trans-acting factor recruited during ribosome stalling. As noted above, this endonuclease appears to differ from the major E. coli endonucleases (RNase E, RNase P, RNase III) or toxins (7, 108, 124).

To my knowledge, only the Moseley group has elaborated a systematic strategy for identifying possible trans-acting factors involved in cleavage (124). To this end, they have devised a genetic screen in which the inhibition of the daaP mRNA cleavage stimulates the expression of a reporter gene. This elegant approach led to the identification of HrpA, a very large putative RNA helicase of unknown function, as one of the factors involved in the cleavage of the daaP mRNA. Cleavage was greatly reduced in its absence, but not eliminated. HrpA belongs to the DEAH family of RNA helicases (125), several of which are involved in RNA catabolism (e.g., the yeast proteins Prp2, Prp16, or Prp22, are involved in splicing; see (126)). Interactome studies revealed that HrpA interacts with several proteins of the small ribosomal subunit, and thus it may bind this subunit (127). Now, what does HrpA do once bound? Several motifs involved in the ATPase and helicase activities of DEAH proteins in vitro are important for the cleavage-stimulating activity of HrpA in vivo. This finding led Moseley and colleagues to speculate that the conformation of either the mRNA or the rRNA is altered by the helicase activity of HrpA, thereby facilitating cleavage (124). However, precise information on the role of HrpA will presumably have to wait until the cleavage reaction can be reconstituted in vitro. One essential pending question concerns the generality of the role of HprA. Is it involved in all cleavage reactions related to ribosome stalling, or only in a subset of them? Finally, the genetic screen of Koo et al. was not exhaustive: it remains to be seen whether other essential trans-acting factors are involved.

E. Ambiguous Lessons from Gram-Positive Organisms

Whereas all the above observations refer to E. coli, apparently similar findings have been reported with Gram-positive organisms such as B. subtilis. However, whether they really constitute the counterpart of the E. coli "killer ribosome" phenomenon remains unclear.

The ermA and ermC genes from Staphylococcus aureus encode rRNA methylases that confer resistance to macrolide antibiotics (including erythromycin) in S. aureus or B. subtilis. Like in the secM–secA case (see above), the ermA and ermC genes are co-transcribed with short upstream ORFs (two in the case of ermA, a single one in ermC). In the presence of sublethal doses of erythromycin, an antibiotic that, like the SecM arrest sequence, interacts with the ribosome exit tunnel (see (*118*) and references therein), ribosomes stall on these small ORFs. This stalling, in turn, favors a conformational change in the mRNA that facilitates the translation of the downstream *ermA* and *ermC* genes ("translational attenuation"). In all these respects, the similarity with the *secM*–*secA* case is striking. Not unexpectedly, then, stalling was found associated with the appearance of a 5′-truncated mRNA fragment extending from the stalling region down to the *ermC* or *ermA* genes; its 5′ extremity was mapped to a few nucleotides upstream of the putative P-site of the stalled ribosome (*128–130*), as in the *secM* or *daaP* cases. Very recently, D. Bechhofer and colleagues (*130*) went a step further by demonstrating that the appearance of the *ermC* 5′-truncated mRNA was dependent on RNase J1, an essential and phylogenetically conserved *B. subtilis* endonuclease that has recently been shown to also possess a 5′→3′ exonucleolytic activity ((*131–133*); for details, see Section V). Interestingly, RNase J1, which has no *E. coli* ortholog, was initially isolated from high-salt ribosome washes (*131*). Is RNase J1, then, the *B. subtilis* equivalent of the (hitherto unidentified) *E. coli* endonuclease responsible for the "killer activity" of stalled ribosomes?

It would be reassuring if the answer were positive, but it seems unlikely. Although this is not the current view of the Bechhofer group, the truncation of the *erm* mRNAs in the stalling region seems to fit better with exonucleolytic trimming from the 5′ end that is halted by the stalled ribosome ("a barricade to 5′→3′ nucleolytic cleavage," (*128*)), than an endonucleolytic cleavage (Fig. 4). First, fragments corresponding to regions upstream of the stalling sequence were looked for but never found, even in a strain deficient in the major 3′→5′ exonuclease PNPase (*129*). Second, processing of the *ermA* mRNA was much slower when the stalling site was located close to 5′ end of the mRNA than further downstream; similarly, in the *ermC* case, mRNA processing could be slowed by appending a 5′ hairpin. These findings are consistent with a 5′-dependent processing enzyme, which fits the characterized exonucleolytic activity of RNase J1 better than its endonucleolytic activity (*132, 133*). Third, the main argument of the Bechhofer group in favor of endonucleolytic processing seems debatable. These authors appended an extra copy of the upstream ORF, complete with its stalling sequence, to the 5′ end of the *ermC* mRNA. Efficient processing was observed at both the normal position and the ectopic position, which, the authors argue, is consistent with two independent (that is, endonucleolytic) cleavages within the two ORFs. However, their observation is just as consistent with the downstream product arising from the upstream one by exonucleolytic trimming after the obstructing ribosome has been removed. In the *ermA* gene, which genuinely carries two erm-sensitive ORFs and thus two processing sites, this precursor–product relationship is clearly observed (*128*).

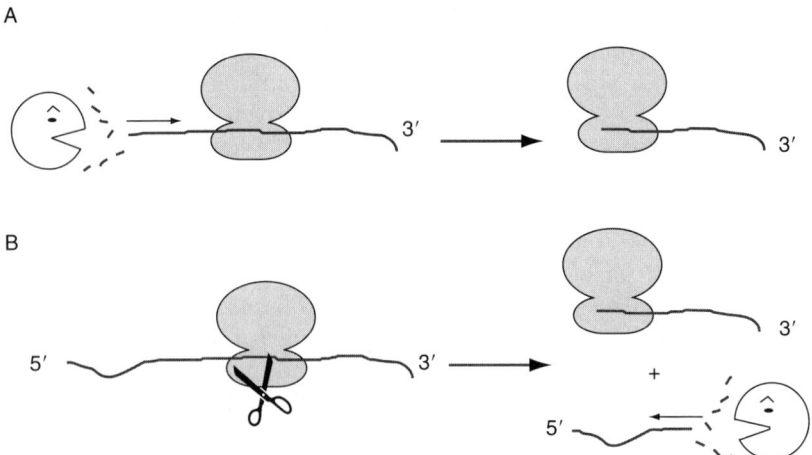

FIG. 4. Killer or protective ribosomes? Ambiguous lessons from the *erm* genes. In erythromycin-treated *B. subtilis* cells, the ribosome stalls on the *erm* leader ORF, resulting in mRNA processing at the stalling site. Only the downstream processed fragment is detected. In model (A), the processing reflects the activity of a $5' \rightarrow 3'$ exonuclease (presumably RNase J1) that is halted by the ribosome, whereas in model (B) it reflects the activity of a ribosome-associated endonuclease (again presumably RNase J1), followed by extremely rapid $3' \rightarrow 5'$ trimming of the upstream fragment. Available evidence seems more supportive of model (A) than (B) (see text).

In addition to ribosome stalling on coding sequences, other circumstances exist where mRNAs get 5′ truncated in *B. subtilis* but not in *E. coli*, supporting the case for a distinct processing mechanism. In the *cryIIIA* mRNA, the mere binding of a 30S ribosomal subunit to a strong SD sequence without an associated start codon stabilizes the downstream sequence but not the upstream portion of the transcript (STAB-SD; (*134*)). The 5′ end of the protected region was mapped to four nt upstream of the STAB-SD, regardless of the local sequence. This processing reaction, which has no equivalent in *E. coli* (*89*), presumably reflects the blocking of RNase J1 exonuclease activity by the bound 30S subunit. Indeed, it is no longer observed in cells deprived of RNase J1, and it can be reproduced *in vitro* with purified components: in this case, the mRNA region upstream of the roadblock is apparently converted to mononucleotides (*132*).

The cleavage of mRNAs during ribosome stalling is conserved in eukaryotes ("no-go decay"; see (*135*)), where it presumably plays both a general scavenging function and a dedicated regulatory role (*136*), as in *E. coli*. It seems very unlikely, then, that a similar phenomenon does not exist in *B. subtilis*, even though the details of the cleavage mechanism may differ from those of *E. coli* (for instance, the *hrpA* gene has no ortholog in *B. subtilis*). The difficulty in observing ribosome-mediated cleavage in the *ermC* and *ermA* mRNAs may be simply circumstantial, reflecting the fact that, in these particular cases, ribosome-mediated endonucleolytic cleavage is slower than 5′→3′ exonucleolytic trimming.

F. Some Pending Questions

As noted above, after ribosome-mediated cleavage, the upstream fragment is usually rapidly degraded unless tmRNA is inactivated (100) (Fig. 5). In contrast, the effect of cleavage on the stability of the downstream fragment is less clear. Whenever the 5′ end of the downstream fragment has been mapped, it invariably corresponds to a position located at the trailing edge of the stalled ribosome, even in cases where cleavage occurs not only in this region but also in the A site or at the leading edge of the ribosome (e.g., in the *secM* case; (7)) (Fig. 5). Presumably, downstream fragments corresponding to these latter cleavages are unstable, whereas downstream fragments starting at the trailing ribosome edge are often very stable. The former fragments are likely to be stripped of ribosomes by *trans*-translation (Fig. 5, cleavages II and III). An

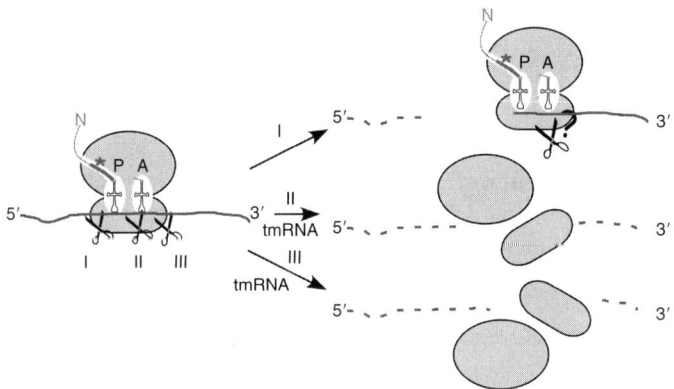

FIG. 5. Killer ribosomes: the fate of fragments. In several cases, cleavage in the stalling region occurs not only at the A-site (cleavage II), but also upstream or downstream of it (cleavages I and III, respectively). However, only downstream fragments corresponding to cleavage I have been detected. Cleavages II an III result in nonstop mRNAs that are targets for *trans*-translation. The upstream fragments are then liberated and subsequently quickly degraded with the help of tmRNA and presumably RNase R; as for the downstream fragments, they are presumably degraded by 5′-dependent nucleases (see text). In contrast, cleavage I does not result in a nonstop mRNA. The upstream fragment, being only loosely bound to the ribosome, may be released and degraded, as for cleavages II and III. In contrast, I propose that the ribosome remains bound at the 5′ end of the downstream fragment, thereby stabilizing it. The reason why this stalled ribosome is not capable of further cleavages that would result in *trans*-translation and ultimately RNA release is unknown (scissors with question mark). Unstable mRNA fragments are shown as dotted lines.

interesting question is whether they carry a 5'OH or in a 5'p extremity, since only the latter would facilitate the action of 5' dependent nucleases (e.g., RNase E/G in *E. coli*; see Section V). Intriguingly in this respect, the bacteriophage T4 RegB endonuclease, which is thought to be ribosome associated (*11*), produces 5'OH extremities (*137*). As concerns the more stable downstream fragments starting within the trailing edge, I speculate that they remain ribosome-bound, which protects them from 5' dependent degradation (Fig. 5, cleavage I). One may wonder, however, why the stalled ribosome is not rapidly removed by further cleavage in the A-site or in the leading edge, followed by *trans*-translation (question mark in Fig. 5, cleavage I). One possibility is that cleavage is slow in the absence of a minimal mRNA length 5' to the ribosome. In principle, this question could be addressed *in vitro*, using *trans*-translation as an assay for cleavage (*123*). In any case, it is possible that the slow removal of the protective ribosome by *trans*-translation determines the stability of the downstream fragment.

Another question concerns the effect of cleavage or processing on the expression of downstream genes. This issue is particularly important when ribosome stalling plays a regulatory role for the expression of these genes (e.g., *secA*, *ermA*, or *ermC*). How does cleavage affects translation attenuation in these cases? To my knowledge, this problem has not been addressed.

Finally, an open question concerns the impact of ribosome-mediated cleavage on bulk mRNA stability in *E. coli*. Since this cleavage occurs under a variety of circumstances—the unifying parameter being ribosome stalling, that is, slowed translation—one may wonder whether it cannot also occasionally take place during normal, unrestricted translation. Even if the probability of cleavage at individual codons is low, it may not be negligible when summed over entire mRNAs, particularly longer ones. It may be worth comparing bulk mRNA stability in $hrpA^-$ and $hrpA^+$ cells, if the HrpA helicase proves to have a general role in ribosome-mediated mRNA cleavage.

V. Protective Ribosomes

Most of the following discussion concerns work on *E. coli*, for which information is most complete. Results obtained with *B. subtilis* are analyzed at the end of the section.

A. mRNAs are Generally Protected by Ribosomes

As noted above, there is ample evidence that the presence of ribosomes over coding sequences generally protect mRNAs against degradation. Indeed, a variety of situations that decrease ribosome packing over individual mRNAs have been found to destabilize them. These include (a) the introduction of point mutations that decrease the efficiency of the RBS (*41, 56, 57, 60*), (b) the introduction of premature stop codons (*58, 61, 138, 139*), (c) the desynchronization of transcription and translation; in this case, a naked mRNA region is created between the RNA polymerase and the leading ribosome (*8, 9, 140, 141*), and (d) the binding of a translational repressor (*12, 24, 68, 69, 71*). As noted above, in the case of Ref. (*12*), mRNA destruction occurs after translation has ceased rather than in competition with it (see "non-nucleolytic inactivation", Section II(B)), but the efficiency of the scavenging process is another illustration of the lability of untranslated mRNAs. Indeed, as already mentioned, ribosome-free mRNA can be so unstable that it is simply undetectable even though there is evidence that it is synthesized (*8, 12*). All these examples support the case for a protective effect of ribosomes. Because the breakdown of most mRNAs is thought to start with an RNase E cleavage (see chapter by A.J. Carpousis and colleagues in this volume), translation must somehow prevent this initial cleavage. How this is achieved is tentatively discussed below (see Section V(C)). It should be kept in mind, however, that in many cases a direct implication of RNase E in mRNA decay has not been demonstrated and when it has, the position of the initial cleavage site is often unknown.

Finally, it should be stressed that although most mRNAs are stabilized by translation, exceptions exist. For example, the translation of the monocistronic *rpsA* mRNA is repressed by its product, ribosomal protein S1. This mRNA accumulates even under conditions of strong repression, indicating that its stability in unaffected by the translation shutoff (*142*).

B. Mechanism of RNase E Cleavage: The 5′ Tethering and Internal Entry Pathways

Before discussing how RNase E activity is affected by translation, we describe some relevant information on the enzyme itself and on its mechanism of cleavage. This section is kept to a minimum as the field is extensively reviewed elsewhere (*143–146*) (see chapter by A.J. Carpousis and colleagues in this volume).

As stated above, *E. coli* mRNAs are stabilized in bulk after the inactivation of RNase E (*46, 47*); hence the belief that this endonuclease controls the decay of many or most individual mRNAs. Other endonucleases (RNase III, P, G, and Z) participate in the decay of a few specific mRNAs, whereas 3′–5′ exonucleolytic degradation presumably constitutes a default pathway for the degradation of intact mRNAs (*146*). Fragments resulting from RNase E cleavage can either be further cleaved by this enzyme or degraded by 3′→5′ exonucleases, particularly RNase II, RNase R, and polynucleotide phosphorylase (PNPase) (see chapter by C. Arraiano and colleagues in this volume). Significantly, RNase E and PNPase are associated in a multienzymatic complex, the *E. coli* "degradosome" (*147–149*) (see (*146*) for review), which presumably helps co-ordinate their activities. For polycistronic mRNAs, the fragments resulting from the initial cleavage are often processed into smaller (often monocistronic) mRNAs which are temporarily protected from further endo- or exonucleolytic attack by secondary structures at their 5' or 3' ends (see, e.g., (*150, 151*)). In contrast, for monocistronic mRNAs (whether synthesized as such or processed from polycistronic mRNAs), fragments are usually not detected in strains that are wild type for exonucleases, suggesting that the initial cleavage is slow compared to subsequent degradation steps.

Even though RNase E is an endonuclease, its activity is sensitive to the nature of the substrate 5' end. *In vitro*, mRNA molecules carrying a free 5' monophosphate end (5'p) are cleaved much faster than their counterparts that carry a 5' triphosphate (5'ppp) or a base-paired 5'p end, or substrates that have been circularized prior to assay (*152*). Further work showed that the preference for 5'p only appeared for concentrations of the enzyme high enough to allow its dimerization (*153*). The X-ray structure of the N-terminal half of RNase E has illuminated these observations (*154*). The enzyme crystallizes as a dimer of dimers (the "principal" dimers), with each protomer carrying a 5'p binding pocket distinct from its active site; the folding of the active site resembles that of DNase I. RNA substrates carrying a free 5'p can span from the 5'p binding pocket of one protomer to the active site of the other protomer within the same principal dimer, explaining both why the enzyme prefers 5'p substrates, and why dimerization is necessary to express this preference. The structure also tells us that the enzyme can accommodate long, even structured RNA stretches between the 5'p binding pocket and the active site. Thus, RNase E does not need to first recognize the 5'p end and then scan the RNA or migrate along it to reach cleavage sites; rather, both binding events can take place simultaneously. Subsequently, I refer to this mode of action of RNase E as the "5'-tethering cleavage pathway." It should be noted that the substrate must then adopt a conformation in which both the 5'p and the cleavage site are single stranded with a separation in space that fits the enzyme geometry (Fig. 6A, left). Presumably, not all substrates can satisfy these constraints (*154*), explaining why some mRNAs decay mainly via other routes (see "internal entry" below).

Intriguingly, *in vivo* work has revealed additional substrate hierarchies that are not seen *in vitro*. In particular, genuine mRNAs, which carry a 5'ppp, are cleaved faster than their circular counterparts *in vivo* (*60*). Yet, the 5'p binding pocket, as revealed by the X-ray structure, cannot accommodate a 5'ppp extremity. This paradox led Belasco and colleagues to hypothesize the existence of an enzyme that converts a 5'ppp to a 5'p extremity *in vivo* (*155*). This enzyme has now been identified and called RNA pyrophosphohydrolase (RppH). Removal of RppH from the cell stabilizes hundreds of mRNAs (*55*). These

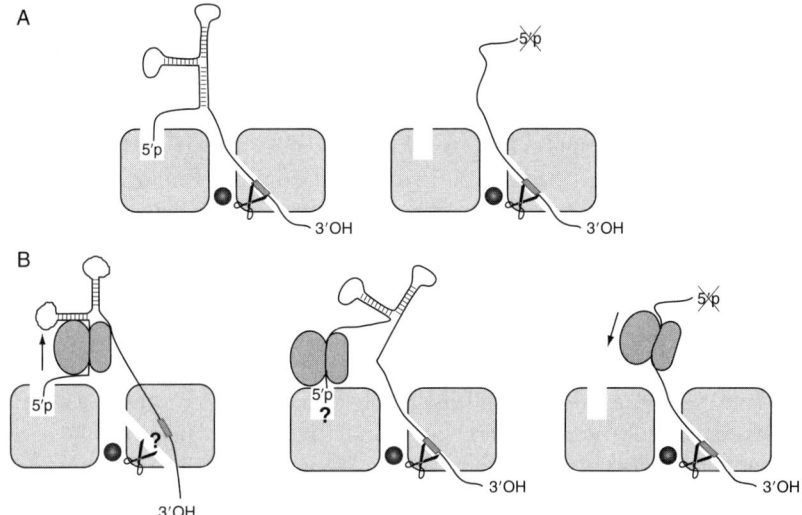

FIG. 6. The 5′ tethering *versus* internal entry pathways: protection at a distance. (A) The two RNase E protomers that constitute the "principal dimer" seen in the X-ray structure are schematized here as two gray squares, with the zinc ion that organizes their interface as a gray sphere. Each protomer possesses a 5′p binding pocket (open rectangle) and an active site (scissors). In the X-ray structure, the free 5′p extremity of the RNA is seen interacting with the 5′p binding pocket of one protomer, whereas the cleavage site (rectangular box) interacts with the active site of the other protomer. This situation corresponds to a cleavage mechanism referred to as the "5′ tethering pathway" (left). In the absence of a free 5′p end (right), RNase E must resort to a default pathway where the substrate interacts with the active site only ("internal entry pathway"). Internal entry is often (but not invariably) less efficient than the 5′ tethering pathway, explaining why many mRNAs are stabilized by appending structures at their 5′ end: in these cases, the extent of the stabilization presumably reflects the relative efficiencies of the two pathways. (B) The presence of ribosomes upstream of the cleavage site (left), and particularly near the mRNA 5′ end (middle), may interfere with the structural constraints of the 5′ tethering pathway (question marks), reducing its efficiency. Compared to the untranslated situation (A, left), the mRNA will then be more stable even if the ribosomes do not cross the cleavage site (protection at a distance). In contrast, the internal entry pathway is structurally less demanding and will not be inhibited by ribosomes unless they directly shield the cleavage site (right). A light cross indicates that no free 5′p extremily is available. Arrows refer to ribosome movement.

mRNAs must undergo pyrophosphate removal as a first and presumably irreversible step in their decay process. In the absence of RppH, the decay of these mRNAs must take alternative, less efficient routes. Interestingly, secondary structures sequestering the 5′ppp end can prevent RppH action. Thus, such structures hinder the 5′-tethering cleavage pathway at two levels *in vivo*, first by inhibiting pyrophosphate removal, then by preventing 5′p binding to RNase E. *In vivo* stabilization of mRNAs by 5′ secondary structures have been amply documented, particularly by the Belasco group (53, 56, 78, 156).

The 5' tethering pathway is obviously not the only route to RNase E cleavage. Circular mRNAs, which lack a 5' extremity, are still cleaved by RNase E at a significant rate (60). In this case, the initial binding of the enzyme to the mRNA must occur internally. Presumably this binding involves only the DNase I-like active site (Fig. 6A, right). Because there is no evidence that RNase E can migrate along the RNA, I assume that the enzyme can bind directly to the cleavage site ("internal entry" pathway). In contrast to the 5'-tethering pathway, this pathway does not require the dimerization of the enzyme: *in vitro*, substrates carrying a 5'OH extremity can be cleaved—presumably via the internal entry pathway—irrespective of the enzyme concentration (153). *In vivo*, there is evidence that internal entry can prevail not only with circular, but also with certain genuine, linear mRNA molecules. For example, the RNase E-mediated cleavage of the *rne* mRNA itself, is largely dependent upon the presence of conserved hairpin structures ("hp2" and "hp3" in (157)) that presumably serve as 5'-independent docking sites for RNase E. More generally, linear mRNAs that cannot be protected from RNase E by appending a 5' hairpin (89, 158) or that are cleaved by RNase E despite the presence of a 5' hairpin (54, 56) are likely to be cleaved principally *via* internal entry.

C. Local *Versus* Distal Protection

The facts. Although the presence of ribosomes usually protects *E. coli* mRNAs from RNase E attack, a direct shielding of the cleavage sites is not always required, that is, ribosomes can sometimes provide protection "at a distance." For instance, the translation of the first 20% of the coding sequence of the *bla* mRNA exerts a protection over the remaining 80%, which would otherwise be labile (the RNase—possibly RNase E—that controls decay in this case has not been characterized) (58). Another example is the *rpsT* mRNA. This mRNA decays *via* two non-exclusive pathways which may both be mediated by the degradosome, that is, PNPase trimming from the 3' end or RNase E cleavage within the coding sequence, at around two-thirds of the mRNA length (143, 159). The presence of ribosomes over the first 15 codons, far away from the targets of RNase E and PNPase, confers some protection to the whole mRNA by inhibiting both pathways (160). Protection at a distance has also been observed in Gram-negative bacteria other than *E. coli*. In the *Rhodobacter capsulatus puf* operon, the stabilization of the mRNA encoding the promoter–distal *pufL* and *pufM* cistrons requires the presence of ribosomes over the two promoter–proximal cistrons, *pufA* and *pufB* and the beginning of *pufL*, but not over decay-promoting endonuclease cleavage sites that have been located further downstream (161). Interestingly, the decay of the *puf* mRNA is controlled by a *R. capsulatus* enzyme that is closely related to RNase E and also assembles into a degradosome (162, 163). Less precise observations, in which translation is not interrupted prematurely but decreased by mutations in the RBS region, are also suggestive of a protection at a distance. Indeed, in several cases, the presence of only a few, if not just one translating ribosome, can significantly protect a whole coding region against RNase E (56, 57). Although the exact location of the RNase E-sensitive sites is not known in these cases, it seems unlikely that protection can be explained by the direct shielding of these sites by the (rare) translating ribosomes: rather, protection at a distance must occur.

In contrast, other cases are known where mRNA stability obviously requires the direct shielding of the cleavage sites by ribosomes. In the *rpsO* mRNA, the major RNase E cleavage site is located only 10 nt downstream from the stop codon and thus is presumably directly shielded by terminating ribosomes. Increasing the spacing between the stop and the cleavage site to 20 nt is enough to substantially accelerate cleavage, indicating that ribosome shielding is required for protection (61). Another particularly clear example is provided by a variant of the *rpsT* mRNA (54). This mRNA can be markedly (but not completely) stabilized against RNase E by appending a hairpin at its 5′ end. However, besides the presence of the hairpin, stabilization also requires translation across the major RNase E cleavage site: the introduction of a stop codon 16 nt upstream of this site dramatically destabilizes the mRNA. Moreover, the introduction of an ectopic RNase E cleavage site in the 5′UTR also accelerates mRNA decay significantly, whereas its introduction within the coding sequence has a milder effect, particularly when translation is efficient (54).

The *lacZ* mRNA is yet another example where ribosomes directly protect cleavage sites ((89); see Fig. 7). To study the effect of prematurely interrupting translation on the stability of this mRNA, we resorted to T7 RNA polymerase (T7 RNAP) to drive transcription, since this enzyme is resistant to polarity (88). The resulting transcript is prone to RNase E attack even when translation is uninterrupted, because T7 RNP outpaces ribosomes (see Section V(A)). However, the complete absence of translation renders the mRNA even more sensitive to RNase E. This sensitivity is not relieved by the presence of an immobile ribosome stably bound near the 5′ end; only the region surrounding the bound ribosome is stable (89). When translation is progressively allowed to proceed further downstream, only the regions that are traversed by the ribosomes get stabilized (Fig. 7). Obviously, in these three cases, ribosomes must be present at the cleavage site to achieve protection.

Possible interpretations. How can ribosomes protect the *bla*, *rpsT*, or *puf* mRNAs without directly shielding endonuclease cleavage sites, whereas in the case of the *rpsO* mRNA, the *rpsT* mRNA carrying a 5′ hairpin, or the *lacZ* mRNA synthesized by T7 RNP they must be present at the cleavage site to provide protection? I believe that this difference reflects the different way RNase E interacts with these two classes of mRNAs—mainly 5′ tethering

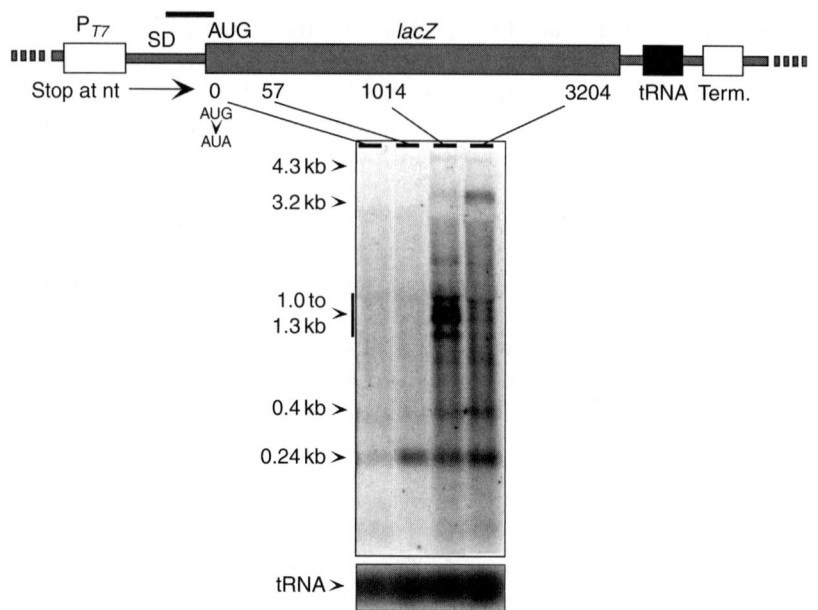

FIG. 7. Protection of the *lacZ* mRNA requires direct shielding of the cleavage sites by translating ribosomes. (Upper scheme) An ectopic copy of the *lacZ* gene, preceded by a T7 late promoter (P_{T7}) and followed by a tRNA gene that is used here as a transcriptional reporter (tRNA; see (74)), has been inserted into the chromosome (dotted line) of a Lac⁻ *E. coli* strain expressing T7 RNA polymerase. Different constructs were made in which translation is interrupted by inserting stop codons at various positions of the 3204 nt *lacZ* coding sequence; "stop at nt 0" corresponds to a change in the initiation codon (AUG→AUA). (Lower panels): Northern blots showing the *lacZ* mRNA in the different situations described above, as well as the reporter tRNA. Note that the latter is equally expressed whatever the transcript, indicating similar levels of transcription. The *lacZ* mRNA was probed with a 5'UTR probe (horizontal bar in the upper scheme). When translation is uninterrupted (last lane), the primary 4.3 kb transcript is rapidly processed into a 3.2 kb species corresponding to the full-length *lacZ* mRNA. In the case of premature translation interruption, no full length transcript is seen, but only fragments thereof: the more distal the interruption, the longer these fragments. The inactivation of RNase E stabilizes the mRNA and yields a similar profile regardless of the position of the stop codon (not illustrated). This experiment indicates that in this system, ribosomes provide only local protection against RNase E cleavage (Susan A. Joyce and MD, unpublished).

pathway for the first class, and internal entry for the second. The *bla* and *rpsT* mRNAs can both be stabilized by appending a 5' hairpin (54, 156), which is indicative of a predominantly 5' tethering pathway. In contrast, the *rpsT* mRNA carrying a 5' hairpin, the *rpsO* mRNA, and the *lacZ* mRNA synthesized by T7 RNP are good candidates for the internal entry pathway. The former lacks a free 5' end altogether, whereas the latter two cannot be stabilized by appending 5' hairpins (89, 158). It is then tempting to speculate that RNase E does not interact with the 5' mRNA end in these cases.

Why, then, would the 5' tethering and the internal entry pathways respond differently to the presence of ribosomes? It should be noted that with an RNase E dimer attached to two defined positions on the same mRNA, the 5'-tethering pathway is more demanding in term of mRNA conformation than the internal entry pathway (compare left and right in Fig. 6A). The presence of ribosomes anywhere between the two sites may perturb their spatial arrangement and hinder cleavage even though they do not directly shield the cleavage site (Fig. 6B, left). In addition, ribosomes lingering near the 5' mRNA end may interfere with RppH activity or hinder the mRNA interaction with the 5'p binding pocket of RNase E (Fig. 6B, middle). In all these situations, the 5' tethering pathway will be less efficient than it is in the absence of ribosomes, resulting in protection at a distance. In contrast, the simpler internal entry pathway appears more robust, and translating ribosomes may not interfere unless sterically shielding the cleavage sites (Fig. 6B, right).

I end with a word of caution. Protection at a distance is interpreted here in the light of recent work on the mechanism of RNase E cleavage. However, before this information was available, protection at a distance had been given a completely different interpretation (57) (see also (56, 58)). According to this view, only certain conformations of an mRNA would be sensitive to RNase E. Newly synthesized mRNA would be resistant but it would slowly refold into more sensitive conformation. The crossing of even a single translating ribosome would "reset the folding clock" and render the message resistant for another period of time. In this model, the mRNA must stop being translated *before* being degraded (non-nucleolytic inactivation), and the ribosome does not need to cross the mRNA regions that will ultimately be cleaved by RNase E to provide protection (Fig. 8). The observed result is protection at a distance but here the ribosome does not interfere directly with a degradation step—a

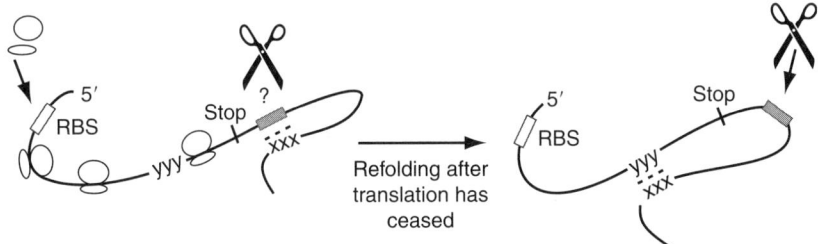

FIG. 8. A possible mechanism for protection "at a distance" based on non-nucleolytic inactivation. When translated (left), the mRNA adopts a conformation in which the RNase E cleavage site is inaccessible (question mark), due to base-pairing with sequence xxx. After translation is interrupted (right), the mRNA refolds into an alternative conformation in which xxx now pairs with yyy, so that the cleavage site becomes accessible. In this model, ribosomes can provide protection without ever traversing the cleavage site itself (see text).

characteristic of non-nucleolytic inactivation (see Section II(B) above). At present, this hypothetical mechanism is an *ad hoc* explanation for protection at a distance, but it certainly cannot be dismissed.

D. A Specific Role of RNase E C-Terminal Half in the Degradation of Untranslated mRNA?

Even though internal entry is a "default" pathway, it can be very efficient, particularly in the absence of translation. Thus, the *rplT* or *bla* mRNAs carrying an appended 5′ hairpin become quite labile when non-translated (54, 56). Similarly the *rpsO* mRNA is very unstable in the absence of translation, as is the *lacZ* mRNA synthesized by T7 RNA polymerase (61, 89). Now, the CTH of RNase E is not essential for catalytic activity: its removal does not affect the maturation of 9S rRNA, a substrate that follows the 5′ tethering mechanism (152), and it has only a moderate effect on bulk mRNA decay (49, 164). In contrast, the CTH is essential for the fast decay of the untranslated *lacZ* mRNA or the *rpsO* mRNA ((49); P. Marujo and P. Regnier, personal communication). The *rne* mRNA, another target for the internal entry pathway, is also stabilized by CTH removal (165). The same holds true for the untranslated *thrS* mRNA, although it is not known whether internal entry is used in this case (12). Put together, these observations lead me to speculate that the CTH plays an important (though undefined) role in the internal entry pathway (166), and that this role explains why the CTH is essential for the rapid breakdown of many untranslated mRNA (167). Intriguingly, several mRNAs that are targeted by sRNAs also seem to require the CTH for fast decay (see Section II(C)). The significance of this similarity, if any, remains to be seen.

E. Transcription, Translation, or RNase E Preferences: The Source of Directionality

It is often stated that *E. coli* mRNAs decay with an overall 5′–3′ direction. Since decay generally consists of endonucleolytic cleavages followed by exonucleolytic trimming of the upstream fragment, this notion requires clarification. By "decay in the 5′→3′ direction," we refer here to the specific situation in which the successive cleavages occur in an orderly 5′–3′ wave, with the first one taking place near the 5′ mRNA end. A 5′→3′ direction of decay has been occasionally observed in polycistronic mRNAs (139, 168), but obviously it is not generally valid in this case (151, 169). In contrast, for monocistronic mRNAs (whether synthesized as such or processed from polycistronic mRNAs), directionality is likely to be the rule, if only because of biological efficiency. Indeed, it is the only decay mode that avoids the wasteful synthesis of incomplete polypeptides from truncated mRNAs (100, 169). Consistent with this idea, the 5′ end starts decaying before mRNA synthesis is completed for several long mRNAs (*39–41*). Some monocistronic mRNAs appear not to conform to this rule, as exemplified by the *rpsO*, *rpsT*, and *cspA* mRNAs which are initially attacked by RNase E near their 3′ ends (*143, 159, 170, 171*), but in the first two cases these exceptions should be relativized given the small size of the mRNAs.

What is the molecular basis for directionality? There is probably more than one answer to this question. It has long been thought that directionality simply reflects the vectorial nature of transcription and translation, which themselves proceed in the 5′→3′ direction. Thus, the 5′ region is synthesized first, and if RNase E susceptibility were uniform over the mRNA, this region should also statistically disappear first. This argument is particularly relevant for long mRNAs for which the synthesis of the 3′ end lags significantly behind that of the 5′ end. As far as translation is concerned, it has been proposed that once the RBS has been removed, preventing further ribosome loading, the naked mRNA region lagging behind the last ribosome is rapidly chopped off by endonucleases (e.g., RNase E), resulting in an orderly 5′→3′ degradation (*169*). While these phenomena—and particularly the first—almost certainly contribute to directionality, it is likely that the mechanism of degradation itself also plays its role. First, whereas the mode of degradation in *E. coli*—endonucleolytic hits followed by rapid trimming of upstream fragments—does not in itself guarantee that the first hit occurs near the 5′ end, it will in a sense contribute to directionality, insofar as regions located downstream of the hits are more stable than upstream ones. This will be particularly true for the mRNA fragment located downstream of the most distal endonucleolytic hit; this fragment is usually genuinely resistant to exonucleases and would therefore be degraded last (for an illustration see (*139*)). Second, the 5′ region of the mRNA, encompassing the 5′UTR and the RBS, has distinctive features which may favor RNase E cleavage so that the first hit occurs there. Cleavage within the 5′UTR may be facilitated by the absence of ribosomes (see e.g., (*54*)). As for the RBS, it is known that efficient ribosome binding usually requires that this region be unstructured (*172*) and statistical analysis has indeed confirmed this point (*173*). As such, the RBS is expected to be intrinsically prone to cleavage by single-strand endonucleases such as RNase E (see (*66, 85*) for discussion). The third and most intriguing possibility is that the affinity of RNase E for 5′ ends directly contributes to directionality. RNase E engaged in the 5′ tethering pathway may have a higher probability of cleaving at 5′-proximal than 5′-distal sites. To my knowledge, this question has not yet been addressed experimentally.

F. Protective Ribosomes in *B. subtilis*

By far the most impressive examples of a protection at a distance have been observed in Gram-positive bacteria, notably *Bacilli*. The stalling of ribosomes over the upstream ORFs of the *erm* genes does not solely cause processing of the mRNA in the stalling region ((128, 130); see Section IV(E)); it also markedly stabilizes the downstream region (18, 174). The same result can be achieved through the mere binding of 30S subunits to the long SD sequence of the *cryIIIA* gene leader region ("STAB-SD"; (134)). In some other cases, the SD is not sufficient for maximal stabilization unless it is associated with a 5′ hairpin structure (e.g., in the *atpE* gene; (175)), or with an initiation codon (even if immediately followed by a stop (176)), or with both (177) (see chapter by D. Bechhofer in this volume). A variety of sequences can be stabilized in this way, including very long untranslated regions such as the *E. coli lacZ* mRNA lacking an RBS (134, 175, 178). Finally, stalled or bound ribosomes can exert their protecting effect not only when present near the 5′ end of the mRNA, but also when located internally (128, 134), although stabilization may be somewhat less efficient in this case (179, 180). Only regions located downstream of the internally located ribosome are stabilized, whereas upstream regions remain labile.

In comparison with these effects, protection at a distance in *E. coli* is far less impressive. For instance, the mere presence of a ribosome at the 5′ end of the *lacZ* mRNA, even if very stably bound, cannot protect the whole message against RNase E (89) (see Section V(C)). Moreover, the STAB-SD stabilizers from SP82 or of *cryIIIA* do not work efficiently in *E. coli* (134, 179). As seen above, protection at a distance in *E. coli* presumably reflects the interference by ribosomes, and in particular ribosomes lingering in the 5′ region of the mRNA, with the 5′ tethering mechanism. This protection cannot be absolute because of the existence of alternative decay pathways, most notably the internal entry pathway, and possibly degradation pathways mediated by 3′→5′ exonucleases (146). The absence of an RNase E ortholog in *B. subtilis* (181) suggests a different interpretation for protection at a distance in this organism.

Only three RNases are known to play a general role in *B. subtilis* mRNA decay, that is, PNPase, and the pair of closely related paralogs RNases J1 and J2 (see chapter by D. Bechhofer in this volume for details). Indeed, the removal of PNPase (182), or of both RNase J1 and J2 (131), significantly increases the half-life of bulk mRNA, whereas the removal of RNase J2 alone has no effect (the technique used does not allow to distinguish between intact mRNAs and fragments). PNPase and RNase J2 are not essential in *B. subtilis*, but RNase J1 is. Both RNases J1 and J2 are representatives of a phylogenetically conserved family present in many classes of Bacteria and Archaea (131). As mentioned above, they were first identified as endonucleases (131), but were later found to also possess 5′→3′ exonucleolytic activity (132), a hitherto unrecognized activity in prokaryotes. Interestingly, the exonuclease activity has a preference for an accessible (single stranded) 5′p end, whereas the endonuclease activity lacks such a preference (132, 133). The recently published structure of RNase J1 explains this 5′p requirement, and suggests how the enzyme may shift efficiently from endo- to exonuclease activity (133). Whether B. subtilis possess an enzyme similar to RppH that would facilitate exonucleolytic attack of 5′ppp extremities is unknown.

Currently, the simplest explanation for the large protection at a distance observed in B. subtilis is that the corresponding mRNAs are degraded by the newly discovered 5′→3′ exonucleolytic activity of RNase J1 or a similar enzyme, with alternative decay pathways playing only a minor role. Depending upon whether they are located close to the 5′ end or at a more distant position, the bound ribosomes would either prevent the exonuclease from accessing the 5′ end or act as an efficient barricade against its 5′→3′ progression. However, the protection of downstream sequences is not absolute: it is likely that 3′→5′ exonucleases participate in the residual degradation observed in this case. For example, in the erythromycin-stabilized *ermC* mRNA, fragments that have been 3′-trimmed up to the leading edge of the stalled ribosome can be detected (129, 183). It is also possible that the obstructing ribosome can be removed over time (see Section IV (F)). On the other hand, there is currently no evidence that the endonuclease activity of RNase J1 participates significantly in the decay of the ribosome-protected mRNAs considered here. If it did, it would probably mitigate the protection at a distance offered by the ribosome, since it is insensitive to the nature (and presumably accessibility) of the mRNA 5′ end. In this sense, the endonucleolytic activity of RNase J1/2 would be functionally equivalent to the internal entry of RNase E (see above). Possibly, the endonucleolytic activity of RNase J1 has sequence or structure requirements that are uncommon within mRNAs. Interestingly in this regard, the introduction of an ectopic Bs-RNase III cleavage site into the *ermC* mRNA is enough to negate the protection offered by the stalled ribosome (178).

VI. Concluding Remarks

Our understanding of the relationship between translation and degradation in *E. coli* and *B. subtilis* has expanded tremendously in the last 10 years, due in particular to a better understanding of the decay process itself. As noted in several parts of the text, many loose ends remain, however. My ambition here has been to present a snapshot of this rapidly moving field, with a few suggestions for future work.

Acknowledgments

Given the many facets of the field, this essay on the translation–degradation interplay cannot be comprehensive and I apologize to the many colleagues whose important contributions are not discussed or even cited. I am much indebted to Dr. Susan A. Joyce for permitting me to show an unpublished experiment (Fig. 7), and to Drs. C. Condon and H. Aiba for comments on the manuscript. I thank Landes Bioscience (Georgetown TX, USA) for the permission of using some items from our former review. Work in my Laboratory is funded by CNRS, by Ecole Normale Supérieure, and by the Agence Nationale de la Recherche (Programme Blanc, grants CARMa and mRNases).

References

1. Jacob F, Monod J. Genetic regulatory mechanisms in the synthesis of proteins. *J Mol Biol* 1961;**3**:318–56.
2. Lu P, Vogel C, Wang R, Yao X, Marcotte EM. Absolute protein expression profiling estimates the relative contributions of transcriptional and translational regulation. *Nat Biotechnol* 2007;**25**:117–24.
3. Dreyfus M, Joyce SA. The Interplay between translation and mRNA decay in procaryotes: A discussion on current paradigms. In: J Lapointe and L Brakier-Gingras, editors. *Translation mechanisms*. Georgetown TX: Landes Bioscience; 2003.
4. Deana A, Belasco JG. Lost in translation: the influence of ribosomes on bacterial mRNA decay. *Genes Dev* 2005;**19**:2526–33.
5. Kaberdin VR, Blasi U. Translation initiation and the fate of bacterial mRNAs. *FEMS Microbiol Rev* 2006;**30**:967–79.
6. Loomis WP, Koo JT, Cheung TP, Moseley SL. A tripeptide sequence within the nascent DaaP protein is required for mRNA processing of a fimbrial operon in *Escherichia coli*. *Mol Microbiol* 2001;**39**:693–707.
7. Sunohara T, Jojima K, Tagami H, Inada T, Aiba H. Ribosome stalling during translation elongation induces cleavage of mRNA being translated in *Escherichia coli*. *J Biol Chem* 2004;**279**:15368–75.
8. Iost I, Dreyfus M. The stability of *Escherichia coli lacZ* mRNA depends upon the simultaneity of its synthesis and translation. *EMBO J* 1995;**14**:3252–61.
9. Makarova OV, Makarov EM, Sousa R, Dreyfus M. Transcribing of *Escherichia coli* genes with mutant T7 RNA polymerases: Stability of *lacZ* mRNA inversely correlates with polymerase speed. *Proc Natl Acad Sci USA* 1995;**92**:12250–4.
10. Sanson B, Uzan M. Dual role of the sequence-specific bacteriophage T4 endoribonuclease RegB: mRNA inactivation and mRNA destabilization. *J Mol Biol* 1993;**233**:429–46.
11. Odaert B, Saida F, Aliprandi P, Durand S, Crechet JB, Guerois R, et al. Structural and functional studies of RegB, a new member of a family of sequence-specific ribonucleases involved in mRNA inactivation on the ribosome. *J Biol Chem* 2007;**282**:2019–28.
12. Nogueira T, de Smit M, Graffe M, Springer M. The relationship between translational control and mRNA degradation for the *Escherichia coli* threonyl-tRNA synthetase gene. *J Mol Biol* 2001;**310**:709–22.
13. Poot RA, Tsareva NV, Boni IV, van Duin J. RNA folding kinetics regulates translation of phage MS2 maturation gene. *Proc Natl Acad Sci USA* 1997;**94**:10110–5.
14. Nierlich DP, Murakawa GJ. The decay of bacterial messenger RNA. *Prog Nucl Acids Res Mol Biol* 1996;**52**:153–216.
15. Schwartz T, Craig E, Kennell D. Inactivation and degradation of messenger ribonucleic acid from the lactose operon of *Escherichia coli*. *J Mol Biol* 1970;**54**:299–311.
16. Puga A, Borras MT, Tessman ES, Tessman I. Difference between functional and structural integrity of mRNA. *Proc Natl Acad Sci USA* 1973;**70**:2171–5.
17. Yamamoto T, Imamoto F. Differential stability of *trp* messenger RNA synthesized originating at the *trp* promoter and P_L promoter of lambda *trp* phage. *J Mol Biol* 1975;**92**:289–304.
18. Sandler P, Weisblum B. Erythromycin-induced stabilization of *ermA* messenger RNA in *Staphylococcus aureus* and *Bacillus subtilis*. *J Mol Biol* 1988;**203**:905–15.
19. Hue KK, Bechhofer DH. Effect of *ermC* leader region mutations on induced mRNA stability. *J Bacteriol* 1991;**173**:3732–40.
20. Jenner L, Romby P, Rees B, Schulze-Briese C, Springer M, Ehresmann C, et al. Translational operator of mRNA on the ribosome: How repressor proteins exclude ribosome binding. *Science* 2005;**308**:120–3.

21. Gottesman S. The small RNA regulators of *Escherichia coli*: Roles and mechanisms. *Annu Rev Microbiol* 2004;**58**:303–28.
22. Marzi S, Fechter P, Chevalier C, Romby P, Geissmann T. RNA switches regulate initiation of translation in bacteria. *Biol Chem* 2008;**389**:585–98.
23. Urban JH, Vogel J. Translational control and target recognition by *Escherichia coli* small RNAs *in vivo*. *Nucleic Acids Res* 2007;**35**:1018–37.
24. Morita T, Mochizuki Y, Aiba H. Translational repression is sufficient for gene silencing by bacterial small noncoding RNAs in the absence of mRNA destruction. *Proc Natl Acad Sci USA* 2006;**103**:4858–63.
25. Aiba H. Mechanism of RNA silencing by Hfq-binding small RNAs. *Curr Opin Microbiol* 2007;**10**:134–9.
26. Morita T, Maki K, Aiba H. RNase E-based ribonucleoprotein complexes: Mechanical basis of mRNA destabilization mediated by bacterial noncoding RNAs. *Genes Dev* 2005;**19**:2176–86.
27. Masse E, Escorcia FE, Gottesman S. Coupled degradation of a small regulatory RNA and its mRNA targets in *Escherichia coli*. *Genes Dev* 2003;**17**:2374–83.
28. Liou GG, Jane WN, Cohen SN, Lin NS, Lin-Chao S. RNA degradosomes exist *in vivo* in *Escherichia coli* as multicomponent complexes associated with the cytoplasmic membrane via the N-terminal region of ribonuclease E. *Proc Natl Acad Sci USA* 2001;**98**:63–8.
29. Taghbalout A, Rothfield L. RNaseE and the other constituents of the RNA degradosome are components of the bacterial cytoskeleton. *Proc Natl Acad Sci USA* 2007;**104**:1667–72.
30. Khemici V, Poljak L, Luisi BF, Carpousis AJ. The RNase E of *Escherichia coli* is a membrane binding protein. *Mol Microbiol* 2008;**70**(4), 799–813.
31. Miller OL, Jr, Hamkalo BA, Thomas CA, Jr. Visualization of bacterial genes in action. *Science* 1970;**169**:392–5.
32. Landick R, Carey J, Yanofsky C. Detection of transcription-pausing *in vivo* in the *trp* operon leader region. *Proc Natl Acad Sci USA* 1987;**84**:1507–11.
33. Landick R, Carey J, Yanofsky C. Translation activates the paused transcription complex and restores transcription of the trp operon leader region. *Proc Natl Acad Sci USA* 1985;**82**:4663–7.
34. de Smit MH, Verlaan PWG, van Duin J, Pleij WA. Intracistronic transcriptional polarity enhances translational repression: A new role for Rho. *Mol Microbiol* 2008;**60**(5), 1278–89.
35. Cardinale CJ, Washburn RS, Tadigotla VR, Brown LM, Gottesman ME, Nudler E. Termination factor Rho and its cofactors NusA and NusG silence foreign DNA in *E. coli*. *Science* 2008;**320**:935–8.
36. Lewis PJ, Thaker SD, Errington J. Compartmentalization of transcription and translation in *Bacillus subtilis*. *EMBO J* 2000;**19**:710–8.
37. Bremer H, Dennis P, Ehrenberg M. Free RNA polymerase and modeling global transcription in *Escherichia coli*. *Biochimie* 2003;**85**:597–609.
38. Sheth U, Parker R. Decapping and decay of messenger RNA occur in cytoplasmic processing bodies. *Science* 2003;**300**:805–8.
39. Cannistraro VJ, Kennell D. Evidence that the 5′ end of *lac* mRNA starts to decay as soon as it is synthesised. *J Bacteriol* 1985;**161**:820–2.
40. Jain C, Belasco JG. RNase E autoregulates its synthesis by controlling the degradation rate of its own mRNA in *E. coli*: Unusual sensitivity of the *rne* transcript to RNase E activity. *Genes Dev* 1995;**9**:84–96.
41. Yarchuk O, Jacques N, Guillerez J, Dreyfus M. Interdependence of translation, transcription and mRNA degradation in the *lacZ* gene. *J Mol Biol* 1992;**226**:581–96.
42. Yarchuk O, Iost I, Dreyfus M. The relation between translation and mRNA degradation in the *lacZ* gene. *Biochimie* 1991;**73**:1533–41.

43. Liphardt J, Onoa B, Smith SB, Tinoco I, JR, Bustamante C. Reversible unfolding of single RNA molecules by mechanical force. *Science* 2001;**292**:733–7.
44. Sousa S, Marchand I, Dreyfus M. Autoregulation allows *Escherichia coli* RNase E to adjust continuously its synthesis to that of its substrates. *Mol Microbiol* 2001;**42**:867–78.
45. Casarégola S, Jacq A, Laoudj D, McGurk G, Margarson S, Tempête M, et al. Cloning and analysis of the entire *Escherichia coli ams* gene. *J Mol Biol* 1992;**228**:30–40.
46. Ono M, Kuwano M. A conditional lethal mutation in an *E. coli* strain with a longer chemical lifetime of messenger RNA. *J Mol Biol* 1979;**129**:343–57.
47. Mudd EA, Krisch HM, Higgins CF. RNase E, an endoribonuclease, has a general role in the chemical decay of *Escherichia coli* mRNA: Evidence that *rne* and *ams* are the same genetic locus. *Mol Microbiol* 1990;**4**:2127–35.
48. Kido M, Yamanaka K, Mitani T, Niki H, Ogura T, Hiraga S. RNase E polypeptides lacking a carboxyl-terminal half supress a *mukB* mutation in *Escherichia coli*. *J Bacteriol* 1996;**178**:3917–25.
49. Lopez PJ, Marchand I, Joyce SA, Dreyfus M. The C-terminal half of RNase E, which organizes the *Escherichia coli* degradosome, participates in mRNA degradation but not rRNA processing *in vivo*. *Mol Microbiol* 1999;**33**:188–99.
50. Bernstein JA, Lin PH, Cohen SN, Lin-Chao S. Global analysis of *Escherichia coli* RNA degradosome function using DNA microarrays. *Proc Natl Acad Sci USA* 2004;**101**:2758–63.
51. Petersen C. Translation and mRNA stability in bacteria: a complex relationship. In: JG Belasco and G Brawerman, editors. *Control of mRNA stability*, San Diego, CA: Academic Press; 1993. p. 117–45.
52. Guillerez J, Gazeau M, Dreyfus M. In the *Escherichia coli lacZ* gene the spacing between the translating ribosomes is insensitive to the efficiency of translation initiation. *Nucl Acids Res* 1991;**19**:6743–50.
53. Hansen MJ, Chen L-H, Fejzo ML, Belasco JG. The *ompA* 5' untranlated region impedes a major pathway for mRNA degradation in *Escherichia coli*. *Mol Microbiol* 1994;**12**:707–16.
54. Baker KE, Mackie GA. Ectopic RNase E sites promote bypass of 5'-end-dependent mRNA decay in *Escherichia coli*. *Mol Microbiol* 2003;**47**:75–88.
55. Deana A, Celesnik H, Belasco JG. The bacterial enzyme RppH triggers messenger RNA degradation by 5' pyrophosphate removal. *Nature* 2008;**451**:355–8.
56. Arnold TE, Yu J, Belasco JG. mRNA stabilization by the *ompA* 5' untranslated region: Two protective elements hinder distinct pathways for mRNA degradation. *RNA* 1998;**4**:319–30.
57. Jain C, Kleckner N. IS*10* mRNA stability and steady state levels in *Escherichia coli*: Indirect effects of translation and role of *rne* function. *Mol Microbiol* 1993;**9**:233–47.
58. Nilsson G, Belasco JG, Cohen SN, von Gabain A. Effect of premature termination of translation on mRNA stability depends on the site of ribosome release. *Proc Natl Acad Sci USA* 1987;**84**:4890–4.
59. Baumeister R, Flache P, Melefors O, von Gabain A, Hillen W. Lack of a 5' non-coding region in Tn1721 encoded *TetR* mRNA is associated with a low efficiency of translation and a short half-life in *Escherichia coli*. *Nucl Acids Res* 1991;**19**:4595–600.
60. Mackie GA. Stabilization of circular *rpsT* mRNA demonstrates the 5' end dependence of RNase E action *in vivo*. *J Biol Chem* 2000;**275**:25069–72.
61. Braun F, Le Derout J, Régnier P. Ribosomes inhibit an RNase E cleavage which induces the decay of the *rpsO* mRNA of *Escherichia coli*. *EMBO J* 1998;**17**:4790–7.
62. Cho KO, Yanofsky C. Sequence changes preceding a Shine–Dalgarno region influence *trpE* mRNA translation and decay. *J Mol Biol* 1988;**204**:51–60.
63. Petersen C. The functional stability of the *lacZ* transcript is sensitive towards sequence alterations immediately downstream of the ribosome binding site. *Mol Gen Genet* 1987;**209**:179–87.

64. Petersen C. Multiple determinants of functional mRNA stability: Sequence alterations at either end of the lacZ gene affect the rate of mRNA inactivation. *J Bacteriol* 1991;**173**:2167–72.
65. Lundberg U, Nilsson G, von Gabain A. The differential stability of the *E. coli* ompA and bla mRNA at various growth rates is not correlated to the efficiency of translation. *Gene* 1988;**72**:141–9.
66. Vytvytska O, Moll I, Kaberdin VR, von Gabain A, Bläsi U. Hfq (HF1) stimulates *ompA* mRNA decay by interfering with ribosome binding. *Genes Dev* 2000;**14**:1109–18.
67. Udekwu KI, Darfeuille F, Vogel J, Reimegard J, Holmqvist E, Wagner EG. Hfq-dependent regulation of OmpA synthesis is mediated by an antisense RNA. *Genes Dev* 2005;**19**:2355–66.
68. Cole JR, Nomura M. Changes in the half-life of ribosomal protein messenger RNA caused by translational repression. *J Mol Biol* 1986;**188**:383–92.
69. Singer P, Nomura M. Stability of ribosomal protein mRNA and translational feedback regulation in *Escherichia coli*. *Mol Gen Genet* 1985;**199**:543–6.
70. Yates JL, Nomura M. Feedback regulation of ribosomal protein synthesis in *Escherichia coli*: Localization of the mRNA target sites for repressor action of ribosomal protein L1. *Cell* 1981;**24**:243–9.
71. Vecerek B, Moll I, Blasi U. Translational autocontrol of the *Escherichia coli* hfq RNA chaperone gene. *RNA* 2005;**11**:976–84.
72. McCormick JR, Zengel JM, Lindahl L. Correlation of translation efficiency with the decay of lacZ mRNA in *E. coli*. *J Mol Biol* 1994;**239**:608–22.
73. Matsunaga J, Simons EL, Simons RW. *Escherichia coli* RNase III (*rnc*) autoregulation occurs independently of *rnc* translation. *Mol Microbiol* 1997;**26**:1125–35.
74. Lopez PJ, Iost I, Dreyfus M. The use of a tRNA as a transcriptional reporter: The T7 late promoter is extremely efficient in *Escherichia coli* but its transcripts are poorly expressed. *Nucleic Acids Res* 1994;**22**:1186–93.
75. Liang ST, Ehrenberg M, Dennis P, Bremer H. Decay of *rplN* and *lacZ* mRNA in *Escherichia coli*. *J Mol Biol* 1999;**288**:521–38.
76. Bernstein JA, Khodursky AB, Lin PH, Lin-Chao S, Cohen SN. Global analysis of mRNA decay and abundance in *Escherichia coli* at single-gene resolution using two-color fluorescent DNA microarrays. *Proc Natl Acad Sci USA* 2002;**99**:9697–702.
77. Pease AJ, Wolf REJ. Determination of the growth rate-regulated steps in expression of the *Escherichia coli* K-12 gnd gene. *J Bacteriol* 1994;**176**:115–22.
78. Bouvet P, Belasco JG. Control of RNase E-mediated RNA degradation by 5′-terminal base pairing in *E. coli*. *Nature* 1992;**360**:488–91.
79. Xu F, Lin-Chao S, Cohen SN. The *Escherichia coli* pcnB gene promotes adenylylation of antisense RNAI of ColE1-type plasmids *in vivo* and degradation of RNAI decay intermediates. *Proc Natl Acad Sci USA* 1993;**90**:6756–60.
80. Jarrige A-C, Mathy N, Portier C. PNPase autocontrols its expression by degrading a double stranded structure in the *pnp* mRNA leader. *EMBO J* 2001;**20**:6845–55.
81. Shen V, Bremer H. Chloramphenicol-induced changes in the synthesis of ribosomal, transfer, and messenger RNA in *Escherichia coli* B/r. *J Bacteriol* 1977;**130**:1098–108.
82. Bessarab DA, Kaberdin VR, Wei C-L, Liou G-G, Lin-Chao S. RNA component of *Escherichia coli* degradosome: Evidence for rRNA decay. *Proc Natl Acad Sci USA* 1998;**95**:3157–61.
83. Lopez PJ, Marchand I, Yarchuk O, Dreyfus M. Translation inhibitors stabilize *Escherichia coli* mRNAs independently of ribosome protection. *Proc Natl Acad Sci USA* 1998;**95**:6067–72.
84. Wagner LA, Gesteland RF, Dayhuff TJ, Weiss RB. An efficient Shine–Dalgarno sequence but not translation is necessary for *lacZ* mRNA stability in *Escherichia coli*. *J Bacteriol* 1994;**176**:1683–8.

85. Komarova AV, Tchufistova LS, Dreyfus M, Boni IV. A/U-rich sequences within 5'-untranslated leaders enhance translation and stabilize mRNA in *Escherichia coli*. *J Bacteriol* 2005;**187**:1344–9.
86. Jacques N, Guillerez J, Dreyfus M. Culture conditions differentially affect the translation of individual *Escherichia coli* mRNAs. *J Mol Biol* 1992;**226**:597–608.
87. Stanssens P, Remaut E, Fiers W. Inefficient translation initiation causes premature transcription termination in the *lacZ* gene. *Cell* 1986;**44**:711–8.
88. Chevrier-Miller M, Jacques N, Raibaud O, Dreyfus M. Transcription of single-copy hybrid *lacZ* genes by T7 RNA polymerase in *Escherichia coli*: mRNA synthesis and degradation can be uncoupled from translation. *Nucleic Acids Res* 1990;**18**:5787–92.
89. Joyce SA, Dreyfus M. In the absence of translation, RNase E can bypass 5' mRNA stabilizers in *Escherichia coli*. *J Mol Biol* 1998;**282**:241–54.
90. Kuwano M., Kwan C. N., Apirion D., Schlessinger D.. Ribonuclease V of *Escherichia coli*: I. Dependance of ribosomes and translocation. *Proc Natl Acad Sci USA* 1969;**64**:693–700.
91. Kuwano M, Schlessinger D, Apirion D. Ribonuclease V of *Escherichia coli*. IV. Exonucleolytic cleavage in the 5' to 3' direction with production of 5'-nucleotide monophosphates. *J Mol Biol* 1970;**51**:75–82.
92. Pato ML, Bennett PM, von Meyerburg K. mRNA synthesis and degradation in *Escherichia coli* during inhibition of translation. *J Bacteriol* 1973;**116**:710–8.
93. Holmes RK, Singer MF. Inability to detect RNase V in *Escherichia coli* and comparison of other ribonucleases before and after infection with coliphage T7. *Biochem Biophys Res Commun* 1971;**44**:837–43.
94. Bothwell AL, Apirion D. Is RNase V a manifestation of RNase II? *Biochem Biophys Res Commun* 1971;**44**:844–51.
95. Schneider E, Blundell M, Kennell D. Translation and mRNA decay. *Mol Gen Genet* 1978;**160**:121–9.
96. Bilge SS, Apostol JM, Jr, Aldape MA, Moseley SL. mRNA processing independent of RNase III and RNase E in the expression of the F1845 fimbrial adhesin of *Escherichia coli*. *Proc Natl Acad Sci USA* 1993;**90**:1455–9.
97. Loomis WP, Moseley SL. Translational control of mRNA processing in the F1845 fimbrial operon of *Escherichia coli*. *Mol Microbiol* 1998;**30**:843–53.
98. Keiler KC, Waller PR, Sauer RT. Role of a peptide tagging system in degradation of proteins synthesized from damaged messenger RNA. *Science* 1996;**271**:990–3.
99. Abo T, Inada T, Ogawa K, Aiba H. SsrA-mediated tagging and proteolysis of LacI and its role in the regulation of lac operon. *EMBO J* 2000;**19**:3762–9.
100. Yamamoto Y., Sunohara T., Jojima K., Inada T., Aiba H. SsrA-mediated *trans*-translation plays a role in mRNA quality control by facilitating degradation of truncated mRNAs. *RNA* 2003;**9**:408–18.
101. Dulebohn D, Choy J, Sundermeier T, Okan N, Karzai AW. *trans*-Translation: The tmRNA-mediated surveillance mechanism for ribosome rescue, directed protein degradation, and nonstop mRNA decay. *Biochemistry* 2007;**46**:4681–93.
102. Keiler KC. Physiology of tmRNA: What gets tagged and why? *Curr Opin Microbiol* 2007;**10**:169–75.
103. Roche ED, Sauer RT. SsrA-mediated peptide tagging caused by rare codons and tRNA scarity. *EMBO J*. 1999;**18**:4579–89.
104. Collier J, Binet E, Bouloc P. Competition between SsrA tagging and translational termination at weak stop codons in *Escherichia coli*. *Mol Microbiol* 2002;**45**:745–54.
105. Hayes CS, Bose B, Sauer RT. Stop codons preceded by rare arginine codons are efficient determinants of SsrA tagging in *Escherichia coli*. *Proc Natl Acad Sci USA* 2002;**99**:3440–5.

106. Hayes CS, Bose B, Sauer RT. Proline residues at the C terminus of nascent chains induce SsrA tagging during translation termination. *J Biol Chem* 2002;**277**:33825–32.
107. Sunohara T, Abo T, Inada T, Aiba H. The C-terminal amino acid sequence of nascent peptide is a major determinant of SsrA tagging at all three stop codons. *RNA* 2002;**8**:1416–27.
108. Hayes CS, Sauer RT. Cleavage of the A site mRNA codon during ribosome pausing provides a mechanism for translational quality control. *Mol Cell* 2003;**12**:903–11.
109. Pedersen K, Zavialov AV, Pavlov MY, Elf J, Gerdes K, Ehrenberg M. The bacterial toxin RelE displays codon-specific cleavage of mRNAs in the ribosomal A site. *Cell* 2003;**112**:131–40.
110. Richards J, Mehta P, Karzai AW. RNase R degrades non-stop mRNAs selectively in an SmpB-tmRNA-dependent manner. *Mol Microbiol* 2006;**62**:1700–12.
111. Karzai AW, Sauer RT. Protein factors associated with the SsrA.SmpB tagging and ribosome rescue complex. *Proc Natl Acad Sci USA* 2001;**98**:3040–4.
112. Sunohara T, Jojima K, Yamamoto Y, Inada T, Aiba H. Nascent-peptide-mediated ribosome stalling at a stop codon induces mRNA cleavage resulting in nonstop mRNA that is recognized by tmRNA. *RNA* 2004;**10**:378–86.
113. Li X, Yokota T, Ito K, Nakamura Y, Aiba H. Reduced action of polypeptide release factors induces mRNA cleavage and tmRNA tagging at stop codons in *Escherichia coli*. *Mol Microbiol* 2007;**63**:116–26.
114. Björnsson A, Isaksson LA. Accumulation of an mRNA decay intermediate by ribosomal pausing at a stop codon. *Nucl Acids Res* 1996;**24**:1753–7.
115. Li X, Hirano R, Tagami H, Aiba H. Protein tagging at rare codons is caused by tmRNA action at the 3′ end of nonstop mRNA generated in response to ribosome stalling. *RNA* 2006;**12**:248–55.
116. McNicholas P, Salavati R, Oliver D. Dual regulation of *Escherichia coli* secA translation by distinct upstream elements. *J Mol Biol* 1997;**265**:128–41.
117. Nakatogawa H, Ito K. Secretion monitor, SecM, undergoes self-translation arrest in the cytosol. *Mol Cell* 2001;**7**:185–92.
118. Nakatogawa H, Ito K. The ribosomal exit tunnel functions as a discriminating gate. *Cell* 2002;**108**:629–36.
119. Collier J, Bohn C, Bouloc P. SsrA tagging of *Escherichia coli* SecM at its translation arrest sequence. *J Biol Chem* 2004;**279**:54193–201.
120. Li X, Yagi M, Morita T, Aiba H. Cleavage of mRNAs and role of tmRNA system under amino acid starvation in *Escherichia coli*. *Mol Microbiol* 2008;**68**:462–73.
121. Christensen SK, Pedersen K, Hansen FG, Gerdes K. Toxin-antitoxin loci as stress-response-elements: ChpAK/MazF and ChpBK cleave translated RNAs and are counteracted by tmRNA. *J Mol Biol* 2003;**332**:809–19.
122. Yusupov MM, Yusupova GZ, Baucom A, Lieberman K, Earnest TN, Cate JH, et al. Crystal structure of the ribosome at 5.5 A resolution. *Science* 2001;**292**:883–96.
123. Ivanova N, Pavlov MY, Felden B, Ehrenberg M. Ribosome rescue by tmRNA requires truncated mRNAs. *J Mol Biol* 2004;**338**:33–41.
124. Koo JT, Choe J, Moseley SL. HrpA, a DEAH-box RNA helicase, is involved in mRNA processing of a fimbrial operon in *Escherichia coli*. *Mol Microbiol* 2004;**52**:1813–26.
125. Tanner NK, Linder P. DExD/H box RNA helicases: From generic motors to specific dissociation functions. *Mol. Cell* 2001;**8**:251–62.
126. Silverman E, Edwalds-Gilbert G, Lin R-L. DExD/H-box proteins and their partners: Helping RNA helicases unwind. *Gene* 2003;**312**:1–16.
127. Butland G, Peregrin-Alvarez JM, Li J, Yang W, Yang X, Canadien V, et al. Interaction network containing conserved and essential protein complexes in *Escherichia coli*. *Nature* 2005;**433**:531–7.

128. Sandler P, Weisblum B. Erythromycin-induced ribosome stall in the ermA leader: A barricade to 5′-to-3′ nucleolytic cleavage of the ermA transcript. *J Bacteriol* 1989;**171**:6680–8.
129. Drider D, DiChiara JM, Wei J, Sharp JS, Bechhofer DH. Endonuclease cleavage of messenger RNA in *Bacillus subtilis*. *Mol Microbiol* 2002;**43**:1319–29.
130. Yao S, Blaustein JB, Bechhofer DH. Erythromycin-induced ribosome stalling and RNase J1-mediated mRNA processing in *Bacillus subtilis*. *Mol Microbiol* 2008;**69**(6), 1439–49.
131. Even S, Pellegrini O, Zig L, Labas V, Vinh J, Brechemmier-Baey D, et al. Ribonucleases J1 and J2: Two novel endoribonucleases in *B. subtilis* with functional homology to *E. coli* RNase E. *Nucleic Acids Res* 2005;**33**:2141–52.
132. Mathy N, Benard L, Pellegrini O, Daou R, Wen T, Condon C. 5′-to-3′ Exoribonuclease activity in bacteria: Role of RNase J1 in rRNA maturation and 5′ stability of mRNA. *Cell* 2007;**129**:681–92.
133. de la Sierra-Gallay IL, Zig L, Jamalli A, Putzer H. Structural insights into the dual activity of RNase J. *Nat Struct Mol Biol* 2008;**15**:206–12.
134. Agaisse H, Lereclus D. STAB-SD: A Shine–Dalgarno sequence in the 5′ untranslated region is a determinant of stability. *Mol Microbiol* 1996;**20**:633–43.
135. Doma MK, Parker R. Endonucleolytic cleavage of eukaryotic mRNAs with stalls in translation elongation. *Nature* 2006;**440**:561–4.
136. Onouchi H, Nagami Y, Haraguchi Y, Nakamoto M, Nishimura Y, Sakurai R, et al. Nascent peptide-mediated translation elongation arrest coupled with mRNA degradation in the CGS1 gene of Arabidopsis. *Genes Dev* 2005;**19**:1799–810.
137. Uzan M, Favre R, Brody E. A nuclease that cuts specifically in the ribosome binding site of some T4 mRNAs. *Proc Natl Acad Sci USA* 1988;**85**:8895–9.
138. Morse DE, Yanofsky C. Polarity and the degradation of mRNA. *Nature* 1969;**224**:329–31.
139. Goodrich AF, Steege DA. Roles of polyadenylation and nucleolytic cleavage in the filamentous phage mRNA processing and decay pathways in *Escherichia coli*. *RNA* 1999;**5**:972–85.
140. Deana A, Ehrlich R, Reiss C. Synonymous codon selection controls in vivo turnover and amount of mRNA in *Escherichia coli* bla and ompA genes. *J Bacteriol* 1996;**178**:2718–20.
141. Deana A, Ehrlich R, Reiss C. Silent mutations in the *Escherichia coli* ompA leader peptide region strongly affect transcription and translation *in vivo*. *Nucleic Acids Res* 1998;**26**:4778–82.
142. Boni IV, Artamonova VS, Dreyfus M. The last RNA-binding repeat of the *Escherichia coli* ribosomal protein S1 is specifically involved in autogenous control. *J Bacteriol* 2000;**182**:5872–9.
143. Coburn GA, Mackie GA. Degradation of mRNA in *Escherichia coli*: An old problem with some new twists. *Prog Nucleic Acid Res Mol Biol* 1999;**62**:55–108.
144. Regnier P, Arraiano CM. Degradation of mRNA in bacteria: Emergence of ubiquitous features. *Bioessays* 2000;**22**:235–44.
145. Kushner SR. mRNA decay in *Escherichia coli* comes out of age. *J Bacteriol* 2002;**184**:4658–65.
146. Carpousis AJ. The RNA degradosome of *Escherichia coli*: An mRNA-degrading machine assembled on RNase E. *Annu Rev Microbiol* 2007;**61**:71–87.
147. Carpousis AJ, Van Houwe G, Ehretsmann C, Krisch HM. Copurification of *E. coli* RNase E and PNPase: Evidence for a specific association between two enzymes important in RNA processing and degradation. *Cell* 1994;**76**:889–900.
148. Miczak A, Kaberdin VR, Wei CL, Lin-Chao S. Proteins associated with RNase E in a multicomponent ribonucleolytic complex. *Proc Natl Acad Sci USA* 1996;**93**:3865–9.
149. Py B, Higgins CF, Krish HM, Carpousis AJ. A DEAD-box RNA helicase in the *Escherichia coli* RNA degradosome. *Nature* 1996;**381**:169–72.

150. Bricker AL, Belasco JG. Importance of a 5′ stem-loop for longevity of papA mRNA in Eschericchia coli. *J Bacteriol* 1999;**181**:3587–90.
151. Klug G, Cohen SN. Combined actions of multiple hairpin loop structures and sites of rate-limiting endonucleolytic cleavage determine differential degrdation rates of individual segments within polycistronic puf operon mRNA. *J Bacteriol* 1990;**172**:5140–6.
152. Mackie G. Ribonuclease E is a 5′-end-dependent endonuclease. *Nature* 1998;**395**:720–3.
153. Jiang X, Belasco JG. Catalytic activation of multimeric RNase E and RNase G by 5′-monophosphorylated RNA. *Proc Natl Acad Sci USA* 2004;**101**:9211–6.
154. Callaghan AJ, Marcaida MJ, Stead JA, McDowall KJ, Scott WG, Luisi BF. Structure of Escherichia coli RNase E catalytic domain and implications for RNA turnover. *Nature* 2005;**437**:1187–91.
155. Celesnik H, Deana A, Belasco JG. Initiation of RNA decay in Escherichia coli by 5′ pyrophosphate removal. *Mol Cell* 2007;**27**:79–90.
156. Emory SA, Bouvet P, Belasco JG. A 5′-terminal stem-loop structure can stabilize mRNA in Escherichia coli. *Genes Dev* 1992;**6**:135–48.
157. Diwa A, Bricker AL, Jain C, Belasco JG. An evolutionary conserved RNA stem-loop functions as a sensor that directs feedback regulation of RNase E gene expression. *Genes Dev* 2000;**14**:1249–60.
158. Marujo PE, Braun F, Haugel-Nielsen J, Le Derout J, Arraiano CM, Regnier P. Inactivation of the decay pathway initiated at an internal site by RNase E promotes poly(A)-dependent degradation of the rpsO mRNA in Escherichia coli. *Mol Microbiol* 2003;**50**:1283–94.
159. Mackie GA. Stabilization of the 3′ one third of Escherichia coli ribosomal protein S20 mRNA in mutants lacking polynucleotide phosphorylase. *J Bacteriol* 1989;**171**:4112–20.
160. Rapaport LR, Mackie GA. Influence of translational efficiency on the stability of the mRNA for ribosomal protein S20 in Escherichia coli. *J Bacteriol* 1994;**176**:992–8.
161. Klug G, Cohen SN. Effects of translation on degradation of mRNA segments transcribed from the polycistronic puf operon of Rhodobacter capsulatus. *J Bacteriol* 1991;**173**:1478–84.
162. Jäger S, Fuhrmann O, Heck C, Hebermehl M, Schiltz E, Rauhut R, et al. An mRNA degrading complex in Rhodobacter capsulatus. *Nucl Acids Res* 2001;**29**:4581–8.
163. Fritsch J, Rothfuchs R, Rauhut R, Klug G. Indentification of an mRNA element promoting rate-limiting cleavage of the polycistronic puf mRNA in Rhodobacter capsulatus by an enzyme similar to RNase E. *Mol Microbiol* 1995;**15**:1017–29.
164. Ow MC, Liu Q, Kushner SR. Analysis of mRNA decay and rRNA processing in Escherichia coli in the absence of RNase E-based degradosome assembly. *Mol Microbiol* 2000;**38**:854–66.
165. Jiang X, Diwa A, Belasco JG. Regions of RNase E important for 5′-end dependent RNA cleavage and autoregulated synthesis. *J Bacteriol* 2000;**182**:2468–75.
166. Marchand I, Nicholson AW, Dreyfus M. Bacteriophage T7 protein kinase phosphorylates RNase E and stabilizes mRNAs synthesized by T7 RNA polymerase. *Mol Microbiol* 2001;**42**:767–76.
167. Leroy A, Vanzo NF, Sousa S, Dreyfus M, Carpousis AJ. Function in Escherichia coli of the non-catalytic part of RNaseE: Role in the degradation of ribosome-free mRNA. *Mol Microbiol* 2002;**45**:1231–43.
168. Morse DE, Mosteller R, Baker RF, Yanofsky C. Direction of in vivo degradation of tryptophan messenger RNA – a correction. *Nature* 1969;**223**:40–3.
169. Kennell DE. The instability of messenger RNA in bacteria. In: W Reznikoff and L Gold, editors. *Maximizing gene expression*. Boston: Butterworth; 1986. P. 101–42.
170. Régnier P, Hajnsdorf E. Decay of mRNA encoding ribosomal protein S15 of Escherichia coli is initiated by an RNase E-dependent endonucleolytic cleavage that removes the 3′ stabilizing stem and loop structure. *J Mol Biol* 1991;**217**:283–92.

171. Hankins JS, Zappavigna C, Prud'homme-Genereux A, Mackie GA. Role of RNA structure and susceptibility to RNase E in regulation of a cold shock mRNA, cspA mRNA. *J Bacteriol* 2007;**189**:4353–8.
172. de Smit MH, van Duin J. Secondary structure of the ribosome binding site determines translational efficiency: A quantitative analysis. *Proc Natl Acad Sci USA* 1990;**87**:7668–72.
173. Ganoza MC, Kofoid EC, Marlière P, Louis BG. Potential secondary structure at translation-initiation sites. *Nucl Acids Res* 1987;**15**:345–60.
174. Bechhofer DH, Dubnau D. Induced mRNA stability in *B. subtilis*. *Proc Natl Acad Sci USA* 1987;**84**:498–502.
175. Hambraeus G, Karhumaa K, Rutberg B. A 5′ stem-loop and ribosome binding but not translation are important for the stability of *Bacillus subtilis* aprE leader mRNA. *Microbiology* 2002;**148**:1795–803.
176. Sharp JS, Bechhofer DH. Effect of translational signals on mRNA decay in *Bacillus subtilis*. *J Bacteriol* 2003;**185**:5372–9.
177. Sharp JS, Bechhofer DH. Effect of 5′-proximal elements on decay of a model mRNA in *Bacillus subtilis*. *Mol Microbiol* 2005;**57**:484–95.
178. DiMari JF, Bechhofer DH. Initiation of mRNA decay in *Bacillus subtilis*. *Mol Microbiol* 1993;**7**:705–17.
179. Hue KK, Cohen SD, Bechhofer DH. A polypurine sequence that acts as a 5′ mRNA stabilizer in *Bacillus subtilis*. *J Bacteriol* 1995;**177**:3465–71.
180. Bechhofer DH, Zen KH. Mechanism of erythromycin-induced *ermC* mRNA stability in *Bacillus subtilis*. *J Bacteriol* 1989;**171**:5803–11.
181. Condon C, Putzer H. The phylogenetic distribution of bacterial ribonucleases. *Nucleic Acids Res* 2002;**30**:5339–46.
182. Wang W, Bechhofer DH. Properties of a *Bacillus subtilis* polynucleotide phosphorylase deletion strain. *J Bacteriol* 1996;**178**:2375–82.
183. Bechhofer DH, Wang W. Decay of ermC mRNA in a polynucleotide phosphorylase mutant of *Bacillus subtilis*. *J Bacteriol* 1998;**180**:5968–77.

mRNA Interferases, Sequence-Specific Endoribonucleases from the Toxin–Antitoxin Systems

YOSHIHIRO YAMAGUCHI AND
MASAYORI INOUYE

Department of Biochemistry, Robert Wood Johnson Medical School, 675 Hoes Lane, Piscataway, New Jersey 08854

I. Introduction	468
A. Toxin–Antitoxin (TA) Systems in Bacteria	468
II. MazF: An mRNA Interferase	471
A. The MazE–MazF System	471
B. Determination of The Target of MazF	471
C. Sequence Specificity of MazF	473
D. *In Vitro* Studies of Cleavage Specificities of MazF Using Synthesized RNAs	475
E. In-Depth Analysis of MazF Cleavage Products	476
F. Effect of MazF Induction in Eukaryotic Cells	479
G. MazE and MazF Structure and Function in *E. coli*	479
III. MazF Homologues and Other mRNA Interferases	480
A. MazF Homologues from *Mycobacterium tuberculosis*	480
B. *Myxococcus xanthus* MazF	484
C. *Staphylococcus aureus* MazF (MazFsa)	486
D. *E. coli* PemK	486
E. *E. coli* ChpBK	488
IV. Other Toxins in TA Systems	489
A. *E. coli* RelE	489
B. *E. coli* YoeB	489
C. *E. coli* YhaV	490
D. *Bacillus subtilis* EndoA	490
V. The Regulation of MazF Expression Under Stress Conditions	491
A. Growth Arrest and Cell Death in *E. coli*	491
B. MazF-Mediated Programmed Cell Death During Multicellular Development in *Myxococcus xanthus*	492
VI. Concluding Remarks	494
References	496

Escherichia coli contains a large number of suicide or toxin genes, whose expression leads to cell growth arrest and eventual cell death. One such toxin, MazF, is an ACA-specific endoribonuclease, termed "mRNA interferase." *E. coli* contains other mRNA interferases with different sequence specificities, which are considered to play important roles in growth regulation under stress conditions, and also in eliminating stress-damaged cells from a population. Recently, MazF homologues with 5-base recognition sequences have been identified, for example, those from *Mycobacterium tuberculosis*. These sequences are significantly underrepresented in the genes for protein families playing a role in the immunity and pathogenesis of *M. tuberculosis*. An mRNA interferase in *Myxococcus xanthus* is essential for programmed cell death during fruiting body formation. We propose that mRNA interferases play roles not only in cell growth regulation and programmed cell death, but also in regulation of specific gene expression (either positively or negatively) in bacteria.

I. Introduction
A. Toxin–Antitoxin (TA) Systems in Bacteria

Amazingly, all free-living bacteria examined so far contain a number of suicide or toxin genes in their genomes (1). The toxins produced from these genes are not aimed to kill other bacteria in their habitats or to kill animal cells in the process of infection, but are produced intracellularly and are toxic to their host cells, causing cell growth arrest and eventual cell death or suicide. Recent developments in this new field of bacterial intrinsic toxins have provided intriguing insights into the role of these toxins in bacterial physiology, persistence in multidrug resistance, bacterial pathogenicity, biofilm formation, and bacterial evolution. Therefore, the study of these toxins has very important implications in medical sciences, especially for treating infectious diseases. Furthermore, it also has become evident that these toxins have great potential in biotechnology, not only because one may develop novel antibiotics which induce the toxin genes or enhance their cytotoxicity, but also because they may be used as a novel tool for developing innovative biotechnological methods such as the single-protein production system (2, 3). Since most of these toxins are cotranscribed with their cognate antitoxins in an operon (thus termed as toxin–antitoxin or TA operons), and they form a stable complex in the cell, their toxicity is not exerted under normal growth conditions (4–6). However, the stability of antitoxins is substantially lower than that of their cognate toxins, and any stress causing cellular damage or growth inhibition-inducing proteases affects the balance between toxin and antitoxin, leading to toxin release in the cell. Although much debated, it is most reasonable to consider that TA toxins function in two different ways depending upon the nature of stress. One way is to regulate growth rate by inhibiting a specific cellular function, such as DNA replication or protein synthesis. Under extensive stresses, which release most of toxins from their cognate antitoxins, cell growth may be completely arrested.

This role in growth regulation is likely to be the primary mission of TA toxins. However, the second role of TA toxins is suicidal, that is, to kill their own host cells. Furthermore, under certain conditions, TA toxins may eliminate highly damaged cells (e.g., through DNA damage or by phage infection) to maintain a healthy population. It is possible that more than one TA system is involved in these processes. Some bacteria including *Escherichia coli* contain a large number of TA systems, suggesting that there may exist a network of TA systems that operates under various stress and physiological conditions.

The TA operons are often located on plasmids, and thus play a role in killing those cells which have lost the plasmids after cell division, a phenomenon called postsegregational killing (7–9). Therefore, TA toxins have been proposed to be primarily bacteriostatic and not bacteriocidal (6). This definitely appears to be applicable for many toxins, since cell growth may be recovered after short expose to a toxin. However, prolonged induction of the same toxin may cause cells to reach a "point of no return," resulting in death (10). This is conceptually very important for defining the function of bacterial toxins. As we will discuss in a latter section, toxins appear to function both ways, as in some bacteria they are primarily used for growth regulation and in others for programmed cell death.

To date, a number of TA modules have been studied in some detailÑthe bacteriophage-encoded *phd–doc* (11) and plasmid-encoded *kis–kid* (12), *pemI–pemK* (13), and *ccdA–ccdB* (14) modules. Six TA systems have been reported on the *E. coli* genome: *relB–relE* (15, 16), *chpBI–chpBK* (17), *mazE–mazF* (18–20), *yefM–yoeB* (21), *dinJ–yafQ* (22), and *hipB–hipA*. The *hipB–hipA* module has been implicated in persistence in multidrug resistance (23, 24). In addition to these six TA systems on the *E. coli* chromosome, there seems to be at least seven more TA systems, *yeeU–yeeV*, *yafW–ykfI*, *yfjZ–ypjF*, *yafN–yafO*, *yfjG–yfjF*, *ydgE–ydgF*, and *pspB–pspC*, which have to be more carefully characterized to identify their toxic function (25). Recently, three new TA systems *hicA–hicB* (26), *prlF–yhaV* (27), and *ybaJ–hha* (28) (Table I) were reported. Interestingly, all of the TA operons appear to use similar modes of regulation–autoregulation by the antitoxins and the formation of complexes between antitoxins and their cognate toxins to neutralize the toxin activity. Furthermore, guanosine 3′,5′-bispyrophosphate (ppGpp), which is known to be produced during various stresses, has been proposed to play an important role in induction of the TA operons (6), although the direct involvement of ppGpp in the regulation of the TA operons is a subject of debate (2) (also see Section II). The cellular targets of some toxins have been studied: CcdB directly interacts with gyrase A and blocks DNA replication (29, 30); RelE, which by itself has no endoribonuclease activity, appears to act as a ribosome-associating factor that promotes mRNA cleavage at the ribosome A-site (31, 32). This cleavage activity was suggested to be due to the intrinsic

TABLE I
TA Systems Present on Chromosomal DNA in *E. coli*

Function of toxin	Operon	Antitoxin	Toxin
RNA interferase	*mazEF*	MazE	MazF
	chpBIK	ChpBI	ChpBK
	prlFyhaV	PrlF	YhaV
	dinJyafQ	DinJ	YafQ
Inhibition of translation	*relBE*	RelB	RelE
	yefMyoeB	YefM	YoeB
Unknown	*hipBA*	HipB	HipA
	yeeUV	YeeU	YeeV
	yafWykfI	YafW	YkfI
	yfjZypjF	YfjZ	YpjF
	hicAB	HicA	HicB
	yafNO	YafN	YafO
	yfjGF	YfjG	YfjF
	ybaJhha	YbaJ	Hha
Unknown (membrane protein)	*ydgEF*	YdgE	YdgF
	pspBC	PspB	PspC

endoribonuclease activity of ribosomes, which is enhanced in the presence of RelE (32). Finally, PemK (13), ChpBK (17), and MazF (33) are unique among toxins, since they target cellular mRNAs for degradation by functioning as sequence-specific endoribonucleases to effectively inhibit protein synthesis and thereby cell growth. In addition to these TA systems described above, there are a few other TA systems that are not present in *E. coli*: *vapB–vapC* (*vapC* is the toxin) (34), *higB–higA* (*higB* is the toxin) (35), and o—ζ (ζ is the toxin) (36). The *vapB–vapC* system is widespread in both Gram-positive and Gram-negative bacteria, and surprisingly *Mycobacterium tuberculosis* contains 23 *vapB–vapC* operons on its genome (1). Since its nonpathogenic fast-growing counterpart, *Mycobacterium smegmatis* has no *vapB–vapC* operons and only one *mazE–mazF* operon in contrast to seven *mazE–mazF* operons in *M. tuberculosis*, it has been speculated that these TA operons may play important roles in the pathogenicity and, particularly, in the extremely long dormancy in human tissues of this devastating pathogen. It is also noteworthy that various TA systems are found in a large number of Archaea, but not in eukaryotes.

MazF, ChpBK, and PemK have now been characterized as sequence-specific endoribonucleases for single-stranded RNA. They are quite distinct from other known endoribonucleases such as RNases E, A, and T1, because

MazF, ChpBK, and PemK function as general protein synthesis inhibitors by interfering with the function of cellular mRNAs. It is well known that the small RNAs, such as micRNA (mRNA-interfering complementary RNA) (37), miRNA (38), and siRNA (39), interfere with the function of specific target RNAs. These small RNAs bind to specific mRNAs to inhibit their expression. Ribozymes also act on their target RNAs specifically and interfere with their function (40). Clearly, MazF, ChpBK, and PemK homologues form a novel endoribonuclease family with a new mRNA-interfering mechanism by cleaving mRNAs at specific sequences. They are thus termed "mRNA interferases" (2).

II. MazF: An mRNA Interferase

A. The MazE–MazF System

In the MazE–MazF system (18, 19, 41, 42), the MazF toxin is stable while the MazE antitoxin/antidote is labile (Fig. 1). The short half-life of MazE is due to its degradation by an ATP-dependent serine protease ClpAP (41). The operon is negatively autoregulated by MazE and a MazE–MazF complex (19, 42). Because of its lability, MazE has to be constantly synthesized in normally growing cells to neutralize MazF toxicity. Therefore, the TA systems are sometime called addiction modules, as in the case of the MazE–MazF system, where the cells are "addicted" to MazE. The proposed regulation of mazE–mazF by ppGpp (41) has been much disputed and it seems likely that ppGpp does not directly regulate mazE–mazF transcription, but indirectly regulates the activation of MazF (e.g., through Lon protease) (6). MazE–MazF-mediated cell growth arrest occurs when transcription of the TA module and/or translation of the mazE–mazF mRNA are inhibited. This is because of MazE's instability compared to MazF in the cell, freeing MazF from its complex with MazE as depicted in Fig. 1. Thus, MazF activation occurs by severe amino acid or thymine starvation (43), certain antibiotics such as rifampicin and chloramphenicol (44), the toxic protein Doc (45), or other stress conditions such as high temperature, oxidative stress, or DNA damage (46). Interestingly, the mazE gene was named from the Hebrew ma-ze, meaning "what is it?." In the following sections, we will present a historical overview of the experiments that helped characterize this and other TA systems.

B. Determination of The Target of MazF

To identify the cellular target of MazF, the mazF gene was cloned into an arabinose-inducible pBAD plasmid (47). E. coli BW25113 (48) carrying pBAD–MazF could not grow on a glycerol–M9 plate in the presence of arabinose. Using a cell-free system prepared from E. coli BW25113 cells

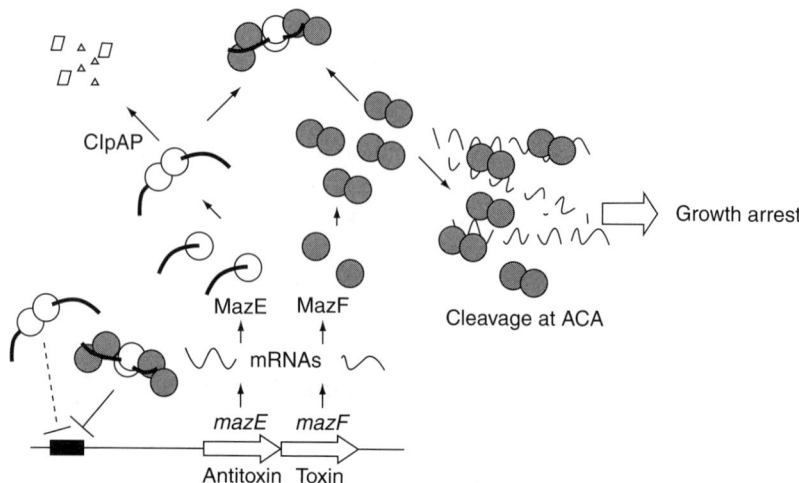

FIG. 1. Regulation of the *mazE–mazF* operon, the structure of the MazE–MazF complex and the cleavage specificity of the MazF mRNA interferase. MazE is represented by white circles with black tail, to represent the C-terminal unstructured region that binds to the MazF dimer (gray circles) to inhibit its endoribonuclease activity. MazE and MazF mRNAs are synthesized from the same operon; one MazE dimer can bind to two MazF dimers to inhibit MazF endoribonuclease activity and to autoregulate the TA module. MazE dimers are subjected to cleavage under stress conditions by ClpAP. MazE dimers can also autoregulate the addiction module, but more weakly than the MazE–MazF heterohexamer complex. MazF dimers, when not bound by MazE, are free to cleave mRNAs. This MazF endoribonuclease activity leads to bacterial cell growth arrest and eventual cell death.

carrying pBAD–MazF permeabilized by toluene treatment (49, 50), ATP-dependent [^{35}S]methionine incorporation was completely inhibited when cells were preincubated for 10 min in the presence of arabinose before toluene treatment. However, the incorporation of [α-^{32}P]dTTP and [α-^{32}P]UTP was not significantly affected under similar conditions (32), demonstrating that MazF inhibits protein synthesis, but not DNA replication or RNA synthesis. From the analysis of the polysome pattern of *E. coli* BW25113 cells carrying the pBAD–MazF plasmid, it was found that MazF disrupts polysomes, either by inhibiting translation initiation or by degrading mRNAs. Furthermore, purified MazF(His)$_6$ inhibited protein synthesis in an *E. coli* cell-free RNA/protein synthesis system (51). Importantly, the addition of the antitoxin (His)$_6$MazE rescued protein synthesis in a dose-dependent manner. MazF(His)$_6$ was also able to inhibit eukaryotic cell-free protein synthesis, which was also recovered upon coaddition of (His)$_6$MazE. Subsequently, toeprinting (TP) analysis was carried out with 70S ribosomes and the *mazG* mRNA in the presence of MazF (His)$_6$ (52). When increasing amounts of MazF(His)$_6$ were added together with

70S ribosomes, a new band called TP(F) appeared, with concomitant reduction of the normal band due to ribosome binding to the Shine–Dalgarno (SD) sequence [TP(r)]. The TP(F) band was mapped to the region between the SD sequence and the initiation codon. Surprisingly, it was found that the TP(F) band could be detected even in the absence of 70S ribosomes, indicating that MazF is able to bind to the mRNA independent of 70S ribosomes, or alternatively that MazF is an endoribonuclease which cleaves between A- and C-residues located between the SD sequence and the initiation codon.

To further determine whether MazF(His)$_6$ indeed cleaved the *mazG* mRNA, the *mazG* mRNA was incubated with MazF(His)$_6$, phenol extracted to remove protein, and assayed by primer extension (*19*). The TP(F) band was observed even after phenol extraction, indicating that MazF(His)$_6$ indeed cleaved the *mazG* mRNA directly. Cleavage of the *mazG* mRNA was blocked when (His)$_6$MazE was added together with MazF(His)$_6$, indicating that MazE exerts its antitoxic effect by blocking the endoribonuclease activity of MazF. It is important to note that the ability of MazF to cleave mRNA in the absence of ribosomes is distinctly different from the function of RelE, whose toxic function requires ribosomes (*15, 31*).

C. Sequence Specificity of MazF

To determine the recognition sequence for MazF, all possible single-base mutations in the UACAU sequence around the MazF cleavage site were examined. It was found that both U-residues at the 5′- and 3′-ends of this motif could be replaced with any other residue (G, A, and C), while any point mutations in the ACA sequence (GCA, CCA, UCA, AGA, AUA, AAA, ACG, ACC, and ACU) made the mRNA resistant to MazF cleavage, demonstrating that the ACA sequence is absolutely required for MazF activity.

To further characterize the specificity of the MazF endoribonuclease activity, two other mRNAs, *lacZ* and *yeeW*, were also examined for the presence of MazF cleavage sites. By comparing the cleavage sites in the *mazG*, *lacZ*, and *yeeW* mRNAs, it became evident that an ACA sequence was indeed present in all of the cleavage sites. Except for the site in *lacZ* mRNA, the mRNAs were cleaved between the first A- and the second C-residues in the ACA sequence, suggesting that MazF is an endoribonuclease that specifically recognizes the ACA sequence and preferentially cleaves 3′ to the first A-residue in ACA. In the case of the *lacZ* mRNA, the cleavage occurred 5′ to the first A-residue in an ACA sequence. The reason for this alternative cleavage site is not well understood at present but is discussed further below.

On the basis of the above observations, Northern blot analysis of total cellular RNA extracted at different time intervals after arabinose induction of MazF was carried out. Among three genes tested, the *ompA* mRNA (*53*) was observed only at the zero time point, and the *secE* mRNA was detected only up to 2.5 min after induction. On the other hand, the *lpp* mRNA (54) was found to be relatively stable. Its band intensity was significantly reduced at 10 min, but was still detectable 20 min after MazF induction. It appears that the observed differences in the stabilities of these mRNAs correlated at least in part with the total number of the ACA sequences present in the mRNA and with the mRNA length. The earlier observation that the synthesis of larger proteins was more sensitive to MazF toxicity is probably due to length-dependent susceptibility of their mRNAs to MazF. The *lpp* mRNA is among the shortest mRNAs in *E. coli* consisting of 322 bases with only four ACA motifs (54), while the *ompA* mRNA consists of 1229 bases with 24 ACA motifs (53). In contrast to the *lpp* and *ompA* genes, the *secE* gene is a member of the essential *secE*–*nusG* operon and cotranscribed with the *nusG* gene (55). The *secE* plus *nusG* ORFs (927 bp) contain a total of seven ACA sequences. Importantly, 16S and 23S rRNA were found to be very stable *in vivo* during the 20-min period of MazF induction, as no significant change was observed in their band intensities, indicating that rRNAs are protected from MazF cleavage *in vivo* even if they contain a large number of ACA sequences. This is likely due to the protection of rRNAs by ribosomal proteins from MazF cleavage. Additional protection is conferred by the extensive secondary structures present in these rRNAs, which are resistant to the single-strand-specific MazF (see below).

In vivo cleavage of cellular mRNAs at ACA sequences upon induction of MazF was confirmed by primer extension analysis of the *mazG* mRNA using total RNA extracted at different time points after MazF induction (20). The same ACA triplet in the *mazG* mRNA identified in the *in vitro* toeprinting experiment described above was tested for the *in vivo* cleavage. A distinct band appeared 5 min after MazF induction, whose intensity further increased at 10 and 20 min. This band was hardly detected at 0 and 2.5 min. Importantly, the cleavage occurred between the first A- and second C-residues in the ACA sequence as in the *in vitro* experiment, indicating that MazF cleaves the same site both *in vivo* and *in vitro*. The ACA sequence is formed between two adjacent codons, AAC for Asn107 and AGU for Ser108, indicating that MazF is able to cleave mRNA at ACA sequences *in vivo* regardless of their reading frames. In addition, the region upstream of the ACA site above was examined using a different primer, and two new ACA cleavage sites were found; one was in the coding frame, while the other was out of frame.

A 30-base RNA, with the ACA sequence in the center, was synthesized to identify the region around the MazF cleavage site necessary for cleavage of *mazG* mRNA. The synthetic RNA was completely cleaved, resulting in a shorter distinct fragment. When the RNA substrate was annealed with an antisense DNA before the addition of MazF(His)$_6$, the RNA cleavage was inhibited. A similar result was obtained when the substrate RNA was hybridized with an antisense RNA, indicating that MazF cannot cleave the

ACA sequence in RNA/DNA or RNA/RNA duplexes. It was also found that MazF cannot cleave single-stranded DNA with the same base sequence as the RNA substrate. Thus, it was concluded that MazF is an endoribonuclease highly specific for single-stranded RNA (33).

Different factors may contribute to the resistance of a particular mRNA against cleavage by MazF, as seen in the case of *lpp* mRNA. Despite its short length, the *lpp* mRNA is relatively stable during MazF induction. Interestingly, three out of four ACA sequences are located in the proposed secondary structures of the *lpp* mRNA (54). The remaining ACA sequence is located very close to the 5'-end of the transcript so that only a 4-base fragment is removed from the 5'-end. The resistance of the *lpp* mRNA to MazF-mediated degradation may also be related to the efficiency of translation initiation, as an mRNA having a better translation initiation efficiency is likely to be more protected by ribosomes bound to the mRNA. The observed resistance of the *lpp* mRNA to MazF degradation may therefore also be due to its high translation efficiency, as the *lpp* product, lipoprotein, is the most abundant protein in *E. coli*. Although computer analysis of the *E. coli* genome reveals that *E. coli* genes contain on average 11 ACA sequences per gene, there are 51 genes of the *E. coli* genome which do not contain ACA sequences. It remains to be determined whether these genes play important roles in cells producing MazF. It is interesting to note that MazF itself contains several ACA sequences, a total of 9 within the *mazF* ORF (only 333 bases long), among which four ACA sequences are clustered in the middle of the ORF, while the *mazE* ORF (249 bases long) contains only two ACA motifs. Indeed, the *mazF* mRNA can be cleaved by MazF *in vitro* (33), suggesting that *mazF* expression may be negatively autoregulated by its own gene product. It is notable that when the ACA sequences in the *mazF* ORF were replaced with MazF-resistant sequences (without changing the amino acid sequence of MazF), the MazF toxicity increased significantly (Suzuki M. and Inouye M., unpublished data).

D. *In Vitro* Studies of Cleavage Specificities of MazF Using Synthesized RNAs

Using a synthesized 15-base RNA, it was shown that the hydrolysis reaction mediated by MazF can occur at either the 5'- or 3'-phosphodiester linkage of the first A-residue in the ACA sequence, yielding a free 5'-OH group on the 3'-cleavage product and thus a 3'-phosphate (or a 2',3'-cyclic phosphate) on the 5'-cleavage product. These data were further substantiated by MALDI-mass spectrometric analysis of the MazF-digested 15-mer RNA. MALDI-mass spectrometric analysis after MazF digestion of shorter (13, 11, and 7 base) RNAs revealed that (i) each was cleaved 5' to the first A-residue of the ACA sequence, (ii) the resulting 5'-products had almost exclusively a 2',3'-cyclic phosphate at their 3′-end, and (iii) all of the resulting 3′-products contained a free 5′-OH group. These results clearly indicate that MazF is an endoribonuclease that cleaves a phosphodiester linkage at the 5′-side of a phosphodiester bond. Using a 12-base RNA, it has shown that the identity of the residue 5′ to the ACA sequence is not important for hydrolysis by MazF.

All of the 5′-cleavage products contain a 2′,3′-cyclic phosphate at their 3′-end, which suggests that the 2′-OH group at the cleavage site plays an essential role for the hydrolysis reaction as in the case of ribonuclease A. This was confirmed using RNAs in which the 2′-OH group was modified with a methyl group at various residues.

E. In-Depth Analysis of MazF Cleavage Products

Two 5′-cleavage products were observed in all cases where the substrates were hydrolyzed by MazF; one contains a 2′,3′-cyclic phosphate at the 3′-end and the other a 3′-phosphate. The ratio of 3′-phosphate to 2′,3′-cyclic phosphate product increased when the reaction was performed with higher concentrations of MazF, suggesting that MazF cleavage reaction of a phosphodiester linkage may occur in a manner similar to ribonuclease A. The hydrolysis reaction by ribonuclease A occurs in two discrete steps: in the first step, the nucleophilicity of the ribose 2′-OH plays a key role to form a 2′,3′-cyclic phosphate, and in the second step, the cyclic intermediate is resolved to form a 3′-phosphate. It was observed that the 2′-OH group at the cleavage site is the only requirement for MazF, that is, it can cleave a 13-mer single-stranded DNA oligo if a single ribonucleotide (rU) is added 5′ to the dAdCdA sequence (Fig. 2). Importantly, the 3′-products of this chimeric substrate from both ribonuclease A and MazF digestions had identical masses, consistent with their bearing the same 5′-OH group (20).

When the upstream MazF product of the 5′-end labeled 13-mer chimeric substrate was treated either with T4 polynucleotide kinase at pH 6.0 [known to function as 3′-phosphatase under these conditions (56)] or with ribonuclease A, the 2′,3′-cyclic phosphate product was converted to 3′-phosphate in both cases. These results provide additional evidence that MazF cleavage of a phosphodiester linkage yields primarily a 2′,3′-cyclic phosphate on one side and a 5′-OH group on the other side. MazF was still able to effectively cleave the chimeric 13-mer substrate when the rU-residue was replaced with rG. The products from MazF treatment migrated at the identical position to the alkaline hydrolysis and to ribonuclease T1 digest. The formation of a 2′,3′-cyclic phosphate at the rG-residue and a 5′-OH group at the dA-residue of the cleavage site was confirmed by MALDI-mass spectrometry. These data indicate that MazF enzymatically functions in a manner similar to ribonuclease A for the cleavage of RNA, where the 2′-OH group at the cleavage site plays a key role in the reaction. One of histidine residues in the active site of ribonuclease A enhances

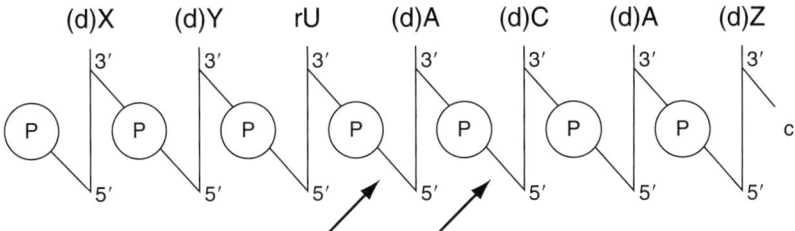

FIG. 2. Cleavage specificity of the MazF mRNA interferase (20). Arrows indicate the cleavage sites, resulting in the formation of a 3'-phosphate group on the upstream fragment and a 5'-OH group on the downstream fragment. rU can be any residue, but cannot be replaced with a deoxyribonucleotide residue, indicating that the 2'-OH at this position is absolutely required for MazF cleavage. All the other residues may be deoxyribonucleotides. For cleavage on the 3'-side of the A-residue, that is, between the A- and C-residues, the A-residue has to be a ribonucleotide.

the nucleophilicity of the ribose 2'-OH group of the U-residue at the cleavage site, resulting in the formation of a 2',3'-cyclic phosphate. It will be useful to determine which residues in the structure of the MazF dimer (18) play a role in the reaction. Although MazF does not contain a conserved histidine residue, it does contain highly conserved basic residues, which may perform nucleophilic attack on the 2'-OH group.

Curiously, the MazF cleavage site within an RNA sequence derived from the *mazG* mRNA was shifted by 1-base upstream when RNA was shortened from 30 to 15 bases. A similar shift in cleavage site was observed in an ACA sequence derived from the *era* mRNA (bases 236–238) when the cleavage conditions were altered. *In vivo* primer extension experiments revealed that the *era*-derived ACA sequence was cleaved between the first A- and C-residue. This result was also confirmed by an *in vitro* primer extension experiment using the full-length (~1 kb) *era* mRNA. However, when only a 98-base portion of the *era* mRNA was digested with MazF in an *in vitro* reaction, cleavage occurred at both the 5'- and 3'-sides of the first A-residue of the ACA sequence. The reason for this cleavage site shifting is unclear, but might indicate that an amino acid residue at the active site of MazF involved in the nucleophilic attack of the 2'-OH group of the cleavage site residue is flexible, enabling the RNA–MazF–dimer complex to attack the 2'-OH group on either side of the first adenine of the ACA sequence.

Although it seems to be evident that the MazF recognition sequence is ACA on the basis of the results described above, a report has been published claiming that the MazF recognition triplet is NAC (N is preferentially A and U) rather than ACA (57). In this report, the authors used three short RNAs synthesized by T7 RNA polymerase, corresponding to the HIV-1 TAR RNA, the CopT RNA of plasmid R1, and its antisense, the CopA RNA. From the analysis of MazF-cleaved products by primer extension experiments, the authors concluded that MazF cleaved these RNAs between N and A of the NAC sequences. The authors speculated that the difference between their results and the results from our laboratory may be due to His-tag, as they used non-His-tagged MazF. However, this seems unlikely, since identical results were observed in in vitro (with use of $(His)_6MazF$) and in vivo (with non-His-tagged MazF) in our experiments (20, 33). It seems likely that these authors were detecting minor cleavage sites with these three small RNAs in vitro. Notably, among the three RNAs, there was only one site (in the CopT RNA) that was cleaved with the least amount of MazF used in their experiment and that one contained an ACA sequence. At this concentration, none of the other cleavage sites were detected.

That MazF cannot efficiently cleave triplet sequences other than ACA was further shown with use of pentaoligonucleotides: the $U1A2C3A4U5$ sequence was mutated to all possible bases at each position (20). Any base change of the ACA sequence (A2, C3, and A4, respectively) completely blocked cleavage of the RNA by MazF, whereas base changes at residues U1 and U5 did not block the RNA cleavage. These results strongly support that the MazF recognition sequence is ACA rather than UAC.

The ultimate confirmation that MazF cleaves mRNA specifically at ACA and not NAC motifs (57) came from an examination of whether a protein could still be produced in MazF-expressing cells from an mRNA where all ACA sequences were removed, leaving the NAC sequences unchanged. It was shown that when an mRNA was engineered to remove all the ACA sequences without altering the amino acid sequence of the protein encoded, the protein was still produced in the presence of MazF, to the exclusion of all the other cellular proteins (3). This system allows the production of only a single protein of interest in living E. coli cells and thus is termed the *single-protein production*, or SPP system. In this report, one human protein, two yeast proteins and two E. coli proteins were shown to be produced exclusively, by simply altering all the ACA sequences in their mRNAs to MazF-resistant sequences. When the ACA sequences were not altered, a significant reduction in protein production was observed. Most importantly, all these mRNAs still contain a large number of NAC sequences (e.g., the ACA-less *envZ* mRNA still contains as many as 23 NAC sequences). These results unambiguously demonstrate that the primary recognition sequence of MazF is ACA.

Gerdes and coworkers (58) reported that MazF inhibits translation by a ribosome- and codon-dependent mechanism very similar to RelE. However, the study presented earlier (33) shows that MazF cleaves mRNAs in a totally codon-independent manner, since MazF cleaves mRNAs at the same site both *in vivo* and *in vitro*. Furthermore, small synthetic RNAs, even as short as 7 bases, are able to serve as MazF substrates (20). Although not pointed out by

Christensen *et al.* (58), a distinct cleavage of the *lpp* mRNA by MazF very close to its 5′-end was evident in their report. On the basis of the sequencing ladder provided, it clearly corresponded to the ACA sequence in the 5′-untranslated region (54). Taken together, MazF functions as a ribosome-independent endoribonuclease that cleaves mRNA at ACA sequences.

F. Effect of MazF Induction in Eukaryotic Cells

As one can imagine, MazF functions as an mRNA interferase not only in *E. coli* but also in yeast and mammalian cells. It has been demonstrated that MazF induces striking degradation of cellular mRNAs and inhibition of protein synthesis leading to apoptosis in mammalian cells (59). MazF expression in mammalian cells caused caspase-3 activation and poly(ADP-ribose) polymerase cleavage, which are hallmarks of apoptotic cell death, all of which were blocked by the antitoxin MazE. A critical point in apoptosis regulation is controlled by members of the BCL-2 family. The BCL-2 family of proteins can be divided into three different subclasses based on conservation of BCL-2 homology (BH1–4) domains: multidomain antiapoptotic proteins (BCL-2, BCL-X_L, MCL-1, BCL-W, and Bfl-1/A1), multidomain proapoptotic proteins (BAX and BAK), and BH3-only proapoptotic proteins (BID, BAD, BIM, PUMA, NOXA, and NBK/BIK). Interestingly, expression of MazF in immortalized baby mouse kidney (iBMK) cells deficient for *bax* and/or *bak*, or BH3-only proapoptotic genes (*puma*, *bim*, *noxa*, and *nbk/bik*), revealed that NBK/BIK and BAK are required for MazF-induced apoptosis. In addition, NBK/BIK bound both MCL-1 and BCL-X_L, thereby promoting the release of BAK from the MCL-1 and BCL-X_L inhibitory complexes in response to MazF induction. Moreover, BAX and BAK, BAK, or NBK/BIK deficiency confer resistance to cell death induced by pharmacological inhibition of protein synthesis and by shutoff of protein synthesis induced by viral infection. This result raises an intriguing possibility that MazF or any other mRNA interferase may be used as a tool for gene therapy dealing with cancer and AIDS.

G. MazE and MazF Structure and Function in *E. coli*

The crystal structure of the MazF–MazE complex has been determined (18, 56). In the structure, MazE and MazF form a 2:4 heterohexmer composed of alternating MazE and MazF homodimers (MazF2–MazE2–MazF2; Fig. 3). It is interesting to note that the unstructured C-terminal region of MazE is highly negatively charged and covers the entire cleft of one of the symmetrical concave interfaces created between two MazF molecules. Since this complex formation completely blocks the MazF endoribonuclease activity, the highly acidic or negatively charged C-terminal MazE extension likely mimics RNA substrates and binds to the active center of the MazF dimer (20, 33). NMR studies on the interaction of MazF dimers with uncleavable RNA substrate

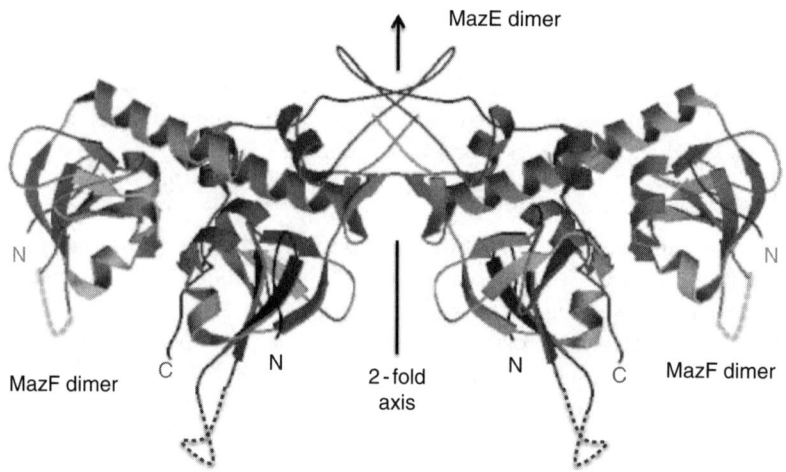

FIG. 3. X-ray structure of the MazE–MazF complex (18). MazE homodimers are in red/orange; MazF homodimers in yellow/yellow green and blue/emerald green/cyan.

analogs demonstrated that the binding of one substrate molecule to one of the two identical interfaces created in a symmetrical MazF dimer prevented binding of a second RNA molecule to other interface through negative cooperativity (60). This result explains why the binding of only a single C-terminal extension of MazE to a MazF dimer is enough to inhibit MazF endoribonuclease activity. At present, it is not known how the ACA triplet sequence is specifically recognized by a MazF dimer.

III. MazF Homologues and Other mRNA Interferases

A. MazF Homologues from *Mycobacterium tuberculosis*

As described earlier, *M. tuberculosis* contains seven MazF homologues, and at least four of them show toxicity, as they cause cell growth arrest when induced in *E. coli* (61). These MazF homologues have been shown to be sequence-specific mRNA interferases (61, 62).

1. MazF–mt1

When expression of MazF–mt1 was induced in *E. coli in vivo*, a specific cleavage of the *era* mRNA between a U and an A-residue was detected by primer extension in a time-dependent manner (61). The identical cleavage site was detected *in vitro* using the *era* mRNA synthesized by T7 RNA polymerase and purified MazF–mt1, tagged with His$_6$ at the N-terminal end. In addition to the site reported for the *in vivo* cleavage (CU!ACC), a weak cleavage site (UU! ACA) was detected *in vitro*. It was confirmed that the cleavage activity was not due to contaminating enzymes using a purified His$_6$MazF–mt1 (E19A) mutant protein, which showed a substantially reduced cleavage activity. This mutation in *E. coli* MazF has recently been shown to dramatically reduce the MazF mRNA interferase activity.

For further biochemical characterization, a synthetic 15-base RNA was used, which was cleaved between U and A of the UAC sequence by purified His$_6$MazF–mt1. A 19-base DNA with a sequence identical to that of the RNA substrate in the center could not be cleaved. When the DNA and RNA substrates were mixed and treated with the enzyme, only the RNA substrate was cleaved, indicating that MazF–mt1 is an endoribonuclease. To further test the cleavage specificity, five 11-base RNA substrates were synthesized, one with the UACA sequence in the center (5'-AUAUACAUAUG-3') and the others with a G-residue in each of the positions defined by the UACA sequence. The first U and the second A-residues could not be replaced with G. The third C-residue also appears to be important, as the cleavage between U and A was significantly reduced when this C was replaced with G. On the other hand, the fourth A-residue was replaceable with G, demonstrating that MazF–mt1 is an endoribonuclease that preferentially cleaves UAC sequences (*61*).

2. MazF–mt3

In the earlier work with MazF (*33*), ChpBK (*17*), PemK (*13*) and MazF–mt1 and MazF–mt6 (*61*), *in vivo* and *in vivo* primer extension was sufficient to determine their cleavage site specificities, since all these mRNA interferases recognize only 3-base sequences in mRNAs. On the other hand, it was not possible to determine the cleavage site specificity for MazF–mt3 by *in vivo* primer extension experiments, since very few cleavage sites were found. It was speculated that this might be due to a longer recognition sequence for MazF–mt3 cleavage. Therefore, a general method had to be developed for the determination of longer recognition sequences of mRNA interferases. For this purpose, the 3.5-kb MS2 phage RNA was used in combination with CspA, an RNA chaperone (*62*). Using the MS2 RNA, several MazF–mt3 cleavage sites indeed became evident. The use of CspA in the cleavage reaction is very important as it removes the secondary structures in the RNA allowing the mRNA interferases access to the entire 3.5-kb RNA. To cover the entire MS2 RNA sequence, a total of 22 primers were synthesized for the primer extension experiments. In most cases the addition of CspA, significantly enhanced RNA cleavage. As a result, eight cleavage sites were identified whose consensus sequence was CUCCU. MazF–mt3 cleaves between the U-residue in the second position and the C-residue in the third position (CU!CCU) (62). In addition to the eight CUCCU sites, four other cleavage sites were observed whose consensus sequence was UUCCU. MazF–mt3 cleaves these sites between the U-residue in the second position and the C-residue in the third position (UU!CCU).

We speculated that the pentad sequences identified as the cleavage sites for MazF–mt3 (CU!CCU and UU!CCU) might allow more stringent control of the synthesis of specific proteins under particular conditions, either by protecting mRNAs from being cleaved or by being more sensitive to MazF–mt3 endoribonuclease activity, depending on cleavage site distribution (62). Indeed, it was found that there are certain genes which have the MazF–mt3 pentad recognition sequences at a much lower frequency than expected. Of 10 genes with the least number of pentad sequences, four were identified as members of the PPE gene family, which has been proposed to play a role in the immunopathogenicity of *M. tuberculosis* (63). Three of these four genes, PPE54, PPE5553, and PPE56, are in a single locus in the chromosome of *M. tuberculosis* although they do not appear to be part of a single operon. All four PPE genes encode members of the major polymorphic tandem repeat subfamily of PPE proteins (PPE_MPTR) (64). The underrepresentation of MazF–mt3 pentad recognition sequences in specific members of the PE and PPE family suggests that the mRNAs for these genes are relatively resistant to MazF–mt3 compared with most mRNAs encoded in the *M. tuberculosis* genome.

3. MazF–mt6

MazF–mt6 preferentially cleaves mRNA in U-rich regions and after a U-residue. (U/C)U!(A/U)C(U/C) has been assigned as a consensus cleavage sequence for MazF–mt6. Cleavage also occurs after G or A-residues in some cases, although all cleavage sites contain UU, UC, or CU pairs (61).

4. MazF–mt7

Specific cleavage sites for MazF–mt7 were also identified using the MS2 RNA-CspA system. Most of the cleavage sites contain the sequence UCGCU, where MazF–mt7 cleaves between the first U- and the second C-residues (U!CGCU) (62). Some cleavage sites have a 1-base mismatch from the UCGCU consensus and a few have a 2-base mismatch. All these cleavage sites, however, share the central G-residue and most of them also have a following C-residue. MazF–mt7 is thus an mRNA interferase that also recognizes a pentad sequence; however, it appears to be less stringent than MazF–mt3. Interestingly, MazF–mt3 has a putative 11-residue S1–S2 loop similar to *E. coli* MazF (Fig. 3), whereas the MazF–mt7 has a much shorter (only four residues) S1–S2 loop. As the S1–S2 loop has been implicated in stabilizing the interaction of

MazF with its mRNA substrates, the S1–S2 loop structures of MazF–mt3 and MazF–mt7 are consistent with our observation of their relative flexibilities in recognition sequence.

Similar bioinformatics analysis to that described for the MazF–mt3 recognition sequence was carried out to search for *M. tuberculosis* mRNAs that are either resistant or hypersusceptible to cleavage by MazF–mt7. Of the top 10 predicted MazF–mt7-resistant genes, four belong to the PE or PPE family. Three of 10 are PPE_MPTR and one is a PE_PGRS family member. Interestingly, both PPE55 and PPE56 are resistant, not only to MazF–mt7, but also to MazF–mt3. PPE55 is known to be highly immunogenic, and is expressed during incipient, subclinical *M. tuberculosis* infection. PPE56 is highly homologous (67%) to PPE55 (65). Based these data, it has been proposed that the expression of different PE and PPE family proteins may be regulated at certain stages of infection by mRNA interferases during *M. tuberculosis* pathogenesis (61, 62).

The function of most genes in the PE/PPE family is not well known at present; however, a number of these proteins appear to localize to the mycobacterial cell wall or cell surface, and several have been shown to provoke an immune response *in vivo* or play a role in host–pathogen interactions (63, 66–68). In addition, members of these families have been shown to be differentially expressed in different tissues or at distinct times during growth *in vitro* (69, 70). Thus, to the extent that MazF–mt3 and MazF–mt7 can affect the production of specific PE or PPE family proteins *in vivo*, these mRNA interferases could play a role in the pathogenesis of tuberculosis. Members of the PPE_MPTR superfamily are present in multiple copies only in *M. tuberculosis* and its close relatives, and are totally absent in the fast-growing mycobacteria, including *M. smegmatis*. This finding suggests that these genes may have been positively selected during the evolution of these slow-growing pathogens, raising an interesting possibility that the target specificity of these mRNA interferases might have played a role in the evolutionary process to expand the PE/PPE family in the slow-growing pathogenic mycobacteria (62). It has been speculated that, since PE family proteins (notably the PE_PGRS group) are rich in glycine and alanine, the low frequency of MazF–mt3 and MazF–mt7 sequences could be secondary to codon usage, driven by processes independent of MazF–mt3 and MazF–mt7 targeting (62).

Interestingly, all these toxin genes appear to be cotranscribed with an overlapping upstream gene, as in the case of the *E. coli mazE–mazF* operon. However, none of these upstream gene products are homologous to MazE. Therefore, it remains to be determined whether these are antitoxins against their cognate toxins. The answer to this question may provide important insights into the role of *M. tuberculosis* MazF homologues in the pathogenicity of this human pathogen.

B. *Myxococcus xanthus* MazF

1. Solitary *mazF* Gene in *M. xanthus*

Surprisingly, not all *mazF* genes are cotranscribed with a cognate antitoxin gene. In some bacteria, the gene encoding a MazF homologue exists as a single gene without an upstream (or downstream) antitoxin gene. It has been speculated that the expression of these solitary toxin genes has to be tightly regulated in the cells and/or their cognate antitoxin gene(s) must exist somewhere else on the genome. A diversity of putative antitoxins for MazF has been observed in some prokaryotes, suggesting that the flexibility of the MazF-antitoxin system with highly diverse antitoxin partners may endow them with unique regulatory capabilities under various stresses. Recently, it has been shown that *M. xanthus* MazF (MazF–mx) is encoded by a monocistronic operon without any cognate antitoxin gene (*71*). Genomic analysis for the eleven known TA families using TBLASTN-Search, Pfam, and COG lists on the *M. xanthus* genomic database (TIGR) suggests there is only a single MazF homolog (MazF–mx; MAXN1659) with no identifiable MazE homolog. MazF–mx (122 aa) has 24% identity and 58% similarity to *E. coli* MazF (111 aa).

2. The Potential Antitoxin of MazF–mx

Using a yeast two-hybrid screen, with MazF–mx as a bait and an *M. xanthus* genomic library (*72*), 32 protein fragments were found to associate with MazF–mx, of which 15 were encoded by *mazF–mx* and 17 were encoded by *mrpC*. This indicates that MazF–mx forms an oligomer (dimer) and that MrpC is a likely candidate antitoxin for MazF–mx. Interestingly, MrpC is a member of the CRP transcription regulator family and its gene is located 4.44-Mbp downstream of the *mazF–mx* gene on the chromosome. Importantly, the *mrpC* gene is essential for *M. xanthus* development (*73*) and is a key early developmental transcription activator of the gene encoding FruA, another essential developmental regulator (*74*). Additionally, phosphorylation of MrpC by a Ser/Thr protein kinase cascade is involved in the regulation of MrpC function (below) (*75*). The MrpC and MazF interaction was also demonstrated by pull-down assays using purified N-terminal histidine-tagged MrpC and nontagged MazF–mx expressed in the soluble fraction of an *E. coli* cell extract (*71*). Furthermore, in vegetative DZF1 cells, chromosomally encoded MazF–mx coimmunoprecipitated with MrpC, using anti-MrpC IgG, demonstrating that MazF–mx forms a complex with MrpC *in vivo*.

3. MazF–mx Is Essential for Cell Death During Development

While vegetative growth of a Δ*mazF* strain was normal, development was profoundly affected (*71*). The cells in the fruiting bodies formed by the Δ*mazF* strain appeared to be very loose and relatively translucent with poor spore yields (15–18%) compared to the parental cells, DZF1. After 12 h of development, DZF1 cell numbers dramatically decreased to 18% due to autolysis. At the 24-h time point, the conversion of surviving wild-type cells to myxospores started. In contrast, the Δ*mazF* strain cell numbers were only slightly reduced to 77% after 12 h of development and were maintained at that level even after 48 h. Interestingly, DZF1 cell viability was substantially reduced (less than 1%) after 24 h of development, while over 30% of Δ*mazF* cells were found to be able to form colonies. A LIVE/DEAD stain was also applied to detect cell death during *M. xanthus* development. At 18 h of development, approximately 63% of DZF1 cells lysed, and could not be detected by staining, but the remaining 37% of the cells were stainable, of which 54.3% were stained as dead cells. On the other hand, only 9.2% of stainable Δ*mazF* cells were found to be dead. From these observations it was concluded that MazF–mx is required for developmental autolysis or programmed cell death to achieve effective fruiting body formation and sporulation (71).

4. Regulation of MazF–mx Expression by MrpC

The level of the *mazF–mx* transcript was significantly found to increase upon nutritional starvation, indicating that expression of *mazF–mx* is developmentally induced (71). This was further confirmed using a *lacZ–mazF–mx* fusion that was introduced into the DZF1 strain at the original chromosomal location. β-galactosidase assay of this strain (*mazF–mx–lacZ*/DZF1) showed that *mazF–mx–lacZ* was expressed during vegetative growth and steadily increased after 6 h, at the onset of development.

Gel-shift assay using purified MrpC and the *mazF–mx* promoter region (P*mazF*) showed that MrpC binds to at least two sites in the *mazF–mx* promoter region. Binding of MrpC to the promoter region was inhibited when MrpC was preincubated with MazF–mx. This is different from other toxin/antitoxin complexes, which usually bind their operators better than the antitoxin alone. Furthermore, *mazF–mx* expression, analyzed by primer extension, became undetectable in a Δ*mrpC* strain during both vegetative growth and the development phase, indicating that MrpC is a transcription activator for developmental *mazF–mx* expression. The regulation of *mazF* expression by MrpC during development will be discussed in more detail in Section V.

5. MazF Toxicity in *M. xanthus*

Severe cell toxicity by MazF–mx was observed in the Δ*mrpC* strain when induced in *M. xanthus* (71). In particular, MazF–mx expression is toxic in the absence of MrpC expression, confirming the idea that MrpC functions as an antitoxin to MazF–mx.

Interestingly, however, MazF–mx expression did not exhibit strong cellular toxicity in *E. coli*, suggesting that MazF–mx may cleave mRNAs at more specific sites than *E. coli* MazF. Purified MazF–mx exhibited endoribonuclease activity against *M. xanthus* total RNAs yielding free 5′-OH groups. When MS2 phage RNA was used as a substrate, it was cleaved into two major bands with many minor bands between them, suggesting that MS2 RNA may contain a preferential cleavage site for MazF–mx. At a low concentration of MazF–mx, a preferential cleavage was observed at 3′ to U-residue 724 (GAGU!UGCA), which yields a 0.8-kb fragment. Preincubation of MazF–mx with MrpC resulted in a complete inhibition of the cleavage. When higher concentrations of MazF–mx were used, new cleavage sites appeared at residues 1087 (AUGU!CAGG), 1106 (ACGU!AAUA), and 1241 (ACGU!AAAG), with several other minor cleavage sites that were detected after prolonged autoradiography of the gel. Interestingly, all cleavage sites are located in predicted single-stranded regions (76), consistent with the previous finding that *E. coli* MazF cleaves only single-stranded RNAs (*19, 33*). From the alignment of all cleavage sites identified, the preferred cleavage sequence for MazF–mx is GU!UGC, in which the first G-residue can be replaced with an A-residue.

C. *Staphylococcus aureus* MazF (MazFsa)

The MazEF antitoxin–toxin module in *S. aureus* consists of a pair of genes, *SA2059* and *SA2058*, encoding antitoxin (MazEsa) and toxin (MazFsa), respectively. MazFsa consists of 120 amino acid residues, and its sequence shows 20% identity and 40% similarity to the *E. coli* MazF toxin. MazFsa functions as an mRNA interferase with a cleavage specificity for a U-rich consensus sequence of VUUV′ (where V and V′ are A, C, or G and may or may not be identical) both *in vivo* and *in vitro* (*77*).

D. *E. coli* PemK

The *pemI–pemK* system and the *kis–kid* system are involved in the stable maintenance of two closely related incFII low copy plasmids, plasmid R100 (*78, 79*) and plasmid R1 (*80, 81*), respectively. These two systems turned out to be identical (*82*). Kid (PemK) has been shown to inhibit ColE1 replication, acting on the initiation of DNA synthesis, but does not inhibit P4 DNA replication *in vitro* (*83*). Toxin Kid (PemK) and antitoxin Kis (PemI) not only function in bacteria but also function efficiently in a wide range of eukaryotes. Kid (PemK) inhibits cell proliferation in yeast, *Xenopus laevis*, and human cells, and the inhibition is released by Kis (PemI) (*84*). These results suggest that there is a common target for Kid (PemK) in both prokaryotes and eukaryotes.

A primer extension assay performed on *mazG* mRNA and *era* mRNA partially digested by PemK showed that PemK is a ribonuclease that cleaves primarily at the 5'- or 3'-side of the A-residue in the UAH sequence (where H is C, A, or U). A single exception is known in which the cleavage occurs between U- and G-residues in the sequence UGC (13). The sequence UAC appears in 11 of the 18 cleavage sites determined. This was further confirmed with a synthetic 30-base RNA (5'-UAAGAAGGAGAUA<u>UAC</u>AUAUGAAUCAAAUC-3'), where the RNA was cleaved equally at either the 5'- or 3'-side of the A-residue in the UAC sequence. However, in the primer extension experiment using the full-length *mazG* mRNA, this UAC sequence was cleaved only at the 5'-side of the A-residue. It is interesting to note that PemK cleavages can occur 3' to the A-residue at some cleavage sites in the full-length *mazG* RNA, whereas at other cleavage sites, cleavages can occur at the 5'-side. The PemK cleavage of the 30-base RNA substrate is blocked by either antisense RNA or DNA, indicating that PemK is another sequence-specific endoribonuclease that cleaves single-stranded RNA.

Primer extension experiments designed to determine PemK cleavage sites in mRNAs *in vivo* revealed that the *mazG* and *era* mRNAs were cleaved at identical sites *in vivo* as *in vitro* (13). The cleaved UAC sequences in the *mazG* and the *era* mRNAs correspond to Tyr codon 41 in MazG and Tyr-7 in Era, whereas the cleaved UAC sequence in the *lpp* mRNA is between two adjacent codons, GCU for Ala73 and ACU for Thr74. The *in vivo* mRNA cleavages by PemK were very specific because no other cleavages were detected. Therefore, unlike RelE, which stimulates codon-specific mRNA cleavage at the A-site on ribosomes (15), PemK is a sequence-specific endoribonuclease or an mRNA interferase that works in a manner independent of ribosomes and codon reading frames.

Based on the fact that Kid (PemK) inhibits ColE1 replication at the initiation stage *in vitro*, but has no significant effect on the P4 DNA replication, DnaB was proposed as the target for the inhibitory action of Kid (PemK) (83). However, since the ColE1 replication is known to be initiated by RNA II and inhibited by RNA I (85, 86), whereas the P4 DNA replication is mainly regulated by protein (87), ribonucleases are expected to play a role in the control of the ColE1 plasmid replication (88). Indeed, RNA II contains several UAC sequences, two of which exist in the loop regions of the first and second stem-loop structures (89, 90). Therefore, the observed inhibition of ColE1 DNA replication by Kid (PemK) is probably due to the degradation of RNA II by its endoribonuclease activity. Furthermore, the fact that Kid (PemK) inhibits the growth of various eukaryotic cells (84) can be readily explained by its mRNA interferase activity against cellular mRNAs.

E. *E. coli* ChpBK

ChpBK is a toxin encoded by the ChpBI–ChpBK antitoxin–toxin module. This module consists of a pair of genes, *chpBI* and *chpBK* encoding antitoxin ChpBI and toxin ChpBK, respectively. ChpBK consists of 116 amino acid residues, and its sequence shows 35% identity and 52% similarity to *E. coli* MazF (*17*).

Interestingly when ChpBK expression is induced, the total protein synthesis is reduced by only ~60%, and a low level of cellular protein synthesis is observed even 60 min after ChpBK induction (*17*). In contrast, induction of MazF under similar conditions almost completely blocks protein synthesis *in vivo* (*33*), indicating that ChpBK is less detrimental to the cells than MazF, although the induction of both ChpBK and MazF results in inhibition of colony formation on agar plates.

Identification of the ChpBK-mediated cleavage sites in the *era* and *mazG* mRNAs *in vivo* revealed that it recognizes XACY, where the cleavage occurs either before or after the A-residue. C-residues were notably absent from the X and Y positions in these *in vivo* cleavage sites. The same cleavage sites were identified by *in vitro* experiments with purified ChpBK.

To further pinpoint the cleavage specificity of ChpBK, a U1A2C3A4 sequence in the *mazG* mRNA was mutated to all possible bases at each position (*17*). Any base change of the AC sequence (A2 and C3) completely blocked cleavage of the RNA by ChpBK. On the other hand, any base could be at the U1 position, although U or C was preferred over A or G. When A4 was mutated to a U-residue, RNA cleavage was observed. When A4 was mutated to a G-residue, RNA cleavage also occurred but at a significantly reduced level. However, when A4 was mutated to a C-residue, no cleavage was observed. Therefore, ChpBK is a sequence-specific endoribonuclease that cleaves mRNAs at ACY sequences (where Y is G, A, or U, but not C). Furthermore, using a synthetic 11-base RNA, ChpBK was found to cleave the substrate yielding a 2′,3′-cyclic phosphate at the 3′-end of the 5′-product and a 5′-OH group at the 5′-end of the 3′-product, in a manner identical to that demonstrated for MazF (*20*).

It is interesting to note that complete cleavage of the *mazG* mRNA by MazF was observed by 5 min after induction, as judged from the density of the band of the full-length mRNA, while the cleavage by ChpBK was much slower and a substantial amount of the full-length mRNA was still visible even 90 min after induction (*17*). This again supports the notion that ChpBK is intrinsically a less potent toxin than MazF. Interestingly, the pI value of ChpBK is 5.2, in contrast to basic pI values of MazF (8.3) and PemK (11.0). The weaker toxic effect of ChpBK may be at least partially attributed to its acidic pI and important for its physiological role in the cells, since ChpBK may arrest cell growth in a less damaging manner than that of MazF.

IV. Other Toxins in TA Systems

A. *E. coli* RelE

The RelBE system is one of the most extensively investigated TA systems. In vitro studies have documented that an association of RelE with the ribosome A-site promotes a novel ribonucleolytic activity that cleaves mRNA codons, preferentially between the second and the third nucleotides of the termination codon (*15, 31*). This activity leads to global inhibition of protein synthesis. Homologs of the RelBE system have been found in other bacterial species and Archaea, in both chromosomes and plasmids, and shown to be functionally active (*6*). RelE by itself has no endoribonuclease activity (*15*) and has been proposed to be a ribosome-associated factor that stimulates an endogenous ribonuclease activity of ribosomes (*15, 32*).

The crystal structure of the aRelB2–aRelE2 complex from the hyperthermophilic archaea *Pyrococcus horikoshii* and an NMR-derived structure of *E. coli* RelB have been determined (*17, 91*). In the tetrameric complex, a molecule of aRelB wraps around a compact aRelE, forming a tight heterodimer. The sequence similarity between the archaeal *P. horikoshii* aRelB and bacterial *E. coli* RelB is relatively low, with only 24% identity and 48% similarity. A high level of homology resides only within the C-terminal part of these antitoxins.

B. *E. coli* YoeB

It has been shown that the YefM antitoxin is highly unstable in the cell and purified YefM exists as a denatured form without secondary structure (*92*). However, when YefM is coexpressed with YoeB, a dimer of YefM forms a stable complex with a single molecule of YoeB. The X-ray structure of this heterotrimer has been determined (*21*). Under normal growth conditions, YoeB presumably exists as heterotrimer and is thus unable to exert its cellular toxicity. The ATP-dependent protease Lon, whose expression is induced under various stress conditions, has been implicated in the degradation of YefM and release of YoeB. This consequently leads to inhibition of protein synthesis (*93*). These authors demonstrated that induction of YoeB expression leads to cleavage of the *lpp* mRNA *in vivo* at a few sites and concluded that YoeB functions similarly to the other *E. coli* toxins, RelE, MazF, and ChpBK, by cleaving translated mRNAs (*93*). However, RelE by itself has no endoribonuclease activity (*15*), while MazF and ChpBK are ribosome-independent endoribonucleases that cleave mRNAs specifically at ACA and ACY sequences, respectively (*17, 33*).

YoeB is a potent inhibitor of protein synthesis, which functions in a manner that is distinctly different from that of MazF (Zhang Y. and Inouye M., unpublished results). Upon induction of YoeB expression, [^{35}S]methionine incorporation into cellular proteins is abruptly stopped, in less than 5 min. This is in sharp contrast to MazF, which causes inhibition of [^{35}S]methionine incorporation only 15–20 min after induction (33). Another important difference between YoeB and MazF is that significant amounts of full-length cellular mRNAs remain in the YoeB-induced cells even 30 min after its induction, while they rapidly disappear after 10 min of MazF induction. In addition, MazF inhibits protein synthesis in both prokaryotes and eukaryotes as expected from its function as an mRNA interferase (33), while YoeB was found to be a prokaryote-specific inhibitor. On the basis of these results, we suspect that YoeB is not a sequence-specific endoribonuclease as originally proposed by Christensen et al. (93) and Kamada and Hanaoka (21). Instead, we believe that YoeB is a 50S ribosome-associating factor, which blocks translation initiation (Zhang and Inouye unpublished results).

Interestingly, the overproduction of Lon ATP-dependent protease has been shown to activate YoeB-dependent mRNA cleavage (93). This mRNA cleavage may not be due to primary effects of YoeB (as it has no mRNA interferase activity) but rather due to the induction of another toxin(s) with mRNA interferase activity. In this regard, it will be interesting to determine the exact cleavage sites of mRNA caused by Lon overproduction, which may provide insights into possible TA networks in E. coli.

C. E. coli YhaV

Recently a new TA system, prlF–yhaV, was found on the E. coli chromosome (27). PrlF belongs to the AbrB superfamily of bacterial transcription factors, which includes the antitoxins of the TA systems mazE–mazF and pemI–pemK (kis/kid). YhaV is a member of the RelE superfamily, which comprises the toxins of the TA systems relBE, yefM–yoeB, and parDE. It has been demonstrated that YhaV has ribonuclease activity and PrlF inhibits this activity in vitro. Toxins of the RelE superfamily contain two highly conserved arginine residues and their mutation strongly reduces the effect of RelE toxicity (16). Mutations in the corresponding arginine residues (either R85A or R94A) of YhaV caused reduced toxicity, but still inhibited cell growth. However, the double mutant (R85A and R94A) was not toxic in an in vivo experiment and the purified double mutant protein showed significantly lower ribonuclease activity than the wild-type YhaV. It remains to be determined whether YhaV is a sequence-specific mRNA interferase.

D. Bacillus subtilis EndoA

EndoA (YdcE) is 25% identical to E. coli MazF, and encoded by the ndoAI–ndoA TA system from B. subtilis, a Gram-positive bacterium (94). This protein also functions as an mRNA interferase having cleavage specificity for UAC sequences, somewhat similar to PemK. Interestingly, its antitoxin EndoAI (YdcD) has little homology to MazE, but nevertheless it effectively inhibits the EndoA mRNA interferase activity both *in vivo* and *in vitro*.

V. The Regulation of MazF Expression Under Stress Conditions

A. Growth Arrest and Cell Death in *E. coli*

In *E. coli*, MazF expression may be induced under various stresses, resulting in inhibition of cell growth and eventually cell death (95, 96). Induction of MazF expression causes dormancy in cell growth, termed "quasi-dormancy," since MazF induced cells still fully retain various cellular metabolic pathways (33). The quasi-dormancy state induced by MazF is considered quite important for cellular physiology, as cellular damage caused by MazF induction appears to be kept to a minimal level to allow a recovery of cell growth upon neutralization of MazF toxicity by MazE. It has been shown that MazF-induced quasi-dormant cells are fully capable of producing a protein in the total absence of other cellular protein synthesis if its mRNA is engineered to be devoid of ACA sequences (3). However, prolonged quasi-dormancy likely causes an unrecoverable damage in the cell, which has been defined as a "point of no return" (10). One of the best ways to study the role of the *mazE–mazF* operon in bacterial physiology in *E. coli* is to isolate a *mazE–mazF* deletion strain and to compare its physiology with that of the wild-type strain. Interestingly, a *mazE–mazF* deletion strain infected with bacteriophage P1 produces significantly more phage particles than a similarly infected wild-type strain, suggesting that the MazE–MazF system may be involved in P1 phage exclusion from the bacterial population (4). MazF seems to play an active role in eliminating a tainted fraction of cells from the entire population to protect the healthy, uninfected population. This type of role for MazF may be crucial for maintaining a healthy bacterial population and MazF, together with other TA systems, may therefore have played an important role in bacterial evolution. A *mazE–mazF* deletion strain has also been shown to be quite resistant to various short-term stress conditions such as (i) extreme amino acid starvation, leading to the production of the starvation-signaling molecule ppGpp (37, 97); (ii) inhibition of transcription and/or translation by antibiotics including rifampicin, chloramphenicol, and spectinomycin (44); (iii) DNA damage caused by thymine starvation (43) as well as by mitomycin C, nalidixic acid, and UV irradiation (45, 98); and (iv) oxidative stress (H_2O_2) (45). Furthermore, the postsegregational killing caused by the *phd–doc* TA system of the plasmid prophage P requires the *mazE–mazF* system. Doc protein is an effective inhibitor of peptide elongation that blocks protein synthesis (99).

The role of the MazE–MazF TA system under various stresses is, however, controversial at present, as an *E. coli* strain with a deletion of five TA systems ($\Delta mazE$–$mazF$, $\Delta relB$–$relE$, $\Delta chpBI$–$chpBK$, $\Delta yefM$–$yoeB$, and $\Delta dinJ$–$yafQ$) did not show a significant difference in growth compared to the wild-type strain under various stress conditions (amino acid starvation, rifampicin treatment, acidic stress, and nutritional downshift) (*100*). The difference between the observations from these two groups maybe attributed to their experimental conditions, as cell density may be a key factor in MazF-dependent cell death, as described below.

Recently, Engelberg-Kulka and colleagues (*101, 102*) showed that *E. coli mazEF*-mediated cell death was dependent on the density of the bacterial population. Adding rifampicin for a short period to inhibit transcription led to *mazEF*-mediated cell death at densities of 3×10^8 or 3×10^7 cells/ml, but not at 3×10^5 or 3×10^4 cells/ml. Moreover, they found that *mazEF*-mediated cell death requires an "extracellular death factor" (EDF), which is a linear peptide with the amino acid sequence Asn–Asn–Trp–Asn–Asn (NNWNN). At a wide range of concentrations (2.5–200 ng/ml), the killing activity of the chemically synthesized EDF was *mazEF* dependent and no EDF activity was observed in a *mazEF* deletion strain. Using database analysis, they found five open reading frames containing the same or similar peptide sequence, which may be the source of EDF. From the analysis of deletion strains of each gene, these authors found that the deletion of *zwf*, encoding glucose 6-phosphate-1-dehydrogenase, which contains the sequence NNWDN (D, Asp), eliminated the production of an active EDF (*102*). They speculated that the *zwf* pentapeptide may be the precursor of EDF, and a subsequent amidation step at the Asp-residue may generate the full NNWNN sequence. These authors argue that *mazEF*-mediated cell death is a quorum-sensing process in which the EDF peptide functions as an autoinducer and that a cellular component, yet to be identified, may directly interact with EDF at the specific stage in *mazEF*-mediated cell death to trigger the process (*101, 102*).

These authors further argue that *mazEF*-mediated cell death is absolutely dependent on EDF (*101, 102*), so that for any studies on *E. coli mazEF*-mediated cell death in *E. coli* a protocol should be used in which an active EDF is produced at the required concentrations. There are clearly more questions to be solved than available answers, including the identification of the protease that generates the pentapeptide from the full-length protein, whose expression or activity should be regulated in a growth-dependent manner. I have changed this sentence a little, since the likely amidase was identified in the Science paper as AsnA.

B. MazF-Mediated Programmed Cell Death During Multicellular Development in *Myxococcus xanthus*

As discussed earlier, *M. xanthus* is a developmental Gram-negative bacterium, which forms multicellular fruiting bodies upon nutrient starvation. Uniquely, 80–90% of total cell population undergoes lysis during the course of the fruiting body formation and these lysed cells are considered to be utilized as nutrients for conversion of the remaining surviving cells to highly stress-resistant spores (*103*). This cell death process during development, which is essential for normal fruiting body formation, can be defined as programmed cell death comparable to the programmed cell death observed in mammalian cells. The solitary *mazF* gene in *M. xanthus* has been shown to play an essential role in programmed cell death and this is the first unambiguous example which demonstrates that a toxin in the TA systems is involved in programmed cell death in bacteria. Note that the role of MazF in PCD in *E. coli* has been debated (*101, 102*).

In vegetative cells, MazF–mx appears to form a complex with nonphosphorylated MrpC (*73*). Importantly, MrpC that is not in a complex with MazF–mx is phosphorylated by Pkn14, a Ser/Thr protein kinase, inactivating MrpC as a transcriptional activator of *mazF–mx* and other early developmental genes (*104*). Consequently, MrpC is not active as a transcription activator, or MazF–mx as a toxin, in vegetative cells. In developmental cells, MrpC exists in two forms: one, nonphosphorylated MrpC in a complex with MazF–mx and the other, as a shorter variant termed MrpC2. MrpC2 activates the transcription of *mrpC* and *mazF–mx*. During early development, MazF–mx is kept inactive in a complex with MrpC, while later in development, both MrpC and MrpC2 are thought to be degraded, leaving MazF–mx free to exert its endoribonuclease activity leading to cell death.

An intriguing question is how 20% of the population is selected to survive avoiding autolysis. It has been speculated that since *M. xanthus* development does not occur uniformly, the seemingly altruistic autolysis may be a matter of timing, and the subpopulation in which the onset of the developmental program is delayed (perhaps because of their position in the cell cycle at the time of nutritional deprivation) may be triggered by the transient release of nutrition from autolyzed cells to initiate the late developmental process (*73*). In this selected population, MazF–mx function may be subdued by a mechanism yet to be determined. It is possible that the expression level of MrpC in this subpopulation is relatively higher than that in the lysing cells, effectively neutralizing MazF–mx toxicity for successful development. Interestingly, a *pkn8–pkn14* cascade deletion strain expresses a high level of MrpC even during vegetative growth, and that fruiting body development progresses significantly faster than in the parental strain (*75*) due to greater expression of MazF–mx. It is interesting to note that the involvement of a Ser/Thr protein kinase cascade in programmed cell death in *M. xanthus* is reminiscent of eukaryotic programmed cell death. It also remains to be elucidated whether MazF–mx triggers programmed cell death in *M. xanthus* through the cleavage of a specific mRNA(s) or by inflicting general damage to the cells as is thought to be the case in *E. coli* (*105*). In this regard, it should be noted that MazF induction in mammalian cells induces BAK-dependent programmed cell death (*59*).

VI. Concluding Remarks

Most bacteria and a large number of Archaea are equipped with TA systems that are considered to function as defense systems against various environmental stresses, such as changes in temperature, pH, radiation, heavy metal ions, toxic chemicals, and antibiotics. Even if many bacteria live as unicellular organisms in their natural habitats, the TA systems appear to regulate cell growth under various environmental stresses and ensure the welfare of the entire bacterial population by inhibiting cell growth of unhealthy or damaged fractions of the population. Furthermore, as seen in *M. xanthus*, TA systems play an essential role in the life cycle of developmental bacteria, causing obligatory cell death or programmed cell death during development, functionally similar to eukaryotic apoptotic systems. It should be noted that TA systems are found only in free-living bacteria, but not symbiotic bacteria such as *Chlamydia muridarum* or *Mycoplasma gallisepticum* (*1*). It was speculated that the number of TA systems in a bacterium may be correlated with its growth rate, that is, TA systems may be beneficial for those bacteria characterized by slow growth like *M. tuberculosis* (*1*). It is also possible that symbiotic bacteria may not require any TA systems as they live in eukaryotic cells that may provide a relatively constant stress-free living environment.

It is quite surprising to find that *E. coli* contains a large number of TA systems on its genome and that their cellular targets appear to be highly diverse. This clearly indicates that TA systems play important roles in basic bacterial physiology and in maintaining genomic stability during the course of evolution. They allow organisms to find the best niche for survival in severe natural habitats by regulating cell growth under stress conditions.

Among the TA systems, a number of toxins have been identified as sequence-specific endoribonucleases or mRNA interferases. All mRNA interferases so far identified are listed in Table II together with their specific recognition sites. It is likely that many more mRNA interferases will be identified from various bacterial species, with different cleavage specificities, a situation somewhat reminiscent of DNA restriction enzymes. However, as the recognition sequences become longer than 5 bases, new technologies will have to be developed to unambiguously identify them. If the recognition sequence is 6 bases long, it would theoretically exist only once in every 4096-base sequence, corresponding to a protein of approximately 150 kDa. As was seen with MazF–mt3 and MazF–mt7 from *M. tuberculosis* (*62*), their recognition sites are underrepresented in specific mRNAs associated with immunity and pathogenesis. It is also possible that some mRNA interferases may be designed to degrade specific groups of mRNAs by having an unusually large number of cleavage sites. Indeed, we have recently found that a MazF homolog from *Streptococcus aureus* appears to target a specific group of genes involved in its

TABLE II
mRNA Interferases in Bacteria

Strain	mRNA interferase	Cutting site
E. coli	MazF	ACA
	PemK	UAC
	ChpBK	ACY (Y is G, A, or U)
M. tuberculosis	MazF–mt1	UAC
	MazF–mt3	CUCCU and UUCCU
	MazF–mt6	(U/C)U(A/U)C(U/C)
	MazF–mt7	UCGCU
M. xanthus	MazF–mx	GUUGC
S. aureus	MazFsa	VUUV' (V and V' are A, C, or G)
B. subtilis	EndoA	UAC

pathogenicity (Zhu L., Yoshizumi S., Kato F., Sugai M., Ouyang M., and Inouye M., unpublished results). Therefore, the identification of the cellular targets of each mRNA interferase is fascinating and important, given their potential role in bacterial pathogenicity. Not only will the study of mRNA interferases provide new insights into the underlying mechanisms of pathogenicity, but will undoubtedly also disclose new ways to control various human pathogens including *M. tuberculosis*.

The use of mRNA interferases also opens new avenues in biotechnology. The single-protein production system (3) has been shown to be highly useful in protein structure biology. Furthermore, taking advantage of the high cytotoxicity of MazF in mammalian cells, the *mazF* gene can be also used for gene therapy against cancer and AIDS. The small size of MazF (111 amino acids) and availability of its cognate antitoxin provide a very useful feature for this application. The development of novel antibiotics inhibiting the MazE–MazF complex formation is also very attractive, as such antibiotics would have two synergistic effects: one to release free mRNA interferase in the cell leading to cell growth arrest and cell death, and the other to derepress the *mazE–mazF* operon expression, as the TA complex is a strong repressor for its own operon.

Acknowledgments

We are grateful to S. Phadtare and S. Baik for critical reading of this manuscript. This study was supported by a research fund from Takara-Bio, Inc., Japan.

References

1. Pandey DP, Gerdes K. Toxin–antitoxin loci are highly abundant in free-living but lost from host-associated prokaryotes. *Nucleic Acids Res* 2005;**33**:966–76.
2. Inouye M. The discovery of mRNA interferases: Implication in bacterial physiology and application to biotechnology. *J Cell Physiol* 2006;**209**:670–6.
3. Suzuki M, Zhang J, Liu M, Woychik NA, Inouye M. Single protein production in living cells facilitated by an mRNA interferase. *Mol Cell* 2005;**18**:253–61.
4. Engelberg-Kulka H, Sat B, Reches M, Amitai S, Hazan R. Bacterial programmed cell death systems as targets for antibiotics. *Trends Microbiol* 2004;**12**:66–71.
5. Buts L, Lah J, Dao-Thi MH, Wyns L, Loris R. Toxin–antitoxin modules as bacterial metabolic stress managers. *Trends Biochem Sci* 2005;**30**:672–9.
6. Gerdes K, Christensen SK, Lobner-Olesen A. Prokaryotic toxin–antitoxin stress response loci. *Nat Rev Microbiol* 2005;**3**:371–82.
7. Bernard P, Couturier M. The 41 carboxy-terminal residues of the miniF plasmid CcdA protein are sufficient to antagonize the killer activity of the CcdB protein. *Mol Gen Genet* 1991;**226**:297–304.
8. Thisted T, Nielsen AK, Gerdes K. Mechanism of post-segregational killing: Translation of Hok, SrnB and Pnd mRNAs of plasmids R1, F and R483 is activated by 3′-end processing. *EMBO J* 1994;**13**:1950–9.
9. Thisted T, Sørensen NS, Wagner EG, Gerdes K. Mechanism of post-segregational killing: Sok antisense RNA interacts with Hok mRNA via its 5′-end single-stranded leader and competes with the 3′-end of Hok mRNA for binding to the mok translational initiation region. *EMBO J* 1994;**13**:1950–9.
10. Amitai S, Yassin Y, Engelberg-Kulka H. MazF-mediated cell death in *Escherichia coli*: A point of no return. *J Bacteriol* 2004;**186**:8295–300.
11. Gazit E, Sauer RT. The Doc toxin and Phd antidote proteins of the bacteriophage P1 plasmid addiction system form a heterotrimeric complex. *J Biol Chem* 1999;**274**:16813–8.
12. Hargreaves D, Santos-Sierra S, Giraldo R, Sabariegos-Jareno R, de la Cueva-Mendez G, Boelens R, et al. Structural and functional analysis of the *kid* toxin protein from *E. coli* plasmid R1. *Structure* 2002;**10**:1425–33.
13. Zhang J, Zhang Y, Zhu L, Suzuki M, Inouye M. Interference of mRNA function by sequence-specific endoribonuclease PemK. *J Biol Chem* 2004;**279**:20678–84.
14. Loris R, Dao-Thi MH, Bahassi EM, Van Melderen L, Poortmans F, Liddington R, et al. Crystal structure of CcdB, a topoisomerase poison from *E. coli*. *J Mol Biol* 1999;**285**:1667–77.
15. Pedersen K, Zavialov AV, Pavlov MY, Elf J, Gerdes K, Ehrenberg M. The bacterial toxin RelE displays codon-specific cleavage of mRNAs in the ribosomal A site. *Cell* 2003;**112**:131–40.
16. Takagi H, Kakuta Y, Okada T, Yao M, Tanaka I, Kimura M. Crystal structure of archaeal toxin–antitoxin RelE–RelB complex with implications for toxin activity and antitoxin effects. *Nat Struct Mol Biol* 2005;**12**:327–31.
17. Zhang Y, Zhu L, Zhang J, Inouye M. Characterization of ChpBK, an mRNA interferase from *Escherichia coli*. *J Biol Chem* 2005;**280**:26080–8.
18. Kamada K, Hanaoka F, Burley SK. Crystal structure of the MazE/MazF complex: Molecular bases of antidote–toxin recognition. *Mol Cell* 2003;**11**:875–84.
19. Zhang J, Zhang Y, Inouye M. Characterization of the interactions within the *mazEF* addiction module of *Escherichia coli*. *J Biol Chem* 2003;**278**:32300–6.
20. Zhang Y, Zhang J, Hara H, Kato I, Inouye M. Insights into the mRNA cleavage mechanism by MazF, an mRNA interferase. *J Biol Chem* 2005;**280**:3143–50.

21. Kamada K, Hanaoka F. Conformational change in the catalytic site of the ribonuclease YoeB toxin by YefM antitoxin. *Mol Cell* 2005;**19**:497–509.
22. Motiejunaite R, Armalyte J, Markuckas A, Suziedeliene E. *Escherichia coli dinJ–yafQ* genes act as a toxin–antitoxin module. *FEMS Microbiol Lett* 2007;**268**:112–9.
23. Korch SB, Henderson TA, Hill TM. Characterization of the *hipA7* allele of *Escherichia coli* and evidence that high persistence is governed by (p)ppGpp synthesis. *Mol Microbiol* 2003;**50**:1199–213.
24. Keren I, Shah D, Spoering A, Kaldalu N, Lewis K. Specialized persister cells and the mechanism of multidrug tolerance in *Escherichia coli*. *J Bacteriol* 2004;**186**:8172–80.
25. Brown JM, Shaw KJ. A novel family of *Escherichia coli* toxin–antitoxin gene pairs. *J Bacteriol* 2003;**185**:6600–8.
26. Makarova KS, Grishin NV, Koonin EV. The HicAB cassette, a putative novel, RNA-targeting toxin–antitoxin system in archaea and bacteria. *Bioinformatics* 2006;**22**:2581–4.
27. Schmidt O, Schuenemann VJ, Hand NJ, Silhavy TJ, Martin J, Lupas AN, et al. prlF and yhaV encode a new toxin–antitoxin system in *Escherichia coli*. *J Mol Biol* 2007;**372**:894–905.
28. Garc'a-Contreras R, Zhang XS, Kim Y, Wood TK. Protein translation and cell death: The role of rare tRNAs in biofilm formation and in activating dormant phage killer genes. *PLoS ONE* 2008;**3**:e2364.
29. Bahassi EM, O'Dea MH, Allali N, Messens J, Gellert M, Couturier M. Interactions of CcdB with DNA gyrase. Inactivation of Gyra, poisoning of the gyrase–DNA complex, and the antidote action of CcdA. *J Biol Chem* 1999;**274**:10936–44.
30. Kampranis SC, Howells AJ, Maxwell A. The interaction of DNA gyrase with the bacterial toxin CcdB: Evidence for the existence of two gyrase–CcdB complexes. *J Mol Biol* 1999;**293**:733–44.
31. Christensen SK, Gerdes K. RelE toxins from bacteria and Archaea cleave mRNAs on translating ribosomes, which are rescued by tmRNA. *Mol Microbiol* 2003;**48**:1389–400.
32. Hayes CS, Sauer RT. Cleavage of the A site mRNA codon during ribosome pausing provides a mechanism for translational quality control. *Mol Cell* 2003;**12**:903–11.
33. Zhang Y, Zhang J, Hoeflich KP, Ikura M, Qing G, Inouye M. MazF cleaves cellular mRNAs specifically at ACA to block protein synthesis in *Escherichia coli*. *Mol Cell* 2003;**12**:913–23.
34. Daines DA, Wu MH, Yuan SY. VapC-1 of nontypeable *Haemophilus influenzae* is a ribonuclease. *J Bacteriol* 2007;**189**:5041–8.
35. Budde PP, Davis BM, Yuan J, Waldor MK. Characterization of a *higBA* toxin–antitoxin locus in *Vibrio cholerae*. *J Bacteriol* 2007;**189**:491–500.
36. Zielenkiewicz U, Ceglowski P. The toxin–antitoxin system of the streptococcal plasmid pSM19035. *J Bacteriol* 2005;**187**:6094–105.
37. Mizuno T, Chou MY, Inouye M. A unique mechanism regulating gene expression: Translational inhibition by a complementary RNA transcript (micRNA). *Proc Natl Acad Sci USA* 1984;**81**:1966–70.
38. Ambros V. microRNAs: Tiny regulators with great potential. *Cell* 2001;**107**:823–6.
39. Billy E, Brondani V, Zhang H, Muller U, Filipowicz W. Specific interference with gene expression induced by long, double-stranded RNA in mouse embryonal teratocarcinoma cell lines. *Proc Natl Acad Sci USA* 2001;**98**:14428–33.
40. Puerta-Fernandez E, Romero-Lopez C, Barroso-del Jesus A, Berzal-Herranz A. Ribozymes: Recent advances in the development of RNA tools. *FEMS Microbiol Rev* 2003;**27**:75–97.
41. Aizenman E, Engelberg-Kulka H, Glaser G. An *Escherichia coli* chromosomal "addiction module" regulated by guanosine 3′,5′-bispyrophosphate: A model for programmed bacterial cell death. *Proc Natl Acad Sci USA* 1996;**93**:6059–63.
42. Marianovsky I, Aizenman E, Engelberg-Kulka H, Glaser G. The regulation of the *Escherichia coli mazEF* promoter involves an unusual alternating palindrome. *J Biol Chem* 2001;**276**:5975–84.
43. Sat B, Reches M, Engelberg-Kulka H. The *Escherichia coli mazEF* suicide module mediates thymineless death. *J Bacteriol* 2003;**185**:1803–7.

44. Sat B, Hazan R, Fisher T, Khaner H, Glaser G, Engelberg-Kulka H. Programmed cell death in *Escherichia coli*: Some antibiotics can trigger *mazEF* lethality. *J Bacteriol* 2001;**183**:2041–5.
45. Hazan R, Sat B, Reches M, Engelberg-Kulka H. Postsegregational killing mediated by the P1 phage addiction module *phd–doc* requires the *Escherichia coli* programmed cell death system *mazEF*. *J Bacteriol* 2001;**183**:2046–50.
46. Hazan R, Sat B, Engelberg-Kulka H. *Escherichia coli mazEF*-mediated cell death is triggered by various stressful conditions. *J Bacteriol* 2004;**186**:3663–9.
47. Guzman LM, Belin D, Carson MJ, Beckwith J. Tight regulation, modulation, and high-level expression by vectors containing the arabinose PBAD promoter. *J Bacteriol* 1995;**177**:4121–30.
48. Datsenko KA, Wanner BL. One-step inactivation of chromosomal genes in *Escherichia coli* K-12 using PCR products. *Proc Natl Acad Sci USA* 2000;**97**:6640–5.
49. Halegoua S, Hirashima A, Inouye M. Puromycin-resistant biosynthesis of a specific outer-membrane lipoprotein of *Escherichia coli*. *J Bacteriol* 1976;**126**:183–91.
50. Halegoua S, Hirashima A, Sekizawa J, Inouye M. Protein synthesis in toluene-treated *Escherichia coli*. Exclusive synthesis of membrane proteins. *Eur J Biochem* 1976;**69**:163–7.
51. Zhang J, Inouye M. MazG, a nucleoside triphosphate pyrophosphohydrolase, interacts with Era, an essential GTPase in *Escherichia coli*. *J Bacteriol* 2002;**184**:5323–9.
52. Moll I, Blasi U. Differential inhibition of 30S and 70S translation initiation complexes on leaderless mRNA by kasugamycin. *Biochem Biophys Res Commun* 2002;**297**:1021–6.
53. Movva NR, Nakamura K, Inouye M. Gene structure of the OmpA protein, a major surface protein of *Escherichia coli* required for cell–cell interaction. *J Mol Biol* 1980;**143**:317–28.
54. Nakamura K, Inouye M. DNA sequence of the gene for the outer membrane lipoprotein of *E. coli*: An extremely AT-rich promoter. *Cell* 1979;**18**:1109–17.
55. Downing WL, Sullivan SL, Gottesman ME, Dennis PP. Sequence and transcriptional pattern of the essential *Escherichia coli secE–nusG* operon. *J Bacteriol* 1990;**172**:1621–7.
56. de la Cueva-MŽndez G. Distressing bacteria: Structure of a prokaryotic detox program. *Mol Cell* 2003;**11**:848–50.
57. Munoz-Gomez AJ, Santos-Sierra S, Berzal-Herranz A, Lemonnier M, Diaz-Orejas R. Insights into the specificity of RNA cleavage by the *Escherichia coli* MazF toxin. *FEBS Lett* 2004;**567**:316–20.
58. Christensen SK, Pedersen K, Hansen FG, Gerdes K. Toxin–antitoxin loci as stress-response-elements: ChpAK/MazF and ChpBK cleave translated RNAs and are counteracted by tmRNA. *J Mol Biol* 2003;**332**:809–19.
59. Shimazu T, Degenhardt K, Nur-E-Kamal A, Zhang J, Yoshida T, Zhang Y, et al. NBK/BIK antagonizes MCL-1 and BCL-XL and activates BAK-mediated apoptosis in response to protein synthesis inhibition. *Genes Dev* 2007;**21**:929–41.
60. Li GY, Zhang Y, Chan MC, Mal TK, Hoeflich KP, Inouye M, et al. Characterization of dual substrate binding sites in the homodimeric structure of *Escherichia coli* mRNA interferase MazF. *J Mol Biol* 2005;**357**:139–50.
61. Zhu L, Zhang Y, Teh JS, Zhang J, Connell N, Rubin H, et al. Characterization of mRNA interferases from *Mycobacterium tuberculosis*. *J Biol Chem* 2006;**281**:18638–43.
62. Zhu L, Phadtare S, Nariya H, Ouyang M, Husson RN, Inouye M. The mRNA interferases, MazF–mt3 and MazF–mt7 from *Mycobacterium tuberculosis* target unique pentad sequences in single-stranded RNA. *Mol Microbiol* 2008;**69**:559–69.
63. Brennan MJ, Delogu G. The PE multigene family: A "molecular mantra" for mycobacteria. *Trends Microbiol* 2002;**10**:246–9.
64. Gey van Pittius NC, Sampson SL, Lee H, Kim Y, van Helden PD, Warren RM. Evolution and expansion of the *Mycobacterium tuberculosis* PE and PPE multigene families and their association with the duplication of the ESAT-6 (*esx*) gene cluster regions. *BMC Evol Biol* 2006;**6**:95.
65. Singh KK, Dong Y, Patibandla SA, McMurray DN, Arora VK, Laal S. Immunogenicity of the *Mycobacterium tuberculosis* PPE55 (Rv3347c) protein during incipient and clinical tuberculosis. *Infect Immun* 2005;**73**:5004–14.

66. Denny PW, Smith DF. Rafts and sphingolipid biosynthesis in the kinetoplastid parasitic protozoa. *Mol Microbiol* 2004;**53**:725–33.
67. Basu S, Pathak SK, Banerjee A, Pathak S, Bhattacharyya A, Yang Z, et al. Execution of macrophage apoptosis by PE_PGRS33 of Mycobacterium tuberculosis is mediated by Toll-like receptor 2-dependent release of tumor necrosis factor-alpha. *J Biol Chem* 2007;**282**:1039–50.
68. Mishra KC, de Chastellier C, Narayana Y, Bifani P, Brown AK, Besra GS, et al. Functional role of the PE domain and immunogenicity of the *Mycobacterium tuberculosis* triacylglycerol hydrolase LipY. *Infect Immun* 2008;**76**:127–40.
69. Delogu G, Sanguinetti M, Pusceddu C, Bua A, Brennan MJ, Zanetti S, et al. PE_PGRS proteins are differentially expressed by *Mycobacterium tuberculosis* in host tissues. *Microbes Infect* 2006;**8**:2061–7.
70. Dheenadhayalan V, Delogu G, Sanguinetti M, Fadda G, Brennan MJ. Variable expression patterns of *Mycobacterium tuberculosis* PE_PGRS genes: Evidence that PE_PGRS16 and PE_PGRS26 are inversely regulated *in vivo*. *J Bacteriol* 2006;**188**:3721–5.
71. Nariya H, Inouye M. MazF, an mRNA interferase, mediates programmed cell death during multicellular *Myxococcus* development. *Cell* 2008;**132**:55–66.
72. Nariya H, Inouye S. Modulating factors for the Pkn4 kinase cascade in regulating 6-phosphofructokinase in *Myxococcus xanthus*. *Mol Microbiol* 2005;**56**:1314–28.
73. Sun H, Shi W. Genetic studies of *mrp*, a locus essential for cellular aggregation and sporulation of *Myxococcus xanthus*. *J Bacteriol* 2001;**183**:4786–95.
74. Ueki T, Inouye S. Identification of an activator protein required for the induction of *fruA*, a gene essential for fruiting body development in *Myxococcus xanthus*. *Proc Natl Acad Sci USA* 2003;**100**:8782–7.
75. Nariya H, Inouye S. A protein Ser/Thr kinase cascade negatively regulates the DNA-binding activity of MrpC, a smaller form of which may be necessary for the *Myxococcus xanthus* development. *Mol Microbiol* 2006;**60**:1205–17.
76. Zuker M. Computer prediction of RNA structure. *Methods Enzymol* 1989;**180**:62–88.
77. Fu Z, Donegan NP, Memmi G, Cheung AL. Characterization of MazFSa, an endoribonuclease from *Staphylococcus aureus*. *J Bacteriol* 2007;**189**:8871–9.
78. Tsuchimoto S, Nishimura Y, Ohtsubo E. The stable maintenance system *pem* of plasmid R100: Degradation of PemI protein may allow PemK protein to inhibit cell growth. *J Bacteriol* 1992;**174**:4205–11.
79. Tsuchimoto S, Ohtsubo H, Ohtsubo E. Two genes, *pemK* and *pemI*, responsible for stable maintenance of resistance plasmid R100. *J Bacteriol* 1988;**170**:1461–6.
80. Ruiz-Echevarria MJ, Berzal-Herranz A, Gerdes K, Diaz-Orejas R. The *kis* and *kid* genes of the *parD* maintenance system of plasmid R1 form an operon that is autoregulated at the level of transcription by the co-ordinated action of the Kis and Kid proteins. *Mol Microbiol* 1991;**5**:2685–93.
81. Bravo A, de Torrontegui G, Diaz R. Identification of components of a new stability system of plasmid R1, ParD, that is close to the origin of replication of this plasmid. *Mol Gen Genet* 1987;**210**:101–10.
82. Engelberg-Kulka H, Glaser G. Addiction modules and programmed cell death and antideath in bacterial cultures. *Annu Rev Microbiol* 1999;**53**:43–70.
83. Ruiz-Echevarria MJ, Gimenez-Gallego G, Sabariegos-Jareno R, Diaz-Orejas R. Kid, a small protein of the *parD* stability system of plasmid R1, is an inhibitor of DNA replication acting at the initiation of DNA synthesis. *J Mol Biol* 1995;**247**:568–77.
84. de la Cueva-Mendez G, Mills AD, Clay-Farrace L, Diaz-Orejas R, Laskey RA. Regulatable killing of eukaryotic cells by the prokaryotic proteins Kid and Kis. *EMBO J* 2003;**22**:246–51.

85. Cesareni G, Helmer-Citterich M, Castagnoli L. Control of ColE1 plasmid replication by antisense RNA. *Trends Genet* 1991;**7**:230–5.
86. Davison J. Mechanism of control of DNA replication and incompatibility in ColE1-type plasmids. *Gene* 1984;**28**:1–15.
87. Briani F, Deho G, Forti F, Ghisotti D. The plasmid status of satellite bacteriophage P4. *Plasmid* 2001;**45**:1–17.
88. Jung YH, Lee Y. RNases in ColE1 DNA metabolism. *Mol Biol Rep* 1995;**22**:195–200.
89. Tomizawa JI, Itoh T. The importance of RNA secondary structure in CoIE1 primer formation. *Cell* 1982;**31**:575–83.
90. Tomizawa J. Control of ColE1 plasmid replication: The process of binding of RNA I to the primer transcript. *Cell* 1984;**38**:861–70.
91. Li GY, Zhang Y, Inouye M, Ikura M. Structural mechanism of transcriptional autorepression of the *Escherichia coli* RelB/RelE antitoxin/toxin module. *J Mol Biol* 2008;**380**:107–19.
92. Cherny I, Gazit E. The YefM antitoxin defines a family of natively unfolded proteins: Implications as a novel antibacterial target. *J Biol Chem* 2004;**279**:8252–61.
93. Christensen SK, Maenhaut-Michel G, Mine N, Gottesman S, Gerdes K, Van Melderen L. Overproduction of the Lon protease triggers inhibition of translation in *Escherichia coli*: Involvement of the *yefM–yoeB* toxin–antitoxin system. *Mol Microbiol* 2004;**51**:1705–17.
94. Pellegrini O, Mathy N, Gogos A, Shapiro L, Condon C. The *Bacillus subtilis ydcDE* operon encodes an endoribonuclease of the MazF/PemK family and its inhibitor. *Mol Microbiol* 2005;**56**:1139–48.
95. Bayles KW. Are the molecular strategies that control apoptosis conserved in bacteria? *Trends Microbiol* 2003;**11**:306–11.
96. Hayes CS, Sauer RT. Toxin–antitoxin pairs in bacteria: Killers or stress regulators? *Cell* 2003;**112**:2–4.
97. Engelberg-Kulka H, Reches M, Narasimhan S, Schoulaker-Schwarz R, Klemes Y, Aizenman E, et al. rexB of bacteriophage lambda is an anti-cell death gene. *Proc Natl Acad Sci USA* 1998;**95**:15481–6.
98. Engelberg-Kulka H, Amitai S, Kolodkin-Gal I, Hazan R. Programmed cell death and multicellular behavior in bacteria. *PLoS Genet* 2006;**2**:1518–26.
99. Liu M, Zhang Y, Inouye M, Woychik NA. Bacterial addiction module toxin Doc inhibits translation elongation through its association with the 30S ribosomal subunit. *Proc Natl Acad Sci USA* 2008;**105**:5885–90.
100. Tsilibaris V, Maenhaut-Michel G, Mine N, Van Melderen L. What is the benefit to *Escherichia coli* of having multiple toxin–antitoxin systems in its genome? *J Bacteriol* 2007;**189**:6101–8.
101. Kolodkin-Gal I, Hazan R, Gaathon A, Carmeli S, Engelberg-Kulka H. A linear penta-peptide is a quorum sensing factor required for *mazEF*-mediated cell death in *Escherichia coli*. *Science* 2007;**318**:652–5.
102. Kolodkin-Gal I, Engelberg-Kulka H. The extracellular death factor: Physiological and genetic factors influencing its production and response in *Escherichia coli*. *J Bacteriol* 2008;**190**:3169–75.
103. Shimkets LJ. Intercellular signaling during fruiting-body development of *Myxococcus xanthus*. *Annu Rev Microbiol* 1999;**53**:525–49.
104. Nariya H, Inouye S. Identification of a protein Ser/Thr kinase cascade that regulates essential transcriptional activators in *Myxococcus xanthus* development. *Mol Microbiol* 2005;**58**:367–79.
105. Engelberg-Kulka H, Hazan R, Amitai S. *mazEF*: A chromosomal toxin–antitoxin module that triggers programmed cell death in bacteria. *J Cell Sci* 2005;**118**:4327–32.

Index

A

Actinobacterium sp., 7
Acyl-CoA dehydrogenase, 295
Addiction modules, 471
Adenosylcobalamin, 166
adhE gene, 120
AdoCbl. *See* Adenosylcobalamin
Aeromonas hydrophila, 68, 209
Aeromonas salmonicida, 68
Aeropyrum pernix, 322
7-Aminoethyl 7-deazaguanine, 166
ams gene, 99
Animal mitochondria, 412–413
Apoptosis, MazF role in, 479
aprE gene, 139
Aquifex aeolicus, 4, 18, 331, 380
Aquificales, phylogenetic tree of, 7
Arabidopsis, RNase E role in, 401
Arabidopsis thaliana, 112
Archaea
  mRNA characteristics
    environmental stimuli, response to, 291–293
    half-lives of, 288–291
    structural features of, 287–288
  nine-subunit exosome of, 305–308
  RNA degradation
    exosome-harboring
      S. solfataricus, 308–309
    without exosome, 309–310
Archaeal exosome
  composition of, 300–302
  crystallographic analysis of, 199–200
  structure and mechanism of, 302–308
Archaeal proteins
  endoribonucleolytic
    activities, 293–294
    H. salinarum, archaeal IF5A from, 297–298
    RNase activity and *Sulfolobus*
  dehydrogenases, 295–297
  Sulfolobus small DNA-binding proteins and RNase activity, 294–295
  RNA degradation, 299–300
Archaeal RNase P RNAs, classes of, 325
Archaeal tRNase Z proteins, 354–355. *See also* tRNA 3′-trailer, removal of
Archaeoglobus fulgidus, 305–306, 322
Arcobacter butzleri, 19
AREs. *See* AU-rich elements
ASD. *See* Aspartate-semialdehyde dehydrogenase
Aspartate-semialdehyde dehydrogenase, 295
AU-rich elements, 278
Azotobacter vinelandi, 404

B

Bacillus amyloliquifaciens, 28
Bacillus cereus, 15, 19
Bacillus halodurans, 344
Bacillus pumilus, 19, 28–29
Bacillus subtilis, 8, 25, 124, 138, 190, 195, 211, 232, 277, 320, 371
  and *E. coli*, RNA degradation, 278–283
  EndoA (YdcE), 490–491
  protective ribosomes in, 455–457
  ribosome-mediated cleavage in, 442–445
  RNase M5 activity in, 372
  RNase R, in mRNA degradation, 262–263
  5S rRNA maturation, 381–382
  16S rRNA maturation, 375–377
  23S rRNA maturation, 378–380
  tRNA, 320
  tRNA 3′-end maturation, 357
  tRNase Z-tRNA complex, 349–353
Bacillus subtilis mRNA decay
  3′ end activity
    exoribonucleases, 262–263
    PNPase, 259–262
    polyadenylation, 263

Bacillus subtilis mRNA decay (cont.)
  terminal fragment turnover, 264
  transcription terminator
    structure, 258–259
  5′ end activity
    destabilizers, 251
    mRNA prediction, stability of, 245–250
    pyrophosphatase activity, 250–251
    RNase J, 237–242
    RNase J1 activity, 243–244
    5′ stabilizers, 244–245
  historical perspectives of, 232–234
  translation of body of message
    Bs-RNase III, 254–255
    messenger RNA length, 253–254
    polycistronic mRNA, diVerential stability
      of, 256–258
    RNase M5, RNase P, RNase Z,
      255–256
  translation and decay, 251–253
Bacillus thuringiensis cry gene, 259
Bacterial 3′–5′ exoribonucleases, RNA
  degradation, 190–193
Bacterial genomes, 5
Bacterial P RNAs. See also Bacterial
    ribonucleases and Bacterial RNase P
  role of, 7
  types of, 6
Bacterial ribonucleases
  of cenome, 26–28
    exported ribonucleases, 28–29
    plasmid and phage ribonucleases, 29–30
  of paleome
    energy-dependent degradation of
      RNA, 8–17
    hydrolytic nucleases, 17–26
    RNase P, 4–8
  role of, 2–4
Bacterial RNase P
  non-tRNA substrates of, 338–342
  processing P RNA transcripts, 334–336
  protein, expression of, 333–334
  P RNAs, subtypes of, 325
  P RNA-tRNA complexes, models of, 328
  substrate recognition by, 328–330
Bacterial toxins. See Toxin-antitoxin (TA)
    systems
Bacterial tRNase Z enzyme, 353–354
Bacteria
  RNA degradation

mechanisms in, 278–287
  stabilizing and destabilizing, 278
  toxin-antitoxin (TA) systems
    EndoA (YdcE), 490–491
    RelBE system, 489
    YhaV, 490
    YoeB expression, 489–490
Bacteriophage T4, development of, 43–44
  mRNA degradation, mechanisms and
    regulation of, 45–47
  early stage, bacteriophage T4
    endoribonuclease RegB in, 50–59
  host RNase E in, 47–50
  in middle and late stages, 59–66
  RNase III role in, 66
transfer RNA
  genetic organization and transcription
    of, 68–71
  physiology and distribution of, 66–68
  precursors, processing of, 71–78
BCL-2 family, of proteins, 479
bla gene, 64
bolA gene, 143
Borrelia burgdorferi, 19, 322
Brucella abortus, 209
Bs-RNase III, 254–255. See also mRNA
  degradation
BUMMER strain, 16S rRNA
  maturation, 374–375

C

Campylobacter jejuni, 24
Candidatus Korarchaeum cryptofilum, 355
Caulobacter crescentus, 110, 166, 208
cca gene, 263
Cef phage factor, 43
Cellular components, degradation of, 2
Central β-CASP domain, 240, 349
Chlamydia muridarum, 494
Chlamydomonas reinhardtii, 405
Chloramphenicol, 434, 436
Chloroflexus aurantiacus, 25
Chloroplasts
  and mitochondria, 394
  RNA degradation and polyadenylation
    in, 408–409
ChpBK, 488
Chromobacterium violaceum, 28

INDEX

Cleavage and polyadenylation specificity factor (CPSF), 154, 349
Codon triplets, role of, 2
Cold shock domain (CSD), 202, 280
Colwellia maris, 142
Controlling inverted repeat of chaperone expression (CIRCE), 251
cry1Aa gene, 143
CSP41 enzymes, 402–403
CsrA protein, 164
Cyanophora paradoxa, 332

D

daa operon, 437–438
DEAD-box RhlB helicase, 196
3′–5′ degradative exoribonucleases, 190–193
Deinococcus radiodurans, 15, 19
Desulfotalea psychrophila, 30
dicF gene, 121
dmd gene, 60–61
dp1 gene, 143

E

EcoTrz, processing activity of, 354
EDF. *See* Extracellular death factor
EndoA, in *B. subtilis*, 490–491
Endonucleolytic cleavage, 395–396
Endoribonucleases
  in *E. coli*
    RNase E and RNase G, functions of, 120–121
    RNase G, 119–120
    RNase III, 121–122
    RNase P, RNase Z, and RNase, 122–123, 319
  in RNA processing and degradation, 94
eno gene, 120
Enzymes, RNA degradation
  endoribonucleases
    CSP41, 402–403
    RNase E, 399–401
    RNase J, 402
    RNase P, 319, 403
    RNase Z, 404
  exoribonucleases
    PNPase, 404–406

RNase II/R, 406–408
poly(A)-adding enzymes family, 408
Erwinia cartovora, 108, 142
Erythromycin-treated *B. subtilis* cells, 442–444
Escherichia coli, 6, 44, 138, 232
  and *B. subtilis*, expression of RNase P RNA, 333
  and *B. subtilis*, RNA degradation in, 278–283
  ChpBK, 488
  growth arrest and cell death in, 491–492
  growth, PNPase role in, 197–198
  lacZ gene, and translation–degradation interplay, 435–436
  MazE–MazF complex structure in, 479–480
  mechanisms of
    chemical and functional inactivation of mRNA, 96–97
    enzymes in, 93–95
    mRNA structure in *E. coli*, 95–96
    stable and unstable RNA, 92–93
  mRNA degradation characteristics
    endonucleolytic initiation of, 100–101
    rapid turnover of mRNA, 97–98
    RNA degradosome, 101
    RNA 5′ end on RNase E activity, 101–102
    RNase E, 98–99
  mRNA degradation pathways, 102
    limitations of endonucleolytic initiation of, 102–103
    rpsO mRNA degradation, 103–104
    rpsT mRNA degradation, 104–105
    translation and cellular localization, 105–106
PemK, 486–487
PNPase, structure of, 196
RelE, 56, 489
ribosome stalling, and mRNA cleavage, 438–442, 445–446
RNA degradation pathways, 213
RNaseE activity, 399–400
  catalytic domain quaternary structure of, 110–112
  in degradosome, 196
  functional domain organization of, 106–110
  modulators and adaptors of, 118–119
  in mRNA decay, 148
  regulators of, 158–159

504 INDEX

Escherichia coli, (cont.)
  and RNA degradosome, 123–125
  structure of catalytic domain, 110
  substrate specificity and catalytic
    mechanism of, 112–118
  in T4 mRNA degradation, 47–50
  RNase II/R, activities of, 407
  RNase II, structure of, 203
  rRNA maturation process, 370–372
    5S rRNA, 380–381
    16S rRNA, 375–376
    23S rRNA, 378–379
  TA systems, on chromosomal DNA,
    469–470
  tRNA, 320–321
  tRNA 3′-end maturation, 355–356
  tRNA 5′-end maturation, 324
  YhaV, 490
  YoeB, 489–490
Eukarya, RNA degradation in, 277–285
  mechanisms in, 285–287
  stabilizing and destabilizing, 278
Evolution of life, 2–4
3′–5′ Exoribonucleases
  abundance of, 189–190
  in RNA degradation, 191
Extracellular death factor, 492

F

Flavin mononucleotide (FMN), 166
Frequency of translation initiation (FTI), 431
Fusidic acid, 434, 436

G

gadX gene, 163–164
gadY gene, 163–164
GAPDH. *See* Glyceraldehyde 3-phosphate
  dehydrogenase
Gemmata sp., 7
Gene expression. *See also* RNA degradation
  environmental factors aVecting, 140–142
  growth-phase and growth-rate, 139–140
Geobacillus kaustophilus, 19, 29
Geobacter metallidurens, 29
Glucosamine-6-phosphate synthase
  (GlmS), 157

Glyceraldehyde 3-phosphate
  dehydrogenase, 296
groESL genes, 284
Guanosine 3′,5′-bispyrophosphate
  (ppGpp), 469

H

Haemophilus influenzae, 123, 165
hag gene, 143
Halobacterium salinarum, 289
Haloferax mediterranei, 140, 288
Haloferax volcanii, 292, 353, 355
Halophilic archaea, RNA degradation, 398
Helicobacter pylori, 165
Hfq protein, 139
HrpA, 442
Human cancer cells, 404–405
Hydrolytic ribonucleases, 93–94

I

Ignicoccus hospitalis, 331
Integral inner membrane protein
  (IIMP), 22

K

Kasugamycin treatment, 434, 437
Killer ribosome, 436–437, 445–446
  daa operon, 437–438
  HrpA, role in cleavage, 442
  ssrA-tagging, 438–441
Klebsiella pneumoniae, 140–141

L

Lactococcus lactis, 140, 331
lacZ gene, 143
lacZ mRNA, 451–452
Legionella pneumophila, 28, 123, 165
leuU gene, 121
Light-harvesting complex I (LHI), 283
Listeria innocua, 19
Lon ATP-dependent protease, 489–490
lpp gene, 140

INDEX

M

*mal*EF, 215. *See also* RNA degradation
Marinobacter aquaeolei, 28
MazF mRNA interferase
  cellular target of, 471–473
  cleavage specificity of, 475–479
  expression, regulation under stress conditions, 491–493
  induction in eukaryotic cells, 479
  MazE–MazF system, 471–472, 479–480
  sequence specificity of, 473–475
MazFsa. *See Staphylococcus aureus*
mbhA gene, 140
Metallo-β-lactamase (MBL), 20, 240, 348
Methanocaldococcus jannaschii, 322, 354
Methanococcus vannielii, 288
Methanothermobacter thermautotrophicus, 301, 322
Methylococcus capsulatus, 28
metY gene, 121
micRNA (mRNA-interfering complementary RNA), 471
Mini-III enzyme, 379–380
motA gene, 51
Motor proteins, 193
mRNA
  in archaea
    environmental stimuli, 291–293
    half-lives of, 288–291
    structural features of, 287–288
  interferases, 471, 494–495
  stability
    direct protection of cleavage sites, by ribosomes, 451–453
    measurement, methods for, 433–435
    protection at a distance, ribosomes role in, 450–454
  structure, in *E. coli*, 95–96
mRNA degradation
  3' end activity in
    exoribonucleases, 262–263
    PNPase, 259–262
    polyadenylation, 263
    terminal fragment turnover, 264
    transcription terminator structure, 258–259
  5' end activity in
    destabilizers, 251
    focus on, 234–236
    mRNA prediction, stability of, 245–250
    pyrophosphatase activity, 250–251
    RNase J, 237–242
    RNase J1 activity, 243–244
    5' stabilizers, 244–245
  factor, 190
  mechanisms and regulation of, 45–47
    early stage, bacteriophage T4 endoribonuclease RegB in, 50–59
    host RNase E in, 47–50
    in middle and late stages, 59–66
    RNase III role in, 66
  model for, 243–244
  regulation of, 264–267
  in *Rhodobacter capsulatus*, 283–285
  translation of body of message in Bs-RNase III, 254–255
  messenger RNA length, 253–254
  polycistronic mRNA, 256–258
  RNase M5, RNase P, RNase Z, 255–256
  translation and decay, 251–253
Mycobacterium smegmatis, 470
Mycobacterium tuberculosis, 470
Mycobacterium tuberculosis, MazF homologues
  MazF–mt1, 480–481
  MazF–mt3, 481–482
  MazF–mt6, 482
  MazF–mt7, 482–483
Mycoplasma gallisepticum, 494
Myxococcus xanthus MazF
  MazF mx, 484
  antitoxin of, 484
  cell toxicity by, 485–486
  expression regulation, by MrpC, 485
  role in development, 484–485
  and programmed cell death, 492–493

N

Nanoarchaeum equitans, 4, 322, 331, 346
NanoRNases distribution, 27
Nocardia farcinica, 28
Noncoding RNAs, 138, 281
N-terminal β-lactamase domain, 240
Nucleolytic *vs.* non-nucleolytic inactivation
  non-nucleolytic inactivation, 426, 430
  nucleolytic inactivation, 426
  sRNAs action, 427–428

Nucleolytic vs. non-nucleolytic inactivation (cont.)
  translation and mRNA decay, 428–429
  translation and non-nucleolytic inactivation, 427
Nucleoside diphosphokinase (NDK), 16
Nucleotide diphosphate (NDP), 94
Nucleotide monophosphate (NMP), 93
NUDIX proteins, 251
nusA gene, 121

O

Oceanobacillus iheyensis, 29
Oenococcus oeni, 142
Oligomer-binding (OB), 298
Oligoribonuclease (Orn), 95, 190, 210–212
OmpA gene, 64, 101, 215–216
OmpA mRNA, growth-phase and growth-rate of, 139
Open reading frames (ORFs), 291, 439, 443–444, 475

P

PAP I. See Poly(A) polymerase I
P-bodies, 429
pcnB gene, 66, 144, 153, 337
PDX. See Phosphate-dependent exoribonuclease
PemK, 486–487
Pentatricopeptide repeat (PPR) proteins, 411
PE/PPE family genes, 482–483
Persephonella marina, 7
Phosphate-dependent exoribonuclease, 259
Phosphoribosyl pyrophosphate (PRPP), 344
Phosphorlytic ribonucleases, 93–94
Photobacterium profondum, 28, 30
Photorhabdus luminescens, 29
Planctomycetes, 7
Plant mitochondria, 409–410
Plant RNase E proteins, 400–401
PNK. See 5' polynucleotide kinase/3' phosphatase
PNPase. See Polynucleotide phosphorylase
pnp gene, 121, 157, 193
Polyadenylation/poly(A)-dependent RNA decay pathway
  characterization of, 143–144
  functions of, 142
  heterogeneous tails and RNA stability, 157
  poly(A)-metabolism enzymes, 145–148
  regulation and cofactors of, 153–156
  RNA stability, polyadenylation aVecting, 144–145
  roles of, 156–157
5' Polynucleotide kinase/3' phosphatase, 54
Polynucleotide phosphorylase (PNPase), 45, 141, 233, 404–406, 409, 413, 435
  activities, 195
  complexes, 195–196
  exosome-like enzyme, 199–200
  expression, control of, 193–194
  in poly(A)-dependent RNA degradation, 151–152
  role for, 199
  rRNA degradation, 387
  structure and function, 196–199
Polyphosphate kinase, 15
Poly(A) polymerase I, 16
Poly(A) tails, 397
Postsegregational killing, 469
PPK. See Polyphosphate kinase
preQ1. See 7-Aminoethyl 7-deazaguanine
Protective ribosomes
  in B. subtilis, 455–457
  directionality, source of, 454–455
  local vs. distal protection, 450–454
  mRNAs protection, by ribosomes, 446–447
  RNaseE cleavage mechanism, 447
  internal entry pathways, 449–450
  5'-tethering cleavage pathway, 448–450
Proteus mirabilis, 165
PRPP. See Phosphoribosyl pyrophosphate
pseT gene, 54–55
Pseudomonas aeruginosa, 15, 165
Pseudomonas fluorescens, 165
Pseudomonas putida, 25
Pseudomonas sp., 24
Pseudomonas syringae, 23, 124, 206, 276
Puromycin treatment, 434, 437
Pyrobaculum aerophilum, 331–332, 355, 358
Pyrococcus abyssi, 308
Pyrococcus furiosus, 327, 352, 355
Pyrococcus horikoshii, 308, 345, 489

INDEX 507

Q

Quantitative reverse-transcriptase PCR, 289
Quasi-dormancy, 491

R

regB gene, 50–51
Repetitive extragenic palindromic (REP), 144, 193
Reverse transcription PCR, 287
RhlB. See RNA helicase B
Rhodobacter capsulatus, 124, 141, 276
  mRNA degradation, 283–285
  and puf mRNA, 450
Ribonuclease E, 18–19. See also Escherichia coli
  activity, RNA 5' end role in, 101–102
  modulators and adaptors of, 118–119
  in mRNA decay, 148
  regulators of, 158–159
  and RNA degradosome, phylogenetic distribution of, 123–125
  role, in degradosome, 196
  structure of
    catalytic domain quaternary structure of, 110–112
    functional domain organization of, 106–110
    structure of catalytic domain, 110
    substrate specificity and catalytic mechanism of, 112–118
Ribonuclease G, 17–18, 119–120
Ribonuclease M5, 19, 255–256
Ribonuclease P (RNase P), 324, 403. See also Transfer-RNA
  association with cellular components and oligomeric state, 345–346
  in E. coli, 122–123
  family tree, 326
  role of, 4–8
Ribonucleotidyl transferase (rNTr) family, 408
Ribosomal RNAs (rRNAs), 369–370
Ribosomes. See also Killer ribosome; Protective ribosomes
  binding sites, 13, 95, 426–428, 432–434, 440, 447, 451, 453, 455–456
  stalling, 438–442, 445–446
Rifampicin treatment, and mRNA stability, 433–434

RISC. See RNA-induced silencing complex
rmf gene, 143
RNA
  helicase B, 193
  and hexameric ring interaction, 302–305
  interference, 286
  polyadenylation
    polyadenylation-stimulated RNA degradation pathway, 395–399
    stable poly(A) tail, of nucleus-encoded mRNA, 394–395
  stability
    Gmr regulator, 159–160
    Hfq, CsrA, and ncRNAs proteins, 160–165
    riboswitches destabilization, 166–167
    RNase E regulators, 158–159
    tmRNA, 165–166
RNA degradation. See also Bacterial ribonucleases
  in archaea
    exosome-harboring S. solfataricus, 308–309
    without exosome, 309–310
  archaeal proteins, 299–300
  energy evolution in, 15–16
  in eukarya and bacteria, 277–287
  3'–5' exoribonucleases, 191
  gene expression regulation
    environmental factors aVecting, 140–142
    growth-phase and growth rate, 139–140
  hydrolytic endonucleases, 17–23
  hydrolytic exonucleases, 23–26
  machinery, dynamics of, 188–190
  mechanism of, 212–217
  pathways of, 190
  phosphorolysis, 10–14
  polyadenylation/poly(A), 138, 148–153
    characterization of, 143–144
    functions of, 142
    heterogeneous tails and RNA stability, 157
    poly(A)-metabolism enzymes, 145–148
    regulation and cofactors of, 153–156
    RNA stability, polyadenylation aVecting, 144–145
    roles of, 156–157
  RNase II activity and, 201–202
  (See also RNase II)
  without polyadenylation, 398

RNAi. *See* RNA interference
RNA-induced silencing complex, 286
RNA 3′ polyadenylation, in *B. subtilis*, 263
RNA pyrophosphohydrolase (RppH), 448–449
RNase BN, 71–73
RNase E, 47, 399–401, 425–426, 429, 434–435, 447–454
RNase exoribonucleases, 192
RNase I, 373, 386
RNase II
  activity and RNA degradation, 201–202
  expression, control of, 200–201
  Rrp44/Dis3, 204–205
  structure and function, 202–204
RNase III, 371
  in *E. coli*, 121–122
  in T4 mRNAs degradation, 66
RNase J, 237–242, 402. *See also Bacillus subtilis* mRNA decay
RNase J1, 190
RNase LS, 61–64
RNase P
  architecture, 324–328
  association with cellular components and oligomeric state, 345–346
  exceptions in bacteria and archaea, 330–332
  expression, 332–337
  life with, 346–347
  organelles, 332
  substrate recognition by, 328–330
  and tRNase Z for cell viability, 322–324
  versatility of, 337–344
RNase PH domains, 280
RNase R, 205
  expression, control of, 206
  protein quality control, 207–208
  RNA
    degradation and mode of action, 206–207
    quality control, 207
  role for, 209
  stress and stationary-phase, 208–209
  structure and function of, 209–210
RNase V, 436
RNase Z, 20, 71–72, 122–123, 255–256, 404. *See also* tRNase Z
rnb gene, 200–201
rne gene, 46, 99, 157
rnlA gene, 61
rnpA gene, 6, 331, 336–337

rnpB gene, 4, 143, 331, 336–337
rnr gene, 206
Rnz. *See* RNase Z
RPDs. *See* RNase PH domains
rplL gene, 121
rpoB gene, 121
rpoS gene, 141, 143
rppH gene, 105
rps gene, 143
rpsO gene, 48, 64, 121, 148, 193, 214
rpsO mRNA, degradation of, 103–104
rpsT gene, 102, 214–215
rpsT mRNA, 104–105
rraA gene, 158
rraB gene, 158
rRNA
  degradation, 383–384
    early studies of, 373
    environmental conditions and, 386–388
    quality control process and, 384–386
    RNase I-mediated degradation, 386
  maturation, 382–383
    early studies of, 370–372
    role of RNase, in *E. coli* and *B. subtilis*, 374
    5S rRNA maturation, 380–382
    16S rRNA maturation, 374–378
    23S rRNA maturation, 378–380
Rrp44/Dis3 protein, 204–205. *See also* RNase II
RT-PCR. *See* Reverse transcription PCR
Rubrobacter xylanophilus, 19, 25

S

Saccharomyces cerevisiae, 290, 332, 410
S-adenosylmethionine, 166
Salmonella dublin, 139
Salmonella enterica, 404
Salmonella typhi, 123
Salmonella typhimurium, 108, 144, 165, 216, 337, 343
SAM. *See* S-adenosylmethionine
Schizosaccharomyces pombe, 410
sdh gene, 139
S-D-lactoylglutathione, 355
secG gene, 121
Secretion monitor (SecM) polypeptide, 440
SELEX. *See* Systematic evolution of ligands by exponential enrichment

INDEX

Serine, and mRNA cleavage, 441
Serratia marcescens, 165
Serratia proteomaculans, 29
Shewanella baltica, 28
Shewanella oneidensis, 28
Shigella flexneri, 108, 209
Single-protein production system, 478
siRNA. *See* Small interfering RNAs
SLG. *See* S-D-lactoylglutathione
Small interfering RNAs, 286
Small noncoding RNAs, 193, 216–217
Small regulatory RNAs (sRNAs), 427–428
soc gene, 63
sRNAs. *See* Small noncoding RNAs
5S rRNA maturation
  in *B. subtilis*, 381–382
  in *E. coli*, 380–381
16S rRNA maturation
  in *B. subtilis*, 375–377
  BUMMER strain and, 374–375
  in *E. coli*, 375–376
  P. syringae and, 377–378
23S rRNA maturation
  in *B. subtilis*, 378–380
  in *E. coli*, 378–379
  in *Salmonella*, 380
Stability of mRNA prediction (STOMP), 247
Staphylococcus aureus, 8, 142, 161, 442
Staphylococcus epidermidis, 15
STOMP. *See* Stability of mRNA prediction
Streptococcus pyogenes, 140, 331
Streptomyces antibioticus, 12, 196, 406
Streptomyces coelicolor, 19, 106, 110, 124, 195
Sulfolobus small DNA-binding proteins and RNase activity, 294–295. *See also* Archaeal proteins
Sulfolobus solfataricus, 15, 288–289, 294–295, 300–301
Sulfurihydrogenibium azorense, 7
Synechocystis sp., 19
Systematic evolution of ligands by exponential enrichment, 52

T

T4-encoded RegB endoribonuclease, 43
5′-Tethering cleavage pathway, 448–450, 453
Tetracycline, 434, 436
Thermomicrobium roseum, 6, 325
Thermoplasma acidophilum, 355

Thermotoga maritima, 8, 15, 349, 354, 357
Thermus thermophilus, 6, 19, 240
Thiamine pyrophosphate, 166
Thiobacillus denitrificans, 28
Thymidine-5′-*p*-nitrophenyl phosphate, 353
tmRNA. *See* Transfer-messenger RNA
T4 mRNA and *E. coli* mRNA, difference, 46–47
Toeprinting (TP) analysis, 472–474
Topoisomerase-primase (TOPRIM), 301
Toxin-antitoxin (TA) systems
  in bacteria, 56, 57, 468–471, 494
  EndoA (YdcE), 490–491
  RelBE system, 489
  YhaV, 490
  YoeB, 489–490
  and mRNA cleavage, 441
TpNPP. *See* Thymidine-5′-*p*-nitrophenyl phosphate
T4 polynucleotide kinase, 43
TPP. *See* Thiamine pyrophosphate
TRAMP molecule, 286
Transfer-messenger RNA, 261
Transfer-RNA, 319–320
  genes encoding CCA, 322
  number and organization of, 320–322
  RNase P
    architecture, 324–328
    association with cellular components and oligomeric state, 345–346
    exceptions in bacteria and archaea, 330–332
    expression, 332–337
    life with, 346–347
    organelles, 332
    substrate recognition by, 328–330
    and tRNase Z for cell viability, 322–324
    versatility of, 337–344
  role of, 3
Translational block and mRNA stabilization, 434–435
Translation–degradation interplay
  direct/translation-mediated changes, 430–433
  lacZ gene and, 435–436
  mRNA stability, measurement of, 433–435
Translation inhibitors, 434, 437
Translation initiation
  mutation eVects, 431–432
  regions, 94–95, 233

TRAP. *See trp* operon regulatory protein
Trichodesmium erythreum, 28
T4 tRNAs, 66–78
tRNA. *See* Transfer-RNA
tRNA 3′-processing, ribonucleases in, 323–324
tRNase Z, 72
   crystal structure, 349–352
   features, 348–349
   inhibitory effects, 352–353
   substrates, 353
tRNA 3′-trailer, 72
   concerted processing by endo-and
      exonucleases, 355–357
   endonucleolytic removal of, 348–355
   tRNase Z, 72
trp operon regulatory protein, 260
trxA gene, 212–214. *See also* RNA degradation
Trypanosoma brucei, 411
Trypanosome mitochondria, 411
Type three secretion systems (TTSS), 199

U

Untranslated regions (UTR), 95, 158, 192

V

Vibrio anguillarum, 164
Vibrio cholerae, 15, 123, 161

X

Xanthomonas campestris, 211
Xenopus laevis, 486

Y

Yeast mitochondria, 410
Yersinia enterocolitica, 198
Yersinia pestis, 29, 108, 123
Yersinia pseudotuberculosis, 29, 108, 117
YmdA phosphohydrolase, 22

Z

Zinc-dependent phosphodiesterase
   (ZiPD), 354